Modern Food Analysis

Modern Food Analysis

F. Leslie Hart, A.M.
Director, Boston District
U.S. Food and Drug Administration (*retired*)

and

Harry Johnstone Fisher, Ph.D.
Chemist Emeritus
The Connecticut Agricultural Experiment Station
New Haven, Conn.

SPRINGER-VERLAG NEW YORK HEIDELBERG BERLIN

1971

Library of Congress Catalog Card Number 72-83661.

Printed in the United States of America

ISBN 0-387-05126-0 Springer-Verlag New York · Heidelberg · Berlin
ISBN 3-540-05126-0 Springer-Verlag Berlin · Heidelberg · New York

To Louise Matherly Fisher, whose death on June 8, 1969 ended 34 years of companionship and inspiration to her husband, the junior author of this book.

Preface

When the present authors entered governmental service, food chemists looked for guidance to one book, Albert E. Leach's *Food Inspection and Analysis*, of which the fourth revision by Andrew L. Winton had appeared in 1920. Twenty-one years later the fourth (and last) edition of A. G. Woodman's *Food Analysis*, which was a somewhat condensed text along the same lines, was published.

In the 27 years that have elapsed since the appearance of Woodman's book, no American text has been published covering the same field to the same completeness. Of course, editions of *Official Methods of Analysis of the Association of Official Agricultural Chemists* have regularly succeeded each other every five years, as have somewhat similar publications of the American Public Health Association in the more limited field of dairy products. However, these books have strictly confined themselves to methods, eschewing interpretations and the tables of authentic analyses so important to the food chemist in deciding whether his sample is what it should be. There have also been textbooks intended for use in elementary courses in food chemistry, encyclopedic three- and four-volume disquisitions on the composition and manufacture of food products, and monographs on single foods. But no text such as those of Leach and Woodman.

In presenting the present book, it has been the intent of the authors to offer what will be in essence a modern version of "Leach". It differs from that book in that familiarity with the everyday practices of analytical chemistry, and the equipment of a modern food laboratory, is assumed. We have endeavored to bring it up-to-date both by including newer methods where these were believed to be superior, and by assembling much new analytical data on the composition of authentic samples of the various classes of foods. Many of the methods described herein were tested in the laboratory of one of the authors, and several originated in that laboratory. In many cases methods are accompanied by notes on points calling for special attention when these methods are used. A particular attempt has been made to give proper representation to the newer techniques such as gas chromatography and atomic absorption spectrophotometry.

We have sought to include all types of foods likely to be encountered by the food analyst, together with the methods we have anticipated would be useful in the examination of such foods, insofar as that could be done and still keep the book of marketable size.

Space will not permit our listing all persons whose help we have received in compiling this book, but special mention must be made of a few. Our thanks are due to Dr. James G. Horsfall, director of The Connecticut Agricultural Experiment Station, for offering office space and secretarial help, as well as the facilities of the station libraries, to one of the authors (H.J.F.). If it had not been for the kindness of the Executive Committee of the Association of Official Analytical Chemists in permitting us to make free use of AOAC methods, the assemblage of a book such as this would scarcely have been possible. Our thanks are also due to Dr. William Horwitz and Mr. Luther Ensminger of that association for their personal help. Mr. George P. Larrick, then U.S. Food and Drug Commissioner, opened government files to us for methods of analysis and data on the composition of foods that would otherwise have been unavailable. Mrs. F. Leslie Hart drew most of the illustrations. The American Public Health Association, Inc., permitted us to reproduce a phosphatase method, and H. E. Cox and David Pearson allowed us to copy a method for the separation of artificial colors. We wish to express our gratitude to Gordon Wood, Director of the Los Angeles Laboratory of the U.S. Food and Drug Administration, for permission to use his library and unpublished analytical data on foods. Thanks should also be given to Gentry Corporation, Glendale, California, for use of unpublished data on the composition of spices, and for facilities of their library and laboratory.

We must also express grateful appreciation to the American Association of Cereal Chemists and the American Oil Chemists Society for permission to include many of their official methods in this book and to Director Eugene H. Holeman, Director and State Chemist, Tennessee Department of Agriculture, for valuable suggestions on the interpretation of analyses of fruit beverages.

F. Leslie Hart
Harry J. Fisher

Definitions of Terms, etc.

1 "Water" means distilled water except where otherwise specified.

2 "Alcohol" or "Ethanol" unqualified means 95% ethyl alcohol.

3 Where ammonium hydroxide or an acid is called for in a method, and the strength is not specified, the usual concentrated reagent is understood to be meant.

4 Where an expression such as (1 +2), (5 +4), etc., is used in a method, the first numeral indicates the number of parts by volume of the first reagent, and the second numeral the number of parts by volume of the second reagent (water unless otherwise specified).

5 Temperatures not otherwise specified are understood to be centigrade.

6 *Perchloric acid*—Persons using this reagent should familiarize themselves with the precautions necessary to prevent fire or explosion. *Perchloric acid should not be evaporated in wooden hoods.*

7 Journal abbreviations employed in the Text References and Selected References are for the most part those of *Chemical Abstracts*. An exception in some chapters is the use of *JAOAC* for *Journal of the Association of Official Analytical Chemists*. (This journal was known as the *Journal of the Association of Official Agricultural Chemists* prior to 1966.)

Contents

CHAPTER 1

Introduction—General methods for proximate and mineral analysis

Proximate analysis is defined by H. Bennett in the *Concise Chemical and Technical Dictionary* as the "determination of a group of closely related components together, e.g. total protein, fat." It conventionally includes determinations of the amount of water, protein, fat (ether extract), ash and fiber, with nitrogen-free extract (sometimes termed Nifext) being estimated by subtracting the sum of these five percentages from 100. In order to emphasize the group nature of the percentage of protein, fat and fiber, many chemists use the word "crude" before these three terms.

Since all of these determinations are empirical, conditions of analysis must be precisely stated and followed. Results obtained in ash and moisture determinations are governed primarily by the temperature and the time of subjection to heat. Protein, fat and fiber figures do not represent single constituents. Any errors made in these five determinations are arithmetically cumulative in the figure obtained for nitrogen-free extract.

Water

All foods, no matter what the method of processing, contain more or less water or moisture. It comprises from 60 to 95 percent of all natural foods. In plant or animal tissues, it may be said to exist in two general forms: "free water" or "bound water". Free or absorbed water, the most prevalent form, is readily liberated and may be determined by most of the methods used for the determination of moisture.

Bound water is combined or adsorbed water. This may be present as water of crystallization in hydrates or as water even more firmly bound with protein or saccharide molecules, or adsorbed on the surface of colloidal particles. These forms of water require different degrees of heat to remove—some water remaining at temperatures up to charring. This means that the phrase "percent moisture" is meaningless unless the method of determination is stated.

Determination of Water

A. Oven–Drying to Constant Weight

While there are a few methods, notably the Karl Fischer titration, that determine water content by a stoichiometric reaction, most methods commonly used involve the direct application of heat by drying the sample to constant weight at specified temperatures and atmospheric pressures. During such procedures, hydrates may form through concentration rendering the water in such compounds more difficult to remove by volatilization. Other methods involve distillation, either directly or by azeotropic distillation with an immiscible solvent. An apparatus house summarized the methods for moisture adopted by the AOAC in their 1950 (7th Edition) *Official Methods of Analysis of the Association of Official Agricultural Chemists.*[1]

Method 1-1. *Moisture, Air Oven Method.*

DETERMINATION

Weigh to 1 mg accuracy 2 to 10 g, depending on solids content, into a weighed metallic flat-bottomed dish equipped with a tight fitting cover, previously dried at 90°–100°. Loosen cover and dry in an air oven, with vents open, for 2–3 hours at 98°–100°. Remove dish, cover and cool in a desiccator; weigh soon after it reaches room temperature. Return dish to oven, redry one hour and weigh. Repeat process until change in weight between successive dryings does not exceed 2 mg. The loss of weight is calculated as percent moisture.

Method 1-2. *Moisture, Vacuum Oven Method.*

DETERMINATION

Weigh to 1 mg accuracy 2 to 10 g, depending on solids content, into a weighed, metallic flat-bottomed dish equipped with a tight-fitting cover, previously dried at 98°–100°. Loosen cover and dry in a vacuum oven connected to a pump capable of maintaining a partial vacuum equivalent to 100 mm or less of mercury in the oven chamber and equipped with a thermometer extending into the oven chamber and near the sample dishes. Crack the stopcock of the vacuum oven to admit a current of air (about two bubbles a second) dried by passing through a sulfuric acid drying tower. Dry 4 to 6 hours at 60°–70°. Stop the suction and carefully admit dried air into the chamber. Insert cover, remove the dish and cool in a desiccator. Weigh soon after dish reaches room temperature. Return dish to oven, loosen lid, redry 1 hour. Repeat process until loss between successive dryings does not exceed 2–3 mg. Calculate the loss of weight as moisture.

NOTES ON OVEN MOISTURE DETERMINATIONS

1. Products with high sugar content and meats with high fat content should be dried in the vacuum oven at a temperature not exceeding 70°. Some sugars, e.g. levulose, decompose about 70°, liberating water.

2. Oven methods are not suitable for substances, e.g. spices, containing appreciable amounts of volatile constituents other than water.

3. The removal of water from a sample requires that the partial pressure of water in the vapor phase be lower than that of moisture in the sample. Hence some air movement is necessary. This is accomplished in an air oven by partially opening the vents and in vacuum ovens by a slow stream of dried air.

4. The oven temperature varies at different sites on the oven plate, hence the precaution of placing the sample near the thermometer. The variation may be as much as 3° or even more in older types of air or gravity convection ovens. The more modern ovens of this type are equipped with efficient thermostats and sensors and claim a variant within 1°.

5. After drying, many substances are quite hygroscopic; the cover should be placed tightly on the drying dish immediately upon opening the oven, and the dish should be weighed as soon as it reaches room temperature. This may require as long as an hour if a glass desiccator is used.

Smith and Mitchell[2] have pointed out that certain dried products are better desiccants than those used in the desiccator itself and will pick up moisture even in a closed dish. Some desiccants can be regenerated by following directions on the label or in the catalog. The choice of desiccant is important. Concentrated sulfuric acid, anhydrous copper sulfate, and granular calcium chloride, which are commonly used, are relatively inefficient. Anhydrous calcium sulfate, anhydrous magnesium perchlorate, anhydrous barium perchlorate, freshly ignited calcium oxide, anhydrous phosphorus pentoxide are all more efficient. Boiled, concentrated sulfuric acid (10th ed. AOAC **22.006**) is one of the best, but its corrosive properties add an element of danger.

6. The familiar browning reaction between amino acids and reducing sugars liberates moisture during drying, accelerated at higher temperatures. Thus, foods with high protein and accompanying high reducing sugars must be dried with caution, preferably in a vacuum oven at 60°.

7. Forced draft ovens are occasionally used to lessen the time required for drying to constant weight. The AOAC uses this as a screening method for moisture in cheese (10th ed. Method **15.158**), marine products (10th ed. Method **18.006**) and meat (10th ed. Method **23.003**).

8. The dried residue from the moisture determination may be used for fat (ether extract determination) and, if the original sample were 2 g, for the crude fiber determination.

B. Drying in a Desiccator at Room Temperature

Drying in a vacuum desiccator over sulfuric acid at ambient temperature has been used at times, particularly on heat-labile materials. It is quite time consuming. Windham[3] found that a one week drying period was necessary for meats and that the method did not always give reproducible results. The AOAC recommends it for determination of moisture in plant materials and in grain and stock feeds.

Method 1-3. *Moisture: Vacuum Desiccator Method (AOAC Method* ***22.006–22.007****).*

This method, drying over H_2SO_4 without heat for 24 hours or more, is seldomly used. See AOAC method for details.

C. Distillation with an Immiscible Solvent

The most frequently used distillation method (the Bidwell-Sterling method)[4] measures the volume of water released from the sample by continuous distillation with an immiscible solvent. The water is received into a specially designed receiver which allows it to settle into a graduated section where it will be measured, while the solvent overflows into the distilling flask. It has one disadvantage common to all moisture methods involving heat treatment in that any water formed at the temperature of distillation by decomposition of any constituents of the sample will be included, and measured in the receiver as water.

Xylene was first used as the solvent but it was found that its boiling point (137°–140°) was high enough to decompose several normal constituents of foods and feeds. Fetzer and Kirst[5] compared toluene and benzene distillations of various food stuffs and recommended use of benzene for heat-sensitive products.

The AOAC and ASTA have adopted this method for the determination of moisture in spices, using toluene (ASTA uses benzene for spices high in sugars). In spite of its limitations, the method has several advantages, particularly if good judgment is used in selection of the solvent: (1) A constant temperature, that of the boiling point of the solvent, is maintained. (2) The rate of distillation can be followed visually; when the upper layer of solvent in the receiver is clear the distillation is complete. (3) It is faster than most drying techniques. (4) No complicated apparatus is required.

Method 1-4. *Moisture: Solvent Distillation* (Adapted from official ASTA Method 2).

APPARATUS

All-glass distilling apparatus.

a. *Short-necked flask* or *Erlenmeyer flask*, 250 or 500 ml, with standard taper joints 24/40.

b. *Bidwell-Sterling trap*, with standard taper joint 24/40 or its equivalent.

c. *West condenser*, 400 mm long, with standard taper joint 24/40. Hot plate equipped with a magnetic stirrer and teflon-coated stirring bar.

DETERMINATION

Transfer 10 to 40 g of sample (depending on amount of moisture expected and the capacity in ml of the receiving trap) into the distilling flask. Add sufficient toluene, about 75–100 ml, to completely cover the sample. Insert the magnetic stirrer, and connect the receiving trap and condenser. Fill the trap with toluene by pouring it through the top of the condenser. Place the assembled apparatus on an electric hot plate, and slowly bring the toluene to a boil. Distill at the rate of 1 or 2 drops per second until most of the water passes into the trap, then increase distillation rate to about 4 drops per second.

When all moisture is apparently over, as shown by a layer of clear toluene in the upper section of the trap, wash down the condenser by pouring toluene in the top. Dislodge any water droplets in the condenser tube by brushing with a long handled nylon burette brush or a copper wire spiral. Rinse the brush or spiral with a few ml of toluene poured through the condenser. Continue the distillation about five minutes longer, cool the trap to room temperature, and read the volume of water, estimating to nearest 0.01 ml.

NOTES ON SOLVENT DISTILLATION PROCEDURE

1. Other solvents recommended by various workers include:

Carbon Tetrachloride	B.P. 77°
Benzene	B.P. 80°
Methyl Cyclohexane	B.P. 100°
Toluene	B.P. 111°
Tetrachloroethylene	B.P. 121°
Xylene	B.P. 137°–140°

2. The American Spice Trade Association, in their Official Analytical Methods, recommends the use of benzene instead of toluene for spices such as red peppers (this includes cayenne, chili pepper, chili powder, and paprika), dehydrated onion and dehydrated garlic, which contain large proportions of sugars and other materials which may decompose, liberating water at the temperature of boiling toluene. For example, levulose begins to decompose at about 70°.[6]

3. The entire apparatus should be cleaned each time before use with sulfuric acid—dichromate solution, rinsed thoroughly with distilled water, then with alcohol, and dried.

4. The receiving trap should be calibrated by distilling successive accurately measured quantities of water into the calibrated portion of the trap with toluene. Readings should be estimated to the nearest 0.01 milliliter.

5. Various modifications of the original receiving trap are available. Choice depends on the expected volume of water to be distilled, degree of calibration required, ease of flow-back and other factors.

D. Chemical Methods

The only chemical method commonly used for foods is that which employs the Karl Fischer reagent. This reagent (discovered in 1936 by Karl Fischer[7]) is composed of iodine, sulfur dioxide, pyridine and a solvent that may be methanol or methyl cellosolve. The reactions involved are:

a. $C_5H_5N.I_2 + C_5H_5N.SO_2 + C_5H_5N + H_2O \rightarrow 2C_5H_5N.HI + C_5H_5N.SO_3.$

b. $C_5H_5N.SO_3 + CH_3OH \rightarrow C_5H_5N(SO_4)CH_3.$

Normally an excess of sulfur dioxide, pyridine and methanol (or other hydroxyl-containing compound) is used so the effective strength of the reagent is established by the iodine concentration. The most widely used reagent is a methanolic solution containing the other components in the ratio:

$$I_2 : 3\ SO_2 : 10\ C_5H_5N.$$

Because the Karl Fischer reagent is in effect a powerful desiccant, the sample and reagent must be protected from atmospheric moisture in all procedures. Both visual and electrometric titration procedures are possible and in the simplest form of the method the reagent acts as its own indicator. The sample solution remains canary yellow as long as water is present, changing to chromate yellow and finally to the brown color of unused iodine at the end point. The visual titration is, however, less precise than procedures employing electrometric means of determining the end point. For this reason most Karl Fischer determinations are now made using various forms of market apparatus based on electrometric principles. Most of these instruments use the "dead stop" technique[8] employing platinum electrodes.

In the simplest form, (Fig. 1-1), apparatus requirements are a battery, variable resistor, galvanometer or microammeter and platinum electrodes. A potential is applied across the electrodes to just balance the system, that is, to the point where the galvanometer is not deflected. During titration, as long as water is present, the anode is depolarized and the cathode polarized. At the end point, the small excess of iodine depolarizes the cathode, resulting in a surge of current.

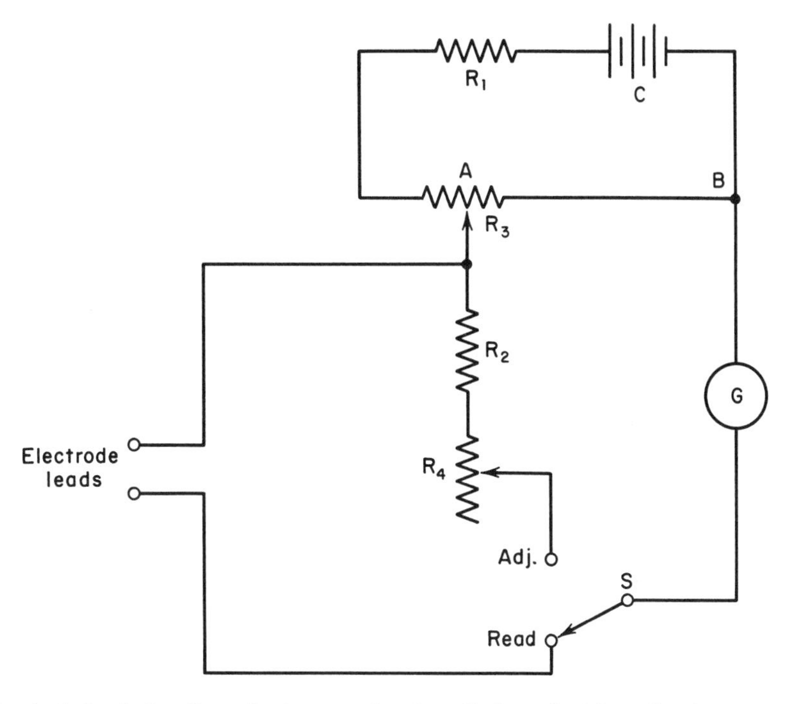

Fig. 1-1. Electrical circuit for direct dead-stop end point: *C*, dry cell, 1.5 v; *G*, galvanometer (Leeds & Northrup No. 2330-C) or microammeter; R_1, resistor, wire-wound, 9000 ohms; R_2, resistor, wire-wound, 1500 ohms; R_3, potentiometer, 1000 ohms; R_4,resistor, 400 ohms, variable; *S*, switch, single pole, double throw.

It is obvious that, like any titration method, the Karl Fischer method will not work unless there is contact between the reagent and the ingredient being titrated—in this case, water. If the food is one that is almost completely soluble in anhydrous methanol or methyl cellosolve there is no problem. Most solid foods, however, are only very partially soluble in these solvents and in such cases techniques such as Soxhlet extraction with methanol followed by titration of the extract have been resorted to. Some foods do not yield all of their moisture rapidly under such extraction procedures, and Johnson[9] has shown that for sweet and white potatoes soaking for 4 to 6 hours in methanol is required to give results comparable to those of vacuum oven drying. McComb and McCready[10] found that if formamide was substituted for methanol as the extracting medium, all of the moisture was removed from ground dehydrated potatoes, sweet potatoes, carrots and peas in 15 minutes.

In spite of the fact that the Karl Fischer method is widely used for the rapid determination of moisture in foods and many other materials, few collaborative comparisons with results by oven drying have been made. For this reason and because this chapter is a general one, only the following two modifications of the method are described:

Method 1-5. *Moisture in Molasses, Karl Fischer Reagent.*[11] (This is an AOAC first action official method).

APPARATUS

a. *Buret with automatic zero reservoir for reagent.*

b. *Magnetic stirring device.*

c. *Titration vessel* (300 ml Berzelius beaker with stopcock attached to side at bottom for withdrawing excess solution is recommended.

d. *Electrodes.*

e. *Circuitry for deadstop end point detection.*

All openings must be tightly closed or protected with drying tubes to prevent contamination from atmospheric H_2O. Various titration assemblies may be obtained from laboratory supply houses, or may be assembled.

Assemble titration apparatus and follow manufacturer's instructions; set for direct titration. Set timer for 30 second end point. Add enough dry methanol to cover electrodes on electrode probes and turn on stirrer. Adjust speed to obtain good stirring without splashing. Do not let stirrer bar contact electrodes. Titrate until satisfactory end point is reached. Newly assembled or not recently used apparatus may require repetition of this step to dry out system.

REAGENTS

a. *Karl Fischer Reagent*—Available from laboratory supply houses or prepare as follows: Dissolve 133 g I_2 in 425 ml dry pyridine in dry glass-stoppered bottle. Add 425 ml dry methanol or ethylene glycol monomethyl ether. Cool to 4° in ice bath and bubble in 102–105 g SO_2. Mix well and let stand 12 hours. (There is less trouble with stopcock leakage when ethylene glycol monomethyl ether is used.) The reagent is reasonably stable, but restandardize for each series of determinations.

b. *Anhydrous Methanol* — Reagent grade CH_3OH containing $<0.1\%$ H_2O. Prepare by distilling over Mg.

DETERMINATION

Add about 120 mg H_2O from weighing pipet or other suitable device and titrate with Karl Fischer reagent.

Calculate C = mg H_2O/ml reagent.

For titration of molasses, C = about 5 mg/ml. Weigh quantity of molasses estimated to give 20–40 ml titer into titration apparatus and titrate. Percent $H_2O = (C \times \text{ml reagent})/(\text{g sample} \times 10)$.

Drain excess liquid and repeat with succeeding samples. If time lapse occurs between titration of samples, adjust liquid in titration vessel to end point by titration with reagent before adding next sample.

Method 1-6. *Moisture in Dehydrated Vegetables, Karl Fischer Reagent.*[12]

APPARATUS. See Method 1–5.

REAGENTS

a. *Formamide*—Practical grade.

b. *Karl Fischer Reagent*—Obtain ready made from a laboratory supply house or prepare as follows: Dissolve 84.7 g resublimed I_2 in 269 ml reagent grade pyridine ($<0.1\%$ H_2O) in 1-liter Pyrex glass-stoppered bottle. Add 667 ml of dried absolute methanol (preferably $<0.05\%$ H_2O). To this add 64 g of sulfur dioxide gas. To avoid appreciable heating add the SO_2 slowly at rate of about 40 g/hour. Stopper solution tightly and store 2–3 days before use.

Some commercial methanol is suitably dry for use in the reagent. If it is found necessary to dry the available supply, add 5 g of Mg turnings to each liter and after the initial vigorous reaction subsides, distill off the alcohol. Take care to keep the distillation system free from contamination by atmospheric moisture.

DETERMINATION

Determine the water equivalent of the Karl Fischer reagent and the blank titer of the formamide daily, or each time a series of determinations is made, because parasitic reactions decrease the reagent's effective strength. To determine the blank, titrate 10 ml of formamide in a 250 ml Erlenmeyer flask. To the same flask add by means of a weight buret 70–100 mg of water and titrate. Calculate the water equivalent of the reagent—mg H_2O/ml reagent. (It is advisable to average at least three blanks and water equivalent values for each standardization. The net titer of the flasks should check within 0.1 ml and the range of the water equivalent values should not exceed 6 parts per thousand). Carry out all titrations drop by drop near the end point until an end point constant for 30 seconds is obtained. Grind the samples of dehydrated vegetables to pass a 40 mesh sieve and store in small tightly sealed containers. Weigh a sample of approximately 500 mg of the ground material into a dry, glass-stoppered 250 ml Erlenmeyer flask which contains a stirrer. Add 10 ml of formamide with slight agitation to disperse the sample and prevent clumping. Heat 40 seconds on a hot plate held at $150 \pm 10°$. Release the stopper slightly and gently rotate the flask during the heating; cool to room temperature and titrate. (Because of the viscous nature of some of the sample-formamide mixtures, it is sometimes expedient to rotate or agitate the flask slightly during the titration in addition to using the magnetic stirrer.)

Calculate the percent water as follows:

$$\%H_20 = \frac{100\ (\text{sample titer–blank titer})\ (H_2O\ \text{equivalent})}{(\text{sample weight in milligrams})}$$

NOTES

1. Besides methanol and formamide listed above, pyridine, dioxane and dimethyl formamide have been employed as sample solvents.

2. Direct titration usually gives total water, that is, free plus hydrated water. When a suitable water-miscible liquid can be found in which the sample is insoluble, free water can be determined by extraction with this liquid and titration of the extract.

3. The method is obviously not applicable without modification to materials containing substances that react with iodine. For other interferences refer to Selected Reference Q.

4. Instead of weighing out portions of water for standardization of the Karl Fischer reagent, finely ground sodium tartrate dihydrate (1 – 1.5 g) dispersed in 50 ml of pretitrated methanol may be used as a standard.

5. Some dehydrated plant material, e.g. spices, contain active aldehydes and ketones that react with the methanol in Karl Fischer reagent to form water. Rader[12] has proposed a method to the AOAC that substitutes methyl cellosolve (ethylene glycol monomethyl ether) for methanol in the Karl Fischer reagent and formamide as the sample solvent.

E. Special Instrumentation Methods

Many specialized instruments have been developed for the direct and rapid determination of moisture. One of the earliest was the Brown-Duvel Moisture Tester devised in 1907 by U.S. Department of Agriculture chemists for the determination of moisture in grains. Water in the sample is distilled along with a hydrocarbon oil and the volume of water distilled is measured.

The Brabender Moisture Tester is a combination drying oven, desiccator, and analytical balance. The Ohaus Moisture Balance is similarly designed. The percentage of moisture is read directly on a scale. Another instrument, the Speedy Moisture Tester, depends on the reaction of calcium carbide with the water of the sample. The pressure of the evolved acetylene is read on a dial directly calibrated in percent of moisture.

There are other instruments on the market which measure the conductivity of the sample. One is the Tag-Heppenstahl Moisture Tester used for the determination of moisture in grains. Others, such as the Tag Dielectric instrument, measure the dielectric properties of the sample. All of these are speedy and moderately precise if instructions are followed exactly. Many of them must be calibrated for each particular food product tested. They are thus suitable for routine in-process or finished product testing of a particular food.

During recent years, instruments have been devised using nuclear magnetic resonance principles, radio frequency power absorption, microwaves as sensors, and other properties of matter. The use of all of these specialized instruments is considered to be beyond the scope of this text. Their manufacturers issue detailed instructions for these instruments.

In 1966 the AOAC adopted a near-infrared spectrophotometric method for moisture in dehydrated vegetables and spices. See Method 17–10.

Reference Method for Moisture Determination

Makower[13] of the USDA Western Regional Research Laboratory has pointed out the need for a "reference method" for the determination of moisture. It would be a means of comparing or calibrating other methods that are more or less empirical but are used extensively because of their convenience. He and his co-workers have developed a procedure based on lyophilization. Briefly stated, the ground sample along with an eight-to-ten-fold addition of water is held over night at about 5° and then frozen by immersion in a solid carbon dioxide-ethanol slurry. The frozen mixture is transferred to a lyophilization apparatus and dried *in vacuo* while still in the frozen state to a moisture level of 2 to 3%. The lyophilized sample is dried in a vacuum oven at 60° or 70° and a pressure of 0.05 to 0.15 mm of mercury, or at room temperature in an evacuated desiccator over an efficient desiccant such as magnesium perchlorate. The original papers should be consulted for a description of the lyophilization apparatus and the details of the method.

Nitrogen

Nitrogen, expressed either as total nitrogen or "protein" (N × 6.25), is almost always determined by a form of wet combustion in which the nitrogen present is converted to ammonium sulfate and finally to ammonia. The ammonia formed is distilled and titrated with a standard acid solution. This method, originally devised by J. Kjeldahl in 1883[14], has been modified by many investigators. The modifications that have become most widely accepted are in what is now known as the Kjeldahl-Gunning-Arnold (KGA) Method.

H. B. Vickery has written an interesting historical review of this method, which is briefly abstracted here (Selected Reference DD). Kjeldahl originally digested the sample in fuming sulfuric acid fortified with phosphorus pentoxide; then potassium permanganate was added to complete the oxidation to ammonium sulfate. He diluted the oxidized mixture, added excess of sodium hydroxide and zinc granules and distilled the ammonia thus formed.

In 1885, H. Wilfarth used a metallic catalyst to shorten the time of oxidation (mercury or copper as oxides were found to be most effective). A few years later, J. W. Gunning suggested the use of potassium sulfate to hasten the removal of water in order to increase the rate of digestion. About this same time, C. Arnold confirmed Kjeldahl's and Wilfarth's techniques and recommended mercury as the most efficient catalyst. Later he suggested addition of benzoic acid and sugar to digest aromatic substances more difficult to analyze, and advocated the combined use of copper and mercury. It might be noted here that A. E. Paul and E. H. Berry showed that the use of the two catalysts gave no faster digestion than did mercury alone. Arnold's main contribution appears to be the accumulation of much analytical data on the method and use of improved apparatus, particularly the trap bulb. Vickery maintains, with considerable justification, that credit is due to Wilfarth for his fundamental contribution of a metallic catalyst, and that the method known since 1912 as the KGA method should be called the Kjeldahl-Wilfarth-Gunning method. In 1955, a joint Association of Official Agricultural Chemists—American Oil Chemists' Society Committee[15] recommended that use of copper as a catalyst in official methods be dropped. This was approved by the AOAC, and the official method given in *Methods of Analysis*, 10th ed. (1965) was revised accordingly. This is given as Method 1-7 in this text:

Method 1-7. *Total Nitrogen in Nitrate-Free Products.*

REAGENTS

a. *Sulfuric acid*, 93–98% H_2SO_4, N-free.

b. *Mercuric oxide*, HgO, or metallic mercury, reagent grade, N-free.

c. *Potassium sulfate*, powdered or anhydrous sodium sulfate reagent grade, N-free.

d. *Sulfide or thiosulfate solution*, Dissolve 40 g commercial potassium sulfide, K_2S, in 1 L water (a solution of 40 g sodium sulfide, Na_2S, or 80 g sodium thiosulfate, $Na_2S_2O_3.5H_2O$, in 1 L water may be used instead).

e. *Sodium hydroxide pellets* or *solution*, N-free; for solution dissolve about 450 g solid NaOH in water and dilute to 1 L (sp g should be 1.36 or higher).

f. *Zinc granules*, reagent grade.

g. *Methyl red indicator*, Dissolve 1 g methyl red in 200 ml alcohol.

h. *Standard* 0.5 *N* or 0.1 *N hydrochloric* or *sulfuric acid.* (Use 0.1 *N* standard acid solution when amount of N is small.)

i. *Standard* 0.5 *N* or 0.1 *N sodium hydroxide solution.*

NOTE: Reagents **h** and **i** should each be standardized against a primary standard, and then checked one against the other. A blank determination should be made on all reagents by substituting 2 g of sucrose for the weighed sample, and proceeding according to Method 1–7.

APPARATUS

a. 500 *ml* or 800 *ml Kjeldahl flask.*

b. *Kjeldahl distilling bulb.*

c. *Straight-tube condenser.*

d. *Heater* (gas or electric) for digestion, adjusted to bring 250 ml water at 25° to a rolling boil in about 5 minutes if electric. To test heaters, preheat 10 minutes if gas, or 30 minutes if electric. Add 3 or 4 boiling chips to prevent superheating.

The Kjeldahl digestion and distillation racks illustrated in most apparatus supply catalogs are suitable and convenient for the purpose. These may be obtained for either gas or electricity as a source of heat.

DETERMINATION

Place weighed sample (0.7–2.2 g, depending upon amount of N) in a Kjeldahl digestion flask. Add 0.7 g of mercuric oxide or 0.65 g metallic mercury, 15 g of powdered potassium sulfate or anhydrous sodium sulfate, and 25 ml sulfuric acid. (If larger sample than 2.2 g is used, increase amount of sulfuric acid 10 ml for each gram of sample.) Place flask in inclined position and heat gently until frothing ceases. (If necessary, add a small amount of paraffin to reduce frothing.) Boil briskly until solution clears and then for at least 30 minutes longer (2 hours for samples containing organic material).

Cool, add about 200 ml water, cool below 25°, add 25 ml of the sulfide or thiosulfate solution, and mix to precipitate mercury. Add a few zinc granules to prevent bumping, tilt flask, and gently add along the side of the neck, without agitation, 25 g sodium hydroxide pellets, or equivalent water solution, to make contents strongly alkaline. (If desired, the sulfide or thiosulfate solution may be mixed with the sodium hydroxide solution before addition to the flask.) Immediately connect the flask to the distilling bulb on the container. Place under the condenser a 500 ml Erlenmeyer flask containing 25 to 50 ml of the standard acid solution, reagent **h**, accurately measured by a pipette or burette. The tip of the condenser should extend beneath the surface of the acid in the Erlenmeyer flask.

Ignite the burner under the distilling flask and rotate flask to mix contents thoroughly. Heat until all the ammonia has been distilled (at least 150 ml of distillate). After distillation has been completed, lower the receiving flask until the condenser tip is above the liquid in the flask and wash off the condenser tip with distilled water. Titrate the excess standard acid in distillate with the standard alkali solution, reagent **i**, using methyl red as indicator. Correct the titration for the blank determination on reagents used and calculate percent of nitrogen, as N, in the sample. If the percent of crude protein is desired, multiply the percent of N by 6.25. This is the factor used conventionally for all proteins. Occasionally other factors are used: 5.7 for wheat, 6.38 for milk, 5.55 for gelatin, 5.95 for rice, 5.77 for soybeans.

Total Nitrogen in Samples Containing Nitrates

The unmodified Kjeldahl digestion of Method 1–7 cannot be relied on to convert all of the nitrate nitrogen in a sample to ammonium sulphate, so this method is not satisfactory for determining total nitrogen in samples containing nitrates. Because Method 1–7 does transform an uncertain proportion of the nitrate to ammonium salts, it must be emphasized on the other hand that it is not a measure of the non-nitrate nitrogen in mixtures containing nitrate.

Since nearly all human foods are essentially nitrate-free, Method 1–7 can be used with little chance of error to determine total nitrogen in foods. However, cured meats contain small proportions of nitrate (and nitrite), and tobacco (if this can be classified as a food) is high in nitrate. When it is necessary to take nitrate nitrogen into consideration in determining total nitrogen, there are several modifications of the Kjeldahl method that may be used. The best known of these are those employing salicylic acid to fix the nitrate, and sodium thiosulfate or zinc dust to reduce it[16]. However, in samples containing high proportions of chloride (ratio of chloride ion to nitrate ion 1:3 or greater), these methods yield low results because the nitric and hydrochloric acids in the digestion mixture react to form nitrosyl chloride (NOCl), which is volatilized and lost. Several methods have been devised to eliminate this error, among which are those employing reduced iron[17] and a special Raney nickel alloy (containing 40% Ni, 10% Co and 50% Al)[18] as reducing agents. Both of these methods have yielded excellent results in our laboratory, but the reduced iron method is slightly shorter.

However, a chromium reduction method recently introduced by Gehrke *et al.*[19] is claimed not only to be even shorter than the iron method but to be a truly universal method in that it gives accurate results with samples containing high chloride/nitrate ratios and with difficultly digestible materials like nicotinic acid and other pyridine derivatives as well as with the more tractable substances that can be handled satisfactorily by Method 1–7. The method is:

Method 1-8. *Total Nitrogen in Samples Containing Nitrates and/or Difficultly-Digestible Compounds.*

REAGENTS

a. *Chromium metal powder, 100 mesh*—Fisher C–318 or Sargent SC–11432.

b. *Norton alundum 14x.*

c. *Silicone antifoam*—General Electric No. 66 or Dow Corning Antifoam Q is suitable.

d. *Sodium thiosulfate or potassium sulfide solution*—Respectively 200 g $Na_2S_2O_3.5H_2O$ or 100 g K_2S per liter.

DETERMINATION

Transfer a sample containing not more than 80 mg of nitrate N to a 500 ml Kjeldahl flask; add 1.2 g of Cr powder (**a**) and 35 ml of water, and let stand 10 minutes with occasional swirling. Then add 7 ml of HCl and two drops of silicone antifoam (**c**) and let stand until visible reaction takes place. Then place on an electric burner adjusted to cause boiling in 7–7.5 minutes, and let stay there until the contents of the flask come to a rolling boil. Remove from the heat and allow to cool.

Add 22 g of K_2SO_4, 1.0 g of HgO, 25 ml of H_2SO_4 and 1.5 g of the alundum (**b**). (If the sample contains considerable organic matter add 0.5 ml of silicone antifoam.) Place on burner regulated to cause boiling in 5.0–5.5 minutes, allow about 15–20 minutes for the copious white fumes to clear out of the bulb of the flask, swirl gently, and digest an additional 30 minutes. If material containing refractory nitrogen is present, continue the digestion for a total of 60 minutes after the copious white fumes clear out of the bulb of the flask. Remove from the heat and cool.

Add 20 ml of the $Na_2S_2O_3.5H_2O$ or K_2S solution (**d**) and complete the analysis as directed in the second paragraph of Method 1–7.

NOTE: Instead of employing a separate solution of $Na_2S_2O_3.5H_2O$ or K_2S, 200 g of $Na_2S_2O_3.5H_2O$ or 100 g of K_2S may be added to the 450 g of NaOH in preparing the sodium hydroxide solution, Method 1–7, (**e**). This eliminates the addition of one solution to the digestion mixture.

Ash

Mineral elements are present in foods as both organic and inorganic compounds. Their exact composition as they exist in the food is often difficult to determine. Ashing the food to destroy all organic matter changes its nature; metallic salts of organic acids are converted to oxides or carbonates or may react during ashing to form phosphates, sulfates or halides. Some elements, such as sulfur and the halogens, may not remain completely in the ash, but may be volatilized.

The determination of total ash in foods is empirical, as are the other determinations discussed in this chapter. It is therefore essential that the detailed instructions in a method of analysis be followed exactly, and that the pertinent factors, such as time, temperature and method of ashing, be recorded by the analyst.

Ash determinations were originally made over an alcohol burner, a gas burner, or a gas-fired muffle. Directions varied from "low red heat" on up. The advent of the electric muffle with indicating pyrometer and thermostatically controlled heat made control of ashing temperature possible for the first time.

Wichmann, in his classic series of papers on ashing technique, made decomposition curves of the ash of typical agricultural products by plotting weight loss against temperatures. (Selected Reference HH). Potassium compounds predominate in plant ash, and sodium compounds in animal ash. Potassium carbonate volatilizes appreciably at 700° and almost entirely at 900°. Sodium carbonate is unchanged at 700°, but suffers considerable loss at 900°. Reactions between carbonates and phosphates also occur (HH).

Klemm, in his survey of approximately 50 ashing procedures listed in *Methods of Analysis, AOAC*, 9th ed. (1960), remarked that it is virtually impossible to incorporate every applicable factor governing ashing of different types of food into one general method. He has devised a table listing sample size, ashing temperature, and ashing time for over 40 classes of food products, and proposed "general directions" for the determination of total ash. The method given below is that proposed by Klemm (Selected Reference FF):

Method 1-9. *Total Ash, General Method.*

Accurately weigh an amount equivalent to about 2–5 g of solids from the well-mixed sample (or measure by means of a pipet, if a free-flowing liquid) into a tared, shallow, rather wide, flat-bottomed ashing dish that has been previously ignited. Use a Pt dish unless otherwise directed.

If considerable moisture is present, dry on a steam bath to apparent dryness, then ignite to constant weight in an electric muffle equipped with an indicating pyrometer and a thermostatic heat control. The control should be set at the temperature recommended in the method given for the food product concerned. If no temperature is recommended, set the control at 525°–550°. Ignite to a white or gray ash, transfer directly to a desiccator (single desiccator for each dish is preferable), cool to room temperature and weigh immediately. Calculate as percent ash.

NOTE: Bring the muffle to the stated temperature slowly, without flaming. Too active combustion may cause loss of ash, or the ash may fuse and cause enclosed carbon to escape ignition. Take care to avoid loss of light, fluffy ash. Cover the dish with a small watch glass even while in the desiccator.

If a carbon-free ash cannot be obtained by following this procedure, remove the dish from the muffle, cool and leach the mass with hot water. Filter through an ashless paper, evaporate the filtrate to dryness and dry the residue and paper at 150°–200°. Then ignite at the stated temperature to a white or gray ash. If a small amount of carbon still remains, moisten the ash in the dish with several drops of water, dry at 150°–200° and re-ignite.

Method 1-10. *Acid-Insoluble Ash.*

This determination is of value in detecting such adulterants as sand or soil in spices, talc in confectionery, etc.

DETERMINATION

To the ashed sample obtained from Method 1–9, add 25–30 ml of diluted hydrochloric acid (1 + 2.5 by volume). Cover the platinum dish with a watch glass to prevent spattering and boil for 5 minutes. Filter through an ashless filter paper and wash with hot water until washings are acid free. Transfer the filter paper and contents to the original dish, dry and ignite at a temperature of 625–650°to a white ash. Cool in a desiccator and weigh.

NOTES ON ASH DETERMINATION

1. Since temperatures may vary in different parts of the muffle, the sample dishes should be placed as close to the pyrometer terminals as possible. Many analysts prefer to place the dishes on a silica plate elevated 1 to 2 cm above the muffle floor on 4 small porcelain crucibles. This provides a more even temperature distribution.

2. See Note 5 in Notes on Oven-drying Methods, moisture, for precautions on choice and use of desiccants.

Crude fiber

The term "crude fiber" is a measure of the material in a food of vegetable origin that has no appreciable food value other than roughage. It consists largely of cellulose, lignin and pentosans, which comprise the cellular structure of the plant along with small amounts of nitrogenous matter. By definition of a combined liaison committee of the AOCS and the AOAC[20], "crude fiber is loss on ignition of dried residue remaining after digestion of sample with 1.25% H_2SO_4 and 1.25% NaOH solutions under specific conditions." The principle of the method of determination has suffered no essential change since its introduction by German agricultural chemists in 1864[21].

The AOCS-AOAC Crude Fiber Liaison Committee of 15 members was established in 1958. After 4 years of intensive study, the Committee recommended a completely revised method based on the original acid-alkali digestion but defining more precisely the equipment and reagents used and the conditions of the test. They also devised new filtering equipment since filtering seemed to be the greatest potential source of error.

Method 1–11 is the method recommended by the combined liaison committee and adopted by the two societies:

Method 1-11. *Crude Fiber.*

REAGENTS

a. *Sulfuric acid solution*—0.255 *N.* 1.25 g of H_2SO_4/100 ml.

b. *Sodium hydroxide solution*—0.313 *N.* 1.25 g of NaOH/100 ml, free or, nearly so, from Na_2CO_3.

(Concentrations of these solutions must be checked by titration.)

c. *Prepared asbestos*—Spread a thin layer of acid-washed, medium or long fiber asbestos in an evaporating dish and heat 16 hours at 600° in a muffle. Boil 30 minutes with 1.25% H_2SO_4, filter, wash thoroughly with water, and boil 30 minutes with 1.25% NaOH. Filter, wash once with 1.25% H_2SO_4, wash thoroughly with water, dry and ignite 2 hours at 600°.

Determine the blank by treating 1.0 g of prepared asbestos with acid and alkali as in *Determination.* Correct crude fiber results for any blank, which should be negligible (about 1 mg). Asbestos recovered from the determination may be used in subsequent determinations.

d. *Alcohol*—Methanol, isopropyl alcohol, or 95% ethyl alcohol.

e. *Antifoam*—Dow Corning Antifoam A Compound diluted 1 + 4 with mineral spirits or petroleum ether, or Antifoam A Emulsion diluted 1 + 4 with water.

f. *Bumping chips or granules*—Broken Alundum crucibles or equivalent. Granules of RR Alundum 90 Mesh, manufactured by Norton Co., Worcester, Mass., are satisfactory.

APPARATUS

a. *Digestion apparatus*—With a condenser to fit a 600 ml beaker and a temperature-adjustable hot plate that will bring 200 ml of water at 25° to a rolling boil in 15 ± 2 minutes (may be obtained from Laboratory Construction Co., 8811 Prospect Avenue, Kansas City, Mo.).

b. *Ashing dishes*—Silica (Vitreosil) 70 × 15 mm;

or porcelain, Coors No. 450, size 1; or the equivalent.

c. *Desiccator*—With efficient desiccant such as 4–8 mesh Drierite. (Calcium chloride is not satisfactory.)

d. *Filtering device*—With No. 200 Type 304 or 316 stainless steel screen (W. S. Taylor Co., 3165 Superior Ave., Cleveland, Ohio), easily washed or digested residue. Either the Oklahoma State filter screen (*see* Figure 1-2; available from Laboratory Construction Co.) or California modified polyethylene Büchner funnel (*see* Figure 1-3; consists of two-piece polyethylene funnel manufactured by Nalge Co., Inc., 75 Panorama Creek Drive, Rochester, New York, as their Catalogue No. J1060, item 4280, 70 mm, without No. 200 screen, or an equivalent. Seal the screen to the filtering surface of the funnel, using a small-tip soldering iron.)

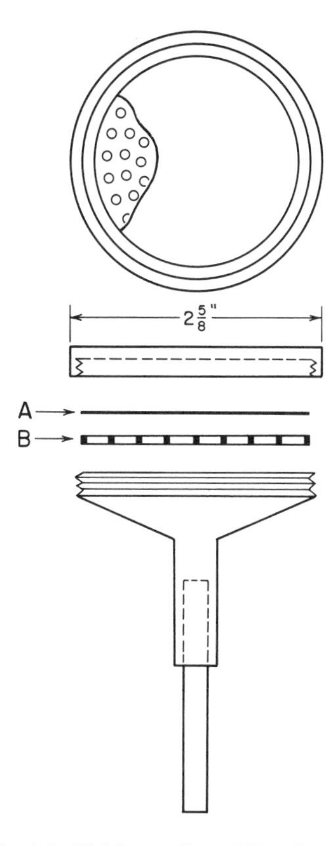

Fig. 1-2. Oklahoma State Filter Screen.

e. *Suction filter*—To accommodate the filtering devices. Attach a suction flask to a trap in line with an aspirator or some other source of vacuum with a valve to break the vacuum.

f. *Liquid preheater*—(*See* Figure 1-4.) A sheet copper tank with 3 coils of $\frac{3}{8}''$ Cu tubing 12.5 ft long. Solder inlets and outlets where the tubing passes through the tank walls. Connect to a reflux condenser and fill with water. Keep the water boiling with two 750 watt thermostatically controlled hot plates. Use Tygon for inlet leads to reservoirs of water, acid and alkali; use gum rubber outlet tubing. The capacity of the preheater is adequate for 60 analyses in 8 hours.

Preparation of Sample

Reduce the sample (a riffle is suitable) to 100 g and place a portion in a sealed container for the moisture determination. Determine moisture immediately. Grind the remainder to uniform fineness (a Webber mill with a 0.033–0.040″ screen, a Mikro mill with a 1/25–1/16″ screen, and a Wiley mill with a 1 mm screen, give comparable fineness.) Since most materials lose moisture during grinding, determine moisture on the ground sample at the same time a sample is taken for the crude fiber determination.

Determination

Extract 2 g of the ground material with ether or petroleum ether (Method 1–13). If the fat content is less than 1%, extraction may be omitted. Transfer the residue to a 600 ml beaker, avoiding fiber contamination from paper or a brush. Add about 1 g of prepared asbestos, 200 ml of boiling 1.25% H_2SO_4, and one drop of diluted antifoam. (An excess of antifoam may give high results; use only if necessary to control foaming.) Bumping chips or granules may also be added. Place the beaker on the digestion apparatus with preadjusted hot plate, and boil exactly 30 minutes, rotating the beaker periodically to keep the solids from adhering to the sides. Remove the beaker and filter as in **a** or **b**.

a. *Using the Oklahoma filter screen*—Turn on the suction and insert the screen (precoated with asbestos if extremely fine materials are being analyzed) into the beaker, keeping the face of the screen just under the surface of the liquid until all liquid is removed. Without breaking the suction or raising the filter add 50–75 ml of boiling water. After this wash is removed, repeat with three 50 ml washings. Work rapidly to keep the mat from

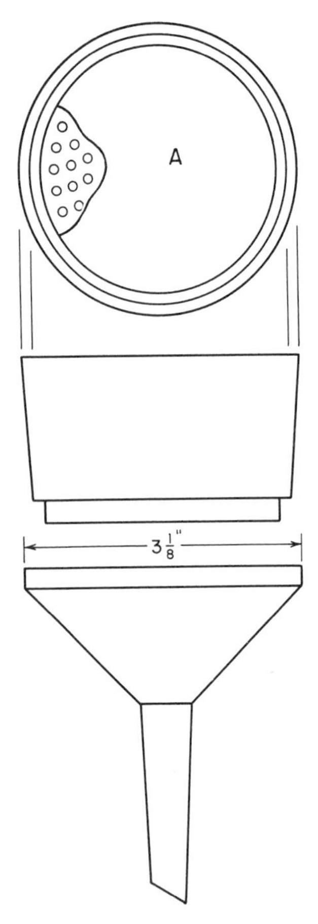

Fig. 1-3. Modified California State Büchner Funnel, 2-piece, polyethylene. Covered with 200-mesh screen, A, heat-sealed to edge of filtering surface.

becoming dry. Remove the filter from the beaker and drain all water from the line by raising above the trap level. Return the mat and residue to the beaker by breaking the suction and blowing back. Add 200 ml of boiling 1.25% NaOH and boil exactly 30 minutes. Remove the beaker and filter as above. Without breaking suction, wash with 25 ml of boiling 1.25% H_2SO_4 and three 50 ml portions of boiling water. Drain free of excess water by raising the filter. Lower the filter into the beaker and wash with 25 ml of alcohol. Drain the line, break the suction and remove the mat by blowing back through the filter screen into the ashing dish. Proceed as in the final paragraph below.

b. *Using the California Büchner funnel*— Filter the contents of the beaker through the Büchner funnel (precoated with asbestos if extremely fine materials are being analyzed), rinse the beaker with 50–75 ml of boiling water and wash through the Büchner funnel. Repeat with three 50 ml portions of water and suck dry. Remove the mat and residue by snapping the bottom of the Büchner funnel against the top of the beaker while covering the stem with the thumb or forefinger and replace in the beaker. Add 200 ml of boiling 1.25% NaOH and boil exactly 30 minutes. Remove the beaker and filter as above. Wash with 25 ml of boiling 1.25% H_2SO_4, three 50 ml portions of water and 25 ml of alcohol. Remove the mat and residue, and transfer to an ashing dish.

Dry the mat and residue 2 hours at $130 \pm 2°$. Cool in a desiccator and weigh. Ignite 30 minutes at $600 \pm 15°$. Cool in the desiccator and reweigh. Percent crude fiber in ground sample = C = (Loss in wt on ignition − loss in wt of asbestos blank × 100/wt sample. Percent crude fiber on desired moisture basis = $C \times (100 - \%$ moisture desired)/ $100 - \%$ moisture in ground sample). Report to 0.1%.

NOTES ON CRUDE FIBER DETERMINATION

1. Chemists using this method should study the four reports of the Crude Fiber Liaison Committee listed under R. E. Holt, in Selected Reference LL. The reports discuss in detail the possible sources of error, choice of filtering equipment and the necessity for following directions exactly as written.

2. Provided a 2 g sample was taken, the residue from ether extraction may be used as the sample for crude fiber. Some laboratories routinely determine moisture by oven drying, ether extract, and crude fiber, successively on the same 2 g sample.

Van Soest and Wine[22] of the Agricultural Research Service of USDA have introduced a new concept of the meaning of the term "crude fiber". They define fiber on a nutritional basis as "insoluble vegetable matter which is indigestible by proteolytic and diastatic enzymes and which cannot be utilized except by microbial fermentation in the digestive tract of animals". These authors point out that the classic acid-alkali digestion to obtain the fibrous portion of plant material

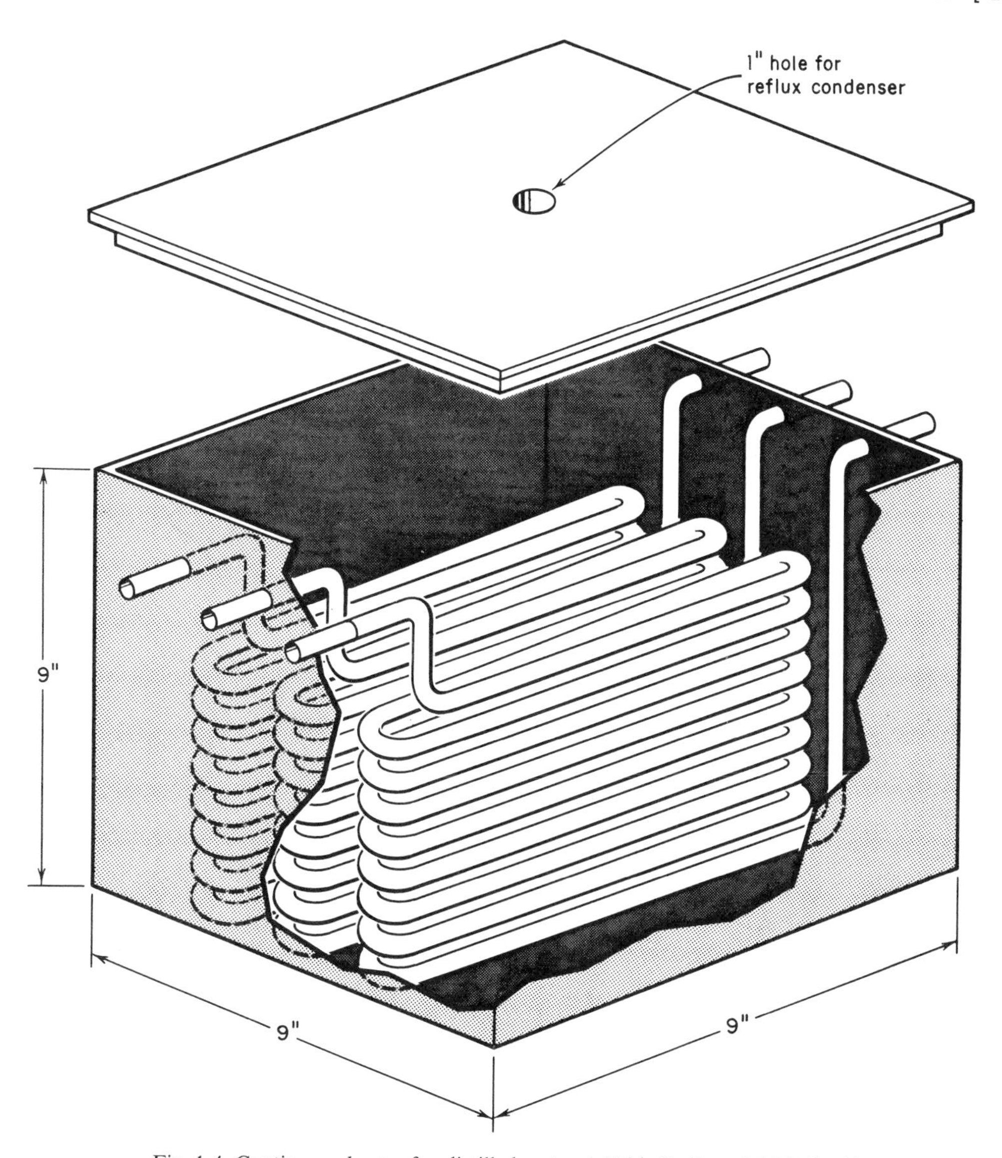

Fig. 1-4. Continuous heater for distilled water, 1.25% alkali, and 1.25% acid.

reported as "crude fiber" gives a figure that has an uncertain and variable relationship to the nutritive value of the "fiber" so obtained. The ideal method is one that will separate lignin, cellulose and hemicellulose with a minimum of nitrogenous material. The residue obtained by acid-alkali digestion retains a considerable portion of plant protein, and part of the lignin is gelatinized or dissolved and is lost.

Van Soest obtained fiber residues of low nitrogen content by the use of neutral detergents[23] in 1963. Further studies led to the development of a method that yields a residue representing the cellulose and lignin portions of the cell-wall, including insoluble ash (mainly silica). It is included in this text as Method 1–12.

This method cannot be considered a replacement for the acid-alkali digestion Method

1–11, especially when it is desired to compare the results with the many recorded analyses of authentic samples of plant material in the literature which were obtained by the established "crude fiber" technique. It does, however, give a truer picture of the nutritional availability for evaluation of foods and feeds of plant origin, particularly for non-ruminant animals. In this connection see a critique of this method by Kimm, Gillingham and Loadholt: *J. Assoc. Offic. Agr. Chemists* **50**; 340 (1967).

Method 1-12. *Plant Cell-Wall Constituents. Fiber Insoluble in Neutral Detergents.*

REAGENTS

a. *Neutral-detergent solution.* To 1 liter distilled water add 30 g of sodium lauryl sulfate, USP, 18.61 g disodium dihydrogen ethylenediamine-tetraäcetate dihydrate, reagent grade, 6.81 g sodium borate decahydrate, reagent grade; 4.56 g disodium hydrogen phosphate, anhydrous, reagent grade and 10 ml 2-ethoxyethanol (ethylene glycol monoethyl ether), purified grade. Agitate to dissolve. Adjust pH to a range of 6.9–7.1.

b. *Decahydronaphthalene.* Technical grade.

c. *Acetone.* Use a grade that is free from color and leaves no residue upon evaporation.

d. *Sodium sulfite.* Anhydrous, reagent grade.

APPARATUS

a. *Refluxing apparatus.* Use any conventional apparatus suitable for crude fiber determinations. Berzelius beakers (600 ml) and condensers made from 500 ml round bottomed flasks are preferred.

b. *Sintered glass crucibles.* Tall form, coarse porosity with plate 40 mm in diameter and large enough to hold 40–50 ml liquid.

DETERMINATION

Weigh 0.5–1.0 g air-dried sample (ground to pass 40 mesh) into refluxing apparatus. Add, in order, 100 ml cold (room temperature) neutral-detergent solution, 2 ml decahydronaphthalene, and 0.5 g sodium sulfite with a scoop. Heat to boiling in 5–10 minutes. Reduce heat as boiling begins, to avoid foaming. Adjust boiling to an even level and reflux 60 minutes, timed from onset of boiling. Swirl beaker and fill previously tared crucible. Admit vacuum only after crucible has been filled. Use low vacuum at first, increasing it only as more force is needed. Rinse sample into crucible with minimum of hot (80–90°) water. Remove vacuum, break up mat, and fill crucible with hot water. Filter liquid and repeat washing procedure. Wash twice with acetone in same manner and suck dry. Dry crucibles at 100° 8 hours or overnight. Cool in efficient desiccator and weigh. Report yield of recovered neutral-detergent fiber as cell-wall constituents. Estimate noncell-wall material by subtracting this value from 100. Ash for 3 hours at 500–550°. Cool in desiccator and weigh. Report ash content of neutral-detergent fiber.

NOTES

1. With concentrated plant material much gelatinous material may form which may tend to clog the filter. As an alternative the Oklahoma State Filter Screen described in Method 1–11 (Figure 1-2) may be used for the initial filtration before the sample is washed into the crucible.

2. Crucibles must be clean for efficient filtration. Used crucibles may be cleaned by ashing at 500° and then forcing water in reverse flow through the filter plate. When crucibles become clogged with ash particles after repeated use, they may be cleaned by forcing in reverse flow a hot solution of 20% KOH, 5% Na_3PO_4 and 0.5% disodium EDTA. Overuse of this cleaning solution should be avoided as it tends to erode the glass.

3. For many purposes, it will be important to know the quantity of cell wall organic matter; therefore an ashing procedure is included in the method. For non-ruminants with little fiber-utilizing capacity, the neutral detergent fiber will be the only fiber value required.

Ether extract (crude fat)

The term "ether extract" embraces all substances extracted by ethyl ether. In addition to fats, it includes phospholipides, lecithins, sterols, waxes, fatty acids, carotenoids, chlorophyll and other pigments. This determination, like that of crude fiber, goes back to work done by early researchers in the German agricultural experiment stations, and later at experiment stations in this country and in England. These workers used anhydrous ethyl ether, and many of the results in literature on the fat content of foods and feeds were obtained by use of this solvent.

Solvents other than ethyl ether have been used but the yield and composition of their resultant extracts differ somewhat from that obtained by ethyl ether. In order to avoid confusion, it is essential that the solvent be named. The determination is made on a sample that has previously been dried in an oven to remove moisture.

Two types of extractors are in general use: the continuous extractor, exemplified by the Underwriters, Knorr, Goldfisch or Bailey-Walker, and the intermittent extractor, such as the Soxhlet and its many modifications. The latter is more efficient. It has the one disadvantage of using a relatively large amount of solvent.

Method 1-13. *Ether Extract* (*Crude Fat*). (This is the official AOAC method for grains, feeds, flour, meats, etc.).

REAGENT

Anhydrous ethyl ether. Wash commercial ether two or three times with water, add several pieces or pellets of sodium or potassium hydroxide, and let stand overnight or longer until most of the water has been absorbed. Decant into a dry bottle, add small pieces of cleaned, freshly cut metallic sodium and let stand until evolution of gas ceases. Keep over metallic sodium in a bottle, protected by a calcium chloride tube.

DETERMINATION

Dry a 2 g sample, preferably in a vacuum oven at 70°, to remove moisture (or use residue from the oven determination of moisture). Transfer dried material to an extraction thimble with porosity permitting a rapid flow of ether. Extract in a Soxhlet extractor at a rate of 5 or 6 drops per second condensation for about 4 hours, or overnight at a rate of 2 or 3 drops per second. Remove ether by cautious evaporation of the contents of the Soxhlet flask, dry in air oven at 100° for 30 minutes, cool and weigh.

NOTES ON ETHER EXTRACT DETERMINATION

1. Some methods for specific foods call for use of some other solvent, for example, the AOAC method for meats permits use of either ether or petroleum ether, whereas certain federal specifications, such as that for pork sausage PP-S-91, require use of petroleum ether.

2. For powdery substances, such as flour, the weighed samples may be mixed with an equal weight of clean, dry sand before drying.

3. The residue remaining after extraction may be dried, and quantitatively transferred to a 600 ml beaker for use for crude fiber determination, provided a 2 g sample was used.

4. The approximate fat content of various classes of basic foods is given below as a guide to sample size:

fresh fruit (other than olives or avocados), leaf or root vegetables, immature corn, beans and peas	up to 1%
mature or dried beans or peas (other than soy)	1 to 5%
cereals and grains	1 to 5%
nuts (other than coconut)	45 to 75%

Nitrogen-free extract

This term is applied to the figure obtained by subtracting the sum of the percentages of moisture, crude protein (N × 6.25), ash, ether extract and crude fiber from 100. It is sometimes called "carbohydrates by difference", or "total carbohydrates", although this latter term is used by most authors to include crude fiber. For example, the authors of USDA *Agriculture Handbook* No. 8, Composition of Foods (revised 1963) define "total carbohydrates" as "the remainder after the sum of the fat, protein, ash, and moisture have been deducted from 100".

It must again be pointed out that any errors

made in laboratory determinations of fat, protein, ash, moisture, and crude fiber are reflected in the figure reported for nitrogen-free extract.

This concludes our discussion on general methods for proximate analysis of foods. We are not including certain long established methods, such as the classical combustion method for carbon and hydrogen, the Dumas combustion method for nitrogen, nor some determinations of physical characteristics, for example, melting points and boiling points. These are given in all books on general chemical analysis and need not be repeated here.

Mineral elements

The term "mineral elements" is, of course, inexact because organic elements such as carbon, hydrogen, nitrogen, oxygen, phosphorus and sulfur do occur in minerals, Nevertheless, it serves as a convenient group classification for those elements, mostly metallic, that occur in relatively minor proportions in foods and are usually determined as elements rather than as specific compounds or groups of compounds.

The number of elements that may occur in traces in foods is very large, including besides the relatively major group composed of Si, Ca, Mg, Na, K, P, S and Cl, such elements as Fe, Al, Mn, F, As, B, Co, Cu, Hg, Mo, Pb, Se, Sn, Zn, and I. While As, Hg, Pb and Sn probably always represent contamination (with the possible exception of As in shellfish), Fe, Mn, Cu, Mo, Zn and I occur as traces as integral parts of enzymes in animal tissues, and these elements and F, B and Se are natural ingredients of plant foods grown on certain soils. It was originally intended to include in the present text methods for all of these elements, but it was impossible to do this without sacrifice of space needed for discussion of the various types of foods, if the book were to be kept of reasonable size. We are therefore listing in this chapter methods for only those elements we have classed as "relatively major". Methods for calcium, copper, iron and iodine specially adapted for dairy products are given in Chapter 6. Analysts interested in determining the minor elements are referred to the texts listed under "Mineral Elements" in the Selected References.

Silicon

Method 1-14. *Silicon* (*AOAC Method* ***6.005***).

Ignite a 10–50 g sample in a flat-bottomed Pt dish in a muffle at 500–550° until the residue is white or nearly so. (Pt dishes must be used with caution in ashing plant materials high in Fe; for such materials it may be advisable to use a well-glazed porcelain or a Vycor dish and run a blank determination.) Moisten with 5–10 ml of HCl, boil about 2 minutes, evaporate to dryness, and heat on a steam-bath 3 hours to render the SiO_2 insoluble. Moisten the residue with 5 ml of HCl, boil 2 minutes, add about 50 ml of water, heat a few minutes on a water-bath, filter through a hardened paper, and wash thoroughly. To this filtrate add the filtrate and washings from the alkali-soluble SiO_2 determination **b**, and dilute to 200 ml. Designate as Solution A. Save for Ca determination.

a. *Sand*—Wash the residue from the filter into a Pt dish, and boil about 5 minutes with about 20 ml of saturated Na_2CO_3 solution; add a few drops of 10% NaOH solution, let the mixture settle, and decant through an ignited and weighed Gooch crucible. Boil the residue in the dish with another 20 ml portion of the Na_2CO_3 solution, and decant as before. Repeat the process. Transfer the residue to the crucible and wash thoroughly, first with hot water, then with a little HCl (1+4), and finally with hot water until Cl-free. Dry the filter and contents, ignite at 500–550°, and weigh as sand. Confirm by microscopic examination.

b. *Alkali-soluble* SiO_2—Combine the alkaline filtrates and washings, acidify with HCl, again evaporate, and dehydrate by heating 2 hours at 110–120°. Moisten the residue with 5–10 ml of

HCl, boil about 2 minutes, add about 50 ml of water, heat on a water-bath 10–15 minutes, filter through an ashless filter or an ignited and weighed Gooch crucible, wash with hot water, ignite at 500–550°, and weigh as SiO_2. Add the filtrate to Solution A.

NOTES

1. The above separation of sand from alkali-soluble SiO_2 is necessary when information on the constitutional ("normal") Si content of vegetable or other plant material (as distinguished from the Si coming from adhering sand) is desired—or, conversely, when contamination of a food with excessive sand must be proved. In those foods where the "normal" Si content is low enough to be ignored, this separation may be omitted.

2. When the filtrate from the above procedure is not being used in other determinations, or the presence of perchloric acid will not interfere in such determinations, $HClO_4$ may be substituted for HCl with some saving in time, because with this acid evaporation to fumes is sufficient to dehydrate the SiO_2, and evaporating to dryness and heating the residue on the steam bath may be omitted.

3. If the quantity of the SiO_2 appears to be excessive, or it is dark-colored, its percentage can be checked by evaporation with H_2F_2 and a few drops of H_2SO_4, ignition, and determination of the loss in weight, provided the SiO_2 was originally collected on an ashless paper and ashed in a Pt dish.

Calcium

Method 1-15. *Calcium by EDTA Titration.*[24]

APPARATUS

a. *Titration stand*—Fluorescent illuminated, such as the "Titra Lite" of the Precision Scientific Co., Chicago, Ill.

b. *Ion-exchange column*—Approximately 20 × 600mm, fitted with a coarse porosity sintered glass disk and a Teflon stopcock. Place 30–40 g from a fresh bottle of Amberlite IR-4B resin (obtainable from Mallinckrodt as their No. 3327) in a 600 ml beaker, and exhaust with three 250 ml portions of 5% Na_2CO_3 or NaOH. Wash with water until excess base is removed. Treat the resin with three 250 ml portions of HCl (3+22), mixing thoroughly after each treatment. Rinse with water until color is removed, and then transfer to the column with water. The column is ready for use after the water has drained to the top of the resin. [The exchange capacity for phosphate is about 1500 mg, so a number of aliquots can be passed through the column before regeneration is necessary. Rinse the column until the eluate is colorless (about 250 ml of water) before each use.]

REAGENTS

a. *Buffer solution*—pH 10. Dissolve 67.5 g of NH_4Cl in 200 ml of water, add 570 ml of NH_4OH, and dilute to 1 liter with water.

b. *Potassium hydroxide—potassium cyanide solution*—Dissolve 280 g of KOH and 66 g of KCN in 1 liter of water.

c. *Potassium cyanide solution*—2%. Dissolve 2 g of KCN in 100 ml of water.

d. *Calcium carbonate*—Primary standard grade, dried 2 hours at 285°.

e. *Hydroxy naphthol blue*—Calcium indicator. Obtainable from Mallinckrodt Chemical Works, 2nd & Mallinckrodt Sts., St. Louis, Mo. 63160, in a dispenser bottle as their No. 5630.

f. *Calmagite [1-(1-hydroxy-4-methyl-3-phenylazo)-2-naphthol—4-sulfonic acid]*—Calcium and magnesium indicator. Obtainable from Mallinckrodt as their No. 4283, or from G. Frederick Smith Chemical Co., Station D, Box 5906, Columbus, Ohio, as their No. 278.

g. *Disodium dihydrogen ethylenediamine tetraäcetate (EDTA) standard solution*—0.01 *M*. Dissolve 3.72 g of EDTA (99+% purity) in water in a liter volumetric flask, and dilute to volume. Accurately weigh enough $CaCO_3$ to give a titration of about 40 ml with the 0.01 *M* EDTA solution, and transfer to a 400 ml beaker. Add 50 ml of water and enough 10% HCl to dissolve the $CaCO_3$. Dilute with water to about 150 ml and add 15 ml of 1 *N* NaOH (disregard any precipitate or turbidity). Add about 200 mg of the hydroxy naphthol blue indicator, and titrate the pink solution to a deep blue end point, using a magnetic stirrer. Add the last few ml of the EDTA solution dropwise. Molarity of the EDTA solution equals

$$\frac{\text{mg } CaCO_3}{\text{ml EDTA} \times 100.09}$$

PREPARATION OF SAMPLE

Drain the liquid from canned whole tomatoes, centrifuge and filter through a fast filter paper. Weigh 100 g of the filtrate into a Pt or porcelain dish, and evaporate to dryness using a forced-draft oven, infrared radiation or other convenient means. Ash at not more than 525° until apparently carbon-free (ash gray to brown). Cool, add 20 ml of water and then cautiously add 10 ml of HCl under a watch-glass while stirring with a glass rod. Rinse the watch-glass off into the dish and evaporate to dryness on a steam bath. Add 50 ml of HCl (1 + 9), heat 15 minutes on a steam bath and filter through a quantitative paper into a 200 ml volumetric flask. Thoroughly wash the paper and dish with hot water. Cool the filtrate, dilute to the mark and mix. For other foods, see Note 2.

DETERMINATION

a. *Removal of phosphate*—Transfer a 50 or 100 ml aliquot of the prepared sample to a 250 ml beaker, and adjust the pH to 3.5 with 10% KOH solution (added drop by drop), using a pH meter and magnetic stirrer. Pass the adjusted solution through the resin column and collect the effluent in a 250 ml volumetric flask—adjusting the flow rate to 2–3 ml/minute. Wash the column thoroughly with two 50 ml portions of water, passing the first 50 ml through the column at the same rate as the sample solution and the second 50 ml at a rate of 6–7 ml/minute. Finally freely pass enough water through the column to make to the mark. Mix thoroughly.

b. *Titration*—Pipet a 100 ml aliquot into a 400 ml beaker and adjust the pH to 12.5–13.0 with the KOH-KCN solution (about 10 ml) using a pH meter and magnetic stirrer. Add 0.100 g of ascorbic acid and 200–300 mg of the hydroxy-naphthol blue indicator. Titrate immediately with the 0.01 *M* EDTA solution through a pink to a deep blue end point, using a magnetic stirrer. Ml EDTA solution consumed = *Titer B*.

$$\%Ca = Titer\ B \times 0.4008 \times 10 \times 100/\text{mg sample}.$$

NOTES

1. *Never pipet any KCN solution by mouth.*
2. The directions for preparation of the sample are specifically directed to canned whole tomatoes, and call for sampling only the drained liquid, because the author of Method 1–15 was interested in the question of whether market canned tomatoes contained more than the 0.026% Ca permitted by Federal regulations. He found by experiment that analysis of the liquid was sufficient since its Ca content was identical with that of the solid portion. The directions should apply without alteration to canned lima beans and potatoes for which the regulations set Ca limits of 0.026 and 0.051% respectively. For other foods it would be advisable to ash an appropriate quantity of the whole material and proceed as directed above. Solution A, Method 1–14, may be used.

Magnesium

Method 1-16. *Magnesium by EDTA Titration.*[24]

Apparatus, reagents, preparation of sample and removal of phosphate are the same as in Method 1–15.

Pipet a 100 ml aliquot of the 250 ml of effluent into a 400 ml beaker and adjust its pH to 10 with the buffer solution, using a pH meter and magnetic stirrer (about 5 ml are required). Add 2 ml of the 2% KCN solution and 200 mg of Calmagite indicator; titrate immediately with the 0.01 *M* EDTA solution, through red to deep blue, using a magnetic stirrer. Ml EDTA solution consumed = *Titer A*

$$\%Mg = (Titer\ A - Titer B) \times 0.2432 \times 10 \times 100/\text{mg sample}$$

Potassium and sodium

The classical way to determine potassium is by the Lindo-Gladding chloroplatinate method.[25] There are also several variations of a method depending on the insolubility of the perchlorate in alcohol or other organic solvents.[26] With careful handling the chloroplatinate method can give very precise results, but lately it has tended to be superseded by methods based on the recent discovery of the insolubility of potassium tetraphenylborate[27], or by the use of flame photometry. Details of a flame photometric method applicable to both potassium and sodium are given below:

Method 1-17. *Flame Photometric Method for K and Na.* (*AOAC Method* ***6.016–6.019***).

REAGENTS

a. *Potassium stock solution*—1,000 ppm K. Dissolve 1.907 g of dried KCl in water and dilute to 1 liter.

b. *Sodium stock solution*—1,000 ppm Na. Dissolve 2.542 g of dried NaCl in water and dilute to 1 liter.

c. *Lithium stock solution*—1,000 ppm Li. Dissolve 6.109 g of dry LiCl in water and dilute to 1 liter.

d. *Ammonium oxalate solution*—0.24 *N*. Dissolve 17.0 g of $(NH_4)_2$ $C_2O_4.H_2O$ in water and dilute to 1 liter.

e. *Extracting solution*—To 250 ml of the NH_4 oxalate solution add enough of the Li stock solution to bring the resulting intensity of the Li line within the range of the instrument being used (*see Preparation of Sample*), and dilute to 1 liter with water.

PREPARATION OF STANDARD SOLUTIONS

Dilute mixtures of appropriate aliquots of the K and Na stock solutions in order to prepare a series of standards containing K and Na in stepped amounts (including 0) covering the instrument range, together with Li and NH_4 oxalate in the same concentrations that will be employed in the extracting solutions.

PREPARATION OF SAMPLE

a. *Dried plant materials* (*including vegetables*)—Transfer a weighed portion of the finely ground and well mixed sample to an Erlenmeyer flask large enough to contain twice the volume of extracting solution to be used. Add a measured volume of extracting solution, stopper the flask and during 15 minutes shake vigorously at frequent intervals. Filter through dry, fast paper. If the paper clogs, pour the contents onto additional fresh paper and combine the filtrates. Use the filtrates for the determination.

NOTE: Do not make extracts more concentrated than required for the instrument because there is a tendency toward incomplete extraction as the ratio of sample weight to volume of extracting solution increases. Prepare separate extracts for Na and K when their concentrations in the sample greatly differ. For K, use a sample weight not exceeding 0.1 g/50 ml extracting solution; for low Na concentrations, use at least 1 g sample/50 ml extracting solution. For higher concentrations, prepare weaker extracts by reducing the ratio of sample to extracting solution rather than by diluting stronger extracts.

b. *Fruit juices and fruit flavored beverages*—Evaporate an aliquot of the sample, ash at low temperature and take the residue up in a measured volume of the extracting solution.

c. *Other foods*[28]—Weigh 1–50 g (according to the expected Na and K contents) of the sample into a shallow Vycor dish. If damp, dry in an oven just below 100°. Then place in a cold muffle furnace and heat to 550°, maintaining this temperature until a gray ash is formed. Allow the furnace to cool and remove the dish. Moisten the ash with redistilled, constant-boiling HCl, evaporate on a steam bath and again heat the residue in the furnace to 550°, maintaining this temperature until the ash is white (generally 1–2 hours). If the ash fails to become white after several hours at 550°, cool the dish and contents, moisten the residue with a 25% solution of redistilled HNO_3, dry and heat again to 550°. (Any color still remaining is usually due to Fe, and cannot be eliminated.) Finally take up the residue in a measured volume of the extracting solution.

DETERMINATION

Rinse all glassware with diluted HNO_3 followed by several portions of water. Protect solutions from airborne Na (or K) contamination. Operate the instrument according to the instructions of the manufacturer. Allow the instrument to reach operating equilibrium before using. Atomize portions of the standard solutions toward the end of the warm-up period until reproducible readings for the series are obtained.

Run standards covering the concentration range of the samples involved at frequent intervals during atomization of a series of sample solutions. Repeat this operation with both standard and sample solutions enough times to result in a reliable average reading for each solution. Plot the analysis curves (K/Li and Na/Li ratios) from readings of the standards and calculate the percentages of K and Na in the samples from these curves.

Phosphorus

Most of the methods for phosphorus depend at some point on conversion of this element to the phosphomolybdate. After the ammonium phosphomolybdate is filtered off, it may be dissolved in an excess of standard alkali which is back-titrated with acid, or it may be dissolved in an excess of ammonia and the phosphorus precipitated as magnesium ammonium phosphate, which is ignited and weighed as magnesium pyrophosphate. For traces of phosphorus, the phosphomolybdate can be reduced to a blue molybdenum compound which is estimated colorimetrically.

Two relatively recent variations are a colorimetric method depending on formation of the yellow phosphomolybdovanadate[29], and a gravimetric method involving precipitation of quinoline phosphomolybdate. It is this latter method and one of the Mo blue colorimetric methods that we have chosen to present:

Method 1-18. *Quinoline Phosphomolybdate Method.* (*AOAC Method* ***2.023–2.025***).

REAGENTS

Quimociac reagent—Dissolve 70 g of $Na_2MoO_4.2H_2O$ in 150 ml of water. Dissolve 60 g of citric acid in a mixture of 85 ml of HNO_3 and 150 ml of water and cool. Gradually add the molybdate solution to the citric-nitric acid mixture with stirring. Dissolve 5 ml of synthetic quinoline in a mixture of 35 ml of HNO_3 and 100 ml of water. Gradually add this solution to the molybdate-citric acid-HNO_3 solution, mix and let stand for 24 hours. Filter, add 280 ml of acetone, dilute to 1 liter with water and mix. Store in a polyethylene bottle.

DETERMINATION

Pipet into a 500 ml Erlenmeyer flask a 50 ml aliquot of Solution A, Method 1–14, and dilute to about 100 ml with water. Add 50 ml of the quimociac reagent, cover with a watch-glass, place on a hot plate in a well ventilated hood and boil one minute. Cool to room temperature, swirling carefully 3–4 times during the cooling. Filter onto a Gooch crucible, lined with a glass fiber filter paper that has been dried at 250° and weighed. Wash five times with 25 ml portions of water. Dry the crucible and contents for 30 minutes at 250°, cool in a desiccator to constant weight and weigh as $(C_9H_7N)_3[PO_4.12MoO_3]$. Subtract a reagent blank. Multiply by 0.01400 to obtain the weight of P.

NOTE: Drying the precipitate at the high temperature specified is necessary to remove variable water of crystallization and obtain stoichiometric results.

Method 1-19. *Molybdenum Blue Micro Method.* (*AOAC Method* ***6.062–6.064***).

REAGENTS

a. *Phosphorus standard solution*—Dissolve 0.4394 g of pure dry KH_2PO_4 in water and dilute to 1 liter. Dilute 50 ml of this solution to 200 ml (2 ml = 0.05 mg P).

b. *Ammonium molybdate solution*—Dissolve 25 g of NH_4 molybdate in 300 ml of water. Dilute 75 ml of H_2SO_4 to 200 ml and add this to the NH_4 molybdate solution.

c. *Hydroquinone solution*—Dissolve 0.5 g of hydroquinone in 100 ml of water and add 1 drop of H_2SO_4 to retard oxidation.

d. *Sodium sulfite solution*—Dissolve 200 g of Na_2SO_3 in water, dilute to 1 liter and filter. Either keep the solution well stoppered or prepare fresh each time.

e. *Magnesium nitrate solution*—Dissolve 950 g of P-free $Mg(NO_3)_2.6H_2O$ in water and dilute to 1 liter.

PREPARATION OF SOLUTION

To a 1 or 2 g sample in a small porcelain crucible add 1 ml of the $Mg(NO_3)_2$ solution and place on a steam bath. After a few minutes, cautiously add a few drops of HCl, taking care that gas evolution does not push portions of the sample over the edge of the crucible. Make 2 or 3 further additions of a few drops of HCl while the sample is on the bath, so that as it approaches dryness it tends to char. If the contents of the crucible become too viscous for further drying on the bath, complete drying on a hot plate. Cover the crucible, transfer to a cold muffle and ignite at 500° for 6 hours until an even gray ash is obtained. (If necessary, cool the crucible, dissolve the ash in a little water or alcohol-glycerol, evaporate to dryness and return to the muffle uncovered for 4–5 hours longer.) Cool, take up in HCl (1+4), and transfer to a 100 ml beaker. Add 5 ml of HCl and evaporate to dryness on a steam bath to dehydrate SiO_2. Moisten the residue with 2 ml of HCl, add about 50 ml of water and heat a few minutes on the bath. Transfer to a 100 ml volumetric flask, cool immediately, dilute to volume and filter, discarding the first portion of the filtrate.

DETERMINATION

To a 5 ml aliquot of the filtrate in a 10 ml volumetric flask, add 1 ml of the NH_4 molybdate solution, rotate the flask to mix, and let stand a few seconds. Add 1 ml of the hydroquinone solution, again rotate the flask and add 1 ml of the Na_2SO_3 solution. (The last three additions may be made with a Mohr pipet.) Dilute to volume with water, stopper the flask with a thumb or forefinger and shake to mix thoroughly. Let stand 30 minutes. Then in a photoelectric colorimeter (equipped with a filter, maximum transmittance 625–675 mμ) immediately compare with 2 ml of the standard KH_2PO_4 solution that has been treated simultaneously and identically with the sample. Report as percent P.

Sulfur

Sulfur occurs in natural foods chiefly as an ingredient of proteins or amino acids such as cystine and methionine. Trace amounts are present in the vitamin thiamine, and in spices such as mustard and vegetables such as cabbage and onions, in the form of organic sulfides. Therefore, while some natural sulfate may be present, the fact that unoxidized forms of this element predominate means that provision must be made for some form of oxidation when the sample is ashed prior to determination of total sulfur as barium sulfate. The usual oxidizing agents are sodium peroxide, magnesium nitrate or perchloric acid. The following method employs the second of these:

Method 1-20. *Total Sulfur.* (*AOAC Method* ***6.059–6.060***).

PREPARATION OF SOLUTION

Weigh a 1 g sample into a large porcelain crucible. Add 7.5 ml of $Mg(NO_3)_2$ solution, Method 1–19e, in such a way that all of the material comes in contact with the solution. (It is important that enough $Mg(NO_3)_2$ solution be added to ensure complete oxidation and fixation of all S present. For larger samples and for samples with a high S content, proportionally larger quantities of this solution must be used.) Heat on an electric hot plate at 180° until no further action occurs. Transfer the crucible while hot to an electric muffle and let it remain at low heat (not over 500°) until the charge is thoroughly oxidized. (No black particles should remain. If necessary, break up the charge and return to the muffle.) Remove the crucible from the muffle and let it cool. First add water, then HCl in excess. Bring the solution to a boil, filter and wash thoroughly. If preferred, transfer the solution to a 250 ml volumetric flask before filtering and dilute to the mark with water.

DETERMINATION

Either dilute the entire filtrate to 200 ml or take a 100 ml aliquot of the solution in the 250 ml volumetric flask (= 0.4 g sample) and dilute to 200 ml. Neutralize with HCl and add 5 ml in excess. Heat to boiling and add 10 ml of 10% $BaCl_2$ solution drop by drop stirring constantly. Continue boiling for about 5 minutes and let stand 5 hours or longer in a warm place. Decant through an ashless paper or an ignited and weighed

Gooch crucible. Add 15–20 ml of boiling water to the precipitate, transfer to the filter or crucible and wash with boiling water until the filtrate is Cl free. Dry the precipitate and filter or crucible, ignite and weigh as $BaSO_4$.Wt. $BaSO_4 \times 0.1374 = S$.

Chlorine

The chlorine content of natural foods is low and, while traces of organic chlorine exist in some of them, essentially all of this element is present in the form of chloride. The amount of chloride is high in some manufactured foods because salt has been added as a seasoning or preservative.

The following method is simple and should suffice for samples containing more than just traces of Cl:

Method 1-21. *Volhard Method. (AOAC Methods **6.065** and **6.067–6.068**).*

REAGENTS

a. *Silver nitrate standard solution*—Prepare a solution slightly stronger than 0.1*N* and standardize by thiocyanate titration against KCl that has been recrystallized three times from water, dried at 100° and heated to constant weight at 500°. Adjust to exactly 0.1*N*. 1 ml = 0.0355 g Cl.

b. *Ammonium or potassium thiocyanate standard solution*—0.1*N*. Prepare a solution slightly stronger than 0.1*N*. Standardize against Solution **a**, and adjust to exactly 0.1*N*.

c. *Ferric indicator*—A saturated solution of $FeNH_4(SO_4)_2 \cdot 12\,H_2O$.

d. *Nitric acid*—Free from lower oxides of N by diluting the usual pure acid with about one-fourth its volume of water. Boil until colorless.

PREPARATION OF SOLUTION

Moisten a 5 g sample in a Pt dish with 20 ml of 5% Na_2CO_3 solution, evaporate to dryness and ignite as thoroughly as possible at 500°. Extract with hot water, filter and wash. Return the residue to the dish and ignite to ash; dissolve in HNO_3 (1+4), filter, wash thoroughly and add this solution to the previous aqueous extract.

DETERMINATION

To the prepared solution add a known volume of the $AgNO_3$ solution (a) in slight excess. Stir well; filter and wash the AgCl precipitate thoroughly. To the combined filtrate and washings, add 5 ml of the ferric indicator and a few ml of the HNO_3. Titrate the excess of Ag with the thiocyanate to a permanent light brown. From the ml of $AgNO_3$ used, calculate the quantity of Cl in the sample.

TEXT REFERENCES

MOISTURE

1. *Central Scientific Co., Chicago:* Pamphlet FM-2 (March 19, 1954).
2. Smith, E. R. and Mitchell, L. C.: *Ind. Eng. Chem.*, **17,** 180 (1925).
3. Windham, E. S.: *J. Assoc. Offic. Agr. Chemists*, **36,** 279 (1953).
4. Bidwell, G. L. and Sterling, W. A.: *Ind. Eng. Chem.*, **17,** 147 (1925).
5. Fetzer, W. R. and Kirst, L. C.: *J. Assoc. Offic. Agr. Chemists*, **38,** 130 (1955).
6. *Official Analytical Methods of the American Spice Trade Association*, American Spice Trade Association, New York (1960).
7. Fischer, K.: *Angew. Chem.*, **48,** 394 (1935).
8. Faulk, C. W. and Bawden, A. T.: *J. Am. Chem. Soc.*, **48,** 2045 (1926); (a) Kao and Hsu: *Acta Chim. Sinica*, **24,** 1 (1958).
9. Johnson, C. M.: *Ind. Eng. Chem., Anal. Ed.*, **17,** 312 (1945).
10. McComb, E. A., McCready, R. M.: *J. Assoc. Offic. Agr. Chemists*, **35,** 437 (1952).
11. Epps, E. A., Jr.: *J. Assoc. Offic. Agr. Chemists*, **49,** 551 (1966).
12. Rader, B. R.: *J. Assoc. Offic. Agr. Chemists*, **49,** 726 (1966).
13. Makower, B.: *Advances in Chemistry*, **3,** 37 (1950)

(a) Makower, B. and Nielsen, E.: *Z. Anal. Chem.*, **20,** 856 (1948).

NITROGEN

14. Kjeldahl, J.: *Z. Anal. Chem.*, **22,** 366 (1883).
15. Report of the Committee on joint collaborative work of the AOAC—American Oil Chemists' Society on total Nitrogen in feeds and fertilizers, *J. Assoc. Offic. Agr. Chemists*, **38,** 56 (1955).
16. Horwitz, W. (ed.): *Official Methods of Analysis of the Association of Official Agricultural Chemists*, 10th ed. Method **2.045.** Association of Official Agricultural Chemists, Washington (1965).
17. Gehrke, C. W., Beal, B. M., and Johnson, F. J.: *J. Assoc. Offic. Agr. Chemists*, **44,** 239 (1961); Gehrke, C. W., and Johnson, F. J.: *J. Assoc. Offic. Agr. Chemists*, **45,** 46 (1962).
18. Burch, W. G., Jr. and Brabson, J. A.: *J. Assoc. Offic. Agr. Chemists*, **48,** 1111 (1965).
19. Gehrke, C. W., Ussary, J. P., Perrin, C. N., Rexroad, P. R. and Spangler, W.: *J. Assoc. Offic. Agr. Chemists*, **50,** 965 (1967).

CRUDE FIBER

20. Holt, K. E.: *J. Assoc. Offic. Agr. Chemists*, **45,** 578 (1962).
21. Henneberg, *Landw. Versuch Stat.*, **6,** 497 (1864).
22. Van Soest, P. J. and Wine, R. H.: *J. Assoc. Offic. Agr. Chemists*, **50,** 50 (1967).
23. Van Soest, P. J.: *J. Assoc. Offic. Agr. Chemists*, **46,** 829 (1963).

MINERAL ELEMENTS

24. Steagall, E. F.: *J. Assoc. Offic. Agr. Chemists*, **49,** 287 (1966); Ibid., **50,** 195, 219 (1967).
25. Ref. 16, Method **2.069–2.071.**
26. Ref. 16, Method **6.021.**
27. Wittig, G. Keicher, G., Ruchert, A. and Raff, P.: *Ann.*, **563,** 100, 126 (1949); Wittig, G.: *Angew. Chem.*, **62A,** 231 (1950); Wittig, G. and Raff, P.: *Ann.*, **573,** 195 (1951).
28. Bills, C. E., McDonald, F. G., Niedermeier, W. and Schwarts, M. C.: *Anal. Chem.*, **21,** 1078 (1949).
29. Ref. 16, Methods **2.018–2.022** and **22.074–22.077.**

SELECTED REFERENCES

PROXIMATE COMPOSITION OF FOODS

A Bailey, E. M.: Analyses of common foods. Conn. Agr. Exp. Sta. Bul. **373** (1934).

B Bureau of Human Nutrition and Home Economics: Tables of food composition in terms of eleven nutrients. U.S.D.A. Misc. pub. **572,** Rev. (1949).

C Canada Dept. of National Health and Welfare, Nutrition Division: Tables of Food Values Recommended for Use in Canada (1951).

D Chatfield, C. and Adams, G.: Proximate composition of American food materials. U.S.D.A. Circular **549** (1940).

E Consumer and Food Economics Research Division, Agricultural Research Service: Nutritive value of foods. U.S.D.A. Home and Garden Bulletin **72** (1960).

F Leach, A. E. and Winton, A. L.: *Food Inspection and Analysis*, 4th ed. New York: John Wiley and Sons, Inc. (1920).

G McCance, R. A. and Widdowson, E. M.: *The Composition of Foods*, 3rd ed. London: H. M. Stationery Office (1960).

H Merrill, A. L. and Watt, B. K.: Energy value of foods, U.S.D.A. Agriculture Handbook **74** (1955).

I Sherman, H. C.: *Chemistry of Food and Nutrition*, 8th ed., New York: Macmillan Co. (1952).

J Watt, B. K. and Merrill, A. L.: Composition of foods—raw, processed, prepared. U.S.D.A. Agriculture Handbook **8** Rev. (1963).

K Winton, A. L. and Winton, K. B.: *The Structure and Composition of Foods.* Vols I–IV. New York: John Wiley and Sons, Inc. (1932–1939).

L Woodman, A. G.: *Food Analysis.* 4th ed. New York: McGraw-Hill Book Co. (1941).

M Wooster, H. A., Jr.: *Nutritional Data.* 2nd ed. Pittsburgh: H. D. Heinz Co. (1954).

MOISTURE

N *Moisture Determination by the Karl Fischer Reagent*, 2nd ed. Dorset: British Drug Houses, Ltd. (1966).

O Cleland, J. E. and Fetzer, W. R.: Historical review of distillation methods for moisture. *Ind. Eng. Chem., Anal. Ed.*, **14,** 442 (1942).

P Fetzer, W. R. and Kirst, L. C., Review of recommended moisture methods for feeds, *J. Assoc. Offic. Agr. Chemists*, **38,** 130 (1955).

Q Kolthoff, J. M. and Elving, F. J.: Extensive review of moisture methods,—*Treatise on*

Analytical Chemistry, Part 2, Vol. 1, New York: Interscience Publishers (1961).

R Kuprianoff, J.: "Bound water" in foods. "Fundamental Aspects of the Dehydration of Foodstuffs." *J. Soc. Chem. Ind.*, **1958,** 14–23.

S Makower, B., *et al.*: Discussion of vacuum oven methods. *Ind. Eng. Chem.*, **38,** 725 (1946).

T McComb, E. A. and McCready, R. M.: Comparison of Karl Fischer and vacuum oven methods. *J. Assoc. Offic. Agr. Chemists*, **35,** 437 (1952).

U McComb, E. A. and Wright, H. M.: Formamide as a Karl Fischer solvent. *Food Technology*, **8,** 73 (1954).

V Mitchell, J. Jr. and Smith, D. M.: Discussion of desiccants. *Ind. Eng. Chem.*, **17,** 180 (1925); *Aquametry*, New York: Interscience Publishers (1948).

W Sair, L. and Fetzer, W. R.: Summary of moisture methods used in the wet milling industry. *Cereal Chem.*, **19,** 633 (1942).

X Windham, E. S.: Comparison of various vacuum oven methods for moisture in meat. *J. Assoc. Offic. Agr. Chemists*, **36,** 279 (1953).

NITROGEN

Y Bradstreet, R. B.: *The Kjeldahl Method for Organic Nitrogen*, New York: Academic Press, (1965).

Z Bradstreet, R. B.: A review of Kjeldahl methods, with extensive bibliography, *Chem. Reviews*, **27,** 331 (1942).

AA Jones, D. B., *et al.*: Specific protein factors. *J. Assoc. Offic. Agr. Chemists*, **25,** 118 (1942).

BB Paul, A. E. and Berry, E. H.: A study of the Kjeldahl-Gunning-Arnold method. *J. Assoc. Offic. Agr. Chemists*, **8,** 108 (1921).

CC Ranker, E. R.: Function of salicylic acid in the Kjeldahl-Gunning-Arnold method. *J. Assoc. Offic. Agr. Chemists*, **10,** 230 (1927).

DD Vickery, H. D.: Historical review of the Kjeldahl-Gunning-Arnold method. *J. Assoc. Offic. Agr. Chemists*, **29,** 358 (1946).

ASH

EE Blade, E.: Survey of AOAC ash methods, *Ind. Eng. Chem., Anal. Ed.*, **12,** 330 (1940).

FF Klemm, G. C., General methods for ashing. *J. Assoc. Offic. Agr. Chemists*, **47,** 40 (1964).

GG St. John, J. L. and Midgley, M. C.: Standardization of the ash procedure for foodstuffs. *J. Assoc. Offic. Agr. Chemists*, **22,** 628 (1939), **23,** 620 (1940); **24,** 848, 932 (1941); **25,** 857, 969 (1942).

HH Wichmann, H. J.: Six papers on factors influencing the ash determination. *J. Assoc. Offic. Agr. Chemists*, **23,** 630 (1940).

CRUDE FIBER

II Bidwell, G. L. and Bopst, L. E.: Report on crude fiber. *J. Assoc. Offic. Agr. Chemists*, **5,** 58 (1921).

JJ Entwistle, V. P. and Hunter, W. L.: Report on crude fiber. *J. Assoc. Offic. Agr. Chemists*, **32,** 651 (1949).

KK Hallab, A. H. and Epps, E. A., Jr.: Variations affecting the determination of crude fiber. *J. Assoc. Offic. Agr. Chemists*, **46,** 1006 (1963).

LL Holt, K. E.: Reports of Crude Fiber Liaison Committee, *J. Assoc. Offic. Agr. Chemists*, **42,** 222 (1959); **43,** 335 (1960); **44,** 567 (1961); **45,** 578 (1962).

FAT (ETHER EXTRACT)

MM Hoffman, H. H.: Review of AOAC investigations and official state methods. *J. Assoc. Offic. Agr. Chemists*, **34,** 558 (1951); **38,** 255 (1955).

NN Taylor, J. J.: Comparison of ethyl ether and petroleum ether extracts. *J. Assoc. Offic. Agr. Chemists*, **30,** 597 (1947).

OO Windham, E. S.: Comparison of ethyl ether and petroleum ether extracts. *J. Assoc. Offic. Agr. Chemists*, **36,** 288 (1953).

MINERAL ELEMENTS

PP Brewer, H. C.: *Bibliography of the Literature on the Minor Elements and Their Relation to Plant and Animal Nutrition.* 4th ed., Vols. I–IV. New York: Chilean Nitrate Educational Bureau, Inc. (1948–1955).

QQ Charlot, G.: *Colorimetric Determination of Elements.* 2nd ed. New York: Elsevier (1964).

RR Diehl, H. and Smith, G. F.: *The Copper Reagents: Cuproine, Neocuproine, Bathocuproine.* Columbus: G. Frederick Smith Chemical Co. (1958).

SS Furman, N. H. and Welcher, F. J. (eds.): *Standard Methods of Chemical Analysis.* 6th ed., Vols. 1–3. Princeton: D. Van Nostrand Co., Inc. (1962–1966).

TT Horwitz, W. A. (ed.): *Official Methods of Analysis of the Association of Official Agricultural Chemists.* 10th ed., Washington: Association of Official Analytical Chemists. (1965).

UU Koch, O. G. and Koch-Dedic, G. A.: *Handbuch der Spurenanalyse.* Berlin-Göttingen-Heidelberg-New York: Springer-Verlag (1964).

VV Kolthoff, I. M. and Elving, P. J.: *Treatise on Analytical Chemistry*, Part II. Vols. 1–12. New York: Interscience Publishers, Inc. (1961–1965).

WW Lundell, G. E. F., Bright, H. A. and Hoffman, J. I.: *Applied Inorganic Analysis.* 2nd ed. New York: John Wiley and Sons, Inc. (1953).

XX Monier-Williams, G. W.: *Trace Elements in Food.* New York: John Wiley and Sons, Inc. (1949).

YY Rosenfeld, I. and Beath, O. A.: *Selenium, Geobotany, Biochemistry, Toxicity, and Nutrition.* New York: Academic Press (1964).

ZZ Sandell, E. B.: *Colorimetric Determination of Traces of Metals*, 3rd ed. New York: Interscience Publishers, Inc. (1959).

AAA Yoe, J. H. and Koch, H. J., Jr. (eds.): *Trace Analysis.* New York: John Wiley and Sons Inc. (1957).

CHAPTER 2

Alcoholic Beverages

Descriptive

Non-Fruit Based Undistilled Beverages

The most important beverages in this category are the malt beverages: ale, beer, porter and stout. Some form of beer has been drunk for at least 6,000 years and there are numerous references to its use in the records of ancient Mesopotamia, China, and Egypt. Next to wine, it is no doubt the oldest of prepared food drinks.[1]

The principal ingredient of beer is barley which has been malted (i.e., germinated) to produce the enzyme (diastase) required to break down the starch of the cereals into maltose and dextrose. The second indispensable ingredient for brewing beer, as it is known today, is hops; these impart the characteristic bitterness and aroma, help to clarify the wort and assist in preserving the beer and improving its foam-holding ability. The other ingredients are the so-called "cereal adjuncts" and, finally, water. Many grains have been used as cereal adjuncts but the most common are corn and rice. The ground raw cereals, unless previously processed to sugar or syrup, are boiled with malt to gelatinize their starch before being added to the main mash. These adjuncts are primarily employed to produce a beer that is paler and more stable. Beer can be produced without the use of cereal adjuncts but the presence of malt, hops and water is essential.

While a more detailed description of the brewing process may be found in Selected Reference C, briefly these are the steps:

Barley is steeped in water to germinate and then dried. The ground malt and cereal adjuncts are heated with water to complete saccharification ("mashing"). The "wort" thus obtained is boiled with hops, filtered, cooled, yeast added and fermentation allowed to proceed for 1–2 weeks at 38–40° F. After fermentation the beer is stored for 3–6 weeks or longer at 32°F., carbonated, filtered and bottled.

The above description applies mainly to lager beer, the type most commonly consumed. Ale is distinguished from lager beer in that a top instead of a bottom fermenting yeast is used. It also usually has a more pronounced hop flavor and is fermented to a higher alcoholic content. Comparative average compositions of beer and ale are as follows:[2]

Ingredient	Beer	Ale
Alcohol, % by volume	4.47	5.32
Extract, %	5.19	4.95
Reducing sugars, %	1.48	1.58
Acidity (as lactic acid), %	0.15	0.17
Protein, %	0.36	0.37
pH	4.35	4.19
Carbon dioxide, %	0.47	0.46

Porter and stout are top-fermented like ale. Porter is a heavier and darker ale made from longer dried, roasted or caramel malt with less hops. It is usually sweeter than beer or regular ale. Stout has a heavier malt flavor than porter and is much darker and sweeter than any other malt liquor; it has a stronger hop taste than porter. Bock beer is a heavy-brewed, dark beer usually coming on the market only around Easter. Weiss beer is primarily a German product and brewed principally from wheat malt. It has a high CO_2 content.

Canada has established quantitative requirements for certain ingredients of malt beverages.[2a] Ale, beer, porter and stout all must contain at least 3.2% of alcohol by volume. Minimum requirements for extract and ash are:

	Ale	Beer	Porter	Stout
Extract, g/100 ml	3.5	3.5	4.0	5.0
Ash, g/100 ml	0.12	0.12	0.13	0.15

An exception is a special "light beer," whose alcohol content is set at 0.96–2.0%, and for which no minimum ash is required.

There are a number of other non-distilled alcoholic beverages not derived from fruit, but most of these have limited geographic distribution. One such beverage is saké which is a Japanese beer made wholly from rice and fermented with a culture of a yeast-like fungus or mold. Saké may contain 17% or more of alcohol by volume, and is sometimes called "rice wine".

Pulque is a sourish beer-like liquid containing 6% alcohol by volume. It is produced by natural fermentation of Maguey (Agave) sap; its consumption is limited to Mexico and some Central and South American countries.

Mead is a beverage made by fermentation of honey whose use dates back to the Biblical era. There is little or no commercial production at the present time, but directions for home manufacture of "honey wine" call for diluting the honey to a 22% sugar content, adding yeast nutrients and fermenting with a wine yeast starter.[3] Canadian regulations[2a] permit the addition of "caramel, natural botanical flavors and honey spirit".

There are two relatively weak alcoholic beverages made from milk, both of which apparently originated in southeastern Russia and the Caucasus. *Kefir* is made by fermenting the milk of various animals with a starter made from Kefir grains and *koumiss* is mares' milk fermented with milk-souring bacteria and yeasts. Kefir contains only up to 1.1% of alcohol, while koumiss may contain 3%.[4]

Fruit-Based Undistilled Beverages

All such beverages may be classified as "wines" in the broader sense of that word, but the unqualified term "wine" when no specific fruit is mentioned is understood to mean grape wine only. In the following discussion of grape wines, the word "grape" will therefore be omitted. Beverages made from other fruits will be described briefly afterwards.

There are two general types of wines, red and white*, and either type may be carbonated, although carbonated red wines do not enjoy a very high reputation among connoisseurs. It is possible to produce wine by simply crushing grapes, permitting the natural yeasts present on the skins to ferment the sugars and filtering the fermented juice. It was probably in this way that wine was made in prehistoric times, but the product of such a primitive process was both too highly variable in flavor and too unstable, because acetic fermentation soon changed it to vinegar, in this fashion of manufacturing to survive once wine-making became an industry.

Modern wine-making starts with selection of the grapes to be used. While a number of species of grapes are known, most wines are made from some variety of the wine grape, *Vitis vinifera*. Some eastern American wines

*There are a few *rosé* or pink wines.

are produced from varieties of the native species, *Vitis labrusca* (Concord, Catawba, Delaware, etc.), which are distinguished by their content of methyl anthranilate. At one time "Scuppernong" wines were produced in the South Atlantic states from *Vitis rotundifolia* varieties.

In large-scale commercial wineries, the decision as to when the grapes should be picked is governed by degree of ripeness. This is determined by measuring the density (i.e., sugar content) of the juice; it should be 21–23° Brix for red wines and 20–23° for white wines. After picking, the grapes are crushed. Whether the juice ("must") is separated from the skins at this point depends on whether white or red wine is being produced. Because the color of red wines is derived from the anthocyanins in the skins, the skins are left in contact with the "must" during fermentation to increase extraction of the coloring matter. White wines may be produced from either white grapes or the colored varieties but if the latter are used, it is necessary to separate the "must" from the skins before fermentation to prevent take-up of color. Fermenting with the skins also introduces more tannin into the wine.

The crushed and stemmed grapes are treated with sufficient sulfur dioxide (added as a solution of SO_2, a metabisulfite or bisulfite or as liquid SO_2) to give a total concentration of 75–150 p.p.m., to prevent growth of wild yeasts and spoilage bacteria. The mixture is then inoculated (after filtration in the case of white wines) with a starter of pure yeast—some strain of *Saccharomyces cerevisiae* var. *ellipsoideus* such as the Champagne or Burgundy type. Cooling during fermentation is required for the wine to be of good quality; if the temperature reaches 100–105°F., the yeast may even be killed.

Within six weeks after crushing, the wine is completely fermented (0.20% or less sugar), and is pumped from the sediment into another vat where 100 p.p.m. of SO_2 are added. It may then be blended with other wines (clarifying agents may be added), after which it is aged about three years in wooden barrels and held in bottles for another year before sale. Wine will improve in the bottle for another 5–10 years.

Particularly in Europe, the primary fermentation caused by yeast conversion of the sugar to alcohol and CO_2 is followed by a malo-lactic fermentation, which is produced by certain *Lactobacilli* that convert malic acid to lactic acid and CO_2. This fermentation reduces the titratable acidity and raises the pH, which is desirable for high acid wines; but in California, where the acidity is already too low and the pH too high, it may lead to actual spoilage of the wine. To prevent the malo-lactic fermentation, early racking, cool storage and maintenance of the SO_2 concentration at 100 p.p.m. or higher, are sufficient.

The above processes yield beverages that are called "table wines" because they are usually consumed during a meal. Such wines also furnish the basis for the carbonated and fortified wines that will be discussed below. However, before proceeding to these special wines, some explanation of the nomenclature of table wines is desirable.

Because the wines of the Bordeaux and Burgundy regions of France and the Rhine Valley in Germany have always been particularly esteemed by connoisseurs, there has been a tendency to label wines of other countries with names derived from these regions. In the United States, this has been particularly noticeable in the case of California table wines. These names (particularly the French ones) properly designate localities (or even individual vineyards) rather than types of wine, and the American wines to which they are applied frequently bear no resemblance in flavor to the originals. One of the most flagrant cases is that of California "sauterne", which is a dry wine whereas the original sauternes of France is sweet. Fortunately, the better-grade California wines are now being sold under the names of the varieties of grapes from which they are made, such as Cabernet, Pinot noir, Riesling, Sémillon, etc.

Average analyses of wines and the musts from which they are made are given in Table 2-1:

Canadian food and drug regulations[2a] carry certain quantitative requirements for wine sold in that Dominion. Volatile acidity (expressed as acetic acid) may not exceed 0.13 g/100 ml, and not more than 70 p.p.m. of

TABLE 2-1
Composition of Musts and Wines[a]

	Must	Wine
	Per cent	Per cent
1. Water	70–85	80–90
2. Carbohydrates	15–25	0.1–0.3
Dextrose	8–13	0.5–0.1
Levulose	7–12	0.05–0.1
Pentoses	0.08–0.20	0.08–0.20
Arabinose	0.05–0.15	0.05–0.10
Rhamnose	0.02–0.04	0.02–0.04
Xylose	T[d]	T
Pectin	0.01–0.10	T
Inositol	0.02–0.08	0.03–0.05
3. Alcohols and related compounds		
Ethyl	T	8.0–15.0
Methyl	0.0	0.01–0.02
Higher	0.0	0.008–0.012
2,3-Butylene glycol	0.0	0.01–0.15
Acetylmethylcarbinol	0.0	0.000–0.003
Glycerol[b]	0[b]	0.30–1.40
Sorbitol	T	T
Diacetyl	0.0	T–0.0006
4. Aldehyde	T	0.001–0.050
5. Organic acids	0.3–1.5	0.3–1.1
Tartaric	0.2–1.0	0.1–0.6
Malic	0.1–0.8	0.0–0.6
Citric	0.01–0.05	0.0–0.05
Succinic	0	0.05–0.15
Lactic	0	0.1–0.5
Acetic	0.00–0.02	0.03–0.05
Formic	0	T
Propionic	0	In spoiled wines
Butyric	0	In spoiled wines
Gluconic		From botrytised grapes only
Glucuronic		From botrytised grapes only
Glyoxylic	?	0.00012
Mesoxalic	?	0.0001–0.0003
Glyceric	. . .	T
Saccharic	. . .	T
Amino	0.01–0.08	0.01–0.20
Pantothenic	. . .	T
Quinic	0	T
p-Coumaric	T	?
Shikimic	T	?
Sulfurous	0	0.00–0.05
Carbonic	T	Various

TABLE 2-1—(continued)
Composition of Musts and Wines

	Must	Wine
	Per cent	Per cent
6. Polyphenol and related compounds		
Anthocyans	T	T
Chlorophyll	T	0–T
Xanthophyl	T	?
Carotene	T	?
Flavonol		
Quercetin	T	T
Quercetrin	T	T
Rutin	?	?
Tannins	0.01–0.10	0.01–0.30
Catechol	T	T
Gallocatechol	T	T
Epicatechol gallate	T	T
Gallic acid	T	T
Ellagic acid	T	T
Chlorogenic acid	T	T
Isochlorogenic acid	T	T
Caffeic acid	T	T
7. Nitrogenous compounds		
Total	0.03–0.17	0.01–0.09
Protein	0.001–0.01	0.001–0.003
Amino	0.017–0.110	0.010–0.200
Humin	0.001–0.002	0.001–0.002
Amide	0.001–0.004	0.001–0.008
Ammonia	0.001–0.012	0.00–0.071
Residual	0.01–0.02	0.005–0.020
8. Mineral compounds	0.3–0.5	0.15–0.40
Potassium	0.15–0.25	0.045–0.175
Magnesium	0.01–0.025	0.01–0.020
Calcium	0.004–0.025	0.001–0.021
Sodium	T–0.020	T–0.044
Iron	T–0.003	T–0.002
Aluminum	T–0.003	T–0.001
Manganese	T–0.0051	T–0.005
Copper	T–0.0003	T–0.0005
Boron	T–0.007	T–0.004
Rubidium	T–0.0001	T–0.0004
Phosphate	0.02–0.05	0.003–0.090
Sulfate	0.003–0.035	0.003–0.22
Silicic acid	0.0002–0.005	0.0002–0.005
Chloride	0.001–0.010	0.001–0.060
Fluoride	T	0.0001–0.001
Iodide	T	T–0.001
Carbon dioxide	0	0.01–0.05[c]
Oxygen	T	T–0.00006

[a]Source of data: Hennig (1958), Amerine (1954, 1958), Ribéreau-Gayon and Peynaud (1958), Eschnauer (1959). [b]Except for botrytised grapes. [c]In normal still wine. About 0.1 is the beginning of gasiness. [d]Trace.

free SO_2 or 350 p.p.m. of combined SO_2 are permitted. Residues of fining agents are restricted to 2 p.p.m. Up to 200 p.p.m. of diethyl pyrocarbonate may be added during manufacture, but none of this compound is permitted to remain more than 120 hours after bottling.

Space will not permit further discussion of the many table wines made throughout the world, with one exception, champagne. True champagne comes from the Champagne region of France—and in that country carbonated wines originating outside that region are not permitted to be called "champagne" but must be labelled "mousseux".

For the details of champagne manufacture, the Selected References should be consulted. Briefly, the process is as follows: A white wine is first produced. After filtration, this undergoes a secondary fermentation in the bottle. After aging 2–3 years, the bottles are gradually inverted and the sediment allowed to collect in the neck. The neck is then frozen and the cork removed long enough for the plug of sediment to be blown out. Enough wine or liqueur, plus an amount of syrup gauged to the final sweetness desired, are added to refill the bottle, which is then corked and wired down.

This is the fermented-in-the-bottle process. Some American champagnes are made by a bulk process in which the product is manufactured in tanks and then bottled.

The terms *brut* or *nature* on a label are supposed to indicate very dry champagne; medium-sweet champagnes are marked *sec* or *demi-sec*; while the sweetest types are labeled *doux*.

We will close the subject of fruit-based undistilled beverages by referring briefly to the "wines" made from fruits other than the grape. Of these, the most important is the product of fermenting apple juice, *hard cider*. Historically the unqualified term *cider* meant *fermented* apple juice, as it still does in Great Britain and France (cidre). At the present time in the United States "cider" unqualified by "hard" is usually understood to mean *sweet* cider, i.e., unfermented apple juice. This is a popular beverage in the late fall, where it is largely produced for quick sale by pressing local apples in small countryside mills. Because the only preservative permitted in sweet cider is sodium benzoate, and this compound is relatively inefficient in preventing alcoholic fermentation, the cider cannot be stored unless it is pasteurized. Once it becomes fermented, it cannot legally be sold unless it complies with all the State and Federal tax requirements for alcoholic beverages. The only American hard cider the present authors have encountered has been a home-made product, but Amerine and Cruess report that commercial hard cider is produced on the West Coast.[5] Sparkling apple wines are produced to some extent, either by treatment of fermented apple wine with carbon dioxide or by tank or bottle fermentation in the manner employed in making champagne; the product resembles champagne in appearance and flavor.

In Canada as in Europe the unqualified term *cider* is legally defined as fermented apple juice; it is required by regulation to contain between 2.5 and 13.0% of alcohol by volume. Canadian regulations permit the addition of up to 10 g/100 ml of sucrose, dextrose or invert sugar to the juice before fermentation. The finished cider must contain not over 8 g/100 ml of total sugars (expressed as invert sugar), 2–12 g/100 ml total solids and 0.2–0.4 g/100 ml ash.

Perry is a fermented drink made from pears in the same manner as cider is made from apples. In France, Switzerland and Germany special varieties of pears of high tannin content are grown especially for making this beverage.[6]

Wines have been produced from a number of other fruits, such as berries, cherries, plums, pomegranates, pineapples, oranges, grapefruit and dried figs, dates and raisins. None of these—with the possible exception of blackberry wine—has an extensive sale. Elderberry wine used to be a common homemade beverage.

Fortified Wines

These are usually called "dessert wines" in order to distinguish them from the "table wines" previously described. The three types of wines most commonly encountered in this category are *Sherry* from Spain, *Port* from Portugal and *Madeira* from the Portuguese island of that name. Sherry and Madeira are white wines only; Port is usually red, but white types exist.

The original Sherry is produced only in a region in Spain around Jerez de la Frontera, and the name "Sherry" arose as a corruption of "Jerez". For a complete description of the process of Spanish sherry manufacture the reader is referred to Text Reference 3, pages 407–421.

Unique to production of Spanish sherry is use of the *Solera* system. This system consists of a series of butts of sherry in the process of aging, arranged in stages so that as wine is withdrawn from the oldest stage the amount withdrawn is replenished from the next-oldest stage, and so on through from 3 stages to 8 or 9 for the best *fino* wines. Spanish sherry as it goes on the market is therefore the result of blending many crops of different years. If necessary to obtain the desired alcohol content (15% or better), very high proof brandy is added, but the concentration from evaporation of water in the solera usually suffices.

There are six classes of Spanish sherries: *Fino*, a very pale, dry sherry of slightly pungent bouquet and flavor; *vino de pasto*, similar to *fino* but somewhat milder; *amontillado*, a *fino* that has been aged for a long while in wood, darker than a *fino*, very dry to slightly sweet; *manzanilla*, a dry *fino* of light color and delicate flavor, often not fortified; *oloroso*, darker than *fino*, usually fairly sweet, not aged under flor yeast film, sometimes called "golden" or "East India" sherry; and *amoroso*, similar to *oloroso* but quite dark in color.

One distinctive feature of sherry manufacture is the use of the so-called *flor* yeasts which form a film over the wine within a few weeks to a year or more after completion of fermentation. This film is permitted to grow undisturbed for years on the wines in all stages of the solera, and its autolysis is undoubtedly responsible for much of the characteristic flavor and bouquet of the wine.

Most California sherries are made to acquire their particular flavor by a procedure that is faster and, therefore, less expensive: the "baking" process. The filtered and fortified wine is heated in tanks at 130–140°F for 9 to 20 weeks—the longer periods of heating being at 130° and the shortest at 140°. California regulations require fortification to between 19.5 and 21.0% alcohol. Such sherries have flavors that somewhat resemble that of the true sherry from Spain, but the resemblance is not so close that the two cannot be readily distinguished by taste. California and European ports resemble each other much more closely than do the corresponding sherries.

Port acquired its name from the town of Oporto in Portugal whence it was shipped to English markets. The original port is grown in a region mainly bordering the Douro River. It is primarily a red sweet wine, although some white port is also made in Portugal from white grapes.

Wines of similar color and flavor are produced and sold as port in several other countries, including Australia, Chile, South Africa and the United States. Most of the port made in this country originates in the San Joaquin Valley in California. Several processes are used; details will be found in Text Reference 3, pages 440–447. The port is adjusted by blending to about 20% alcohol and 12–14% total extract before bottling.

Madeira is a term covering several white dessert wines produced on the Portuguese island of that name some 600 miles off the north coast of Africa. It formerly had a high repute but the introduction of various plant diseases to the island affected the ripening of the grapes and prevented long natural aging. There are three types of Madeira, called respectively *Sercial*, *Boal* (or *Bual*) and

Malmsey in order of increasing sugar content.

Other fortified "dessert" wines are *Angelica*, *Malaga*, *Marsala*, *Muscatel* and *Tokay*. *Angelica* is a smooth, fruity-flavored wine of full body and light to dark amber color of California origin. *Malaga* is a Spanish wine made from partially dried Muscat of Alexandria grapes; it is a dark amber wine with a caramel odor, containing 16–18% alcohol. It has a sugar content of 12–20%, which makes it one of the sweetest wines in the world. *Marsala* is produced in western Sicily; it is a manufactured rather than a natural wine, made by blending dry wine with grape concentrate, unfermented grape juice and reduced must.

At least 51% of California *Muscatel* is required by law to be derived from Muscat grapes; practically the only variety of these grapes used is Muscat of Alexandria. The finished wine is fortified to 20% alcohol and contains over 10% of sugar.

The original *Tokay* came from Hungary, where it was produced from a particular grape, the *Furmint*; the sweeter varieties (*Aszu* and *Tokay essence*) were made from grapes shriveled by *Botrytis* attack. Such Tokay was not a fortified wine. The wine sold as California Tokay, on the other hand, is usually a blend of one-third each of port, sherry and angelica.

Distilled Liquors

Numerous beverages have been made by fermenting fruits and other plant products and raising the alcohol content by distillation, but those having the widest distribution are brandy, gin, rum, whiskey and vodka.

Brandy

Brandy is the beverage resulting from the distillation of wine. As is the case with the term "wine", the product of the distillation of grape wine is called "brandy" unqualified, while similar products obtained from distillation of apricot and blackberry wines are known as "apricot brandy" and "blackberry brandy". Brandy was probably the first distilled liquor known to man, but nevertheless, whereas wine-making goes back to prehistoric times, it is probable brandy was unknown before the fourteenth century. The original idea of brandy as a medicine rather than a beverage is perpetuated in the pharmaceutical term *aqua vitae* and the French name *eau de vie*.

The most esteemed (and most expensive) of the brandies comes from the Charente and Charente Maritime departments of France, and is the only brandy that can legally be sold under the name *Cognac*. Another French brandy that has a reputation second only to Cognac is *Armagnac*, which is produced in an area in south-western France mainly in the department of Gers.

Brandy is made in many countries other than France. Next to France, Spain is the chief exporter to the United States. California produces several million gallons a year. Some of this brandy is aged in bond for four or more years and bottled at not under 100° proof; other portions are bottled at 84 to 88° proof, and may contain up to 2 per cent of flavoring material. A little *grappa* (pomace brandy) is also made.

Canada requires all brandy to be aged 2 years in wood.[2a]

Of the brandies made from fruits other than grapes, the greatest production is probably of apple brandy, but the total quantity that is consumed of even this beverage is far less than that of grape brandy. Apple brandy has been made in America since colonial times; it was originally manufactured by a double pot-still distillation (first distillation to 60° proof, second to 110–130° proof with head and tail cuts) from hard cider, but continuous column stills have largely superseded pot stills. The American product is usually called *applejack*; it is aged for a relatively short period in wood.

Gin

Gin is a colorless distilled beverage with an odor and flavor derived primarily from juniper. The original gin was produced in Holland by distilling a fermented mash of grain (primarily barley) with juniper berries and other aromatics; it had the character of new whiskey plus the aromatic flavor of the juniper and other flavoring ingredients. This type of gin is still sold on a relatively small scale under the names of "Holland gin" or "Schiedam gin".

Most of the gin now sold is of the type called "distilled gin", "English gin" or "London dry gin". It differs from Holland gin in that it is made by distilling pure alcohol diluted with water through a head containing trays, bags or other receptacles laden with juniper berries, coriander seeds, angelica and other aromatics; the vapors passing through these aromatics carry the flavoring principles into the condensate, which is then reduced with distilled water to the desired proof and bottled. There is no aging of the water-white product. Millions of gallons of such gin are sold annually, but very little is now consumed straight, most of it going into cocktails and other mixed drinks. (The original cocktail, the Martini, is a mixture of gin and Vermouth.)

Canadian regulations permit Holland gin to contain up to 2% of sugar; no sugar is permitted in dry gin.[2a]

Sloe gin is strictly speaking not a gin, but a sweetened reddish liqueur made from alcohol flavored with sloe berries from the blackthorn.

Rum

Rum is a beverage distilled directly from sugar-cane products; it has no other source. Rum is made chiefly in the islands of the West Indies, particularly Cuba and Puerto Rico, but also St. Croix in the Virgin Islands and Jamaica. Some rum is produced in the United States, and in colonial times the town of Medford in Massachusetts was famous for its rum. Jamaica is claimed to be the earliest commercial producer of rum, and the oldest rums may be found there.

Rum sold in Canada must be aged 2 years in wood.[2a]

Like gin, very little rum is consumed straight; most of it goes into mixed drinks such as the Bacardi, Cuba Libre, Daiquiri and Rum Collins.

Vodka

Vodka is a distilled liquor originating in Russia but also produced in Poland. Its principal source is wheat, but rye, barley, corn and potatoes have been used in its production. Malt is added to the cooked cereal mash to convert the starches to sugar, which is then fermented as in the production of whiskey. The resulting beer is distilled in column stills to produce pure or nearly pure neutral spirits, which are then reduced to about 100° proof. The resulting beverage has almost no odor or flavor beyond that of the alcohol it contains; it is not aged.

American vodkas are simply made by diluting neutral spirits and filtering through activated carbon; very small amounts of flavoring ingredients are added to some of them to give a weak anise-like flavor.

In recent years, the popularity of vodka in the United States has increased because of a belief that it cannot be detected on the breath as can whiskey or gin. In this country, it is consumed either as a substitute for other distilled beverages in mixed drinks, or simply "on the rocks", while in Russia it is drunk "straight".

Whiskey

Whiskey (or "whisky") is by far the most popular distilled beverage in the United States, and is approached only by gin as an ingredient of mixed drinks. Most whiskey is drunk either "straight" or mixed with plain or carbonated water or ginger ale. There are four main varieties; two of these (*Bourbon* and *Rye*) are made in the United States, while *Scotch* comes from Scotland and *Irish* whiskey from Ireland.

Scotch is no doubt the original whiskey, but it has undergone a number of changes since its origin. Up to about 1831 it was solely a pot still product, but since that time the continuous

Coffey still has been used to produce a distillate of 90–94% alcohol content (called "Scotch grain spirits") that is blended with pot still distillates to yield a liquor of less marked taste and aroma that is marketed (after aging 4 years or more) as "blended Scotch whisky". Practically all of the Scotch imported into the United States is of this type.

The smoky flavor is produced by allowing the green or partially dried malt to absorb some of the smoke from the smoldering peat used in drying the malt. This flavor is dissolved by the mash and passes over into the distillate. The Scotch grain spirit used for blending is essentially a very light-bodied corn whiskey (see under *Bourbon* below).

Irish whiskey is largely an all-malt product, but small amounts of rye, wheat and oats are used. It is mashed and fermented as in Scotland, but because no peat is used for drying the malt, there is no smoky flavor. Irish whiskey is bonded at 143° U.S. proof (71.5% alcohol by volume).

American whiskeys are made both in the United States and Canada. The legal definition is in part as follows:[7]

"Rye, bourbon, wheat, malt, rye malt, and corn are whiskies which have been distilled at not exceeding 160° proof from a mash containing not less than 51% of the grain which bears its name, and are called straight whiskies. The minimum age required in order to be classed as straight whisky is 2 years.

"The aging must be carried out in new, charred, white oak casks, except for corn whisky, which must be aged in plain, seasoned white oak barrels. The minimum proof for bottling whisky is 80°.

"Bottled-in-bond whisky . . . must be aged in charred barrels in bond for not less than 4 nor more than 8 years. It must be bottled at 100° proof and cannot consist of two or more different whiskies."

While this definition provides for six different types of straight whiskey, most of the American whiskeys actually encountered on the market are either straight rye, straight bourbon (both of these must legally be aged at least 2 years) or a blend of rye or bourbon with neutral spirits. Blended whiskeys are usually made from several lots of whiskey of varying ages.

Canadian regulations provide for a "Highland whisky", which is defined as a "whisky manufactured and blended in Canada consisting of a blend of malt whisky distilled in Canada or Scotland with grain whisky". This is required to contain at proof at least 25% of malt whisky.

Historically the early colonists in Maryland and Pennsylvania raised rye as the principal grain crop, so rye became the raw material for the first American whiskey. As a result of the "Whiskey Rebellion" against the taxes levied on this rye whiskey, some distillers and growers migrated to Kentucky and Indiana, where they started making whiskey from the principal grain crop of that region, which happened to be corn. Because corn whiskey was first distilled in Bourbon county, Kentucky, it came to be called Bourbon whiskey. Such is the political-economic foundation for the growth of the two main types of whiskey made in the United States. Rye whiskey, as its name indicates, is made primarily from rye, while Bourbon is simply corn whiskey that has been aged in charred oak barrels.

The process of whiskey manufacture is as follows: The grain is mashed into boiling water and cooked (frequently under pressure) until the starch granules are loosened and ruptured. The mash is then cooled and about 10% malt worked in to convert the starch to sugar. The resulting "sweet mash" is cooled to the optimum temperature for yeast action, a selected pure yeast culture is added and the mixture is allowed to ferment into a "distilling beer". (This process results in a "sweet mash" whiskey; "sour mash" whiskey is made by using spent beer or slop and barm from previous runs instead of yeast to start fermentation.) The fermented mash is usually distilled in continuous stills. After distillation, the whiskey is adjusted with water to the proper proof and placed in new or re-used charred white oak barrels to age. After the desired period of aging, the proof is adjusted with water and the product is bottled for sale.

Much has been made in advertising lately about whether a whiskey is "light-bodied" or "heavy-bodied". The term "light-bodied" (or just plain "light") has nothing to do with the alcoholic content; a "light" whiskey is one that is relatively low in flavoring ingredients ("congenerics"), color and extractive matter. If most of the by-products of fermentation are removed by distillation and rectification and the whiskey is stored in re-used, charred or plain casks in unheated warehouses, the resulting product is a "light" whiskey. If, on the other hand, most of the congenerics remain in the whiskey and it is stored in new charred casks in a heated warehouse, the product will be heavy-bodied and have more aroma and flavor than a "light" whiskey. Blending any straight whiskey with neutral spirits will, of course, also make the resulting product lighter bodied.[8]

Average analyses of American whiskeys are given in Table 2-2.[9]

In late years gas chromatographic analyses by several authors have identified some 50 individual constituents of the whiskey components listed as "esters, aldehydes and fusel-oil" in the table below.[9a,9b]

Other Distilled Beverages

There are several other distilled beverages made in various parts of the world, but these have only restricted regional distribution. The best known is *tequila*, which has a limited sale in the United States as well as in its country of origin. Production of tequila is limited to the State of Jalisco, Mexico, and concentrated in the town of Tequila, from which it takes its name. It is produced by fermenting a mash from the cooked central head and heart portion of the Tequila cactus (*Agave tequilana*), which is cultivated for that purpose; each head produces about 2 gallons of juice. A week or longer after fermentation, the mash is distilled in pot stills and rectified by re-

TABLE 2-2

Characteristics of American Whiskeys at Various Ages[a]

Age Yr.	Age Mon.	Proof	Total Acids	Fixed Acids	Esters	Aldehydes	Furfural	Fusel Oil	Solids	Color (Density)	Tannins	pH
	0	101.8	5.9	0.8	16.7	1.4	0.2	111	8.6	0.032	0.7	4.92
	1	101.4	20.4	3.7	17.2	2.1	1.2	123	44.1	0.156	12	4.62
	3	101.3	32.2	5.3	18.5	2.8	1.5	131	66.6	0.205	21	4.46
	6	101.4	42.5	6.6	21.8	3.3	1.6	131	87.7	0.243	28	4.38
1	12	102.0	53.4	8.3	26.8	4.1	1.7	132	111.1	0.282	35	4.38
	18	102.5	58.1	9.0	31.1	4.8	1.8	132	127.6	0.308	39	4.29
2	24	103.1	61.8	9.2	35.5	5.5	1.8	134	137.5	0.328	42	4.29
	30	103.6	64.1	9.3	38.9	5.8	1.9	136	147.7	0.341	44	4.28
3	36	104.1	65.8	9.3	41.8	6.0	1.8	135	152.7	0.352	47	4.27
	42	104.7	67.8	9.4	44.7	6.0	1.9	137	157.7	0.360	48	4.26
4	48	105.2	69.2	9.4	47.6	6.1	1.8	138	165.9	0.365	48	4.26
	54	105.5	69.7	9.4	48.0	6.1	1.7	...	166.0	0.367	49	4.26
5	60	106.0	70.2	9.5	51.9	6.2	1.7	...	173.0	0.368	49	4.26
	66	106.7	72.0	9.5	55.6	6.3	1.8	...	174.2	0.369	49	4.26
6	72	107.4	71.6	9.5	57.6	6.5	1.8	...	181.5	0.380	49	4.24
	78	107.9	74.4	9.6	61.2	7.0	1.8	...	186.0	0.385	50	4.34
7	84	108.6	76.2	9.7	62.0	7.0	1.8	...	198.6	0.389	50	4.23
	90	108.9	79.4	9.7	64.4	7.0	2.0	...	198.9	0.413	50	4.22
8	96	109.3	81.9	9.7	64.8	7.0	2.0	...	209.6	0.449	53	4.20

[a] All figures represent average values and are expressed as grams per 100 litres at 100 proof, except proof (expressed as degrees proof), color (expressed as density), and pH.

distillation. Tequila is usually sold colorless and unaged; it has an herbaceous weedy flavor.

A similar Mexican beverage of lesser quality is *Mescal*, which is made similarly to tequila but from another variety of *Agave* cactus; mescal is not redistilled.

Arrack is a Ceylonese beverage made by distillation of *Toddy*, which is the fermented juice of the flowers of several varieties of palm.

Okolehao is a Hawaiian beverage originally made by distillation of a mash made from the ti root, but the name has come to mean little more than "any distilled liquor produced in Hawaii".

Liqueurs (Cordials)

These are beverages produced by mixing or redistilling neutral spirits, brandy, gin or other distilled spirits with or over fruit, flowers, plants or their juices or other natural flavoring materials, extracts derived from infusions, percolations or macerations of such materials, and adding not less than 2.5% of sucrose and/or dextrose. (Liqueurs sold in Canada are also required to contain a minimum of 23% of alcohol by volume.) They are not supposed to contain synthetic or imitation flavoring materials, but they may be colored with caramel or a certified food dye if this fact is declared on the label. Most liqueurs are very sweet, often syrupy. As the name "cordial" indicates, they were originally produced centuries ago as "heart" remedies, but nowadays they are simply beverages, consumed either straight at the end of a dinner or as ingredients of certain cocktails.

One liqueur that is somewhat in a class by itself is *absinthe*, whose original chief flavoring ingredient was wormwood (*Artemisia absinthium*). Because of the deleterious effects of this liqueur on those who drank it regularly (presumably due to the thujone in the wormwood), its sale was forbidden in a number of countries. The green liqueur with an anise flavor now on the market under the name of "absinthe" is free of wormwood.

Methods of analysis

Odor and Flavor

At such time as all of the factors influencing the odor and flavor of alcoholic beverages are identified and the effects of various proportions of these factors on the overall odor and flavor are determined, it will be possible to establish the relative qualities of two wines (or brandies or whiskeys) by running the chemical determinations necessary to solve for x, y, z, etc., in a mathematical equation. Even with the modern help of gas chromatography, this time is still in the remote future.

Nevertheless, the purchaser who does not buy his alcohol solely for the "glow" resulting from its consumption is governed more by odor and flavor than by any other characteristic in deciding what brand he chooses. Consequently judgments on these factors must be made by manufacturers. Since chemical tests are unreliable bases for more than rough comparisons, taste tests by panels of expert tasters are resorted to. The details of such tests as they relate to wines are outlined in Amerine, Berg and Cruess's *The Technology of Wine Making*, 2nd Ed., pp. 678–684 (Westport: Avi Publishing Co. Inc., 1967).

Total Solids (Extract)

Amerine, Berg and Cruess[10] state that the extract (alcohol-free soluble solids) of a wine "is made up mainly of cream of tartar, malic, lactic, and succinic acids, glycerin, tannin, protein, and other nitrogenous compounds, and sugar. In low acid (high pH) wines there may be considerable neutral potassium tartrate," while "in sweet wines sugar predominates over the other constituents".

There are two common methods for determining extract. One of these, employing a Balling or Brix hydrometer (both of which are graduated in terms of percent by weight of sucrose), is relatively inaccurate but widely used in wineries. The other depends on evaporating a sample *in vacuo* and weighing the residue:

Method 2-1. *Extract by Specific Gravity.*[10]

Evaporate 100 ml of wine in a 250 ml beaker to 20 ml or slightly less on a hot plate or steam bath, taking care to prevent spattering or burning. Transfer to a 100 ml volumetric flask, cool, and make to volume with water. Pour into a hydrometer cylinder, float a Balling hydrometer in the liquid and take the hydrometer reading and the temperature. Correct the reading to 20° by Table 2-3:

This reading is in terms of percent by weight. To obtain the extract as g/100 ml, multiply by the specific gravity of the sample.

NOTE: Instead of evaporating a separate sample as above, the residue from an alcohol determination may be used if the sample was not made alkaline before distillation.

Method 2-2. *Extract by Weighing.*[10]

Pipet 25 ml of dry table wine or 10 ml of sweet dessert wine or sauternes into a weighed flat-bottomed evaporating dish about 65 mm in diameter, and evaporate on a steam bath to a syrupy consistency. Dry at 70° *in vacuo* (28–29″ vacuum) for at least 6 hours, cool in a desiccator and weigh. Subtract the weight of the evaporating dish and multiply the difference by 4 or 10,

TABLE 2-3

Corrections for Brix or Balling Hydrometers[a] Calibrated at 68°F. (20°C.)

Temperature of Solution		Observed Percentage of Sugar						
		0	5	10	15	20	25	30
Below Calibration								
°C.	°F.	Subtract						
15	59.0	0.20	0.22	0.24	0.26	0.28	0.30	0.32
15.56	60.0	0.18	0.20	0.22	0.24	0.26	0.28	0.29
16	60.8	0.17	0.18	0.20	0.22	0.23	0.25	0.26
17	62.6	0.13	0.14	0.15	0.16	0.18	0.19	0.20
18	64.4	0.09	0.10	0.11	0.12	0.13	0.13	0.14
19	66.2	0.05	0.05	0.06	0.06	0.06	0.07	0.07
Above Calibration		Add						
21	69.8	0.04	0.05	0.06	0.06	0.07	0.07	0.07
22	71.6	0.10	0.10	0.11	0.12	0.13	0.14	0.14
23	73.4	0.16	0.16	0.17	0.17	0.20	0.21	0.21
24	75.2	0.21	0.22	0.23	0.24	0.27	0.28	0.29
25	77.0	0.27	0.28	0.30	0.31	0.34	0.35	0.36
26	78.8	0.33	0.34	0.36	0.37	0.40	0.42	0.44
27	80.6	0.40	0.41	0.42	0.44	0.48	0.52	0.52
28	82.4	0.46	0.47	0.49	0.51	0.56	0.58	0.60
29	84.2	0.54	0.55	0.56	0.59	0.63	0.66	0.68
30	86.0	0.61	0.62	0.63	0.66	0.71	0.73	0.76
35	95.0	0.99	1.01	1.02	1.06	1.13	1.16	1.18

[a]Source of data: Association of Official Agricultural Chemists (1955).

according to whether a 25 or 10 ml sample was taken.

NOTE: For whiskeys use a 25 to 100 ml sample and dry the residue for 30 minutes at 100° without the use of vacuum.

Alcohol

Alcohol may be determined with an ebullioscope (which measures the boiling points of alcohol-water mixtures), by the specific gravity or index of refraction of the distillate, or by a chemical method such as dichromate titration of the distillate. These methods are as follows:

***Method 2-3.** Alcohol by Ebullioscope.*[11]

The most common type of ebullioscope is the Dujardin-Salleron, which may be obtained from Arthur H. Thomas Co., Philadelphia, Pa. or from Van Waters and Rogers, Los Angeles and San Francisco, Cal. This nickel plated copper apparatus consists of: (1) a boiling chamber into which is inserted the bulb of a Centigrade thermometer; (2) a metal reflux condenser; (3) an alcohol lamp; (4) a measuring cylinder; and (5) a slide rule with adjustable scales showing the relation between the boiling point and alcohol content. Directions for analyzing dry wine with this apparatus are as follows:

Place 25 ml of water in the boiling chamber, insert the thermometer and screw the reflux condenser into place without filling with water. Place the lighted alcohol lamp or a micro gas burner under the boiling spout of the ebullioscope and heat the water to boiling. Continue boiling until the Hg column in the thermometer remains constant. (This will be about 100° in regions near sea level.) Set the slide rule so that 0.0% alcohol is opposite the observed temperature. Repeat this adjustment once or twice a day, particularly if the barometric pressure is changing rapidly.

Now empty the ebullioscope, rinse it with a few ml of the sample and discard the rinsings. Pipet 50 ml of the sample into the boiling chamber, reinsert the thermometer, screw the condenser into place and fill it with cold water. Light the burner and heat until the boiling point registered on the thermometer becomes constant. Read the temperature and remove the flame. Read from the slide marked "Wine" the percentage of alcohol opposite the observed boiling point.

NOTES

1. When used at maximum efficiency, the above method can give results within ±0.25% of the true alcohol content. If the wine is first distilled and the distillate run in the ebullioscope, readings on the scale marked "Alcohol" can be accurate to ±0.15%.

2. With fermenting musts or sweet wines, the sample should be diluted to 5% alcohol or less in a volumetric flask before running on the ebullioscope; the result obtained is then multiplied by the appropriate factor.

***Method 2-4.** Alcohol from Specific Gravity of the Distillate.*[12]

APPARATUS

a. *Distillation apparatus*—A 500 ml round-bottomed (Kjeldahl) flask attached by an Iowa State type connecting bulb (Fisher No. 13–188 or equivalent) to a Liebig condenser with a jacket at least 400 mm long and inner tube of 9 ± 1 mm i.d. The condenser is mounted vertically.

b. *Pycnometer*—Of 25 or 50 ml capacity, preferably of the type with attached thermometer, such as Kimble No. 15123. This may be obtained as No. J–6456 from Will Scientific Inc., Rochester, N.Y. Standardize by weighing empty and filled with water at 15.56° (60°F).

DETERMINATION

For carbonated beverages such as beer or champagne, first expel the free CO_2 by pouring back and forth between two beakers until the liquid is comparatively quiet and free of foam. Other beverages need no previous treatment.

Pipet an appropriate quantity of the prepared sample (25 ml for distilled liquors and cordials, 100 ml for wines, ciders and malt beverages) into the distilling flask and dilute to 150 ml with water. Neutralize abnormally acid wines with 1*N* NaOH. Add a little antifoam if foaming is anticipated. Add a few glass beads and distil into a 100 ml volumetric flask nearly to the neck; make to volume with water and mix. Fill the pycnometer with the distillate and cool to 10–12° in a refrigerator. Then wipe off external moisture with a cloth and warm between the hands until the thermometer registers exactly 15.56°. At this point remove the drop of liquid at the top of the pycnometer with filter paper, and weigh to 0.1 mg.

$$\text{Sp.g. of distillate} = \frac{\text{Wt. distillate in pycnometer.}}{\text{Wt. water in pycnometer}}$$

Obtain from Table 23–7 the percentage of alcohol by volume in the distillate at 15.56° (60°F) corresponding to this specific gravity. If a 100 ml sample was taken, this is the percentage of alcohol in the sample; for a 25 ml sample it is necessary to multiply by 4.

NOTES

1. This is generally considered to be the most accurate method of determining alcohol. Preferably the figure so obtained should be checked by the refraction of the distillate (Method 2–5); if values by the two methods do not closely agree, the presence of methyl and isopropyl alcohols should be tested for.

2. If the distillate is not clear, essential oils are probably present and must be separated before alcohol can be determined. This can be done by saturating the distillate with NaCl, extracting with petroleum ether and redistilling the aqueous layer after sufficient dilution to prevent the salt from causing bumping. [See A. G. Woodman: *Food Analysis*. 4th ed. (1941) p. 477.]

Method 2-5. *Alcohol by Refraction.*

Measure the immersion refractometer reading at 15.56° (60°F) of a portion of the distillate from Method 2–4, and determine the corresponding percentage of alcohol from Table 23–8. *See* Note 1 above.

Method 2-6. *Alcohol in Wine by Dichromate Oxidation.*[13]

REAGENTS

a. *Standard potassium dichromate solution*—Dissolve 33.768 g of reagent grade $K_2Cr_2O_7$ in water and dilute to 1000 ml at 20°. (One ml ≎ 1% alcohol/ml wine.)

b. *Ferrous ammonium sulfate solution*—Dissolve 135.1 g of $Fe(NH_4)_2(SO_4)_2.6H_2O$ and 20 ml of H_2SO_4 in water and dilute to 1000 ml at 20°. Two ml of this solution are equivalent to 1 ml of the $K_2Cr_2O_7$ solution when first made up but because its titer slowly decreases, it must be compared daily with the $K_2Cr_2O_7$ solution.

c. *Indicator*—Dissolve 0.5 g of Ba p-diphenylamine sulfonate (Eastman No. 3104) in 100 ml of water, and filter.

DETERMINATION

a. *Distillation*—Place 25 ml of wine in a distilling flask such as a 300 ml Kjeldahl, add 50 ml of water and slowly distil at least 50 ml into a 250 or 500 ml volumetric flask containing about 100 ml of water. [Use a 250 ml flask for table wines (10–14% alcohol) and a 500 ml flask for dessert wines (20% alcohol).] Attach to the condenser a delivery tube that dips into the water in the receiver. When distillation is complete, dilute the distillate to volume with water at 20°. (If the wine contains more than 300 ppm of SO_2, neutralize before distilling.)

b. *Oxidation and titration*—Pipet 20 ml of the standard $K_2Cr_2O_7$ solution into a 500 ml glass-stoppered flask and add 10 ml of H_2SO_4. Agitate and cool under running water to about room temperature, add 10 ml of the distillate (**a**) with gentle swirling, stopper tightly and let stand 15 minutes.

Add 325 ml of water, 35 ml of 85% H_3PO_4 and 1 ml of the indicator, mix and titrate with the $Fe(NH_4)_2(SO_4)_2$ solution to the color change from blue violet to light green. Calculate the percentage of alcohol by volume in the wine as follows:

Vol. % alcohol = $20 \times (1 - n/N) \times$ (ml distillate collected)/ml distillate oxidized × ml wine distilled), where n = volume reducing solution used, and N = volume reducing solution equivalent to 20 ml of the standard $K_2Cr_2O_7$ solution.

Where 25 ml of wine have been distilled to 250 ml and 10 ml of the distillate have been oxidized, this reduces to the formula:

$$\text{Vol \% alcohol} = 20\left(1 - \frac{n}{N}\right).$$

Other Methods. A field test for determining alcohol in distilled liquors that depends on the distribution of this compound between immiscible solvents is given in AOAC Method ***9.018–9.021.***

The "Proof" of Alcoholic Beverages

This term originated in England in the days before exact chemical analysis was common, when a simple test was desired to identify distilled beverages for tax purposes. "Proof spirits" were then defined as liquids of such alcoholic concentration that gunpowder wet with them would just burn. It is because of this historical definition that British law defines proof spirits as containing 57.10% of alcohol by volume. The United States, for the sake of simplicity, has abandoned any attempt to perpetuate an exact relation between the alcoholic concentration of proof spirits and their ability to wet gunpowder and has established 50% of alcohol by volume at 60°F as being 100 proof. Therefore, the "proof" of any beverage can be obtained by doubling its alcoholic concentration, expressed as percent by volume at the standard temperature.

Acidity

Method 2-7. *Total Acidity.*[14]

a. *Distilled liquors*—Neutralize about 250 ml of boiled water in a porcelain evaporating dish (about 7½" dia.) with 0.1*N* NaOH, using 2 ml of phenolphthalein indicator (0.5% in alcohol). Add 25 ml of sample and titrate with the 0.1*N* NaOH. Express as g/100 ml of acetic acid (1 ml 0.1*N* NaOH = 0.006005 g of CH_3COOH).

b. *Malt beverages*—Proceed as (**a**) except to heat the 250 ml of water to boiling, add 25 ml of the previously-decarbonated and filtered beer sample from a fast-flowing pipet, heat 30 seconds more and cool to room temperature before titrating. Then add 0.5 ml of phenolphthalein indicator and titrate against a white background to the first faint pink. Report as lactic acid to the nearest 0.01% (1 ml 0.1*N* alkali = 0.0090 g lactic acid).

c. *Wines*—Remove CO_2 if present by placing 25 ml of sample in a flask and either agitating 1–2 minutes while under vacuum from a water aspirator, or heating to incipient boiling for 30 seconds, swirling and cooling. Then to 200 ml of hot, boiled water in a wide-mouthed Erlemmeyer flask add 1 ml of phenolphthalein indicator and neutralize with 0.1*N* NaOH. Pipet in 5.00 ml of the degassed sample and titrate with the standard alkali. Report as g/100 ml of tartaric acid (1 ml 0.1*N* NaOH = 0.00750 g tartaric acid).

NOTE: Beers too dark to give a good endpoint with phenolphthalein even when diluted should be titrated potentiometrically (*see* AOAC Method **10.024**).

Method 2-8. *Volatile Acidity.*[15]

APPARATUS

Cash still. See Figure 2-1. This all Pyrex apparatus consists of outer and inner chambers, a trap, a two-way stopcock, an electric coil heater and a "T" inlet-outlet for water. After distillation has been completed, the residue in the inner chamber is flushed out automatically by vacuum action when the current is shut off. The device may be obtained from Braun-Knecht-Heimann Co., 1400 Sixteenth Street, San Francisco, California.

DETERMINATION

Connect the Cash still to a vertically mounted Liebig condenser. Place 500–600 ml of water in the boiling flask, open the sidearm and heat the water to boiling for 2–3 minutes. Reduce the heat and place a 250 ml Erlenmeyer flask (marked at the 100 ml level) under the outlet of the condenser. Pipet 10 ml of wine or degassed beer into the inner tube and insert the stopper connecting the still to the condenser. Increase the heat, bring the water to boiling and while it is boiling vigorously close the sidearm. Continue boiling until 100 ml of distillate have been collected. Open the sidearm and turn off the heat.

To the distillate add 3–4 drops of phenolphthalein indicator and titrate to a definite pink with 0.025*N* NaOH. Ml 0.025*N* NaOH × 0.015 = g acetic acid/100 ml sample.

NOTES

1. For distilled liquors, volatile acidity is usually determined as the difference between total acidity and the fixed acidity obtained by evaporating the sample to dryness, taking the residue up in water and titrating. (*See* AOAC Method **9.031**.)

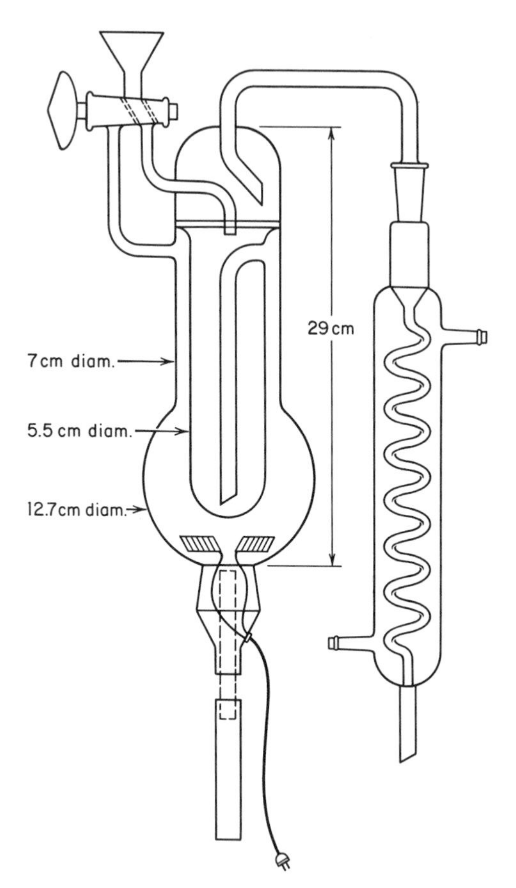

Fig. 2-1. Volatile Acid Still (Cash Still).

2. If it is necessary to correct for the acidity due to SO_2, AOAC Method **11.037** (which depends on the insolubility of $BaSO_3$) can be followed.

3. The volatile acidity is a measure of the soundness of fermentation of new wine and the keeping quality of old wines. An appreciable rise in volatile acidity during storage indicates bacterial spoilage, and calls for immediate pasteurization or the addition of 100–150 p.p.m. of SO_2, or both. Sound new wine should show less than 0.05 g/100 ml of volatile acidity; the corresponding figure for sound aged wine is 0.08 g/100 ml.[16]

Method 2-9. *pH.* (*AOAC Method **11.032**.*)

Let the pH meter and the glass and calomel electrodes warm up before use according to the manufacturer's instructions. Check the meter with a freshly prepared saturated aqueous solution of K bitartrate. Adjust the meter to read 3.55 at 20°, 3.56 at 25°, or 3.55 at 30°.

Rinse the electrodes free of bitartrate by dipping in H_2O and then in the sample. Place the electrodes in a fresh portion of the sample, determine the temperature and read the pH to the nearest 0.01 unit.

Esters and Aldehydes

Method 2-10. *Esters and Aldehydes in Whiskey.* (Determination of esters follows Liebmann and Scherl: *Ind. Eng. Chem.* **41,** 535 (1949); the method for aldehydes is the same as *AOAC Method **9.036***).

Reagents

a. *Sodium thiosulfate standard solution*—0.05*N*. Prepare by diluting with boiled water just before use a 0.1*N* solution made as follows: Dissolve about 25 g of $Na_2S_2O_3.5H_2O$ in 1 liter of water, boil gently 5 minutes and transfer while still hot to a storage bottle previously cleaned with hot H_2SO_4-$K_2Cr_2O_7$ solution and rinsed with warm boiled water. Store in a dark, cool place. Standardize as follows: Accurately weigh 0.20–0.23 g of NBS Standard Sample $K_2Cr_2O_7$ (dried 2 hours at 100°) and place in a glass-stoppered iodine flask. Dissolve in 80 ml of water containing 2 g of KI. Add, swirling, 20 ml of approximately 1*N* HCl (104 ml/liter), immediately place in the dark, and let stand 10 minutes. Titrate with the $Na_2S_2O_3$ solution, adding starch solution after most of the I has been consumed.

$$\text{Normality} = \frac{\text{g } K_2Cr_2O_7 \times 100}{\text{ml } Na_2S_2O_3 \times 49.032}$$

b. *Iodine solution*—Approximately 0.05*N*. Weigh 6.35 g of I_2 and 10 g of KI into a glass-stoppered flask, dissolve in a small amount of water, wash into a liter volumetric flask and make to volume.

c. *Sodium bisulfite solution*—Approximately 0.05*N*. Dissolve 2.6 g of $NaHSO_3$ in water, add 100 ml of alcohol to retard deterioration and dilute to 1 liter with water. Discard after one week.

Preparation of Sample

To 200 ml of sample in a 500 ml all-glass distilling apparatus, add about 35 ml of water and a few grains of carborundum. Distill slowly into a 200 ml volumetric flask until the distillate is nearly at the mark. Dilute to the mark and mix.

DETERMINATION OF ESTERS

Transfer 100 ml of the distillate to a 500 ml flask, add a measured excess (2–10 ml) of $0.1N$ NaOH, and let stand 24 hours at room temperature. Titrate the excess of alkali with $0.1N$ H_2SO_4. Express the results as g ethyl acetate/100 ml sample at 100 proof. 1 ml $0.1N$ $H_2SO_4 \approx 0.008806$ g ethyl acetate.

DETERMINATION OF ALDEHYDES

Place the remaining 100 ml of distillate in a 500 ml flask, add about 100 ml of water and an excess of the $NaHSO_3$ solution, and let stand 30 minutes, shaking occasionally. (The excess of $NaHSO_3$ should be equivalent to about 25 ml of the I solution.) Add an excess of the I solution, and titrate this excess with the standard $Na_2S_2O_3$ solution. Run a blank containing the same quantities of I solution and $NaHSO_3$ solution as were used with the sample. The difference between the titrations in ml $Na_2S_2O_3$ solution $\times 1.1$ = mg acetaldehyde in 100 ml of sample. Report as g acetaldehyde in 100 liters at 100 proof.

Furfural in Whiskey and Hydroxymethylfurfural in Wine

For these determinations see respectively AOAC Method **9.046–9.047** and Text Reference 3, pp. 724 and 715.

The presence of hydroxymethylfurfural in dessert wines is an indication either of heat treatment or of the presence of grape concentrate.

Fusel Oil

Method 2-11. *Fusel Oil in Whiskey.*[17]

REAGENTS

a. *Standard fusel oil*—Mix 4 volumes of *iso*amyl alcohol with 1 volume of *iso*butyl alcohol.

b. *Standard fusel oil solution*—Weigh 1.00 g of the standard fusel oil into a liter volumetric flask and make to volume with 15% alcohol. 1 ml = 1 mg fusel oil.

c. *p-Dimethylaminobenzaldehyde solution*—Dissolve 1 g of p-dimethylaminobenzaldehyde in 100 ml of alcohol.

PREPARATION OF SAMPLE

Place 25 ml of sample in a 500 ml round bottomed flask, add 0.5 g of Ag_2SO_4 and 1 ml of H_2SO_4 (1+1), and dilute to 110 ml with water. Reflux gently 15 minutes, then make the solution alkaline with 5 ml of NaOH solution (1+1), and reflux 30 minutes more. (Bumping can be prevented by the addition of a little granulated Zn; if foaming occurs, add 15 g of NaCl.)

After saponification distil the liquid and collect 75 ml of distillate. This distillate contains all of the higher alcohols at one third of the original concentration. (The extreme sensitivity of the color reagent makes this reduction in concentration necessary.)

DETERMINATION

Place 2.00 ml of the distillate in a 125 ml flask and add 20.0 ml of H_2SO_4, swirling the flask in a bath of cold water during the addition. Then add 2.00 ml of the p-dimethylaminobenzaldehyde solution while swirling in the cold bath.

Prepare a similar flask containing 2.00 ml of the standard fusel oil solution, acid and reagent.

Place the flasks simultaneously in a bath of vigorously boiling water. After 20 minutes transfer the flasks to the cold bath. When cool, add 25 ml of H_2SO_4 (1+1) to each flask and mix thoroughly by swirling. Read the transmittances at 541 $m\mu$ against a reagent blank.

If x = mg fusel oil in the 2 ml of distillate, T = transmittance of the distillate and T' transmittance of the standard solution, since the 2 ml of standard contains 2 mg of fusel oil,

$$x = \frac{2T'}{T}, \text{ and mg fusel oil/ml distillate} = \frac{T'}{T}.$$

Because of the 1:3 dilution in preparation of the sample, mg fusel oil/ml sample = $3T'/T$. This value $\times$ 100 = g fusel oil/100 liters of sample.

NOTES

1. This method is similar to AOAC Method **9.037–9.039**.

2. The term "fusel oil" is applied to the combined alcohols higher than ethyl alcohol. While natural fusel oil contains other alcohols than those

used in preparing the above standard, these two are the predominant ones and occur in approximately the proportions employed for the standard.

3. Fusel oil values in Table 2-2 were obtained by the old Allen-Marquardt method involving extraction with CCl_4 and oxidation with H_2CrO_4, which yields figures about 65% as high as those of Method 2–11. This should be taken into consideration when using Table 2-2 to interpret results on whiskeys analyzed by Method 2–11.

Sugars

See Chapter 16. Alternatively, wines may be analyzed by the Lane-Eynon method as described in Text Reference 3, pp. 703–704.

Tannin

Method 2-12. *Tannin in Wine and Spirits.*[18]

REAGENTS

a. *Folin-Denis reagent*—To a mixture of 100 g of $Na_2WO_2.2H_2O$ (Folin grade), 20 g of phosphomolybdic acid and 50 ml of H_3PO_4 add 750 ml of water, reflux 2 hours, cool and dilute to 1 liter.

b. *Sodium carbonate solution*—Saturated. Add 35 g of Na_2CO_3 to 100 ml of water at about 80°. Allow to cool over night, and seed with a few crystals of $Na_2CO_3.10H_2O$.

c. *Standard tannic acid solution*—100 mg/liter of water. Prepare daily.

DETERMINATION

Pipet 1 ml of sample into a 100 ml Nessler tube containing 80 ml of water, add 5 ml of the Folin-Denis reagent and 10 ml of the Na_2CO_3 solution, shake well and make to the mark. After 30 minutes, compare the color with those developed from 0.0, 0.2, 0.4, 0.6, 0.8, 1.0, 1.2, 1.4, 1.6, 1.8, 2.0 and 2.4 ml portions of the standard tannic acid solution treated in the same way as the sample.

Fixed Acids

Method 2-13. *Tartaric Acid in Wine.*[19]

APPARATUS

a. *Gas Chromatograph*—Micro Tek Model 220 or equivalent, with a flame ionization detector.

b. *Column*—6′ × $\frac{1}{4}$″ o.d. glass column containing 3.8% SE-30 on Diatoport S, operated isothermally at 130°, with inlet and detector temperatures set at 180°.

c. *Syringe*—Hamilton No. 705 microliter syringe.

REAGENTS

a. *Lead acetate solution*—Dilute 75 g of Pb acetate and 1 ml of acetic acid to 250 ml with water.

b. *Internal standard solution*—Dissolve 0.8 g of undecanoic acid (Eastman No. 1238) in pyridine and dilute to 200 ml (20 mg/5 ml). Add a few chips of Drierite to this solution.

c. *Sulfuric acid*—1*N*. Dilute 28.4 ml of H_2SO_4 to 1 liter with water.

d. *Hexamethyldisilazane and trimethylchlorosilane*—*See* Method 16–11, Reagent (**c**), (**i**) and (**ii**), and note.

DETERMINATION

Pipet 15 ml of wine into a 250 ml centrifuge bottle, and dilute with water to about 100 ml. Add 3 ml of the 1*N* H_2SO_4 and 20 ml of the Pb acetate solution. Then add about 0.2 g of Celite 545, shake vigorously about 2 minutes, and centrifuge. Decant and discard the supernatant liquid. Wash the Pb salt in the centrifuge bottle twice with 100 ml portions of 80% alcohol, shaking, centrifuging, decanting and draining each time. Heat the residue in the bottle at 100° until moisture-free (about 2 hours).

Suspend the dried Pb salt in 5 ml of the internal standard solution, (**b**), and break up the precipitate with a glass rod. To this suspension add, in order: A few chips of Drierite, 1 ml of trimethylchlorosilane and 1 ml of hexamethyldisilazane. Stir about 2 minutes to complete the reaction, and then decant the mixture into a small 5 ml polyethylene-stoppered vial, and centrifuge. Chromatograph 20μl of the supernatant in a 50 ml/minute current of helium.

Carry out the same procedure for 5, 10, 15 and 20 mg of known fixed acids (tartaric, malic, citric and succinic), and plot peak height ratios of TMS derivatives to internal standard *vs.* concentrations of the fixed acids.

Inject the wine samples twice and the standards three times, and identify the fixed acids in the sample by comparison with the retention times of the known acids.

NOTES

1. To confirm the identities of the fixed acids in their wine samples, the authors of this method ran additional chromatograms on a similar sized glass column containing 20% Apiezon L on 60–80 mesh Chromport XXX. This column was operated isothermally at 160°, with inlet and detector temperatures both 225°.

2. The method was tested on grape wines and those made from a number of other fruits (apple, blackberry, grapefruit, honey, orange, peach, plum and strawberry). All were found to contain malic acid (0.008–0.102 g/100 ml), and many contained succinic and citric acids, but only the grape wines contained any tartaric acid (0.051–0.190 g/100 ml). The presence of tartaric acid in a wine sold as blackberry wine, for instance, is therefore a definite indication of adulteration with grape wine.

3. The Drierite is used to remove traces of water that would prevent the trimethylsilylation reaction from going to completion.

4. By this method it is possible to determine concentrations of fixed acids as low as 1 mg/100 ml.

Thujone

For a qualitative test for this compound as an indication of the presence of wormwood in absinthe, *see* AOAC Method **9.091**.

Sulfur Dioxide

Method 2-14. *Total Sulfur Dioxide.*

Employ Method 14-2, using 100–300 ml of wine. Report as mg SO_2/liter. (Because the SO_2 in wine is unstable, give the sample no preliminary degassing and keep exposure to air to a minimum.)

Method 2-15. *Free Sulfur Dioxide in Wine.*[20]

To 50 ml of wine add 10 ml of H_2SO_4 (1+3). Quickly add a few drops of starch indicator and titrate to the first permanent blue color with 0.02N iodine. Ml iodine consumed × normality of the iodine × 32 × 20 = p.p.m. free SO_2.

NOTE: This determination is of some value in the wine industry. In champagne manufacture a free SO_2 content exceeding 10 p.p.m. may interfere with secondary fermentation. When *Botrytis cinerea* develops extensively on grapes, it secretes a polyphenol oxidase that will cause browning of white wines; in such cases undue darkening is prevented by maintaining a fairly high level of free SO_2. It is also the level of free rather than total SO_2 that affects the flavor of a wine.

Carbon Dioxide

Method 2-16. *Carbon Dioxide in Wine.* (*AOAC Method* **11.056–11.057**).

REAGENT

Carbonic anhydrase solution—Prepare an aqueous solution containing about 1 mg enzyme/ml. This solution is stable about 2 weeks in the refrigerator.

Carbonic anhydrase may be obtained from Calbiochem, Box 54282, Los Angeles, California 90054, as their No. 2157.

DETERMINATION

Cool the sample to 0° or below, so that it can be pipetted without loss of CO_2. With an automatic 25 or 30 ml pipet with a Teflon stopcock, dispense an aliquot of 0.1N NaOH into a beaker. Rinse a 20 ml pipet with the sample to prevent warming of sample and possible loss of CO_2. Then pipet the sample into the beaker with the tip of the pipet submerged just below the surface of the NaOH solution. Add 3–4 drops of the enzyme solution and place the beaker under glass and calomel electrodes (Beckman Nos. 41263 and 40463 are satisfactory). Titrate to pH 8.45 with 0.1N H_2SO_4 from a 5 ml burette graduated in 0.01 ml.

To correct for the presence of acids other than H_2CO_3, place 50 ml of the cooled wine in a 500 ml heavy-walled flask and agitate at room temperature under a vacuum of about 27″ for 1 minute. Then titrate 20 ml to pH 7.75 with 0.1N NaOH as above. Subtract ml used in this titration from that used in the first titration.

(Net ml NaOH × normality − ml H_2SO_4 × normality) × 100 × 44/ml sample = mg CO_2/100 ml wine.

Caramel

Method 2-17. *Mathers Test.* (*AOAC Method* **11.045–11.046**).

REAGENTS

a. *Pectin solution*—Dissolve 1 g of pectin in 75 ml of water and add 25 ml of alcohol as a preservative. Shake well before using.

b. 2,4-*Dinitrophenylhydrazine solution*—Dissolve 1 g of 2,4-dinitrophenylhydrazine in 7.5 ml of H_2SO_4 and dilute to 75 ml with alcohol. (This solution will remain clear and stable several months if kept in a glass-stoppered bottle.)

TEST

Place 10 ml of a filtered sample of beer, cordial, distilled liquor or wine in a Babcock cream bottle or a centrifuge tube. Add 1 ml of the pectin solution and mix; add 3–5 drops of HCl and mix; fill the bottle or tube with alcohol (about 50 ml), mix, centrifuge and decant. Dissolve the precipitate in 10 ml of water and repeat the additions of HCl and alcohol; shake well, centrifuge and decant. Repeat until the alcoholic liquid is colorless. Finally, dissolve the gelatinous residue in 10 ml of hot water. If the solution is colorless, caramel is absent; if it is clear brown, caramel may be present. Confirm by adding 1 ml of the 2,4-dinitrophenylhydrazine solution, mixing and heating 30 minutes in boiling water. If a precipitate forms, caramel is present.

NOTE: For many years the usual test for caramel in distilled liquors was the Marsh test. This consisted of shaking the sample with *iso*amyl alcohol containing a little dilute H_3PO_4; color in the lower layer indicated caramel.[21] False positives were, however, obtained in the presence of extractives from uncharred oak chips (as well as coal-tar dye), and this led to the devising of several more specific tests of which the above Mathers test is considered the most reliable and most widely applicable.

Methyl and Isopropyl Alcohols and Acetone

Qualitative Tests. Methanol and isopropanol may be tested for by a number of methods. Those found particularly convenient by one of the present authors are the method of the U.S.P., X (1925), page 39, for methanol, and the method of Leffman and Pines, *Am. J. Pharm.*, **102**, 41 (1930), for isopropanol. This latter method with omission of the $KMnO_4$ solution serves as a test for acetone. These tests had their chief use in Prohibition days when the composition of alcoholic beverages on the "bootleg" market was always suspect but they still find occasional use in checking possible contamination with denatured alcohol.

The methanol test is designed not to give positive tests for the traces naturally occurring in some wines.

Method 2-18. *Methanol Determination.* (*AOAC Immersion Refractometer Method* ***9.055***).

Determine the Zeiss immersion refractometer reading at 17.5° of the distillate obtained in the determination of alcohol, Method 2–4. If, upon reference to Table 2-4, the refractometer reading corresponds to the specific gravity obtained by Method 2–4, it may be assumed that no methyl alcohol is present. A low refractometer reading indicates the presence of an appreciable proportion of methyl alcohol. If absence from the solution of refractive substances other than water and the alcohols is assured, such a difference in refraction is conclusive evidence of the presence of methanol.

Addition of methanol to ethyl alcohol decreases the refractive index in direct proportion to the quantity added; hence a quantitative calculation can be made by interpolation in Table 2–4 of the figures for pure ethyl and methyl alcohols of the same specific gravity as the sample.

NOTE: The AOAC *Methods of Analysis* formerly contained an accurate gravimetric method for methanol that was based on the fact that tetramethylammonium iodide is insoluble in absolute alcohol while the corresponding quaternary derivative of ethyl alcohol is relatively soluble.[22] This method was dropped from *Methods of Analysis* for reasons unrelated to its accuracy, and while it requires special apparatus, it might be well for analysts interested in the accurate determination of low percentages of methanol in mixtures of the two alcohols to consul the paper cited in Text Reference 22. AOAC Method **9.051–9.054** gives a colorimetric method for this determination.

Diethyl Pyrocarbonate

Method 2.19. *Diethyl Pyrocarbonate in Wine.*[23]

REAGENTS

a. *Pentane-ether mixture*—Mix in the proportion of 105 volumes of n-pentane to 75 volumes of ether.

b. 4-*Aminoazobenzene*—Eastman No. 1375 p-phenylazoaniline.

c. *Di-isobutylamine solution*—0.1N in monochlorobenzene. Dissolve 12.92 g of di-isobutylamine (Eastman No. 1388) in chlorobenzene and make to 1 liter with this solvent.

d. *Bromophenol blue indicator*—0.04%. Grind 0.1 g of bromophenol blue (Eastman No. 752 3′, 3″, 5′, 5″ tetrabromophenolsulfonphthalein) in a mortar with 3 ml of 0.05N NaOH, transfer to a volumetric flask, and dilute to 250 ml with water.

QUALITATIVE TEST (by paper chromatography)

Extract 500 ml of the wine twice with 75 ml portions of the pentane-ether mixture. To the combined extracts, add 50 mg of 4-aminoazobenzene and 0.1 ml of acetic acid; concentrate to about 1 ml *in vacuo*. Apply one drop of the orange concentrated solution to the starting point of a Whatman No. 1 filter paper and develop over night with 15% acetic acid (*see* Method 16–53 (e).) If 10 p.p.m. or more of diethyl pyrocarbonate are

TABLE 2-4

Scale Readings on Zeiss Immersion Refractometer at 17.5°, Corresponding to specific Gravities of Ethyl and Methyl Alcohol Solutions

Sp. Gr. 15.56°/15.56°	Scale Readings		Differences	Sp. Gr. 15.56°/15.56°	Scale Readings		Differences
	Ethyl Alcohol	Methyl Alcohol			Ethyl Alcohol	Methyl Alcohol	
1.0000	15.0	15.0	0.0	.9720	51.5	27.0	24.5
.9990	15.8	15.3	0.5	.9710	53.0	27.5	25.5
.9980	16.6	15.6	1.0	.9700	54.6	28.1	26.5
.9970	17.5	15.9	1.6	.9690	56.1	28.7	27.4
.9960	18.5	16.2	2.3	.9680	57.6	29.2	28.4
.9950	19.4	16.5	2.9	.9670	59.1	29.6	29.5
.9940	20.4	16.9	3.5	.9660	60.6	30.1	30.5
.9930	21.4	17.2	4.2	.9650	62.0	30.6	31.4
.9920	22.5	17.5	5.0	.9640	63.3	31.0	32.3
.9910	23.6	17.9	5.7	.9630	64.6	31.5	33.1
.9900	24.7	18.2	6.5	.9620	65.8	31.9	33.9
.9890	25.9	18.6	7.3	.9610	67.0	32.4	34.6
.9880	27.1	19.0	8.1	.9600	68.1	32.8	35.3
.9870	28.4	19.5	8.9	.9590	69.2	33.3	35.9
.9860	29.6	19.9	9.7	.9580	70.2	33.7	36.5
.9850	31.0	20.4	10.6	.9570	71.2	34.1	37.1
.9840	32.4	20.8	11.6	.9560	72.1	34.5	37.6
.9830	33.8	21.3	12.5	.9550	73.0	34.9	38.1
.9820	35.2	21.8	13.4	.9540	73.8	35.3	38.5
.9810	36.7	22.3	14.4	.9530	74.6	35.6	39.0
.9800	38.3	22.8	15.5	.9520	75.4	35.9	39.5
.9790	39.9	23.4	16.5	.9510	76.2	36.2	40.0
.9780	41.5	24.0	17.5	.9500	76.9	36.5	40.4
.9770	43.1	24.5	18.6	.9490	77.6	36.8	40.8
.9760	44.8	25.0	19.8	.9480	78.3	37.0	41.3
.9750	46.5	25.5	21.0	.9470	79.0	37.3	41.7
.9740	48.2	26.0	22.2	.9460	79.7	37.6	42.1
.9730	49.8	26.5	23.3				

Scale readings are applicable only to instruments calibrated in arbitrary scale units proposed by Pulfrich, *Z. angew. Chem.*, 1899, p. 1168. According to this scale, 14.5 = 1.33300, 50.0 = 1.34650, and 100.0 = 1.36464. If instrument used is calibrated in other arbitrary units, refractive index corresponding to observed reading can be converted into equivalent Zeiss reading by referring to Table 23-8.

present, a yellow spot of 4-azobenzene ethyl carbonate is left at the origin; unchanged aminoazobenzene rises upward to an Rf value of 0.54.

DETERMINATION (By titration)

Extract 250 g sample with three 25 ml portions of the pentane-ether mixture. To the combined extracts add 3 g of NaCl, 200 ml of acetone and 25 ml of the 0.1*N* di-isobutylamine, and titrate the excess base with 0.1*N* HCl to a bromophenol blue endpoint, shaking during the titration.

P.p.m diethyl pyrocarbonate = 16,200,000 V/E, where E = mg sample, and V = ml 0.1*N* di-isobutylamine consumed.

NOTES

1. This method is suitable for determining up to 20 p.p.m. of diethyl pyrocarbonate.

2. U.S. Food Additive Regulation 121.1117 permits the addition of up to 200 p.p.m. of diethyl pyrocarbonate to still wines as a fermentation inhibitor before or during bottling, but any wine so treated must be free of this compound within 5 days of bottling. As noted on page 33, the requirements of the Canadian regulations, while somewhat differently expressed, are identical in their effect.

The Age of Whiskey

A problem that may present itself to the analyst is verification of the labeled age of a whiskey. No chemical analysis can answer this question with precision, but it may be possible to identify cases of gross misrepresentation. Many blended whiskeys are labeled as containing mixtures of several straight whiskeys of different ages with grain spirits; for such whiskeys the most that may be done is to estimate what may be called the "diluted average age", i.e., the figure that would be obtained by multiplying the age of each straight whiskey by the fraction of that whiskey present in the sample and adding the results.

The latest published analyses of authentic American whiskeys of known age that can be used as a basis for estimating the age of a suspected sample are those of Liebmann and Scherl: *Ind. Eng. Chem.* **41**: 534 (1949); Table 2-2 is taken from that article. These analyses represent 469 different barrels of American straight bourbon and rye whiskeys.

The analyst wishing to check the age of a sample of whiskey is advised to study both Table 2-2 and the separate curves showing variation with age for each constituent that can be found in the original article. Note particularly that Liebmann and Scherl state "The increase in esters with age is probably the most regular and consistent of all the characteristics and therefore may be regarded as the most reliable index for determination of age."

Recent gas chromatographic analyses of de Becze *et al.* have confirmed that 8-year-old Bourbon contains about 19 times as much ethyl acetate and 10 times as much ethyl formate as the unaged article (Text Reference 9a, Table 3). These authors also found 3-4 times as much *iso*butyl alcohol in rye whiskey as in Bourbon.

TEXT REFERENCES

1. Jacobs, M.B. (ed.): *The Chemistry and Technology of Food and Food Products*. 2nd ed. New York: Interscience Publishers, Inc. (1951) III 2382.
2. Ibid. 2390.

2a. *Canadian Food and Drug Regulations, Division* 2, *Alcoholic Beverages*, *B*.02.001—*B*.02.135 (3rd September 1964).

3. Amerine, M. A., Berg, H. W. and Cruess, W. V.: *The Technology of Wine Making*, 2nd ed., Westport: Avi Publishing Company (1967) 544.
4. Jacobs: Op. cit. II, 892.
5. Amerine, M. A., Berg, H. W. and Cruess, W. V.: Op. cit. 527–531.
6. Ibid. 540.
7. Jacobs: Op. cit. 2431. Vol. III.
8. Ibid. 2430–2431. Vol. III.
9. Liebman, A. J. and Scherl, B.: *Ind. Eng. Chem.* **41,** 536 (1949).

9a. De Becze, G. I., Smith, H. F. and Vaughn, T. E.: *J. Assoc. Offic. Anal. Chem.*, **50,** 311 (1967).
9b. Kahn, J. H., La Roe, E. G. and Conner, H. A.: *J. Food Science*, **33,** 395 (1968).
10. Amerine, M. A., Berg, H. W. and Cruess, W. V.: Op. cit. 701 and 690.
11. Ibid. 696.
12. Leach, A. E. and Winton, A. L.: *Food Inspection and Analysis.* 4th ed. New York: John Wiley and Sons Inc. (1920) 47, 687.
13. Guymon, J. F. and Crowell, E. A.: *J. Assoc. Offic. Agr. Chemists*, **42,** 393 (1959).
14. Horwitz, W. (ed.): *Official Methods of Analysis of the Association of Official Agricultural Chemists.* 10th ed. Methods **9.030, 10.023, 11.033.**
15. Amerine, M. A., Berg, H. W. and Cruess, W. V.: Op. cit. 693–695.
16. Ibid. 693.
17. Penniman, W. B. D., Smith, D. C. and Lawshe, E. I.: *Ind. Eng. Chem. Anal. Ed.* **9,** 91 (1937).
18. Amerine, M. A., Berg, H. W. and Cruess, W. V.: Op. cit. 709.
19. Brunelle, R. L., Schoeneman, R. L. and Martin, G.E.: *J. Assoc. Offic. Anal. Chemists*, **50,** 329 (1967).
20. Amerine, M. A., Berg, H. W. and Cruess, W. V.: Op. cit. 708.
21. Horwitz: Op. cit. Method **9.057.**
22. Wilson, J. B.: *J. Assoc. Offic. Agr. Chemists*, **18,** 477 (1935).
Bailey, E. M.: *J. Assoc. Offic. Agr. Chemists*, **19,** 192 (1936).
23. Dept. of Health, Education and Welfare: Food Additives Analytical Manual. Washington: Government Printing Office.

SELECTED REFERENCES

A Amerine, M. A., Berg, H. W. and Cruess, W. V.: *Technology of Wine Making.* 2nd ed. Westport: Avi Publishing Company Inc. (1967).
B Don, R. S.: *Teach Yourself Wine.* New York: Dover Publications Inc. (1968).
C Jacobs, M. B. (ed.): *The Chemistry and Technology of Food and Food Products.* 2nd ed.: New York: Interscience Publishers Inc., (1951). III. Chap. 51.
D Johnson, H.: *Wine.* New York: Simon and Schuster (1967).
E Lichine, A., Fifield, W., Bartlett, J. and Stockwood, J.: *Alexis Lichine's Encyclopedia of Wines and Spirits.* New York: Alfred A. Knopf. (1967).
F Simon, A. (ed.): *Wines of the World.* New York: McGraw-Hill Book Company (1967).
G Street, J.: *Wines, Their Selection, Care and Service.* New York: Alfred A. Knopf (1933).
H Underkofler, L. A. and Hickey, R. J.: *Industrial Fermentations.* New York: Chemical Publishing Company Inc. (1954). I.

CHAPTER 3

Non-Alcoholic Beverages and Concentrates

Carbonated beverages constitute by far the largest volume of foods sold in this category, excluding fruit juices and concentrates, which are discussed in Chapter 11. Many of these are sweetened with artificial sweeteners instead of sugar and sold as low calorie or dietetic drinks. They are made by drawing a measured amount of the combined ingredients (called a "throw") into a container, filling with chilled carbonated water and capping or sealing.

The finished beverage contains from 2.0 to 5.0 volumes of carbon dioxide, as measured at 760 mm pressure and 60°F. Under these conditions, water will absorb 1 volume of CO_2. Club soda and ginger ale usually have the highest carbonation (up to 4.5 or 5.0 volumes), while the fruit-flavored soft drinks may be carbonated to only 2 or 3 volumes. The sugar content of those beverages sweetened with sugar ranges from 8.5° to around 14° Brix.

The U.S. Food and Drug Administration has promulgated a definition and standard of identity for "soda water".[1] This requires a CO_2 content of at least 1 volume at 760 mm and 60°F. If it is a "cola" or "Pepper" beverage it shall contain caffeine not to exceed 0.02%. It may be sweetened with one or more of the following nutritive sweeteners: sugar, dextrose, corn syrup, glucose syrup and sorbitol.

The permitted acidulants are acetic, adipic, citric, fumaric, lactic, malic, phosphoric and tartaric acids. Certain buffering salts and emulsifying, stabilizing or viscosity-producing and foaming agents are permitted. Up to 83 parts quinine per million in the finished beverage may be added. Preservatives are permitted under proper labeling, and in the case of canned soda water, stannous chloride may be used as a preservative in amounts up to 11 parts per million, calculated as tin (Sn). Artificial and/or natural flavors, colors and artificial sweeteners may be used if declared on the label.

Cola drinks characteristically contain extract of the kola nut (*cola acuminata*) along with various other flavor ingredients such as citrus oils, spice oils, vanilla, etc. Some contain only the caffeine normally found in kola extract; in others, additional caffeine is obviously present. Quinine has also been found in some cola drinks. Beattie[2] gives several formulations, as does Merory[3].

Beverage bases in the form of a dry powder are sold in consumer-size packages for "instant use." These may produce a still or a carbonated finished beverage. If the latter, sodium bicarbonate, calcium carbonate or other carbonates are used as the source of CO_2 in conjunction with an excess of powdered acid such as citric, tartaric, fumaric, malic, hexose diphosphoric, etc. Some require sugar to be added, others are artificially sweetened. A mixture of cyclamate and saccharin in the ratio of 8 or 10:1 seems to be preferred over either sweetener alone.

Analysis

A preliminary examination is always helpful. Note and record whether the beverage is clear or cloudy, or if there is any sediment, characteristic odor or taste that can be recognized, or any synthetic flavor or sweetener that can be identified organoleptically.

Method 3-1. *Preparation of Sample.*

If the beverage is carbonated, chill before opening; then pour it back and forth several times between two large beakers to remove CO_2 before proceeding with the analysis.

If a dry beverage base is under examination, weigh the dry sample, and then dilute according to label directions for making the finished beverage. Report on either the basis of the dry product or the finished beverage, as circumstances dictate.

Methods for analysis for many constituents of beverages may be found in Chapters 1 and 11. Total solids are usually determined by use of a Brix spindle on the degassed liquid; ash is determined by ignition at 525–550°. This analysis is usually of value in fruit-base beverages, as is the determination of potassium and phosphorus (Methods 1–17, 1–18 and 1–19). A convenient method for potassium determination is by a flame photometer. Estimation of the fruit juice content may be obtained by comparison with analyses of authentic fruits given in Chapter 11, Section III and a review of "Interpretation of Analytical Results" in Section II.

Total acidity may be determined by titration, either to phenolphthalein end point or electrometrically (Method 11–18). Unless it is known that only one acidulant is present, a determination of pH is more significant than acidity by titration. Individual fruit acids may be determined by Methods 11–8, 11–9, 11–10. Phosphoric acid, when in combination with other acids, may be estimated by adding excess NaOH solution, ashing and determining phosphates by Method 1–18, 1–19 or by the classic phosphomolybdate (yellow precipitate) method. A method for determining fumaric acid will be given later in this chapter.

Sugars: As a general rule, total sugar after inversion is the only significant determination, since, if sucrose had been used initially, it will have been totally or partially inverted by the acids present. Sugars after inversion may be determined by polarization or by Munson-Walker chemical methods. Consult Chapter 16. Qualitative tests should be made for artificial sweeteners, and if present, they can be determined quantitatively according to methods given in Chapter 14.

An indication of the use of pectin, gums, carboxymethylcellulose for thickening purposes (especially if the beverage is artificially sweetened) may be obtained by determining alcohol-insoluble precipitate by Method 11–17.

Volatile oils may be isolated from citrus-flavored beverages by steam distillation by Method 11–5. Terpeneless citrus oils are frequently used in citrus-flavored soda waters. Grape beverages may be flavored either with a grape concentrate or artificially with methyl anthranilate. This compound is present in Concord grape juice itself only in amounts averaging 2 mg/liter or up to about 3.8 mg/liter. Therefore, amounts above 2 mg/liter or thereabouts in a grape beverage are an indication of artificiality. One of us[4] has found as high as 13 mg/liter of methyl anthranilate in grape sodas. Method 11–11 may be used for its estimation.

Stannous chloride may be used as a preservative for carbonated beverages packed in tins. It may be determined by Method 11-18. Tests should be made for other chemical preservatives. See Chapter 14.

The degree of carbonation, measured as volumes of CO_2 dissolved in 1 volume of carbonated beverage, is an important index of quality. This may be measured by a special gauge called a "gas volume tester", devised by the laboratory of the National Soft Drink Association.[5] This device clamps the bottle in a frame, and punctures the cap, and the gauge pressure is read in lbs/sq. in. This is converted to volumes of CO_2 from a table accompanying the instrument. Details may be obtained from the Technical Director of the Association.

Fumaric Acid

This acid has been approved by FDA as "generally recognized as safe" in amounts reasonably required to accomplish its intended effect. It is included as an optional acidulant in the FDA standard for soda water. Its rather low solubility, 0.5% at 20° and 0.33% at 10° in water, may limit its use. Two forms are available; one is virtually 100% fumaric acid, the other contains 0.3% dioctyl sodium sulfosuccinate, DSS, to make it more rapidly soluble. DSS has been approved by FDA as a food additive in fruit juice drinks and dry beverage bases in amounts that will not furnish more than 10 parts per million DSS in the finished beverage, which means that a finished beverage should not contain more than 0.33% modified fumaric acid.[6]

Fumaric acid may be determined in fruit beverages and beverage bases by Method 2–13, substituting a 5 ml or 5 g sample for the 15 ml of wine, and running standards on fumaric acid instead of the specified tartaric, malic, citric and succinic acids.

See also the polarographic method by Smith and Gagan.[7, 8]

Caffeine

Caffeine is permitted in cola beverages up to 0.2%. The following method was developed by Vessiny, of Truesdail Laboratories, Inc., Los Angeles, California:[9]

Method 3-2. *Caffeine in Cola Beverages.*

Standard Preparation

Dissolve a suitable quantity of previously dried caffeine in water and dilute quantitatively to prepare a solution containing 10 micrograms of caffeine in each ml.

Determination

Place 25 ml of the beverage in a 100 ml beaker and warm for 5 minutes in a steam bath, stirring occasionally. Cool to room temperature. Pipet 5.0 ml into a 50 ml volumetric flask and make to volume with distilled water. Pipet 5.0 ml of this solution into a 60 ml separatory funnel containing 10 ml of chloroform. Shake vigorously for 1 minute and allow to settle. Transfer the solvent layer to another 60 ml separatory funnel. Repeat extraction twice more. Wash the combined extracts with 5 ml of water by shaking vigorously in the separatory funnel. Transfer the solvent to a 50 ml Erlenmeyer flask by passing through a pledget of cotton, add 5 drops of HCl-acidified alcohol (1+9) and evaporate to dryness on a steam bath. Extract a 5.0 ml aliquot of the standard preparation in a similar manner. Dissolve the residues in 5.0 ml of distilled water and use as the sample and the standard solution. Determine, concurrently, the absorbance of the sample and standard solutions at 252, 272, and 292 millimicrons in 1 cm cells with a suitable spectrophotometer using distilled water as the blank. Designate the absorbance from the sample solution as Au and that from the standard solution as As.

Calculate the quantity in micrograms of caffeine in the portion of the beverage taken by the formula:

$$5C \times \frac{Au\,272 - \frac{1}{2}(Au252 + Au292)}{As\,272 - \frac{1}{2}(As252 + As292) \times 2V},$$

where

C = micrograms/ml of the standard preparation, and

V = volume in ml of the beverage taken before dilution (in this case 5 ml).

Note: This formula uses the base-line technique. It allows for UV determinations in the presence of background absorption. It is assumed that this background absorption is a relatively straight line between the wavelength extremes—in this case 252 and 292 mμ.

In essence this technique is a means of measuring the height of a peak for comparison with a standard, in the presence of background. Without such use one would have to measure absorbance at the maximum in order to determine concentration, and this procedure would yield a high result when UV-absorbing impurities were present.

Quinine

Method 3-3. *Quinine in Carbonated Beverages.*

This method, developed by MacLean[10], is included through the courtesy of the Canada Dry Corporation. They use the method successfully on both "quinine water" and "bitter lemon" beverages.

Apparatus

Beckman DU spectrophotometer or equivalent capable of measuring absorption at 347.5 mμ.

Determination

Degas the beverage as directed in Method 3–1. Pipette 50 ml of degassed liquid into a 100 ml volumetric flask. Add 20 ml of 0.5*N* HCl and bring to mark with water, Mix, and determine absorption at 347.5 mμ using a 1 cm path length cell. Determine mg of quinine present corresponding to the observed optical density by reference to a standard graph.

Preparation of Standard Graph: Weigh accurately 100 mg of reagent grade quinine (dried for 3 hours at 100°) into a 100 ml volumetric flask, dissolve in 20 ml of 0.5*N* HCl, bring to mark and mix. (1 ml = 1.0 mg quinine alkaloid.) Transfer separately 0.5, 1.0, 2.0, 3.0, 4.0, 5.0 and 6.0 ml into seven 100 ml volumetric flasks and proceed as above, beginning "Add 20 ml 0.5*N* HCl and bring to mark. . . ." Plot absorbance against mg quinine on semi-log paper. A straight line should result.

The usual concentration of quinine in commercial beverages is up to 120 p.p.m. (6 mg/50 ml), so this procedure should accommodate all commercial carbonated beverages. The standard graph should be established by the analyst for the equipment and reagent used. It should then be checked periodically.

Safrole

Method 3-4. *Safrole in Carbonated Beverages.*[11]

Safrole, now prohibited in this country as being carcinogenic, is a major constituent (up to 85%) of oil of sassafras. It and methyl salicylate are the principal flavoring ingredients of root beer, sarsaparilla and birch beer. We have found the following method by Wilson to be quite satisfactory:

Determination

Decarbonate the beverage as directed in Method 3–1. Transfer 250 ml to the distilling flask of an all-glass steam distillation assembly, and add 10 ml of ethanol. Collect 150 ml of distillate in a receiving flask containing 20 ml of ethanol. The delivery end of the condenser should extend below the surface of the alcohol. The receiver should be marked at the 170 ml level.

Disconnect the steam and the condenser, and wash down the condenser tube with three 5 ml portions of ethanol, letting the tube drain completely between washings. Transfer the distillate and washings into a 200 ml volumetric flask, wash out the receiver with a little alcohol, bring to room temperature and fill to mark with water.

Pipette a 50 ml aliquot into a 125 ml glass-stoppered Erlenmeyer flask fitted with a reflux air condenser, add 5 ml of about 0.1*N* NaOH solution and reflux on a steam bath for 20 minutes, rotating the flask occasionally to bring the alkaline solution in contact with all inner surfaces of the flask. Remove from the bath, and cool to room temperature. Transfer contents to a 125 ml separatory funnel, and wash out the air condenser and flask with about 25 ml of $CHCl_3$, adding the $CHCl_3$ washings to the separator.

Stopper, and shake gently for 1 minute, using a horizontal rotary motion to avoid formation of emulsions. After separation of the layers, draw off the lower $CHCl_3$ layer through a small filter paper in a funnel into a 100 ml volumetric flask. Wash out the stem of the separatory funnel with a few ml of $CHCl_3$, and add to the volumetric flask through the filter. Extract with 3 more portions of $CHCl_3$ (25, 20, 20 ml) and add these in turn to the flask through the filter. Bring to the mark with $CHCl_3$. Retain the extracted aqueous portion for determination of methyl salicylate, if desired, as directed later.

STANDARD ABSORPTION DATA

Prepare a standard solution containing 50 mg of safrole in 100 ml of $CHCl_3$, and transfer 1, 2, 4 and 6 ml into a series of 4 volumetric flasks (0.5, 1, 2 and 3 mg safrole). Bring each flask to the mark with $CHCl_3$, mix and read the absorbances in a UV spectrophotometer at 287 and 308 mμ using matched quartz cells against a $CHCl_3$ blank. Record the difference between the readings at each wave length for each solution. Calculate the difference value corresponding to 1 mg safrole/100 ml solution for each, and calculate the reciprocal of the average value.

Estimation of Quantity of Safrole in Sample. Read off the $CHCl_3$ extract of the distillate at 287 and 308 mμ against a $CHCl_3$ blank. Multiply the difference between the two readings by the reciprocal value obtained above and calculate to p.p.m. safrole in the original sample.

NOTE: Wilson reported the safrole content of commercial root beers to be within the 10–20 p.p.m. range.

Determination of Methyl Salicylate

This determination is seldom required. To conserve space, we give it by reference only. It may be determined on the aqueous portion remaining after $CHCl_3$ extraction of safrole. This portion is evaporated to dryness, and salicylate determined colorimetrically by reaction with a ferric salt. See Text Reference 11.

Composition of Dry Beverage Bases

Purchasing specification MIL-B-35023C, Beverage Base, Powdered, issued by Department of Defense February, 1967, typifies the various types of these products now on the market. This specification covers 3 classes:

(A) Without full complement of sugar, (B) with full complement of sugar, (C) artificially sweetened.

These may be flavored with dehydrated fruit juice (orange or lemon) or with encapsulated natural or imitation flavors. They may contain added ascorbic acid and vegetable colors or U.S. certified food colors.

These products are blends of the flavor base, food acids, gums, corn sirup solids, sorbitol or sugars and at times tricalcium phosphates, which the Federal Specification permits up to 1.0%. In Class C, sugars and other carbohydrates are replaced in whole or part by artificial sweeteners.

TABLE 3-1
Analytical Requirements—Dry Beverage Bases

Ingredient	Class A	Class B	Class C
Moisture, maximum	7.0[a]	7.0[a]	4.5[a]
Essential oil (in orange juice bases) minimum	0.66[b]	0.082[b]	—
Essential oil (in lemon juice bases) minimum	0.25[b]	0.032[b]	—

[a] Per cent. [b] ml/100 g.

The acidity requirements determined on the finished beverage, diluted according to directions, range from 0.26% to 0.48% for the fruit juice based type to 0.2% to 0.4%, expressed as citric acid, for those products containing no fruit juice.

The method used for these determinations are described elsewhere in this text:

Essential Oils	Method 15–6
Moisture	Method 16–2
Acidity (glass electrode titration)	Method 11–8

Dietetic Beverages

One other class of non-alcoholic beverages that should be considered in this chapter is the dietetic beverage known under various names such as "900 calorie", "Instant Meal", etc., designed to furnish a balanced nutritional diet but with minimum caloric intake. These are complex mixtures fortified with vitamins and minerals and, possibly, other nutritional elements, but with minimal carbohydrates. A typical product consists of dehydrated skim milk, whey solids, milk proteins or other dairy concentrate, yeast and/or hydrolyzed vegetable protein, some form of fat, such as partially hydrogenated soy oil or peanut oil, a sugar plus an artificial non-nutritive sweetener, flavor (cocoa, vanillin, etc.), a thickener such as Irish moss, carboxymethyl cellulose, etc. and other ingredients to make it palatable. It is practically impossible to duplicate such a product from a quantitative analysis alone. Considerable trial-and-error formulation must follow.

"900 Calorie" Beverages

The Connecticut Agricultural Experiment Station made a market survey in 1962 of these products on sale in that state[12]. Both complete beverages and dry bases were sampled.

The liquid products declared the ingredients on a "per quart" basis. The analyses are therefore reported on that basis. The water content of all 7 liquid products tested was close to 800 g/quart, and the calorie declaration ranged from 720 to 1024/quart. These, except for one 13% below labeled calories, were all within 4% of the labeled figure.

This Experiment Station also examined 15 "low calorie" dry mix bases. All but one of these, which claimed to provide 1440 calories/8 oz. but actually furnished 1080, claimed to provide 900/8 oz. and did so. The moisture content of these dry mixes ranged from 5.7 to 8.4 g/8 oz.

These products varied somewhat in composition, but the general pattern showed them to be mixtures of non-fat milk solids and/or soy protein, yeast, milk protein, fat (at times partially hydrogenated), sugars, along with flavor, artificial sweeteners and added vitamins.

TABLE 3-2
Composition of "900 Calorie" Beverages

	Liquids, g/quart	Dry Mixes, g/8 oz.
Moisture	789–826	5.7– 8.4
Ash	10.0–17.0	12.6–19.5
Protein (NX6.25)	57.6–73.6	66.5–75.5
Fiber	0.2– 1.8	0.0– 2.6
Lactose	53.4–107	32.1–110
Total Carbohydrate	87.4–145	108–119
Fat	5.9–35.9	15.7–26.2
Calories	813–1050	880–1081

These products were also fortified with added vitamins. Vitamin D was present in declared amounts in both liquid and solid products. There was a tendency to under-declare the amount of Vitamin A present in both types. This amounted to as high as 240% excess of the declared amount, usually 5000 units of Vitamin A. Water-soluble vitamins in general were closer to label declaration, with the exception of one liquid product which contained only 33% of the declared thiamine content. We are assuming here that an excess of water-soluble vitamins, which may run as high as 100%, presents no health hazard, but is an economic factor only.

Fruit Beverages (See also Chapter 11)

Definitions and Standards: Standards of identity for fruit juices and juice concentrates are discussed in Section 1 of Chapter 11. No Federal standards have been issued for the various diluted fruit juices sold as nectars, ades, etc. Historically, this area of legislation in this country has been left to the individual states.

The Association of Food and Drug Officials of the United States (AFDOUS), in an effort to promote uniformity among the states, has suggested standards of identity for the more common fruit juice beverages. These were incorporated into a "Statement of Policy," published by AFDOUS.[13] The standards have been officially adopted by several states, and are used by all states as guide lines in enforcing state food laws.

AFDOUS recognizes the commercial practice of using canned or frozen juice concentrates, along with single-strength juice and water, to produce the finished beverage. To these fruit constituents may be added organic acids, ascorbic acid up to 40 mg/100 ml, sweeteners, true fruit flavors, preservatives and coloring which do not simulate the color of the natural fruit juice. As a guide to manufacturers in diluting these concentrates to natural strength juices, AFDOUS has defined "average concentration of single-strength juices in degrees Brix" as follows:

TABLE 3-3
Average Concentration of Single-strength Juices in Degrees Brix

Name of Fruit	Degrees Brix
Apple	13.3
Apricot	14.3
Blackberry	10.0
Boysenberry	10.0
Cherry	14.3
Grape (Labrusca)	14.3
Grape (Vinifera)	22.0
Guava	7.7
Loganberry	10.5
Orange	11.8
Peach	11.8
Pineapple	14.3
Plum	14.3
Cranberry	6.5

Lemon is defined as containing 5.7 g anhydrous citric acid per 100 ml.

AFDOUS has defined these various fruit beverages in the same Statement of Policy.[13] All of these definitions permit addition of the non-fruit ingredients (organic acids, ascorbic acid, etc.) mentioned above. These will not be repeated in the definition. The term "fruit juice" in these definitions means the natural strength juice, fresh or reconstituted, with soluble fruit solids as listed in Table 3–3.

AFDOUS Definition of Fruit Beverages

A. *Nectar*: A blend of pureed fruit pulp or whole fruit with fruit juice and water such that the finished product contains not less than 40% fruit ingredient. Exceptions: Apricot 35%, guava 25%, orange 50%, papaya $33\frac{1}{3}$%. In addition, orange nectar shall have at least 0.5145 lb of juice soluble solids per gallon, 11.8° Brix basis.

B. *Juice Drink*: A blend of juice with not more than an equal volume of water, except that cranberry juice may be calculated at twice its volume in determining if the product satisfies the 50% by volume requirement.

C. *Ade*: A blend of juice and water such that the finished product contains not less than 25% juice by volume, except that cranberry-, lemon- or lime-ade shall not be less than 12.5% by volume. Orange ade shall contain at least 0.257 lb of orange juice soluble solids per gallon, 11.8° Brix basis.

D. *Punch*: A blend of 2 or more juices with water, such that the finished product contains not less than 10% juice by volume.

E. *Drink*: Same as Ade except that cranberry-, lemon- or lime-drink shall contain not less than 6.0% juice. (0.342 g anhydrous citric acid per 100 ml obtained solely from lemon or lime juice.) Orange drink shall contain not less than 0.103 lb orange juice solids per gallon, 11.8° Brix basis.

F. *Flavored Drink*: Same as "Drink."

Methods and Interpretation of Analysis

The methods given in Chapter 11, Sections I and II, are in general applicable to these fruit beverages. The suggestions on interpretation given in Section II may also be helpful. Calculation of fruit content from the ash and ash constituents (K_2O, P_2O_5) alone are usually not sufficient, since these may be sophisticated by adding appropriate salts. Total acidity, determination of individual fruit acids and ascorbic acid give valuable information.

Determination of chlorides and sulfates is helpful in detecting the addition of mineral salts, since pure fruit juices contain very small amounts of these ingredients. Fruit ash in general contains 50–55% K_2O and 6–12% P_2O_5. If K_2O is significantly below 50% of the ash, addition of non-fruit, inorganic salts are indicated (other than potassium salts). Corn syrup made by acid hydrolysis is sometimes used as a sweetener; it may contain 0.3–0.4% ash, usually NaCl. Chlorides in pure juices are not over 5 or 6 mg/100 ml.

The formol titration for amino acids and the determination of polyphenolics, as developed by a group at the U.S.D.A. Western Regional Laboratories, are helpful in detecting adulteration of lemon juice. The ratio of citric acid to amino acids should be around 40 to 50. A higher figure indicates addition of citric acid. We include in this text a modified version of the formol titration for total amino acids as used by the Tennessee Division of Food and Drugs.[14]

Method 3-5. *Free Amino Acids.*

PREPARATION OF SAMPLE

Mix about 5 or 6 g Celite analytical filter aid with 175 ml of the sample (about 80–100 milliequiv. acid/100 ml sample). Filter with suction and, if not completely clear, refilter through a fresh Celite pad. Store in a glass-stoppered flask.

DETERMINATION

Pipete a 20 ml aliquot of the prepared sample into a 150 ml beaker. Partially neutralize by adding NaOH solution (1+1) drop by drop to pH 6–7, and continue the titration potentiometrically to pH 8.4. Adjust 37% formaldehyde to pH 8.4 by titrating potentiometrically not more than 1 hour before being used.

Add 20 ml of the adjusted 37% formaldehyde to the 20 ml sample and titrate the resulting acidity back to pH 8.4 as before. Calculate the formol index by the equation:

$$\text{Formol index} = \frac{\text{ml } 0.1N \text{ NaOH} \times 100}{\text{ml sample}}$$

Authentic lemon juice and orange juice both have an average formol index of 20. The Tennessee regulatory officials consider any figure materially below 20 as suspect and to be confirmed by other analyses.

Smotherman has also furnished the following limiting values by which Tennessee state officials judge these listed juices or beverages[14]:

TABLE 3-4
"Scale of Values" for Interpretation of Analyses of Juices or Beverages

	Orange	Lemon	Lime	Grape	Cranberry
Ash, %	0.400	0.290	0.300	0.280	0.190
K_2O, mg/100 g	210	155	150	150	96.8
P_2O_5 mg/100 g	38	24.6	27	26.7	9.5
Acidity as citric, %	0.810	6.06	5.60		
Brix Degrees	11.8		8.6		10.7
Brix/acid ratio	10–1				
Formol Index	20	20			

Vandercook and co-workers at the USDA Western Regional Laboratories developed a method that has been adopted by AOAC for the determination of total polyphenolics in lemon juice. They also showed a relationship between l-malic acid, the citric acid content and the formol titration for amino acids. Since the exact quantitative composition of the polyphenolics is not known the assay is expressed as absorbance.[15, 16, 17, 18, 19]

Method 3-6. *Total Polyphenolics in Lemon Juice. (AOAC Method **20.077**).*

PREPARATION OF SAMPLE: See Method 3–1.

APPARATUS: *Recording spectrophotometer.*

CALIBRATION OF SPECTROPHOTOMETER

Weigh accurately two 1.4–1.5 g portions of KNO_3 into 250 ml volumetric flasks. Dissolve in water and dilute to the mark. Zero the instrument at 302 mμ and measure the absorbance of each solution, A, in a 1 cm cell. Calculate the standard absorbance for each solution:

$$A' = \frac{a \times \text{molarity } KNO_3\ (= 6.99 \times \text{g } KNO_3)}{101.11 \times 0.25}$$
$$= 0.2765 \times \text{g } KNO_3$$

Divide average A′ by average measured absorbance, A, at 302 mμ to obtain correction factor a.

DETERMINATION

Pipette 0.5 ml of prepared sample into a 10 ml volumetric flask and dilute to the mark with ethanol. Transfer to a centrifuge tube, cover with aluminum foil to prevent evaporation and centrifuge. Measure the UV spectrum of the supernatant liquid with a recording spectrophotometer from 300 to 400 mμ (or with a manual instrument at 2 mμ intervals from 325 to 335 mμ). Multiply the absorbance of the 323–335 mμ peak by the correction factor (a) and report as absorbance of total polyphenolics.

Rolle and Vandercook[16] gave the following characteristics of California-Arizona lemon juice, as compiled from analysis of 61 authentic juices:

Total amino acids	2.01 meq/100 ml
1-malic acid	3.41 meq/ml
citric acid	89.38 meq/100 ml
polyphenolics	0.645 absorbance units

TEXT REFERENCES

1. *Federal Register* **33**: 8593 (June 12, 1968).
2. Beattie, G. B.: *Perfumery and Essential Oil Record* **47**: (No. 12) 437 (1956).
3. Selected Reference D.
4. Fisher, H. J.: *Conn. Agri. Experiment Station Bull.* **617** (1955), **629** (1959).
5. Levine, M., Toulose, J. H. and Sharp, J. F.: *Product Uniformity in Bottled Carbonated Beverage Manufacture.* Washington D.C.: National Soft Drink Assn. (rev. 1957).
6. *Code of Federal Regulations, Title* 21—*Food and Drugs, Parts* 1 *to* 129. Washington, D.C.: Office of Federal Register, General Services Adm., (1966).
7. Smith, H. R. and Gagan, R. J.: *J. Assoc. Offic. Agr. Chemists*, **48**: 699 (1965).
8. Smith, H. R.: Ibid., **49**: 701 (1966).
9. Vessiny, R. E. (Truesdail Laboratories Inc.): Private communication (1967).
10. MacLean, J. S. (Canada Dry Corporation): Private communication (1967).
11. Wilson, J. B.: *J. Assoc. Offic. Agr. Chemists*, **42**, 696 (1959).
12. Fisher, H. J.: Unpublished report (1962).
13. *Quarterly Bull. Assoc. Food and Drug Officials of the U.S.* Proceedings Issue (Oct. 1963).
14. Smotherman, T. (Tenn. Div. of Food & Drugs): Private communication (1968).
15. Vandercook, C. E., and Rolle, L. A.: *J. Assoc. Offic. Agr. Chemists*, **46**: 359 (1963).
16. Vandercook, C. E., Rolle, L. A. and Ikeda, R. M.: *J. Assoc. Offic. Agr. Chemists*, **46**: 353 (1963).
17. Rolle, L. A. and Vandercook, C. E.: *J. Assoc. Offic. Agr. Chemists*, **46**: 362 (1963).
18. Yokayama, H.: *J. Assoc. Offic. Agr. Chemists*, **48**: 530 (1965).
19. Rockland, L. B. and Underwood, J. C.: *Anal. Chem.*, **26**: 1557 (1954).

SELECTED REFERENCES

A Bennett, H., ed.: *Chemical Formulary* (various eddition). New York: Chemical Publishing Co.

B Gardner, W. H.: *Food Acidulants.* New York: Allied Chemical Corporation (1966).

C Jacobs, M. B.: *Manufacture and Analysis of Carbonated Beverages.* New York: Chemical Publishing Co. (1959).

D Merory, J.: *Food Flavorings, Composition, Manufacture and Use.* 2nd ed. Westport: Avi Publishing Co. (1968).

E Sand, F. E. M. J.: *A Study in Beverage Shelflife.* Baltimore: Naarden-Flavorex Inc. (undated).

CHAPTER 4

Cereal Foods

Cereals are agricultural grains, usually of the grass family (*Gramineae*) although commercially buckwheat is included, grown for their edible seed. The cultivation of these grains predates historical record. Corn, the only cereal indigenous to this hemisphere, was introduced to the old world shortly after the discovery of America. Wheat seems to have developed from a wild grass native to Asia Minor. Rice is an oriental plant, cultivated in China as far back as 3000 B.C. Rye and oats are believed to have originated in Central Europe, barley in Central Asia and buckwheat in Northern Asia.

With the possible exception of rice, whole grains per se have a limited use as food and, in general, their analysis is considered to be beyond the scope of this text. Proximate analyses of grains and milled cereal products may be made by following the procedures in Chapter 1, and sugar determinations by the methods in Chapter 16. The AOAC directs that grains be ground to pass through a No. 20 sieve before analysis. Cereal Laboratory Methods of the American Association of Cereal Chemists (AACC) directs grinding wheat to a floury consistency, or so that at least 50% passes through a 36 grits gauze sieve prior to analysis. Most of the cereal-producing countries have issued official grades or standards for grains. The United States has issued official standards under the U.S. Grain Standards Act administered by USDA, and Canada has done the same under the Canada Grain Act, administered by the Canada Board of Grain Commissioners, Winnipeg, Manitoba.

Flours

Modern milling of cereal grains is a very complex operation. Its aim for wheat is the separation of the less desirable germ and kernel coats (bran) from the endosperm. This is accomplished by passing the tempered wheat through alternate series of break rolls and sifters and results in a series of flour streams containing progressively larger quantities of bran and germ. In practice, certain streams are combined to make the various grades and types of flour in demand for different purposes. Air-separation techniques are also used by some mills to fractionate the flour. Interested readers are referred to the books on milling technology listed in "Selected References" at the end of this chapter.

There are two general types of commercial wheats, hard (including durum) and soft. Hard wheat comprises about 70% of the United States production and almost 100% of the Canadian output. The hard wheats are most suitable for manufacture of bread flour; the soft wheats are used for the most part in production of flour for cakes, crackers, pastries, etc. Durum wheat is used almost entirely for the manufacture of semolina for pasta products (macaroni, spaghetti, noodles, etc.). Farina, the corresponding fraction from wheat other than durum, is also used for these purposes. Semolina and farina are more coarsely ground than other flours.

Millers use the term "extraction" to denote percent by weight of flours obtained from the wheat milled. Some confusion exists as to the base of this calculation, i.e., cleaned wheat of natural moisture content, 14% moisture content, tempered wheat, or some other basis. Wheat consists of about 84% endosperm, 14% bran and 2% germ; hence theoretical extraction is 84%. This is never obtained. Current yields are 70–84%. During war time, millers have been required by government fiat to extract as high as 80% flour. This lowered the quality of the flour but did conserve the food supply.

Official Standards

Standards of Identity have been issued for the various wheat flours by FDA.[1] These are given in Table 4-1. In addition, standards have been promulgated for cracked and whole wheat. These vary only in the granulation. They are milled so that the proportions of the natural constituents of the wheat remain unaltered, and contain not more than 15% moisture. Cracked wheat is milled so that 40% or more passes through a No. 8 standard sieve and less than 50% through a No. 20 sieve. Crushed wheat is milled so that not less than 90% passes through a No. 8 sieve and not more than 20% through a No. 20 sieve.

In 1965 a standard of identity was promulgated for Instantized Flours (instant blending flours, quick-mixing flours). This covered flour, enriched flour, enriched bromated flour, self-rising flour, enriched self-rising flour and phosphate flour, agglomerated so as to be readily pourable. All such flour shall pass through a No. 20 sieve, and not more than 20% shall pass through a No. 200 sieve.[1]

The Canadian standards for flour[2] are mainly quite similar to those of the United States. White flour must be bolted through a cloth equivalent to a No. 100 woven wire screen (149 microns). The ash content shall not exceed 1.20% moisture-free basis, and moisture shall not exceed 15%. They allow the same bleaching and aging agents, and in addition, they permit potassium bromate up to 50 p.p.m., ammonium persulfate up to 250 p.p.m. and ammonium chloride up to 2000 p.p.m. The limit for benzoyl peroxide is 150 p.p.m.

The list of permitted enrichments for enriched flour does not include Vitamin D; otherwise, the scale of enrichment is the same as in the United States. They also have defined a Vitamin B White Flour and Enriched Vitamin B White Flour.

The Canadian standard for Whole Wheat Flour includes an ash requirement of not less than 1.25% nor more than 2.25%. Otherwise, it is the same as the U.S. standard except for the additives permitted, as in white flour. Canada distinguishes between whole wheat flour and Graham flour by the method of preparation. The former is milled wheat from which a part of the bran has been removed, while Graham flour is flour to which part of the bran has been added. The ash limits are the same, 1.5% to 2.5%, moisture-free. Canada has one flour, gluten flour, not standardized in the United States. This is white flour from which part of the starch has been removed. It contains not more than 10% moisture and not more than 44% starch, moisture-free basis. In general, the Canada Food and Drug Directorate uses AOAC Methods of Analysis for the examination of flours.

TABLE 4-1
U.S. Standards of Identity for Wheat Flours

Official Name of Product	Source	Granulation (referred to by NBS Standard Sieve No.)	Maximum Moisture (vac. oven 90°–100°)	Ash, Moisture-Free Basis (550°)	Optional Ingredients
Wheat flour, White flour, Flour, Plain flour	Wheat other than Durum or Red Durum	Not less than 98% through No. 70	15%	1/20 the % of protein (NX5.7) on moisture-free basis +0.35	Nitrogen oxides, nitrosyl chloride, chlorine, chlorine dioxide, 1 part benzoyl chloride to not more than 6 parts by weight potas. alum, magnesium carbonate, sodium alum. sulfate, calcium carbonate, calcium sulfate, di- or tri-calcium phosphate and/or starch. Acetone peroxide. Azodicarbonamide 45 ppm max. Not more than 5% defatted wheat germ. (Ash should be corrected for that contributed by added minerals.)
Enriched flour	Same as wheat flour except for added specified vitamins and minerals				
Bromated flour	Same as wheat flour except for addition of 50 ppm max. potassium bromate				
Enriched bromated flour	Same as bromated flour plus enrichment				
Durum flour	Durum wheat	Not less than 98% through No. 70	15%	1.5%	
Self-rising flour, self-rising white flour, self-rising wheat flour	Flour plus $NaHCO_3$ and acidic ingredients	—	—	—	Mixture of flour, sodium bicarbonate and the acid-reacting ingredients monocalcium phosphate, sodium acid pyrophosphate and/or sodium alum, phosphate at rate of 4.5 parts max./100 parts flour. Not less than 0.5% CO_2 evolved.
Enriched self-rising flour	Same as self-rising flour plus enrichment				
Whole wheat flour, graham flour, entire wheat flour	Wheat other than Durum, proportion of natural constituents unaltered	Not less than 90% through No. 8 and not less than 50% through No. 70	15%	—	Malted wheat or malted barley to correct natural enzyme deficiency, but not more than 0.75% may be added. Azocarbonamide (max. 75 ppm), chlorine or chlorine dioxide may be added
Whole durum wheat flour	Durum wheat	Same as whole wheat flour			
Bromated whole durum wheat flour	Same as whole durum wheat flour				75 ppm max. potassium bromate
Farina	Wheat other than durum	Through No. 20, not more than 3% through No. 100	15%	0.6%	
Enriched Farina	Same as farina plus enrichment				
Semolina	Durum	Through No. 20, not more than 3% through No. 100	15%	0.92%	
Phosphated flour	Same as wheat flour except for addition of a phosphate				Monocalcium phosphate, not less than 0.25% or more than 0.75%

Method 4-1. Fineness of Flour. Official Method of Canada Food and Drug Laboratories.

Fit a No. 8 sieve (2380 microns) into a No. 20 sieve (840 microns). Attach a bottom pan to the No. 20 sieve. Pour 100 g of the sample into the No. 8 sieve. Attach the cover and hold the assembly in a slightly inclined position with one hand. Shake the sieves by striking the sides against the other hand with an upward stroke, at the rate of about 150 times per minute. Turn the sieves about one-sixth of a revolution each time in the same direction after each 25 strokes. Continue shaking for 2 minutes. Weigh the material which fails to pass through the No.8 sieve and the material which passes through the No. 20 sieve.

Method 4.2. Moisture, Vacuum Oven Method.

Follow Method 1–2, using a 2 g sample, a vacuum of 25 mm of Hg and a temperature of 98–100°.

Method 4-3. 2-Stage Air Oven Method.

This has been the official method of the AACC for wheat and other grains. In March 1967, the Proximate Analysis Committee of AACC proposed that it be revised and extended to include flour, farina, semolina, bread, grains, corn grits, rolled oats, rolled wheat, bulgur and breakfast foods.[3]

DETERMINATION

a. *Samples containing less than 16% water (except rice containing less than 13% water), unless the product is already ground to a flour or meal.*—Grind through a Wiley mill equipped with an 18- or 20-mesh screen, or any other mill that will grind to this degree of fineness without undue exposure to the atmosphere and appreciable heating. Grind 30 to 40 g sample, leaving as little as possible in the mill. Mix the sample rapidly and immediately transfer 2- to 3-g portions to each of 2 or more dried, tared metal dishes about 55 mm in diameter (described in Method 1–2).

Place uncovered dishes in an oven provided with ventilation ports, and set to maintain a temperature of 130 ± 1°. The thermometer should be placed so that the tip of the bulb is even with the top of the dishes, but not directly over any dish. Dry for exactly 1 hour, starting timing when oven temperature regains 130°. After the 1 hour has elapsed, cover the dishes while still in the oven, transfer to a desiccator, and weigh after the dishes have reached room temperature (45–60 minutes).

NOTE: AACC permits use of either a forced draft or a mechanical convection oven.

b. *Samples often containing more than 16% water (except rice containing more than 13% water).*—Loss of moisture incident to grinding such grains may be excessive, hence the following 2-stage drying should be used:

Fill 2 or more tared moisture dishes nearly full with the unground grain (*see* NOTE for procedure for bread). Cover the dishes and weigh.

Place the uncovered dishes in a warm, well-ventilated area free from dust (on top of heated oven is suitable) until the grain is air-dried (about 14–16 hours). Cover the dishes, cool and weigh. Record percent of loss.

Grind the air-dried sample, and proceed as directed in **a**. Calculate the percentage moisture in the original sample by the equation:

$$\text{T.M.} = A - \frac{(100 - A)B}{100}, \text{ in which}$$

T.M. = % total moisture
A = % moisture lost in air-drying
B = % moisture in air-dry sample as determined by oven-drying at 130°.

NOTES

1. The oven should regain its temperature of 130° within 15–20 minutes after insertion of a full load (24 moisture dishes). If the oven requires a longer time, it should not be used.

2. To determine moisture in bread, weigh a representative loaf to nearest 0.2 g. If not sliced, place loaf on a large sheet of smooth paper and slices being careful not to lose any crumbs. Let cut slices dry out at room temperature until they are in approximate equilibrium with the moisture in the air (15–20 hours). Weigh the dried slices and proceed as in **b** beginning "Grind the air-dried sample . . ." If sample contains raisins or other fruit, comminute by passing twice through a food chopper instead of grinding.

The AOAC directs that air-dried samples of bread containing pieces of fruit be dried in an uncovered dish for about 16 hours at 70° in a vacuum under pressure not exceeding 50 mm Hg, instead of the 130° air oven. See also Method 4–23.

Method 4-4. *Ash in Flour, Ground Grains, Pastas and Similar Products.*

Weight 3–5 g, and follow general Method 1–9, starting at 425° and finally ashing at 550°. This is the basic method of the American Association of Cereal Chemists (AACC).

NOTE: AACC permits final ashing of hard wheat flours at 575°–590°.

The AOAC and AACC manuals include methods for determining "Original Ash in Phosphated and Self-rising Flour" (AOAC **13.009**) and "Added Inorganic Material in Phosphated Flour" (AOAC **13.010**).[4] These methods are of limited use and are given by reference only.

Crude Fiber in Flour and Grain

Follow Method 1–11, using a 2 g sample. The residue from the extraction of fat, Method 4–11 (without use of sand), may be used if the sample weight was 2 g. If fat content is less than 1%, the ether extraction may be omitted.

Crude fiber, along with ash, is a measure of the quantity of bran, hence the degree of extraction of the flour.

Method 4-5. *Iron and Calcium in Enriched, Self-Rising and Phosphated Flour, and Bread.* (Adapted from AACC and AOAC Official Methods.)

Ash 10 g of flour or air-dried bread in a shallow, relatively broad porcelain or silica dish as directed in Method 4–4. (Platinum should not be used if chlorides are present.) Weigh, if the percentage of ash is desired. Cool, dissolve ash in about 5ml HCl, rinse down sides of dish, and evaporate to dryness on steam bath. Dissolve residue in 2.0 ml of accurately measured HCl reagent grade. Place a watch-glass over the dish, heat 5 minutes on a steam bath, rinse watch-glass with water, and filter into a 100 ml volumetric flask. Cool and dilute to mark.

Determine iron by AOAC Method **13.011–13.013** on a 10 ml aliquot of the ash solution and calcium on a 50 ml aliquot by Method 1–15.

NOTE: If difficulty occurs in obtaining an ash that is practically carbon free, the ash may be moistened with magnesium nitrate solution (50 g $Mg(NO_3)_2.6H_2O$ in water, diluted to 100 ml), dried and reignited carefully in a muffle. This ash aid should not be used on self-rising flour, or on products containing chlorides if ashing is done in a platinum dish.

Total Nitrogen (Crude Protein) in Flour or Grain

Follow Method 1–7, using a 1 g sample of flour or grain ground to pass a 20-mesh sieve. Calculate the percentage of crude protein in any wheat product by multiplying the percentage of nitrogen by 5.70. For products of other grains, use the conventional factor 6.25. In every case, report the factor used in converting from nitrogen to crude protein.

Early attempts to separate and identify the individual proteins of wheat and other grains were for the most part empirical in methodology. These depended on fractionation of the proteins by specific metallic salt solvents at varying concentrations. In more recent times, the use of electrophoresis techniques has proved promising. These are reviewed in Chapter 6 of Hylnka's text: *Wheat Chemistry and Technology* (Selected Reference F).

NOTE: The FDA definition and standard for wheat flours requires that the protein content be calculated by multiplying the percentage of total nitrogen, moisture-free basis, by the factor 5.7. Grains other than wheat and flours made from wheat are conventionally calculated by multiplying the percentage of N by 6.25.

Method 4-6. *Amino Nitrogen in Wheat Flour and Semolina.*

This modified Sørensen method has been used in one form or another for at least 50 years. It has been adopted as a laboratory method by AACC.

REAGENTS

a. *Sodium hydroxide* N/14, made from carbonate-free concentrated NaOH solution.

b. *Buffer solution*, pH 8.0. Combine 50 ml 0.2 M monopotassium phosphate solution (27.232 g anhydrous KH_2PO_4/L) and 46.85 ml 0.2 M NaOH made from carbonate-free concentrated NaOH solution. Dilute to 200 ml. (Standard buffer solutions can be purchased from laboratory suppliers.)

DETERMINATION

Add exactly 100 ml of water to 20.0 g of flour in a centrifuge bottle, shake thoroughly, and let stand for an hour, shaking every 10–15 minutes. Centrifuge, and filter if necessary, to obtain 40–60 ml clear extract.

Transfer 10 ml of extract to a 125 ml Erlenmeyer flask and titrate with 0.14N NaOH to pH 8 electrometrically using glass electrodes and a micro-burette calibrated to 0.01 ml. Record as titration A. Add 8 ml of commercial 40% HCHO and titrate again to pH 8.0. Correct for acidity of the HCHO and record as B. B − A represents the amino acid value of the flour. Calculate as mg amino N.

Method 4-7. *Crude Fat.* (Acid hydrolysis method—applicable to flour, semolina, alimentary pastes, bread and baked products not containing fruit). Official AACC method, and has been official in AOAC methods since 1925.

DETERMINATION

Place a 2 g sample of flour or similar product, or the equivalent of 2 g of air-dried bread or baked product in a 50 ml beaker. Moisten with 2 to 3 ml of ethanol and stir to prevent subsequent lumping. Add 25 ml HCl (25 + 11), mix, and set the beaker in a water bath held at 70–80° for 30–40 minutes, stirring frequently. Add 10 ml of ethanol and cool.

Transfer the mixture to a Mojonnier extraction flask, and rinse the beaker into the extraction flask with successive 10 10 and 5 ml portions of ethyl ether. Stopper with a cork or neoprene stopper (not rubber) and shake vigorously for 1 minute. Add 25 ml of redistilled petroleum ether and shake vigorously for 1 minute. Let stand until the solvents and aqueous solution separate.

Bring the aqueous layer up to the center of the neck, and decant as much of the solvent layer as possible through a funnel containing a pledget of cotton into a weighed 125 ml wide-mouthed Erlenmeyer flask (beaker-flask) to which a few porcelain chips or broken glass have been added. (The flask with the porcelain chips is more accurately tared by oven-drying at 100°, then air-drying to constant weight against a similarly treated flask as a counterpoise.)

Re-extract the aqueous layer in the extraction flask twice, each time with 15 ml each of ethyl ether and petroleum ether. Allow the layers to separate, and filter the combined ether layer into the same tared flask as before. Wash off the stopper and rim of the extraction flask and the funnel and funnel stem with a few ml of a mixture of the two ethers in equal volumes.

Evaporate the ethers slowly on a steam bath, then dry in a 100° oven to constant weight (60 to 90 minutes) along with the counterpoise flask. Remove the flasks from the oven, let stand in air to constant weight (about 30 minutes) and weigh.

Run a blank extraction on reagents used and report as: "Percentage fat by acid hydrolysis."

NOTE: A Röhrig fat extraction apparatus may be used instead of a Mojonnier flask. The Mojonnier flasks may be centrifuged instead of allowing the ether and aqueous layers to separate by gravity.

Method 4-8. *Crude Fat.* (Ether Extraction Method).

Follow the procedure given in Method 1—13, using a 3 to 5 g sample. If the sample is a fine flour, an equal weight of clean, dry sand may be mixed with the flour.

The AACC and AOAC direct use of petroleum ether for extraction of soy flour. Extraction may be accomplished in a Soxhlet, Butt or Goldfisch extractor. The solvent should drop on the center of the thimble at a rate of at least 150 drops/minute. Extraction will be complete in 5 hours.

Method 4-9. *Acidity of Flour (pH).*

Suspend 10 g of flour in 100 ml freshly boiled, distilled water at 25°. Let stand for 30 minutes, shaking occasionally. Let stand for 10 minutes, then decant supernatant liquid and determine pH electrometically at 25°. Standardize the pH meter at pH 4.00 and 9.18 against standard acid potassium phthalate and borax solutions respectively [*Methods of Analysis*, AOAC, 10th ed., **42.007(c), (f)**]. The pH of flour is normally about 6.1. Lower values are considered an indication of bleaching with a chlorine compound.

Method 4-10. *Gasoline Color Value of Flour.*[5]

The color extracted from flour has long been used as an indication of bleaching. The method is no longer in general use, having been replaced by a method made official by AACC involving extraction with water-saturated butyl alcohol, which was later adopted by AOAC. The gasoline color value method, dropped by AOAC in 1950, is given here because it is still used by some laboratories. It is a rough measure of the remaining pigments in flour after bleaching.

DETERMINATION

Add 100 ml of colorless gasoline to 20 g of flour in a wide-mouth, glass-stoppered bottle. Stopper and shake vigorously for 5 minutes. Allow to stand overnight, shake again for a few seconds to loosen the flour from bottom of the bottle, and filter immediately through a dry, 11 cm paper into an Erlenmeyer flask. Cover the funnel with a watch glass during filtration. If filtrate is not clear, return it through the filter until clear.

Determine the color value of the filtrate against a 0.005% K_2CrO_4 solution, using a Schreiner or similar colorimeter or Nessler tubes. The 0.005% K_2CrO_4 solution is arbitrarily assigned a gasoline number of 1.0. It is conveniently made by diluting 10 ml of a 0.5% solution to 1 liter.

Adjust colorimeter tube containing the gasoline solution to read 50 mm and raise or lower the tube containing the K_2CrO_4 standard solution until the yellow shades in both tubes match.

$$\text{Gasoline Color Value} = \frac{K_2CrO_4 \text{ tube reading}}{\text{Gasoline extract tube reading}}$$

Method 4-11. *Pigments in Flour* (*AOAC Method* ***13.050***).

This method has largely replaced the gasoline color value determination. It has been made official by both AACC and AOAC.

DETERMINATION

Add 50 ml of water-saturated n-butyl alcohol to 10.00 g of flour in a 125 ml glass-stoppered flask. Stopper tightly, shake thoroughly, and let stand 15 minutes in the dark. Reshake well, and filter through a 12.5 cm folded filter. Refilter if extract is not entirely clear.

Fill a 1 cm spectrophotometer cell with the flour extract, and fill a duplicate cell with the filtered solvent. Read absorbance at 435.8 mμ, taking 3 readings. Prepare a standard curve from pure beta-carotene for calculation, or, in the absence of carotene, calculate pigment value (C) as p.p.m. carotene from formula. $C = 5.0 \times \text{absorbance}/bK = 30.1 \times \text{absorbance}$, where b is cell length (1 cm) and K (0.16632) is absorptivity in mg/L for carotene.

Maturing and Bleaching Agents

Whiteness is a primary characteristic of bread and cake flour, although flours used for pastas (semolina and farina) should have a yellow cast for best acceptance. (This is doubtless because the color suggests the presence of egg.) For this reason, bleaching is practiced extensively in the milling of the white flours. The Alsop patents for generating nitrogen peroxide (U.S. Patents 758,883 and 758,884 of 1904) were the basis for the earliest bleaching used extensively in this country. This agent is seldom used now.

In 1921, nitrogen trichloride came into use under the name "Agene". Its use was banned by FDA in 1949 when it was found to have toxic effects on dogs. Chlorine dioxide took its place. Benzoyl peroxide, mixed with a carrier, was patented by Sutherland in 1925 (U.S. Patent 1,539,701) and soon became popular under the name "Novadel". This is probably the most widely used additive at present for bleaching alone. Chlorine, nitrosyl chloride and chlorine dioxide have both bleaching and maturing actions, the latter being an artificial aging effect due to inhibition of the active proteolytic enzymes present in flour.

Harrel[6] published a review of bleaching and maturing agents used in flour in 1952. Other agents have been introduced since that date.

Maturing agents are not used indiscriminately in present-day milling practice. They are only added if baking tests show that the flour requires such treatment. In some crop years, no maturing treatment is necessary. This is particularly true in the more highly refined flour stream, known as "patents" and used largely for bread-making. The maturing effect is, according to Harrel, more marked in the clear flours used along with rye and whole wheat flours for baking those types of bread.

Method 4-12. *Qualitative Tests for Additives (Oxidizers and Improvers) in Flour.*

With the exception of potassum bromate and azodicarbonamide, flour standards in the United States have no quantitative restrictions for these additives other than that only quantities sufficient to effect the purpose should be used. Canada permits benzoyl peroxide up to 150 p.p.m., ammonium persulfate up to 250 p.p.m. and ammonium chloride up to 2000 p.p.m. Analysts are more often interested in the proof of the presence of these additives than in their quantitative estimation.

A. *Chlorine Bleaches*: Extract about 30 g of flour with 50 ml of petroleum ether. Decant through a filter, catching the solvent in a porcelain evaporating dish. Remove the petroleum ether by evaporation. A small amount of oil will remain. Heat a piece of copper wire in a nonluminous Bunsen flame until it is fully oxidized (black) and no longer colors the flame green. Dip the hot end of the wire into the residual oil and again bring it into the flame. A green color of the flame indicates presence of chlorine bleaching agents.

Bromine bleaching agents give the same or a blue color. These are not permitted bleaches. Potassium bromate is permitted in the United States and Canada in bromated flour as a maturing agent. It is without action in dry flour, and only reacts when the flour is made into dough.

B. *Oxides of Nitrogen*: This is a modification of a widely-used test for nitrites in water.

REAGENTS

a. *Sulfanilic acid solution*—0.5 g sulfanilic acid in 150 ml of 20% acetic acid.

b. *Alpha-naphthylamine hydrochloride solution*—0.2 g of the salt in 150 ml of 20% acetic acid. Dissolve by heating, if necessary.

NOTE: The hydrochloride should be completely colorless. If not, add about 5–10 g of the salt and 2 g of decolorizing carbon in 100 ml water and boil for 10 minutes. Filter rapidly through a Büchner funnel. If the solution is not clear and colorless, add 2 g of charcoal and repeat the process. Add 25 ml conc. HCl to the filtrate, cool to 0° in an ice bath, filter and air-dry the crystals on a porous plate in the dark. Keep in a tightly-closed amber bottle.

TEST

To about 20 g of flour in an Erlenmeyer flask, add 20 ml nitrite-free water at 40°. Stopper, and shake vigorously for 5 minutes, then digest at 40° for 1 hour, shaking every 10 minutes. Filter on a washed and dried, nitrite-free, pleated filter, refiltering until the solution is clear.

Transfer 50 ml of the filtrate to a small flask. Add 2 ml each of the sulfanilic acid and the alpha-naphthylamine solutions, shake, and allow to stand 1 hour to develop color. A pink to reddish purple color denotes the presence of nitrite nitrogen.

This method has been made quantitative by the AOAC (Method **13.043**, *Methods of Analysis* 10th ed.,) by comparing the color against a series of known amounts of nitrites. Others have compared the colors against a series of standards in Nessler tubes or a colorimeter, or the absorbance at 520 mμ may be measured in a spectrophotometer.[7]

C. *Oxidizing Agents*: This is an AACC qualitative test. This method detects all oxidizing agents except perchlorates and benzoyl peroxide.

REAGENT

Benzidine—0.5 g in 500 ml of HCl (1 + 1).

TEST

Transfer 10 to 25 ml of acid-benzidine solution into a test tube. Shake small portions of the flour under test into the benzidine solution with a spatula. As the flour particles settle, brownish streaks indicate presence of oxidizing agents.

Confirmatory test: Add 200 ml of water to 50 g of flour in a 500 ml Erlenmeyer flask. Stopper, shake vigorously, and allow to stand for about an hour, shaking frequently. Filter or centrifuge. To 5 ml of the clear liquid, add 5 ml of 10% KI solution and 5 ml of H_2SO_4 (1 + 10). A yellow or brown color indicates presence of oxidizing agents.

D. *Benzoyl Peroxide*: AACC qualitative test.

REAGENT

Rotherfusser's reagent: Triturate 1.0 g of diparadiaminophenylamine sulfate with several ml of ethanol and grind in a mortar. Dilute to 100 ml with ethanol, added little by little. Refluxing on a water-bath may be necessary to complete solution.

TESTS

Shake about 1 g of flour into a test tube with 3.5 ml of petroleum ether. After settling, add 1.5 ml of Rotherfusser's reagent. A green color in the petroleum ether indicates the presence of benzoyl peroxide. Blue crystals in the sediment indicate the presence of persulfates.

E. *Persulfates*: This is an AACC and AOAC qualitative test. Place about 10 g of flour on a glass plate. Slick the surface of the flour with a spatula, and spray or pour gently over the surface a solution of 0.5 g of benzidine in 100 ml ethanol. If persulfates are present bluish-black specks appear.

F. *Bromates*: This is an AACC and AOAC qualitative test. Sieve flour over the surface of a dry pan and spray a freshly prepared mixture of equal volumes of 1% KI solution and HCl (1 + 7) solution. Black or purple specks indicate the presence of bromates. Iodates and persulfates will give the same result.

Confirmatory test for iodates: Spray a layer of flour with a mixture of 1 volume of 1% KSCN solution and 4 volumes of HCl (1 + 32) solution. Black or purple specks indicate presence of iodates.

Method 4-13. *Bromates in Bromated Flour, Quantitative.*[8] (This method has been adopted as official by AACC and AOAC).

REAGENTS

a. *Zinc sulfate solution*—20 g $ZnSO_4.7H_2O$ per liter.

b. NaOH, 0.4 *N*, 17 g/liter. Adjust to 0.4 ± 0.01 *N*.

c. NaOH, 0.5 *N*, 21 g/liter. Adjust to 0.5 ± 0.01 *N*.

d. H_2SO_4, dilute. Add 112 ml H_2SO_4 to 800 ml. of water. Dilute to 1 liter.

e. KI *solution*. 25 g in water, diluted to 50 ml. Discard if solution has a yellow tinge.

f. *Ammonium molybdate solution.* 3 g $(NH_4)_6Mo_7O_{24}.4H_2O$ in water, diluted to 100 ml.

g. $KBrO_3$ *stock solution.* 5 g $KBrO_3$, dried at 110° for 1 hour, in water, diluted to 1 liter.

h. $KBrO_3$ *standard solution.* 25 ml stock solution (f), diluted to 250 ml. Prepare daily.

i. KIO_3 *stock solution.* Weigh 3.204 g of KIO_3, dried 1 hour at 110°. Dissolve in water and dilute to 1 liter.

j. KIO_3 *standard solution*, 0.00359*N*. 10 ml of stock solution (**i**) diluted to 250 ml. Prepare daily.

k. $Na_2S_2O_3$ *stock solution.* 22.5 g $Na_2S_2O_3.5H_2O$ and 0.06 g of Na_2CO_3 in water. Dilute to 1 liter.

l. $Na_2S_2O_3$ *standard solution*, 0.00359 *N*. Dilute 10 ml stock solution (**k**) to 250 ml. Prepare daily and check normality.

TEST

Transfer quantitatively 200 ml of $ZnSO_4$ reagent (**a**) to an 800 ml beaker and stir with a variable stirrer or with a glass paddle or glass-covered magnetic bar so that the vortex does not extend quite to the bottom. Add 50 ± 0.1 g of flour sample to the stirred solution in small portions (2 to 5 g). Stir 5 minutes or until all dry flour on the surface is uniformly dispersed. While stirring, add 50 ml 0.4 NaOH solution (**b**). Decrease stirrer speed and continue stirring for 5 minutes more.

Filter or centrifuge to obtain a clear solution and transfer 50.0 ml to a 200 ml Erlenmeyer flask (or, if smaller aliquot is taken dilute to about 50 ml). Add 10 ml of H_2SO_4 solution (**d**), 1 ml KI solution (**e**), 1 drop of molybdate solution (**f**) and 50 ml water. With continuous agitation, add excess (5–10 ml) of 0.00359 *N* $Na_2S_2O_3$ solution (**l**). Now add 5 ml of freshly-prepared, 1% starch solution, and titrate excess thiosulfate with 0.00359 *N* KIO_3 solution (**j**). As end point nears, titrate drop by drop, swirling, and viewing flask against a white surface after each addition. The first reddish or purplish tinge is the true end point Read the burette and add a drop or two to confirm. Add another 1 ml of 0.00359 *N* thiosulfate solution and again titrate to the same end point. Average the differences between the amounts of standard 0.00359 *N* solutions of thiosulfate and iodate used in these titrations.

$KBrO_3$ in p.p.m. = 10 times the difference in titration, provided the solutions are exactly 0.00359 *N*.

Correct results by a "recovery factor" determination as follows: dilute a known volume (x), more than 3.0 ml and less than 10.0 ml, of standard bromate solution (**h**) to 250 ml. Proceed as in second paragraph under "Determination" beginning ". . . and transfer 50.0 ml of the clear solution. . . . ," ending " . . . and add a drop or two to confirm."

Suspend 50 g portions of nonbromated flour in 2 separate 200 ml portions of $ZnSO_4$ solution by stirring as directed in first paragraph of "Determination." To one (blank) suspension add 10 ml water; to the others (recovery) add the same quantity, "x" ml, of standard bromate solution (**h**) and (10 − x) ml of water. Continue as in the first paragraph of "Determination," beginning "While stirring add 50 ml 0.4 *N* NaOH . . . ," except this time add 40 ml of 0.5 *N* NaOH (**c**) stirring continuously. Add 10 ml of H_2SO_4 solution (**d**), 1 ml KI solution (**e**), 1 drop molybdate solution (**f**) and 50 ml water to each suspension. Now add 5 ml of standard $Na_2S_2O_3$ standard solution (**l**) to "blank" and 10 ml of the same to "recovery" suspension, stirring constantly during the addition. Titrate excess thiosulfate with standard KIO_3 solution (**j**) as before. Deduct blank value, if any, from the value of $KBrO_3$ found in "recovery" determination and multiply by 10 to obtain p.p.m. "recovered bromate."

$$\text{Recovery factor} = \frac{\text{added bromate}}{\text{recovered bromate}}$$

Ammonium persulfate (quantitative method):

There is little call for a quantitative method. Reference is directed to AACC Method **48–62** and to the method of Auerbach *et al.*[9]

Method 4.14. *Starch in Flour*. (This is the official AACC Method applicable to flour and semolina; also, with slight modification, to corn, rye, barley, rice, grain sorghum and buckwheat.) *See* NOTE.

REAGENTS

a. *Alcohol solvent.* Dissolve 1 g mercuric chloride ($HgCl_2$) in 900 ml water, and add 100 ml 95% ethyl alcohol. Mix thoroughly.

b. *Acid calcium chloride solution.* Dissolve sufficient $CaCl_2.6H_2O$ (or $CaCl_2.2H_2O$) in water to bring density to 1.30 at 20°. Acidify with acetic acid to pH 2.2–2.5 and make further adjustment of density if necessary. Filter with suction through several layers of paper until reagent is crystal-clear.

c. *Stannic chloride*, 4%. Dissolve 4.0 g $SnCl_4.5H_2O$ in 100 ml reagent (**b**).

DETERMINATION

Grind the sample finely and accurately weigh 2.463 g into 50-ml round-bottomed, lipped centrifuge tube. Add 10 ml of alcohol solvent, and stir vigorously for about 2 minutes. Filter with vacuum through 9 cm of hard paper supported on filter cone in a 60° funnel. Rinse the tube, transferring the residue quantitatively, and wash the residue twice with about 25 ml of alcohol solvent. Apply vacuum until the residue is substantially dry.

Quantitatively transfer to a 400 ml Berzelius beaker, add 10 ml water, and stir with a rubber-tipped glass rod until smooth suspension is obtained. Add 60 ml of acid $CaCl_2$ reagent (**b**) and mark liquid height with wax pencil or other suitable means.

Bring to boiling in 4–5 minutes with frequent stirring; avoid overheating sides. Let boiling proceed rather briskly for 15 minutes, stirring periodically. Add water during heating to maintain liquid level (rate of evaporation loss 1–2 ml per minute). If frothing is troublesome, add few drops of octanol.

Cool to room temperature, add 2.5–5.0 ml $SnCl_4$ reagent (**c**), and quantitatively transfer to a 100 ml volumetric flask, rinsing with successive portions of reagent (**b**). Destroy any foam in the neck of the flask by adding a drop or two of 95% ethanol, and shaking as the alcohol reaches the froth. Bring accurately to volume with reagent (**b**), and mix thoroughly.

Filter through Whatman No. 12 fluted paper or sintered glass filter. Discard the first 20 ml of filtrate, and collect 30–50 ml for polarization. If filtration is slow, cover the funnel with a watch-glass.

Polarize in a 2-dm tube, using sodium light.

CALCULATION

Starch, % = polarimeter reading in degrees × 10, if 2.463 g sample is used.

NOTE: By substituting 10 ml of a solution of 5% uranyl acetate dihydrate for $SnCl_4$ (**c**) as a protein precipitant, this procedure is satisfactory for use with corn, rye, barley, rice, grain sorghum and buckwheat.

Sugars in Flour

Early reports on the composition of flour show a wide divergence in the quantity of sugar(s) present in flour. More recent reports show that total sugars are present in the magnitude of 1–3%. The quantity of non-reducing sugars is usually arbitrarily expressed as mg sucrose/10 g flour and reducing sugars as mg maltose/10 g flour. The true identity of these sugars was not determined until the techniques of chromatography and electrophoresis became available. Raffinose, sucrose, maltose, dextrose and possibly others have been reported in flour.

Method 4-15. *Reducing and Non-Reducing Sugars in Flour.*

This method was first proposed by Sandstet[10] in 1937. It has since been adopted by both AACS and AOAC.

REAGENTS

a. *Acid buffer solution*—Dilute 3 ml of acetic acid, 4.1 g of anhydrous sodium acetate and 4.5 ml of sulfuric acid to 1 liter.

b. *Tungstate solution*—12.0 g of sodium tungstate, $Na_2WO_3.2H_2O$, to 100 ml.

c. 0.1 *N alkaline ferricyanide solution*—33.0 g $K_4Fe(CN)_6$ and 44.0 g Na_2CO_3 diluted to 1 liter.

d. *Acetic acid-salt solution*—Dissolve 70.0 g KCl and 44.0 g $ZnSO_4.7H_2O$ in about 750 ml water. Slowly add 200 ml of acetic acid and dilute to 1 liter.

e. *Soluble starch-iodide solution*—Suspend 2 g of soluble starch in a small quantity of cold water and pour slowly into about 50 ml boiling water with constant stirring. Cool thoroughly (or resultant mixture will turn dark); add 50 g KI and dilute to 100 ml. Add 1 drop NaOH solution (1 + 1).

f. 0.1 *N thiosulfate solution*—24.82 g $Na_2S_2O_3.5H_2O$ and 3.8 g $Na_2B_4O_7.10H_2O$. Dilute to 1 liter.

Make blank determination each day to guard against deterioration in reagent (**c**) and to correct for reducing impurities in the reagents as follows: Combine 5 ml ethanol, 50 ml of buffer solution (**a**) and 2 ml of tungstate solution (**b**). Transfer 5 ml of this mixture (to replace 5 ml of the flour extract) to a 1″ × 8″ pyrex test tube. Add 10 ml of 0.1 *N* ferricyanide solution, mix and proceed as under "reducing sugars" below, beginning, "... and immerse test tube..." Ten ml of the 0.1 *N* thiosulfate solution should discharge the blue starch-iodine color. If titration falls within 10 ± 0.05 ml do not discard reagents but correct subsequent sugar titrations by using the thiosulfate equivalent of 10 ml ferricyanide solution instead of 10 as basis for subtraction.

DETERMINATION

Transfer 5.675 g sample to a 125 ml Erlenmeyer flask, tipping the flask so that all the flour is on one side. Wet the flour with 5.0 ml ethanol. Tip flask so that the wet flour is on the upper side and add 50 ml acid buffer solution (**a**), keeping this solution from coming into contact with the flour until all has been added. Shake the flask to bring the flour into suspension, then add immediately 2 ml of 12% tungstate solution (**b**) and mix thoroughly. Filter at once through a rapid filter paper (Whatman No. 4 or equivalent), discarding first 8–10 drops of filtrate.

Reducing sugars—Pipette 5 ml of the filtered extract into a 1 × 8″ pyrex test tube, add exactly 10 ml of 0.1 *N* ferricyanide solution (**c**), mix and immerse test tube into vigorously boiling water so that liquid in the test tube is 3–4 cm below surface of the boiling water (delay between filtering and treatment in boiling water should not exceed 15–20 minutes). Let test tube remain in boiling water exactly 20 minutes, remove and cool tube and contents under running water. At once pour the contents into a 125 ml Erlenmeyer flask, rinse test tube with 25 ml acetic acid-salt solution (**d**), adding rinsings to the flask. Mix and add 1 ml of starch-iodide indicator solution (**e**).

Titrate with 0.1 *N* thiosulfate solution (**f**), preferably with a micro burette, until complete disappearance of the blue color. Subtract the ml 0.1 *N* thiosulfate solution from 10 (or from the "thiosulfate equivalent" of the 0.1 *N* ferricyanide solution). Compute reducing sugars as mg maltose/10 g flour by reference to Table 4–2.

Non-reducing sugars—Pipette 5 ml of the filtered flour extract into a 1 × 8″ pyrex test tube and immerse in vigorously boiling water for 15 minutes. Cool test tube and contents under running water. Pour contents of test tube into a 125 ml Erlenmeyer flask, rinse out test tube with 25 ml acetic acid-salt solution (**d**) and proceed with the reduction and thiosulfate titration as under "reducing sugars". Calculation for non-reducing sugars:

Ml ferricyanide consumed after above hydrolysis, less ml ferricyanide reduced by reducing sugars in flour, equals ml equivalent to non-reducing sugars present. Report as mg sucrose/10 g flour, from Table 4-2.

Pentosans

The amount of pentosans present in flour is one of the indices of the quantity of bran left in the flour, along with crude fiber and ash determinations. Wheat flour contains 4% or less pentosans while the bran contains about 25%. Cleaned, white rice and rice flour contain no more than 1 or 2% pentosans, while rice hulls contain around 18% and rice bran around 12–14%. It is claimed that use of pentosan-rich flour decreases the spreading of cookie dough.

Method 4-16. *Pentosans in Cereal Flours.*[11] (This is an AACC Official Method).

APPARATUS

a. *Distillation assembly*, including a 500 ml separatory funnel and a thermometer attached to stopper of distilling flask.

b. *Glycerol bath.*

REAGENT

Bromide-bromate solution—50 g KBr and 3.0 g $KBrO_3$/liter.

DETERMINATION

Weigh 0.500 g sample into a 500 ml distilling flask containing a few glass beads or boiling chips and add 125 ml HCl (1+2). Place 360 ml of HCl (1+2) in the separatory funnel and attach it to the distilling flask so that acid can be introduced into the flask during distillation. Heat the flask on a glycerol bath, distilling at the rate of 30 ml per 10 minutes at 150°. After 30 ml have distilled over, slowly add 30 ml acid from the separatory flask, allowing 1–2 minutes for the addition. Again distil at the same rate, adding another 30 ml increment of acid each time 30 ml distillate has been collected, until 360 ml have been collected. Stopper receiving flask, place immediately in an ice bath and cool to 0°.

Add 50 ml of the bromide-bromate solution, using a volumetric pipet, to the cooled distillate, stopper immediately, swirl and let react for *exactly* 4 minutes. Add 10 ml 10% KI solution, re-stopper, and shake gently. Remove flask from ice bath and titrate liberated I with 0.1 *N* standard sodium thiosulfate solution, using starch solution as indicator. Determine a blank on the bromide-bromate solution, adding 5 ml conc. HCl before adding the 10% KI solution.

1 ml thiosulfate solution = 0.0082 g pentosans.

Method 4-17. *Sorbitol in Bakery Products.*[12]

Sorbitol is used as a humectant or emulsifier in many bakery products. It also has sweetening properties. This method, developed by analysts of the Kansas City District of FDA, extracts the sorbitol with water, inverts non-reducing sugars with acid, and degrades sugars other than sugar alcohols to salts of organic acids which are removed on a basic ion exchange column; the separated sugar alcohols are then determined by periodate oxidation.

APPARATUS

a. *Thin-layer chromatography apparatus.*

b. *Chromatographic columns*—glass tubing 12–14 mm diameter and 45 cm high, fitted with stopcock or adjustable pinch clamp.

REAGENTS

a. *Standard acid* KIO_4 *solution*—dissolve 0.60 g KIO_4 in 400 ml water plus 20 ml H_2SO_4, and dilute to 1 liter.

b. *Standard* $Na_2S_2O_3$ *solution*—0.02 *N*. Prepare fresh before use by dilution and standardization of 0.1 *N* $Na_2S_2O_3$ solution.

c. *Strongly basic ion exchange resin*—Amberlite IRA-400, 20–50 mesh.

d. *Silica gel G, specially prepared for TLC use.*

TABLE 4-2

Ferricyanide-Maltose-Sucrose Conversion Table[a]

0.1 *N* Ferricyanide reduced	Maltose per 10 g flour	Sucrose per 10 g flour	0.1 *N* Ferricyanide reduced	Maltose per 10 g flour	Sucrose per 10 g flour
ML	MG	MG	ML	MG	MG
0.10	5	5	4.50	237	214
0.20	10	10	4.60	244	218
0.30	15	15	4.70	251	223
0.40	20	19	4.80	257	228
0.50	25	24	4.90	264	233
0.60	31	29	5.00	270	238
0.70	36	34	5.10	276	242
0.80	41	38	5.20	282	247
0.90	46	43	5.30	288	251
1.00	51	48	5.40	295	256
1.10	56	52	5.50	302	261
1.20	60	57	5.60	308	266
1.30	65	62	5.70	315	270
1.40	71	67	5.80	322	275
1.50	76	71	5.90	328	280
1.60	80	76	6.00	334	285
1.70	85	81	6.10	341	290
1.80	90	86	6.20	347	294
1.90	96	91	6.30	353	299
2.00	101	95	6.40	360	304
2.10	106	100	6.50	367	309
2.20	111	104	6.60	373	313
2.30	116	109	6,70	379	318
2.40	121	114	6.80	385	323
2.50	126	119	6.90	392	328
2.60	130	123	7.00	398	333
2.70	135	128	7.10	406	337
2.80	140	133	7.20	412	342
2.9C	145	138	7.30	418	347
3.00	151	143	7.40	425	352
3.10	156	148	7.50	431	357
3.20	161	152	7.60	438	362
3.30	166	157	7.70	445	367
3.40	171	161	7.80	451	372
3.50	176	166	7.90	458	377
3.60	182	171	8.00	465	382
3.70	188	176	8.10	472	387
3.80	195	181	8.20	478	392
3.90	201	185	8.30	485	397
4.00	207	190	8.40	492	402
4.10	213	195	8.50	499	407
4.20	218	200	8.60	505	...
4.30	225	204	8.70	512	...
4.40	231	209	8.80	519	...

[a]These values are arbitrarily given for 10 g of flour although determination is made on only 0.5 g of flour.

e. *Mobile solvent*—isopropyl alcohol—acetone-water (4:2:1 by volume).

f. *Chromogenic agent*—0.5% $KMnO_4$ in 1 *N* NaOH solution.

g. 1% *solutions of sorbitol, mannitol, and any other sugars or sugar alcohols which may be present.*

PREPARATION OF CHROMATOGRAPHIC COLUMN

Rinse enough Amberlite IRA-400 resin (**c**) into the column with water to form an 8″ layer. Add more water and invert several times to remove all air bubbles and redistribute the resin according to particle size. After resin settles, add a small glass wool plug, and pass 150 ml 5% NaOH solution through the column at the rate of about 5 ml/minute. Wash with about 200 ml water or until the eluate is neutral to pH paper.

NOTE: Do *not* let the column become dry. Do not retain the column in the hydroxide form for more than a few hours.

DETERMINATION

Chop a representative sample in a Hobart food chopper. Weigh a portion of the well-mixed sample containing about 400 mg sorbitol (usually about 10 g) into a 250 ml centrifuge bottle. Add 50 ml water, stopper and shake for 2 minutes. Centrifuge and decant into a 200 ml volumetric flask. Repeat this extraction until the volume is 200 ml. Mix well.

Pipette 5 ml extract into a 250 ml flask fitted with a reflux condenser. Add 10 ml 0.1 *N* HCl and reflux for 30 minutes. Add 25 ml 0.1 *N* NaOH through the top of the condenser, without disconnecting, and reflux at a rapid boil for 1.5 hours. Transfer the cooled solution quantitatively to a 100 ml volumetric flask, and dilute to mark with water. Pipette 25 ml of this solution onto the prepared ion exchange column and let it pass into the resin bed at about 3 ml/minute. When the solution has just passed into the column, rinse its sides with about 10 ml water and let this flow into the resin. Complete the elution with 200 ml water at about 3 ml/minute, collecting the eluate in a 500 ml Erlenmeyer flask.

Add 50 ml acid periodate solution (**a**) to the eluate and heat on a steam bath for 15 minutes. Cool, and add 2 g potassium iodide. Let solution stand for 5 minutes and titrate liberated iodine with 0.02 *N* thiosulfate solution (**b**) using starch solution as an indicator. Run a blank determination with 50 ml acid periodate solution plus amount of water equal to the volume of eluate collected.

1 ml 0.02 N thiosulfate = 0.3644 mg d-sorbitol ($C_6H_{14}O_6$).

NOTE: The titration is not valid for over 5 mg sorbitol. When the calculated result of the titration is 4.5 mg or more, take a smaller aliquot from the alkali-degraded solution in the 100 ml volumetric flask, pass through the ion exchange column, and titrate. Recovery obtained by these authors ranged from 97.1 to 99.1%. The method was tested on cakes and cookies.

There are other recent methods reported in the literature that are of interest. Among these is one developed by the Canada Food and Drug Directorate[13] using gas-liquid partition chromatography, and one by Graham[14], using the same acid and alkali hydrolysis procedure followed by a spectrophotometric determination.

Method 4-18. *Gluten in Flour.*

This term is loosely used. It is not a constituent of wheat, but is thought to be formed by a combination of glutenin and gliadin. Wheat is the only grain whose flour yields significant quantities of gluten, and damaged wheat kernels may yield little or none. For this reason bakers have used the gluten content as an index of the quality of flour. The USDA developed an empirical method for gluten:

DETERMINATION

Add about 15 ml water to 25 g of flour in a mortar or bowl, and work up into a dough by use of a spatula. Avoid adherence to the container. Let stand for an hour, then knead gently in a stream of cold tap water, letting the washings pass through a fine sieve until all starch and soluble matter are removed. Test for removal of starch by squeezing a little water from the ball of gluten into a beaker of clear, cold water, A cloudiness indicates that starch is still present. Gather up any bits of gluten caught by the sieve. Add them to the ball of gluten. Let the gluten stand in cold water for an hour, then work with the hands to squeeze out as much water as possible. Place the ball in a tared, flat-bottomed dish, and weigh as "moist gluten". Transfer to a 100° oven, and dry for 24 hours or to a constant weight. Weigh and report as "dry gluten".

Method 4-19. Detection of Rye Flour in Wheat Flour. (This is an AACC Method).

REAGENTS

a. 70% *ethyl alcohol.*

b. 1.0 *N sodium hydroxide in* 70% *ethyl alcohol.*

TEST

Weight 5 g of flour into a 50 ml centrifuge tube, add 20 ml of 70% alcohol, (**a**), and shake continuously for 15 minutes. Cool in an ice-salt bath at −3° for 10 minutes, stirring the viscous mass with a glass rod repeatedly during cooling.

Centrifuge for 5 minutes, then decant the supernatant liquid, which, if cloudy, must be filtered clear. To 10 ml of the filtrate add 0.5 ml of the 1.0 *N* alcoholic NaOH solution (**b**). Wheat flour gives a clear solution, or only slight turbidity. In the presence of rye flour there is a pronounced cloudiness or even a precipitate. As little as 10% rye flour can be detected in this manner.

Method 4-20. Detection of Soy Flour in Wheat Flour or Macaroni. (Not applicable to cooked cereal products. This is included in both AACC and AOAC Methods).

REAGENT

2% *urea.*

TEST

Mix an approximately 0.5 g sample with 5 ml of 2% urea solution in a test tube. Partly immerse a strip of red litmus paper in the liquid. Stopper the test tube and heat in a water bath held at 40° for 3 hours. If soy flour is present in more than traces, the litmus paper will turn blue from evolution of NH_3 liberated from the urea by urease present in soy flour. Urease has not been found in wheat flour.

Physical Properties of Dough

The rheological properties of dough, as measured by instruments designed to evaluate certain physical attributes, are useful indices of consumer acceptance of the finished, baked product. Such tests are used by bakery laboratories with realization that conclusions drawn from such data must at times be confirmed by subsequent test bakings.

Detailed directions for the use of these instruments are, we feel, beyond the purpose of this book. A brief description of the more widely used devices will suffice. Further information may be found in other texts.[15, 16]

These instruments may be classified in principle as:

1. *Recording dough mixers.* These are sold under various names—Farinograph, Mixograph, Rheograph, and others. These record changes in rheological properties during mixing time. They give an indication of such factors as dough development time (peak time), stability, gluten development, mixing tolerance, moisture absorption.

2. *Extensograph or alveograph,* which measures or records the extensibility, or breaking point, of dough under a uniformly applied stress. It is useful in measuring the effect of proposed additives on flour.

3. *Recording viscometer.* One such is the amylograph which is used to measure the effect of alpha-amylase on viscosity of flour as a function of temperature.

4. *Fermentograph,* to make and record gasometric measurements on fermenting dough.

Bread and rolls

FDA standards of identity have been promulgated for white bread, enriched white bread, milk bread, raisin bread and whole wheat (graham) bread.[17] The corresponding rolls and buns are included in these definitions. "Bread" is defined as the baked product in units weighing 0.5 lb or more after cooling, while "roll" and "bun" are defined as units weighing less than 0.5 lb after cooling.[17]

This definition and standard is too voluminous to be reproduced here (for example, the standard lists 94 optional ingredients for white bread alone). A summary may suffice:

All breads, rolls and buns shall contain not less than 62% solids.

The sole moistening ingredient of milk bread is milk, or, in lieu of milk, a combination (with or without water) of milk, concentrated, evaporated, dried or sweetened condensed milk, butter or cream, in quantities containing not less than 8.2 parts by weight of milk solids for each 100 parts by weight of flour used. In addition, the combination of such dairy ingredients shall be in such proportion that the weight of nonfat milk solids therein is not more than 2.3 times and not less than 1.2 times the milk fat therein.

Raisin bread shall contain not less than 50 parts by weight of seeded or seedless raisins for each 100 parts by weight of flour used.

Whole wheat (graham) bread shall be made from whole wheat flour as the sole flour ingredient.

Enriched bread is white bread containing the following mandatory enrichment ingredients per pound: (a) not less than 1.1 mg and not more than 1.8 mg thiamin, (b) not less than 0.7 mg and not more than 1.6 mg riboflavin, (c) not less than 10 mg and not more than 15 mg niacin or niacinamide, and (d) not less than 8 mg and not more than 12.5 mg iron (Fe).

Enriched bread may also contain per pound: (a) not less than 150 and not more than 750 USP units Vitamin D, and (b) not less than 300 mg and not more than 800 mg calcium (Ca).

The Canadian standards for bread[18] are generally similar to the United States standard. There are no minimum solids requirements and the enrichment values are slightly different. Enriched bread in Canada must be made from enriched flour as the only wheat flour ingredient, and shall contain not less than 2 parts by weight of skim milk solids for each 100 parts of flour used.

Federal Specification EE-B-671c, February 1, 1962, covers purchases by all Federal agencies for 9 varieties each of bread and rolls: white, white enriched, milk, whole wheat, raisin, rye, part whole wheat, French and Vienna. FDA standards of identity have been promulgated for the first 5 varieties and the Federal specification conforms to these standards. In all cases except for rye bread and rolls, the flour used must have been milled from hard wheat (other than durum). Total solids of all varieties shall not be less than 62%.

The flour ingredient used for "part whole wheat bread and rolls" shall consist of not less than 25% nor more than 50% whole wheat flour; that for "rye bread and rolls" shall consist of not less than 20% nor more than 40% rye flour.

There is also a military specification for canned bread, MIL-B-1070D, April 17, 1961. This is bread made by the usual straight dough, or the sponge and dough method, baked in the can, then sealed. The dough formula is given in the specification. Enriched hard wheat flour (or hard wheat flour plus enrichment) must be used. The baked loaf must occupy at least 95% of the volume of the can. Analytical requirements are:

Moisture not more than 35% or less than 32%; pH of crumb not to exceed 4.8, as determined by Method 4–9.

The Department of Defense has issued a specification for Instant Bread Mix—MIL-B-35092, September 28, 1962. This is a free-flowing powder containing not more than 5.5% moisture as determined by Method 4–2, Vacuum Oven Method, and a total fat content of not less than 4.3% as determined by Method 4–8, Ether Extraction Method.

Chemical leavening agents glucono-delta-lactone and sodium bicarbonate, both pre-coated with shortening, are required by this specification, and flour containing not more than 6.0% moisture must be used to ensure a finished product containing not more than 5.5% moisture. The specification should be consulted for processing and packaging directions.

Alimentary pastes (Pasta)

This is a generic term comprising macaroni products and noodle products. The trade names—macaroni, spaghetti, vermicelli, cannelloni, etc., designate a specific size or shape. The term "noodle" always denotes a product containing egg yolk; macaroni products by definition contain no egg yolk, although egg white may be used. Thus, we can have on the market macaroni (no egg yolk) or egg macaroni, which is a tubular shaped noodle. There are well over 100 sizes and shapes of alimentary pastes on the market.

FDA has promulgated standards of identity for certain of these products.[19] They define "macaroni" as 0.11″ to 0.27″ tubes, "spaghetti" as 0.06″ to 0.11″ tubes and "vermicelli" as cord-shaped (not tubular), not more than 0.06″ in diameter. "Noodles" are defined as ribbon-shaped. Egg white may be used to prevent collapse of the pasta tube during cooking; disodium phosphate promotes quick cooking; gum gluten makes the dough more tenacious; and monostearate improves the texture. See Tables 4-3 and 4-4 for summary of standards.

Canada has only one regulation defining

TABLE 4-3
FDA Standards for Macaroni Products
All foods listed shall contain not less than 87% solids

Product	Mandatory Ingredients	Optional Ingredients
Macaroni Products	Semolina, durum flour, farina, flour, or mixtures thereof with water.	Egg white[a], not less than 0.5% or more than 2.0% of the finished product. Disodium phosphate not less than 0.5% or more than 1.0%. Gum gluten so that the total protein content of the finished product is not more than 13%. Salt, onion, garlic, celery, bay leaf. Concentrated glyceryl monostearate not exceeding 2%.
Milk Macaroni	Same, except milk[b] is the only moistening substance.	Same, except egg white and disodium phosphate prohibited.
Whole Wheat Macaroni	Whole wheat flour or whole wheat durum only wheat ingredient permitted.	Same, except egg white, gum gluten and disodium phosphate not permitted.
Wheat and Soy Macaroni	Same as macaroni products except not less than 12.5% of the total flour shall be soy flour.	Same, except egg white and disodium phosphate not permitted. Gluten content same as for macaroni products.
Vegetable Macaroni	Same, except not less than 3% of the finished product shall be vegetable solids from tomato, artichoke, beet, carrot, parsley, spinach.	Same optional ingredients as wheat and soy macaroni.
Enriched Macaroni and enriched vegetable macaroni	Same as unenriched, except for enrichment/lb: 4.0–5.0 mg thiamine, 1.7–2.2 mg riboflavin, 27–34 mg niacin, 13–16.5 mg iron (Fe).	250–1000 USP units Vitamin D and/or 500–625 mg calcium (Ca)/lb. Partly defatted wheat germ not to exceed 5% of the finished food.

NOTE: [a]"Egg white" includes liquid, frozen or dried egg white, with or without water.
[b]"Milk" includes milk, concentrated, evaporated or dried milk and a mixture of butter with skim milk, concentrated, evaporated or dried skim milk, with or without water, in such proportion that the weight of nonfat milk solids is not more than 2.275 times the weight of milk fat therein.

TABLE 4-4

FDA Standards for Noodle Products

All foods listed shall contain not less than 87% solids and the noodle solids shall contain not less than 5.5% egg or egg yolk solids

Product	Mandatory Ingredients	Optional Ingredients
Noodles (Egg Noodles), Egg Macaroni Egg Spaghetti, Egg Vermicelli	Semolina, farina, durum flour, or mixtures thereof with liquid, frozen or dried eggs, liquid, frozen or dried egg yolk with or without water.	Onions, celery, garlic and/or bay leaf, salt to season. Gum gluten so that total protein content does not exceed 13%. Concentrated glyceryl monostearate not exceeding 2%.
Wheat and Soy Noodles	Same as noodles except that not less than 12.5% of the total flour shall be soy flour.	Same as noodles.
Vegetable Noodles	Same as noodles except that not less than 3% of the finished product shall be solids from tomato, artichoke, beet, carrots, parsley or spinach.	Same as noodles.
Enriched Noodles and enriched vegetable Noodles	Same as the unenriched product except for enrichment/lb: 4.0–5.0 mg thiamine, 1.7–2.2 mg riboflavin, 27–34 mg niacin, 13–16.5 mg iron (Fe). Carrots, because they impart an egg-yolk color, not permitted in enriched vegetable noodles.	250 to 1000 USP units of Vitamin D and/or 500 to 625 mg calcium (Ca)/lb. Partly defatted wheat germ not to exceed 5% of the finished food.

alimentary pastes: pastes sold as "egg noodles", "egg macaroni", etc., shall contain on the dry basis not less than 4% egg yolk solids.[20]

Federal Purchase Specification N-M-51d, January 14, 1963 (Macaroni, Spaghetti and Vermicelli) governs all government purchases of the foods listed in Table 4-3. It details the physical forms and sizes of the pieces. The permissible ingredients follow the FDA standards except that they must be made from semolina, durum flour, farina or hard wheat flour other than durum flour (except for whole wheat products, which may be made of whole wheat flour, whole durum wheat flour or both). Maximum acceptable moisture content is 12.0%. Limits for protein and ash vary with the type.

Chemical Requirements, Moisture-Free Basis

Type	Protein–% (NX5.7)	Ash–%
Plain	12.2	0.80[a,c]
		0.60[b,c]
Milk	12.7	0.98[a]
		0.80[b]
Whole Wheat	13.2	2.5
Wheat and Soy	15.2	1.7
Vegetable	12.1	1.5

[a]Applies to products in which the principal farinaceous ingredients are semolina and/or durum flour.
[b]Applies to products in which the principal farinaceous ingredients are farina and/or hard wheat flour other than durum flour.
[c]Exclusive of ash from Na_2HPO_4 in quick-cooking products and of added calcium in enriched products.

The Federal specification for wheat and soy macaroni products sets an upper limit of 15.0% soy flour in the total farinaceous ingredients. The FDA standard establishes only a minimum of 12.5% of the total flours as soy flour.

This specification also sets a carotenoid pigment color score of 2.5 p.p.m. on all of these products, as determined by Method 4–11.

The Department of Defense has issued

Military Specification MIL-M-35067, September 13, 1961, for Instant Macaroni. This product shall be made from semolina and water, the resultant dough extruded through a die, then precooked and dried to meet these analytical requirements:

Moisture	6% maximum
Ash, moisture-free basis	0.90 maximum
Protein (NX5.7), moisture-free basis	12.2 minimum

General Services Administration has issued Federal Specification N-N-591e, January 7, 1966, governing purchases of egg noodles by all Federal agencies. This permits egg noodles to be made from either semolina or durum wheat flour, as defined by the FDA Standard, provided that the ash of the semolina shall not exceed 0.78% and the ash of the flour shall not exceed 0.85% (both on a 14.0% moisture basis). The egg solids content of the 3 types of noodles covered, 5.5% minimum egg yolk solids, is controlled by the minimum lipid phosphorus content. The full chemical requirements are:

Type	Minimum Limits of		
	Lipoid P[a] (P_2O_5)%	Protein[a] (NX5.7)%	Moisture %
Plain or enriched	0.136	13.1	12.0
Wheat and soy	0.136	17.3	12.0
Vegetable	0.136	13.1	12.0

[a]Moisture-Free Basis.

Setting a lipid phosphorus limit as a measure of egg content judiciously avoids any controversy over the ratio of lipid phosphorus to egg yolk solids. Similarly, analytical interpretation of the soy flour requirement (not less than 12.5% nor more than 15.0% of the combined wheat and soy ingredient) is avoided by the protein minimum set for wheat and soy egg noodles. The specification further requires that the soy flour used shall be low fat flour as produced by solvent extraction. Vegetable egg noodles shall contain not less than 3.0% nor more than 5.0% vegetable solids (red tomato, artichoke, beet, parsley, carrot or spinach). Since this specification was issued in 1966 these minimum limits can be said to represent modern commercial practice.

Breakfast foods

Ready-to-eat, dry, packaged cereals, familiarly known as "breakfast foods", had their origin around 1880 in Battle Creek Sanitarium, Battle Creek, Michigan as a health food, although there is record of a short-lived product called "Granola" in Danville, New York, in 1863. The Battle Creek product, developed by Dr. J. H. Kellogg, Medical Superintendent, was first served in the Sanitarium. Later, C. W. Post saw the possibilities as a convenience food, and it became big business.

Cereals requiring cooking before serving are either granular meals like farina, flaked wheat and corn meal, or flakes like rolled oats. Some of these may have an initial steam-cook, as in Instant Oatmeal, before drying. Others are simply milled and sieved, or flaked by rolling or pressing. There is no FDA definition or standards for these cereals, except standards for farina (see Table 4-1) and corn meal.[21]

Federal Specifications for Breakfast Cereals

The Federal Supply Service and Department of Defense have issued purchase specifications for breakfast cereals. Federal Specification N-C-196e, July 27, 1962, covers 24 ready-to-eat cereals made from wheat, corn, rice or oats. A list of these products, and their analytical requirements, are given in Table 4-6.

Federal Specification N-C-201d, May 20, 1966, includes regular and quick-cooking uncooked wheat cereals. The chemical requirements laid down by the specification are as in Table 4-5.

TABLE 4-5
Chemical Requirements for Uncooked Wheat Cereals

Type and Class Regular and Quick-Cooking	Protein (NX5.7) Minimum	Ash Maximum	Crude Fiber		Moisture Maximum
			Minimum	Maximum	
	%	%	%	%	%
Farina: plain and enriched	8.5	0.60	...	...	12.0
with bran and germ	10.5	...	0.7	1.5	12.0
with malt	10.0	0.60	...	...	12.0
Whole Wheat Meal	11.0	...	2.0	3.0	12.0
Rolled or Flaked Wheat	11.0	...	2.0	3.0	12.0
Cracked Wheat	11.0	...	2.0	3.0	12.0

NOTE: Ash and crude fiber limiting values are on a moisture-free basis; ash value is exclusive of ash contributed by added calcium in enriched farina or added Na_2HPO_4 in quick cooking farina.

"Quick-cooking wheat cereals" are defined as products that have been so prepared that they can be sufficiently cooked for table use by boiling in water at 99–100° in an open vessel for 5 minutes. In the case of farinas cooking time is reduced by incorporating from 0.5% to 1.0% of disodium phosphate into the farina.

Two government purchase specifications for Rolled Oats Cereal have been issued. Federal Specification N-C-195, September 17, 1965, describes the 3 types included as:

Type I—Regular: reasonably thin flakes. Not less than 70% shall be retained on a U.S. No. 7 standard sieve and not more than 5% shall pass through a U.S. No. 25 standard sieve.

Type II—Quick-cooking: very thin flakes. Not less than 40% shall be retained on a No. 7 sieve and not more than 6% shall pass through a No. 25 sieve.

Type III—Steam Table: reasonably thick flakes. Not less than 90% shall be retained on a No. 7 sieve and not more than 1% shall pass through a No. 25 sieve.

The chemical requirements of all 3 types are:

Moisture	12.0% maximum
Protein (NX6.25)	15.0% minimim, moisture-free basis
Crude Fiber	1.8% maximum, moisture-free basis

"Quick-cooking" means a product that can be sufficiently cooked for table use by boiling in an open vessel for 1 minute, covering the pan and allowing to stand for 3 minutes (or cooking for 3 minutes without additional standing). "Steam table" type means a product that can be sufficiently cooked for table use by adding to boiling water, continuing to boil for 5 to 6 minutes, transferring to a covered container on a steam table and holding at least 30 minutes.

Instant Rolled Oats, Military Specification MIL-C-35070, December 27, 1961, covers purchases by the armed forces. This is made by steel-cutting the groats, steam-cooking, rolling and drying. Not more than 10% of the finished product shall pass through a No. 20 U.S. Standard sieve.

Chemical requirements

Moisture	6% maximum (vacuum oven 100°)
Protein (NX6.25)	15.0% minimum[a]
Ash	1.8% maximum[a]
Fat	7.0% maximum, Method 4-8[a]
Fat acidity	30 mg KOH/100 g sample maximum

[a]On moisture-free basis.

Composition of Breakfast Foods

The moisture content of commercial ready-to-eat cereals has decreased somewhat since

TABLE 4-6

Chemical Requirements for Breakfast Cereals

Federal Specification N-C-196e

Name of products	Minimum limits of			Maximum limits of		
					Moisture	
	Crude fiber[a]	Protein[a] (Nx6.25)	Reducing sugars[a,b]	Crude fiber[a]	Process inspection prior to packaging	Inspection 72 hours or more after packaging
	PERCENT	PERCENT	PERCENT	PERCENT	PERCENT	PERCENT
Shredded wheat biscuits, large	1.8	—	—	3.0	8.0	11.0
Shredded wheat biscuits, small	1.8	—	—	3.0	4.0	6.0
Loose shredded wheat	1.8	—	—	3.0	4.0	6.0
Pressed-flake whole wheat biscuits	1.8	—	—	3.0	4.0	6.0
Wheat flakes	1.5	—	—	3.0	4.0	6.0
Malted wheat flakes	1.5	—	12.0	3.0	4.0	6.0
Bran flakes (25 to 40 percent)	3.5	—	—	4.7	4.0	6.0
Bran flakes (25 to 40 percent) and raisins	3.5[c]	—	—	4.7[c]	7.0[c]	9.0[c]
Wheat bran (prepared)	7.5	—	—	—	4.0	6.0
Puffed wheat (gun-puffed)	—	—	—	—	4.0	6.0
Puffed wheat (gun-puffed), coated	—	—	—	—	4.0	6.0
Malted cereal granules	—	—	—	—	4.0	6.0
Corn flakes	—	—	—	—	4.0	6.0
Corn flakes, coated	—	—	—	—	4.0	6.0
Corn cereal (gun-puffed)	—	—	—	—	4.0	6.0
Corn cereal (gun-puffed), coated	—	—	—	—	4.0	6.0
Shredded corn with soya	—	18.0	—	—	4.0	6.0
Shredded corn, small biscuits (oven-puffed)	—	—	—	—	4.0	6.0
Puffed rice (gun-puffed)	—	—	—	—	4.0	6.0
Rice cereal (oven-puffed)	—	—	—	—	4.0	6.0
Rice cereal (oven-puffed), coated	—	—	—	—	4.0	6.0
Rice flakes	—	20.0[d]	—	—	4.0	6.0
Shredded rice, small biscuits (oven-puffed)	—	—	—	—	4.0	6.0
Oat cereal (oven- or gun-puffed)	—	—	—	—	4.0	6.0

[a]These limiting values are on a moisture free basis.
[b]Calculated as maltose monohydrate.
[c]This limiting values applies only to the bran flakes ingredient.
[d]When specified.

their first inception. Bailey[22] compiled analyses of many common foods in 1935 in connection with a study of diets for diabetics. He reported the moisture range in these products as:

Corn flakes	7.7–12.1%
Wheat flakes	5.2–10.8%
Shredded Wheat	5.8–11.3%
Bran flakes	9.6–11.6%

Contrast these figures with the moisture

limits given in Table 4-5 for government purchases of ready-to-eat cereals and those given for Table 4-6 below for the composition of modern breakfast cereals, as compiled from USDA Agriculture Handbook No. 8.

One point is worth mentioning in considering the nutritive value of these cereals, particularly that contributed by the protein. Sure[23] has pointed out that heat processing may impair the nutritive value of cereal proteins. He compared the "protein efficiency ratio" (gain in body weight/gram of protein intake) against the calculated protein content as determined from the analysis for nitrogen to arrive at the biological nutritive value.

Composition of Whole Grains

The composition of whole grains, except possibly rice, is of little interest to the food chemist, except for those factors that affect the quality of the flours and other food ingredients milled therefrom. USDA Handbook No. 8[24] and National Research Council Bulletin on Composition of Cereal Grains (Selected Reference K) are good sources for such information.

Juliano and his co-workers at the International Rice Research Institute, Manila Philippines, have published analyses of 16 varieties of rice grown at the Institute in 1961–62.[26]

Composition of Flours

Most commercial wheat flours are blends of several mill streams, the blend depending on the end-use of the flour. Workers at Kansas Agricultural Experiment Station have made extensive studies of both wheat and flour made therefrom. Analyses of modern American flours and millstreams are included in two 1967 reports from this Station[27, 28] and their Bulletin 392, published in 1957 contains further information. See also the Selected References at the end of this chapter.

The analytical requirements for wheat flours purchased by all Federal agencies, given in Federal Specification for Wheat Flour, N-F-481h, December 20, 1965, and that for whole wheat flour, NF-485a, August 17, 1965, give a rather accurate picture of modern flours. These requirements are as follows for

TABLE 4-7
Composition of American Breakfast Cereals

Product	Water—%	Ash—%	Protein—%[a]	Fiber—%	Other carbo-hydrates—%	Fat—%
Corn flakes	3.8	0.7	7.9	0.7	84.6	0.2
Puffed corn	3.6	0.4	8.1	0.4	80.4	4.2
Shredded oats	3.9	3.2	18.8[b]	1.8	70.2	2.1
Puffed oats	1.9	2.4	6.7	0.7	84.9	3.4
Rice flakes	3.2	0.4	5.9	0.6	87.7	0.3
Puffed rice	3.7	2.9	5.9	0.6	89.5	0.4
Wheat flakes	3.5	4.2	10.2	1.6	78.9	1.6
Puffed wheat	3.4	1.6	15.0	2.0	76.5	1.5
Shredded wheat	6.6	1.6	9.9	2.3	77.6	2.0
Bran flakes 40%	3.0	4.4	10.2	3.6	77.0	1.8
Farina	10.0	0.4	11.4	0.4	76.6	0.9
Oatmeal	8.3	1.9	14.2	1.2	67.0	7.4
Rolled wheat	10.1	1.8	9.9	2.2	74.0	2.0

[a]Handbook No. 8, p. 159 used the following protein factors to convert N to protein: corn—6.25, oats—5.83 rice—5.95, wheat—5.70.
[b]With added protein.

plain, enriched, bleached or unbleached hard-wheat, soft wheat or clear flours, and for bleached, unbleached or bromated whole wheat flour:

	Moisture—% Maximum	Protein (Nx5.7)—%[a] Min.	Max.	Ash—%[a,c] Maximum
Hard-wheat flours	13.5	11.0	—	0.46
Soft wheat flours	13.5	6.2	9.2	0.38
Clear flours	13.5	13.0	—	0.75
Whole wheat flour[b]	13.5	11.0	—	1.9

[a]Basis of 14.0% moisture.
[b]Also shall contain not more than 2.8% crude fiber on 14.0% moisture basis.
[c]Exclusive of ash contributed by any compound of calcium used for enrichment in enriched flour. The extractable fat content of these flours usually range from 0.8% to 1.2% and crude fiber from 0.2% to 0.4%.

Composition of Commercial Bread

The FDA Standards and Definition of Identity for Breads are, generally, not very restrictive as to composition, other than the total solids requirement of 62.0%. Despite this, analyses of the standard breads (white, whole wheat, raisin) over the last 35 years or so do not show much difference in quantities of the main nutrients. This is shown in comparing analyses of market breads in 1934 and again in 1956–57 by the Connecticut Agricultural Experiment Station:[29]

Average Composition of Bread 1956-1957 and 1934

	White		Whole Wheat		Raisin	
	1956–57	1934	1956–57	1934	1956–57	1934
Water—%	36.28	35.3	38.49	38.4	33.16	None Analyzed
Ash—%	1.89	1.1	2.50	1.3	1.76	
Protein (N × 5.7)%	10.31	9.2	9.86	9.7	8.25	
Fiber—%	0.34	0.5	1.41	1.2	0.35	
Other carbohydrates	48.54	52.66	44.78	48.5	53.04	
Fat—%	2.56	1.3	2.95	0.9	3.45	
Calories/100 g	258	259	245	241	270	

Agriculture Handbook No. 8 (revised December 1963), lists representative composition of American breads. These should indicate current bakery practices:

Representative Composition of American Bread[24]

Bread	Water—%	Ash—%	Protein—%	Fiber—%	Other Carbo-hydrates—%	Fat—%
White, 2–4% MSNF[a]	35.8	2.0	8.7	0.2	50.0	3.2
Whole wheat 2% MSNF	36.4	2.4	10.5	1.6	46.1	3.0
Rye—1/3 rye, 2/3 clear flour	35.5	2.2	9.1	0.4	51.7	1.1
Pumpernickel	34.0	2.6	9.1	1.1	52.0	1.2
Raisin	35.3	1.7	6.6	0.9	52.7	2.8
Salt-rising	36.5	1.0	7.9	0.2	52.0	2.4

[a]Milk Solids Non Fat

Harris has published analyses of 44 typical wheat and rye breads from 14 countries. These include proximate analysis, vitamins, minerals and certain amino acids. He also gives these breads a "nutritional rating".[30]

Other cereal products

Corn

Definitions and Standards of Identity have been established by the Food and Drug Administration for 15 corn food products. Their analytical requirements are given in Table 4-7, and the text following the table.[31]

TABLE 4-8

U.S. Standards of Identity for Milled Corn Products[a]

Product	Granulation Woven wire U.S. Standard sieves	On Moisture-Free Basis Crude fiber	 Fat
White corn meal and Yellow corn meal	95% minimum through No. 12 45% minimum through No. 25 35% maximum through No. 72[b]	12% minimum	Difference ± 0.30% from fat content of original cleaned corn
Bolted white corn meal and Bolted yellow corn meal	95% minimum through No. 20 45% minimum through No. 45 25% maximum through No. 72[b]	less than 1.2%	not less than 2.25%
White corn flour and Yellow corn flour	98% minimum through No. 50[c]	Not to exceed that of cleaned corn from which made	
Grits (corn grits, hominy grits) and Yellow Grits (yellow corn grits, yellow hominy grits)	95% minimum through No. 10 20% maximum through No. 25	1.2% maximum	2.25% maximum
Quick Grits	Same as grits and yellow grits, except in preparation grits are lightly steamed and slightly compressed.		

[a]Maximum moisture limit 15% for corn meals, bolted corn meals, degermed corn meals, corn flours, and enriched corn meals. No moisture limit set on other products.

[b]Through a No. 72 XXX grits gauze, by manual shaking method, p. 85.

[c]As determined by Method 4-22.

The definition and standard for Self-rising White Corn Meal and self-rising Yellow Corn Meal defines these products as "an intimate mixture of white (or yellow) corn meal, sodium bicarbonate, and one or both of the acid-reacting substances monocalcium phosphate and sodium aluminum phosphate, seasoned with salt." The acid-reacting substance is added in sufficient quantity to neutralize the bicarbonate, and the combined weight of the acid-reacting substances and the bicarbonate is not more than 4.5 parts to each 100 parts by weight of corn meal used.

The product shall not yield less than 0.5% CO_2 when tested with the Chittick apparatus by AOAC Method **13.017**.

Enriched corn meals and corn grits shall conform to the standards for the unenriched foods, and in addition shall contain as enrichment per pound: 2.0–3.0 mg thiamine, 1.2–1.8 mg riboflavin, 16–24 mg niacin or niacinamide, and 13–26 mg iron (Fe). It may contain 250–1000 USP units Vitamin D and 500–750 mg calcium (Ca). Dried yeast up to 1.5% may be used as one source of enrichment.

Enriched Rice, Enriched Parboiled Rice

These are the only cereal or bakery products not yet mentioned for which an FDA standard of identity has been promulgated.[32] This standard (which does not apply to "coated rice" bearing a film of talc and glucose) gives requirements only for enrichment. These are: each pound shall contain 2.0–4.0 mg thiamine, 1.2–2.4 mg riboflavin, 16–32 mg niacin or niacinamide and 13–26 mg iron (Fe). It may also contain 500–1000 mg calcium (Ca) and 250–1000 USP units Vitamin D. Enriched parboiled rice may contain not more than 0.0033% butylated hydroxytoluene as an antioxidant.

Tests for Compliance with Granulation, or Sieving Requirements, for Meals and Flours

Several methods for determining granulation of meals, flours, and other particulate substances appear in many specifications and standards. The basic apparatus is the same in most of these, viz. a nested set of standard sieves (either U.S. or Tyler series) with a bottom pan, and a cover that will fit all the sieves. Usually the directions call for woven-wire, 8″, full-height sieves, or woven silk cloth fitted over a frame of the same dimensions as the wire-cloth sieves.

Shaking the nested sieves may be done as directed by the specification, either by hand or by mechanical means. Both are empirical, so that the prescribed procedure must be followed exactly.

Method 4-21. *Granulation, Mechanical Shaking.*

This method is prescribed by the FDA standard for wheat flours.[34] See also Method 4–1.

APPARATUS

a. *Nested set of standard sieves* as prescribed by the specification involved.

b. *Ro-Tap mechanical sifter*, or equivalent.

DETERMINATION

Assemble the sieves in order, with the finest mesh sieve fitted into the pan, followed by the next finest, and so on. Place a ping pong ball or other sieving aid in each sieve and insert the assembly into the Ro-Tap sifter. Use a No. 70 sieve alone for white flours and durum flour; for whole wheat flour use Nos. 8 and 20 sieves and for instantized flour, Nos. 20 and 200 sieves.

Weigh a 100 g sample into the top sieve, place the cover on the sieve and shake in the Ro-Tap sifter for exactly 5 minutes. Weigh residues on each sieve and report as cumulative percents passing through each sieve.

Granulation, Manual Shaking for Corn Meal

This method is prescribed by the FDA Standard for corn meal. It is given in Code of Federal Regulations.[35,36]

Method 4-22. *Granulation by Solvent Extraction.*

APPARATUS

a. *Truncated metal cone* (top diameter 5 cm, bottom 2 cm, height 4 cm), fitted at the bottom with 70-mesh wire cloth complying with the specification for a U.S. No. 70 standard sieve.

b. *U.S. No. 50 standard seive*, 8″ diameter, full height.

DETERMINATION

Weigh a 5 g sample into the tared truncated cone. Attach the cone to a suction flask. Wash the sample with 150 ml of petroleum ether, without suction, while gently stirring the sample with a small glass rod. Apply suction for 2 minutes after washing is completed, then shake the cone for 2 minutes with a vigorous horizontal motion, striking the side against the hand. Weigh the cone and contents. The loss in weight, calculated to percent by weight, shall be considered the percent passing through No. 70 woven wire cloth.

Transfer the residue from the cone to the No. 50 sieve. Shake for 2 minutes as described above for the cone, remove and weigh the residue. Calculate this as percent by weight and subtract from 100 to obtain the percent passing through the No. 50 sieve.

Analytical methods for bread, alimentary pastes and breakfast foods

The general methods given in Chapter 1, and the methods listed under Flour in this chapter, are applicable to the analysis of bread, alimentary pastes and breakfast foods. Methods devised primarily for these foods alone will be given in the following pages, under the appropriate headings.

Method 4-23. *Preparation of Sample of Bread and Determination of Total Solids.*

The FDA definition and standard of identity directs that this method be used[37] for determination of total solids of bread. It is an AOAC method.

DETERMINATION

Use 1 entire unit if the baked unit weighs 1 lb or more. If the baked unit weigh less than 1 lb, use such number of units as weigh 1 pound or more.

a. *All types of bread not containing fruit*: Accurately weigh the sample to the nearest 0.2 g immediately upon receipt. Designate this weight as "A". Cut bread into slices 2–3 mm thick ($\frac{1}{2}$ loaf may be used). So that no crumbs are lost, spread slices on paper. Let dry in a warm room (15–20 hours) and when apparently dry, break into fragments. If bread is not entirely crisp and brittle, let it dry longer until it is in equilibrium with the air. Quantitatively transfer the air-dried bread to scale pan and accurately weigh. Designate this weight as "B". Grind to pass a 20-mesh sieve, mix well and keep in air-tight container. Determine percent of total solids (C) in the dried bread by Method 4–2, using either the vacuum oven method at 98–100° or the air oven method at 130°.

Percent total solids in sample = $(B \times C/A)$.

b. *Raisin bread and bread containing raisins and fruit*:

Proceed as in (**a**) except to comminute the air-dried sample twice through a food chopper instead of a grinder. Dry the 2 g air-dried sample in an uncovered dish in a vacuum oven at 70° and not more than 50 mm mercury pressure for about 16 hours.

NOTE: The air-dried sample, if preserved in an air-tight container, may be used as the sample for other determinations, such as protein, fat, ash, minerals, crude fiber, etc. Percentages should be calculated on the basis of the original bread sample.

Method 4-24. *Lactose and Non-Fat Dry Milk Solids in Bread.* (This is an Official AACC Method).

REAGENTS

a. *Pancreatic amylase.* The activity should be checked by standardizing with a known amount of lactose.

b. *Baker's yeast.* Fleischmann's yeast or equivalent is satisfactory. Wash 25 g yeast 5 times with 4 times its volume of water, centrifuging and decanting off the supernatant liquid between each washing. The last washing should be clear. Suspend the washed yeast with 100 ml of water and store at 0 to 4°. Prepare 24 hours before using and do not store longer than 4 days.

c. *Neutral lead acetate*, saturated solution.

d. *Mercuric chloride*, 5% solution $HgCl_2$.

e. *Phosphotungstic acid*, 20% solution.

f. *Hydrogen sulfide*

DETERMINATION

Weigh 15 g air-dried, ground sample (Method 4–23) into 300 ml volumetric flask, add 200 ml water and digest in boiling-water-bath for 30 minutes, with occasional shaking. Cool, make to volume, mix well and centrifuge. Pipette 150 ml of the supernatant liquid into 300 ml volumetric flask. Make to volume with ethanol, shake, let stand 10–15 minutes, and centrifuge.

Pipette 250 ml of the supernatant liquid into a beaker, and boil until no odor of ethanol is perceptible. Transfer to a 200 ml volumetric flask, keeping volume below 150 ml. Cool to 55°, add 0.25 g pancreatic amylase (**a**) and digest at 55° in a thermostatically controlled bath for 30 minutes. Place flask in boiling-water-bath for 15 minutes, cool, add another 0.25 g of pancreatic amylase and digest at 53–55° for 30 minutes.

Place flask in boiling-water-bath for 15 minutes, cool to room temperature, add 10 ml washed baker's yeast suspension (**b**), plug with cotton and ferment at 26.5–29° for 17–18 hours. Make to volume with water, mix and centrifuge. Transfer 190 ml of the supernatant liquid to beaker, evaporate to 25–50 ml volume, and transfer into a 100 ml volumetric flask by washing with hot water. Add 10 ml saturated lead acetate solution (**c**), cool, make to volume and centrifuge. Transfer 75 ml of this supernatant liquid to 100 ml volumetric flask, add 2.5 ml $HgCl_2$ solution (**d**), let stand 25–30 minutes with repeated shaking, add 10 ml phosphotungstic acid solution (**e**), make to volume, mix and centrifuge. Decant and, if solution is not clear, filter through a dry filter paper.

Saturate the decanted liquid with H_2S, filter and pipette 50 ml of the clear, colorless filtrate into a 300 ml Erlenmeyer flask. Boil until free from H_2S. Make the volume to 50 ml and determine lactose by the Munson-Walker gravimetric Methods 16–10 and 16–6, determination (**b**), copper by direct weighing.

Obtain the equivalent weight of lactose from Hammond Table 23-6, and calculate the percent of lactose from the following equation:

Weight of sample represented in final aliquot is 2.327 g, correcting for volume of precipitate in original 300 ml volumetric flask (13 ml).

$$15\text{ g} \times \frac{150}{(300-13)} \times \frac{250}{300} \times \frac{190}{200} \times \frac{75}{100} \times \frac{50}{100} = 2.327\text{ g}$$

$$\%\text{ Lactose in air-dried sample} = \frac{(\text{g lactose} - 0.006)100}{0.97 \times 2.327}$$

In this formula 0.006 is correction for a blank and 0.97 corrects for lactose fermentation loss. % Lactose × 2 = % milk solids not fat.

Express % lactose and milk solids not fat on basis of fresh bread, or on any desired basis.

Method 4-25. *Preparation of Sample of Alimentary Pastes.*

Select from the sample on hand sufficient strips or units to ensure a representative sample. These should be drawn from representative unopened packages when available. Break into small fragments and grind ½ pound through a mill until all material passes through a 20-mesh sieve. Preserve the sample in a sealed, air-tight container to prevent loss of moisture.

For years regulatory chemists had been estimating the egg content of noodles and other baked products by determining the lipid content, and the phosphoric acid content of the lipid expressed as P_2O_5, by AOAC Methods **13.033** and **13.034**. The quantity of egg was then calculated from formulae derived from analysis of authentic samples of egg products, flour, semolina and farina.[38, 39]

Mitchell[40] uncovered a weakness in the estimation of egg content of cereal products from their lipid P_2O_5 content when he showed the possibility of loss of phosphorus during storage of noodles or by incipient decomposition of the egg constituent. We therefore give the digitonin method 4-26 as our method of choice. For other approaches to the problem see text references 41 and 42.

Method 4-26. *Sterols Egg in Noodles. AOAC Method* ***13.131****).*

Weigh 5 g prepared sample (Method 4–25) into 300 ml Erlenmeyer flask, and add, with shaking, 15 ml HCl (1 + 1) so that particles adhering to the side are kept at a minimum. Heat 30 minutes on steam bath, shaking frequently to break lumps and insure complete hydrolysis. While cooling the inclined flask under tap water, carefully add 15 g KOH pellets at such a rate that liquid may boil, but not so violently as to cause loss. Cool, add 20 ml ethanol, rinsing down sides of flask, insert an air condenser and heat on steam bath 45 minutes.

Add 25 ml of water, rinsing sides of flask, mix, and cool. Add 50 ml ether, swirl vigorously 1 minute and transfer to a 500 ml separatory funnel. Wash flask with 25 and 10 ml portions of ether, then with 50 ml 1% KOH, adding washings to separatory funnel in a slow stream while gently swirling the liquid. Continue swirling 10–15 seconds. Let liquids separate and slowly draw lower soap solution into a 250 ml separatory funnel. Do not drain off any small quantity of emulsion or insoluble matter at the interface. Rinse the sides of the 500 ml separatory funnel with 5 ml 1% KOH, let separate and drain this into the 250 ml separatory funnel. Add 25 ml ether to the KOH solution in the 250 ml separatory funnel shake vigorously 1 minute, let separate and discard lower layer. Transfer remaining ether solution to the 500 ml separatory funnel, rinsing the the smaller one with 10 ml ether, adding rinsings to the larger separatory funnel; wash ether as before with 3 additional 50 ml portions of 1% KOH, still avoiding any draining off of emulsion or insoluble matter. Wash ether solution twice by swirling with 50 ml water, and drain as much of aqueous layer as possible without losing any ether.

Transfer the ether to a 300 ml Erlenmeyer flask, add a porcelain chip, rinse separatory funnel with three 5 ml portions of ether, and rinse stem of separatory funnel with ether, adding all rinsings into the flask. Evaporate off ether on steam bath.

Dissolve residue in 5 ml acetone. Filter, using suction if necessary, through a Knorr extraction tube containing a medium porosity fritted glass disk (Ace Glass Co., Vineland, N.J. No. 8571D or equivalent, covered with a few grams of washed and ignited sand), into a 100ml centrifuge tube or test tube under a bell jar. Wash flask and Knorr tube 3 times with 4 ml acetone each time and rinse tube and stem with a little acetone, making a total volume of about 20 ml.

Dissolve 40 mg digitonin in 5 ml of 80% alcohol, warming if necessary under hot water tap at 40–50°, and add to the acetone solution. (Products containing over 6% egg yolk solids, moisture free basis, require additional digitonin solution or use of an aliquot portion for precipitation.)

Mix by rotating, place a porcelain chip in the tube, and suspend the tube in steam bath with a small amount of steam to avoid boiling or spattering. Evaporate nearly to dryness, add 50 ml hot water (near boiling) and stir well with a glass rod to disperse the precipitate and dissolve excess digitonin. Place tube in boiling water bath for several minutes, stirring frequently. Cool to about 60°, and add 25 ml acetone. Mix well and cool to room temperature in a beaker of cold water.

When precipitate has nearly all settled (about 15 minutes) remove glass rod, rinsing it off with acetone. Decant into a previously dried and weighed 10 ml capacity Gooch crucible containing an asbestos pad covered with about 1 g washed and ignited sand. Transfer all the precipitate to the Gooch crucible by means of acetone from a wash bottle (avoid transfer of any particle of chips). Finally, rinse the crucible with acetone, then 5 ml ether, dry 30 minutes at 100° and weigh. Check weight after a second 30 minute drying.

Weight of residue × 0.243 = weight of sterol.

Report percent of sterol, moisture-free basis.

NOTE: Formula for Calculating Percent Egg Yolk (or Whole Egg) Solids in Noodles:

$$\% \text{ yolk solids} = \frac{\% \text{ sterols (moisture-free basis)} - 0.052}{2.87 - 0.052}$$

$$\% \text{ whole egg solids} = \frac{\% \text{ sterols (moisture-free basis)} - 0.052}{2.08 - 0.052}$$

in which 0.052 is average percent sterol in durum flour or semolina, 2.87 is average percent sterol in egg yolk solids and 2.08 is average percent sterol in dried whole egg as found by Munsey.[43]

Studies made by the AOAC in 1967 reveal significant changes in sterol content of durum flour, semolina and eggs. See Table 8-2.

TEXT REFERENCES

1. *Code of Federal Regulations—Title* 21 *Food and Drugs, Parts* 1 *to* 129. Section 15.1 to 15.150, inclusive. Washington, D.C.: Government Printing Office, (Jan. 1, 1966).
2. *Office Consolidation of the Food and Drugs Act and of the Food and Drug Regulations.* Regulation B.13.001—B.13.013. Ottawa: Queen's Printer (1966).
3. Halverson, J. C., chairman: *Cereal Science Today* 12:93 (1967).
4. Collins, F. H.: *J. Assoc. Offic. Agr. Chemists*, 32: 257 (1949).
5. U.S.D.A., Bureau of Chemistry: *Bulletin*, **137**: 144 (1911); *Methods of Analysis A.O.A.C.* 6th ed. **20**: 54 (1945).
6. Harrel, C. J.: *Ind. Eng. Chem.*, **44**: 95 (1952).
7. Snell, F. D. and Cornelia T.: *Colorimetric Methods of Analysis.* 3rd ed. New York: D. Van Nostrand & Co. Inc. (1949). II, 803.
8. Rainey, W. L.: *J. Assoc. Offic. Agr. Chemists*, **37**: 395 (1954).
9. Auerbach, M. F., Eckert, H. W. and Angell, E.: *Cereal Chemistry*, **26**: 490 (1949).
10. Sandstet, R. M.: *Cereal Chemists*, **14**: 767 (1937).
11. Hughes, E. E. and Acree, S. F.: *Ind. Eng. Chem. Anal. Ed.* **6**:123 (1934), **9**:318 (1937).
12. Hundley, H. K. and Hughes, D. D.: *J. Assoc. Offic. Agr. Chemists*, **49**: 1180 (1966).
13. Jones, H. G., Smith, D. M. and Sahasrrabudhe, M.: *J. Assoc. Offic. Agr. Chemists*, **49**: 1183 (1966).
14. Graham, H. D.: *J. Food Science*, **28**: 440 (1963).
15. Selected Reference I.
16. Selected Reference F, Chap. 5.
17. Ref. 1, Sec. 17.1—17.5.
18. Ref. 2, Reg. B.13.021—B.13.030, inclusive.
19. Ref. 1, Sec. 16.1—16.12, inclusive.
20. Ref. 2, Reg. B.13.051.
21. Ref. 1, Sec. 15.500—15.505, inclusive.
22. Bailey, E. M.: *Conn. Agric. Experiment Station Bulletin* 373, New Haven (1935).
23. Sure, B.: *Food Research*, **16**: 161 (1951).
24. Watt, B. K. and Merrill, A. L.: "Composition of Foods." *U.S.D.A. Agri. Handbook No.* 8. (revised Dec. 1963).
25. Selected Reference M.
26. Juliano, B., Bautista, G. M., Lugan, J. C. and Reyes, A. C.: *J. Agr. Food Chem.*, **12**: 131 (1964).
27. Farrell, E. P., Ward, A., Miller, G. D. and Lovett, L. A.: *Cereal Chemistry*, **44**: 39 (1967).
28. Waggle, D. H., Lambert, M. A., Miller, G. D. and Deyoe, C. W.: *Cereal Chemistry*, **44**: 48 (1967).
29. Fisher, H. J.: *Bull.* 629, *Conn. Agricultural Experiment Station*, New Haven (1959).
30. Harris, R. S.: *J. Amer. Dietetic Assoc.*, **38**: 26 (1961)
31. Ref. 1, Sec. 15.500—15.514.
32. Ref. 1, Sec. 15.525.
33. Ref. 2, Reg. B.13.010
34. Ref. 1, Sec. 15.1 (c) (4)
35. Ref. 1, Sec. 15.500 (b) (2)
36. Ref. 1, Sec. 15.508 (b) (2)
37. Ref. 1, Sec. 17.1 (a) (16)
38. Buchanan, R.: *J. Assoc. Offic. Agr. Chemists*, **7**: 407 (1923–24).
39. Munsey, V. E., *J. Assoc. Offic. Agr. Chemists*, **35**: 693 (1952) **36**: 760 (1953).
40. Mitchell, L. C.: *J. Assoc. Offic. Agr. Chemists*, **15**: 282 (1932).
41. Munsey, V. E.: *J. Assoc. Offic. Agr. Chemists*, **36**: 766 (1953).
42. Haenni, E. O.: *J. Assoc. Offic. Agr. Chemists*, **24**: 114, 143 (1941), **25**: 365, 639 (1942).
43. Munsey, V. E.: *J. Assoc. Offic. Agr. Chemists*, **37**: 408 (1954).

SELECTED REFERENCES

SELECTED REFERENCES

A Anderson, J. A.: *Enzymes and Their Role in Wheat Technology.* New York: Interscience Publishers, Inc. (1940).

B Bailey, C. N.: *The Constituents of Wheat and Wheat Products.* New York: Reinhold Publishing Co. (1944).

C Furia, T. E. (ed.): *C.R.C. Handbook of Food Additives, ch.* 9, Cleveland: Chemical Rubber Co. (1968).

D Grist, D. H.: *Rice*, 4th ed. London: Longmans, Green & Co. Ltd. (1965).

E Hummel, C.: *Marcaroni Products, Manufacturing Processing and Packing.* London: Food Trade Press, Ltd. (1950).

F Hylnka, I. (ed.): *Wheat Chemistry and Technology.* St. Paul: Amer. Assoc. of Cereal Chemists, Inc., (1964).

G Jacobs, M. B.: *Chemistry and Technology of Foods and Food Products.* New York: Interscience Publishers, Inc. (1944) I, chap. 17; II chap. 15.

H Kent-Jones, D. W.: *Modern Cereal Chemistry.* 3rd ed. Liverpool: Northern Publishing Co. (1939).

I Knight, J. W.: *Chemistry of Wheat Starch and Gluten.* Cleveland: Chemical Rubber Co. (1965).

J McMasters, M. M.: chairman Revision Committee: *Cereal Laboratory Methods.* 7th ed., St. Paul: American Assoc. of Cereal Chemists, Inc. (1962).

K Matz, S. L. (ed.): *Chemistry and Technology of Cereals.* Westport: Avi Publishing Co. (1959).

L Matz, S. L.: *Bakery Technology and Engineering.* Westport: Avi Publishing Co. (1960).

M Miller, D. F.: "Composition of Cereal Grains and Forages." *Publication* 585, Washington D.C.: National Academy of Sciences (1958).

N Storck, J. and Teague, W. D.: *Flour for Man's Bread.* Minneapolis: Univ. of Minn. Press (1952).

CHAPTER 5

Cocoa, Coffee, Tea and Yerba Maté

Descriptive

These beverages have certain characteristics in common: all are natural products that have been heat-treated to develop their characteristic flavors; all contain caffeine; the flavoring ingredients of all four are complex and partially due to tannins; and all are drunk as aqueous infusions (or emulsions in the case of cocoa).

Cocoa

Chocolate consists of the ground cotyledons or nibs of the seeds of the fruit of the cacao tree (*Theobroma cacao* L.) that have been fermented and roasted to develop the characteristic flavor. Cocoa is chocolate from which a portion of the fat has been removed by pressure while hot. True chocolate (that is, the undefatted nibs) is rarely used to prepare a beverage, despite the advertising of "hot chocolate" in restaurants; its use in confectionery is discussed in Chapter 16.

Cacao beans flourish in a warm, moist climate within an area of about 20 degrees north and south of the equator, and are cultivated in a number of South and Central American countries where the tree originated: the Amazon and Plata river areas of Brazil, in Ecuador, Venezuela and Costa Rica. The tree is also grown in Mexico and Trinidad, Granada, Jamaica, Haiti, Santo Domingo and Cuba as well as in Ghana and Nigeria, Ceylon and Java. A discussion of the flavor qualities of the beans from these different regions will be found in Selected Reference K, pages 114–115.

The cacao beans when removed from the fruit or "pod" possess neither the color nor flavor of chocolate or cocoa, and do not develop these properties by drying and "roasting" alone. For conversion to chocolate the beans are first allowed to undergo natural fermentation for from 2 to 12 days. After that, they are dried and then roasted.[1]

The fermentation and subsequent drying are necessary to destroy viability of the beans and to convert a substance (or complex of substances) "A" into a substance or complex "B" which on roasting will produce the chocolate flavor. What "A" and "B" are, and the path by which one produces the other, have not been definitely established, but there is evidence that a reaction between amino acids and sugars is involved.[2] There is evidence that a substance "C" present in unfermented beans that affects the aroma adversely on roasting may be a protein; it is known that there is extensive proteolysis during fermentation.[3]

The development of the chocolate odor and flavor takes place during the roasting process. Several types of machines are used for roasting; some of these involve counter-current circulation of air heated to 95–134°.

The heat must be so regulated that it penetrates the center of the beans without causing the outer portion to acquire a burnt flavor.

Roasting accomplishes four purposes. It develops the flavor and aroma, produces the characteristic brown color, renders the nibs sufficiently plastic for easy grinding, and changes the seed coat into an easily-removable shell.[4]

The roasted beans are then winnowed to separate the nibs from the 11.6–13.4% of shell and about 0.8% of germ that they contain. The separated nibs form a plastic mass, known as "chocolate liquor" or simply "chocolate"; it is from this that cocoa is made by removing a portion of the fat in a press and pulverizing the remaining press cake.

Federal regulations provide for three general grades of cocoa, differing in the amount of cacao fat that they contain:

Grade	Required Fat Percentage
Breakfast cocoa (high fat cocoa)	22 minimum
Cocoa (medium fat cocoa)	Less than 22; not less than 10
Low-fat cocoa	Less than 10

Any of these types may be flavored with ground spice, ground vanilla beans, a natural food-flavoring oil, oleoresin or extract, vanillin, ethyl vanillin, other artificial food flavoring or salt.

The so-called "Dutch process" cocoa is a special type which has been processed for increased solubility by heating with one or more of the following alkaline substances: sodium, ammonium or potassium bicarbonate, carbonate or hydroxide, or magnesium carbonate or oxide. Regulations require that the total quantity of such alkalies be equivalent in neutralizing value to not more than 3% of anhydrous potassium carbonate.

Cocoas sold to food manufactures are divided by the trade into a larger number of grades according to their fat content, ranging from a "semi-cocoa" containing 32–45% fat to a "defatted cocoa" that contains only 0.5–3.0% fat.[5] Most of these special grades do not reach the public as such, and need not be discussed here.

Regulations under the U.S. Food and Drug Act of 1906 preceding the present law, which were merely advisory and did not have the full force and effect of law, set the following maximum limits for non-fat ingredients of cocoa (on a moisture- and fat-free basis):[6] total ash, 8%; acid-insoluble ash, 0.4%; and crude fiber, 7%.

The average composition of cocoa on a moisture- and fat-free basis is: protein, 25.69%; theobromine, 2.21%; caffeine, 0.86%; starch, 17.10%; fiber, 5.61%; ash, 7.04%, and undetermined (mostly tannin), 41.49%.[7]

Coffee

There are several species of the genus *Coffea*, but the original species to be cultivated was *Coffea arabica*, which is native to Ethiopia and neighboring regions of Africa. From there cultivation spread to Arabia, and about 1700 the tree was introduced by the Dutch into Java and Surinam, and later throughout the East and West Indies and South America. The Arabian or short berry Mocha coffee produced in Yemen and the Abyssinian or long berry Mocha from Ethiopia are still considered the finest grades, but total production is small. At the present time the largest coffee producer is Brazil.[8] While 90% of the coffee grown is still *Coffea arabica*, the "beans" of other species, particularly *C. liberica* and *C. robusta*, are found in commerce.[9]

The coffee fruit or berry resembles a small cherry in appearance, and like the cherry is red when ripe. The fruit contains two seeds which are the "coffee beans" from which the beverage is made. The berries are prepared for market by two different processes ("wet" and "dry"), both of which involve separating the husks from the seeds ("beans"), which are then dried. Some natural fermentation may take place.

At this stage the seeds or "beans" possess none of the aroma associated with the finished beverage. Like the cacao bean, this must be developed by roasting, but the preparation of coffee differs from that of chocolate and cocoa in that fermentation is not a factor in flavor development. After cleaning, the green coffee is roasted in gas-heated, revolving, perforated metal cylinders. Both the temperature and air supply are carefully regulated; the roasting temperature is between 300 and 415°F, and the average time of roasting is 16–17 minutes. Because conditions required to produce the best roast vary with different lots of beans, roasting is more of an art than a science. After roasting, different coffees are mixed to give the blend desired, and the blend is then ground and packaged. Vacuum packing may be used to preserve the flavor by preventing oxidation on standing.

At one time there was much confusion in the trade with regard to the meanings of the various terms employed to express the type of grind, but the National Coffee Association in coöperation with the National Bureau of Standards on December 1, 1947 finally developed standards defining the three grinds of *regular*, *drip* and *fine* in terms of the proportions that would pass certain standard sieves. When 100 grams of coffee were agitated in a sieve shaker for 5 minutes, the three grinds were required to meet the following specifications (in terms of U.S. Standard sieves):[10]

	Percent coffee retained on		Percent coffee passing through
GRADE DESIGNATION	12 AND 16 MESH SIEVES	20 AND 30 MESH SIEVES	30 MESH SIEVE
Regular	33	55	9–15
Drip	7	73	16–24
Fine	0	70	25–40

More recently Lockhard[11] has questioned this system; he considers the terms for the grinds to be inappropriate and the number of grinds to be too limited. The original article should be consulted.

Selected Reference Q, pages 1674–1678, should be consulted for information on the grading of coffee, the regional designations used in the trade and the differing characteristics of the coffees from these regions.

While coffee has been known as a beverage for 700 years (and in the Western world since the 17th century), until comparatively recent times almost all of it reached the consumer either as the intact beans for extemporaneous grinding or in the preground form. The present era of "convenience foods" has seen a large portion of this market taken over by the so-called instant coffees, which are actually coffee extracts. (For a while certain brands were in fact truthfully labeled as "coffee extracts", but this more accurate designation eventually disappeared.) "Instant coffee" is essentially a dried aqueous extract of coffee. It is most often made by extracting ground roasted coffee with water in a countercurrent extractor and evaporating the extract (containing 10–20% solids) in a spray drier with a stream of hot air.[12]

The fact that there were many people who enjoyed the flavor of coffee but were troubled with its keeping them awake at night led to the development of decaffeinated coffee. In general decaffeinated coffees are made as follows: The green (unroasted) beans are first softened by steam under pressure, the caffeine is extracted with benzene, chloroform or alcohol, and the retained solvents are then driven out by resteaming. The extracted beans are then roasted, blended, ground and packed in the same manner as ordinary coffee. The resulting product contains from 0.07 to 0.30% of caffeine as against the 1.20% in regular coffee.[13]

The adulteration of coffee with foreign substances such as roasted and ground cereals, peas, beans, etc., is probably extremely rare at the present time; such adulteration, where it does exist, is most readily detected by microscopic examination. The most likely adulteration at the present time is with exhausted coffee from "instant" coffee manufacture, i.e., coffee grounds. There is one of the former adulterants that is in a class by itself, however: chicory. While the sophistication of coffee with roasted chicory root (*Cichorum intybus* L.) has probably disappeared, some coffee drinkers are fond of the dark color and somewhat bitter flavor that a little chicory imparts to black coffee. As a result, coffee-chicory mixtures in which the presence of chicory is plainly declared on the label are on the market.

Jacobs[14] gives the following average analyses of raw and roasted coffee:

	Raw, percent	Roasted, percent
Moisture	10.73	2.16
Sugar	8.62	0.75
Caffeine	1.07	1.20
Crude fiber	24.00	13.03
Ether extract	11.08	13.75
Water extract	30.35	12.62
Ash	3.00	4.03
Dextrin	0.86	—
Tannic acid	9.02	—

Five market brands of "instant" coffee were analyzed by the Connecticut Agricultural Experiment Station in 1951.[15] Before reporting the results of these analyses, it is necessary to note that three of these brands declared the presence of dextrin, maltose and dextrose added for the specified purpose of improving the flavor, while the other two were labeled as "100% Pure Coffee" and therefore were supposedly straight coffee extracts. Average analyses for the two groups were as follows:

	Straight "instant" coffees	"Instant" coffee containing added carbohydrates
Moisture, percent	3.98	2.70
Ash	11.21	5.70
Protein	14.01	6.88
Crude fiber	0.00	0.00
Dextrose	2.13	0.75
Maltose	4.44	31.04
Dextrins, etc., by difference	60.79	51.23
Fat	0.06	0.06
Caffeine	3.40	1.65

In 1950 the same source[16] analyzed two types of decaffeinated coffee; the *Drip Grind* was a straight decaffeinated coffee, while the *Instant Sanka* was an extract of decaffeinated coffee containing added dextrose and (probably) dextrin. Results were as follows:

Drip grind Sanka: Moisture, 3.84%; ash, 4.76%; water-soluble ash, 3.56%; water-insoluble ash, 1.20%; soluble solids, 21.46%; fat, 15.29%; protein, 12.06%; crude fiber, 13.22%; total carbohydrate (by difference), 50.83%; caffeine, 0.08%, and dextrose, 0.05%.

Instant Sanka: Moisture, 3.45%; ash, 11.15%; water-soluble ash, 8.66%; water insoluble ash, 2.49%; soluble solids, 95.32%; fat, 0.04%; protein, 12.31%; crude fiber, 0.06%; total carbohydrate (by difference), 72.98%; caffeine, 0.19%, and dextrose, 4.44%.

Federal purchasing specifications for instant coffee call for two types—with and without added ascorbic acid. Details may be found in Specification HHH-C-5756 (May 4, 1964).

Tea

Tea is the leaf of a shrub of the *Camellia* family to which also belong the camellias that are cultivated as flowers. Cultivated teas are largely hybrids of the two varieties of China tea (*C. sinsensis* L.) and Assam tea (*C. assamica*), plus a possible third variety called the "Southern form". It has been stated that "aside from water, tea is the world's most popular beverage. In 1959, world tea output reached a record 1,671,400,000 pounds."[17] Tea is known to have been grown in China for at least 4000 years; the first tea reached England and America about 1650. The leading tea export nations are now India and Ceylon.[18]

Teas fall into the three general classifications of black, green and oolong leaf tea, which depend respectively on whether the leaves are fermented, steamed, or partially fermented:

In preparing black tea, the withered leaf is made to undergo an enzymic oxidation inaccurately called "fermentation"; it is then heated ("fired") to stop enzymatic action and remove moisture. At this point the originally green leaf has turned black. Black tea is screened or graded according to size into eight major classifications: orange pekoe, pekoe, pekoe souchong, souchong, broken orange pekoe, broken pekoe, fannings, and dust.

In the making of green tea, withering and "fermentation" are completely omitted. Green tea is often "polished" in a rotating drum that is charged with a mixture of the tea with soapstone or French chalk. The common grades are: Young Hyson, Twankay, Fannings (or Soumee) and Dust. The first grade, which forms nearly 90% of the total, reaches the consumer in the teas sold as such; the next two grades (made mostly of broken fragments) go into tea bags, while the dust is used for beverages in the Orient.[19]

Oolong tea is a semifermented product almost exclusively produced in Formosa; its flavor, somewhere between that of black and green tea, is primarily due to the variety grown and the soil and climatic conditions of Formosa rather than to the manufacturing process, which is similar to that for green tea except that the leaf is slightly withered and lightly fermented before being dried.

A product known as brick tea is used in Russia, Tibet and Central Asia. Black brick tea is made from the waste left over from the manufacture of regular black tea (including siftings, dust and stalks); this is steamed over a boiler until flaccid and soft, and then compressed in molds into bricks in a hydraulic press. Green brick tea is processed in similar fashion, but is made from the leaves only. The user prepares his beverage by whittling off a portion of a brick and grinding this before infusing it in hot water.[20]

Table 5-1[21] shows the proximate composition of tea:

TABLE 5-1
General Analyses of Tea as Given by Kursanov (1956)

	Fresh tea shoots	Black tea
Catechins and catechin-tannins	26.0	18.9
Protein	15.7	16.6
Caffeine	2.7	2.7
Other N-containing compounds	8.7	10.2
Sugar	4.1	4.6
Starch	1.9	0.6
Pectin	12.7	11.9
Cellulose	7.3	7.9
Lignin	6.0	6.1
Lignin (—OCH_3)	(3.0)	(7.6)
Ash	4.9	5.2
Inositol	0.8	—
	90.8	84.7

It will be noted that the major constituent listed is "Catechins and catechin-tannins." These polyphenolic substances include the group of compounds formerly lumped together as "tea tannins." The term "tannin" was originally applied to any natural organic substance that would tan leather. In fact one method that is still in use for determining the tannin content of a vegetable product is based on use of a standard hide powder as a reagent. The best-known general method for tannin, that of Lowenthal[22] (which depends on estimating the material in an aqueous infusion that is oxidized by $KMnO_4$ and precipitable by gelatin), was also designed to measure the substances that had actual tanning properties. Since the general discovery that tannic acid was a depside (i.e., a product of the condensation of the carboxyl group of one polyphenol with a hydroxyl group of another), and the identification by paper chromatography of the presence of over 20 phenolic substances in tea, the use of the nonspecific term "tannin" as the name for a very important group of tea ingredients should probably be abandoned except as purchasing specifications may require it.

Text Reference 17, pp. 207–241, should be consulted for detailed information on the chemistry of fresh green tea leaves and the changes they undergo in processing to black tea. It should be pointed out here that the polyphenols are an important factor in tea flavor (3-monogalloylquinic acid, "Theogallin", is apparently a unique constituent of the tea leaf) and color, which is mostly due to the thearubigens and theaflavins (constituting respectively 10 and 2 percent of the dry weight of black tea). While the taste of tea is due to the polyphenols, the aroma is caused mainly by the volatile oils, although reactions between the tannins and amino acids also appear to be a factor. A large number of components have been isolated from tea essential oil by different investigators, who are unfortunately not in complete agreement. Two substances that do appear to be present in relatively high concentration are cis-3-hexen-1-ol (which by itself has a grassy odor) and 2-hexen-1-al. A constituent of tea which is more important for its physiological effect than for its effect on flavor is caffeine, of which from 2.8 to 4.0% (average 3.30%) is present in India tea; there is little difference between black and green, and between India and China and Turkish, teas in their caffeine contents, but Brazilian teas are apparently lower in caffeine (2.2–2.9%). While the presence of theobromine is usually considered to distinguish cocoa from tea, a trace of theobromine (0.17%) is present in black tea.

The quality of tea sold in the United States is controlled primarily by the supervision over tea imports that is exercised under a special Act.[23] This Act sets up a Board of Tea Examiners that annually prepares a series of standard teas, each of which represents the lowest quality of tea in its class that may be imported. For 1969–1970 there are 5 standard samples representing Formosa Oolong, Ceylon-India-Indonesia black (all black tea except Formosa and Japan black), Formosa black (Formosa black and Japan black), Green tea and Canton type (all Canton type teas including scented Canton and Canton oolong types) respectively.[24] As each lot of tea is offered for import, a licensed tea taster brews a cup of tea from the lot under standard conditions and compares its quality with that of a cup prepared from the corresponding standard. If the taster ("examiner") finds the submitted lot to be of equal or better quality than the standard, entry is permitted; otherwise it is refused.

It is inaccurate to say that a tea taster depends on the taste of the brew alone, since the appearance, twist and smell of the dry leaf, the appearance of the brewed leaf, and the color, "brightness" and aroma of the infusion are also factors in his judgment as to the origin and quality of a particular tea. For a more detailed discussion of this subject the reader is referred to Selected Reference X, pp. 241–245.

Canadian regulations are distinguished by setting limits on the basis of chemical determinations:

Black tea must contain (on the dry basis) not less than 30% water-soluble solids and between 4 and 7% ash. For unblended black tea packaged in the country of origin, the water-soluble solids content may be as little as 25%.

Green tea must contain (dry basis) 33% water-soluble solids and 4–7% ash.

The Tea Board of India in 1958 published required and recommended chemical standards for Indian teas as shown in Table 5–2:[25]

TABLE 5-2
Tentative and Recommended Standards on Indian Tea

Chemical constituents	Present[a] standards	Specifications[b] recommended for adoption
1. Moisture	—	Max. 8.0%
2. Total ash	5.0–8.0%	4.8–7.0%
3. Water-soluble ash	Min. 40.0%	Min. 50.0%
4. HCl-insoluble ash	Max. 1.0%	Max. 0.6%
5. Alkalinity of sol. ash	1.3–2.0%	1.4–2.0%
6. Water-soluble extract	Min. 35.0%	Min. 38.0%
7. Crude fiber	Max. 15.0%	Max. 12.0%
8. Caffeine	—	Min. 2.5%
9. Tannin	—	6.0–12.0%

[a]Tea Board, India (1958).
[b]Sen Gupta, Roy, Mathew, Mallik, Sundararajan (1958).

The same demand for convenience foods that caused "instant" coffee to appear on grocery shelves led to the manufacture and sale of "instant" tea, although this product has never attained the popularity of "instant" coffee—perhaps because tea bags are at least equally convenient as the source of an individual cup of tea—and finds it greatest use in preparing pitchers of iced tea. Such teas are mixtures of a dried aqueous tea extract with dextrine, maltose and dextrose, claimed to be added "solely to protect the really fine tea flavor." A sample analyzed at the Connecticut Agricultural Experiment Station in 1958[26] showed: moisture, 3.46%; ash, 6.76%; protein, 6.38%; crude fiber, 0.13%; dextrose, 4.06%; maltose, 36.67%; dextrin, tannin, etc. (by difference), 39.93%; fat, 0.21%, and caffeine, 2.41%.

Yerba Maté

Tea is a member of the Camellia family, while *Yerba Maté* (or just plain *Maté*), *Ilex paraguayensis* St. Hil., is a holly. Its leaves are used to prepare a beverage which is drunk primarily in Brazil south of the Amazon and in Argentina and Paraguay, and which largely replaces tea and coffee in those regions. The infusion is served not in a cup like tea, but from a small gourd ("maté"), about the size of a large orange, whence it is imbibed through a tube fitted with a strainer at the lower end that is called a *bombilla*.[27] In preparing the drink, sugar and a little hot water are placed in the gourd, the maté is added, and the container is then filled to the brim with boiling water or hot milk. Burnt sugar or lemon juice may be added to the water[28]

The present method of preparing maté leaves for consumption is as follows: The leaves are plunged for a moment into hot water, then dried over brick stoves and finally broken into fragments.[27]

Analyses of 17 samples of maté by Krauze [*Mitt. Lebensm. Hyg.*, **23**, 218 (1932)] as reported by Winton and Winton[29] showed the following range of values:

	Minimum	Maximum
Water, percent	6.90	10.40
Caffeine, percent	0.58	1.64
Tannin, percent	7.80	10.98
Total ash, percent	6.09	7.38
Water extraxt, percent	35.27	49.60

Somewhat older analyses by Hennings [*Ber. deut. pharm. Ges.*, **30**, 22 (1920)] reproduced by the same authors indicated the following averages: water, 9.00%; protein, 9.75%; caffeine, 2.10%; fat, 7.75%; volatile ether extract, 2.05%; resin, 9.10%; tannin, 9.80%; fiber, 15.45%; total ash, 6.62%; soluble ash, 2.26%; sand, 3.26%, and water extract, 33.10%.

A related species, *Ilex cassine* L., var. *myrtifolia*, indigenous to the southeastern United States, has been used to make a beverage called *Cassina*; by varying the method of curing the leaves, *black* and *green cassinas* resembling the corresponding teas, and a *cassina maté* similar to *Yerba maté*, have been prepared. Analyses of black and green cassinas by the Connecticut Agricultural

Experiment Station in 1922 were as follows:[30]

Black cassina: Moisture, 3.15%; fat, 1.68%; hot water extract, 31.00%; protein, 12.88%; crude fiber, 14.13%; caffeine, 0.69%; total ash, 6.00%; water-soluble ash 1.71%, and sand, 1.09%.

Green cassina: Moisture, 3.68%; fat, 1.98%; hot water extract, 40.00%; protein, 13.69%; crude fiber, 12.29%; caffeine, 0.38%; total ash, 6.00%; water-soluble ash, 1.72%, and sand, 1.40%.

Methods of analysis

Method 5-1. *Macroscopic and Microscopic Examination.*

a. *Coffee* (AOAC Method **14.008**)—Pick out and identify microscopically artificial coffee beans, apparent from their regular form, and roasted legumes and lumps of chicory in whole roasted coffee. For ground coffee, sprinkle some of the sample on cold water and stir lightly. If they are not over-roasted, fragments of pure coffee float, while fragments of chicory, legumes, cereals, etc., sink immediately, chicory coloring the water decidely brown. In all cases, use microscopic examination to identify the particles that sink.

b. *Tea* (Method **14.40** of AOAC Methods of Analysis, 7th Ed., 1950).

i. *General*—Examine the ash for mineral pigments. Shake a quantity of the tea with a large volume of water and remove the leaves by sieving. Allow the insoluble matter in the aqueous portion to settle, filter, and examine the residue on filter paper for insoluble pigments. Catechin and other soluble substances, if used, will be in the filtrate.

ii. *Paraffin and waxy substances*—Spread a quantity of the tea between two sheets of unglazed white paper, and place a hot iron on the upper paper. Any greasy substance will stain the paper.

iii. *Pigments used for coloring or facing*—Weigh a 60 g sample onto a 60 mesh, 5–6″ sieve provided with a cover. Sift a small quantity of the dust (about 0.1 g) onto an 8 × 10″ piece of semi-glazed white paper. (To obtain the required quantity of dust, it is sometimes necessary to rub the leaves gently against the sieve.) Place the paper on a plain, firm surface (preferably glass or marble), and crush the dust by pressing firmly upon it with a flat steel spatula about 5″ long. Repeat the crushing process until the dust is ground almost to powder, when particles of coloring matter (if present) become visible as streaks on the paper. Brush off the loose dust and examine the paper under a simple lens magnifying 7.5 diameters. Bright light is essential to distinguish these particles and streaks. In many cases the character of the pigment is indicated by the behavior of these streaks when treated with reagents and examined under the microscope. Crushed particles of leaf of either black or green tea appear in such quantity that there is no chance of mistaking them for coloring or facing material. Repeat this test, using black semi-glazed paper, for facings such as talc, gypsum, $BaSO_4$, or clay.

Method 5-2. *Moisture*

a. *In cocoa* (AOAC Method **12.002**)—Dry a 2 g sample to constant weight in a Pt dish at 100° in an oven. (An Al dish may be used when ash is not determined on the same sample.) Report the loss in weight as water.

b. *In coffee* (AOAC Method **14.010**)—Dry a 5 g sample at the temperature of boiling water under a pressure not over 100 mm of mercury, or at 105–110° under atmospheric pressure, for 5 hours and subsequent periods of 1 hour each until constant weight is reached. For whole coffee, grind rapidly to a coarse powder, and without sifting and unnecessary exposure to air, weigh out portions for the determination. For ground coffee, sample directly without further grinding.

c. *In tea and Yerba Maté—See* Method 1–2.

Method 5-3. *Ash.* (Use Method 16-4 for cocoa, coffee, tea or yerba maté).

Method 5-4. *Alkalinity of Soluble Ash in Cocoa.* (*See* Method 16–17 and Note).

Method 5-5. *Protein.*

Use Method 1–7 for cocoa, coffee, tea or yerba maté. Subtract N due to caffeine from total N, and multiply the difference by 6.25.

Method 5-6. Crude Fiber.

a. *Cocoa* [AOAC Method **12.014(a)** modified]—Treat 7 g of cocoa in a centrifuge with two 100 ml portions of petroleum ether, centrifuging and decanting the supernatant after each addition. Dry the residue in an oven at 100°, and then powder in a bottle with a flat-ended glass rod. If necessary, grind the material in a mortar, extract a third time with petroleum ether, and dry to constant weight in a Pt dish. Weigh 2 g of this residue and treat by Method 1–11. This yields the percentage of crude fiber on a moisture-and-fat-free basis. If crude fiber on an *as is* basis is desired, correct for moisture as determined by Method 5–2(**a**) and fat as determined by Method 5–7(**a**).

b. *Coffee, tea and yerba maté—See* Method 1–11.

NOTE: Instead of following Method 5–6(**a**), crude fiber in cocoa may be determined directly by Method 1–11 on the fat-free residue of Method 5–7(**a**). This procedure gives the percentage of crude fiber in the sample directly; no correction for moisture or fat is required.

Method 5-7. Fat.

a. *In cocoa*[31]—Prepare a Knorr extraction tube by adding a tightly packed mat of asbestos that has been freed of coarse particles. Wash this filter with alcohol, ether and a little petroleum ether (redistilled below 60°). Weigh 2–3 g of sample into the tube. Insert the tube into a rubber stopper in a filtering bell-jar connected to the suction through a two-way stopcock, taking care that no rubber particles adhere to the tip of the stem. Place a weighed 150 ml Erlenmeyer flask at such a height that the tube stem passes through the neck into the flask. (Lengthen the tube stem if necessary.) Fill the tube to about two-thirds of its capacity with the redistilled petroleum ether, and by means of a rod having a flattened end stir the sample thoroughly, taking care to crush all lumps. Let stand one minute and drain by suction. Regulate the suction so that the collected solvent in the flask will not boil violently. Add the solvent from a wash-bottle while turning the tube between the thumb and forefinger so that the sides of the tube are washed down by each addition. Repeat the extractions, with stirring, until the fat is removed. (Ten extractions are usually sufficient.) Remove the tube with stopper from the bell, wash the traces of fat from the end of the stem with petroleum ether, evaporate the solvent, and dry to constant weight at 100°.

b. *In coffee, tea and yerba maté.* (AOAC Method **14.029**)—Dry 2 g of sample, ground to pass a 30 mesh sieve, at 100°, and extract with petroleum ether (b. pt. 35–50°) for 16 hours. Evaporate the solvent, dry the residue at 100°, cool, and weigh.

Method 5-8. Sugars.

a. *In cocoa—See* Methods 16–19(**b**) and 16–20.

b. *In coffee, tea and yerba maté*—Determine dextrose by Method 16–8(**b**) or (**c**). Determine total reducing sugars by Method 16–10, subtract Cu equivalent to the dextrose from the total Cu obtained, and calculate the difference to maltose.

Method 5-9. Tannin in Tea. (*AOAC Method* ***14.048–14.049***).

REAGENTS

a. *Potassium permanganate solution*—Dissolve 1.33 g of $KMnO_4$ in water and dilute to 1 liter. Determine ml 0.1 *N* oxalic acid equivalent to 1 ml of this solution as follows: Dissolve 0.2500 g of Bureau of Standards Na oxalate in 200–250 ml of hot water (80–90°), add 10 ml of H_2SO_4 (1 + 1), and titrate with the $KMnO_4$ solution, added no more rapidly than 10–15 ml/minute, while stirring vigorously and continuously; add the last 0.5–1 ml drop by drop.

$$\text{Ml } KMnO_4 \text{ used} \times 0.0268 = \text{ml } 0.1N\ H_2C_2O_4/\text{ml } KMnO_4.$$

b. *Indigo carmine solution*—Weigh 6 g of indigo carmine (free of indigo blue) (Eastman No. C 1009) into a liter volumetric flask, dissolve in water and 50 ml of H_2SO_4, and make to volume with water.

c. *Gelatin solution*—Soak 25 g of gelatin in saturated NaCl solution for 1 hour, heat until the gelatin dissolves, and dilute to 1 liter with saturated NaCl solution.

d. *Acid sodium chloride solution*—Acidify 975 ml of saturated NaCl solution with 25 ml of H_2SO_4.

DETERMINATION

Boil a 5 g sample 30 minutes in 400 ml of water, cool, transfer to a 500 ml volumetric flask, and dilute to the mark. To 10 ml of this infusion (filtered if not clear) add 25 ml of the indigo carmine solution and about 750 ml of water. Add the $KMnO_4$ solution from a buret, a little at a time while stirring, until the color of the solution changes to light green, then continue the titration dropwise until the solution becomes bright yellow (or faint pink at the rim). Designate ml $KMnO_4$ used as a.

Mix 100 ml of the clear tea infusion in a stoppered flask with 50 ml of the gelatin solution, 100 ml of the acid NaCl solution and 10 g of powdered kaolin, and shake several minutes. Let the mixture settle, and decant through a filter. Mix 25 ml of the filtrate with 25 ml of the indigo carmine solution and 750 ml of water, and titrate with the $KMnO_4$ solution as before. Designate ml $KMnO_4$ used as b.

Then $a-b$ = ml $KMnO_4$ required to oxidize the tannin in the sample. 1 ml $0.1N$ oxalic acid = 0.0042 g tannin (gallotannic acid).

NOTE: *See* the remarks on pp. 95–96 concerning the significance of the term "tannin", particularly as it relates to the above [Lowenthal[22]] method.

Method 5-10. *Chlorogenic Acid in Coffee.* (*AOAC Method* ***14.025***).

REAGENTS

a. *Basic lead acetate solution*—Sp. g. 1.25. *See* Method 16–6, Preparation and Use of Clarifying Agents, (**a**).

b. *Chlorogenic acid*—Obtainable from Calbiochem, Box 54282, Los Angeles, Calif. 90054, as their No. 2207.

PREPARATION OF SAMPLE SOLUTION

a. *Roasted coffee*—Weigh a 1 g sample, ground to pass a 30 mesh sieve, transfer to a 750 ml Erlenmeyer flask, and add 400 ml of boiling water. Reheat quickly to boiling, and continue to boil gently for exactly 15 minutes; then cool quickly to room temperature under the tap. During boiling, swirl the flask frequently to keep the coffee submerged in the solution. Transfer to a 500 ml volumetric flask and dilute to volume. Filter through a retentive paper, discarding the first 25–50 ml of the filtrate. If the filtrate is more than faintly cloudy, refilter through a fine porosity fritted glass filter, using suction. *Do not use filter aids.*

b. *Instant coffee*—Weigh a 0.25 g sample, transfer to a 500 ml volumetric flask, dilute to volume, and filter as under (**a**) if necessary.

DETERMINATION

Transfer 10 ml of the filtrate to a 100 ml volumetric flask and dilute to volume with water. Determine the absorbance at 324 mμ against water. Transfer another 100 ml of the sample solution to a 200 ml Pyrex volumetric or Kohlrausch flask, and add 2 ml of saturated K acetate solution and 10 ml of the basic Pb acetate solution with swirling. Place the flask in a boiling water bath for 5 minutes, swirling occasionally. Remove, cool under the tap, and place in an ice-water bath. Stir mechanically for 1 hour, using a glass rod with a paddle-shaped blade at the lower end that reaches the bottom of the flask, while the flask is immersed in the bath. Remove, wash down the stirrer, warm to room temperature, and dilute to volume with water. Filter through a fluted filter paper, discarding the first 25–50 ml. Immediately determine the absorbance of the solution at 324 mμ.

From the standard curve determine (1), the apparent concentration of chlorogenic acid in the solution taken for absorbance measurement without Pb treatment (C_0); (2), the apparent concentration in the filtrate after Pb treatment (C_1). From the latter value subtract 0.00045 mg/ml to correct for the solubility of Pb chlorogenate.

Corrected concentration = $C_0-[C_1-0.00045/5]$.

PREPARATION OF STANDARD CURVE

Weigh 40.0 mg of dried chlorogenic acid, transfer to a 500 ml volumetric flask, dissolve in water, and dilute to volume. Prepare a series of standards by transferring 5, 10, 15 and 20 ml aliquots to 100 ml volumetric flasks and diluting to volume. Determine the absorbance at 324 mμ of each solution against water. Plot the concentrations of chlorogenic acid, in mg/ml, against absorbances.

NOTES

1. Cells must be carefully cleaned after each

use of Pb-treated solutions because $PbCO_3$ slowly accumulates on the optical surfaces.

2. Chlorogenic acid is a depside in which the 3-carbon atom of a molecule of quinic acid is joined to a molecule of caffeic acid.[32] It was thought to occur in coffee partly as the free acid, partly as the K salt, and partly as the double salt of K and caffeine, but this has been disputed by Griebel,[32] who believes that it is present as Ca, Mg and K salts. Interest in its determination arose partly because of advertising claims that coffee low in this compound made a less bitter beverage[33] and in part due to general studies of the effect on coffee flavor of decomposition of the organic acids brought about by roasting.[32]

Besides chlorogenic acid, coffee contains an isomeric depside, *iso*chlorogenic acid, in which the quinic and caffeic acids are joined through the 5-carbon atom of the quinic acid. Since both chlorogenic and *iso*chlorogenic acids are precipitated by Pb acetate, they are not separated by the above method, and at least one-sixth of the approximately 6% of "chlorogenic acid" determined by Method 5–10 is actually *iso*chlorogenic acid.[32] If determination of the individual depsides is desired, recourse must be had to column and paper chromatographic separations such as are outlined in Text Reference 32.

In addition to chlorogenic and *iso*chlorogenic acids, at least two other caffeylquinic acids (pseudo- and neochlorogenic acids) are known to exist in plant material, including coffee.[35]

Method 5-11. *Caffeine in Coffee, Tea and Yerba Maté, Bailey-Andrew Method.* (*AOAC Methods **14.020*** and ***14.047***).

Into a weighed 1-liter Erlenmeyer flask weigh 10 g of regular coffee, or 5 g of "instant" coffee, tea or yerba maté, ground to pass a 30 mesh sieve. Add about 500 ml of water, swirl, and heat to boiling. Add 10 g of heavy MgO. Boil gently over a low flame 2 hours with occasional shaking, adding water as necessary to prevent frothing, and wash down the sides of the flask. Cool, place the flask on a balance, and add sufficient water that total wt. = tare wt. + sample wt. + 510 g. Filter, collect 200 ml of clear filtrate (equivalent to 0.4 sample wt., or 2 g for "instant" coffee, tea or yerba maté, and 0.8 g for regular coffee), add 20 ml of H_2SO_4 (1 + 9), and transfer to a 500 ml separatory funnel. Shake 6 times with $CHCl_3$, using 25, 20, 15, 10, 10 and 10 ml portions. Treat the combined extracts with 5 ml of 1% KOH solution; when the liquids separate completely, drain the $CHCl_3$ layer into a Kjeldahl flask. Wash the alkaline solution in the separatory funnel with two 10 ml portions of $CHCl_3$ and combine the washings with the remaining bulk of the extract. Evaporate or distil off the $CHCl_3$ to less than 25 ml, and determine N as in Method 1–7. One ml 0.1 *N* H_2SO_4 = 4.85 mg anhydrous caffeine

Method 5-12. *Caffeine in Decaffeinated Coffee, Chromatographic-Spectrophotometric Method.* (*AOAC Method **14.022–14.024***).

PREPARATION OF SAMPLE

Place 1.0 g of decaffeinated coffee ground to pass a 30 mesh sieve, or 0.5 g of decaffeinated "instant" coffee, in a 100 ml beaker. Add 5 ml of NH_4OH (1 + 2), and heat 2 minutes on the steam bath. Add 6 g of Celite 545 and mix thoroughly.

PREPARATION OF COLUMN

a. *Acid column*—Place a fine glass wool plug in the base of a 25 × 250 mm column. Add 2 ml of 4*N* H_2SO_4 (209 g H_2SO_4/liter) to 2.0 g of Celite 545 and mix well by kneading with a spatula blade. Transfer to the column, using a powder funnel, and tamp, using gentle pressure, to a uniform mass. Place a small glass wool wad above the surface. Insert the tip into a 50 ml volumetric flask.

b. *Basic column—Layer A*—Mix 3 g of Celite 545 and 2 ml of 2*N* NaOH (84 g NaOH pellets/liter), and place in a 25 × 250 mm column over a glass wool plug as in (**a**). *Layer B*—Transfer the above sample-Celite mixture to the column directly over Layer *A*. Dry-wash the beaker with about 1 g of dry Celite 545, transfer to the column, and tamp to a uniform mass.

DETERMINATION

Mount the basic column over the acid column. Pass 150 ml of water-saturated ether through the basic column onto and through the acid column, and follow with another 50 ml of the ether. Discard the eluate. Place a 50 ml volumetric flask under the acid column. Measure 50 ml of water-saturated $CHCl_3$ into a graduate, use a little of this to wash the tip of the basic column onto the acid column, and disconnect the basic column.

Pass the balance of the $CHCl_3$ through the acid column and collect the eluate in the volumetric flask. Make the flask to volume with the $CHCl_3$, mix, and read the absorbance at 276 mμ against a $CHCl_3$ blank.

Prepare caffeine standards containing 0.25, 0.5 and 0.75 mg/50 ml in water-saturated $CHCl_3$, read their absorbances, and plot a curve of wts of caffeine vs. absorbances therefrom. Calculate percentage caffeine in the sample from this curve.

NOTE: Both the acid Celite and basic Celite mixtures may be prepared in large batches and used as needed. If this is done, use 4.2 g of acid mixture and 5 g of basic mixture per column.

Method 5-13. *Theobromine and Caffeine in Cocoa.*[36]

A. Theobromine.

REAGENTS AND EQUIPMENT

a. *Zinc acetate solution*—Dissolve 219 g of $Zn(C_2H_3O_2)_2.2H_2O$ and 30 ml of acetic acid in water and dilute to 1 liter.

b. *Potassium ferrocyanide solution*—Dissolve 106 g of $K_4Fe(CN)_6.3H_2O$ in water and dilute to 1 liter.

c. *Silver nitrate*—1*N*. Dissolve 170 g of $AgNO_3$ in water and dilute to 1 liter.

d. *Silver nitrate*—0.1*N*. Dissolve 17 g of $AgNO_3$ in water and dilute to 1 liter.

e. *Sodium hydroxide*—0.1*N* and 0.025*N*. Prepare from a 50% solution of CO_2-free NaOH and standardize against National Bureau of Standards K acid phthalate. (*See* AOAC Method **42.030–42.033.**)

f. *Sulfuric acid*—1*N*. Dissolve 49 g of H_2SO_4 in water and dilute to 1 liter.

g. *Sulfuric acid*—0.1*N*. Dissolve 4.9 g of H_2SO_4 in water and dilute to 1 liter.

h. *Potassium acid phthalate buffer*—0.05*M*. Dissolve 10.20 g of dried (1 hour at 105°) NBS Standard Sample 185 K acid phthalate in water and dilute to 1 liter.

i. *Sodium nitrate solution*—Add 100 g of $NaNO_3$ to 100 ml of water.

j. *Universal indicator paper*—No. 14–839 Test Paper, Alkacid Wide Range, or No. 14-852 Test Paper, Nitrazine, Squibb's, obtainable from Fisher Scientific Co., 633 Greenwich Street, New York, N.Y. 10014.

k. *Mixed indicator*—Dissolve 0.625 g of methyl red in 450 ml of alcohol, and filter through asbestos. Dissolve 0.412 g of methylene blue in 50 ml of water. Mix the two solutions, and dilute to 500 ml with water. This indicator has a pH range of 5.34–5.65 (reddish purple to green).

l. *Fuller's earth—Celite absorption mixture*—Thoroughly mix equal quantities (by weight) of English Superfine XL fuller's earth (obtainable from L. A. Salomon and Bros., 216 Pearl Street, New York, N.Y.) and Celite 535 or 545 (obtainable from Johns-Manville).

Test the fuller's earth for its capacity to adsorb theobromine as follows: Stir 2 g of the fuller's earth with 200 ml of a neutral aqueous solution containing 100 mg of theobromine for 10 minutes. Centrifuge, and determine theobromine in a 100 ml aliquot of the supernatant liquid by the $AgNO_3$ titration procedure described below. At least 75 mg of the theobromine should have been adsorbed.

m. *Extraction tubes*—Attach 100 mm lengths of 8 mm glass tubing to the bottoms of 25 × 200 mm test tubes.

n. *Adsorption tubes*—Attach 80 mm lengths of 6 mm glass tubing to the bottoms of 18 × 150 mm test tubes.

o. *Salt bridge for electrometic titration*—Seal a thread of asbestos into the end of an 18 × 100 mm test tube.

EXTRACTION AND CLARIFICATION

Place a weighed portion of the prepared sample (preferably containing about 40 mg of theobromine; usually 2.0–3.0 g) and one g of heavy MgO in a 200 ml casserole. Add 4 g of Celite 545 and mix thoroughly with sufficient hot water (about 10 ml) to make a smooth paste. Dilute to about 50 ml with hot water, mix thoroughly, and transfer to a 25 × 200 mm extraction tube (containing a glass wool plug and a filter bed of about 1 g of Celite 545) that is attached by a one-hole stopper to a liter filter flask. Attach a vacuum line to the filter flask and percolate boiling water through the tube, regulating the vacuum so that 500 ml of extract is collected in 30–40 minutes. (Do not allow the tube to go dry at any time during the percolation, but make sure the layer of water on the sample does not exceed 2.5 cm, in order to maintain the temperature of the sample bed between 80 and 90°.)

Neutralize the extract to about pH 6 with the $1N$ H_2SO_4, using universal indicator paper. Transfer the extract to a 1-liter beaker and concentrate to about 150 ml. Transfer the concentrate to a 200 ml volumetric flask, using a small amount of wash water and keeping the total volume to about 170 ml. Add 7 ml of the Zn acetate reagent and mix, then immediately add with swirling 7 ml of the $K_4Fe(CN)_6$ solution. Make to volume with water and mix thoroughly by shaking. After standing for a minimum of 3 and a maximum of 5 minutes, filter the solution through a dry 27 cm folded filter paper. Discard the first 5–8 ml of filtrate and collect the remainder. Place 150 ml of the solution in a separatory funnel equipped with a rubber stopper to fit the top of an adsorption tube.

Adsorption of theobromine by Fuller's Earth

Place a plug of glass wool in the bottom of the adsorption tube and press it firmly and evenly in place with a glass rod. Add about 0.5 g of Celite 545 so as to make a bed about 10 mm thick. On top of this Celite bed place about 6 g of the fuller's earth-Celite mixture, (**l**). To obtain a compact uniform column that will not channel, tap the tube to distribute the packing, then compress the bed by drawing air through the dry column. This is conveniently carried out by placing the adsorption tube on a 500 ml suction flask and drawing a vacuum through the flask. To test for channeling, add water to the tube while the vacuum is not on, and then turn the vacuum on slowly.

When a tube with a properly-compacted bed without channeling has been prepared, immediately connect the separatory funnel holding the clarified sample solution containing 40–60 mg of theobromine to the tube, and draw the solution through the column (filtration time should be 20–30 minutes). As soon as only 4–5 ml of solution remain on top of the bed, wash the column with 50 ml of water. (Some liquid should be kept on top of the column at all times to avoid channeling.)

Elution of Theobromine from Fuller's Earth Column

Remove the suction flask containing the alkaloid-free filtrate and wash water, and replace it with a clean 250 ml suction flask. Place 75 ml of $0.1N$ NaOH in the separatory funnel, and elute the theobromine by connecting the funnel to the adsorption tube and drawing the alkali through the column by applying vacuum to the suction flask. The time for the elution step should be 10–20 minutes, making the total time for the adsorption, washing and elution about 40–60 minutes.

Titration of Theobromine

Add 2 drops of mixed indicator to facilitate rough adjustment of the pH, then neutralize the NaOH eluate with the $1N$ H_2SO_4 to a faint red end point. Transfer the neutralized solution to a 250 ml beaker, rinsing the flask with water. In the beaker place a glass electrode and a calomel electrode, making contact with the solution by means of a salt bridge filled with saturated $NaNO_3$ solution. Adjust the solution to pH 6.40 ± 0.05, using $0.025N$ NaOH for the final adjustment. (The total volume should be about 125 ml.) Add 25 ml of $0.1N$ $AgNO_3$ solution from a graduate or automatic pipet, and stir thoroughly. With thorough stirring, titrate to pH 6.40 ± 0.05 with $0.025N$ NaOH. (For best results, this titration should be made without interruptions and should not require more than 5 minutes for its completion.) Record the amount of $0.025N$ NaOH required for the titration.

Blanks and Controls

Carry a sample of water through the Zn acetate-$K_4Fe(CN)_6$ treatment, adsorption, elution with NaOH, neutralization and titration in the presence of $AgNO_3$. Perform each operation in the exact manner prescribed for unknown samples. This blank titration should require not over 0.20 ml of $0.025N$ NaOH.

Carry a sample containing 40 mg of theobromine through the same procedure to establish the quantitative nature of all steps involved.

Calculations

$$\%\ \text{Theobromine} = \frac{100\,(\text{vol. NaOH for sample-vol. NaOH for blank}) \times \text{theobromine factor}}{\text{wt. of sample clarified}}$$

$$\text{Theobromine factor} = \text{normality of NaOH} \times 0.180 \times \frac{\text{total volume clarified}}{\text{aliquot analyzed}}.$$

(The addition of a 3% correction as recommended in the Notes may be made by substituting 103 for 100 in the preceding formula.)

B. Caffeine

REAGENTS

a. *Chloroform*—Redistil USP $CHCl_3$.

b. *Trisodium phosphate*—0.5*M*. Dissolve 190 g of $Na_3PO_4.12H_2O$ in 1 liter of water.

EXTRACTION AND CLARIFICATION

Mix 4 g sample with 2 g of heavy MgO and 8 g of Celite 545 and extract by the same procedure as used for theobromine. Neutralize the extract, concentrate to 150 ml, and clarify with $Zn_2Fe(CN)_6$. Add 10 ml of the 0.5*M* Na_3PO_4 to 150 ml of the filtrate, make up to 200 ml, and filter on fluted paper. Collect 150 ml of filtrate and transfer to a 500 ml separatory funnel.

CHLOROFORM SEPARATION

Shake the clarified water extract five successive times for one minute each with 30 ml portions of $CHCl_3$. After each shaking draw off the $CHCl_3$ solution into a 250 ml separatory funnel. After the $CHCl_3$ extraction is completed, add 5 ml of 1*N* H_2SO_4 to the $CHCl_3$ solution in the 250 ml separatory funnel. Mix thoroughly, let stand about 10 minutes, and draw off the $CHCl_3$ solution through a cotton plug in the stem of the separatory funnel into a 650 ml Kjeldahl flask. Wash the acid with one 30 ml portion of $CHCl_3$ and add the washings to the Kjeldahl flask. Recover most of the $CHCl_3$ by distillation, but do not permit the distillation to proceed too rapidly, and discontinue while there is still 10–15 ml left in the Kjeldahl flask.

NITROGEN DETERMINATION

Determine total N in the $CHCl_3$ extract by Method 1–7. Run a nitrogen blank on the reagents used for the determination starting with 180 ml of $CHCl_3$.

CALCULATIONS

$$\% \text{ Caffeine} = \frac{100(\text{vol. NaOH for blank} - \text{vol. NaOH for sample}) \times \text{normality of NaOH} \times 0.0485}{\text{wt. sample analyzed}}$$

NOTES

1. To determine the theobromine content of the pure or crude alkaloid, proceed as follows: Pulverize the sample with a mortar and pestle and accurately weigh about 0.1 g into a 500 ml Erlenmeyer flask. Add about 250 ml of water, place a small funnel in the flask, and boil about 30 minutes. Filter while still hot through a coarse filter paper into a 400 ml beaker. (Filtration may be omitted for samples containing 90–100% theobromine.) Wash the filter thoroughly with about 50 ml of hot water. Cool the filtrate and washings to room temperature, adjust to pH 6.40, add 5 ml of 1*N* $AgNO_3$, and titrate back to pH 6.40 with 0.025*N* NaOH. Correct for a water blank, using 250 ml of boiled water and 5 ml of 1*N* $AgNO_3$.

2. Cocoas which contain more than 15% fat should be extracted with hexane or pentane prior to the above determinations to facilitate the extraction of theobromine and the subsequent clarification and adsorption steps.

3. The $AgNO_3$ titration method originated by Boie[37] is based on the reaction $C_7H_8N_4O_2 + AgNO_3 \rightarrow AgC_7H_7N_4O_2 + HNO_3$. The HNO_3 formed by the reaction is titrated with standard NaOH solution. Use of a visual endpoint with phenol red indicator for this titration is limited to solutions that are essentially colorless, while by introducing an electrometric titration as in the present procedure colored solutions may be titrated with even greater precision.

4. The customary combination of glass and calomel electrodes placed directly in the solution being titrated cannot be used very long in the presence of $AgNO_3$ because the reaction between $AgNO_3$ and the KCl at the liquid junction of the calomel electrode interferes with the measurement of the potential. A saturated $NaNO_3$ solution possesses the required properties of being a highly dissociated salt with a neutral pH that is not reactive with KCl or $AgNO_3$. The level of the solution in the salt bridge should be above the point to which the bridge is immersed in the titrated solution.

Method 5-14. *Water Extract of Tea.*[38]

To 2 g of the ground sample in a 500 ml graduated flask add 200 ml of hot water and boil over a low flame for 1 hour, rotating the flask occasionally. The flask should be closed with a rubber stopper through which passes a glass tube 30" long for a condenser. Boil very slowly so that no steam escapes from the top of the air condenser. Cool, dilute to volume, mix thoroughly and filter through a dry filter paper. Take an aliquot of 50 ml and evaporate to dryness over a steam bath. Then place in oven and heat to 100° for 1 hour, cool and weigh.

Method 5-15. *Egg in Coffee.*[39] (Method of Wickroski and Agostini).

Grind the sample finely in a *Mikro-Samplmill* or similar mill, and weigh 12.5 g into an Erlenmeyer flask. Add 125 ml of $CHCl_3$—absolute alcohol (1 + 1), stopper, shake $\frac{1}{2}$ hour on a shaking machine, and filter. Evaporate 100 ml of the filtrate (= 10 g sample) on the steam bath in a weighed 125 ml Pyrex beaker, and dry the beaker and contents in an oven at 100° to constant weight (about 90 minutes). Let the beaker stand in air to constant weight (about 30 minutes), weigh, and report the residue as percent lipids.

Add to the residue in the beaker 2 ml of $Mg(NO_3)_2$ solution [Reagent (**e**), Method 1–19] and 8 ml of alcohol, and evaporate to dryness on the steam bath, using care to avoid spattering. Place the beaker in an oven for 30 minutes at 100° to remove any remaining water, then transfer while hot to a muffle heated to 500°, and keep at this temperature 1 hour. Cool, add a few drops of water, and break up the charge with a flat-ended glass rod. Cover the beaker with a watch-glass, slowly add 5 ml of HNO_3 (1 + 3), mix, and wash the watch-glass and filter, collecting the filtrate in a 500 ml Erlenmeyer flask. Thoroughly wash the charred material and filter with water, dilute the filtrate to about 100 ml, and proceed as in Method 1–18. Report as percent lipid P_2O_5.

NOTE: This method was worked out to check the claim of the manufacturer that a brand of coffee sold as "Regular Grind Fireside Egg Coffee" had had 12 lb of whole eggs added to each 400 lb of coffee. Analyses of known samples showed 0.0133–0.0160% (average 0.0148%) of lipid P_2O_5 in straight coffee and 0.0210–0.0260% (average 0.0226%) in coffee treated with the claimed amount of whole egg.

Ascorbic Acid in Instant Coffee and Tea—See Chapter 20.

Method 5-16. *Sediment in Instant Coffee.* (This is Federal Purchasing Specification HHH-C-575b.4.3.7).

Weigh 5.0 g of instant coffee into a 250 ml beaker. Add 200 ml of hot (not below 85°) water. Stir until the coffee is completely dissolved. Filter the hot solution through the sediment apparatus [*See* AOAC Method **36.010(c)(1)**], using Johnson and Johnson cotton lintane sediment testing filter material. (The temperature of the coffee solution must not be less than 60° at the time of filtration. Wash the beaker and sediment tester with small portions of hot water so as to transfer all sediment to the surface of the filter material, and to wash out all of the soluble coffee from the filter. The total wash water must not exceed 200 ml. Remove the filter containing the sediment from the apparatus, dry, and compare with photographs of standard sediment disks in the American Public Health Association sediment chart.

NOTE: The purchasing specification requires that the sediment of an acceptable coffee not exceed that shown on the 2.0 mg disk.

TEXT REFERENCES

1. Roelofsen, P. Q.: *Fermentation, Drying, and Storage of Cacao Beans, Advances in Food Research*, Vol. VIII, pp. 226–228. New York: Academic Press, Inc. (1958).
2. Rohan, T. A.: *J. Sci. Food Agr.*, **14,** 799 (1963); *J. Food Sci.*, **29,** 456 (1964); Rohan, T. A. and Stewart, T.: *J. Food Sci.*, **30,** 416 (1965); **31,** 202 (1966); **32,** 395, 399 (1967).
3. Ref. 1, pp. 278–281.
4. Ibid, pp. 134–137.
5. Merory, J.: *Food Flavorings, Composition, Manufacture and Use*, 118. Westport: Avi Publishing Co., Inc. (1960).
6. *Rules and Regulations Relating to the Food and Drug Law of Connecticut, Revision of July* 1, 1937, pp. 104–106.

7. Winton, A. L. and Winton, K. B.: *The Structure and Composition of Foods*, Vol. IV, p. 121. New York: John Wiley and Sons, Inc. (1939).
8. Ibid, Vol. IV, pp. 140–141.
9. Jacobs, M. B. (ed.): *The Chemistry and Technology of Food and Food Products*, 2nd ed., Vol. II, pp. 1660 and 1665. New York: Interscience Publishers, Inc. (1951).
10. Ibid, Vol. II, pp. 1666–1674.
11. Lockhard, E. E.: *Food Research*, **24,** 91 (1959).
12. Ref. 9, Vol. II, pp. 1680–1681.
13. Ibid, Vol. II, p. 1681.
14. Jacobs, M. B.: *The Chemical Analysis of Foods and Food Products*, 3rd Ed., p. 626. Princeton: D. Van Nostrand Co., Inc. (1958).
15. Fisher, H. J.: *Conn. Agr. Expt. Sta: Bull.* **574,** 12 (1953).
16. Fisher, H. J.: *Conn. Agr. Expt. Sta. Bul.* **558,** 19 (1952).
17. Stahl, W. H.: *The Chemistry of Tea and Tea Manufacture, Advances in Food Research*, Vol. XI, p. 202. New York and London: Academic Press (1962).
18. Ref. 7, Vol. IV, p. 94.
19. Ref. 17, pp. 203–204.
20. Ref. 17, p. 204.
21. Ref. 17, p. 207.
22. Lowenthal: *J. Soc. Chem. Ind.*, **3,** 82 (1884); *See also* A.O.A.C. Method **14.048–14.049.**
23. *Tea Importation Act*, 29 Stat. 604 (1897), as amended 21 U.S.C. §§ 41–50 (1952).
24. Federal Register, **34**, 5987 (April 2, 1969).
25. Ref. 17, p. 247.
26. Fisher, H. J.: *Conn. Agr. Expt. Sta. Bul.* **635,** 18 (1960).
27. Ref. 7, Vol. IV, pp. 87–88.
28. Ref. 9, Vol. II, pp. 1702–1703.
29. Ref. 7, Vol. IV, p. 90.
30. Bailey, E. M.: *Conn. Agr. Expt. Sta. Bul.* **248,** 434 (1923).
31. Woodman, A. G.: *Food Analysis*. 4th ed., p. 349, London and New York: McGraw-Hill Book Co., Inc. (1941).
32. Barnes, H. M., Feldman, J. R. and White, W. V., *J. Am. Chem. Soc.*, **72,** 4178 (1950); Griebel, C.: *Z. Lebensm-Unt. Forsch*, **104,** 173 (1956).
33. Lepper, H. A.: *J. Assoc. Offic. Agr. Chemists*, **33,** 523 (1930).
34. Bulen, W. A., Varner, J. E. and Burrell, R. C.: *Anal. Chem.* **24**, 187 (1952); Mabrouk, A. and Deatherage, F. E.: *Food Technol.*, **10**, 194 (1956); Lentner, C. and Deatherage, F. E., *Food Research* **24**, 483 (1959).
35. Urtani, I. and Migano, M.: *Nature*, **175**, 812 (1955); Corse, J. W., *Nature*, **172,** 771 (1953); Sondheimer, E.: *Arch. Biochem. Biophys.* **74,** 131 (1958).
36. Moores, R. G., and Campbell, H. A., *Anal. Chem.*, **20**, 40 (1948); Chatt, E. M., *Cacao*, p. 245. New York: Interscience Publishers, Inc. (1953).
37. Boie, H.: *Pharm. Ztg.*, **75,** 968 (1930).
38. Bailey, E. M. and Andrew, R. E.: *Conn. Agr. Expt. Sta. Bul.* **248,** 433 (1923).
39. Fisher, H. J.: *Conn. Agr. Expt. Sta. Bul.* **660**, 18 (1963).

SELECTED REFERENCES

COCOA

A Bywaters, H. W.: *Modern Methods of Cocoa and Chocolate Manufacture.* London: Churchill (1930).
B Chatt, E. M.: *Cocoa Cultivation Processing Analysis.* New York: Interscience Publishers, Inc. (1953).
C Clarke, W. T.: *Chocolate and Cocoa, Encyclopedia of Chemical Technology*, Vol. III, p. 8. New York: Interscience Encyclopedia Inc. (1949).
D Fincke, H.: *Handbuch der Kakaoerzeugnisse.* Berlin: Springer-Verlag (1936).
E Hall, C. J. J. van: *Cacao.* New York: Macmillan Co. (1932).
F Jacobs, M. B. (ed.): *The Chemistry and Technology of Food and Food Products*, 2nd ed., Vol. II. pp. 1639–1654. New York: Interscience Publishers, Inc. (1951).
G Jensen, H. R.: *The Chemistry, Flavouring and Manufacture of Chocolate, Confectionery and Cocoa.* London: Churchill (1931).
H Knapp, A. W.: *Cacao Fermentation.* London: Bale Sons and Curnow (1937).
I Knapp, A. W.: *The Cocoa and Chocolate Industry.* London: Sir Isaac Pitman and Sons, Ltd. (1923).
J Lecocq, R., *Cacao.* Paris: Vigot (1926).
K Merory, J.: *Food Flavorings, Composition, Manufacture, and Use*, Chapter 8. Westport: Avi Publishing Co., Inc. (1960).
L Roelofsen, P. A.: *Fermentation, Drying and Storage of Cacao Beans, Advances in Food Research*, Vol. VIII, pp. 225–296. New York and London: Academic Press, Inc. (1958).
M Smith, H. H., (ed.): *The Fermentation of Cacao.* London: Bale Sons and Daniellsen, Ltd. (1913).
N Whymper, R.: *Cocoa and Chocolate, Their*

Chemistry and Manufacture 2nd ed. Philadelphia: P. Blakiston's Son and Co. (1921).

O Winton, A. L. and Winton, K. B.: *The Structure and Composition of Foods*. Vol. IV, pp. 114–136. New York: John Wiley and Sons, Inc. (1939).

P Zipperer, Dr. P.: *The Manufacture of Chocolate and Other Cacao Preparations*, 2nd ed. New York: Spon and Chamberlain (1902).

COFFEE

Q Jacobs, M. B., (ed.): Ref. F, Vol. II, pp. 1656–1682.

R Lindner, M. W., *Warenkunde und Untersuchung von Kaffee*. Berlin: Verlag, A. W. Hayns Erben (1955).

S Ukers, W. H.: *All About Coffee*, 2nd ed. New York: Tea and Coffee Trade Journal Co. (1935).

T Ukers, W. H.: *The Romance of Coffee*. New York: Tea and Coffee Trade Journal Co. (1948).

U Winton, A. L. and Winton, K. B.: Ref. O, Vol. IV, pp. 139–166.

TEA

V Jacobs, M. B., (ed.): Ref. F, Vol. II, pp. 1683–1705.

W Roberts, E. A. H.: *Chemistry of Tea Manufacture*, *J. Sci. Food Agr.*, **9**, 381 (1958).

X Stahl, W. H.: *The Chemistry of Tea and Tea Manufacturing*, *Advances in Food Research*, Vol. XI, pp. 201–262. New York and London: Academic Press (1962).

Y Subramanyan, V., Batia, D. S., Nataranjan, C. P., Venkateawara Rao, R., Nagarajan, R., Chandrasekhara, N., Sebastian, J. and Sankaran, E. N.: *Chemical Composition of Indian Tea*, *J. Proc. Inst. Chemists India*, **31**, 167 (1959).

Z Ukers, W. H.: *All About Tea*. New York: Tea and Coffee Trade Journal Co. (1959).

AA Ukers, W. H.: *The Romance of Tea*. New York: Alfred A. Knopf (1936).

BB Winton, A. L. and Winton, K. B.: Ref. O, Vol. IV, pp. 94–111.

CHAPTER 6

Dairy Products (Including Ice Cream, Ice Milk and Sherbets)

Milk

"Milk notwithstanding that it seemeth to be wholly of one substance, yet it is compact or made of three severall substances, that is to say in effect of Creame, Whey and Curds."[1] It is remarkable that although this statement appeared in a book published in the 17th century, it is still a sound first approximation of the composition of milk. Control of the quality of this "lacteal fluid" differs from that of most other foods in that it is chiefly under the jurisdiction of the individual states rather than of a Federal agency, and for this reason legal definitions are not completely uniform within the United States. The United States Public Health Service does however act as a unifying influence through its recommendations to state and municipal health departments. The following definition from its proposed *Milk Ordinance and Code—1953* may be taken as representative of those of the majority of states:

"Milk is hereby defined to be the lacteal secretion, practically free from colostrum, obtained by the complete milking of one or more healthy cows, which contains not less than $8\frac{1}{4}\%$ milk solids-not-fat, and not less than $3\frac{1}{4}\%$ milk fat."[2]

Such definitions frequently occur in actual statutory form rather than as regulations and may give limits for bacterial content in addition to specifying minimum percentages of fat and solids-not-fat. For instance, Connecticut General Statute 22-152 states that "unpasteurized milk shall not contain more than three hundred thousand standard plate count bacterial colonies per milliliter; pasteurized milk shall not contain more than thirty thousand standard plate count bacteria colonies per milliliter." Statutes that set the above limits for fat and solids-not-fat may permit naturally low-fat milk to be "standardized" by the addition of sufficient butter fat or cream to bring its fat content to 3.25%, as well as by blending with milk of higher fat content. At one time, only this latter form would have been tolerated because milk to which anything had been added would have been considered adulterated.

As is implied in the name, upon parturition females of all mammalian species secrete milk to nourish their young. During the earliest phases of lactation, the secretion is known as *colostrum*. This is a thick, yellow fluid whose composition differs both qualitatively and quantitatively from mature milk, and most strikingly in its nitrogenous consistuents. In bovine colostrum not only is the protein content higher, but the distribution of proteins is very different, being high in albumin and globulin; more than traces of ammonia are also present. Immune globulins have been identified in colostrum and are believed to be responsible for the immunity to certain infections that newborn mammals acquire after suckling.

While colostrum is a very interesting substance, both chemically and physiologically, it is not an article of commerce nor is it a food to any except the newborn of the respective species. Therefore it will not be discussed further in this chapter, which will treat primarily the mature milk of cows and products manufactured therefrom.

Qualitatively the milks of all mammals are at least grossly similar in composition, but they do differ in the relative proportions of their major constituents, as Table 6-1 shows:[2]

TABLE 6-1

Mature Milks of Different Species

Representative values for some major constituents in g per 100 ml whole milk*

Constituent	Human	Cow[a]	Buffalo[b]	Goat	Ewe	Mare	Camel	Reindeer
Solids-Not-Fat	8.9	8.6	9.3	8.7	10.9	8.5	8.7	14.2
Protein (N × 6.38)	1.2	3.2	3.7	3.3	5.6	2.2	3.7	10.3
Lactose (anhydrous)	6.9	4.6	4.9	4.4	4.4	6.0	4.1	2.4
Fat	4.6	3.5	7.4	4.5	7.5	1.6	4.2	22.5
Calcium	0.03	0.12	0.19	0.13	0.20	0.09	[c]	[c]

*Summarized by Bernard L. Oser [*Hawk's Physiological Chemistry*, 14th Ed., p. 369. New York. Blakiston Div., McGraw-Hill Book Co., (1965)], from FAO Nutritional Study No. 17, *Milk and Milk Products in Human Nutrition*, 1959. Courtesy, Food and Agriculture Organization of the United Nations.

[a]Holstein-Friesian.

[b]Indian water buffalo.

[c]Reliable information not available.

Total ash contents also differ; average percentages are 0.71 for cow's milk, 0.77 for goat's milk and 0.21 for human milk.

The one ingredient that most readily distinguishes milk as a class from all other foods* is the carbohydrate *lactose*, which is a dextrose-galactose disaccharide (D-glucopyranose-4-β-D-galactopyranoside). The protein *casein* is also found only in milk; this is actually a complex of phosphoproteins, and minor differences between the caseins of different species can be detected by electrophoresis. Much detailed work has been done in the composition of milk, which is a highly complex fluid, but most of this is primarily of interest to the specialist.

The fat content of cow's milk varies with the breed; it averages 3.5% for Holsteins and 5.2% for Jerseys, but individual variations are greater.[3] Milk fat is a mixture of glycerides of over 16 fatty acids. The distribution of these acids is calculated to be as follows: Butyric, 3.5%; caproic, 2.0%; caprylic, 1.0%; capric, 2.0%; lauric, 2.5%; myristic, 10.0%; palmitic, 25.0%; stearic, 10.5%; arachidic, 5.0%; decenoic, dodecenoic and tetradecenoic, 5.0%; oleic, 33.0%; linoleic, 4.0%; and C_{20}–C_{22} unsaturated acids, 1.0%. Forty-three percent of the total acids are unsaturated, and about half of them are liquid at room temperature. It has been established that the glycerides in milk fat are of the mixed type; a typical formula is:

$$\begin{array}{l} CH_2\ \text{(Oleic)} \\ | \\ CH\ \text{(Palmitic)} \\ | \\ CH_2R \end{array}$$

where R is one of the lower fatty acids.[4]

Of the total nitrogenous compounds in milk, 5% are non-protein and 95% of proteinaceous nature. These latter are divided into:

*Kuhn and Low: *Ber.* **82**: 479 (1949), claim to have found lactose in a plant, but this does not significantly alter the above statement.

Casein, 78.5%; albumin, 2.2%; globulin, 3.3%; and proteose-peptone, 4.0%.[5]

Fresh milk comes to the consumer in several grades that are based primarily on sanitary considerations. Insofar as such grading depends on laboratory examination, the tests involved are bacteriological in nature and therefore beyond the scope of the present text. Suffice it to say that nearly all of the milk reaching the average customer is pasteurized, and that while the pasteurization process used may vary in the temperature employed and the time of exposure to that temperature, in every case it is designed to expose "all the product to a heat treatment which will destroy all pathogenic organisms and nearly all other bacteria, and yet not alter the flavor or composition of the product."[6] Pasteurization is not synonymous with sterilization, because only 95–99% of the bacteria are destroyed.

There are a number of standardized milk products that involve either the removal of an ingredient from natural milk or the addition of a foreign substance to it. Because these are of concern to the dairy chemist, their requirements will be listed herewith:[7]

Skimmed milk must not contain more than 0.5% fat. Connecticut Statute **22–127** also recognizes a *fortified skimmed milk* of the same maximum fat content, but fortified with added milk solids to a total not exceeding 17%.

(The state of California has a standard for *concentrated skim milk* containing 9.9% fat and a minimum solids-not-fat content of 24.0%[7a] that is widely sold on the West Coast.)

Buttermilk must contain not less than 8% of solids-not-fat.

Vitamin D milk is milk "the vitamin D content of which has been increased . . . to at least 400 USP units per quart."

Vitamin mineral fortified milk must contain in each quart not less than: Vitamin A, 4000 and vitamin D, 400 USP units; thiamine 1, riboflavin 2, niacin 10, iron 10, and iodine 0.1, milligrams.

Vitamins A and D skimmed milk must contain 2000 USP units of vitamin A and 400 USP units of vitamin D in each quart, and its fat content must not exceed 0.5%.

Evaporated milk must contain not less than 7.9% fat and 25.9% total milk solids. Disodium phosphate, sodium citrate, or both, or calcium chloride, may be added to a total extent of 0.1%. Vitamin D equivalent to at least 7.5 USP units/oz may be added if declared. "Adjustment" of the fat content with cream, etc. is permitted.

Sweetened condensed milk must contain not less than 28% of total milk solids and 8.5% of milk fat. "The quantity of refined sugar or sucrose or combination of such sugar and refined corn sugar or dextrose used shall be sufficient to prevent spoilage."

California also has a *low-fat milk* (*2% fat milk*) whose fat content is standardized at 1.9–2.1%; this must also contain 10% of solids not fat.[7a]

Methods of Analysis

Both in the laws covering its sale and in practice, fresh milk has always been placed in a special category among foods because it is the primary component of the diet of infants and occupies a unique nutritional place in the food intake of children and adults in the Western world. As noted before, it is the one food whose control has been in the hands of municipal and state agencies rather than of the Federal Government. Even in those states having general food laws, there are usually separate acts that specify in detail the manner in which milk must be handled before it reaches the consumer and that set minimum standards of quality.

Because of these laws, specialized dairy control laboratories are primarily interested in the bacteriological testing of milk, and such chemical determinations as they normally run are designed to detect watering or skimming and to test for proper pasteurization. The better-equipped laboratories may also check for the presence of penicillin and pesticide residues and for the level of contamination with radioactive elements such as strontium 90. Bacteriological methods are beyond the scope of this book, and the reader interested in such methods is referred particularly to *Standard Methods for the Examination of Dairy Products Microbiological and Chemical.* 12th ed. (1967), published by American Public Health Association Inc., 1790 Broadway, New York, N.Y. 10019. The following methods are designed to cover the likely chemical needs of a nonspecialized food laboratory:

Method 6-1. *Collection of Sample.* (*AOAC Method* ***15.001***).

The necessary sample size varies with the analysis required. For a usual analysis, collect a 250–500 ml sample; for a fat determination only, collect 50–60 ml.

For bottled milk, collect one or more containers as prepared for sale. Thoroughly mix bulk milk by pouring from one clean vessel into another 3–4 times, or stir at least 30 seconds with a utensil reaching to the bottom of the container. If cream has formed, detach all of it from the sides of the vessel and stir until the liquid is evenly emulsified, or use a hand homogenizer.

Place samples in non-absorbent, air-tight containers, and keep cold but above freezing temperature until examined. When transporting samples, completely fill the containers. Stopper tightly and identify. Tablets containing $HgCl_2$, $K_2Cr_2O_7$, or another suitable preservative at a concentration of at least 0.5 g active ingredient/8 fl-oz milk, but having total weight not exceeding 1 g, may be used provided the preservative does not interfere in the tests to be made; 0.1 ml (2 drops) of 36% HCHO solution per fl oz may also be used with the same proviso. If a phosphatase test is to be made, $CHCl_3$ is the only permissible preservative and stoppers must be of phenol-free material such as red rubber.

Method 6-2. *Preparation of Sample.* (*AOAC Method* ***15.002***).

Bring the sample to about 20°, mix until homogeneous by pouring into a clean receptacle and back repeatedly. Promptly weigh or measure the test portion. If lumps of cream do not disperse, warm the sample in a water bath to about 38° and keep mixing until homogeneous, using a policeman if necessary to reincorporate any cream adhering to the container or stopper. Where it can be done without interfering with dispersal of the fat, cool warmed samples to about 20° before transferring a test portion.

When fat is determined by the Babcock method (Method 6–9), adjust both fresh and composite samples to about 38°, mix until homogeneous, and immediately pipet portions into the test bottles.

Method 6-3. *Total Acidity.* (*AOAC Method* ***15.004***).

Measure or weigh a suitable quantity of the sample (about 20 g) into a suitable dish and dilute with twice its volume of CO_2—free water. Add 2 ml of phenolphthalein solution (1% in alcohol), and titrate with 0.1*N* NaOH to the first persistent pink. Report the acidity as percent lactic acid by wt (1 ml 0.1*N* NaOH = 0.0090 g lactic acid). If a 17.6 ml Babcock milk pipet is used to measure the sample, ml 0.1*N* NaOH required÷20 = % acid as lactic acid.

Method 6-4. *Total Solids.*

a. *By weighing* (*AOAC Method* ***15.014***)—Weigh 5 g of prepared sample into a weighed flat-bottomed dish not over 5 cm in diameter; a Pt dish of this shape and size is preferable, and essential if ash is to be determined on the same portion. Heat on a steam-bath 10–15 minutes, exposing the maximum surface of the bottom of the dish to live steam. Then heat 3 hours in an air oven at 98–100°, cool in a desiccator, weigh quickly, and report the percentage of residue as total solids.

b. *By lactometer* (*AOAC Method* ***15.015***)—Determine the specific gravity of the milk with a Quévenne lactometer (reading to the top of the meniscus), observe the temperature, and correct reading L to 60°F by Table 23–9 Calculate the. total solids either from the formula $0.25L + 1.2F$, in which F = percentage fat in the milk, or from Table 23–10.

Method 6-5. *Ash.* (*AOAC Method* ***15.016***).

Weigh about 5 g of prepared sample into a suitable Pt dish and evaporate to dryness on a steam bath. Ignite in a muffle at not over 500° until the ash is C-free. Cool in a desiccator, weigh, and calculate the percent of ash.

Method 6-6. *Total Nitrogen.* (*AOAC Method **15.017***).

Transfer a 5 g sample to a Kjeldahl digestion flask and proceed as in Method 1–7. Percent $N \times 6.38$ = percent protein.

Method 6-7. *Lactose—Polarimetric Method* (*AOAC Method **15.026–15.027***).

REAGENTS

a. *Acid mercuric nitrate solution*—Dissolve Hg in twice its weight of HNO_3 and dilute with five volumes of water.

b. *Mercuric iodide solution*—Dissolve 33.2 g of KI and 13.5 g of $HgCl_2$ in 200 ml of acetic acid and 640 ml of water.

DETERMINATION

Weigh 65.8 g of milk into each of two volumetric flasks, 100 and 200 ml respectively. Add to each flask 20 ml of the acid mercuric nitrate solution or 30 ml of the HgI_2 solution. To the 100 ml flask add 5% phosphotungstic acid solution to the mark. To the 200 ml flask add 15 ml of 5% phosphotungstic acid solution and dilute to the mark with water. Shake both flasks frequently during 15 minutes, filter through dry filters, and polarize. (It is preferable to read the solution from the 200 ml flask in a 400 mm tube to reduce the error of reading; the solution from the 100 mm flask may be read in a 200 mm tube.) Calculate the percent of lactose as follows:

1) Subtract the reading of the solution from the 200 ml flask (using a 400 mm tube) from the reading of the solution from the 100 ml flask (using a 200 mm tube); 2) multiply the difference by 2; 3) subtract the result from the reading of the solution from the 100 ml flask; 4) divide the result by 2.

Method 6–8. *Lactose—Gravimetric Method.* (*AOAC Method **15.028***).

Dilute a 25 g sample with 400 ml of water in a 500 ml volumetric flask. Add 10 ml of a 6.9g/100 ml solution of $CuSO_4.5H_2O$, and about 7.5 ml of a KOH solution of such concentration that one volume is just enough to precipitate completely the Cu as $Cu(OH)_2$ from one volume of the $CuSO_4$ solution. (Instead, 8.8 ml of 0.5*N* NaOH may be used.) After addition of the alkali solution, the mixture must still be acid and contain Cu in solution. Dilute to the mark, mix, filter through a dry filter, and determine lactose in an aliquot of the filtrate as in Method 16–6, DETERMINATION (**b**). Obtain from Table 23-6 the weight of lactose equivalent to the weight of Cu_2O.

NOTE: K. K. Fox *et al.*: *Agr. Food Chem.* **10**: 408 (1962), have devised a colorimetric modification of the Munson-Walker method that depends on converting the unreacted Cu to the $Cu(NH_3)_4$ ion and measuring its absorbance at 625 mμ. This method is claimed to be applicable not only to milk but to other dairy products, including sucrose-containing materials such as condensed milk and ice cream.

Method 6-9. *Fat—Babcock Method.* (*AOAC Method **15.030–15.031*** and ***15.087–15.088***).

APPARATUS

a. *Standard Babcock milk and cream test bottles*—8%, 18g, 6″ milk-test bottle, and 50%, 9 g, short neck, 6″, 50%, 9 g, long neck, 9″, or 50%, 18 g, long neck, 9″, cream-test bottle. Detailed specifications for these bottles are given in AOAC *Methods of Analysis*, 10th ed. (1965), Sections **15.030(a)** and **15.087(a)**. Bottles certified by state agencies to meet these specifications are available from laboratory supply houses.

b. *Pipet*—Standard milk pipets graduated to contain 17.6 ml that have been certified by state agencies to comply with the specification of AOAC Section **15.030(b)** are available. Pipets designed to deliver 9 g of cream are not officially approved and will not be certified by the appropriate agencies; cream samples should be weighed instead of pipetted. (*See* **c.**)

c. *Cream weighing scales and weights*—With a sensibility reciprocal of 30 mg, i.e., the addition of 30 mg to either pan of the scale, when loaded to capacity, causes a deflection of at least one subdivision of graduation. Set the scales on a support and protect from drafts. These scales are supplied with 9 g and 18 g weights.

d. *Acid measure*—A device used to measure

H_2SO_4, graduated to deliver 17.5 ml; it may be either a graduated cylinder or a pipet attached to a Swedish acid bottle.

e. *Centrifuge*—A centrifuge capable of being heated to at least 55° during centrifuging and provided with a speed indicator, permanently attached if possible. The proper rate of rotation varies with the diameter of the wheel (i.e., the distance between the inside bottoms of opposite cups measured through the center of rotation while the cups are horizontally extended), as follows:

Diameter of wheel, inches	rpm
14	909
16	848
18	800
20	759
22	724
24	693

f. *Dividers or calipers*—For measuring the fat column.

DETERMINATION

a. *In milk*—With the pipet transfer 18 g (17.6 ml) of the prepared sample to a milk test bottle. Blow out the milk in the pipet tip about 10 seconds after free outflow ceases. Add portionally about 17.5 ml of H_2SO_4 (sp. g. 1.82–1.83 at 20°) tempered at 15–20°, washing all traces of milk into the bulb. Shake until all traces of curd disappear; place the bottle in a heated centrifuge, counterbalance, and after the proper speed is reached, centrifuge 5 minutes. Add water at 60° (or above) until the bulb of the bottle is filled. Centrifuge 2 minutes. Add hot water until the liquid column approaches the top graduation of the scale. Centrifuge 1 minute longer. Immerse it in the water bath at 55–60° to the level of the top of the fat column, and leave until the column is in equilibrium and the lower fat surface assumes its final form (not over 3 minutes). Remove the bottle from the bath, wipe it, and with the aid of the dividers or calipers measure the fat column, in terms of percent by weight, from the lower surface to the highest point of the upper meniscus.

The fat column, at the time of measurement, should be translucent, golden-yellow or amber, and free from visible suspended particles. Reject all tests in which the fat column is milky or shows the presence of curd or charred matter, or in which the reading is indistinct or uncertain; repeat the test, adjusting the quantity of H_2SO_4 added.

b. *In cream*—Weigh 9 g of prepared sample directly into a 9 g cream-test bottle, add 9 ml of water and mix thoroughly; add about 17.5 ml of the H_2SO_4 and shake until all lumps completely disappear. Transfer the bottle to the centrifuge, counterbalance it, and after the proper speed is reached, centrifuge 5 minutes. Fill the bottle to the neck with hot water and centrifuge 2 minutes. Add hot water until the liquid column approaches the top graduation of the scale, and centrifuge one minute longer at 55–60°. Adjust the temperature and with the end of dividers or calipers measure the fat column, in terms of percent by weight, from the lower surface to the bottom of the upper meniscus.

NOTES

1. It is important to note that the H_2SO_4 used in the Babcock method is not the straight concentrated acid of sp. g. 1.84 containing 95–98% H_2SO_4, but a somewhat more dilute acid. It may be purchased as *Sulfuric Acid for Babcock Test*; or it may be prepared by dilution of the reagent or technical grade acid. If the latter (containing a trace of Fe) is used, the acid layer in the Babcock bottle assumes a violet color when the milk is preserved with formaldehyde.

2. If the sample is an homogenized milk, American Public Health Association Method **18.12**[8] recommends the following procedure: Add the 17.5 ml of acid in 3–4 steps, with a first portion of about 7.5 ml and the remainder in 2–3 equal amounts, mixing thoroughly for at least 30 seconds after each addition. Then set the bottle aside for not over 30 minutes before centrifuging, and centrifuge for 10 instead of 5 minutes the first time.

Even using this procedure, the Babcock method is not wholly reliable for homogenized milks, and where accuracy in determining fat in such milks is essential, Method 6–10 should be employed.

Method 6-10. *Fat-Roese-Gottlieb Method.* (*AOAC Method **15.029***).

Transfer a 10 g sample to a Mojonnier fat-extraction flask or a Röhrig tube. Add 1.25 ml of NH_4OH (2 ml if the sample is sour) and mix thoroughly. Add 10 ml of alcohol and mix well. Add 25 ml of peroxide-free ether, stopper with a cork or synthetic rubber stopper, and shake vigorously for 1 minute. Add 25 ml of petroleum ether (boiling below 65°), and repeat the vigorous shaking. Centrifuge the Mojonnier tube at 600 rpm, or let it or the Röhrig tube stand until the upper liquid is practically clear. Decant the ether solution into a suitable flask or metal dish. Wash the lip and stopper of the extraction flask or tube with a mixture of equal parts of the two solvents, and add the washings to the weighing flask or dish. Repeat the extraction twice, using 15 ml of each solvent each time.

Evaporate the solvents completely on a hot plate or steam bath at a temperature that does not cause spattering or bumping. Dry the fat to constant weight in an oven at the temperature of boiling water. Weigh the cooled flask or dish, using as a counterpoise a duplicate container handled similarly; avoid wiping either immediately before weighing. Remove the fat completely from the container with warm petroleum ether, dry, and weigh as before. The loss in weight = weight of fat. Correct by a blank determination on the reagents.

Method 6-11. *Fat—TeSa Method.* (*AOAC Method* ***15.032–15.034***).

APPARATUS AND REAGENTS

This method employs 8% 18 g milk test bottles of a special shape, and a non-acid test reagent prepared from a mixture of urea, Na_2CO_3, EDTA, Na_2HPO_4 and polyoxyethylene ethers of mixed fatty and resin acids. Both the bottle and the prepared "*TeSa Reagent Concentrate*" can be obtained from Technical Industries, 2711 S.W. Second Ave., Fort Lauderdale, Florida. If the user prefers to prepare his own reagent, he is referred to AOAC Method **15.032**.

A *fat test reagent* is prepared by dissolving 156 g of "TeSa Reagent Concentrate" in water, diluting to 1 liter, mixing, and allowing to stand for at least 6 hours. Prepare fresh every 2 weeks. The *milk test reagent* is made from this by mixing one volume of methanol with 5 volumes of *fat test reagent*. Prepare fresh every 2 days.

DETERMINATION

With a 17.6 ml pipet, transfer 18 g prepared sample to a "TeSa" milk test bottle through the side tube. Blow out the milk in the pipet tip about 10 seconds after free outflow ceases. Add 15 ml of milk test reagent through the side tube, washing all traces of milk into the bulb, and immediately swirl to obtain a uniform mixture. Place the bottle in a boiling water bath for 10–12 minutes. Remove the bottle and slowly add hot water through the side arm until the bottle is full. Let stand at room temperature 5–6 minutes. Place the bottle in a tempering bath to the level of the top graduation for 3 minutes. Add water at 55–60° through the side arm to raise the fat level to (or just below) the O graduation. Read the lower meniscus, and report the difference between the upper and lower menisci in percent by wt of fat. (Read the meniscus as in usual volumetric measurements, not as in Babcock tests.)

If a small quantity of foam or undigested material obscures the lower meniscus, add a few drops of 40% methanol through the fat column, retemper, and read as above. If the quantity of undigested material obscuring the lower meniscus is large, repeat the test, using 17–18 ml of milk test reagent, and leaving in the boiling water bath for 15–20 minutes, shaking 3–4 times during the first 10 minutes.

NOTE: This method was designed primarily for field testing where it was desired to avoid the hazard that could arise from the use of strong acid in the Babcock test. It has been adopted by the AOAC as official only for raw milks, but results for other milks do not differ greatly from those by the Babcock method.

Method 6-12. *Fat by the Foss "Milko-Tester."*

This method employs a special Danish apparatus that is obtainable from Foss America Inc., Fishkill, N.Y., 12524. (*See* Figure 6–1.)

A 35 ml sample is sucked into the machine, where it is heated to 60°, and homogenized. An aliquot is then pipetted out, diluted with a reagent that neutralizes the transmission of the protein, and the transmission due to the fat is measured by a photocell that acts as a galvanometer to give a direct reading in percent fat. The whole operation is automatic except for pushing two buttons and is claimed to take only 30 seconds. Detailed directions may be obtained from the supplier of the machine. Results are claimed to show agreement

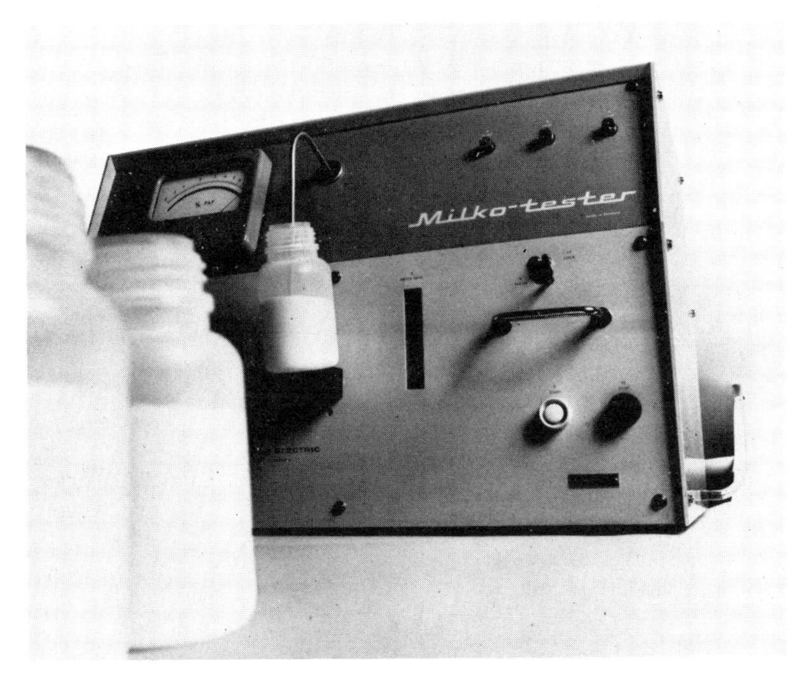

Fig. 6-1. Foss "Milko-Tester."

with Babcock Tests and a probable precision of ± 0.05% on unhomogenized products, with readings on homogenized milks running 0.05% lower than those of the Babcock test.[9]

Because the instrument is quite expensive, this method will find its chief use in large dairies where speedy results are important.

Methods for Detecting Watering

All of these methods depend in principle on the fact that the addition of water to milk will dilute the dissolved substances in the aqueous portion (i.e., serum) of the milk. At one time, the two factors chiefly relied on for the detection of watering were the refraction of the copper serum and the percentage of ash in the serum obtained from natural souring of the milk. Both were largely replaced by the freezing point as determined by the Hortvet cryoscope[10] because this method was capable of determining a smaller percentage of added water. The Fiske Cryoscope Method 6–15 employs a modification of the Hortvet cryoscope that makes use of a more convenient thermistor to measure the freezing point.

Method 6-13. *Copper Serum Refraction.* (*AOAC Method* ***15.036.***)

REAGENT

Dissolve 72.5 g of $CuSO_4.5H_2O$ in water and dilute to 1 liter. Adjust to read 36 at 20° on the Zeiss immersion refractometer, or to a sp. g. 20/4° of 1.0443.

DETERMINATION

To one volume of the $CuSO_4$ solution, add 4 volumes of the milk. Shake well and filter. Determine the immersion refractometer reading of the clear serum at 20°.

NOTES

1. Bausch and Lomb immersion refractometers give readings identical with those of the Zeiss instrument, except those with Serial Nos. 4,000–10,000, whose readings are 0.4 lower.

2. A Cu serum refraction at 20° of less than 36.0 suggests watering. This method will detect 10% of added water.[11]

Method 6-14. *Acetic Serum Ash.* (*AOAC Method* ***15.035b***).

To 100 ml of milk, measured at 20° into a beaker, add 2 ml of 25% acetic acid (sp. g. 1.035).

Cover the beaker with a watch-glass, keep in a water-bath 20 minutes at 70°, and then 10 minutes in ice-water. Separate the curd from the serum by rapid filtration through a small filter. Pipet 25 ml of the serum into a weighed flat-bottomed Pt dish and evaporate to dryness on a water bath. Heat over a low flame (to avoid spattering) until the contents are thoroughly charred, place the dish in a muffle furnace (preferably pyrometer-controlled) and ignite to a white ash at a temperature not exceeding 500°. Cool and weigh. Express the results as g/100 ml.

NOTES

1. This method is essentially a development from the original sour serum ash method in which the serum from naturally soured milk was used. The change became necessary when nearly all market milks were pasteurized before reaching the consumer, because pasteurized milk will not sour normally.

2. An ash content of less than 0.73 g/100 ml indicates watering.[12]

Method 6-15. *Freezing Point by the Fiske Cryoscope.* (*AOAC Method* ***15.040–15.041***).

APPARATUS

The Fiske Cryoscope is obtainable from Advance Instruments, Inc., 45 Kenneth Street, Newton Highlands, Massachusetts 02161. It consists of a cooling bath, an air agitator, a thermistor probe, a seeder rod, a Wheatstone bridge and a galvanometer measuring circuit. Freezing point values are read from a temperature dial that balances the bridge.

DETERMINATION

Operate the instrument in accordance with the instructions of the manufacturer. Observe the following precautions:

The sample tubes (16 mm dia.) and pipets must be clean and dry. Prechill the tubes containing 2 ml of sample by immersing in an ice-bath to two-thirds of their lengths. Locate the thermistor in the center of a tube and 10 mm from the bottom of the tube. (This distance must be kept constant for reproducibility.) Supercool to the same temperature each time, because variable supercooling affects accuracy. After seeding (seed only once), a long steady temperature plateau must be obtained. Take the freezing point as the position where the galvanometer spot ceases to move to the right. Follow the same technique for standard solutions as for milk samples, but note that the time required to reach a plateau is less for standards.

NOTES

1. The presence of added water is indicated if the freezing-point is above −0.530°. (This applies to either raw or pasteurized milk that has not been subjected to vacuum treatment; vacuum treatment may raise the freezing-point by as much as 0.008°.) It should not be assumed that milk with a freezing-point below −0.530° is necessarily free of added water. In fact, samples representing large mixed lots of raw milk will probably have a freezing-point below −0.540°. Such milk with a freezing-point above −0.540° should be regarded with suspicion, as should large fluctuations in the freezing-points of bulk milk from day to day. If desired, the minimum percentage of added water, W, can be calculated as $W = (100 - TS)(T - T')/T$, where $T = -0.530$, T' = f. pt. of sample, and TS = % total solids.

2. If the titratable acidity (Method 6–3) is over 0.18%, results by this method may underestimate the amount of added water.

3. The cryoscope method is capable of detecting as little as 3% of added water.

Detection of Skimming

Skimming milk, i.e., removing a portion of the fat, changes the relation of the fat content to the other ingredients, and the detection of skimming is based on this fact rather than on some special chemical test. One of the factors employed is the protein/fat ratio. Average market milk has a protein/fat ratio of 0.82, and a ratio of 0.90 is pretty good evidence of skimming. In fact, the mixed milk of Guernsey and Jersey cows has a protein/fat ratio as low as 0.6; for the milk of such cows the 0.90 ratio is unduly generous and would permit considerable skimming.

This method is not reliable for milks whose fat content exceeds 3.5%. If a milk has a fat content below 2.2% and the solids-not-fat exceeds 8.5% it is certainly skimmed.[13]

Preservatives

For the detection and determination of preservatives in general, see Chapter 14. The following qualitative tests are suggested for formaldehyde and hydrogen peroxide in milk:

Method 6-16. *Formaldehyde.*[14]

REAGENT

Chromotropic acid solution—Pour 42 ml of H_2SO_4 into water, cool, and dilute to 100 ml. Saturate this solution with 1,8-dihydroxynaptha-lene-3,6-disulfonic acid (about 500 mg). The solution should be a light straw-colored.

TEST

Dilute 100 ml of milk with an equal volume of water, acidify with H_3PO_4, and add 1 ml in excess. Slowly distil through a condenser with a trap, collecting 50 ml of distillate. To 1 ml of the distillate in a test-tube add 5 ml of the chromotropic acid reagent and place in a boiling water-bath for 15 minutes. In the presence of HCHO the solution turns a light to deep purple, the depth of color varying with the amount of HCHO.

NOTE: *See* note 1 under Method 6–9.

Method 6-17. *Hydrogen Peroxide.*[14]

Dissolve 1 g of V_2O_5 in 100 ml of H_2SO_4 (6+94). Add 10–20 drops of this reagent to 10 ml of milk and mix. A pink or red color indicates H_2O_2.

Method 6-18. *Hypochlorites and Chloramines.* (*AOAC Method* **15.044–15.045**).

Unreliable in presence of over 2.5 ppm Cu.

REAGENTS

a. *Potassium iodide solution*—Dissolve 7 g of KI in 100 ml of water. Prepare fresh.

b. *Dilute hydrochloric acid*—To 100 ml of HCl add 200 ml of water.

c. *Starch Solution*—Boil 1 g of starch in 100 ml of water. Cool before using.

TESTS

a. To 5 ml of milk in a test-tube add 1.5 ml of the KI solution, mix thoroughly by shaking, and note the color of the milk.

b. If unaltered, add 4 ml of the HCl, mix thoroughly with a flat-ended stirring rod, and note the color of the curd.

c. Next place the tubes in a large water-bath (previously heated to 85°) and let stand 10 minutes (during this interval the curd rises to the surface); then cool rapidly by placing in cold water. Note the color of the curd and liquid.

d. Then add 0.5–1.0 ml of the starch solution to the liquid below the curd, and note the color. Concentrations of available Cl present (if any) are as indicated in Table 6-2:

Sediment

For a standardized method for estimating the amount of foreign undissolved material (manure, dirt, etc.) in milk and cream, the

TABLE 6-2
Reactions with the Various Tests

Concentration of Available Cl	1:1000	1:2000	1:5000	1:10000	1:25000	1:50000
Test a	Yellowish brown	Deep yellow	Pale yellow, fades	—	—	—
Test b	Yellowish brown	Deep yellow	Light yellow	—	—	—
Test c	Yellowish brown	Deep yellow	Yellow	Yellow	Pale yellow	Yellowish
Test d	Blue purple	Blue purple	Blue purple	Dark red-purple	Red purple	Pale red-purple

reader is referred to AOAC Methods **36.010–36.013** and **36.015**, or to Chapter 15 of the American Public Health Association *Standard Methods for the Examination of Dairy Products Microbiological and Chemical*, 11th ed. (1960).

Residual Phosphatase

The phosphatase tests are chemical tests designed to determine whether a dairy product has been pasteurized and can consequently be assumed to be free of pathogenic bacteria. They depend on the fact that an enzyme, alkaline phosphatase, present in raw milk, is progressively inactivated when exposed to temperatures similar to those employed in pasteurization. In 30 minutes most of it is destroyed at 143°F., and inactivation is complete at 145°F. Consequently the determination of the amount of enzyme present is an indication of whether milk has been underprocessed (in temperature or time) or has had raw milk added to it. Such determinations are based on the hydrolysis of disodium phenyl phosphate to free phenol by the enzyme. The phneol is then easily estimated by any one of several colorimetric reactions.[15]

A number of officially recognized methods for residual phosphatase can be found in Chapter 10 of the American Public Health Association *Standard Methods for the Examination of Dairy Products Microbiological and Chemical*, 11th ed. (1960), and in AOAC Methods **15.049–15.066**, **15.154**, **15.177–15.180** and **15.199**. Some of these are relatively long laboratory methods, designed to measure accurately the amount of enzyme remaining activated; others are short tests, intended to determine only whether a sample has or has not been pasteurized, but capable of field use. There is also a special test for identifying so-called "reactivated phosphatase" that is not associated with underpasteurization, which may develop when pasteurized products are stored without adequate refrigeration—particularly when the temperature of pasteurization was between 220 and 230°F.[16]

The following method, of the "rapid" variety, was designed by Scharer:[17]

Method 6-19. *Residual Phosphatase in Milk. (Scharer Rapid Method).*

APPARATUS

a. *Test tubes*—Matched set of uniform bore and equal inside diameter, preferably 12 × 114 mm, calibrated at 5, 5.5 and 8.5 ml to the top of the meniscus, equipped with phenol-free stoppers.

b. *Pipettes*—5 ml and 1.1 ml graduated at 0.5 ml.

c. *Light source and filter*—Such as the Catalogue No. 335 Fluoro-Dent Illuminator of Wolf X-Ray Products Inc., 93 Underhill Ave., Brooklyn, New York.

d. *Color Standards*—1, 2 and 5 unit standards may be obtained from Applied Research Institute, 27 East 22nd Street, New York, N.Y. 10010, or standards may be prepared from a standard 1 mg/ml phenol solution as directed in Reference 8, page 340.

REAGENTS

a. *Buffer*—Dissolve 100 g of $NaHCO_3.Na_2CO_3.2H_2O$ in water and make up to 1 liter.

b. *Buffer substrate*—Dissolve 0.5 g of phenol-free $Na_2C_6H_5PO_4$ crystals in water, add 25 ml of the buffer, (**a**), and make to 500 ml. Prepare sufficient solution for immediate needs and store under refrigeration. Optionally this substrate may be prepared by dissolving one PHOS-PHAX tablet (new type)* in 50 ml of water. [For the purification of reagents not phenol-free, *see* Reference 8, pp 322–323, Section **19.09** (**b**).]

c. *CQC reagent*—Dissolve 30 mg of crystalline 2,6-dichloroquinone chloroimide (special for phosphatase work) in 10 ml of alcohol or methanol. Do not prepare large quantities. Store under refrigeration in a glass-stoppered bottle. For convenience, transfer a few ml to a brown dropping bottle with a dropper calibrated to deliver about 50 drops/ml. The dropper closure must be phenol-free. Discard the solution after one week, or if it turns brown. The solution may also be prepared by dissolving one INDO-PHAX tablet* in 5 ml of methanol.

*These tablets may be obtained from Applied Research Institute, 27 East 22nd Street, New York, N.Y. 10010.

d. *Catalyst*—Dissolve 200 mg of $CuSO_4.5H_2O$ in 100 ml of water. INDO-PHAX tablets bearing lot or control numbers over 2000 contain Cu, so if tablets of such numbers are used in (**c**), the separate catalyst may be omitted.

e. *n-Butyl alcohol*—Neutralized. To n-butyl alcohol of b.pt. 116–118° add 0.1*N* NaOH until a small portion tested with bromothymol blue indicator gives a green or light blue color. (Shake the sample with an equal volume of neutral water, allow to separate, and test the aqueous layer.)

COLLECTION OF SAMPLES

Collect samples preferably of not less than 10 ml. If these are refrigerated, avoid freezing. If a preservative is necessary, use only 1–3% of $CHCl_3$.

CONTROLS

a. *Negative control*—Place 5 ml of the milk in a tube containing a thermometer, and heat in a beaker of hot water for at least 1 minute after the thermometer registers 80–90°. Stir and mix to assure complete inactivation of phosphatase.

b. *Positive control.* Add 0.1 ml of fresh raw milk to 100 ml of boiled milk. (If raw milk is unavailable, add 1 drop of 0.005% phenol per ml of boiled milk.) This control should yield about 2 μg phenol/ml.

DETERMINATION

Add 5.0 ml of buffer substrate to enough tubes to allow for all samples, duplicate tests and other necessary controls. Transfer 0.5 ml of a well-mixed sample or control, using a clean pipette for each. Label or otherwise identify the tube before or immediately after the transfer. Stopper the tubes and mix by inverting several times.

Place the tubes in a water-bath and warm to 40°, then incubate at 39–41° for 15 minutes. Remove from the water-bath, add 6 drops of the CQC reagent (and 2 drops of catalyst if necessary) to each tube, stopper, mix, and incubate 5 minutes.

Remove the tubes from the water bath, cool in tap water, add 3 ml of the neutralized butyl alcohol, stopper, and extract the indophenol blue by gently inverting the tubes four times through a half-circle. (To avoid emulsions caused by violent shaking, take about 1 second to invert the tube, pause about 1 second to allow bubbles to break and the alcohol to separate, take another second to return the tube to an upright position, pause 1 second and repeat.) Lay the tubes on a flat surface for 2 minutes to permit separation of butyl alcohol. Repeat the extraction and separation step. (Extraction of indophenol blue by n-butanol must be complete and without emulsification. If an emulsion interferes, clarify by centrifuging or repeat the test.)

Stand the tubes erect and compare the colors of the butyl alcohol layers with standards. The negative control should show no blue color; if it does, the cause should be determined.

NOTE

A value of 1 μg or more of phenol per ml of milk is indicative of improper pasteurization or contamination with unpasteurized substances.

Penicillin

Penicillin usually gets into milk as a result of treatment of the cow for a mastitis infection. Milk from such cows is not supposed to be sold until after a lapse of time calculated to be sufficient for complete disappearance of the penicillin. Testing of market milk for this antibiotic is important because of the possibility that sensitization of consumers to penicillin could result, or conversely that people already sensitive to penicillin might suffer allergic reactions.

Those interested in running tests for penicillin in milk are referred to AOAC Methods **15.067–15.071** and **15.072**, sensitive respectively to 0.05 and 0.01 units/ml of penicillin. (Attention should be called to the note of Hankin: *J. Assoc. Offic. Agr. Chemists.* **47**: 692 (1964).)

Mineral Elements in Milk

Unfortunately, space limitations forbid inclusion in this book of methods for all of the mineral elements in milk whose determination is sometimes desirable. We are listing a method for calcium because that is the major metallic ingredient of milk and the method is new, methods for iron and iodine because the specifications for Vitamin Mineral Fortified Milk set limits for those elements, and a method for copper because of the effect of excess copper on the development of "cardboardy" flavor.

Readers interested in determining other elements are referred to the Selected References in Chapter 1.

Method 6-20. *Calcium.*[18]

REAGENTS

a. *Standard calcium solution*—1.00 mg Ca/ml. Dissolve 2.4972 g of reagent grade $CaCO_3$, dried at 110°, in HCl, and make to 1 liter with water.

b. *Glyoxal bis(2-hydroxyanil) solution*—0.5%. Dissolve 0.5 g of Fisher G-147 glyoxal bis(2-hydroxyanil) in 100 ml of redistilled methanol.

c. *Tris buffer solution*—pH 12.7. To 1 liter of 0.1*M* tris(hydroxymethylaminomethane) add 100 ml of 10% KOH solution.

d. *Ammonium oxalate—oxalic acid solution*—pH 5. Dissolve 27 g of $(NH_4)_2C_2O_4$ and 1.26 g of $H_2C_2O_4.H_2O$ in water and dilute to 1 liter.

e. *Congo red indicator*—Dissolve 0.1 g of Congo Red water soluble (Colour Index No. 22120, Fisher Catalogue No. A-795) in water and dilute to 100 ml.

DETERMINATION

Mix 20 ml of milk with an equal volume of 20% trichloracetic acid, let stand 10 minutes, and filter. Pipet 20 ml of the filtrate (=10ml sample) into a 50 ml round-bottomed centrifuge tube, and adjust the pH to greater than 5 with 10% KOH solution, using 1 drop of Congo Red solution as indicator. Then add 10 ml of the $(NH4)_2C_2O_4$–$H_2C_2O_4$ solution, cap the tube, and let stand for 1 hour.

At the end of this time, centrifuge the tube for 15 minutes at 3000 rpm, and carefully pour off the supernatant liquid. Dissolve the precipitate in about 2 ml of 1*N* HCl, and quantitatively transfer the solution to a volumetric flask and dilute to give a Ca concentration within the desired range. (For a skim milk sample, transfer the solution of the precipitate to a 200 ml flask and make to volume.)

After further 1:1 dilution, pipet a 10 ml aliquot into a 125 ml Erlenmeyer flask, and add 5 ml of the Tris buffer solution. Run a blank on 10 ml of water and 5 ml buffer. To each add 0.5 ml of the glyoxal bis(2-hydroxyanil) solution and 10 ml of alcohol, in that order, and mix before proceeding from one flask to the other. After 10 minutes, measure the absorbances of the two solutions at 524 mμ and calculate the concentration of Ca in the sample by reference to a curve prepared by carrying known amounts of the standard Ca solution through the procedure.

NOTES

1. The oxalate precipitation step eliminates phosphate interference; Mg does not interfere unless its concentration is above 40 μg/ml.

2. The Ca concentration of normal milk averages 0.126%.[19]

Method 6-21. *Iron.* (Bipyridine Method).[20]

REAGENTS

a. *2,2′-Bipyridine solution*—Dissolve 0.2 g of the reagent in 5 ml of acetic acid and dilute to 100 ml with water.

b. *Ammonium hydroxide*—About 6*N*. Dilute 420 ml of NH_4OH to 1 liter with water.

c. *Buffer solution*—Dissolve 27.2 ml of acetic acid and 33.4 g of Na acetate in water and dilute to 250 ml.

d. *Hydrochloric acid*—6*N*. Dilute 531 ml of HCl with water to 1 liter.

e. *p-Nitrophenol*—0.1% in water.

f. *Standard iron solution*—Dissolve 0.1 g of Fe wire in dilute HCl, and dilute to 1 liter with water (1 ml = 100 μg Fe). Dilute 10 ml of this solution to 1 liter (1 ml = 1 μg Fe).

g. *Trichloracetic acid solution*—25% in water.

DETERMINATION

To 5 ml of milk in a 15 ml centrifuge tube add 5 drops of mercaptoacetic acid (Eastman No. P2249), 2 ml of the trichloroacetic acid solution, and 1 ml of HCl. Also prepare a blank on reagents with 5 ml of water. Stir well, place for 5 minutes in a water bath heated to 90–95°, and cool in water to room temperature. Stir thoroughly to break up the precipitate and centrifuge 15 minutes at 3500 rpm. Decant the supernatant liquid into a 25 ml volumetric flask. Add to the precipitate in the centrifuge tube a mixture of 2 ml of water, 1 ml of trichloroacetic acid and 1 ml of HCl, and repeat the heating, centrifuging and decantation.

To the combined supernatants, add one drop of the *p*-nitrophenol indicator, and slowly add 6*N* NH_4OH until the solution turns yellow. Make acid with 1–2 drops, or enough to cause the yellow color to disappear, of the $6N$ HCl. Add 1 ml of the buffer solution, dilute to 25 ml, and mix. Pipet a 5 or 10 ml aliquot (estimated to contain 1–2 mg of Fe) into a beaker or flask, and dilute if necessary to 10 ml with water. Add 2 drops of mercaptoacetic acid (to reduce any ferric Fe present) and 1 ml of the 2,2′ bipyridine reagent, mix, and read the absorbance in a spectrophotometer at 522 mμ. Correct for the absorbance of the reagent blank, and determine Fe present by reference to a standard curve. If a 5 ml aliquot was taken, this value = p.p.m. Fe in the milk. For a 10 ml aliquot, divide by 2.

PREPARATION OF STANDARD CURVE

Dilute 0, 1, 2, 3, 4 and 5 ml of the standard Fe solution, (**f**), to 5 ml with water, and treat as under Determination (except that the centrifugation may be omitted), using 5 ml aliquots for the color development. Subtract the reading for the solution to which no Fe was added from the readings for the other solutions, and plot the absorbances of solutions 1–5 against micrograms of Fe.

NOTE: State regulations generally require the presence of 10 mg of iron per quart in vitamin-mineral fortified milk. Buegamer *et al.* found 0.114–0.650 p.p.m. of Fe in unfortified raw and market whole milks.[20]

Method 6-22. *Iron in Skim Milk.* (PPDT Method).[21]

REAGENTS

a. 3-(4-*Phenyl-2-pyridyl*)5,6-*diphenyl*-1,2,4-*triazine* (*PPDT*)—0.005*M*. To 0.193 g of PPDT [prepared by the method of Case, *J. Org. Chem.* **30**: 391 (1965)] add a few drops of HCl and 100 ml of alcohol.

b. *Standard iron solution*—Using electrolytic Fe, proceed as in Method 6–21.

c. *Hydroxylamine hydrochloride solution*—10%. Dissolve 100 g of the salt in 900 ml of water. Remove Fe by adding PPDT solution and extracting with *iso*amyl alcohol.

d. *Acetate buffer*—pH 4–5. Dissolve 60 g of acetic acid and 136 g of $CH_3COONa.3H_2O$ in water and dilute to 1 liter.

e. *Ammonia buffer*—pH 8.5–9.2. Dissolve 53.5 g of NH_4Cl in water, add 70 ml of NH_4OH, and dilute to 1 liter.

f. *Perchloric acid*—70%. Use the doubly distilled grade of G. Frederick Smith Chemical Co., 867 McKinley Ave., Columbus, Ohio.

g. *Nitric acid*—Preferably use acid that has been redistilled from a glass still and stored in a borosilicate glass bottle.

DETERMINATION

Pipet a 5.00 ml sample into a 250 ml Erlenmeyer flask, add a 10 ml mixture of equal volumes of $HClO_4$ and HNO_3, and heat below boiling. (Four stages are observed during the wet ashing process: first, the evolution of brown fumes of NO_2; then a quiescent stage during which the solution becomes increasingly concentrated in $HClO_4$; next a rather vigorous evolution of gaseous reaction products, accompanied by dense white fumes of $HClO_4$, lasting about 10 minutes; and finally a quiescent stage with dense white fumes completely filling the flask. To completely digest the sample, continue heating the reaction mixture for at least 10 minutes after the fourth stage has been reached.) After the flask and contents have cooled to room temperature, add 20–30 ml of water (washing down the sides of the flask), and boil briefly to remove any Cl_2. Add 2 ml of the $NH_2OH.HCl$ solution, 2 ml of the PPDT solution, 2 ml of the acetate buffer, and sufficient NH_4OH (about 4 ml) to adjust to pH 4–7 with indicator paper. Transfer the solution to a 60 ml separatory funnel, add 9 ml of *iso*amyl alcohol, shake 10–20 seconds, and discard the lower aqueous phase after complete separation. Add 20 ml of the NH_3 buffer, shake at least 15 seconds, discard the aqueous layer, and collect the *iso*amyl alcohol layer in a 10 ml volumetric flask. Dilute to volume with alcohol, and measure the absorbance against a blank at 561 mμ. (Make this measurement not later than 30 minutes after shaking with the NH_3 buffer.) Read μg Fe present from a curve prepared by carrying aliquots of the standard Fe solution through the procedure.

NOTES

1. The wet ashing can be performed under a Transite hood, or outside of the hood using the glass fume eradicators of the G. Frederick Smith Chemical Company.

2. If this or any other method is used to deter-

mine Fe, Cu, Zn, etc. in a food that must be comminuted before analysis, it is essential not to employ a metal mill or sieve until the analyst has established that these metals are not being picked up during the grinding process. It is usually safer to use a clean porcelain mortar and pestle. Water employed for dilution or washing should be redistilled from glass.

Method 6-23. *Copper.* (2,2′-*Biquinoline Method.*)[22]

REAGENTS

a. *Redistilled water*—Redistil from glass.

b. *Cuproine reagent*—0.1%. Dissolve 1 g of 2,2′-biquinoline (obtainable from G. Frederick Smith Chemical Co., 867 McKinley Ave., Columbus, Ohio, as their No. 116) in 1 liter of alcohol.

c. *Isoamyl alcohol*—Shake 800 ml of *iso*amyl alcohol with 100 ml of 10% $Na_2S_2O_5$, separate the alcohol layer, dry over anhydrous $MgSO_4$, filter and redistil. Store in a brown bottle.

d. *Hydroxylamine hydrochloride solution*—10%. Dissolve 10 g of $NH_2OH.HCl$ in 100 ml of water, and extract with 10 ml portions of the *iso*amyl alcohol, to which a few ml of the cuproine solution have been added, until free of Cu.

e. *Sodium acetate buffer solution*—1*M*. Dissolve 136 g of $CH_3COONa.3H_2O$ in 1 liter of water, and free of Cu as under (**d**).

f. *Standard copper solution*—Place 0.2000 g of Cu wire or foil in a 125 ml Erlenmeyer flask, add 15 ml of HNO_3 (1+4), cover the flask with a watch-glass, and let the Cu dissolve, warming to complete solution. Boil to expel fumes, cool, transfer to a 200 ml volumetric flask, and make to volume with the redistilled water (1 ml = 1 mg Cu). Dilute 20 ml of this solution to 200 ml with redistilled water (1 ml = 100 μg Cu). Prepare a working standard (2 μg/ml) daily by diluting 5 ml of this solution to 250 ml with 2.0 *N* H_2SO_4 (made by diluting 104 g of H_2SO_4 to 1 liter with redistilled water).

PREPARATION OF SAMPLE

Proceed as in Method 6–22, *Determination*, through "boil briefly to remove any Cl_2". Cool the flask and residue and wash into a separatory tunnel. (If the sample is expected to contain more than 250 μg of Cu, make the solution to a fixed volume and transfer a suitable aliquot to the funnel.)

DETERMINATION

To the solution in the separatory funnel, add 5 ml of the 10% $NH_2OH.HCl$ solution and adjust the pH to 5–6 with NH_4OH (1+3). Add 2 ml of the 2,2′ biquinoline reagent and 10 ml of the *iso*amyl alcohol. Shake 1 minute and allow the layers to separate. Drain the lower layer into a second separatory funnel, and repeat the extraction with another 10 ml of *iso*amyl alcohol. Transfer both alcohol layers to a 25 ml volumetric flask and make to volume with *iso*amyl alcohol. Measure the absorbance of the solution at 546 mμ, and determine the amount of Cu present from a curve plotted by carrying known amounts of the standard Cu solution through the same procedure. Run a blank on the reagents.

Method 6-24. *Copper by Atomic Absorption.*[23]

APPARATUS

Atomic absorption spectrophotometer—Model 303 of the Instrument Division, Perkin-Elmer, Norwalk, Conn., should be operated under standard conditions for Cu as outlined in *Analytical Methods for Atomic Absorption Spectrophotometry* of the manufacturer, except to use 2X scale expansion and added external capacitance. Reduce the flow of acetylene in the air-acetylene flame almost to the point where the flame lifts off the burner. Under these conditions the introduction of the organic solvent produces a clear blue flame.

REAGENTS

a. *Ammonium pyrrolidine dithiocarbamate solution*—1%. Dissolve 1 g of NH_4 pyrrolidine dithiocarbamate (obtainable from K and K Laboratories, Inc., 121 Express St., Engineers Hill, Plainview, N.Y. 11203, as their No. 2061) in water and make to 100 ml.

b. *Hydrochloric acid*—2.4*N*. Dissolve 250 g of HCl in water and dilute to 1 liter.

c. *Hydrochloric acid*—6*N*. Dissolve 625 g of HCl in water and dilute to 1 liter.

d. *Hydrochloric acid*—0.1*N*. Dissolve 10.4 g of HCl in water and dilute to 1 liter.

PREPARATION OF SAMPLE

To 25 ml of whole milk in a Vycor evaporating dish add 4 ml of 2.4*N* HCl and evaporate to dryness. Heat the dried sample slowly in a muffle to 500°, and leave over night at this temperature. This leaves a gray ash. Add 2 ml of HNO_3 to the residue and evaporate to dryness on the steam bath. Flame the dish a few minutes, and return to the muffle at 500° for 1 hour. This produces a white ash. Take the ash up in 5 ml of 6*N* HCl, and dry on the steam bath. Take up again in 10 ml of 0.1*N* HCl, bring to boiling, and transfer to a 50% Babcock cream bottle. Rinse the evaporating dish by adding another 10 ml of 0.1*N* HCl, bringing to a boil, and transferring to the Babcock bottle. Finally add 10 ml of water to the dish and pour into the bottle.

To the combined solution in the Babcock bottle add 1 ml of 1% NH_4 pyrrolidine dithiocarbamate solution and 5 ml of *iso*amyl methyl ketone (Eastman No. P8327 5-methyl-2-hexanone). Shake 1 minute by hand, and centrifuge 10 minutes. Bring the organic phase into the narrow neck of the bottle by adding water, and centrifuge again for 10 minutes.

DETERMINATION

Prepare a series of standards containing 0, 1, 2, 3, 4, 5, 10, 20, and 30 μg of Cu in 5 ml of *iso*amyl methyl ketone by adding appropriate aliquots of Reagent (**f**), Method 6–23, to 25 ml portions of 0.1*N* HCl in 50% Babcock cream bottles, and treating as in the previous paragraph. Run atomic absorption analyses on the organic extracts and plot a curve therefrom. Measure the absorption of the organic phase of the sample, and calculate its Cu concentration from the curve after correction for a blank on reagents that have been carried through the whole procedure.

NOTES

1. All water used should be redistilled from glass or deionized.

2. The determination of Cu is particularly important in milk because abnormally high concentrations of this element result in an off-flavor known as "oxidized" or "cardboardy" flavor. Normal milk from cows not in the early stages of lactation contains 0.02–0.04 p.p.m. of Cu.[24]

Method 6-25. *Iodine.* (*AOAC Method* ***22.087***).

REAGENTS

a. *Sodium thiosulfate solution*—About 0.1*N*. Dissolve about 25 g of $Na_2S_2O_3.5H_2O$ in 1 liter of water, boil gently 5 minutes, and transfer while hot to a storage bottle previously cleaned with hot H_2SO_4-$K_2Cr_2O_7$ solution and rinsed with warm boiled water. (Temper the bottle, if not made of resistance glass, before adding the hot solution.) Store in a dark, cool place.

b. *Standard sodium thiosulfate solution*—0.005*N*. Just before use, dilute 50 ml of the approximately 0.1*N* $Na_2S_2O_3$ solution, (**a**), to 1 liter with boiled water, and standardize as follows:

Pipet 25 ml of a solution containing 0.1308 g/liter of KI into a beaker and add 200 ml of water, 5 ml of 20% $NaHSO_3$ solution, and 2–3 g of NaOH. Neutralize the mixture with syrupy H_3PO_4 and add 1 ml excess, using methyl orange indicator. Add an excess of Br_2 water and boil the solution gently until colorless, then 5 minutes longer. Add a few crystals of salicylic acid and cool the solution to 20°. Add 1 ml more of 85% H_3PO_4 and about 0.5 g of KI, and titrate the I_2 with $Na_2S_2O_3$ solution, adding starch solution when the brown of the liberated I_2 is nearly gone. If X = ml $Na_2S_2O_3$ solution titrated, mg I/ml $Na_2S_2O_3$ solution = $15/X$.

DETERMINATION

Place a sample containing 3–4 mg of I in a 200–300 ml Ni dish. Add about 5 g of Na_2CO_3, 5 ml of NaOH solution (1 + 1), and 10 ml of alcohol, taking care that the entire sample is moist. Dry at about 100° to prevent spattering upon subsequent heating. Thirty minutes usually suffices.

Place the dish and contents in a furnace heated to 500° and keep at that temperature for 15 minutes. (Ignition of the sample at 500° appears to be necessary only to carbonize any soluble organic matter that would be oxidized by Br_2-water if not so treated. A temperature exceeding 500° may be used if necessary.) Cool, add 25 ml of water, cover the dish with a watch glass, and boil gently for 10 minutes. Filter through an 18 cm filter paper and wash with boiling water, catching the filtrate and washings in a 600 ml beaker. The solution should total about 300 ml.

Add an excess of Br_2 water and boil the solution

gently until colorless, then 5 minutes longer. Add a few crystals of salicylic acid and cool the solution to 20°. Add 1 ml of 85% H_3PO_4 and about 0.5 g of KI, and titrate with the 0.005*N* $Na_2S_2O_3$.

Vitamins

The proportions of the vitamins in unfortified milk varies considerably with the cow's diet; notably the milk of cows on green pasture is much higher in the vitamin A precursor, β carotene—a fact that would be visually evident in butter made from such milk if nearly all butter were not artificially colored. At the present time a large proportion, perhaps most, of the fresh milk reaching the market is fortified with vitamin D to a level of 400 International units per quart or greater. As a result, in some states this vitamin is routinely tested for in the market supply. Unfortunately, there is as yet no wholly reliable chemical method for Vitamin D in milk, so laboratories undertaking assays for this vitamin must resort to biological methods based on estimating the curative effect of samples on the bones of rachitic rats.[25]

Products such as Vitamins A and D Skimmed Milk and Vitamin Mineral Fortified Milk must carry definite levels of other vitamins: Vitamin A, thiamine, riboflavin and niacin. Directions for vitamin assays are beyond the scope of the present book. The reader interested in such assays should consult the references in Chapter 21. We should also like to call particular attention to the colorimetric method of Sobell and Rosenberg[26] for vitamin A in milk, and the microbiological method Hankin and Squires[27] for thiamine. Both methods are routinely employed in the laboratory of one of the authors.

Radioactivity

The radioactivity of milk was a subject of little or no interest prior to the advent of the atom bomb, but since the testing of such bombs by the United States, Russia and China resulted in the dissemination of radioactive isotopes (particularly the long-lived strontium 90) that could be carried by cows from pasture into their milk, there has been considerable activity in surveying the levels of radioactive elements in milk, as well as air and water, subsequent to each explosion. The details of such tests, which require expensive special apparatus, are beyond the scope of this book. The interested reader is referred to the American Public Health Association *Standard Methods for the Examination of Dairy Products Microbiological and Chemical.* 11th ed. (1960), pp. 312–322.

Cream

The dictionary defines cream as "The yellowish part of milk . . . that rises to the surface on standing or is separated by centrifugal force"[28], but it is unlikely that any of the cream now on the market is produced by the first of these alternatives. Statutes of the various states go into detail in defining several different grades of cream. Those of Connecticut have been taken as representative; besides listing the maximum permissible numbers of bacterial colonies, they set the following fat limits:

Type of Cream	Required Fat Content, Percent
Extra light	12–15
Light	16–29
Medium (or "all-purpose")	30–35
Heavy (or "whipping")	36 or over

Methods of Analysis

Method 6-26. *Preparation of Sample. (AOAC Method **15.075**).*

Immediately before withdrawing test portions, mix the sample by shaking, pouring or stirring (or by using a hand homogenizer) until it pours readily and a uniform emulsion forms. If the sample is very thick, warm to 30–35° and mix. In case lumps of butter have separated, heat the sample to about 38° by placing in a warm water-bath. (A temperature appreciably higher than 38° may cause the fat to "oil off," especially in the case of thin creams.) Thoroughly mix the portions for analysis and weigh immediately. (In commercial testing for fat by the Babcock method, it may be advisable to warm all samples to about 38° in a water bath before mixing.)

Method 6-27. *Total Solids.* (*AOAC Method* ***15.081***).

Proceed as in Method 6–4 (**a**) except to use a 2–3 g sample.

Ash, Total Nitrogen, and Lactose

Proceed as in Methods 6–5, 6–6, and 6–8.

Fat

For the Babcock method, proceed as directed in Method 6–9 for cream. For the Roese-Gottlieb method, proceed as in Method 6–10 except to use a 5 instead of a 10 g sample and dilute to 10.5 ml before adding the NH_4OH.

Method 6-28. *Water—Insoluble Fatty Acids* (*WIA*). (*AOAC Method* ***15.148–15.150***).

REAGENT

Sodium ethylate—0.05*N*. Dissolve a piece of Na, about 1 ml in volume, in 800 ml of absolute alcohol. Titrate 10 ml of 0.1*N* HCl with this solution and add the calculated volume of absolute alcohol to make the solution 0.05*N*. Standardize against 0.1*N* HCl on the day the solution is used. (Methanol or 95% alcohol may be substituted for absolute alcohol, and K may be substituted for Na.)

PREPARATION OF SAMPLE

Weigh 20 g of cream into a 125 ml glass-stoppered Erlenmeyer flask with a No. 22 $\mathfrak{T}$ stopper. Add 25 ml of ice-cold water, cool to 10°, and shake until the butter fat separates in granular form. Discard if the granular fat conglomerates into one lump.

DETERMINATION

Insert a special filter sieve (obtainable from Clark Dairy Supply Co. Inc., P.O. Box 157, Greenwood, Ind. 46142) into the Erlenmeyer flask and pour off the serum layer. Add 50 ml of ice-cold water, insert the $\mathfrak{T}$ 22 glass stopper, and shake 5 seconds. Remove the stopper, insert the filter sieve, and pour off the liquid portion. Wash three additional times. Dissolve the washed butter in 25 ml of ether, pour into a small separatory funnel, wash the Erlenmeyer flask with a few ml of ether and add this to the separatory funnel. Let settle for a few minutes and drain off the aqueous layer. Drain the ether–fat solution into a 125 ml Erlenmeyer flask, wash the separatory funnel with a few ml of ether, add to the Erlenmeyer flask and titrate with the 0.05*N* $NaOC_2H_5$, using phenolphthalein indicator. Calculate the WIA in mg/100 g fat.

$$1 \text{ ml } 0.05N\ NaOC_2H_5 = 13.5 \text{ mg WIA}.$$

NOTE: With creams classified on a flavor basis into the four classes 0 (sweet or good clean sour), 1 (indefinite off-flavor due to feed, etc.), 2 (definitely recognizable "cheesiness," rancidity or putridity) and 3 (markedly stronger off-flavor), Hillig and Ahlmann[29] found the following ranges for WIA values:

Class	WIA, mg/100 g fat
0	25–68
2	53–837
3	55–396

Classes 2 and 3 creams are considered to be unfit for sale as such or for butter making.

Non-dairy cream substitutes

The Defense Department has issued a purchase specification, MIL-C-43338A, December 1965, for a dry, non-dairy cream substitute for use by the Armed Forces as a component of operational rations. Similar products are in domestic consumption. This product is made from sodium caseinate or other edible protein, water, hydrogenated vegetable fat, starch, lactose or other sweetener, and salt, plus emulsifiers, stabilizers, artificial flavor and color. The product is then pasteurized, homogenized and dried. The fat content shall not be less than 33.0%, moisture not more than 3.0% and free fatty acids not over 0.2% as oleic.

Methods given for the analysis of cream are generally applicable to this product. Protein may be determined by Method 1–7 and ash by Method 1–9. If the question arises as to the origin of the fat- animal or vegetable- Method 13–17 is applicable for detection of animal fat.

Method 6.29. *Peroxide Value and Free Fatty Acids of Fat.*

This method is specified by purchase specification MIL-C-43338A (1965).

PREPARATION OF SAMPLE

Blend 50 g of sample, 100 ml $CHCl_3$ and 15 g anhydrous Na_2SO_4 in a blender for 3 minutes, then filter through a No. 12 Whatman or equivalent filter paper into a 250 ml glass-stoppered flask. Rinse the blender 4 times with 50 ml $CHCl_3$ and pour through the filter. (Between 150 and 200 ml of filtrate is necessary.) Pipette a 25 ml aliquot into a tared dish, evaporate the $CHCl_3$ in a draft oven set at 100°F, cool and weigh. Use this weight as sample weight in subsequent determinations.

DETERMINATIONS

Peroxide Value: Transfer a 25 ml aliquot of the $CHCl_3$ extract into a 250 ml glass-stoppered flask, add 30 ml acetic acid and mix. Continue according to Method 13–8, "Determination," beginning "Add 0.5 ml saturated KI solution . . . ," but titrating with 0.01*N* thiosulfate solution. Run a blank titration on 25 ml $CHCl_3$ and 30 ml acetic acid and correct titration of aliquot accordingly. Report as meq/Kg of fat, using the weight of fat in a 25 ml aliquot as sample weight, as determined under "Preparation of Sample."

Free Fatty Acids: Pipette a 25 ml aliquot of the chloroform extract into a 250 ml glass-stoppered flask. Add 50 ml $CHCl_3$ and titrate with 0.01*N* alcoholic KOH solution, using phenolphthalein as indicator. Conduct a blank determination.

% free fatty acids (as oleic)

$$= \frac{\text{ml } 0.01N \text{ KOH} \times 0.282}{\text{wt. of fat in 25 ml aliquot}} \times 100$$

Butter

Butter is the only food that has been singly defined by Congress (March 4, 1923) for purposes of enforcement of the Food and Drugs Act (now the Food, Drug and Cosmetic Act). It is the "product usually known as butter, made exclusively from milk or cream, or both, with or without common salt and with or without additional coloring matter, and containing not less than 80 per centum by weight of milk fat, all tolerances having been allowed for."

One unforseen development from the passage of the Act has been that, whereas in the early 1920's commercial butter contained an average of around 82.5% fat, since passage the fat content has been lowered, and the average fat content is now around 80.5%.

Under the Food and Drugs Act, Canada requires that butter contains not less than 80% milk fat and not more than 16% moisture. Salt and food color are permitted.

Methods of Analysis

Methods of sampling, preparation of sample for analysis, and analytical procedures included here for butter apply equally well to oleomargarine.

Sampling

The variation in composition of bulk butter from center to perimeter of a cube or tub creates a problem in sampling. Prints cut from cube butter will continue this variation. After considerable study by both state and Federal officials, carried on largely through the facilities of AOAC, that Association adopted the following methods:

Method 6-30. *Sampling and Preparation of Sample.* (*AOAC Method* ***15.129–15.130***).

a. *Tub or cube butter.* Insert regular trough butter trier practically its full length from point near top edge (or corner in case of cube) through center to point at bottom diagonally opposite point of entry. Make one complete turn and withdraw full core. Hold point of trier over mouth of sample container and immediately transfer core of butter in about 3″ sections, working it from trier by aid of spatula fitted to groove. Leave plug about 1″ long to place in hole from which core was removed. Add 2 other trierfuls, taken similarly at points equidistant from first (2 other corners in case of cube), to the jar to constitute subdivision from tub or cube sampled. Do not include moisture adhering to outside of trier. Clean and dry trier before each drawing. Use unwarmed trier for butter stored above freezing point. For harder butter use trier warmed to temperature that may be just borne by hand. Soften butter frozen so hard as to resist trier by storage in tempering room 24 hours.

Containers should be sampled according to the churn numbers, if so marked. Sample 1 unit container from each churn number containing up to 9 units in the lot, 2 containers from each churn containing 10–14 units, and 3 containers from churns of more than 12 units. In no case sample less than 2 units in a lot.

b. *Print butter.* Draw 1 print from each of the number of cases equivalent to the square root of the number of cases in the lot, with a minimum of 3 and a maximum of 25. Each churn or batch mark should be sampled if marked. With prints of 1 lb or over, quarter the print and remove diagonally opposite quarters as the sample.

c. *Sample containers.* Use glass jars of such type as to prevent loss of moisture by evaporation or gain by entrance into the jar. Do not use jar lids containing a liner.

PREPARATION OF SAMPLE[30]

Soften entire sample in sample container by warming in water bath kept at as low a temperature as practicable, not more than 39°. Avoid overheating, which causes visible separation of curd. Shake at frequent intervals during softening process to reincorporate any separated fat and to observe fluidity of sample. Optimum consistency is attained when emulsion is still intact but fluid enough to reveal sample level almost immediately. Remove from bath and shake vigorously at frequent intervals, or place sample container in mechanical shaking machine that simulates hand shaking, with arm 9 inches long set to oscillate at 425 ± 25 times/min through arc of 1.75 inches. Continue shaking until sample cools to thick, creamy consistency and sample level can no longer readily be seen. Weigh portion for analysis promptly.

Method 6-31. *Moisture, Rapid Method.*[31]

This method has been used for some 60 years as a rapid sorting method not requiring an analytical balance. In the hands of a practiced technician, results may be obtained agreeing ±0.1% with those obtained by the official oven-drying method.

APPARATUS

Balance sensitive to 0.05 or 0.10 g.

DETERMINATION

Rapidly weigh 10 g butter into a tared 100 ml pyrex beaker. Heat over a low gas flame, moving the beaker constantly by use of tongs, to avoid spattering and over-heating. After visible evolution of vapor ceases, heat cautiously until the residue *just* begins to discolor. Cool in open air, or desiccator if available, and weigh. Loss of weight, times 10, equals percent moisture.

Method 6-32. *Moisture, Official AOAC Method.* (*AOAC Method* ***15.131***).

This and following procedures determine moisture, fat, casein, ash and salt on one sample weighing of butter.

DETERMINATION

Weigh between 1.5 and 2.5 g sample to the nearest mg into a tared flat-bottomed dish not less than 5 cm diameter. Dry to constant weight in an oven kept at the temperature of boiling water. Calculate the loss in weight as moisture.

Method 6-33. *Fat.* (*AOAC Method* ***15.132***).

Add about 15 ml petroleum ether to the dish containing the dried residue from the moisture determination, Method 6–32. Macerate the residue with a small glass rod with flattened end and transfer it to a weighed asbestos Gooch crucible, fitted into a suction filtering flask. Wash the residue in the dish into the crucible with aid of a wash bottle filled with petroleum ether, and wash dish and crucible free from fat with about 100 ml of the solvent. (Pass the last 25 ml solvent through the crucible without suction.) Dry the crucible and contents in an oven at 100° to constant weight. Repeat washing with 25 ml petroleum ether and drying to constant weight until no further significant loss of weight occurs due to washing. This is known as the "indirect method".

$$\% \text{ fat} = 100 - (\% \text{ moisture} + \% \text{ residue})$$

The residue in the crucible is sometimes reported as percent curd and salt, without further treatment.

Certain military specifications prescribe the "AOAC direct method" for determination of fat. This is simply the extraction of the dry residue from the moisture determination with petroleum ether or anhydrous, alcohol-free ethyl ether, followed by filtration of the extract into a tared flask, evaporation of the solvent, drying of the residue at 100° and weighing.

Method 6-34. *Casein, Ash and Salt.* (*AOAC Method* ***15.135***).

Cover the crucible containing the residue from fat extraction, Method 6–33, and heat, gently at first, then raise the temperature to not over 500°. Remove cover and continue heating until residue is white. Cool and weigh. Calculate the loss in weight as casein and the residue in the crucible as ash.

Transfer the contents of the crucible to a 100 ml flask or beaker with water, add a few drops of HNO_3 free from lower oxides of nitrogen, and determine chlorides by Method 1–21, Volhard titration, beginning "Determination . . .". 1 ml 0.1N $AgNO_3$ equals 0.00584 g NaCl.

Methods 6–32, 6–33, and 6–34 have the advantage that all determinations may be made from one sample weight.

AOAC is now engaged in a collaborative study with the International Dairy Federation, Brussels, Belgium and the Food and Agricultural Organization of the United Nations with the ultimate aim of adoption of international official methods for dairy products.[32, 33] The following method for salt in butter, resulting from one such study, was adopted by AOAC as an official method in 1966.

Method 6-35. *Salt in Butter.*

REAGENTS

a. *Potassium chromate indicator.*—Dissolve 5g K_2CrO_4 in water, add saturated $AgNO_3$ solution until a slight, permanent red precipitate forms. Filter and dilute to 100 ml.

b. *Standard* 0.1N *silver nitrate solution.*—Dissolve slightly more than theoretical quantity of $AgNO_3$ (equivalent weight 169.87) in halogen-free water and dilute to 1 litre. Use thoroughly clean glassware, avoid contact with dust and store in amber glass-stoppered bottles, away from light.

STANDARDIZATION

Weigh accurately about 0.3 g KCl that has been recrystallized 3 times with water and dried at about 500° to constant weight, and transfer to a 250 ml glass-stoppered flask with 40 ml of water. Add 1 ml of the K_2CrO_4 indicator (**a**) and titrate with the $AgNO_3$ solution until the first perceptible pale red-brown color appears. Similarly titrate 75 ml of water containing 1 ml of K_2CrO_4 indicator to the same colored end point. Deduct this titration from the titration of KCl, and calculate normality of the $AgNO_3$ solution as follows:

$$N = \frac{\text{g KCl} \times 1000}{\text{ml AgNO}_3 \times 74.555}$$

DETERMINATION

Weigh 5–10 g sample, ± 10 mg, into a 250 ml flask, add 100 ml of boiling water and let stand 5–10 minutes, swirling occasionally, while cooling to 50–55°. Add 2 ml K_2CrO_4 indicator (**a**) and titrate with $0.1N\ AgNO_3$ to a pale red-orange end point.

$$\% \text{NaCl} = \frac{\text{ml } 0.1N\ \text{AgNO}_3 \times 0.585}{\text{Weight of sample in grams}}$$

Method 6-36. *Examination of Separated Butterfat.*

PREPARATION OF SAMPLE OF BUTTERFAT

Melt the butter and keep in a dry place at about 60° until the moisture and curd separate completely. Filter the clear supernatant oil through a dry filter paper placed in a hot-water funnel. The oil may also be filtered through a dry filter paper placed in an air-oven at about 60°. If filtered oil is not perfectly clear, refilter.

Detection of Margarine in Butter

The critical temperature of dissolution, Method 13–20, of the separated fat will surely distinguish substitution of margarine for butter. Since the difference between the average CTD of butterfat and margarine fat is about 24°, this test may be used as a confirmation of the Reichert-Meissl test to detect substantial amounts of margarine mixed with butter.

Detection of Foreign Fat in Butterfat

The Reichert-Meissl, Polenske and Kirschner values, considered together, serve to distinguish butterfat from other oils in general, and from the palm family (coconut palm, and palm kernel) specifically. (*See* Method 13–5, Interpretation of Oil Analyses, Chapter 13, and Table 13-2.) The Reichert-Meissl value is a measure of water-soluble fatty acids volatile with steam; the Polenske value measures water-insoluble, steam-volatile fatty acids; and the Kirschner value is an index of the characteristic acid of butter fat, butyric acid.

The Reichert-Meissl value of most mixed herd samples or commercial samples of butter fat is around 27 or 28. Values below 24 are suspicious. A Polenske value above 2.5 greatly intensifies this suspicion, indicating presence of one of the palm oils. Frequently, determination of these 2 values alone will establish adulteration. If in doubt, determination of the Kirschner value will establish it with certainty. Pure butter fat from commercial butter has a Kirschner value of 20 or over; with most butters this value is close to 23.

A critical review of the Reichert-Meissl and Polenske determinations, which date back to 1887, was made by Fine in 1955,[34] followed by a further study by Klayder and Fine[35] in 1957. These studies formed the basis for the present AOAC method, which gives detailed specifications of the apparatus and precise instructions for the determination.

Williams[36] has postulated a linear relationship between these 3 values for pure butterfat. Bolton and co-workers give a formula by which the quantity of butterfat in a mixture can be calculated from Kirschner and Polenske values:[37]

$$\% \text{ butterfat} = \frac{K - (0.1P + 0.24)}{0.244}$$

where K = Kirschner value and P = Polenske value.

There are other indications of admixture of foreign oils in butterfat. Refractive index above 1.4555,$^{25°}$ and a saponification value below about 200, indicate presence of vegetable oils other than oils of the palm family (which also have low refractive indices and high saponification values).

The AOAC has substituted the determination of mole percent butyric acid for the Kirschner value determination. This method, developed by Anglin and Mahon, of Canada Food and Drug Directorate[38, 39], separates butyric acid from other fatty acids by chromatographing the fatty acids on a silicic acid column and subsequently titrating separately the butyric acid fraction and the remaining fatty acids, and expressing the butyric acid titration as the percent of the sum of the two titrations. AOAC method **26.034–26.039** should be consulted for details. Normal range is 9.6–11.3.

Another approach to this problem has been made by Windham.[40] He proposes the proportion of tocopherol as an indication of addition of vegetable fat to animal fat. Butterfat contains less than 55 μg tocopherol per 100 g while most of the commonly used vegetable oils, except coconut oil and peanut oil, contain from 300 to 1600 μg tocopherol in 100 g.

Detection of measurable quantities of phytosterols in oils is positive proof of the presence of vegetable oils. This can be done by AOAC Method **26.061–26.063** or **26.065–26.070**.

Still another approach is that of Bhalerao and Kummerow of the Food Technology Department, University of Illinois.[41] They fractionate the butterfat into 3 fractions, 1) absolute alcohol-soluble fraction, 2) absolute alcohol-insoluble, acetone-soluble fraction and 3) absolute alcohol-insoluble, acetone-insoluble fraction, in order to increase the concentration of the adulterant in one of these fractions. They then determine the refractive index of each fraction, then iodinate fractions 1) and 2) with Wijs reagent. From these accumulated data they deduced the type, and at times the amount, of adulterant present.

Authentic Analyses of Butterfat

The literature is replete with reports of Reichert-Meissl, Polenske and Kirschner values of butterfat and of oils and fats that may be used as adulterants. According to Fine,[34] while many changes in apparatus and procedure for these determinations have been incorporated into the methods since they were devised, the reported Reichert-Meissl values for authentic butterfat obtained from butters made from mixed herd milk have not varied more than 2 or 3 units. He believes that the wide ranges found in various reports are due to abnormalities not encountered in truly authentic mixed herd milk. For this reason, we shall refer to only two extensive studies on the subject, one in this country and one in Canada.

Riel[42] of Canada Department of Agriculture made a 1 year survey of milk produced throughout Canada during 1953–1954. He found the Reichert-Meissl value higher in summer than in winter milk. His study revealed ranges as follows:

Reichert-Meissl value—24.6 to 31.4, average 27.8.

Polenske value—1.4 to 3.0, average 2.0.

Refractive Index$^{40°}$—1.4530 to 1.4553, average 1.4542.

The American study was made by Zehran and Jackson [43] of the University of Wisconsin. They made a nation-wide survey of commercial milk suppliers, analyzing 42 samples of butter each month during 1 year (1953–1954).

They reported:

Reichert-Meissl value—24.24–33.55 (96% between 26.01–31.00).

Polenske value—1.12–2.95 (89% between 1.41–2.40).

Refractive Index$^{40°}$—1.4531–1.4557 (90% between 1.4535–1.4550).

These workers sent these samples to Keeney, University of Maryland, who determined mole % butyric acid.[44] They reported a range of mole % butyric acid of 9.6–11.3, with a mean of 10.41, and a standard deviation of 0.27–0.31. They state that in practice "any commercial sample containing less than 9.6 mole % butyric acid is adulterated unless it can be shown that this is the normal value for authentic mixed herd butterfat samples in the area of origin of the sample."

Commercial milk fat

Federal Milk Fat Specifications

There are two purchase specifications for milk fat issued for mandatory use by the Department of Defense. One, MIL-M-1036D (1966), covers pasteurized butterfat for use in recombined milk, cream and other manufactured dairy products. There are two types: plastic cream, containing not less than 78% milk fat, and churned fat, containing not less than 85% milk fat. The other analytical requirements are the same for both types, i.e., peroxide value not more than 1 meq/Kg of fat; copper (Cu), not more than 0.2 p.p.m.; and phosphatase activity not more than 4 μg/g. Plastic cream is an oil in water emulsion, while churned milk fat, like butter, is a water in oil emulsion.

The other specification, MIL-M-3233B (1966), governs Department of Defense purchase of anhydrous milk fat as ingredient for recombined milk, ice cream and other manufactured foods. It shall contain not less than 98.8% fat, not more than 0.1% moisture, not more than 0.30% free fatty acid, not more than 0.10 p.p.m. copper, and the peroxide value shall be not more than 0.50 meq/Kg of fat.

Plastic cream, churned milk fat and anhydrous milk fat are all available for general use as raw materials for many manufactured dairy products and other food products.

Methods of Analysis

The following methods are applicable to the analysis of commercial milk fats:

Moisture, 6–31 *and* 6–32. Moisture in anhydrous milk fat should be determined by the vacuum oven method 1–2.

Fat, 6–33.

Free Fatty Acid, 13–7. Calculated as oleic acid.

Peroxide value, 13–8.

Copper, 6–23.

Federal Standards for Concentrated Milk

Federal Standards of identity have been promulgated by the Food and Drug Administration for concentrated liquid milks. They are listed in Code of Federal Regulations, Title 21, as follows:

Regulation No. and Product	Minimum % Milk Fat	Minimum % Milk Solids	Optional Ingredients
18.520 Evaporated Milk heat processed and packed in sealed container.	7.9	25.9	Disodium phosphate, sodium citrate and/or sodium chloride up to a total of 0.1%. Carrageenin or its salts up to 0.015%. Vitamin D-25 USP units per fluid ounce minimum.
18.525 Concentrated Milk not heat processed or sealed.	7.9	25.9	None
18.530 Sweetened Condensed Milk made by evaporating a mixture of milk and sucrose and/or dextrose	8.5	28.0	None
18.535 Condensed Milk with Corn Syrup. Corn syrup and/or corn sugar instead of sucrose.	8.5	28.0	None

FDA in Federal Register 33, page 6977 (May 9, 1968) proposed Definition and Standard or Identity for Vitamins A and D Fortified Nonfat Dried Milk. This proposes not less than 500 USP units Vitamin A and 100 units Vitamin D per 8 fluid ounces of reconstituted milk.

Concentrated milk products

The Federal Standards for Evaporated Milk and Sweetened Condensed Milk are given just before this section. In general, the methods of analysis for fluid milk are applicable here with some modifications.

Evaporated Milk

Method 6-37. *Preparation of Sample.* (*AOAC Method* ***15.096***).

a. Temper the unopened can by immersing in a water-bath at about 60° for 2 hours, removing the can every 15 minutes and shaking vigorously. Cool to room temperature, remove entire lid and thoroughly mix by stirring. (If fat separates, the sample was not properly prepared.) This is the "undiluted sample."

b. Dilute 40 g of the prepared mix (**a**) with 60 g of water and mix thoroughly. This is the "diluted sample".

NOTE: If sample will not emulsify uniformly, weigh out separate portions of (**a**) for each determination. Analytical results obtained from the diluted sample should of course be corrected for dilution.

Methods of Analysis—Evaporated Milk

Determination	Size of Sample	Method
Total Solids	4–5 g of "diluted sample"	6–4(a)
Ash	dry residue from total solids	6–5
Nitrogen	5 g "undiluted sample"	6–6
Fat	4–5 g "undiluted sample", dilute to 10 ml in extraction flask	6–10
Lactose	"diluted sample" as directed	6–7 or 6–8

Sweetened Condensed Milk

Method 6-38. *Preparation of Sample.* (*AOAC Method* ***15.108***).

a. Temper unopened can in a water bath at 30–35° until warm. Remove entire top and scrape out all milk adhering to interior of can. Transfer to a dish large enough to permit thorough stirring, and mix until the whole mass is homogenized.

b. Weigh 100 g of thoroughly mixed sample into a 500 ml volumetric flask, dilute to mark with water and mix thoroughly. This is the "diluted sample."

Methods of Analysis—Sweetened Condensed Milk

Determination	Size of Sample	Method
Total Solids	10 ml of "diluted sample", heat on steam bath 30 minutes, then in vacuum oven	1–2
Ash	Residue from total solids	6–5
Nitrogen	10 ml of "diluted sample"	6–6
Fat	4–5 g of "diluted sample" dilute to 10 ml in extraction flask	6–10

Method 6-39. *Lactose in Sweetened Condensed Milk.* (*AOAC Method* ***15.114***).

Transfer 100 ml "dilute sample" to a 250 ml volumetric flask, add about 100 ml of water, then 6 ml of a 6.9 g/100 ml solution of $CuSO_4.5H_2O$, followed by 5.3 ml of 0.5*N* NaOH. Continue as in Method 6–8, Lactose, gravimetric method, beginning, "After addition of the alkali . . . ." Correct the result for dilution.

Dried Milk and Dried Buttermilk

Dried milks comprise the ultimate concentrates of milk in commercial production. The American Dry Milk Institute (ADMI) has established standards of grades for dry whole milk, nonfat dry milk and dried buttermilk. These are:

ADMI Standards—Dry Whole Milk

	Spray Dried				Roller Dried	
	Gas Pack		Bulk Pack			
	PREMIUM	EXTRA	EXTRA	STANDARD	EXTRA	STANDARD
Moisture—% max	2.25	2.50	2.50	2.75	3.00	4.00
Butterfat—% min	26.00	26.00	26.00	26.00	26.00	26.00
Acidity—% max[a]	0.15	0.15	0.15	0.17	0.15	0.17

[a]By titration, calculated as lactic acid in the reconstituted product.

ADMI Standards—Nonfat Dry Milk

	Spray Dried		Roller Dried		Instant
	EXTRA	STANDARD	EXTRA	STANDARD	EXTRA
Moisture, % max	4.00	5.00	4.00	5.00	4.50
Butterfat, % max	1.25	1.50	1.25	1.50	1.25
Acidity, % max[a]	0.15	0.17	0.15	0.17	0.15

[a]By titration, calculated as lactic acid in the reconstituted product.

ADMI Standards—Dried Buttermilk

	Extra Grade		Standard Grade	
	SPRAY	ROLLER	SPRAY	ROLLER
Moisture, % max	4.00	4.00	5.00	5.00
Butterfat, % min	4.50	4.50	4.50	4.50
Acidity, % max[a]	0.10–0.18	0.10–0.18	0.10–0.20	0.10–0.20

[a]By titration, calculated as lactic acid in the reconstituted product.

There is a Federal purchase specification for this product, C-B-825a (1965). This defines low acid and high acid sweet cream buttermilk and buttermilk from fresh cultured skim milk. Except for titratable acidity in "high acid" and "fresh cultured" buttermilk, the requirements are the same as the ADMI grade standards. The Federal specification requires not less than 0.50% titratable acidity of the reconstituted buttermilk for the "high acid" and "fresh cultured" products. They also limit the phosphatase activity to 4μg per ml of the reconstituted product (10 g dry buttermilk plus 100 g water at 60–70°F) and alkalinity of the ash to 125 ml 0.1N HCl per 100 g.

Methods of Analysis

In general, the method for analysis of these dried milks are the same as for their fresh counterparts, allowing for the difference in solids content. Use the following methods and sample size as indicated:

Determination	Method	Size of Sample
Moisture	1–2	1.000–1.50 g, heat 5 hours at 100° and 100 mm Hg
Protein	1–7	1.000 g—Use 6.38 factor
Ash	1–9	1 g at 550°

NOTE: ADMI uses toluene distillation Method 1–4 for moisture.

Method 6-40. Total Fat in Dry Milks.

This is essentially the ADMI or AOAC Method **15.125**.

APPARATUS

Mojonnier fat-extraction flask, or Röhrig tube.

DETERMINATION

Quickly weigh 1 to 1.25 g sample to nearest milligram and transfer to the fat extraction flask or Röhrig tube. Add 9 ml hot water, cork the container and shake vigorously until homogeneous. Add 1 ml NH_4OH and heat in a water bath at 60–70°, shaking occasionally. Cool and proceed as in Method 6–10, beginning "Add 10 ml of alcohol. . . ." Report as percent butterfat.

Cheese

Cheese is one of the oldest foods known to man. Its origin goes beyond recorded history. It is mentioned several times in the Bible. The Imperial Romans spread knowledge of its manufacture through their empire, and the early Christian monasteries, more or less by accident, developed special varieties, some of which exist to this day.

Commercial production of cheese on a large scale is a product of the 19th century industrial revolution, and its scientific quality control in this country was fostered by the state experiment stations and state universities that were established during that period.

Cheese Standards

Up to July 1969, some 46 standards of identity for natural cheeses have been promulgated as mandatory standards in the United States. In addition, standards have been issued for certain of these cheeses "for manufacturing". The standards of composition for these are identical with the natural cheese of the same name, except that the milk need not be pasteurized, curing is not required and, in the case of Swiss cheese for manufacturing, the holes or "eyes" have not developed throughout the cheese.[45] Standards of composition of these natural cheeses are listed in Table 6-4.

Mandatory Standards of Identity have also been promulgated for 17 pasteurized process cheeses, cheese foods, cheese spreads and related products.[46] These regulations involve many mandatory and optional ingredients and their requirements for moisture and fat depend in part on the variety of natural cheese used in their manufacture and in part on other ingredients used. The complete specification should be consulted in each case.

The word "process" in the name of a cheese product indicates use of emulsifying agents (phosphates, citrates or tartrates). These are mandatory in pasteurized processed cheese or processed cheese spreads, and optional in processed cheese food or grated American cheese food, and may be used up to 3% of the finished food.

Stabilizers (specified vegetable gums, alginates, cellulose gum, etc.) are optional in pasteurized process cheese spreads, pasteurized cheese spreads, cold-pack cheese foods, grated American cheese food and cream (or pasteurized Neufchatel) cheese with other foods, not to exceed 0.8% of the finished food.

Acidifiers to adjust pH are optional in all of these products except pasteurized blended cheese and cream cheese with other foods.

Other dairy ingredients (cream, milk, skim milk, cheese whey, or albumen from cheese whey) are optional in all cheese foods and cheese spreads. Cream may be used as an ingredient of pasteurized processed cheeses or pasteurized blended cheeses. The USDA has published a summary of state and federal laws affecting the cheese industry.[47]

TABLE 6-4
U.S. Standards of Composition of Natural Cheeses

Name of Cheese and FDA Standard Reg. No.		Max % Moisture	Min % Fat in Solids
Asiago, fresh[c]	19.615	45	50
Asiago, medium[c]	19.620	35	45
Asiago, old[c]	19.625	32	42
Blue	19.565	46	50
Brick[c]	19.545	44	50
Caciocavalle Siciliano[c]	19.591	40	42
Camembert[a]	19.665	—	50
Cheddar[c]	19.500	39	50
Colby[c]	19.510	40	50
Cook	19.635	80	—
Cottage	19.525	80	—
Cream	19.515	55	33[b]
Creamed Cottage	19.530	80	4[b]
Edam[c]	19.555	45	40
Gammelost	19.639	52	—
Gorgonzola	19.567	42	50
Gouda[c]	19.560	45	46
Granular[c]	19.535	39	50
Gruyère[c]	19.543	39	45
Hard Cheeses[c]	19.650	39	50
Hard grating cheese[c]	19.680	34	32
Jack, High Moisture	19.585	44–49	50
Limburger	19.575	50	50
Monterey[c]	19.580	44	50
Mozzarella (Scamoya)	19.600	52–60	45
Mozzarella, part-skim	19,601	52–60	30–44
Mozzarella, low moisture	19.605	45–52	45
Mozzarella, low moisture, part skim[c]	19.606	45–52	30–44
Muenster[c]	19.550	46	50
Neufchatel	19.520	65	20–32[b]
Nuworld	19.569	46	50
Parmesan[c]	19.595	32	32
Provolone[c]	19.590	45	45
Ricotta	19.532	80	11
Ricotta, skim milk	19.533	80	6–10
Romano	19.610	34	38
Roquefort	19.570	45	50
Samsoe	19.544	41	45
Sapsago	19.637	38	—
Semisoft[c]	19.655	39–50	50
Semisoft, part skimmed[c]	19.660	50	45–49
Soft-ripened	See Camembert		
Spiced[c]	19.670	—	50
Spice, part-skimmed[c]	19.675	—	20–49
Swiss,[c] Emmentaler	19.540	41	43
Washed curd[c]	19.505	42	50

NOTE: Salt is a permitted ingredient in all cheeses.

[a]Listed by FDA Standard as "soft, unripened cheese".

[b]On "as is", not "solids" basis.

[c]0.3% sorbic acid or sodium or potassium sorbate optional as an antimycotic agent for consumer sized packages.

Canada has slightly different minimum fat requirements from those of the United States for certain cheeses:[48]

Asiago, Blue, Brick, Camembert, Gouda, Granular, Limburger and Neufchatel are 48%, Gruyere 45%, Romano is 32%, and Part Skim Mozzarella is 30%. Provolone is known as Pasta Filata.

They have also issued minimum fat standards for the following cheeses that have not been standardized in this country:

Alpin, Bel Paese and Feta 48%; Esrom, Havarti, Maribo, Steppe, Tilsiter 45%; Bra, Leyden 40%; Part Skim Pizza 30%; Cream Cheese 65%; Skim Milk Cheese, not more than 15%.

The standards for fat are all on the dry basis.

Canada has not established any maximum requirements for moisture in natural cheeses, except for hard grating cheese (34%), cottage cheese (80%) and cream cheese (55%). There are also some differences in the standards for processed cheeses and cheese spreads. These two products have the same limitations and definition in Canada. The regulations themselves should be consulted for these products.

Government Purchase Specifications for Cheese

Federal and Military Specifications have been issued for several natural and processed cheeses. Their analytical requirements follow the FDA Standards of Identity given above. In the case of cheddar cheese, deliveries must also comply with USDA Standards for Grades of Cheddar Cheese[49], grades AA, A or B as ordered. Processed American cheese must not contain more than 1.5 p.p.m. Cu. The government also buys Pizza cheese for the armed forces; Military Specification MIL-C-35088 (1962) describes two types: whole milk, containing not more than 45% moisture and not less than 45% fat, solids basis, and part-skim milk, containing not more than 45% moisture; the fat content, solids basis, shall be less than 45% but not less than 30%.

Dehydrated Cheese

In recent years dehydrated cheeses and process cheeses have become articles of commerce. The Defense Department has issued purchase specifications for certain dehydrated cheese products. One for dehydrated processed American Cheese, MIL-C-35053A(1964), specifies a product containing not less than 48% milk fat, and not more than 4% salt or 3% water. The copper content should not be more than 1.5 p.p.m. It is made by grinding American (cheddar) cheese; cream, emulsifying agents, color, salt and acidifying agents may be added during the grinding, but not lactose, milk solids, whey or buttermilk. The mixture is then pasteurized and spray dried. There are two types listed, mild and sharp, along with a sharp, seasoned type which contains added salt and black pepper. It is gas-packed in nitrogen, in hermetically sealed tins, and the oxygen content shall not be more than 2% when tested by Method 17–14.

Another Defense Department specification is for dehydrated creamed cottage cheese, MIL-C-43274A (1966). This product shall contain not more than 1.5% moisture, and when rehydrated at the rate of 1 ounce of the dehydrated product plus 4 avoir. ounces of water, shall contain not less than 4.0% milk fat and between 1.0 and 1.5% salt. It is made from freshly made creamed cottage cheese, containing not less than 4% milk fat, frozen, and freeze-dehydrated at an absolute pressure of 1.5 mm mercury. It is then immediately gas-packed over nitrogen into hermetically sealed tins, and the oxygen content shall not be more than 2% when tested by Method 17–14. Not less than 80% of the finished product shall be retained on a U.S. Standard No. 8 sieve.

Various techniques have been used for dehydration of cheeses: lyophilization, spray drying (normal pressures or in vacuum), tray drying and foam-spray drying as developed by U.S.D.A. laboratories. The group at Michigan State University, headed by T. I. Hedrick and Carl W. Hall, has published several papers thus far and more are contemplated. Analytical problems arose on loss of flavor volatiles during the various dehydration processes studied. This work is continuing, and interested persons should write to the University for current developments in analysis.

Flavor Components of Cheese

Considerable attention has been devoted in recent years to the source and identification of the flavor components of cheese, particularly by American State Universities and agricultural experiment stations. These flavor profiles, particularly in cured cheeses, are quite complex, as can be demonstrated by gas chromatography. For example, over 50 flavor elements have been found in cheddar cheese. This subject is not within the scope of this book, but some mention should be made of source material for those interested.

Kristoffersen, of Ohio State University, Dairy Technology Department, has summarized a 10-year study by that group.[50]

They concluded that cheddar cheese flavor was an inter-dependent function of simultaneous proteolytic and lipolytic activities. Since then this group has continued their work, identifying certain carbonyl and sulphydryl compounds.[51, 52]

Methods of Sampling and Analysis

Adequate sampling procedures for cheese have been a problem for many years and are still not solved. It is really a conflict between economics and statistics. Cheeses come in all sizes, from a few ounces to wheels weighing 300–400 pounds or more. The moisture content of a large cheese, like Swiss or cheddar, varies from perimeter to center, and different cheeses made from the same vat lot vary in moisture content among themselves.

The International Dairy Federation (IDF) and the International Standards Organization (ISO), in cooperation with other interested agencies, are engaged in the laudable task of unifying sampling procedures and analytical methods used in dairy products. The AOAC is committed to a program of full cooperation with these organizations.[53, 54] The three organizations agreed to adopt the Karl Fischer method as the reference method for all dairy products for which it is applicable.[55] This method was compared with the two official AOAC methods **15.157** and **15.158**, and a proposed IDF oven method for moisture in cheese.[56]

We include in this text the IDF-ISO sampling procedure for cheese, moisture in cheese and fat in cheese. These all were adopted by AOAC in 1967.

Method 6-41. *Sampling Procedure for Cheese*[57].

a. *By cutting:* This is the preferred method. Using a stainless steel knife with pointed blade, make 2 radial cuts if the cheese has a circular base, or parallel to the sides if rectangular. The size of piece so removed should be such that not less than 50 g of edible portion remains after inedible surface layer has been removed.

b. *By trier:* Acceptable, particularly for hard, large-sized cheese. With a cheese trier suitable to the cheese being sampled, use one of the following techniques, according to shape, weight and type of cheese to be sampled:

1) Insert trier obliquely towards the center of the cheese at one or more points not less than 10 cm from edge.

2) Insert trier perpendicularly into 1 face, through the center to the opposite face.

3) Insert trier horizontally into vertical face, midway between 2 plane faces, toward center of cheese.

4) In case of cheese in bulk containers (barrels, boxes, etc.), or cheese formed into large compact blocks, pass trier obliquely from top to base.

5) For large cheeses, use outer 2 cm or more of plug containing rind for closing the hole left in cheese. Remainder constitutes the sample. If possible, seal over the reinserted plug with sealing compound [mixture of white petrolatum and paraffin $(1+1)$].

Use sample containers with air-tight closures, filling them immediately after withdrawing the sample. Sample may be cut into pieces, but must not be compressed or ground.

6) Sample bulk containers of soft cheeses, such as cottage and similar cheeses, by stirring the contents of the container thoroughly for at least 5 minutes with a dairy stirrer (a perforated concave metal disk about $5\frac{1}{2}''$ across, attached to a metal rod about 27″ long as handle). Fill a pint jar by removing portions from top surface with a small spoon.

Method 6-42. *Preparation of Sample.*

Carefully remove only the inedible surface layer, if any, including moldy or horny portions. Do not remove outer rind or crust of soft cheese sold by piece, e.g., Camembert. Cut wedges of cheese sample into strips and pass 3 times through a food chopper. Pass plug samples through a food chopper or cut or shred finely and mix thoroughly. Blend cottage and similar soft cheeses by placing about 300 g at 15° into the quart cup of a high-speed blender. Stir for a minimum of 2–5 minutes to ensure a homogeneous mix. Final temperature should not be over 25°.

Moisture in Cheese

AOAC now has two official methods. One dries a 2–3 g sample in a vacuum oven at 100° and not over 100 mm Hg; the other partially dries the sample on a steam bath and completes the drying in a forced draft oven for 4 hours at 135°. We prefer to give here the proposed IDF method[58,56] now under study by AOAC. Strange reported that the AOAC 100° vacuum oven method is not suitable for cheeses with a relatively high content of volatile compounds, e.g., Bleu Cheese. He proposed a distillation method, such as Method 1–4, using a 2+1 mixture of xylene and n-amyl alcohol as the immiscible solvent.[56]

Method 6-43. *Moisture in Cheese. (IDF Method).*

Place about 20 g coarse-grained quartz or sea sand previously digested with HCl, washed and ignited, along with a short, flat-ended glass stirrer, in a Ni or Al dish about 6–8 cm in diameter and 2 cm in height, and dry in an oven at 105° to constant weight. Cool in a desiccator containing an efficient desiccant and weigh. Quickly weigh about 3 g of the prepared sample into the dish and reweigh. Mix the cheese and sand with the stirrer. See Note.

Dry in an oven at 105° for 4 hours, cool in a desiccator and weigh. Repeat heating and weighing at ½ hour intervals to constant weight. Accuracy, dry matter = ±0.1%.

NOTE: With cheeses that melt to a horn-like mass at 105°, place the dish containing the cheese and sand mixture in a desiccator at room temperature for 10 hours before drying in the oven. The sand and cheese should be well mixed with the stirrer from time to time to prevent formation of a crust.

Method 6-44. *Fat in Cheese.*

This, the Roese-Gottlieb method, is AOAC Method **15.164**.

In a small, tall beaker rub 1 g of prepared sample to a smooth liquid with 9 ml water and 1 ml NH_4OH, using a small, flattened glass rod as stirrer. Digest at low heat until casein is well softened, neutralize with HCl, using litmus as an indicator, then add 10 ml more HCl. Add a few glass beads or SiC granules, previously digested in HCl, to prevent bumping, and boil gently for 5 minutes, keeping beaker covered with a watch glass.

Cool, add 10 ml alcohol if desired (this is a step included in the FAO method), and transfer to a Mojonnier fat-extraction flask or Röhrig tube and proceed as in Method 6–10, beginning "add 25 ml peroxide-free ether. . . ." Compare with Method 4–7.

Correct the weight of milk fat found by a blank determination on the reagents.

An analyst, unless engaged in a specific research problem, has very little occasion to analyze cheese for permitted additives such as citrates, tartrates, gums and alginates. We are not including these seldom-used methods, in the interest of conservation of space. Methods for these are given in *Official Methods of Analysis*, AOAC, 10th ed. (1965).

Protein Content of Cheese

Follow Method 1-7, using 1 g sample. N × 6.38 = milk proteins.

Ash of Cheese

Follow Method 1-9, burning 1 g sample at 550°.

There is occasional need for the further examination of fat separated from cheese to determine whether it is indeed composed entirely of milk fat:

Method 6-45. *Separation of Fat for Further Examination.*

Weigh about 250 g of cheese, cut into pea-sized bits, into a 1.5 L wide-mouthed flask. Add 700 ml of 5% KOH solution at 20°, shaking vigorously to dissolve casein. (The casein should dissolve in 5–10 minutes and the fat rise to the surface in lumps.) Add cold water to the flask until fat is driven up into the neck of the flask. Scoop out the fat with a fork or perforated spoon and wash with just enough water to remove residual KOH. Melt the fat and hold in a dry place at about 60° until any curd and water separate. Filter the clear fat through a dry filter supported in a hot-water funnel, and refilter if not perfectly clear.

Determine chemical and physical characteristics as given in Chapter 13. Review section on Examination of Separated Butter-Fat in this chapter for suggestions on further examination.

Method 6-46. *Chlorides in Cheese.* (*AOAC Method* ***15.160***).

Weigh about 3 g of the prepared sample into a 200 ml Erlenmeyer flask, and add 25 ml 0.1*N* $AgNO_3$. Add 10 ml halogen-free HNO_3 and 50 ml water. Boil. During boiling add about 15 ml 5% $KMnO_4$ solution in 5 ml portions. Solution becomes yellowish and clear. Cool, filter into a 200 ml volumetric flask. Wash the paper thoroughly with cold water, then dilute to mark. Titrate excess $AgNO_3$ in a 100 ml clear aliquot with 0.1*N* KCNS using 2 ml saturated solution of ferric alum as indicator.

Determine blank on reagents used, except to add sugar to destroy excess $KMnO_4$. 1 ml 0.1*N* $AgNO_3$ = 0.0058 g NaCl.

Wrapper Preservatives for Packaged Cheese

Sorbic and, dehydroacetic acids and their alkali salts are at times impregnated into cheese wrappers as mold inhibitors. Analysts are interested in the penetration of these additives into the food. Sorbic acid and salts are also permitted ingredients in the cheese itself. The FDA Definitions and Standards for Cheeses permit up to 0.3%, calculated as sorbic acid, in the cheese. Sorbic acid and salts are permitted as antimycotics in other standardized foods, e.g., artificially sweetened jams and jellies, soda waters, orange juice and orange concentrates.[59]

Sorbic Acid: Rather than include an analytical method in this text, we believe that chemists interested in this determination should read a series of papers presented at the 1953 and 1955 meetings of the Institute of Food Technologists. The methods for determination of sorbic acid, devised by the Best Foods, Inc. laboratory, are included in these papers.[60] The 1954 paper gives a method specifically devised for cheese and cheese wrappers; the 1955 report gives a method for foods in general.

Method 6-47. *Dehydroacetic Acid in Cheese; Quantitative.*[61]

APPARATUS

Beckman DU spectrophotometer or equivalent.

REAGENT

Dehydroacetic acid, Eastman or equivalent.

DETERMINATION

Weigh a 50–60 g sample of cheese to the nearest 0.1 g, place in Waring blendor cup and comminute (covered) with 80 ml $CHCl_3$ for 3 minutes, scraping down the walls and cover once during the operation. Place a filter paper on a 2–3″ diameter sintered glass Büchner funnel (if a sintered glass funnel is not available, use an ordinary Büchner funnel), transfer the cheese to the funnel with a spatula and filter the $CHCl_3$ with suction. Return the cheese cake and filter paper to the cup, add 80 ml of $CHCl_3$, blend for 1 minute and filter again into the same flask. Use a fresh filter paper for each filtration. Repeat the extraction and filtration for the third time with 80 ml portion of $CHCl_3$. Wash the sides of the filter and cake once with 25 ml of $CHCl_3$. A greater portion of the $CHCl_3$ may be removed if the cheese cake is compressed.

Transfer the combined $CHCl_3$ filtrates to a 500 ml separatory funnel. Rinse the filtering flask with 2 small portions of $CHCl_3$ and add to the funnel. Extract the $CHCl_3$ solution with about 33 ml of 0.5*N* NaOH. Transfer the $CHCl_3$ layer to a 600 ml beaker and the caustic layer to a 300 ml Erlenmeyer flask. Return the $CHCl_3$ to the separatory funnel and repeat the above caustic extraction twice. An emulsion may be formed during extraction, but most of it will break on standing. Transfer this emulsified layer to the alkali solution only in the final extraction. Acidify the caustic extraction with 70 ml of 1*N* HCl, and rapidly aerate for such time as required to remove dissolved $CHCl_3$ (5–10 minutes). To check complete removal of $CHCl_3$, smell the top of the flask while aerating. Be sure to remove all the $CHCl_3$ by aeration or low values will be obtained. Filter the solution through a medium or fine porosity sintered glass funnel fitted with filter paper and make to volume with H_2O in a 500 ml volumetric flask. If this solution is turbid, clarify by refiltering through a fine filter or an asbestos pad.

Prepare a reagent blank by extracting 250 ml of $CHCl_3$ with the 0.5*N* NaOH, adding 1*N* HCl, aerating, and diluting to volume as above. Place a portion of the reagent blank in one cell and a portion of the sample solution in another. Determine absorbance of the solution at 307 mμ. Dilute the sample solution if necessary to obtain a reading in the region covered by the standard curve (ordinary range dilution for absorbance readings is from no dilution to 1+5 dilution).

STANDARD CURVE

To prepare the standard curve, use fresh dehydroacetic acid; low readings are obtained from older solutions. Weigh exactly 100 mg of dehydroacetic acid into a 100 ml volumetric flask. Dissolve in about 50 ml of water plus 4 ml of 0.5*N* NaOH. Dilute to volume with water and mix. Pipette 1.0, 3.0 and 5.0 ml (1.0, 3.0 and 5.0 mg dehydroacetic acid) aliquots into separate 500 ml volumetric flasks. To each add the equivalent of about 100 ml of 0.5*N* NaOH and 70 ml of 1*N* HCl, dilute to volume and mix. Determine absorbance of solutions at 307 mμ, using the reagent blank prepared as above. Plot absorbances against mg dehydroacetic acid per 500 ml prepared solution. Calculate dehydroacetic acid to p.p.m.:

$$\text{p.p.m.} = \frac{(\text{mg}/500\ \text{ml}) \times 1000}{\text{weight of sample}}$$

Method 6.48. *Dehydroacetic Acid in Cheese; Qualitative.* (*AOAC Method* ***27.017***).

REAGENT

Salicylaldehyde solution.—10 ml salicylaldehyde dissolved in 95% alcohol and diluted to 50 ml.

TEST

Transfer the dehydroacetic acid solution remaining in the 500 ml flask after the spectrophotometric quantitative determination to a 1 liter separatory funnel. Add 100–125 ml of ether and shake vigorously. Allow to separate, draw off the aqueous layer, and discard. Draw off the ether into 125 ml Erlenmeyer flask, taking care not to include any emulsion or water. Evaporate the ether extract to dryness on the steam bath and dissolve the residue in 1 ml of 0.5*N* NaOH. Pour the 1 ml of alkaline solution into a test tube (do not rinse flask), add 0.5 ml of alcoholic salicylaldehyde solution and 1 ml NaOH (1+1). Mix and place in a boiling water bath for 5 minutes. Remove tube from bath, add 2 ml of water and observe the color. Included a reagent blank and a control containing 0.2 or 0.3 mg of dehydroacetic acid for comparison. With as little as 10 p.p.m. or less of dehydroacetic acid in cheese, a red or orange color is obtained. The intensity of the color is approximately proportional to the quantity of dehydroacetic acid present.

Residual Phosphatase in Cheese

A discussion of this test for under-pasteurization may be found under "Milk" in this chapter. That section gives the rapid Scharer test as it is applied to milk.

Dr. Kosikowski of Cornell University has developed a simplified phosphatase test in which the sample is incubated in an 8″ section of seamless cellulose tubing immersed in $CuSO_4$ solution. Phenol liberated by residual phosphatase is dialyzed into the $CuSO_4$ solution, the color developed with CQC, and compared visually with standards, or measured spectrophotometrically.[62] The method as applied to fluid milk and cheese is given in APHA Standard Methods.[63] The 10th ed., *Methods of Analysis*, AOAC, gives a longer but more accurate laboratory method for residual phosphatase in cheese (Method **15.177–15.180**). That book (p. 252) gives the test modifications and the phenol equivalent criteria for the various cheeses and cheese foods and spreads.

Test for Sufficiency of Pasteurization

The FDA standards for certain cheeses require that they be made from pasteurized milk, or, in lieu of this, be cured for a stated minimum number of days.[64] The same standards include a rather involved and tedious test method which in reality tests quantitatively for the presence of active alkaline phosphatase. This same test is used for fluid milk and cream.

We choose, instead, to give the rapid Scharer test, as a practical and reliable field test for pasteurization of milk used in the manufacture of cheese. This method, given herein as Method 6–19, is discussed in the section on "Milk" in this chapter.

This modification of the Scharer test, as applied to cheese, is taken from APHA Standard Methods, 11th edition:[65]

Method 6-49. *Test for Pasteurization.* (*APHA Standard Method* ***19.20*** (11th ed).

REAGENT

Diluted neutralized normal butyl alcohol.—7.5 ml neutralized normal butyl alcohol, reagent (**e**), Method 6–19, diluted to 100 ml with water.

TEST

Weigh on a clean balance pan (using waxed paper, foil, a small beaker, watch glass or other convenient container) 5.0 g of sample in duplicate—one for the sample test and the other for control. Transfer to 16 or 18 × 150 mm test tubes. If cheese is sticky, weigh samples on piece of wax paper and insert paper with sample into tube. Grind sample and its control with glass rod. Heat control from 1 to 2 minutes in beaker of boiling water, making sure that all of it has been heated to at least 85°–90°. Cool to room temperature and from this point treat control and blank in same manner.

Add 20 ml of a solution containing 7.5% of neutralized normal butyl alcohol in water. Shake well, macerate if necessary, and allow to stand for about 30 minutes or until the fat has collected at the top while the protein settles to the bottom. Withdraw a 0.5 ml portion from the relatively clear middle portion (with care taken to avoid any inclusion of fat), add to 5 ml of buffer substrate solution reagent (**b**) Method 6–19 and shake. Withdraw a few drops of the mixture and test on a spot plate with thymolphthalein indicator solution. If a blue color is not obtained, adjust the pH of the mixtures by addition of successive small amounts of Na_2CO_3 until a blue reaction is obtained, thus assuring a proper pH condition. Place the tubes in a water bath and continue the determination as for milk, as given in Method 6–19.

Ice Cream

Definition and Standards

The FDA standards for ice cream have been in the making for some years. While other reasons enter into the picture, the primary reason is that traffic in ice cream and other frozen desserts has been primarily a problem for state and local regulation. Different states have had widely varying definitions for ice cream, particularly in the minimum amount of milk fat required for a legal ice cream, which has in the past ranged from 8% to 14%.

As finally promulgated, the standard[66] requires not less than 10% milk fat, and 20% total milk solids, provided that in no case shall the milk solids not fat be less than 6%. In the case of bulky flavoring ingredients (cocoa, fruit, nuts, malted milk, confectionery, cereal), the weight of milk fat and total milk solids (exclusive of those in malted milk) is not less than 10% and 20% respectively, of the remainder obtained by subtracting the weight of these flavoring ingredients from the weight of the finished ice cream; in no case is the weight of the milk fat or total milk solids less than 8% or 16%, respectively, of the weight of the finished ice cream.

This definition of ice cream permits many optional ingredients, including flavorings, nuts, confectionery, various dairy ingredients, eggs in various forms, gums, alginates, glycerides, sorbitan compounds, sugars, etc. To show its complexity, manufacturers have a choice of 13 sugars, 18 gums and 6 egg products. The standard should be consulted for these details.[66]

One controversial segment was decided

when the standard decreed that if both a natural and an artificial flavor are used, e.g., vanilla and vanillin, if the natural flavor predominates, the product should be labeled by the predominating flavor, in this case "Vanilla Flavored". If the artificial flavor predominates or if the artificial flavor be used alone, the ice cream should be labeled as "artificially flavored", e.g., "artificial vanilla". Coloring, including artificial color, may be added. FDA Standards have been promulgated for other frozen desserts. They are equally complicated as to the choice of optional ingredients. Here again those interested should review the standard.[66] Here is a summary of the Frozen Dessert Standards:

	Min % Milk Fat	Min % Milk Solids	Min Wt/Gallon
Ice Cream	10.0	20.0	4.5 lbs.
Chocolate, Nut, etc. Ice Cream	8.0	16.0	4.5 lbs
Frozen Custard, French Ice Cream	10.0	20.0	4.5 lbs
Ice Milk	2.0–7.0	11.0	4.5 lbs
Fruit Sherbet[a]	1.0–2.0	2.0–5.0	6.0 lbs
Water Ice[a]	—	—	6.0 lbs

[a]Titratable acidity—not less than 0.35%, as lactic acid.

The Canadian standards for frozen desserts are similar to those in this country, but expressed somewhat differently:[67]

	% milk fat	% total solids
Ice Cream	10% min.	36% min.
Ice Cream Mix	10% min.	36% min.
Ice Milk	3% min.	33% min.
Ice Milk Mix	3% min.	33% min.
Sherbet	—	5% total milk solids max.

Both countries have requirements for the minimum pounds of milk solids per gallon, as well as (in U.S.) a minimum weight per gallon to control overrun. The respective regulations should be consulted for these details.

Government Purchase Specifications for Frozen Desserts

General Services Administration has issued a purchase specification, EE-1-116b (1953), for use by all Federal agencies, covering ice cream, sherbets and ices.

Ice cream shall be made from milk products, sweetening ingredient and flavoring, with color, stabilizers, emulsifiers, eggs, salt and water optional. Two levels of milk fat content, 12% and 10%, are provided for plain ice cream, and 2 levels, 10% and 8% milk fat, for chocolate, fruit, nut or other bulky flavors. The analytical requirements for the 2 types of plain ice cream are:

	Plain	Chocolate, nuts, fruits, etc.
Milk fat, not less than	12%	10%
Milk solids not fat, not less than	8%	10%
Stabilizers and emulsifiers, not more than	0.5%	0.5%
Weight per gallon, not less than	4.5 lbs	4.5 lbs
Food solids per gallon, not less than	1.6 lbs	1.6 lbs

The requirements for chocolate, fruit or nut ice cream, other than the milk fat content of 10% or 8%, are the same as for plain ice cream. Chocolate ice cream shall contain not less than 2% chocolate, nut ice cream not less than 5% nuts, and fruit (or other bulky flavors) shall not contain less than 10% of such ingredient. Frozen custard is an ice cream containing not less than 1.4% egg yolk solids.

This specification also establishes minimum

requirements for sherbets and water ices. These are:

	Sherbet	Water Ice
Milk fat, not less than	1.0%	—
Total milk solids, not less than	4.0%	—
Stabilizers and emulsifiers, not more than	0.5%	0.5%
Acid content, as lactic, not less than	0.35%	0.35%
Weight per gallon, not less than	6 lbs	6 lbs

The Defense Department has issued a purchase specification for imitation ice cream (MIL-I-35027 1963). Analytical requirements for the finished products are:

	All Flavors	
	Type I	Type II
Edible fat, not less than	10.0%	8.0%
Protein (NX6.38), not less than	3.6%	4.3%
Total solids, not less than	35.5%	35.5%
Weight per gallon, not less than	4.5 lbs	4.5 lbs

Ice Cream Mix is more convenient for use as a military ration item than ice cream itself. The Defense Department issued a mandatory specification for sweetened and flavored Ice Cream Mix in 1965 (MIL-1-7050). Two types are included: dehydrated and paste. The dehydrated mix is made by spray drying, the paste by concentration of the fluid mix by application of heat. The analytical requirements are as follows:

	Dehydrated	Paste
Milk fat, not less than	27.0%	19.0%
Protein (Nx6.38), not less than	9.75%	6.70%
Moisture, not more than	2.0%	28.0%
Titratable acidity*	0.22%	0.22%
Iron, not more than	10 p.p.m.	5 p.p.m.
Copper ,, ,, ,,	1.0 p.p.m.	1.0 p.p.m.
Oxygen in headspace gas, not more than	2.0%	2.0%

*As lactic acid in reconstituted mix.

Label directions for these mixes call for dilution of the dehydrated powder at the rate of $4\frac{1}{4}$ lbs to $3\frac{1}{2}$ quarts of water, and for the paste at the rate of equal volumes of paste and water. These are equivalent to a mixture of 68 g of dehydrated mix plus 112 ml water or 120 g of paste plus 100 ml of water.

Methods of Analysis of Dairy Frozen Desserts. (Also Applicable to Imitation Ice Cream)

Since the various states of the United States had established definitions for frozen desserts long before the promulgation of the Federal Standard, it is natural that official methods of analysis should vary also. This is particularly true in methods of determination of fat. The Babcock method for fat, long established for milk and cream, presented some problems in technique when adapted to ice cream. Consequently some of the states promulgated varying analytical techniques, Babcock bottle specifications, etc., for the determination of fat in ice cream. For example, laboratory supply houses list 9 g and 18 g Babcock bottles, made to the specifications of the several states. It is impossible to include here a Babcock method for fat in ice cream, applicable throughout the United States. The applicable state laws and regulations must be consulted. We limit ourselves here to the Roese-Gottleib method for fat, as adopted by the AOAC, and by many foreign countries.

Method 6-50. *Preparation of Sample.*

Plain products: Let sample soften at room temperature, without application of heat, which would cause separation of melted fat. Mix thoroughly with a spoon or by pouring back and forth between beakers.

Products containing fruit, nuts, candy and other discrete pieces: Fill the cup of a blender or malted milk mixer not more than 1/3 full (4–8 oz). Melt at room temperature or in a closed container set in an oven at 37–40°. Mix until a fine, uniform pulp is obtained (2–5 minutes for fruit desserts, up to 7 minutes for desserts containing nuts or candy). Transfer the blended mixture to a suitable container for convenience in weighing. Stir thoroughly immediately before each weighing.

Method 6-51. *Total Solids.* (*AOAC Method* ***15.184***).

Weigh quickly 1–2 g sample to nearest mg into a tared flat-bottomed dish not less than 5 cm in diameter. Heat on steam bath for 30 minutes, then in an air oven at 100° for 3.5 hours. Cool in a desiccator, and weigh quickly to avoid absorption of moisture.

Method 6-52. *Fat, Roese-Gottlieb Method.*[68] (*AOAC Method* ***15.186***).

Weigh to the nearest mg 4–5 g of the thoroughly mixed sample into a Mojonnier flask or Röhrig tube, using a free-flowing pipette. Dilute with water to about 10 ml, working the sample into the lower chamber and mixing by shaking. Add 2 ml of NH_4OH, mix thoroughly and heat by submerging in a water bath 20 minutes at 60°, with occasional shaking. Cool, and proceed as in Method 6–10, beginning "Add 10 ml of alcohol...."

Method 6-53. *Separation of Fat for Further Examination.* (*AOAC Method* ***15.187***).

Melt about 500 g and pass through a 20-mesh sieve, if necessary, to remove insoluble particles (fruit, nuts, etc.). Transfer about 300 g to a 1L separatory funnel, add 100 ml of water and 50 ml of NH_4OH and shake well. Add 200 ml of ethanol and shake 1 minute. Add 200 ml of ether and shake 1 minute. Add 300 ml of petroleum ether, shake 1 minute, then let stand until the emulsion breaks. Drain off and discard the lower aqueous layer. Add 25 g of anhydrous Na_2SO_4, shake, and decant through a rapid-flowing filter paper. Evaporate the solvents, and dry the residual fat in an oven at 55° overnight. Examine the fat by the methods given in Chapter 13. Review "Interpretation of Oil Analyses" and "Margarine" in Chapter 13.

Nitrogen, ash and ash constituents: Determine by methods given in Chapter 1. Use factor 6.38 to convert percent N to percent milk protein.

The oxygen content of dehydrated ice cream mix, packed in cans, may be determined by Method 17–14.

Method 6-54. *Weight per Unit Volume of Packaged Ice Cream.*[69] (*AOAC Method* ***15.181–15.182***).

This is a measure of "overrun" of ice cream, which is defined by the equation:

$$\frac{A-B}{B} \times 100 = \%\ \text{overrun},$$

in which A = weight of a given volume of ice cream mix and B is the weight of the same volume of frozen dessert. Some laws attempt to control this quality factor by setting a minimum weight per gallon for a frozen dessert.

REAGENTS

Kerosene of known specific gravity, cooled to 5–10° (icebox temperature) before use.

APPARATUS

a. *Overflow can*, Fig. 6–2. This is a No. 10 or gallon can with overflow spout soldered to an opening about ½″ from the bottom of the can and bent upwards, parallel to the sides, to form a bent-over spout. The upper edge of the spout opening should be above, and the lower edge below, the highest point (line "A", Fig. 6–2) of the interior surface of the top bend.

b. (optional) A *thin gauge tinned metal* or *aluminum strip* with a metal rod attached to the center as a handle. The ends should be bent down at a 90° angle to form a "bridge," used to submerge the sample in the overflow can. The bent end should be long enough to submerge the sample below line "A" of Fig. 6–2.

DETERMINATION

Collect samples of packaged ice cream from a freezing cabinet or cold room and immediately place in an insulated container with dry ice, for transportation to the laboratory. Surround

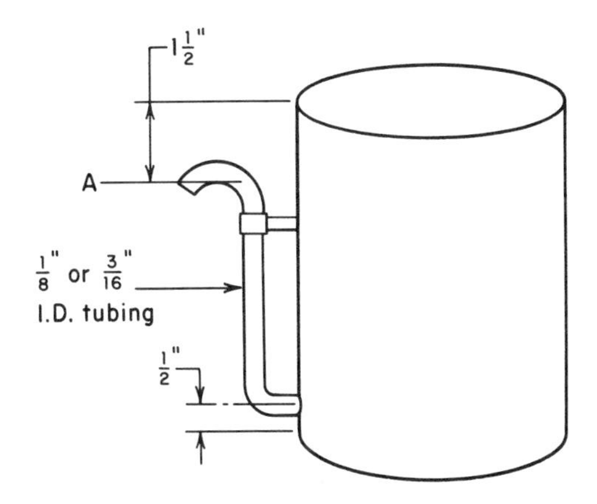

Fig. 6-2. Overflow can.

the samples with pieces of dry ice until frozen solid. Place the overflow can on a level bench so that the overflow will discharge into a container placed in the sink. Fill the can with the cold kerosene until it overflows. When overflow ceases, place a tared 500 ml graduate (or beaker) under the spout.

Remove the frozen brick from the dry ice, quickly remove the carton and weigh to the nearest gram. Designate this weight as W. Slowly immerse the brick in the kerosene, finally immersing it completely by aid of a small spatula or the "bridge" described under "apparatus". Weigh the displaced kerosene to the nearest gram and subtract the tare of the graduate to ascertain the net weight of the displaced kerosene. Divide this net weight by its specific gravity and designate the resultant volume as V.

$$\text{Weight per unit of volume (lbs/gallon)} = \frac{W \times 8.345}{V}$$

A tared graduated cylinder may be used instead of a tared beaker to catch the overflow. The volume reading will then be a check against the calculated volume of the displaced kerosene. The volume as calculated from the weight is more accurate.

For bulk ice cream, cut a wedge-shaped piece with a sharp knife from the bulk container, avoiding side pressure as much as possible. Cut off the edge of the wedge, wrap the remainder in waxed paper, and immediately freeze in dry ice as directed for packaged ice cream.

Remove the waxed paper and determine the weight per unit of volume as described under packaged ice cream. Calculate as weight/gallon as directed. Directions for bulk ice cream are not included in AOAC official method.

Method 6-55. *Sucrose and Lactose in Ice Cream.*

Weigh a 100 g thoroughly-mixed sample into a 500 ml volumetric flask, dilute to the mark with water and mix thoroughly. Dilute 100 ml of the prepared solution in a 250 ml volumetric flask to about 200 ml, add 6 ml of $CuSO_4$ solution [Method 16–6, Determination (**b**), Reagent (**ii**)] and then alkali solution as in Method 6–8. Dilute to the mark, mix thoroughly and filter through a dry filter.

Determine Cu before and after inversion on aliquots of the filtrate by Method 16–6, Determination (**b**). Calculate the wt of lactose in the aliquot from Cu before inversion by Table 23–6. For sucrose, subtract Cu before inversion from Cu after inversion, obtain the weight of invert sugar corresponding to the difference and multiply by 0.95.

Special Dietary Frozen Desserts

Several special diet frozen desserts are now on the market. One is "low-sodium" in which low-sodium fluid, or more often spray-dried, milk, containing about 6–8 mg of sodium per 100 g, is used. Another is "dietary ice cream", carrying calorie claims. In these, sugar is in part replaced by sorbitol or by a non-nutritive sweetener. Some states, for example Connecticut, permit no more than 10% carbohydrates in such a product. A more extreme product is "diabetic ice cream". Lampert gives a typical analysis for this (Selected Reference K):

Fat	12%
Non-fat milk-solids	9%
Sorbitol	15%
Non-nutritive sweetener	0.01%

There are low lactose non-fat milk solids available for the manufacture of these products.

Another frozen dessert becoming increasingly available is one made with vegetable fats instead of milk fat. Originally called Mellorine when it was first introduced in Texas, it must be labeled "Imitation Ice Cream" according to many state laws. Its fat content is the same as the corresponding dairy product, except for the origin of the fat. According to Lampert (Selected Reference K), Texas produced 50% of the 50 million gallons made in this country in 1963. Since then its production has extended into many states, although quite a few either prohibit it, or require it to be labeled as "Imitation". A typical composition, again according to Lampert, is:

Vegetable fat	10.5%
Milk solids non-fat	11.0%
Sucrose	10.5%
Dextrose	8.5%
Stabilizer and emulsifier	0.45%
Flavor	

U.S. Department of Agriculture Market Research Reports Nos. 212 and 296 give further information.

Methods of analysis of ice cream are applicable to all of these products, with, of course, identification of the fat when so indicated.

Sherbets

The Federal Standard and many state standards contain quantitative requirements for milk fat, milk solids and acidity calculated as lactic acid. Milk fat may be determined by Method 6–10 and acidity by 6–3. Also see the Note following Method 6–9. A close approximation of milk solids may be made by determining lactose by method 6–55, then multiplying the percentage of lactose by 2 and adding the percentage of fat.

TEXT REFERENCES

MILK

1. Cogen, T.: *The Haven of Health* (1936), quoted in Ling, E. R.: *A Textbook of Dairy Chemistry*. 3rd rev. ed. New York: Philosophical Library Inc. (1957) I, 1.
2. Oser, B. L.: (ed.), *Hawk's Physiological Chemistry*. 14th ed. New York: Blakeston Div., McGraw-Hill Book Company (1965). 369.
3. Ibid., 371.
4. Ling, E. R.: *A Textbook of Dairy Chemistry*. 3rd ed. New York: Philosophical Library (1957). I, 12, 15.
5. Ibid. 33.
6. Lampert, L. M.: *Modern Dairy Products*. New York: Chemical Publishing Co. Inc. (1965). 170.
7. Bulletin No. 12, 13th Ed.: *Dairy Laws of the State of Connecticut and Rules and Regulations of the Milk Regulation Board*. Rev. May 1, 1954. Hartford: Office of the Commissioner of Agriculture.

7a. Bureau of Dairy Science, California Dept. of Agriculture, *Standards of Dairy Products* 72–118 (Rev.).

8. Black, L. A. (ed.): *Standard Methods for the Examination of Dairy Products Microbiological and Chemical*. 11th ed. New York: American Public Health Assoc. Inc. (1960). 301.
9. Anon.: *Food Engineering*. (May 1957).
10. Hortvet, J.: *Ind. Eng. Chem.*, **13**, 198 (1921).
11. Woodman, A. G.: *Food Analysis*. 4th ed. New York: McGraw-Hill Book Company (1941). 139.
12. Ibid. 140.
13. Ibid. 152–153.
14. Reference 8, 384.
15. Ibid., Ch. 19, 326–327.
16. Ibid. 329–330.
17. Scharer, H. J.: *Dairy Sci.* **21**. 21 (1938). *J. Milk Tech.* 1, **5**, 35 (1938). Reference 8, 331–335.
18. Nickerson, T. A., Moore, E. E. and Zimmer, A. Q.: *Anal. Chem.*, **36**, 1676 (1964).
19. Reference 4, 50.
20. Buegamer, W. R., Michaud, L. and Elvehjem C. Q.: *J. Biol. Chem.*, **158**, 573 (1945).
21. Schilt, A. A. and Hoyle, W. C.: *Anal. Chem.*, **39**: 114 (1967).
22. Diehl, H. and Smith, G. F.: *The Copper Reagents: Cuproine, Neocuproine, Bathocuproine*. Columbus: G. Frederick Smith Chemical Co. (1960) 16–18; Hoste, J., Eckhart, J. and Gillis, J.: *Anal. Chim. Acta*, **9**, 263 (1953); Cheng, K. L. and Bray, R. H.: *Anal. Chem.*, **25**, 655 (1953); Sandell, E. B.: *Colorimetric Determination of Traces of Metals*, 3rd Ed. New York: Interscience Publishers, Inc. (1959) 230.

23. Morgan, M. E.: *Atomic Absorption Newsletter* No. 21 (1964); Allan, J. E., *Spectrochim. Acta*, **17,** 467 (1961).
24. King, R. L. and Dunkley, W. L.: *J. Dairy Sci.*, **42,** 420 (1959).
25. A.O.A.C. Methods **39.116–39.128**; *The Pharmacopeia of the United States of America*, 15th Rev. Easton: Mack Publishing Co. (1955) 889–892.
26. Sobel, A. S. and Rosenberg, A.: *Anal. Chem.*, **21,** 1541 (1949).
27. Hankin, L. and Squires, S.: *Applied Microbiology*, **8**: 209 (1960).

CREAM

28. *Webster's Third New International Dictionary Unabridged*, Springfield, G. and C. Merriam Co., (1961).
29. Hillig, F. and Ahlman, S. W.: *J. Assoc. Offic. Agr. Chemists*, **31**: 748 (1948).

BUTTER

30. Weber, A. L. and Edelson, H.: *J. Assoc. Offic. Agr. Chemists*, **35**: 194 (1952).
31. Patrick, G. E.: *J. Am. Chem. Soc.* **28**: 1611 (1906), **29**: 1126 (1907).
32. Weik, R. L.: *J. Assoc. Offic. Agr. Chemists*, **49**: 518 (1966).
33. Selected Reference B.
34. Fine, S. D.: *J. Assoc. Offic. Agr. Chemists*, **38**: 319 (1955).
35. Klayder, T. J. and Fine, S. D.: *J. Assoc. Offic. Agr. Chemists*, **40**: 509 (1957).
36. Selected Reference W, Chap. 13, p. 218–220.
37. Bolton, E. R., Richmond, H. D. and Revis, C.: *Analyst*, **37**: 183 (1912).
38. Anglon, C. and Mahon, J. H.: *J Assoc. Offic. Agr. Chemists*, **39**: 365 (1956).
39. Mahon, J. H.: *J. Assoc. Offic. Agr. Chemists*, **40**: 531 (1957).
40. Windham, E. S.: *J. Assoc. Offic. Agr. Chemists*, **40**: 522 (1957).
41. Bhalerao, V. R. and Kummerow, F. A.: *J. Dairy Science*, **39**: 947, 956 (1956).
42. Riel, R. H.: *J. Assoc. Offic. Agr. Chemists*, **38**: 495 (1955).
43. Zehran, V. L. and Jackson, H. C.: *J. Assoc. Offic. Agr. Chemists*, **39**: 194 (1956).
44. Keeney, M.: *J. Assoc. Offic. Agr. Chemists*, **39**: 212 (1956).

CHEESE

45. *Code of Federal Regulations, Title* 21, *Food and Drug, Sec.* 19.500–19.685. Washington, D.C.: Office of Federal Register, General Services Adm. (1966).
46. Ibid., Sec. 19.750–19.790.
47. Anon: *Summary of Laws Affecting the Cheese Industry*. Washington, D.C.: U.S. Dept. of Agri., Consumer and Marketing Service, Agriculture Handbook No. 265 (1966).
48. Office Consolidation of the Food and Drugs Act and Regulations to Feb. 16, 1967. *Sec. B*-08.030–*B*-08.054, Ottawa: Queen's Printer.
49. *United States Standards for Grades of Cheddar Cheese*. Washington, D.C.: U.S. Dept. of Agriculture, Consumer and Marketing Service (Federal Register Dec. 5, 1958).
50. Kristoffersen, T.: *Flavor Chemistry Symposium*. Camden: Campbell Soup Company (1961).
51. Bassett, E. W. and Harper, W. J.: *J. Dairy Science* **41**, 1206 (1958).
52. Kristoffersen, T. and Purvis, G. A.: *J. Dairy Science*, **46**: 1135 (1963), **47**: 599 (1964).
53. Anon: *Int'l. Standard FIL-IDF* 4, Int'l. Dairy Federation. Brussels, General Secretariat (1958).
54. Horwitz, W.: *J. Assoc. Offic. Agr. Chemists*, **49**: 56 (1966).
55. Weik, R. W.: *J. Assoc. Offic. Agr. Chemists*, **50**: 531 (1967).
56. Strange, T. E.: *J. Assoc. Offic. Agr. Chemists*, **50**: 547 (1967).
57. Anon: *J. Assoc. Offic. Agr. Chemists*, **50**: 203 (1967).
58. Weik, R. W. and Horwitz, W.: *J. Assoc. Offic. Agr. Chemists*, **49**: 575 (1966).
59. Reference 45, Reg. 121: 101, 121: 2001 (1966).
60. Melnick, D. and Luckmann, F. H.: *Food Research* **19**: 1–60 (1954), **20**: 639, 649 (1955).
61. Ramsey, L. L.: *J. Assoc. Offic. Agr. Chemists*, **36**: 744 (1953).
62. Kosikowski, F. V.: *J. Dairy Science*, **47**: 748 (1964).
63. Reference 8, Ch. 16.
64. Reference 45, part 19.
65. Reference 8, p. 338.

FROZEN DESSERTS

66. Reference 45, Reg. 20.1–20.5 (1966).
67. Reference 48, Reg. B.08.061–B.08.072 (1967).
68. Stegall, E. F.: *J. Assoc. Offic. Agr. Chemists*, **28**: 207 (1945).
69. Hart, F. L.: *J. Assoc. Offic. Agr. Chemists*, **28**: 601 (1945).

SELECTED REFERENCES

A Andersen, G.: *Margarine*. London: Pergamon Press (1954).

B Anon: *Code of Principles Concerning Milk and Milk Products and Associated Standards*. 4th ed. Rome: Food and Agriculture Organization, (1963).

C Anon: *Rapport de la Quatrième Session du Comité d'Experts Gouvernementaux sur le Code de Principes Concernant de Lait et les Produits Laitiers*. Rome: Organization des Nations Unies pour l'Alimentation et Agriculture. (April 1961).

D Arbuckle, W. S.: *Ice Cream*. Westport: Avi Publishing Company. (1966)

E Fisher, R. C.: (ed.), *Laboratory Manual*. Chicago: International Association of Milk Dealers (1943).

F Frandsen, J. H.: Dairy Handbook and Dictionary, Westport: Avi Publishing Company. (1958).

G Hall, C. W. and Hedrick, T. I.: *Drying of Milk and Milk Products*. Westport: Avi Publishing Company. (1966).

H Heineman, P. C.: *Milk*. Philadelphia: W. B. Saunders Co. (1921).

I Hillig, F.: "The Role of Chemistry in the Control of Quality in Dairy Products". *J. Assoc. Offic. Agr. Chemists*, **39**: 49 (1956).

J. Koskikowski, F. V. and Mocquot, G.: *Advances in Cheese Technology*. Rome: Food and Agriculture Organization (1958).

K Lampert, L. M.: Modern Dairy Products. New York: Chemical Publishing Co., Inc. (1965).

L Leach, A. E. and Winton, A. L.: *Food Inspection and Analysis*, 4th ed. New York: John Wiley and Sons Inc. (1920) Chap. VII.

M Ling, E. R.: *A Textbook of Dairy Chemistry*, 3rd. ed. Rev. New York: Philosophical Library, Inc. (1957) I. Theoretical.

N Robertson, P. S.: "Review of Recent Developments in Cheese-Making". (220 references). *J. Dairy Research*, **33**: 343 (1966).

O Sanders, G. P.: *Cheese Varieties and Descriptions*. Agriculture Handbook **No. 54,** Washington, D.C.: U.S. Dept. of Agriculture, (1953).

P Silverman, G. S. and Kosikowski, F. V.: "Observations on Cheese Flavor Production by Pure Chemical Compounds". *J. Dairy Science*, **36**: 574 (1953).

Q Webb, B. H. and Johnson, A. H.: *Fundamentals of Dairy Chemistry*. Westport: Avi Publishing Company (1965).

R Winton, A. L. and K. B.: *The Structure and Composition of Foods*. New York: John Wiley and Sons (1937) III, Part I.

S Woodman, A. G.: *Food Analysis*. 4th ed. New York: McGraw Hill Book Co. Inc. (1941).

CHAPTER 7

Dessert Mixes (Non-Dairy) and Gelatine

About two-thirds of the 50,000,000 or more pounds of gelatin produced in this country goes into edible gelatin. Over half of this goes into gelatin desserts, primarily as dry mixes. This use generally requires gelatin of a high gel strength, blended with sugar, an organic acid, flavor and color. An appropriate specification for gėlatin used for this purpose is that included in the Federal specification for dessert powders and gelatin.[1] This calls for a "high quality", edible ground gelatin of not more than 13% moisture content. It shall be free of preservatives or bleaches, and its hot solution shall be clear, and light colored.

As a practical requirement, gelatin for dessert mixes should have a low calcium content to prevent precipitation of calcium citrate or phosphate upon dilution. The gelatin should have a jelly strength not less than 225 g when tested by the Bloom Gelometer, Method 7–2(**a**).

The Federal Purchase Specification for Gelatin Dessert Powder[1] typifies the quality of most of these powders now on the market. The powder shall not contain more than 2.0% moisture (vacuum oven method) and 1% salt. Buffering salts used (citrates or phosphates) shall not amount to more than the weight of the acids used and must be sufficient to produce the required pH of the gelled dessert. The finished gelled dessert shall have a Bloom jelly strength of not less than 60 g and a pH between 3.0 and 4.5. The remainder of the gelatin dessert powder, other than flavor and color, consists of sucrose or a mixture of sucrose with dextrose not to exceed 18% of the finished weight.

Method 7-1. Preparation of Sample for Analysis.

Ground gelatin—Mix thoroughly.

Sheet gelatin—Crumble into small fragments and mix thoroughly.

Gelatin dessert powders—Mix by sieving through a 30 mesh sieve onto a large sheet of paper, rubbing through the sieve until all material passes through. Repeat twice, mixing each time. Work quickly to avoid absorption of moisture. Keep in an air-tight container.

Method 7-2. Jelly Strength (Bloom Gelometer) of Gelatin and Gelatin Dessert Powders.

This is the standard method used by the industry. It is AOAC Method **21.006** and **21.012**.

APPARATUS

Bloom Gelometer, with 0.5″ plunger and 1″ plunger.[2]

DETERMINATION

a. *Edible Gelatin*—Pipette 105 ml of water at 10–15° into a standard Bloom bottle, add 7.5 g sample and stir. Let stand 1 hour, then bring to

62° in 15 minutes by placing in a water bath regulated at 65° (sample may be swirled several times). Finally mix by inversion, let stand 15 minutes, then place in a water bath at $10 \pm 0.1°$. Chill 17 hours. Determine jelly strength in Bloom Gelometer, adjusted for 4 mm depression and to deliver 200 ± 5 g shot/5 seconds, using the 0.5″ plunger.

b. *Gelatin Dessert Powders*—Weigh 20 g of well-mixed sample into a standard Bloom bottle, and pipette 100 ml of water at 10–15° into the bottle. Let stand 15 minutes, then bring to 62° in 15 minutes in a water bath regulated at 65°. (Sample may be swirled several times.) Mix by inversion, let stand 15 minutes, then place in a water bath controlled at $10 \pm 0.1°$. Chill for 17 hours. Determine jelly strength in Bloom Gelometer, adjusted for 4 mm depression and to deliver 200 g shot/5 seconds, using the 1″ plunger and a light weight (paper or plastic) shot receiver.

The Bloom Gelometer method was adopted by the Edible Gelatin Manufacturers' Society in 1924. Manufacturers of gelatin desserts have found the instrument rather unsatisfactory for measuring the weaker jelly strength of finished gelatin desserts. Borker and coworkers at General Foods Technical Center have modified the Gelometer for more efficient use as a quality control test.[3] This is given by reference only because of the wide acceptance by both buyers and sellers of the present Bloom Gelometer:

Method 7-3. *pH of Gelatin Dessert Powders.*

This method is specified by Federal Specification C-D-221e, Dessert Powders and Gelatin, Sec. 4.3.1.1.

DETERMINATION

Adjust the jelled sample remaining from the Bloom test, Method 7–2, to 25° and determine the pH at that temperature, using a glass-calomel electrode system and a potentiometer.

Protein

Determine nitrogen by Method 1–7, using a 1 g sample. Percent $N \times 5.55$ = percent of anhydrous ash-free gelatin.

Method 7-4. *Total Ash of Gelatin and Gelatin Desserts.*

Follow Method 1–9 using a 5 g sample. Ignite at a temperature of 525°. The sample should be placed in a cold muffle, and the sample should first be charred, without flaming, before the muffle temperature is adjusted to 525°.

Calcium

Determine from the residue from ash determination, Method 7–3, using Methods 1–15 or 1–16, or by flame photometer following directions accompanying the instrument.

Moisture

Follow Method 1–2, vacuum oven method, at not over 70°, using a 2 g sample.

Method 7-5. *Total Acidity of Gelatin Dessert Powder. (AOAC Method **21.011**).*

Dissolve 20 g in 1L of recently boiled water. Titrate a 100 ml aliquot with 0.1*N* NaOH, with phenolphthalein as indicator. Report as % citric acid.

Method 7-6. *Sucrose and Dextrose in Gelatin Desert Powders. (AOAC Methods **21.013, 21.014, 21.015**).*

REAGENTS

a. *Tannin solution*—Dissolve 5 g tannin in 100 ml cold H_2O.

b. *Lead acetate solution*—Dissolve 100 g $Pb(C_2H_3O_2)_2.3H_2O$ in 200 ml of H_2O. (This makes 30° Bé solution.)

DETERMINATION

a. *Sucrose.* Place 13 g sample in 300 or 400 ml beaker, add 2 g $CaCO_3$ and 2 g Filter-Cel and mix well with glass rod. Add 175 ml of boiling water, creaming the mixture with a little of the water at first. Stir thoroughly and let stand a few minutes to insure solution. Cool under cold water to 30°, add slowly, stirring, 25 ml of the tannin solution, and let stand 5 minutes. (This quantity of tannin solution is sufficient for most powders; if 30 ml are required, use 170 ml of water instead of 175 ml.) Add slowly, stirring, 10 ml of the lead acetate solution and filter on an 18.5 cm Whatman No. 2

paper. (Total quantity of liquid used in each case is 210 ml, which yields 200 ml after evaporation and concentration. If precipitation has been conducted properly, the solution will filter readily and the filtrate will be clear.) Polarize this solution in a 200 mm tube at 20°.

If sample contains reducing sugar, delead with $K_2C_2O_4$, add Filter-Cel and filter. Invert by placing 50 ml of the filtrate in a 100 ml volumetric flask with 5 ml of HCl and allowing to stand overnight. After inversion, neutralize with concentrated NaOH solution, using phenolphthalein indicator. Discharge the color of the indicator with 0.1 *N* HCl. Cool to 20°, dilute to volume and polarize in a 200 mm tube. Use the following Clerget formula modified for the percent sucrose in gelatin dessert powders:

$$S = \frac{100(4P-8I)}{142.66+0.0676\,(m-13)-t/2}, \text{ where}$$

S = percent sucrose; P = direct reading; I = invert reading; t = temperature at which readings are made (20°); and m = g total solids from original sample/100 ml invert solution (3.25 g). Simplified—

$$S = 100(4P-8I)/132.$$

b. *Dextrose*. Determine polarization due to dectrose (D) by subtracting percent sucrose (S) as found in **a** from direct reading of polariscope in circular degrees (P) multiplied by 4: $D = 4P - S$,

Calc. percent dextrose (D') from following formula:

$$D' = D \times 66.5/52.5 = 1.267D, \text{ where}$$

D = polarization due to dextrose; 66.5 = specific rotation of sucrose; and 52.5 = specific rotation of dextrose.

Starchy dessert powders

A cursory stroll through a super-market will reveal a large variety of quick-cooking dessert mixes on display. Most of these are designed to be mixed with milk or water. Some require slight cooking; others, the "Instant" type, thicken properly when mixed with cold milk or water. This type contains a pre-cooked or otherwise modified starch to insure rapid thickening in cold liquids.

Defense Department purchase specification MIL-D-350338 (1962) illustrates the composition of these mixes:

To be prepared with milk:

Dextrose—not more than 17%.

Starch—not less than 12 nor more than 35%.

Salt—not more than 3%.

Flavoring ingredients and color—To produce an acceptable product.

Sugar—to make 100%.

Buffering agents, gelatin, gums, vegetable oils and enzymes may be used if they do not adversely affect the quality, stability or utility characteristics of the finished product. If the product is labeled "chocolate", the dry mix shall contain at least 11% cocoa; if "coconut," the mix shall contain at least 10% desiccated coconut.

To be prepared with water: The Military Specification suggests the typical formulations on page 152.

For coconut mix, replace 10 pounds of sugar in the vanilla formulation with 10 pounds of coconut, and reduce the amount of vanilla flavoring.

Either type shall contain not more than 3.5% moisture.

Methods of Analysis

Moisture, ash, protein, sucrose and dextrose may be determined by the methods given in this chapter. Fat may be determined by Method 5–7.

	Weight in pounds		
FORMULATION	VANILLA	BUTTERSCOTCH	CHOCOLATE
Sodium alginate	6.0	6.0	6.0
Calcium gluconate powder	4.0	4.0	4.0
Tetrasodium pyrophosphate (anhydrous powder)	2.3	2.3	1.5
Potassium bitartrate	0.5	0.5	—
Mono-and di-glycerides	0.2	0.2	0.2
Vanilla flavoring powder	0.9	—	0.5
Titanium dioxide	0.4	0.4	0.4
Salt	0.5	0.5	0.5
Starch	5.0	5.0	5.0
Powdered shortening	10.0	10.0	—
Sugar	70.3	70.6	66.4
Nonfat milk	45.4	45.4	45.4
Imitation butterscotch flavor powder	—	0.5	—
Butterscotch color	—	As needed	—
Cocoa	—	—	15.0
Chocolate brown color	—	—	As needed
Total	145.5	145.4	144.9

TEXT REFERENCES

1. *Federal Specification C-D-221F, Federal Supply Service*. Washington, D.C.: General Services Adm. (1968).
2. Bloom, O. T.: U.S. Patent 1,540,979 (1925).
3. Borker, E., Stefanucci A., and Lewis A. A.: *J. Assoc. Offic. Agr. Chemists*, **49**: 528 (1966).

SELECTED REFERENCES

A Alexander, J.: *Glue and Gelatin*. New York: *Chemical Catalog Co.* (1923).

B Bogue, B. H.: *Chemistry and Technology of Gelatin and Glue*. New York: McGraw-Hill, (1922).

CHAPTER 8

Eggs and Egg Products

To a great extent eggs and egg products used in the food industry are bought in either the frozen or dried state. The objective of either drying or freezing is to retain the native characteristics of the raw egg to the greatest possible extent. Egg yolks and whole eggs are spray-dried commercially, while egg whites may be either spray-dried (powder) or pan-dried (flaked).

The production of frozen whole eggs is described in a U.S. Military Purchase Specification.[1] Briefly, eggs are candled, broken, pumped through a flash heat pasteurizer, cooled rapidly to 45°F or less, packed into cans and frozen. Production of stabilized dried eggs is discussed in another military specification.[2]

It is impossible to completely separate yolks and whites commercially. Theoretically, a white-free yolk has about 51% solids. The commercial yolk has about 45% solids. Such a product contains about 15% whites.

Official Standards for Egg Products

No standards other than U.S.D.A. Standards for Grades have been established for shell eggs. Standards of identity for liquid eggs, liquid yolks, frozen eggs, frozen yolks, dried eggs and dried yolks have been established under the Federal Food, Drug and Cosmetic Act:[3]

Liquid whole eggs and frozen whole eggs are defined as hen eggs broken from shells, with yolks and whites in their natural proportions.

Liquid yolks and frozen yolks are yolks so separated from the whites as to contain not less than 43% total egg solids.

Dried whole eggs and dried yolk may have the glucose content reduced before drying by (a) action of a glucose oxidase-catalase preparation and hydrogen peroxide at pH 6.0–7.0 or (b) by fermentation with baker's yeast (S. cerevisiae).[4, 5] Use of the anti-caking agent sodium silicoaluminate in amounts less than 2% is optional.

The moisture content of dried whole eggs shall not exceed 5% if anti-caking agent is used, and 8% if this is not used. The moisture content of dried yolk shall not exceed 3% if anti-caking agent is used, or 5% if not used.*

The only standard promulgated under the Food and Drug Act of Canada defines egg products as the product obtained by shelling

*Amended January, 1969 [34 Fed. Reg., 251 (1969)] to provide that both products shall contain not less than 95% total egg solids.

fresh or stored eggs, and processing them as dried or frozen whole egg, egg-yolk, egg-white, egg-albumen, or mixtures thereof. Salt, sugar and a stabilizing agent are permitted. The only quantitative requirement is that dried eggs may contain 2% anti-caking agent.[6]

Federal Specifications

Military Specification(A) MIL-E-35001A (1963), Egg, Whole Frozen,(B) MIL-E-1037C (amended 1961 and 1965), Egg and Egg Products, Frozen, Federal Specification(C) C-E-230a (1960), Egg and Egg Products, Frozen (for use by all Federal agencies) and Military Specification(D) MIL-E-35062A (amended June 1966), Egg, Whole, Dried, are all in force for U.S. government purchases.

The analytical requirements for these various egg products are as follows:

Specification	Product	Egg Solids, %	Total Solids, %	Moisture %	Sugar %	Salt %
D[b]	Dried, Whole Egg	—	—	2.5	—	—
A, B, C	Frozen Whole Egg	25.5	—	—	—	—
B, C	Frozen Egg White	11.5	—	—	—	—
B, C	Frozen Egg Yolk	43.0	—	—	—	—
B, C	Frozen Sugar Yolk	38.7[a]	48.5	—	10±0.2%	—
B, C	Frozen Salt Yolk	38.7[a]	48.5	—	—	10±0.2%

[a]Equivalent to 43.0% salt-(or sugar-) free basis.
[b]This specification also establishes a pH range of 7.0 to 8.0 and a glucose content of not over 0.03%.

Composition of Egg Products

On the average shell eggs consist of about 58% white, 31% yolk and 11% shell.

Average Analysis of Egg[7]

	% of Total	Water, %	Protein, %	Fat, %	Ash, %
Whole Egg	100	65.5	11.8	11.0	11.7
Yolk	31	48.0	17.5	32.5	2.0
White	58	88.0	11.0	0.2	0.8
Shell	11				

Blanck[8] compiled a table from several sources giving the composition of egg product solids:

Constituent	Whole Egg, %	Yolk, %	Whites, %
Protein (lipid free)	51	32	92
Lipids	43	63	1.5
Free Sugars (95% glucose)	1.1	0.4	3.0
Other N-free organics	1.8	2.4	0.5
Inorganic	2.4	2.2	3.0
Sulfur	0.7	0.35	1.5
Phosphorus	0.8	1.16	0.13

Reproduced by permission of the publisher from: Fred C. Blanck, *Handbook of Food and Agriculture*. New York: Reinhold Book Corp. (1955).

Van Arsdel and Copley report the composition of dried whole egg and dried egg yolk:[9]

	Dried Whole Egg—%		Dried Egg Yolk—%	
	STANDARD[a]	STABILIZED[b]	STANDARD[a]	STABILIZED[b]
Water	5.0	5.0	5.0	5.0
Protein	46.5	46.5	35.3	35.3
Fat	42.5	42.5	57.5	57.5
Glucose	1.17	trace	0.36	trace

[a]Dehydrated without previous removal of glucose.
[b]Glucose removed by fermentation with yeast.

Another table by these same authors gives the composition of separated liquid yolks and whites:

	Egg White % Composition	Egg Yolk % Composition
Water	87.7	49.5
Solids	12.3	50.5
Sodium Chloride	0.3	0.3
Protein	10.7	16.3
Fat	trace	31.9
Glucose	0.38	0.17

In the interpretation of egg analyses there is always occasion to refer to L. C. Mitchell's classical work on the composition of shell eggs. The extensive tables of data in the original publication often make it necessary to refer to several different pages to find the specific values required. Table 8–1 (page 156), summarizing the data, has been found very useful for reference and is published here for the convenience of others who frequently use the Mitchell tables.

Chemical analysis

Method 8-1. *Preparation of Sample.*

a. *Sampling frozen whole egg or egg yolk.* Samples are best taken from a representative number of cans by drilling with an auger. First cut away any ice crystals on the surface, then drill at least 3 equidistant points along an imaginary circle located midway between the center and circumference of the can. Extend the borings as close as possible to the bottom of can. Composite the drillings, totaling 250–500 g, into a glass jar, and keep the contents frozen under a tight seal until ready for use. Warm sample to no more than 50° in a water bath, and mix well before weighing. A weight burette is a convenient method of weighing out successive samples.

b. *Sampling of dried egg products.* If small, use entire package as sample. From drums or other large containers remove about a 6″ top layer and scoop out small quantities from a representative number of containers to total about 500 g. Place in a tightly closed jar or can and keep in a cool place until ready for use. Before beginning weighings, sift 2 or 3 times through a 20-mesh sieve or household flour sifter to break up any lumps.

Method 8-2. *Total Solids.*

Follow Method 1–2, Moisture, Vacuum Oven Method, drying at 98–100°, and at not over 25 mm mercury. Weigh out 5 g of liquid or frozen egg product, or 2 g dried egg product as the sample. First dry liquid or frozen egg to apparent dryness on a steam bath before placing in the oven. Report as percent solids.

Total Nitrogen

Proceed as in Method 1–7, using 2–3 g liquid egg or about 1 g dried egg for sample.

Method 8-3. *Total Fat by Acid Hydrolysis.* (*AOAC Method* ***16.008–16.009***).

Weigh accurately by difference about 2 g of yolk, 3 g whole egg or 5 g whites into a Mojonnier

TABLE 8-1

Composition of Commercial Fresh Eggs

Campiled From Mitchell's Data, J. A.O.A.C., **15,** 310 (1932) by E. O. Haenni (1938).
(Values in parentheses are on the Dry Basis)

	Solids	Chlorine as NaCl	Total P_2O_5	Total Nitrogen	Water-Soluble N	Crude Albumen N	Fat (Acid Hydrolysis)	Dextrose
Whites:								
Max.	12.76	0.31(2.58)	0.05(0.43)	1.85(14.50)	1.71(13.62)	1.48(11.60)	0.05 (0.42)	0.45(3.80)
Min.	11.56	0.27(2.18)	0.03(0.24)	1.62(13.84)	1.52(13.00)	1.29(10.95)	0.02 (0.16)	0.34(2.72)
Ave.	12.21	0.29(2.37)	0.04(0.34)	1.72(14.11)	1.62(13.25)	1.38(11.26)	0.03 (0.28)	0.40(3.31)
Yolks:								
Max.	51.33	0.32(0.63)	1.44(2.82)	2.70 (5.31)	0.54 (1.08)	0.20 (0.40)	32.97(64.53)	0.20(0.40)
Min.	49.37	0.28(0.55)	1.34(2.67)	2.49 (4.95)	0.47 (0.94)	0.11 (0.22)	31.09(62.12)	0.15(0.30)
Ave.	50.47	0.30(0.59)	1.38(2.74)	2.61 (5.16)	0.51 (1.01)	0.16 (0.32)	31.88(63.16)	0.17(0.35)
Whole Eggs:								
Max.	27.16	0.31(1.19)	0.55(2.10)	2.12 (8.21)	1.28 (4.97)	1.01 (3.91)	12.45(47.45)	0.36(1.39)
Min.	25.26	0.28(1.06)	0.48(1.89)	1.99 (7.53)	1.14 (4.30)	0.83 (3.32)	10.88(42.80)	0.28(1.06)
Ave.	26.05	0.29(1.12)	0.52(2.01)	2.04 (7.83)	1.12 (4.64)	0.94 (3.59)	11.55(44.36)	0.32(1.23)

Average weight of egg = 56.7 g
Average % shell = 10.94
Average % white in edible portion = 63.81
Average % yolk in edible portion = 36.19

$$\text{Ratio } \frac{\text{White}}{\text{Yolk}} = \frac{63.81}{36.19} = 1.76$$

$$\text{Ratio } \frac{\text{White Solids}}{\text{Yolk Solids}} = \frac{12.21}{50.47} = 0.24$$

extraction flask or Röhrig tube. Add slowly, stirring vigorously, 10 ml of HCl. Place the flask in a water-bath at 70°, bring bath to boiling, and heat for 30 minutes, carefully shaking at 5-minute intervals. (For dried eggs, weigh 1 g sample into the extraction flask, add 10 ml of HCl (4+1) and proceed as above.)

Add 25 ml of ether and shake. Add 25 ml of redistilled petroleum ether (b.p. below 60°), shake and let stand (or centrifuge if using Mojonnier flask) until solvent layer is clear.

Proceed as in Method 4–10, beginning "Bring aqueous layer. . . ." Omit filtering through cotton.

Run a blank extraction on reagents used. Report as "Percent fat by acid hydrolysis."

Lipids and Lipid Phosphorus.

The lipid and lipid phosphorus content of egg and egg products was once used almost exclusively for estimation of egg content. Determination of sterols, Method 4–26, is now the method of choice. Since there are considerable published data on lipid and lipid phosphorus contents of eggs, chemists may wish to use AOAC Method **16.010–16.012** as confirmation.

Method 8-4. *Total Chlorides.*

Frozen whole egg or egg yolk is at times sold with 10% salt added. (*See* Federal purchase specifications above.)

DETERMINATION

Size of Sample: Frozen egg products (without added salt): 4–5 g of yolk, 7 g whole egg, 10 g whites. Frozen egg products with added salt: 1 to 2 g of yolk or whole egg. Dried whole egg or yolk: 2 g. Dried whites: 1 g.

Method A: Weigh the sample into a 150 ml Pyrex beaker, add 20 ml of 10% Na_2CO_3 solution and proceed as in Method 1–21, beginning "Evaporate to dryness, and ignite. . . ."

Method B: Weigh the sample into a 300 ml Erlenmeyer flask, and proceed as in Method 10–25, beginning "Add a known volume of standard 0.1 *N* $AgNO_3$. . . ."

In each method, run a blank on the reagents used, and correct titration accordingly. For the blank on Method B, add about 0.25 g sucrose to provide organic matter for oxidation. Report as percent sodium chloride after correcting for normal salt content of egg yolk, approximately 0.2%.

Method 8-5. *Dextrose in Dried Egg Products.*[10]

This method, developed by chemists of the Defense Subsistence Supply Center, is the method directed by the purchase specification MIL-E-35062A, Amendment 2. It replaced the previous method using Somogyi reagent, which was found to give high apparent values.

APPARATUS

a. *Chromatographic apparatus*, Mitchell type (*J. Assoc. Offic. Agr. Chemists* **40:** 999 (1957)).

b. *Capillary pipettes*, capable of delivering 1 μl.

REAGENTS

a. *Mobile solvent*—85 ml acetone diluted to 100 ml with water.

b. *Chromogenic reagent*—mix together 0.5 g of benzidine in 5 g of acetic acid, 5 g of monochloroacetic acid, 10 ml of distilled water and 80 ml of acetone.

c. *Standard dextrose solution*—0.0800 g of dextrose in 200 ml of 70% ethanol.

DETERMINATION

Add 50 ml of 70% ethanol to 10 g whole egg or egg albumen solids in a 200 ml centrifuge bottle, and stir for 10 minutes. Centrifuge for 5 minutes at 2000 rpm. Filter through Whatman No. 541 paper; 5–10 ml of filtrate is sufficient to constitute the sample solution. If the concentration of dextrose is high, dilute sample solution and consider dilution factor in final calculation.

Spot 10 μl sample solution on 8 × 8″ Whatman No. 1 chromatographic paper. Spot should be concentrated in an area no greater than 5 mm diameter for maximum sensitivity; this can be done by drying the paper thoroughly after each 1 μl application. (An improvised apparatus useful for drying consists of a 100 watt bulb inside a box having a 2.5 mm diameter hole over which the wet spot can be dried after each application.) Spot 2, 3, 4, 5 and 6 μl of the standard dextrose solution to estimate the concentration of sugar in the sample. The 5 μl spot represents 0.1% dextrose.

Draw a line perpendicular to the starting line with a ball point pen 5 mm from the edge of the chromatographic paper to mark the solvent front. This line marks the advancing solvent front and forms a permanent record.

Develop 1 hour with the mobile solvent, or until the solvent front has traveled about 5″ above the starting line in an ascending column. Dry the chromatogram and spray both sides lightly with the chromogenic reagent. While the chromatogram is still wet, heat in a 100° oven for 5 minutes. Compare the area and intensity of the sample spot with the standard spots for quantitative estimation of the percent dextrose in the sample. A densitometer may be used to give a more exact reading. The R_1 value of the dextrose spot is approximately 0.32.

See also papers by Coleman and co-workers in *J. Agr. Food. Chem.* **10**: 293 (1962) and *J. Assoc. Offic. Agr. Chemists* **45**: 463 (1962).

Residual Oxygen in the Container

Military Specification for Egg, Whole, Dried, MIL-E-35062A, requires that the product be gas-packaged by replacing the air in the container with pure nitrogen or with an 80–20 mixture by volume of nitrogen and carbon dioxide. The oxygen content of the sealed container shall not exceed 2.0%. *See* Method 17–14.

Sterols in Egg Products

Determine sterols by Method 4–26, using as sample weight: not more than 2.5 g whole egg or frozen egg; not more than 1.5 g yolk; not more than 0.7 g dried yolk; and not more than 1.0 g dried whole egg.

The AOAC has found significant differences in the present-day sterol contents of durum flour, semolina and eggs as determined by the digitonin method, when contrasted with earlier reports. Their present study shows higher values for the flours, and lower values for eggs. These new values will be used by AOAC in the derivation of a general formula for estimation of egg content of foods.[11]

These new data are given in Table 8-2:

TABLE 8-2
Percent Sterol (Moisture-Free Basis) in Egg Contents

Source	Year	% Sterol
	DURUM FLOUR[a]	
Minn.	1965	0.058
Minn.	1965	0.057
Minn.	1965	0.063
Minn.	1965	0.065
N.Y.	1964	0.068
	SEMOLINA[a]	
Minn.	1965	0.056
Minn.	1965	0.055
Minn.	1965	0.056
Minn.	1965	0.050
N.Y.	1964	0.049
	DRIED WHOLE EGGS[b]	
Wash.	1964	1.95
Iowa	1964	1.88
Mo.	1964	1.92
Kan.	1964	2.03
Ohio	1964	1.83
Minn.	1965	1.80
Mo.	1965	1.88
Neb.	1965	1.89
	DRIED EGG YOLK[c]	
Mo.	1965	2.71
Mo.	1964	2.59
Kan.	1964	2.70
Ohio	1964	2.59
Neb.	1965	2.71
Wash.	1964	2.65
Iowa	1964	2.76
Minn.	1965	2.54
	FROZEN EGG YOLK[d]	
Ohio	1964	2.60
Kan.	1964	2.67
Wash.	1964	2.54
Minn.	1965	2.44
Mo.	1965	2.30
Mo.	1964	2.48
Calif.	1965	2.45
Minn.	1965	2.56

[a]For 10 samples: maximum 0.065%, minimum 0.049%, average 0.058%.
[b]For 8 samples: maximum 2.03%, minimum 1.80%, average 1.90%.
[c]For 8 samples: maximum 2.76%, minimum 2.54%, average 2.66%.
[d]For 8 samples: maximum 2.67%, minimum 2.30%, average 2.51%.

Method 8-6. Sodium Lauryl Sulfate in Egg White.[12]

Sodium lauryl sulfate is a permitted additive up to 125 p.p.m. in liquid or frozen egg white, and 1000 p.p.m. in egg white solids, as an emulsifier, or a whipping aid.[13]

REAGENTS

a. *Crystal violet indicator*—0.5 g in 100 ml acetic acid.

b. *Bromphenol blue indicator*—Weigh 100 mg bromphenol blue into a mortar and grind with 7.5 ml 0.2 *N* NaOH. Transfer to a 100 ml volumetric flask and dilute to mark with water.

c. *Standard 0.1 N acetous perchloric acid*—Mix 8.5 ml of 72% perchloric acid with 500 ml of glacial acetic acid and 30 ml of acetic anhydride in a 1L volumetric flask. Cool and dilute to the mark with glacial acetic acid. Let stand 24 hours before use. Standardize against 400–500 mg of potassium acid phthalate, previously dried 2 hr, at 105° and dissolved in 80 mg of glacial acetic acid, using 3 drops of crystal violet solution (**a**) as indicator. Titrate to a blue-green endpoint stable for 60 seconds. Run a blank titration on 80 ml of glacial acetic acid and the same amount of indicator, and correct the volume of the standardization titre.

$$\text{Normality} = \frac{\text{mg potassium acid phthalate}}{204.22 \times \text{ml perchloric acid}}$$

d. *Mercuric acetic solution*—6 g in 100 ml of glacial acetic acid. If necessary, warm to dissolve.

e. *Benzethonium chloride*, 99% pure. (Obtainable from Rohm and Haas, Philadelphia, Pa.)—Determine purity as follows: accurately weigh 1 g sample, previously dried in vacuum oven at 80°, into a 250 ml Erlenmeyer flask. Dissolve in 80 ml of glacial acetic acid, add 10 ml of 6% mercuric acetate solution (**d**), and 3 drops of crystal violet indicator, and titrate with 0.1 *N* acetous perchloric acid (**c**) to the same blue-green endpoint as in the above standardization of the perchloric acid. Run a blank on 80 ml of glacial acetic acid and 10 ml of mercuric acetate solution and correct the titration accordingly.

$$\%\ \text{purity} = \frac{\text{ml corrected titrant} \times N \times 44.81}{\text{weight in grams of benzethonium chloride}},$$

where N = normality of standard perchloric acid.

f. *0.002 N benzethonium chloride*. Accurately weigh about 0.9 g of dried benzethonium chloride, and dilute to 1L with water.

$$\text{Normality} = \frac{(\text{weight of sample in grams})\ (\%\ \text{purity})}{448.1}$$

DETERMINATION

a. *Liquid and frozen egg white*—Let frozen egg white thaw at room temperature. Accurately weigh about 40 g of liquid egg white into a beaker and transfer to a 500 ml glass-stoppered Erlenmeyer flask with 40 ml of water. Swirl gently about 1 minute and let stand 1 hour with occasional swirling. Add 400 ml of ethanol and mix by gentle shaking. Heat mixture on steam bath 15 minutes and shake at 3 minute intervals to promote complete precipation. Let cool 45 minutes and filter through a Whatman No. 30 filter paper on an 11 cm Büchner funnel. Wash precipitate in flask and on filter paper with three 50 ml portions of ethanol. Transfer filtrate to beaker with several small portions of alcohol and evaporate to 50–100 ml on steam bath with air current. Refilter if a precipitate forms during the first ½ hour of evaporation. (Note: Low results are obtained if cloudy filtrates are analyzed.) Transfer concentrate to 150 ml beaker with small portions of ethanol. Evaporate to dryness and dissolve residue with 25 ml of water. Let stand 10 minutes and adjust to pH 5.0 ± 0.1 with 0.1 *N* HCl or 0.1 *N* NaOH, using a pH meter. Transfer the solution to a 250 ml glass-stoppered Erlenmeyer flask with several small portions of water. Add 50 ml of $CHCl_3$, 0.5 ml of bromphenol blue indicator, and 250 mg of anhydrous granular Na_2SO_4. (The presence of the solid Na_2SO_4 minimizes emulsions during the titration.) Titrate with 0.002 *N* benzethonium chloride solution by adding it in 0.25 ml increments. Shake vigorously 2–3 seconds after each addition and let layers separate for 15 seconds before next addition. When the end point is approached, the layers will separate quite rapidly. When the $CHCl_3$ layer turns yellow-green, add benzethonium chloride solution dropwise, and shake after each addition until the $CHCl_3$ layer remains blue-green for 15 seconds. (Note: The $CHCl_3$ layer changes from yellow-green to green to blue-green.) Each ml of 0.002 *N* benzethonium chloride is equivalent to 0.5768 mg of sodium lauryl sulfate. Titrate a reagent blank, using 35 ml of water, 50 ml of $CHCl_3$, 0.5 ml bromphenol blue indicator, and 250 mg of Na_2SO_4. (Blank titration requires about 0.5 ml.)

b. *Powdered egg white*—Accurately weigh about 5 g of powdered egg white into a 300 ml Erlenmeyer flask and add 40 ml of water. Let stand 2 hours with occasional swirling. If the egg white is not completely suspended after 1 hour, break up lumps with a stirring rod and swirl at 5-minute intervals until all egg white is suspended. After 2 hours, add 200 ml of ethanol and gently shake 1 minute. Heat mixture on steam bath ½ hour, remove, and let cool ½ hour. Continue as in (**a**) above, beginning with "filter through a Whatman No. 30. . . ."

c. *Flake-dried egg white*—Grind flakes to about 20 mesh in a mortar. Proceed as in (**b**) above.

Blank determinations were made by the author of the method on each type of egg white. These blanks were reported as negative (less 0.006%). The author claims 95–100% recovery.

Chemical indices of decomposition of eggs

In this section, we shall discuss both oxidative deterioration and decomposition caused by the action of microorganisms.

A research team from Hormel Institute and the USDA Western Regional Laboratory studied methods of evaluating oxidative flavor deterioration in powdered yolk.[14] They devised special apparatus for low-temperature collection of volatiles for examination by gas-liquid chromatography and UV absorbancy at 280 mμ. They found the latter correlated better with organoleptic examination. They also compared peroxide, carbonyl and thiobarbituric acid (TBA) values with the GLC and UV values.

These workers concluded that over a wider range of conditions the UV absorbancy method correlated better than did the various chemical techniques. This is particularly true of "standard" dried yolk or whole egg (containing no sugar). The peroxide values were obtained by a modification of Method 13–8, the TBA values by measuring absorbancy of the distillate at 530 mμ, and the carbonyl values by reacting the distillate with 2,4-dinitrophenylhydrazine reagent, and measuring the absorbancy at 435 mμ.

The GLC analyses resulted in a complex series of peaks in which these workers could not associate any particular peak(s) with the onset of rancidity.

The original paper should be consulted for details.

Succinic and Lactic Acids

Succinic acid in any appreciable amount in commercial egg products constitutes convincing evidence of the use of rotten and/or sour eggs in the preparation of the finished product. Lepper and Hillig[15], who made the original investigation for the AOAC, reported that they never found succinic acid in edible, marketable quality eggs or in dried egg prepared therefrom.

Lactic acid in egg products indicates the use of soured eggs in their manufacture or, in the case of frozen eggs, incomplete pasteurization before freezing.

AOAC Methods **16.034** and **15.012—15.013** separate both acids by continuous extraction, followed by spectrophotometric estimation of lactic acid as its colored iron salt. Succinic acid is estimated by partition chromatography and titration of the separated succinic acid with standard $Ba(OH)_2$ solution. AOAC Methods **16.037—16.042** should be consulted for details.

Ammoniacal Nitrogen

Assay of preformed ammonia as an index of protein decomposition has been used by food analysts for many years in testing such foods as fish, meat and eggs. As a rule, determination of NH_3 has been based on an adaptation of the Folin-Denis Method, developed originally for the determination of NH_3 in urine.[16]

The method given here is that used by Bailey in Connecticut.[17] That laboratory found that a strictly fresh egg should not yield more than 1 mg ammonia N/100 g of egg; eggs with an NH_3 content over 2 mg N/100 g begin to show characteristics of staleness, as evidenced by odor and candling (air-space over 1″ in diameter). Some states have settled on 2 mg N/100 g of eggs as the lower limit of ammonia for "passably fresh" eggs;[17] an NH_3 content of 3 mg N/100 g or over is considered definite evidence of decomposition.[18] This conclusion should be fortified by organoleptic or other chemical evidence. The same figures hold for frozen eggs.

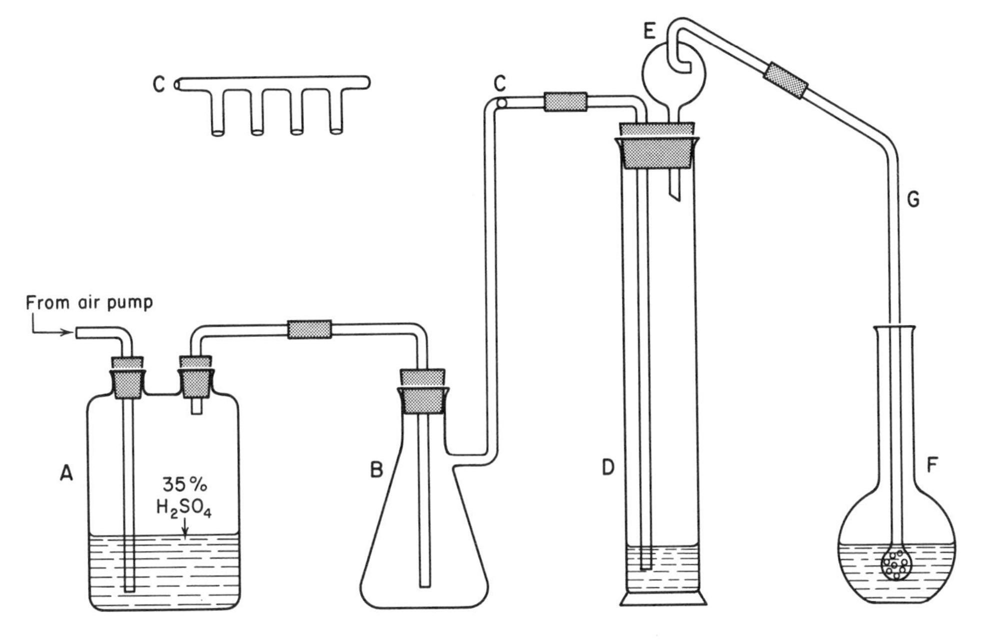

Fig. 8-1.

Method 8-7. *Free Ammonia*[17, 18].

APPARATUS

See Fig. 8-1. This can be set up for a single determination without using the manifold C. It is more convenient to use a four-place manifold. Then one can aerate a sample in duplicate, 1 reagent blank and 1 standardization against a known amount of $(NH_4)_2SO_4$. (Massachusetts Department of Public Health at one time used a 12-place manifold.)

REAGENTS

a. *Pyridine-free ammonium sulfate.* This is obtainable from reagent supply houses, or it may be prepared by adding Na_2CO_3 to an aqueous saturated solution of NH_4Cl and aerating the liberated ammonia into a dilute solution of CP sulfuric acid until the acid is neutralized. The $(NH_4)_2SO_4$ is precipitated by adding an equal volume of ethanol, and the salt is then filtered and dried.

b. *Standard $(NH_4)_2SO_4$ solution.* Dissolve 0.1552 g of the dried $(NH_4)_2SO_4$ in water and dilute to the mark in a 500 ml volumetric flask. 5 ml = 0.4 mg NH_3.

c. *Potassium oxalate*—saturated solution.

d. *Sodium carbonate*—saturated solution.

e. *Nessler reagent*—dissolve 50 g of KI in a minimum quantity of cold water. Add a saturated solution of $HgCl_2$ until a slight permanent precipitate is formed. Add 400 ml of a solution containing 143 g of NaOH, dilute to lL, mix, allow to settle, and decant.

DETERMINATION

Set up the aeration apparatus as shown in Figure 8-1. Add 2 ml 0.1 *N* HCl and about 60 ml of NH_3-free water to the volumetric flask serving as the receiving flask (F).

Break 3 or 4 shell eggs, or transfer 50–100 g of rapidly thawed, well-mixed frozen eggs, into a Waring blendor, blend 1 or 2 minutes and weigh a 20 g sample into the glass cylinder (D) with a minimum amount of NH_3-free water. Add 2 ml of saturated potassium oxalate solution (**c**), 5 ml of mineral oil to minimize foaming, and 5 ml of sodium carbonate solution (**d**).

Start the air pump at a rate of 1 or 2 bubbles a second and continue this slow aeration for one-half hour.

NOTE: It has been our experience that the air should be forced through the glass cylinder shown, and not drawn through by suction. This avoids excessive foaming.

Increase the rate of airflow gradually until the volume of air is as great as possible without undue frothing. Continue at this rate for 2 hours. Also run a reagent blank. At the end of aeration disconnect the receiving flask(s), add 5 ml of Nessler reagent (**e**) and 20 ml of NH_3-free water. Mix to develop the color, dilute to the mark and mix. Compare the color against 5 ml of standard $(NH_4)_2SO_4$ solution (**b**) aerated the same way, using a visual colorimeter.

NOTE: This method is included here because it is being used by several states, particularly in those states that have laws defining "fresh eggs," "cold storage eggs," etc., as offering acceptable supplementary evidence to that from candling or other tests.

Shell Eggs

Methods of analysis for frozen and liquid egg products given here are applicable to shell eggs. U.S.D.A. Poultry Division Regulations (Selected References D-1, D-2) include governing regulations on United States Standards, Grades and Weight classes, and detailed descriptions of candling methods with colored illustrations, along with other criteria of grades and standards.

Tables in the Section on "Composition of Egg Products" are helpful in the evaluation of shell eggs, as is the Section on Chemical Indices of Decomposition of Eggs.

TEXT REFERENCES

1. Military Specification MIL-E-35001A: *Eggs, Whole, Frozen.* Washington, D.C.: U.S. Govt. Printing Office (1963).
2. Ibid. MIL-E-35062A (1961).
3. *Code of Federal Regulations, Title* 21, *Food and Drugs Parts* 1 *to* 129. Washington, D.C.: General Services Adm. (1966). Part 42.
4. Ayres, J. C., and Stewart, G. F.: *Food Technology,* **1**: 519 (1943).
5. Carlin, A. and Ayres, J. C.: *Food Technology,* **5**: 172 (1951) and **7**: 228 (1953).
6. Office Consolidation of the Food and Drugs Act and Food and Drug Regulations, *Reg. B.*22.031. Ottawa: Queen's Printer.
7. Anon.: *Agriculture Handbook No.* 75, Egg Grading Manual. Washington, D.C.: U.S. Dept. of Agriculture (Revised June 1968).
8. Blanck, F. C.: *Handbook of Food and Agriculture.* New York: Reinhold Publishing Co. (1955).
9. Selected Reference D, 684, 653.
10. Coleman, C. H., Despaul, J. E., and Ruiz, S. *J. Assoc. Offic. Agr. Chemists,* **47**, 594 (1964).
11. Stauffer, L. J.: *J. Assoc. Offic. Agr. Chemists,* **50**: 851 (1967).
12. Wiskerchen, J.: *J. Assoc. Offic. Agr. Chemists,* **50**: 847 (1967).
13. Reference 3, section 121.1012
14. Privett, O. S., Romano, O. and Kline, L.: *Food: Technology* **18**: 1485 (1964).
15. Lepper, H. A. and Hillig, F.: *J. Assoc. Offic. Agr. Chemists,* **31**: 734 (1948).
16. Folin, O. and MacCullum, A. B.: *J. Biol. Chem.* **11**: 523 (1912), Folin, O. and Denis, W.: Ibid. **11**: 532 (1912).
17. Bailey, E. M.: *Conn. Agri. Experiment Station Bull.* **255**: 185 (1924), Bull. **341**: 702 (1931).
18. Michael, G. A.: (Mass. Dept. of Public Health) Private communication (1967).
19. Boyce, E. B.: *J. Assoc. Offic. Agr. Chemists,* **33**: 703 (1955).

SELECTED REFERENCES

A Anon: *Findings of Fact*. Federal Register 20:9614 (Dec. 20, 1955).

B Brooks, J. and Taylor, D. J.: *Eggs and Egg Products*. Gr. Brit. Science and Industry Research. Food Investigation Special Report No. 60. London: H. M. Stationery Office (1965).

C Romanoff, A. L. and Romanoff, A. J.: *The Avian Egg*. New York: John Wiley and Sons (1947).

D1 U.S. Dept. of Agriculture, Consumer and Marketing Service, Poultry Division: *Regulations Governing Grading of Shell Eggs, and U.S. Standards* (1967).

D2 Ibid. *Shell Egg Grading Manual* (in press).

E Van Arsdel, W. B. and Copley, M. J.: *Food Dehydration*. Westport: Avi Publishing Co. (1964) II, Chapter 21.

CHAPTER 9

Extracts and Flavors

Today, flavor is attracting more attention from food chemists than ever before. The growth in popularity of ready-to-eat or convenience foods has accelerated this trend, and it has been further enhanced by greatly improved analytical techniques that have enabled chemists not only to isolate predominant flavor constituents, but to reconstitute a flavor profile that simulates to a remarkable degree the natural product.

Modern instrumentation, described in Chapter 21, brings invaluable tools to the command of the chemist.[1] Still, even after an analysis or isolation of a flavoring compound is completed—and there may be many unidentified compounds present—the analyst has the problem of ascertaining which of these many compounds contribute the elements of flavor. With our present knowledge of the physiology of flavor perception, and our meager vocabulary for subjective description of flavor elements, complete analysis or synthesis of a complex flavor may be an extremely difficult, if not impossible, task.

It will suffice here to quote a few references as examples of modern techniques in this field:

1. Bengsstrom and Bosund, in Sweden,[2] used gas chromatography to demonstrate post-harvest changes in stored peas.

2. Teraneshi and co-workers from the USDA Western Regional Research Laboratories isolated over 150 components in volatiles from strawberries by a programmed-temperature capillary gas-liquid chromatograph and a fast-scan mass spectrometer in tandem.[3]

3. Herz and Chang, from Rutgers University, have devised an apparatus that involves the principle of flash evaporation and vaporization from a continuous, thin heated film, with subsequent condensation in a series of specially designed cold-traps. The condensate may then be fractionated by gas chromatography, and the gas fractions isolated by infrared and mass spectrometry.[4]

4. Nawar and Fagerson, of Massachusetts Institute of Technology, have compared direct sampling with an enrichment or concentration technique for gas chromatographic examination of volatiles.[5]

These citations are given as examples of a few of the analytical techniques that are now available for the analysis of natural or synthetic flavors. It must be remembered that artifacts of the original flavor components may form under certain conditions. This tendency must be minimized by judicious choice of apparatus and temperature conditions.

Official Standards for Vanilla Extracts and Flavors[6]

In the jargon of the analytical chemist, the FDA Definition and Standard requires as a minimum per 100 ml of extract the solids extracted by 35% alcohol (by volume) from 10 g of vanilla beans containing not more than 25% moisture. In the language of the Standard, these are the solids extracted by 35% alcohol from 13.35 oz of beans per gallon of finished extract. The standard also defines concentrated extracts equivalent to *X*-fold, *X* being a whole number. In practice, these may be 4-, 10- or 26-fold, as compared with single strength. The extract may also contain glycerol, propylene glycol, sugar, dextrose and/or corn syrup.

There are Standards for "vanilla flavoring" and "concentrated vanilla flavoring" which have the same flavor strength as the corresponding extracts, but contain less than 35% alcohol by volume.

FDA has also issued standards for "vanilla-vanillin extract" and "vanilla-vanillin flavor". These conform to the definition for the extract except that each gallon, containing the extractives from 13.35 oz of beans, may also contain not more than 1 oz of vanillin (equivalent to 0.75 g vanillin in 100 ml). "Vanilla-vanillin Flavoring" may contain less than 35% alcohol by volume.

FDA Standards have been issued for "vanilla powder" and "vanilla-vanillin powder". Eight pounds of these are each equivalent in flavor strength to 1 gallon of vanilla extract. These powders are mixtures of ground vanilla beans and/or oleoresin, with vanillin if declared, and a carrier (a sugar, starch or dried corn syrup) and not more than 2 percent of anticaking agent.

Vanilla preparations are the only food flavor standards thus far promulgated by the United States. Canada has issued standards for 19 flavoring extracts.[7] Their standard for vanilla extract corresponds to that of the United States. The Canadian definition for ginger extract requires that 100 ml contain the alcohol-soluble matter from not less than 20 g of ginger. This usually varies 1 to 2 percent in genuine commercial extracts, depending on the source of the ginger.

The Federal purchasing specification[8] follows the FDA Standard for vanilla flavoring compounds. It shall contain not less than 0.17 percent vanillin and 0.10 percent vanilla resins and shall have a Wichmann lead number of not less than 0.70. In addition, it includes a specification for "Imitation Vanilla Flavor". Each 100 ml shall contain not less than 2.8 g of vanillin or 0.94 of ethyl vanillin, or, alternatively, it may be a blend of not less than 1.4 g of vanillin and 0.47 g of ethyl vanillin. Solvents may be alcohol, glycerin, propylene glycol or any combination thereof,

Minimum Canadian Requirements for Flavor Extracts or Essences—Ml Volatile Oil/100 Ml[7]

Less than 1%[a]	1%	2%	3%	5%
Sweet basil (0.1)	Almond	Cassia[b]	Anise	Orange
Lemon, 0.2% citral	Marjoram	Cinnamon[c]	Peppermint	
Thyme (0.2)		Nutmeg	Spearmint	
Celery Seed (0.3)		Clove	Wintergreen	
Savory (0.35)				
Rose (0.4)				

[a]Figures in ( ) are ml volatile oil
[b]Cassia oil from Cinnamomum Cassia L
[c]Cinnamon oil from Cinnamomum Zeylanicum Nees

in amounts such that crystals are not separated when held at 0° for 8 hours. It shall be colored with caramel.

Composition of Vanilla Extracts

Much of the information on the composition of vanilla and vanilla extract is unpublished or of very limited distribution, such as to members of trade associations or in the files of regulatory agencies. This is unfortunate, since vanilla extract is not only the most widely used flavoring compound, but also the one most subject to sophistication.

One good indication of the composition of genuine commercial vanilla extract is the requirement of the Federal Purchase Specification:[8] minimums of 0.17 percent vanillin and 0.10 percent vanilla resins and a Wichmann lead number of 0.70. Failure to meet these specifications, however, is not alone sufficient to establish sophistication.

Fitelson, a recognized authority on the chemistry of vanilla, states: "Methods of manufacture (of extracts) differ so much that there is rather a wide range in constants, . . . in some cases laboratory extracts do not show the proper constants."[14] It has been established by Blomquist and his co-workers[19] that analyses of extracts made from the same beans by commercial extraction and by laboratory extraction do not agree, although the difference for each constituent analyzed is not great. In each case, the commercial extracts give higher figures.

Analysis of Authentic Extracts

Many of the analyses of authentic samples for vanillin reported in the literature, particularly those made before 1960, were made by the Folin-Denis phosphomolybdotungstic acid reagent, which was an official AOAC method up to 1960. It has been pointed out[10, 14, 20] that this method is not specific for vanillin, and gives high results, sometimes as much as 30–35% high. These earlier reports must therefore be used with caution. The Flavoring Extract Manufacturers Association (FEMA), through their research projects, is accumulating considerable information, not only on composition, but on methods of analysis. For example, Stahl and associates[21] have reported on chromatographic separation of organic acids, analysis by gradient elution for foreign materials by two-dimensional fluorescent chromatography, and amino acid separation by ion exchange chromatography. By these means they were able to identify the two botanical species of vanilla beans, *V. planifolia* and *V. tahitensis*, and to detect sophistication. In all of this work, samples of adulterated extracts were compared against authentic extracts. Much of Fitelson's published work was also under FEMA projects.[11,13]

All this adds up to the conclusion that determination of adulteration or compliance with a given standard is best based on an analysis for several constituents, each compared with the results obtained from an authentic extract that is preferably obtained by surveillance of the commercial extraction throughout the entire process.

Vanilla Extract and Flavor

Alcohol by Volume: Follow the procedure of Method 2–4, using a 100 ml sample for products containing not more than 60% alcohol, and 50 ml for those with higher percentages of alcohol. Confirm by determining the refractive index of the distillate by Method 2–5.

NOTE: This method is also applicable to ginger extract.

Methyl alcohol, isopropyl alcohol and acetone. If disagreement occurs between percent of alcohol by specific gravity and refractive index, other volatile substances should be suspected. Methyl and isopropyl alcohol may be identified by the methods listed in Chapter 2, p. 48. The isopropanol method, omitting $KMnO_4$ oxidation, may be used to identify acetone. Cf. Chapter 2, Method 2–18.

Total solids: In absence of glycerol or a glycol, follow Method 2–3, using a 10 ml sample.

Method 9-1 *Moisture in Vanilla Beans.* (This is the method set by FDA standard for vanilla extract.[9])

PREPARATION OF SAMPLE

Cut beans into pieces not more than $\frac{1}{4}''$ long, using care to avoid moisture change.

DETERMINATION

Follow Method 1–4, except that the toluene used is blended with 20% benzene by volume, and the total distillation time is 4 hours.

Method 9-2. *Qualitative Test for Coumarin.* (*AOAC Method* **19.008.** *FDA Prohibits Coumarin in Foods.*)

TEST

Place 5 ml of vanilla extract, or 2 ml of imitation extract or concentrate, in a 100 ml volumetric flask. Dilute to the mark with water and mix. Pipette 2 ml of this solution into a second 100 ml volumetric flask, add 2 ml of 0.1 *N* NaOH, dilute to the mark with water, and mix. Pour about 20 ml of this solution into a small beaker and place under UV light in a dark room. A brilliant green fluorescence, developing in 5 minutes, indicates coumarin. (Sensitive to 0.01% original extract.) The remainder of this solution may be used for the determination of vanillin by Method 9–3.

For quantitative determination, see: *Official Methods, AOAC,* 10th ed., **19.009-19.011** or **19.012-19.020.**

Method 9-3. *Vanillin in Absence of Coumarin and Ethyl Vanillin, UV Screening Method.* (*AOAC Method* **19.008.**)

Pipette 5 ml of extract, or 2 ml of imitations and concentrates, into a 100 ml volumetric flask, dilute to the mark with water, and mix. Pipette 2 ml of this solution into a second 100 ml volumetric flask, add 2 ml of 0.1 *N* NaOH and dilute to the mark with water. Transfer about 20 ml of this solution and test for coumarin by exposure to UV light, Method 9–2.

If coumarin is not observed, read the absorbance of the remaining alkaline solution at 270, 348 and 380 mμ. Obtain background absorbance as 0.29 × absorbance at 270 mμ + 0.71 × absorbance at 380 mμ. Subtract this value from absorbance at 348 mμ and divide by absorbance of 1 ppm vanillin (about 0.150), determined from a standard solution of 3 ppm of vanillin containing 2 ml of 0.1 *N* NaOH/100 ml and multiply by the dilution factor (1000) to give ppm in original sample. Make all readings within 2 hours of the final dilution. If the background absorbance is greater than 0.15, clarify with isopropyl alcohol as follows:

Pipette a 5 ml sample into a 50 ml volumetric flask and dilute to the mark with isopropyl alcohol. Mix, transfer to a centrifuge bottle, and centrifuge for about 10 minutes at high speed. Carefully pipette 1 ml of the supernatant liquid into a 100 ml volumetric flask. Add 2 ml of 0.1 *N* NaOH, dilute to the mark with water and proceed as above.

Method 9-4. *Determination of Vanillin in Flavoring Materials.*

The most recent method proposed for the estimation of vanillin is based on UV absorption in alkaline solution. This originated in a proposal by Smith[10], modified for collaborative tests by the AOAC. It was recommended for adoption as an official AOAC method by Fitelson in 1967.[11] The reproducibility, as shown by collaborative tests, is excellent.

PREPARATION OF STANDARD CURVE

Dissolve 100 mg of vanillin in 5 ml of ethanol, dilute to 100 ml with water in a volumetric flask, and mix (1 ml = 1 mg vanillin). Transfer 5, 10 and 15 ml respectively to 250 ml volumetric flasks, dilute to the marks with water, and mix (Solutions A). Pipette 10 ml of each Solution A into a 100 ml volumetric flask, add 2 ml of 0.1 *N* NaOH and 80 ml water, mix, dilute to the mark, and mix again. Determine absorbances of the second series (alkaline solutions) at 348 mμ, using the first (neutral) series of dilutions as reference blanks. Plot a standard curve.

DETERMINATION

If the sample contains more than 0.3 g vanillin/100 ml, dilute with 35% alcohol (by volume) to below this concentration. Pipette 10 ml of the sample (or diluted sample) into a 100 ml volumetric flask, dilute to the mark with water, and mix. Pipette 2 ml of this diluted solution into each of 2 100 ml volumetric flasks. Dilute one to the mark with water, and mix. To the other add 80 ml water and 2 ml of 0.1 *N* NaOH solution, dilute to the mark with water, and mix. Determine absorbances at 348 mμ of the alkaline solution, using the diluted neutral solution as a reference blank. Obtain the vanillin content from the standard curve.

NOTE: Fitelson, in the same articles, recommended a quantitative paper chromatographic method for vanillin and ethyl vanillin when occurring together. Collaborative study showed good accuracy and reproducibility. Details are given in the referenced papers.[11] See also AOAC Method **19.012-19.020.**

Method 9-5. *Wichmann Lead Number of Vanilla Extract.*[12]

This method was proposed in 1921 and adopted by AOAC in 1925, Method **19.021–19.022.**

Lead acetate precipitates the lead salts of organic and inorganic acids normally present in the vanilla bean. Imitation flavors colored in imitation of vanilla extract give a very slight, or no, precipitate.

REAGENT

Normal lead acetate solution, 8 g/100 ml, filtered if not clear.

DETERMINATION

Into 175 ml of boiled water in a 1 L round-bottomed flask, pipette 25 ml of 8% lead acetate solution and 50 ml of single-strength vanilla extract (for concentrated extracts, dilute to single strength with 40% ethyl alcohol before pipetting).

Place the flask in a hole in an asbestos board just wide enough so that 50 ml of the liquid in the flask will be even with the top of the board or slightly above it.

Connect flask to a condenser and, using moderate heat, distil 200 ml to the mark in a 200 ml volumetric flask. (The distillate may be reserved for determination of alcohol by specific gravity if the sample is single strength extract.) Transfer the residue in the 1 L flask to a 100 ml volumetric flask with CO_2-free water, using a rubber-tipped policeman to loosen any residue adhering to the flask. Cool, dilute to the mark with CO_2-free water, mix, and filter through a dry filter. Designate the filtrate as Solution A.

Conduct a blank determination, using 5 drops of acetic acid instead of the sample, and distilling 150 ml. Take up the residue with CO_2-free water and determine Pb as below.

DETERMINATION OF LEAD IN THE FILTRATE

Pipette 10 ml of Solution A, and 10 ml of the "blank", into separate 400 ml beakers, add 2 ml of acetic acid, 25 ml of water and 25 ml of approximately 0.1 *N* $K_2Cr_2O_7$, to each beaker. Heat the contents of the 2 beakers immediately, with a moderate flame, until the color of the precipitate changes from yellow to orange. Filter on a tared Gooch crucible, wash well with hot water, then with a few ml each of ethanol and ether. Dry at 100°, cool in a desiccator and weigh. Calculate the Wichmann lead number (*W*) by the equation:

W = (Wt. of $PbCrO_4$ from blank—wt. $PbCrO_4$ from sample) 12.82.

NOTE: The AOAC referee for flavors is now studying the EDTA titration for lead number of vanilla extracts. Interested analysts should watch for progress in this field through referee reports in forthcoming issues of *J. Assoc. Offic. Anal. Chemists*. One such report, by A. R. Johnson, is listed in *J. Assoc. Offic. Anal. Chemists*, **51**: 822 (1968).

Detection of Foreign Plant Extractives in Vanilla Extract

J. Fitelson of Fitelson Laboratories, Inc., New York, extensively investigated the proposition that the chromatographic pattern of the organic acids of vanilla should differ from that of other plant material.[13] He reported some 16 to 18 organic acids present in vanilla, malic acid being the most prominent, and proposed a paper chromatographic method and a gradient elution method to detect foreign plant material in vanilla extract. The methods are also useful in detecting sophistication of vanilla extract by addition of organic acids to enhance the Wichmann lead number. The entire series of articles should be consulted.

One method devised by Fitelson was adopted by AOAC as Method **19.031.** It has also been adopted by the Flavoring Extract Manufacturers Association:

Method 9-6. Vanilla Resins in Vanilla Extract. (*AOAC Method **19.026.***)

This method descended from UDSA Bureau of Chemistry Bull. 107, revised (1907) and, after collaborative study, was adopted with more precise directions by AOAC.

DETERMINATION

Pipette 50 ml of the extract into a 250 ml beaker, add 50 ml of water, and dealcoholize by boiling until the volume is reduced to 50 ml. Cool and make slightly alkaline with NH_4OH (1 + 3), then add 3 drops in excess. Stir vigorously 2minutes to dissolve the resins. Add HC1 (1 + 1) drop by drop until acid to litmus paper, then 2 ml excess. Stir, let stand at least 1 hour, but not over 24 hours. Add 0.5 g of filter-aid (Celite) and filter by means of suction through a long-stemmed 30 ml medium porosity fritted glass Büchner funnel, previously fitted with a pad by pouring an aqueous suspension of 1 g of filter-aid through the funnel and washing with water. If filtration slows, scratch the surface of the pad gently with a pointed glass rod to break the resin film.

Transfer the resins in the beaker to the funnel quantitatively with the aid of a rubber policeman. Reserve a portion of the filtrate for qualitative test below. Wash the beaker and funnel with six 20 ml portions of 0.05*N* HCl, allowing each portion to drain before adding further portions. Dry as thoroughly as possible by suction, transfer the funnel to a small, dry suction flask, and dissolve the resins on the filter with several small portions of boiling alcohol. Use some of this alcohol to rinse out the beaker. Let each portion flow through the funnel before adding the next portion, stirring up the filter-aid in the funnel with a small glass rod. Repeat the extractions until the alcohol filters through colorless. Rinse the tip of the funnel with hot alcohol, and transfer the filtrate quantitatively to a weighed beaker. Evaporate on steam bath, then dry at 100° 1 hour, cool in a desiccator and weigh. The normal quantity in a vanilla extract is 0.08 to 0.11 g/100 ml.

QUALITATIVE TESTS

a. Dissolve a portion of the dried residue in a few ml of 5% NaOH. The residue dissolves to a deep red color.

b. Dissolve a portion in alcohol. To one portion of the alcohol solution add a few drops of 10% $FeCl_3$ solution. To the remainder, add 1 ml of HC1. No marked change of color occurs if the resins are from vanilla.

c. To a portion of the filtrate from the resin precipitate reserved from the quantitative determination above, add a few drops of basic lead acetate solution [*see* Method **16-6** (a)]. A very bulky precipitate is formed and the filtrate from it is almost colorless.

d. Test another portion of the filtrate with a gelatin solution. A very small precipitate of tannin appears.

NOTE: "Vanilla resins" (or "oleoresins") is a rather vague term indicating a concentration of a vanilla extract formed by removal of alcohol and water. According to Fitelson[14], when the resins are diluted back to the single-strength basis, the Wichmann lead number remains unchanged from that of the original extract.

Method 9-7. Qualitative Test for Ethyl Vanillin.[15]

This test converts ethyl vanillin, after separation from vanillin and coumarin, into a salicylate.

TEST

Dealcoholize, if necessary, 25 g of extract or flavor by evaporation to 5 ml, add 5 ml of NH_4OH (1 + 3), and extract with 15 ml of ether. Separate the ether layer and evaporate to dryness on the steam bath. Add 5 drops of 50% KOH solution, evaporate cautiously to dryness, and fuse at the lowest possible temperature that avoids blackening.

Dissolve the residue in a few ml of water, transfer to a test tube and shake well with 5 ml of

$CHCl_3$. Pipette the lower $CHCl_3$ residue through a plug of cotton into another test tube containing 1 or 2 drops of 1% neutral $FeCl_3$ solution (made by adding a few drops of NH_4OH to a 1% $FeCl_3$ solution to incipient precipitation and filtering). Add 2 or 3 drops of water. If salicylic acid was formed, a purple color will develop.

Qualitative Test for Other Flavoring Additives

The AOAC has adopted a thin layer chromatographic method for detection of vanillin, ethyl vanillin, coumarin, p-hydroxybenzaldehyde, veratraldehyde, piperonal and vanitrope. See *Methods of Analysis*, 10th ed., **19.036–19.038.**

Analysis of Vanilla Powders

Fitelson, in the course of his extensive research on the analysis of vanilla flavorings, found that the methods that were applicable to vanilla extract could not be applied without modification to vanilla powders.[16] These are mixtures of ground beans or vanilla oleoresins with a sugar, starch or dried corn syrup.

Method 9-8. *Analysis of Vanilla Powder.*

PREPARATION

Powders with little or no starch: Add about 150 ml of 35% alcohol to 50 g of powder, mix well and place on a steam bath for 1 hour with frequent stirring. Cool, dilute to 200 ml in a volumetric flask with 35% alcohol, mix, and centrifuge. Filter the supernatant liquid through a rapid filter, discarding the first 10-15 ml of filtrate.

Powders with starch: Weigh 50 g of the powder into a beaker, add about 100 ml of 45% alcohol and let stand for 1 hour with frequent stirring. Pour onto a bed of Celite in a coarse porosity sintered glass disc Büchner funnel, using gentle suction for filtering. Shut off the suction, wash the powder with 45% alcohol, mix, and let stand a few minutes before reapplying suction. Repeat this treatment until about 180 ml of filtrate have accumulated. (The last washing should show very little color.) Transfer filtrate to a 200 ml volumetric flask, dilute to the mark with 45% alcohol and mix.

DETERMINATIONS

Lead Number: Follow Method 9-5, but use 100 ml of the prepared solution in place of 50 ml of extract, and 125 ml instead of 175 ml of boiled water.

Vanillin: Use 20 ml of the prepared solution, assuming a single-strength powder and following the UV screening Method 9-4.

Ethyl Vanillin: This may be determined by the AOAC column chromatographic Method **19.012-19.017**, or by the paper chromatographic method of Fitelson.[11]. See Note: Method 9-4.

Organic Acids: To 40 ml of prepared solution, add 100 ml of alcohol, mix, add 2.5 ml of lead acetate solution (8 g/100 ml water), centrifuge, and proceed by AOAC Method **19.031-19.035** or by one of Fitelson's methods listed in reference 13. See *Detection of Foreign Plant Extractives* above.

Method 9-9. *Benzaldehyde in Almond Extract.*

This method was adopted by AOAC in 1966[17] along with a UV spectrophotometric method. We include this gravimetric method as being of more general use.

REAGENT

2, 4-Dinitrophenylhydrazine solution. Add 50 ml of ethanol to 3.0 g of 2,4-dinitrophenylhydrazine. Slowly add 10.0 ml of H_2SO_4 while stirring. After the reagent dissolves, add another 50 ml of ethanol, and filter through a Whatman No. 12 filter paper or equivalent.

DETERMINATION

Measure a sample containing 10 to 50 mg of benzaldehyde (about 5 ml for flavors, 100–200 ml for cordials) into a distillation flask. Add enough ethanol to ensure at least 10% by volume in the distillate, and dilute to about 350 ml with water. Collect about 300 ml of distillate in a 600 ml beaker.

Add 100 ml of alcohol and 25 ml of H_2SO_4 to the distillate, mix well and cool to room temperature. Add 25 ml of the dinitrophenylhydrazine reagent with stirring and continue to stir for about 2 minutes.

Let settle and decant most of the supernatant liquid through a tared Gooch crucible with a thin asbestos mat. Transfer the bulk of the precipitate into the Gooch crucible with a stream of wash water. Wash with water followed by 10 ml of ethanol. (Both water and ethanol should be at not more than room temperature.) Dry at 100° to constant weight (about 2 hours).

Weight of precipitate × 0.3707 = weight of benzaldehyde.

NOTE: Benzaldehyde occurs in "oil of bitter almond" in amounts of 90% or more. Commercially this oil is obtained from bitter almond, peach or apricot kernels by distillation after enzymatic decomposition. The extract contains not less than 1% of the oil by volume.

Almond extract has been occasionally adulterated with nitrobenzene. The incidence of this has been quite rare for several years, hardly enough to justify space for a test. One good test is that of AOAC **19.094**, in which nitrobenzene is reduced by nascent hydrogen to phenylisonitrile, recognizable by its intense disagreeable odor.

A more probable contamination is that of hydrocyanic acid formed from the cyanogenetic glucoside amygdalin by hydrolysis.

Method 9-10. *Test for Cyanide in Almond Extract.*

Add 3 drops of freshly prepared 3% solution of $FeSO_4.7H_2O$ and 1 drop of 1% $FeCl_3$ solution to 5–10 ml of sample. Add 10% NaOH solution drop by drop until no further precipitation occurs, then add HCl(1 + 4) until the precipitate dissolves. If HCN is present, the characteristic Prussian Blue suspension or color develops.

Method 9-11. *Essential Oil Content of Flavoring Extracts, Precipitation Method.* (*AOAC Method **19.109**.*)

This is a general method for determining the percent of essential oil by volume in those flavoring extracts in which the solvent is alcohol or dilute alcohol, such as lemon, clove, mint, nutmeg and others.

APPARATUS

Babcock milk test bottle (*8%, 18 g, 6″*). Each whole percent graduation equals 0.2 ml.

REAGENTS

a. *Solvent*—equal parts of USP mineral oil and water-free kerosene.

b. *Saturated NaCl solution.*

DETERMINATION

Pipette a 10 ml sample (5 ml if oil content is over 5% by volume) into a Babcock milk test bottle and add by pipette or burette 0.5 ml of solvent and 1 ml of HCl (1 + 1). Fill to the shoulder with saturated salt solution. Shake bottle for 3 minutes, then add saturated NaCl solution to bring the column of oil within the graduation of the neck. Centrifuge 10 minutes at high speed and read the volume of oil from extreme bottom to the extreme top of the column. Babcock dividers or calipers are convenient for this purpose. (For allspice or peppermint extracts, read from the extreme bottom of the oil column to the bottom of the meniscus at the top of the column.) To obtain percentage of oil by volume, subtract 2.5 divisions to correct for the 0.5 ml of solvent added, and multiply the remainder by 2 (by 4 if a 5 ml sample is used).

NOTE: We have used this method, but without addition of the 0.5 ml hydrocarbon solvent, as a quick screening method. Some analysts use a glass-stoppered cassia flask instead of the Babcock bottle. In this case, instead of centrifuging, the cassia flask is allowed to stand in a water bath at about 60° until the oil separates.

The minimum oil content of the most used flavoring extracts in the United States, following good commercial practice, is, as expressed as percentage by volume:

Almond	1%
Cinnamon	2%
Clove	2%
Lemon	5%
Orange	5%
Peppermint	3%
Wintergreen	3%

Method 9-12. *Propylene Glycol in Flavoring Extracts and Flavors.*[18] (*AOAC Method **19.004–19.007**.*)

This method is official for vanilla extract and substitutes, but we have used it successfully for other extracts as well. Propylene glycol is quantitatively isolated by codistillation with heptane and finally oxidized by KIO_4.

APPARATUS

All glass distillation assembly with 24/40 joints, 250 ml Erlenmeyer flask, a 20 ml Barrett receiver (Kontes K75125 or equivalent) and West condenser with drip tip.

REAGENTS

a. *Heptane*, b.pt. 96–100°. Eastman practical grade or equivalent.

b. *Potassium arsenite*, standard 0.02*N* solution. Dissolve 4.9460 g of reagent grade As_2O_3, pulverized and dried to constant weight at 100°, in 75 ml of 1 *N* KOH. Add 40 g of $KHCO_3$ dissolved in about 200 ml of water, and dilute with water to 1 L at 25°. Dilute 200 ml of this to 1 L with water to make a 0.02 *N* solution.

c. *0.02 N NaOH solution*, made by diluting 0.1 *N* NaOH.

d. *Potassium periodate*, standard 0.02 *M* solution. Dissolve 4.6 g of KIO_4 in about 500 ml hot water, dilute with water to about 900 ml, cool to 25°, and dilute to 1 L with water. Standardize frequently because this solution decomposes on standing.

e. *Bromocresol purple indicator*, 1 g in 100 ml of ethanol. Filter if necessary.

f. *Propylene glycol*, to meet the following test: Dilute 0.5 ml to 25 ml with water, add 25 ml of 0.02 *M* KIO_4 solution and let stand 10 minutes. Titrate with 0.02 *N* NaOH solution, using 3 drops of bromocresol purple (e) as indicator. Titrate 50 ml of water to the same color as an end-point correction. The volume of 0.02 *N* NaOH consumed by the 0.5 ml propylene glycol, minus the end-point correction, should not be greater than 0.1 ml.

ISOLATION OF PROPYLENE GLYCOL

Place sample containing about 1 g of propylene glycol in a 200 ml ₮ Erlenmeyer flask. Add enough water, if necessary, to make the total vlume 10 ml. Add 60 ml of heptane, and a few glass beads and/or SiC grains. Connect the flask to the receiver attached to the condenser. Fill the receiver with heptane, heat the flask with a variable heat hot plate, and reflux at such a rate that a rapid stream of distillate flows from the tip. Reflux about 8 hours and cool.

Open the stopcock of the receiver and transfer the aqueous layer to a 250 ml (or other convenient size) volumetric flask. Wash the condenser, receiver, and solvent layer by pouring six 10 ml portions down the condenser, collecting each portion in the receiver, and draining it into the volumetric flask. Finally wash with enough water (about 25 ml) to completely fill the receiver, causing the solvent layer to return to the distillation flask. Dilute to volume and mix well.

DETERMINATION

Glycerol absent. Place an aliquot of the aqueous solution containing not more than 45 mg of propylene glycol in a glass-stoppered flask, add 35 ml of the 0.02 *M* KIO_4 solution, dilute to about 100 ml with water, and let stand 1 hour. Add about 1.0 g of $NaHCO_3$, 0.5 g of KI, and 2.5 ml of starch indicator. Titrate with the 0.02*N* $KAsO_2$ solution to the disappearance of blue. Standardize 25 ml of the 0.02*M* KIO_4 solution by the same titration procedure and calculate the amount of KIO_4 reduced by the sample. 1 ml of 0.02*N* $KAsO_2$ = 0.76 mg propylene glycol.

Glycerol present. Proceed as above. If iodine is not liberated on addition of $NaHCO_3$ and KI, insufficient KIO_4 was present. Repeat the determination, using a smaller aliquot or increasing volume of KIO_4 solution.

To determine glycerol in the aqueous solution, place the same volume aliquot used above in a glass-stoppered flask, add 1 drop of bromocresol purple, and add 0.02*N* NaOH until the solution is light purple. Add the same volume KIO_4 solution used above, dilute to about 100 ml, and let stand 1 hour. Add 10 drops of propylene glycol (about 0.5 ml), mix well, wash down the sides of the flask with water, and let stand 10 minutes. Add 3 drops of the indicator and titrate with 0.02*N* NaOH to a light purple end point. Titrate rapidly, but in order to avoid excessive absorption of interfering CO_2 from air do not shake the flask violently. Determine the blank for this determination by repeating the above procedure, using water in place of the sample and omitting the 1 hour standing. Sutbract this blank from the titration obtained for the sample aliquot. 1 ml 0.02*N* $KAsO_2$ = 0.46 mg glycerol; 1 ml 0.02*N* NaOH = 1.84 mg glycerol.

Mg propylene glycol in aliquot = [ml 0.02 *N* $KAsO_2$ — (4 × ml 0.02 *N* NaOH)] × 0.76.

The reactions involved in this method are:

$$CH_2OH.CHOH.CH_2OH + 2\ KIO_4 =$$

$$HCOOH + 2HCHO + 2KIO_3 + H_2O;$$

$$CH_2OH.CHOH.CH_3 + KIO_4 =$$

$$HCHO + CH_3CHO + KIO_3 + H_2O.$$

Method 9-13. *Alcohol, Percentage by Volume, in Flavoring Extracts Containing Volatile Oils.*

Pipette a 25 ml or 50 ml sample, measured at 15.56° (60°F.), into a separatory funnel, add about 100 ml of water, then add fine salt, NaCl, until the solution is saturated. Add 40-50 ml of petroleum ether (b.p. below 60°), shake thoroughly, let stand about ½ hour, and draw off the lower aqueous layer into a second separatory funnel. Repeat the extraction of the aqueous layer twice with 50 ml and 25 ml of petroleum ether. Combine the petroleum ether extractions, and wash the combined extracts twice with saturated salt solution, adding the washings to the original salt solution. This will give a total volume of about 150 ml. Transfer the combined salt solutions to an all-glass distillation assembly.

Distill into a 50 ml or 100 ml volumetric flask or pycnometer. When the pycnometer is almost full, cool to below 15.56° (60°F.) and bring to the mark. From this point follow the procedure in Method **2-4** beginning "Then wipe off external moisture. . . ." Report alcohol as percent by volume.

Method 9-14. *Lemon Oil or Orange Oil in Extracts, Polarization Method.*

This method goes back to USDA *Bureau of Chemistry Bull.* **65** (1902). It was later adopted by AOAC as Method **19.056.**

Polarize directly a sample of the extract at 20°, using a 200 mm tube. Divide the reading on the Ventzke scale (°*S*) by 3.2 for lemon extract and by 5.2 for orange extract, to obtain percentage volatile oil by volume, provided no other optically active compounds are present.

Lemon and orange oils occasionally contain small quantities of sucrose to facilitate solution. This may be established easily by pipetting 10 ml or 20 ml into an evaporating dish, evaporating to dryness, washing several times with ether, and redrying to constant weight. Deduct 0.38°*S* for each 0.1 percent of sucrose before calculating the percentage of oil.

A more exact determination of sucrose may be made by following Method **9-15**:

Method 9-15. *Sucrose in Lemon or Orange Extracts.* (*AOAC Method* ***19.068.***)

Neutralize the normal weight (26.00 g) of the sample, evaporate to dryness, and wash several times with ether. Dissolve the residue in water, dilute to 100 ml in a volumetric flask, and determine sugar by direct polarization or by the Munson–Walker copper reduction method. *See* Chapter 16.

Determination of Citral in Flavoring Extracts

The Hiltner method, which first appeared in *USDA Bureau of Chemistry Bull.* **122** (1908), has been adopted by AOAC, although its shortcomings have been realized. This method is based on the reaction of citral with *m*-phenylenediamine to form a yellow-orange complex which may be measured colorimetrically.

Yokayama *et al*[22] summarized the deficiencies of the method: (1) the color is not stable; (2) such parameters as reaction time and temperature are not precisely stated; (3) the spectrum of the compound has no sharp peaks at which the color may be measured; (4) aldehydes and ketones other than citral react to form yellow, orange or green colors.

Chemists of the Canada Food and Drug Directorate developed a barbituric acid condensation method for citral in essential oils[23] and later applied it to other food and drug substances containing citral.[22, 24] This was later studied collaboratively by Airth and coworkers at the Directorate and adopted by AOAC (Method **19.112–19.114**) for essential oils.[25]

Stanley and others at USDA Western Reginal Laboratory[26] proposed a specific vanillin-piperidine method for citral in oils. Water interferes, therefore the method cannot be applied to most extracts which commonly are made with 80% alcohol. Stanley, in a private communication in 1967, suggested that the method might be modified by separating the water with a hydrophobic solvent such as hexane. He has not had an opportunity to try out this procedure.

These two specific methods for citral are being studied by AOAC for their application to hydro-alcoholic extracts. Until such studies bring forth a tested method, we believe it preferable to give these methods by reference only, since modifications will almost certainly arise from such study.

Onion and Garlic Flavor

An interesting problem that occasionally arises is the chemical identification of onion *vs* garlic flavor. Saghir *et al*[27], from the Davis campus of the University of California, utilized gas chromatography to determine the composition of the volatile aliphatic disulfides that are produced enzymatically in the various allium vegetables when the cells are ruptured. These workers found that the disulfide volatiles from onion (*A. cepa*) are largely n-propyl disulfides (67–96 percent), while garlic (*A. sativum*) produces 80–90 percent of the allyl derivatives. Chives evolved up to 93 percent of the methyl compound in the volatiles.

TEXT REFERENCES

1. Jorysch, D., and Broderick, J. J.: *Cereal Science Today*, **12,** (11): 292 (1967).
2. Bengstrom, B., and Bosund, J.: *Food Technol.*, **18:**773 (1964).
3. Teranishi, R., Corse, J. W., McFadden, W. H., Black, D. R., and Morgan, A. I.: *J. Food Sci.*, **28:**478 (1963).
4. Herz, K. O., and Chang, S. S.: *J. Food Sci.*, **31:**937 (1966).
5. Nawar, W. W., and Fagerson, I. S.: *Food Technol.*, **16** (11):107 (1962).
6. *Code of Federal Regulations, Title 21*, reg. 22.1–22.9. Washington, D.C.: Office of U.S. Fed. Register (1966).
7. *Office Consolidation of the Food and Drugs Act and Regulations:* regulations B.10.003–B.10.027. Ottawa: Queen's Printer.
8. *Federal Specification EE–E–911f, Extracts, Flavorings.* Washington, D.C.: General Services Adm. (1966).
9. Reference 6, regulation 22.1(b).
10. Smith, D. M.: *J. Assoc. Offic. Agr. Chem.*, **48:**509 (1965).
11. Fitelson, J.: *J. Assoc. Offic. Agr. Chem.*, **49:**566 (1966), **50:**859 (1967).
12. Wichmann, H.: *Ind. & Eng. Chem.*, **13:**414 (1921).
13. Fitelson, J.: *J. Assoc. Offic. Agr. Chem.*, **42:**638 (1959), **43:**596 (1960), **45:**246 (1962), **46:**626 (1963), **47:**1161 (1964), **51:**1224 (1968).
14. Fitelson, J. (Fitelson Laboratories Inc.): Private communication (1967).
15. Winton, A. L., and K. B.: *The Analysis of Foods*, New York: John Wiley and Sons Inc. (1945) 921.
16. Fitelson, J.: *J. Assoc. Offic. Agr. Chem.*, **48:**911 (1965).
17. Anon.: *J. Assoc. Offic. Agr. Chem.*, **49:**215 (1966).
18. Bruening, C. F.: *J. Assoc. Offic. Agr. Chem.*, **33:**103 (1950), **38:**726 (1955).
19. Blomquist, V. H., Kovach, E. I., and Johnson, A. R.: *J. Assoc. Offic. Agr. Chem.*, **48:**1202 (1965).
20. Martin, B. E., Feeney, F. J., and Scaringelli, F. P.: *J. Assoc. Offic. Agr. Chem.*, **47:**561 (1964).
21. Stahl, W. H., Voelker, W. A., and Sullivan, J. H.: *J. Assoc. Offic. Agr. Chem.*, **43:**601, 606 (1960), **44:**549 (1961), **45:**108 (1962).
22. Yokayama, F., Levi, L., Laughton, P. M., and Stanley, W. L.: *J. Assoc. Offic. Agr. Chem.*, **44:**535 (1961).
23. Levi, L., and Laughton, P. M.: *J. Agr. Food Chem.*, **7:**850 (1959).
24. Laughton, P. M., Skakum, W., and Levi, L.: *J. Agr. Food Chem.*, **10:**49 (1962).
25. Airth, J. M., Stringer, B., Skakum, W., and Levi, L.: *J. Assoc. Offic. Agr. Chem.*, **45:**475 (1962).
26. Stanley, W. L., Lindwall, L. C., and Vannier, S. H.: *J. Agr. Food Chem.*, **6:**858 (1958).
27. Saghir, A. R., Mann, L. K., Bernhard, R. A., and Jacobsen, J. V.: *Proc. Am. Soc. Hort. Sci.*, **84:**386 (1964).

SELECTED REFERENCES

A. Arctander, S.: *Perfume and Flavor Materials of Natural Origin.* Elizabeth: Steffen Arctander (1960).

B. Guenther, E.: *The Essential Oils*, Vols. 1–6. New York: D. Van Nostrand Co. (1948–1952).

C. Gould, R. F. (ed.): *Flavor Chemistry, A Symposium. In*: Advances In Chemistry, Series 56. Washington: American Chemical Society.

D. Hall, R. L.: *Toxicants Occurring Naturally in Spices and Flavors. In:* Toxicants Naturally Occurring in Foods, Pub. 1354. Washington: National Academy of Sciences (National Research Council) (1964). 164–178.

E. Anon.: "Flavoring Agents as Food Additives", *Food Technol.*, **13**(7):14 (1959); **14**(10):488 (1960).

F. Furia, T. E. (ed.), *CRC Handbook of Food Additives*, Chapters 12–13. Cleveland: Chemical Rubber Co. (1968).

G. Hall, R. L., and Oser, B. L.: *Food Technol.*, **15**(12): 20 (1961), **19**(2), part 2:151 (1965).

H. Hornstein, I.: *Advances in Chemistry*, Series 56. Washington: American Chemical Society (1966).

I. Little, A. D., Inc.: *Flavor Research and Flavor Acceptance.* New York: Reinhold Publishing Co. (1958).

J. Merory, J.: *Food Flavorings: Composition, Manufacture and Use.* Westport: Avi Publishing Co., 2nd ed. (1968).

K. Stahl, W. H.: "Critical Review of Methods of Analysis of Oleoresins", *J. Assoc. Offic. Agr. Chemists*, **48**:515 (1965).

L. Winton, A. L., and K. B.: *The Analysis of Foods.* New York: John Wiley and Sons (1945), 881–927.

CHAPTER 10

Flesh Foods

Beef, Veal, Mutton, Lamb, Pork, Horse Meat, Etc.

Definition

The term "meat" is legally restricted to "the properly dressed flesh derived from cattle, from swine, from sheep, or from goats sufficiently mature and in good health at the time of slaughter", including only "that part of the striated muscle which is skeletal or that which is found in the tongue, in the diaphragm, in the heart, or in the esophagus", and specifically excluding "that found in the lips, snout, or in the ears, with or without the bone, skin, sinew, nerve and blood vessels which normally accompany the flesh and which may not have been separated from it in the process of dressing it for sale." Only when qualified with the name of the respective animal, e.g. "horsemeat", may the term "meat" be applied to the flesh of animals other than cattle, swine, sheep or goats.[1]

Cuts and Grades of Meat

Because most of the recorded analyses of meat refer to specific cuts from the slaughtered animal, Figs. 10–1 to 10–4 are reproduced as a guide to the location of such cuts.[2]

The U.S. Department of Agriculture has established seven grades for beef which are stamped on all U.S. inspected meat. In order of descending quality, these grades are: U.S. Prime, U.S. Choice, U.S. Good, U.S. Commercial, U.S. Utility, U.S. Cutter and U.S. Canner. Because the distinctions between these grades depend to a large extent on subjective factors, proficiency in grading meat can only be obtained by experience, but the following is a general description of the various grades:[3]

U.S. Prime

Only beef from steers or heifers can fall into this grade, which includes the finest finished animals that come to market. These animals will generally be quite young in age. In conformation Prime cattle tend to be low set, compact, thickly fleshed and short of neck and body. The fat covering is firm.

U.S. Choice

Greater variation in appearance is permitted. Too much or too little fat may make an otherwise "Prime" grade only a "Choice".

U.S. Good

This grade permits greater variation and less finish than the "Choice" grade.

U.S. Commercial

This grade has fair quality. Fat covering is small; marbling is seldom present.

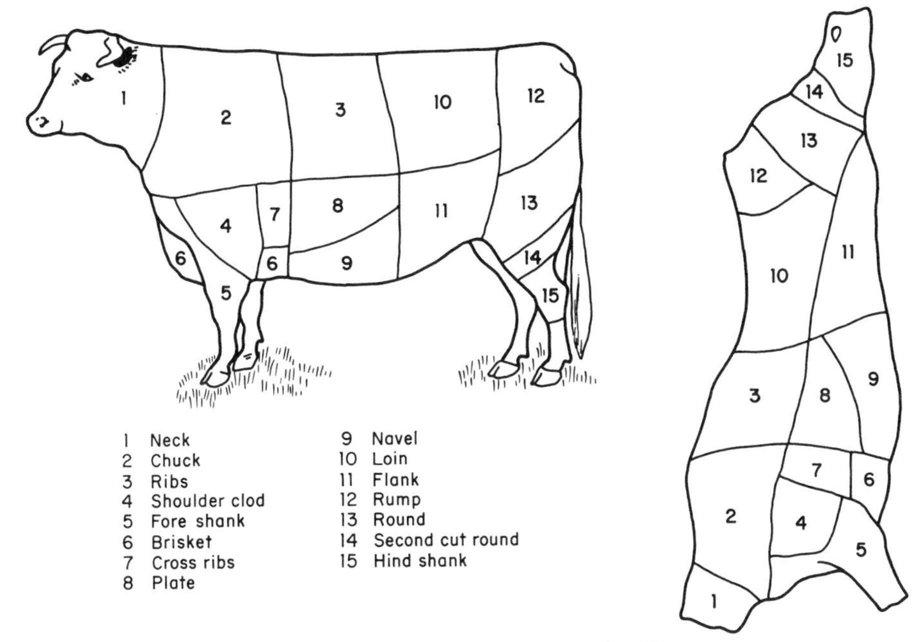

FIG. 10-1. Diagram showing cuts of beef.

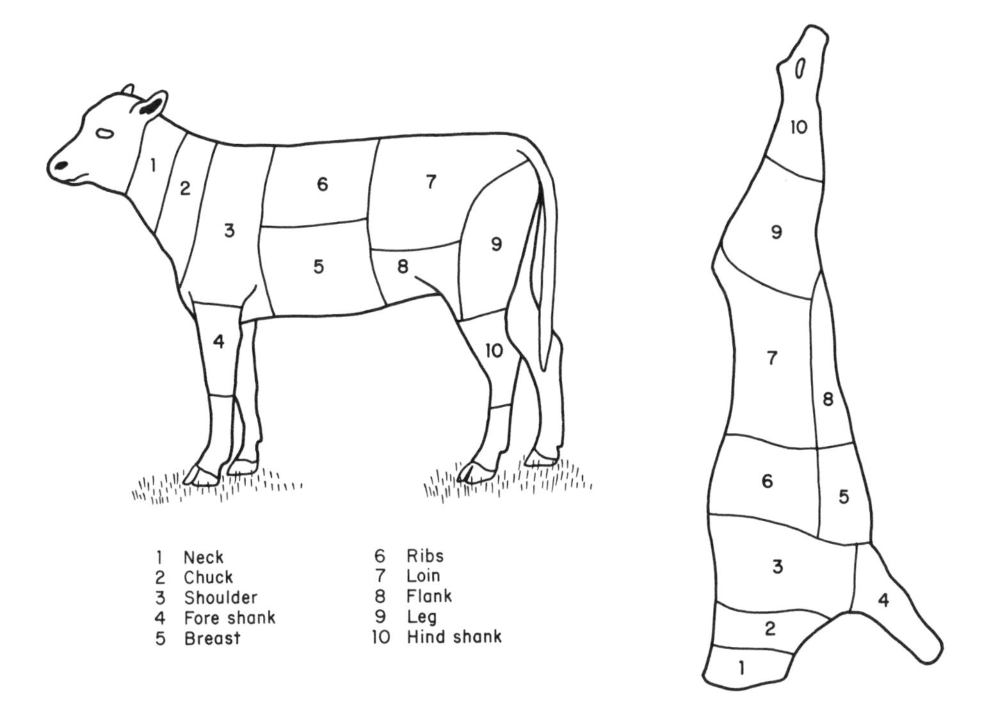

FIG. 10-2. Diagram showing cuts of veal.

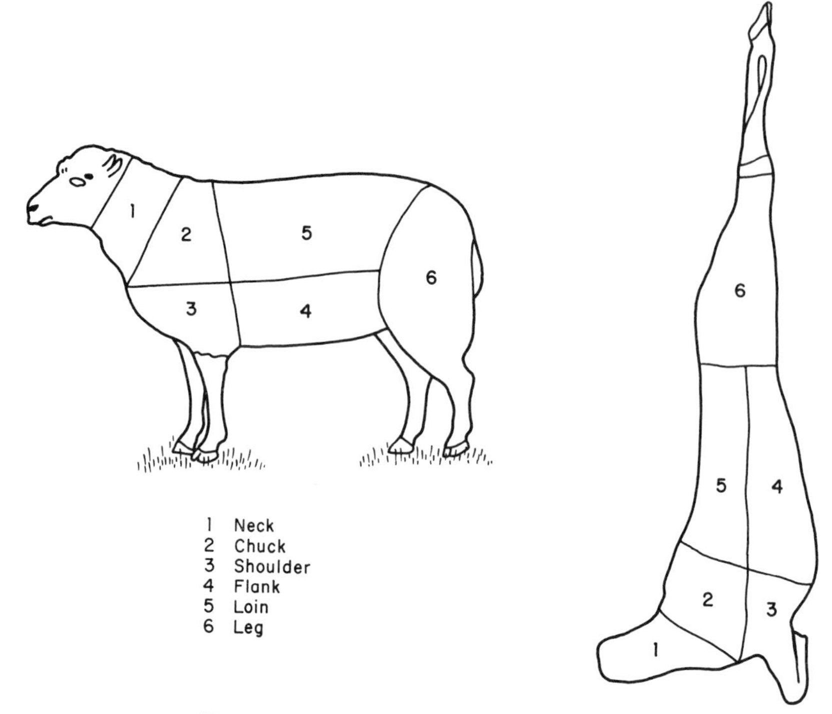

FIG. 10-3. Diagram showing cuts of mutton.

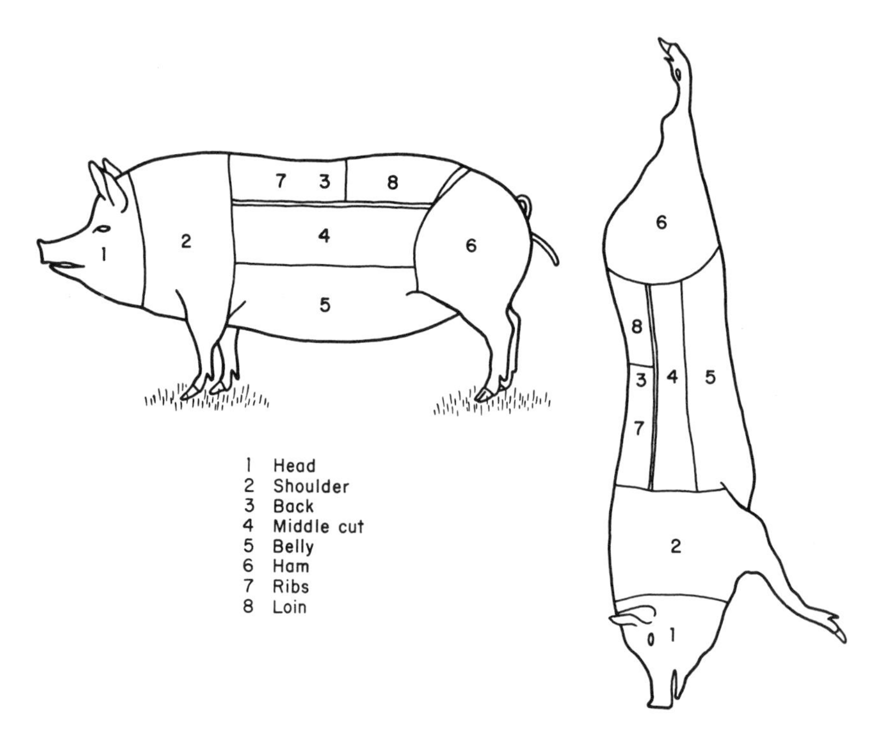

FIG. 10-4. Diagram showing cuts of pork.

U.S. Utility

The animals of this class usually have bony carcasses and little fat covering, which is yellow.

U.S. Cutter

Cows make up most of the animals of this grade. The flesh is very dark.

U.S. Canner

This is the lowest grade and is usually made up of old cows.

Most of the Federally-inspected meat sold to the average consumer is either U.S. Choice or U.S. Good; the Prime grade usually goes to the better restaurants.

There are five U.S. grades of veal: Choice, good, commercial, utility and cull. These grades are relatively unimportant because the animals are all young.

The meat of sheep is divided into two groups on the basis of age: lamb and mutton. Sheep reach maturity at 12 to 14 months, and it is at this age that the change from lamb to mutton occurs. In market parlance, a lamb from 3 to 6 months old is a "genuine spring lamb", while lambs up to one year old (and usually marketed in the fall and winter) are called "spring lamb" unqualified. There are six grades of lamb: Prime, choice, good, commercial, utility and cull.

Most of the pork reaching the market comes from hogs 7 to 12 months old. The three grades are simply designated as: one, two, and three.

Chemical Composition of Fresh Meat

The major constituents of fresh muscle meats are water, protein, fat and ash. The proportions vary with the cut and animal; for beef they average about 18 percent protein, 61 percent water, 20 percent fat and 0.9 percent ash. The most abundant muscle protein is the globulin complex *actomyosin*, which is responsible for the contractile properties of muscle. The respiratory pigment *myoglobin* is the primary source of meat color, since most of the *hemoglobin* in the blood is removed by the bleeding process during slaughter. Both myoglobin and hemoglobin are complex proteins in which the peptide portion of the molecule, the globin, is complexed with a non-peptide portion, the "heme", which is composed of an iron atom joined to a ring of four pyrrole groups. The fats in meat are principally triglycerides of straight-chain fatty acids having an even number of carbon atoms, but small amounts of mono- and diglycerides may be present.

A more detailed discussion of this subject is given in Selected Reference A.

The Proximate Analysis and Mineral and Vitamin Contents of Fresh Meat

Tables 10-1 and 10-2 show respectively the average proximate composition and energy values of the various cuts of fresh muscle, and of the organ meats.

Analyses made at the Connecticut Agricultural Experiment Station in 1956[5] of 63 samples of beef, including the "eye" (a triangular piece cut between the top and bottom round), fat and lean meat of 21 animals, showed the following values:

		Dry matter, %	Ether extract, %	Nitrogen %
Eye ...	Max.	30.68	9.76	3.60
	Min.	22.54	3.10	3.12
	Aver.	27.75	5.80	3.35
Fat ...	Max.	90.75	86.31	0.95
	Min.	84.46	78.96	0.52
	Aver.	88.49	83.73	0.69
Lean ...	Max.	39.19	21.65	3.23
	Min.	30.52	9.67	2.71
	Aver.	34.74	14.89	2.96

TABLE 10-1

The proximate composition and energy value of fresh muscle meats[a]

	Cut of meat, Medium Grade	Protein %	Water %	Fat %	Ash %	Cal/100g[b]
Beef	Chunk	18.6	65	16	0.9	220
	Flank	19.9	61	18	0.9	250
	Loin	16.7	57	25	0.8	290
	Rib	17.4	59	23	0.8	280
	Round	19.5	69	11	1.0	180
	Rump	16.2	55	28	0.8	320
Veal	Cutlet	19.0	70	5	1.3	140
	Leg	19.1	68	12	1.0	190
	Shoulder	19.4	70	10	1.0	170
Pork	Ham	15.2	53	31	0.8	310
	Loin	16.4	58	25	0.9	300
	Shoulder	13.5	49	37	0.7	390
	Spareribs	14.6	53	32	0.8	350
Lamb	Breast	12.8	48	37	—	380
	Leg	18.0	64	18	0.9	240
	Loin	18.6	65	16	—	220
	Rib chop	14.9	52	32	0.8	360
	Shoulder	15.6	58	25	0.8	300

[a] From Chatfield and Adams, 1940; Schweigert and Payne, 1956.

[b] Here, and in subsequent tables, energy values are calculated on the basis of 4 calories/g for protein and carbohydrate, and 9 calories/g for fat, with the calculated values rounded to the nearest 10 calories/100g of meat.

Table 10-3[6] shows the average proportions of the common mineral elements in cooked beef, veal, pork and lamb, while Table 10-4[7] gives the B vitamin contents of such cooked meats. More recent analyses by Assaf and Bratzler[8] showed the following proportions, in ppm, of 12 major and minor mineral elements in raw beef muscle: Potassium, 3,500; phosphorus, 2,400; sodium, 650; magnesium, 210; calcium, 120; iron, 48; zinc, 17; aluminum, 0.9–9.0; copper, 2.0; manganese, 1.7; boron, 0.04–0.1; molybdenum, 0.05–0.90. Dialysis experiments indicated the following percentages of the total amounts of these elements to be in organically bound form: Boron, 20; magnesium, 31; molybdenum, 40; manganese, 41; copper, 50; calcium, 54; aluminum, 55; zinc, 60, and

TABLE 10-2
The proximate composition and energy value of fresh organ meats[a]

	Organ meat	Protein %	Water %	Fat %	Ash %	Carbohydrate %	Cal/100g
Beef	Brain	10.5	78	9	1.4	0.8	130
	Heart	16.9	78	4	1.1	0.7	110
	Kidney	15.0	75	8	1.1	0.9	140
	Liver	19.7	70	3	1.4	6.0	140
	Lung	18.3	79	2	1.0	—	90
	Pancreas	13.5	60	25	1.2	—	280
	Spleen	18.1	77	3	1.4	—	100
	Thymus	11.8	54	33	1.1	—	340
	Tongue	16.4	68	15	0.9	0.4	210
	Calf liver	19.0	71	5	1.3	4.0	140
	Calf pancreas	19.2	70	9	1.4	—	160
	Calf thymus	19.6	75	3	1.9	—	110
Pork	Brain	10.6	78	9	1.5	0.7	130
	Heart	16.9	77	5	1.1	0.4	120
	Kidney	16.3	78	5	1.2	0.8	110
	Liver	19.7	72	5	1.5	1.7	130
	Lung	12.9	84	2	0.8	—	70
	Pancreas	14.5	60	24	1.1	—	270
	Spleen	17.1	77	4	1.0	—	100
	Tongue	16.8	66	16	1.4	0.5	210
Lamb	Brain	11.8	79	8	1.4	1.0	120
	Heart	16.8	72	10	1.0	1.0	160
	Kidney	16.6	78	3	1.3	1.0	110
	Liver	21.0	71	4	1.4	2.9	140
	Lung	17.9	79	2	1.1	—	80
	Spleen	18.8	74	4	1.6	—	110
	Thymus	14.1	80	4	1.3	—	90
	Tongue	13.9	70	15	0.8	0.5	190

[a] From Chatfield and Adams, 1940; Schweigert and Payne, 1956

iron, 83. Potassium, sodium and phosphorus were present almost wholly in ionic form.

U.S.D.A. Agriculture Handbook 8 (*see* Selected Reference F) gives the following average analysis for uncooked, domesticated rabbit meat: Moisture, 70, protein, 21, fat, 8, carbohydrate, 0, and ash, 1, percent; calcium 20, phosphorus, 352, iron, 1.3, sodium, 43, potassium, 385, thiamine, 0.08, riboflavin, 0.06, and niacin, 12.8, mg/100g.

Cured Meats

Historically, meat curing was the addition of salt to meat for the purpose of preservation. Eventually other substances such as nitrates and sugar were added to salt for the purpose of preserving and flavoring meat, and it became known that nitrite resulting from bacterial reduction of nitrate was responsible for production of the thermally stable meat pigment in cured meats; thereafter sodium nitrite (and occasionally potas-

TABLE 10-3
The mineral content of cooked beef, veal, pork and lamb[a]

	Cooked meats	Portion[b]	Calcium (mg/100g)	Phosphorus (mg/100g)	Sodium (mg/100g)	Potassium (mg/100g)	Magnesium (mg/100g)
Beef	Ground beef	LMF	7.3	220	47.4	450	21.3
	Round steak, bottom	LM	12.7	228	44.4	484	24.8
	Round steak, top	LM	8.4	255	41.8	493	26.2
	Sirloin steak	LM	14.8	226	45.9	436	21.0
	Standing rib roast	LM	7.5	164	53.6	413	18.6
	Standing rump roast	LM	8.3	197	54.1	386	20.4
Veal	Cutlet	L	12.1	318	61.6	606	25.2
	Loin chop	L	8.5	261	66.3	475	21.4
	Rump roast	LM	9.1	247	74.3	509	20.6
	Sirloin roast	L	9.1	258	64.1	580	21.4
Pork	Ham, cured, shank end	LM	7.7	186	863	398	19.5
	Ham, fresh	L	7.3	269	71.8	510	28.0
	Loin chop	L	12.1	263	59.6	568	25.3
	Sirloin, fresh	L	5.4	268	55.0	509	23.7
Lamb	Leg	L	8.2	220	84.9	512	23.6
	Loin chop	L	8.2	228	82.9	485	22.0

[a]From Leverton and Odell, 1958.
[b]L = lean portion
LM = lean plus marble portion.
LMF = lean marble plus fat (entire portion).

sium nitrite) as such became a constituent of pickling compounds.

The curing reaction in its simplest form may be expressed by the equation:

Myoglobin + Nitrite → Nitrosomyoglobin.

In addition to salt, nitrite and nitrate, sugars (either sucrose or glucose) are commonly used in curing mixtures for their flavoring effect and to reduce the harshness of the salt. In the cure of bacon, it was found that reducing sugars caused more rapid darkening upon frying (presumably due to a Maillard reaction between the glucose and amino groups), so until recently sucrose was employed almost exclusively in bacon curing.

Until its use was forbidden, the artificial sweetening agent sodium cyclamate was replacing sugars because it was found that due to its lack of reaction with other components of the finished bacon, the bacon could be cooked to any degree of crispness without appreciable darkening.

Formerly cured meats were so heavily salted as to be quite resistant to decomposition, but modern taste preferences have resulted in the use of lower proportions of salt to produce mildly cured hams and bacon that require refrigeration.

Curing may be introduced as a dry rub or by immersion of the meat in a solution ("cover pickle"), both used chiefly for bacon; or by injection ("pumping").

Besides the commonly accepted curing ingredients, a number of adjuncts have been introduced primarily in connection with color problems for use by the meat processor. Ascorbic acid or its sodium salt has been shown to be effective in producing a deeper

TABLE 10-4

The B vitamin content of cooked beef, veal, pork, and lamb[a]

	Cooked meat	Portion[b]	Thiamine (mg/100g)	Riboflavin (mg/100g)	Nicotinic Acid (mg/100g)	Vitamin B_6 (mg/100g)	Pantothenic Acid (mg/100g)	Vitamin B_{12} (μg/100g)
Beef	Ground beef	LMF	0.16	0.18	5.6	0.46	0.44	1.3
	Round steak, bottom	L	0.13	0.33	5.7	0.45	0.63	1.7
	Sirloin steak	L	0.11	0.60	2.9	0.42	—	3.8
		LM	0.10	0.46	3.3	0.41	—	3.0
	Stading rib roast	L	0.08	0.23	5.5	0.48	0.60	2.9
		LM	0.06	0.21	4.0	0.34	0.54	2.3
	Standing rump roast	L	0.11	0.24	4.4	0.41	0.73	1.6
		LM	0.10	0.23	4.3	0.38	0.60	2.2
	T-Bone steak	L	0.13	0.09	6.7	—	0.61	1.1
		LM	0.10	0.12	6.1	—	0.86	1.4
Veal	Cutlet, round	L	0.14	0.37	7.3	0.50	0.50	2.2
	Loin chop	L	0.21	0.32	7.1	0.43	0.50	2.7
	Rump roast	L	0.19	0.22	8.9	0.48	0.71	2.7
		LM	0.16	0.20	8.2	0.48	0.71	2.5
	Sirloin roast	L	0.18	0.26	8.7	0.52	0.84	2.7
Pork	Ham, cured, shank end	L	0.90	0.29	4.6	0.44	0.64	1.3
		LM	0.78	0.24	4.1	0.34	0.57	1.2
	Ham, fresh	L	0.68	0.32	5.3	0.44	0.49	1.4
	Loin chop, center cut	L	1.2	0.19	5.5	0.48	0.40	1.1
	Sirloin roast	L	1.3	0.34	4.6	0.56	0.88	1.2
Lamb	Loin chop	L	0.21	0.33	7.9	0.33	0.59	2.4
	Leg	L	0.23	0.31	7.3	0.32	0.61	3.1

[a]Leverton and Odell, 1958; Lushbough *et al.*, 1959.
[b]L = lean portion.
LM = lean plus marble portion.
LMF = lean plus marble plus fat (entire portion).

red color in hams cured three days prior to processing. Its exact mode of action is uncertain; it may act by reducing metmyoglobin to myoglobin, or it may function by reacting with nitrite to produce higher yields of nitric oxide than would normally occur from nitrous acid alone.

Alkaline phosphates are employed in curing primal cuts chiefly to decrease the amount of shrinkage in smoked products, and to reduce the degree of "purge" or "cook-out" in canned products. Sodium tripolyphosphate, sodium hexametaphosphate, sodium acid pyrophosphate, sodium pyrophosphate and disodium phosphate, either separately or in combination, are employed for this purpose. Regulations restrict the proportion of these salts that may be present in the finished product to 0.5%.

Monosodium glutamate has been used in certain meat products, but because its flavor-enhancing properties are generally more evident in vegetables and mixtures of vegetables and meat, its use is not widespread in the meat industry.

The practice of smoking meat originated

as a means of preservation in the days before refrigeration, but the consumption of smoked meats resulted in the development of a taste for the smoked flavor that has persisted to the present time. Smoking of meat products is usually carried out simultaneously with heat processing. The smoke is commonly generated from hardwood sawdust. Hardwood logs may be used in conjunction with the sawdust in open-pit smokehouses, but for most applications smoke is produced from moist sawdust by controlled combustion in a mechanically-controlled smoke generator. Oak and hickory sawdust are most extensively used.

Federal regulations permit no more than 200 ppm of sodium nitrite in any meat product. Only 10% added water is permitted in hams; such water must be declared.

Sausages

The origin of sausages dates back to ancient history; there are specific references going back to 500 B.C. Early sausage manufacture was confined to home or small commercial operations, and this resulted in a number of distinctive varieties associated with the names of the districts or cities where they originated, e.g., Genoa salami, Thuringer sausage.

Sausages may be loosely classified into three general categories: fresh, cooked or smoked, and dry. Some sausages, especially the dry and semi-dry types, depend on bacterial fermentation for production of their characteristic flavors, while in the manufacture of some of the more common sausages such as frankfurters bacteria play no role.

The meat ingredients used chiefly in preparing sausage are those parts of the animal that do not have a ready sale as such. These are classified according to their "water-binding" properties, i.e., their ability to retain moisture during thermal processing. Meats considered to have the best binding properties are skeletal tissue from the beef animal, and include bull meat, shank meat, chucks and boneless cow meat. "Medium" binders are head meat, cheek meat and lean pork trimmings. The meats with "low" binding properties usually contain large proportions of fat or are non-skeletal or smooth muscle; examples are regular pork trimmings, jowls, ham fat, beef briskets, hearts, hanging tenders, weasand meat, giblets and tongue trimmings. Meat tissues classified as "filler meats" and considered to have little or no binding properties include ox lips and tripe, pork stomachs, skin, snouts, lips, and particularly defatted pork tissue; their use in sausage must be severely limited if the quality of the product is to be maintained.

The addition of moisture (added as ice at the time of chopping) is necessary in the preparation of many smoked and fresh sausages, both because the sausages would be dry and unpalatable without it and because ice is needed to prevent chopping temperatures exceeding 16° (60°F.) that would lead to instability of the "emulsion" and promote bacterial growth. The amount of water that may be added is limited by a Federal regulation that requires that the moisture in the final cooked sausage shall not exceed four times the meat protein plus 10%. For fresh sausage that has not been heat-processed, the limit is four times the protein plus 3%.

The common curing agents for sausage are salt, sodium nitrite and/or nitrate and sugar. Three percent of salt on the basis of the meat ingredients is frequently used in sausages. Sodium nitrite and/or nitrate are usually added along with the salt to the beef portion of the sausage emulsion. The proportion of sodium nitrite may not exceed 1/4 oz per 100 lb of meat; if sodium nitrate is used with the nitrite, not more than 2 oz should be employed per 100 lb of meat. Sugar is added to many sausage products at a level of 0.5–1.0%.

Variation in seasoning is one factor that is responsible for the large number of sausage varieties. These seasonings may be added as ground natural spices, extracted oils and oleoresins, or a mixture of the two. Spices that are employed include allspice, black pepper, cardamom, cinnamon, coriander, garlic, mace, nutmeg, paprika and sage.

Filler or binder materials may be added to improve color, provide better binding properties, improve slicing characteristics, change or improve flavor, or reduce costs. Their value depends on their ability to absorb moisture in the "emulsion" and retain it throughout heat processing.

It is obvious from this fact that use of an excessive proportion of fillers would permit selling water at the price of meat, and for this reason Federal regulations limit the proportion of cereals and milk products in the finished sausage to 3.5%. Corn syrup and corn syrup solids (which are considerably less sweet than sucrose) are also frequently added to sausages such as frankfurters. Their proportion is limited to 2%, but this is not counted as part of the total 3.5% of other non-meat materials.

Federal regulations permit the use of 3/4 oz of ascorbic or d-*iso*ascorbic (erythorbic) acid, or 7/8 oz of their sodium salts, in each 100 lb of sausage products. These are rather commonly used in smoked sausage to assure maximum color development and retention. Monosodium glutamate is occasionally employed in pork sausage at a level of 0.1% to enhance flavor.

Sausage casings may be natural casings prepared from some part of the alimentary tract of cattle, sheep or hogs, or they may be cellulosic casings made from a special grade of cotton linters which are solubilized and then regenerated into casings. These latter are frequently used on frankfurters and may be clear or colored.

Many sausages are both smoked and cooked, and some of these may be eaten without further processing.

It is not possible within the limits of this book to attempt to name and characterize all the different varieties of sausages on the market and describe their manufacture. A concise description of two types must suffice: *Fresh pork sausage* is neither cooked nor smoked; it is comprised of ground and seasoned pork intended to be cooked immediately prior to eating. Pork trimmings are put through a grinder, seasoning is added and mixed, and the product is stuffed into natural casings or sold in bulk.

Most seasonings for pork sausage contain 1.75–2.0 lb of salt, 6–8 oz of dextrose, 4–5 oz of pepper, 2–3 oz of sage, and 0.25 oz of ginger, per 100 lb of meat.

Federal regulations require no more than 50% of "trimmable" fat. In Connecticut the *total* fat must not exceed 50 percent.

More *frankfurters* are consumed than any other type of cooked and smoked sausage; they represent 25% of all sausages sold. The meat formulation of ordinary frankfurters is usually around 40–60% beef and 60–40% pork, but all-beef products are on the market. Bull meat, boneless chuck, plates, hearts and trimmings are the most commonly used beef ingredients; pork trimmings, containing various proportions of trimmable fat, back fat and hearts, comprise the major pork ingredients. Filler meats such as tongues, snouts, lips and other by-products may be found in proportions not exceeding 20%; more will result in unsatisfactory products. Dry skim milk, corn syrup solids, and sometimes cereals are employed as fillers to the extent permitted by federal and state regulations. Usually these regulations set limits of 3.5% total fillers (exclusive of corn syrup solids or other sweetening agents, which may be present to the extent of 2%). Meat used for frankfurter-making was formerly pre-cured, but at the present time the curing agents are normally added at the time of chopping.

Salt, sugar and curing salts are added at levels of 3 lb of salt, 1/2 lb of dextrose, 1/4 oz of sodium nitrite, and 2 oz of sodium nitrate, for each 100 lb of meat. The more common spices and seasonings used are pepper, nutmeg, mace, cinnamon, mustard and garlic.

After chopping and mixing, frankfurters are stuffed into casings and linked. They then are held at refrigerated or ambient temperature for varying lengths of time prior to heat processing to permit completion of the cure. Ascorbic acid may be used to obviate the need for this holding period. The sausages are then heated and smoked.

Immediately after smoking, they are cooked with a spray of hot water at a temperature of 77–82° (170–180°F). Some frankfurters are finished simply by raising the smokehouse temperature; such dry-processed products are usually given a brief hot water shower to plump the sausages and provide better peeling characteristics. The frankfurters are then cooled by cold-water showering to an internal temperature slightly above ambient; the remaining heat is usually sufficient to dry the product prior to its placement in the holding cooler at 2–7° (35–45°F). The total processing time may be as short as 65 minutes (when the cure proceeds rapidly) or as long as 2.5 hours.

Colored cellulose casings may be used to add color, or dye may be mixed with the hot shower water, which must then be recirculated; occasionally frankfurters are colored by dipping. Certified coal-tar colors, alkanet, annatto, carotene, cochineal, chlorophyll, saffron and turmeric have been used. FD & C Orange No. 1 and FD & C Red No. 32, which formerly were frequently employed, have been removed from the certified list and their use is no longer legal. Federal regulations require that the coloring matter shall not penetrate into the frankfurter itself.

Frankfurters processed in cellulosic casings may have their casings peeled from them after processing to produce the product sold as the "skinless" variety.

A more extended discussion of sausage manufacture may be found in Chapter 11 of Selected Reference A.

The Proximate Composition of Cured and Processed Meats

Table 10-5[9] shows the average proximate composition and energy values of several cured and processed meats:

TABLE 10-5
The proximate composition and energy value of cured and processed meats

Cured and/or Processed Meat	Protein %	Water %	Fat %	Ash %	Cal/100 g
Bacon	9.1	20	65	4.3	630
Bologna	14.8	62	16	3.3	220
Braunschweiger	15.2	56	24	2.7	280
Corned beef	15.8	54	25	5.0	290
Dried beef	34.3	48	6	11.6	200
Dutch loaf	15.0	58	14	—	190
Frankfurters	15.2	64	14	3.1	200
Head cheese	15.1	62	20	2.3	240
Kolbase	13.5	51	29	—	310
Liver sausage	16.7	59	21	2.2	260
Salami	23.9	31	37	7.0	430
Salami, cooked	17.1	46	27	—	310
Salt pork	3.9	8	85	3.5	780
Country style sausage	16.2	52	27	3.9	310
Polish sausage	16.4	56	23	3.6	270
Pork sausage, fresh	10.8	42	45	2.1	450
Smoked sausage links	15.4	46	32	—	350
Summer sausage	23.5	34	35	6.8	410
Thuringer	17.7	49	24	—	290

Methods of Analysis

Fresh cuts of meat would normally be examined only to determine whether they were edible and of the claimed grade, and whether they came from the declared species of animal, e.g., whether horse meat was being sold as beef. Chemical analyses would be required only in nutritional studies (proximate and mineral analyses, amino acid distribution) and as a means of identifying the species of animal from which the meat came.

The composition of processed meats and meat products is governed by numerous Federal and state regulations that specify how much moisture and fat may be present, what additives (cereals, dry skim milk, sugars, preservatives, color, etc.) may be used and in what amount. Testing for compliance with these regulations requires various chemical analyses. It is this type of analysis with which the food control laboratory is primarily concerned when examining meat products, and the following methods were selected with the needs of such a laboratory in mind:

Visual and Organoleptic Examination

Any attempt to determine the grade of a cut of meat by visual examination is probably futile for anyone but an expert. The color and odor of a sample of whole or ground beef should however be noted. If the fat is yellow, the beef came either from an older animal or from dairy cattle that had been fed rations high in carotenes; very deep yellow fat would raise a suspicion that the meat might be horse meat. If both the fat and muscle portion of ground beef seem to be nearly uniformly red, the addition of blood should be suspected.

Meat from older animals has a stronger odor than that from younger ones of the same species; the odor of meat from mature male animals may be ammoniacal or "staggy." Such odors in raw meat may carry over when the meat is cooked and even be intensified on heating. Odors like these in ground beef frequently result in consumer complaints that the meat is decomposed, and it is necessary in any examination to distinguish between such odors and that of decomposition. Nevertheless either type of odor is probably sufficient to categorize a sample as unfit for food.

Tenderness

Flavor and tenderness are the two qualities that primarily determine the customer acceptance of the meat of different animals and different cuts from the same animal. This fact is inherent in the U.S. grading system outlined on pages 175 and 179. Because there is greater variation in tenderness between different lots of beef than between lots of other meats, most of the studies of physical methods of estimating tenderness have been made on beef.

Soon after death the muscles become very hard, inflexible, contracted and tough; after aging for a few days they again become soft, pliable, relaxed and tender. The tenderness of meat is usually improved by "aging" or "ripening" at 0–1.5° for about 17 days. These changes are related to changes in composition of the proteins and to the ability of these proteins to bind water.[10]

The impression of tenderness is the resultant of at least three factors: 1) The ease with which the teeth sink into meat when chewing begins; 2) the friability or mealiness, i.e., the ease with which the meat breaks into fragments); and 3) the amount of residue that remains after chewing.

Those muscles containing the least connective tissue are the most tender. During cooking the muscle fibers become tougher, while the connective tissue increases in tenderness. This is the reason why cuts containing relatively large proportions of connective tissue (such as the round) are commonly subjected to a long heating period and a moist atmosphere, while cuts such as rib and loin steaks are cooked for a short time with dry heat, i.e., broiled.[11]

Some work has been done on chemical methods of evaluating the quality of beef by determining the rate of digestion by proteolytic enzymes such as pepsin; for this the references cited in Text Reference 10 should be consulted. It has been more common to depend on taste tests by panels, or on the use of mechanical devices that mimic the physical act of chewing, to compare the relative tenderness of different lots of meat.

The various instruments measure either the force (load) required to cut or shear through a piece of meat, or the amount of work done in cutting or shearing. A review of these instruments with their advantages and disadvantages can be found in an article by H. W. Schultz in *Proceedings of the 10th Annual Reciprocal Meat Conference, National Live Stock and Meat Board*, p. 17 (1957). For tests with such instruments to correlate with taste panel tests, the samples must be of uniform size (with the same orientation of muscle fibers in some cases), the testing must be done at a single temperature (preferably 0–7°), and several replicate tests must be run on each sample. Results obtained on raw beef are of little value as an indication of the tenderness of the meat when cooked.[11]

Method 10-1. *Preparation of Sample for Chemical Analysis.*

Nearly all meat products occur in forms that are so far from homogeneous that the chief obstacle to be surmounted in obtaining duplicate analyses that check is the preparation of subsamples that are both representative and of uniform composition. The following method of Erwin J. Benne *et al*[12] is recommended:

Use a small motor-driven grinder equipped with a fine-mesh cutting plate to which is attached a dividing attachment that splits the material coming from the grinder into two equal portions (Figure 10-5).

Pass the meat through the grinder, collect one portion, and discard the other. If the collected portion does not appear to be homogeneous, pass it through the grinder again, splitting as before. Repeat as often as necessary. (When working with a type of product not previously analyzed, it may be necessary to determine by test analyses just how many passes are required to obtain homogeneity in the final sample.)

Completely fill a small jar or bottle with the prepared sample, close with a tightly-fitting screw cap, and hold at a temperature slightly above freezing until the contents are analyzed. (If long periods of storage are necessary, freezing the meat may be advisable to prevent spoilage.)

Method 10-2. *Moisture.*[12]

Use aluminum moisture dishes approximately 55 mm in diameter and 25 mm deep. Put the dishes in an oven at 100–105° for at least 2 hours, cool in a desiccator, and weigh as soon as they reach room temperature. Place 3-5 g of prepared sample in each dish, close the lid, and weigh immediately. Subtract the tare to obtain the weight of sample taken. When all samples have been weighed, place the dishes in the oven with lids slightly ajar. Heat samples of fat meat 5 hours, and samples of lean meat 24 hours. At end of heating period remove dishes from oven, close the lids, and cool in a desiccator. Weigh as soon as the dishes reach room temperature, and calculate percent moisture from oss in weight.

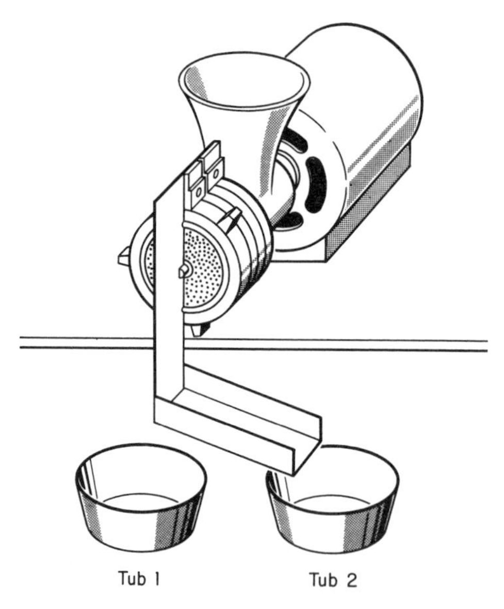

FIG. 10-5. Meat grinder with divider.

Method 10-3. Ash.[12]

Place portions of the prepared samples (3–6 g lean meat or 5–7 g fat meat) in ignited and weighed porcelain crucibles (large enough to contain the increased volume caused by the swelling and frothing that some fresh meats undergo in the initial stages of ashing). Reweigh and subtract the tare to obtain the weight of sample taken. Heat the crucibles in an oven until all moisture is expelled. (A vacuum oven at approximately 75° is convenient, but an ordinary hot air oven at 100–105° will suffice. In a hot air oven at 100–105°, five hours heating for fat and 24 hours for lean samples is enough.)

When the samples are dry, remove them from the oven. (They may be cooled in a desiccator and weighed to determine moisture at this point if desired.) Place the crucibles in a cool electric muffle furnace, raise the temperature gradually to 550°, and heat until ashing is complete, usually overnight. At the beginning of the heating period watch the samples carefully to be sure that losses from the crucibles due to swelling and frothing do not occur. Where frothing is excessive and the contents appear likely to run over, remove the crucibles from the furnace with tongs, permit the frothing to subside, and then return the crucibles to the furnace and continue the gradual heating. (This process must be repeated as often as necessary to insure against loss until the samples are completely charred and frothing ceases. When working with tissues that are known to froth excessively, such as liver and spleen, it may be most convenient to ash them partially over a suitable burner in a well-ventilated hood before placing them in the furnace.)

After ashing is completed in the muffle furnace, remove the crucibles, cool them to room temperature in a good desiccator, reweigh, and calculate the percent of ash.

The color of ash from different tissues varies widely, and may be white, gray, brown, pink or even blue. Hence, whiteness of ash cannot always be relied on as the criterion of complete ashing. However, it is usually possible to tell whether unburned carbon is present. If it is, the crucibles should be cooled, a little water added, the contents dried on a steam bath or in an oven or both, and the crucibles reheated in the muffle furnace.

Method 10-4. Salt (Sodium Chloride)

Questions are occasionally presented as to whether samples of ground beef sold in markets catering to persons who observe the Jewish dietary laws are in fact *Kosher*. Because these laws require treating the meat with salt, the salt content of *Kosher* hamburg is always higher than that of ordinary ground meat, which contains naturally present chloride equivalent to only 0.25% of sodium chloride on the dry basis. Analyses of some 19 samples of, allegedly authentic *Kosher* hamburg on the other hand showed from 1.07 to 4.00 percent, and an average of 2.07%, of salt (also on the dry basis).[13]

Salt may be determined in hamburg by the following AOAC method:[14] Moisten a 1.5–3 g sample in a 300 ml flask with an excess of 0.5*N* $AgNO_3$ (5 ml or more, depending on the NaCl content of the sample). Add 15 ml of HNO_3 and boil until the meat dissolves (10 minutes is usually sufficient). Add concentrated $KMnO_4$ solution in small portions, boiling after each addition until the permanganate color disappears and the solution becomes colorless or nearly so. Add 25 ml of water, boil five minutes, cool, dilute to about 150 ml, add 25 ml of ether, and shake.

Transfer the aqueous layer to an Erlenmeyer flask, add 5 ml of saturated ferric ammonium sulfate, and titrate with standard 0.1 *N* ammonium thiocyanate to a permanent light brown color. Titrate in similar fashion the amount of 0.5*N* $AgNO_3$ used to precipitate the sample. The difference between these two titrations is equivalent to the amount of salt in the sample. Each ml of 0.1*N* $AgNO_3$ — 0.00584 g NaCl.

Method 10-5. Fat—Method I.[12]

Prepare a thin mat of asbestos in a Gooch crucible. Place the crucible in the top of a 30 ml tall-form beaker, marked to match the crucible, which contains a small wad of cotton previously extracted with ether. Dry the crucible and beaker 5 hours at 100–105° in a hot-air oven, cool in a desiccator, and weigh. With a spatula add 4–7 g of prepared sample to the crucible, and spread it as evenly over the bottom as possible. Remove the cotton from the beaker and spread it over the sample in the crucible. Place the crucible in the top of the beaker, and reweigh to obtain the weight of sample by difference. With the crucible still in the top of the beaker, dry in a hot-air oven for approximately 24 hours, remove from oven, cool in a desiccator, and reweigh. If desired, the percent of moisture may be calculated at this point from the loss in weight.

Place the crucible in a Bailey-Walker extraction flask containing approximately 50 ml of anhydrous ether. Connect the flask to a Bailey-Walker extractor and extract for 16 hours. If any fat melted and ran into the beaker during the drying period, remove it by dissolving the fat in ether and wiping the beaker with a dry cloth.

At the end of the 16-hour extraction period, remove the flask from the extractor, discard the used ether, and by use of a suction flask and water aspirator remove any ether that remains in the Gooch crucible. Set the crucible in a warm place away from flames or sparks until the odor of ether can no longer be detected in the sample; then place in the original beaker and dry for 2–3 hours in a hot air oven. Remove the crucible and beaker from the oven, cool in a desiccator, and reweigh. Calculate the loss in weight due to extraction as percent fat.

The arrangement of the crucible, beaker, and cotton before and after the sample is placed in the crucible, is shown in Fig. 10-6:

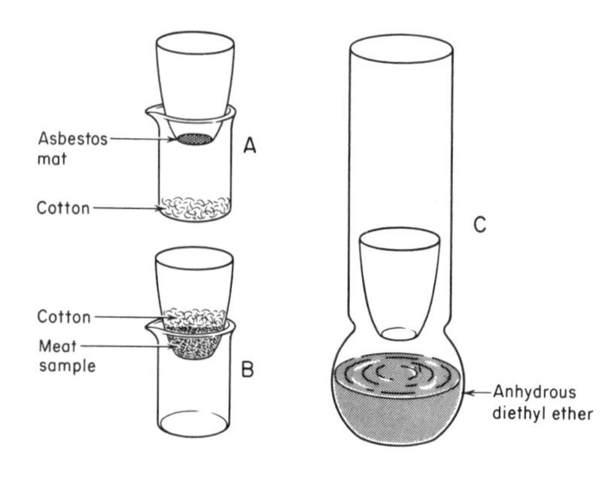

FIG. 10-6. Arrangement of Gooch crucible and beaker for determining fat. A) Crucible and beaker dried and ready to weigh for tare; B) Meat and cotton in place ready to weigh to obtain weight of sample; C) Crucible in place for extraction.

When some kinds of meat dry, they contract into hard compact masses which are not readily penetrated by diethyl ether. This occurrence is evident in poor duplication of results. The authors attempted to insure penetration of ether into the mass of meat by adding extra asbestos or clean sea sand to the Gooch crucible before it was dried and weighed. This was then mixed with the fresh sample so that it would be more easily penetrated when dried. In cases of extreme difficulty in this regard, it may be necessary to dry the sample, grind it finely, and use portions of the ground material for the fat determinations. When this is done it is necessary to evaluate the moisture lost during the drying process so that the result can be calculated to the original moisture basis.

Method 10-6. *Fat—Method II. (Wickroski Method).*[15]

Weigh a two to five gram sample into a beaker, mix with a teaspoonful of Celite 545, and dry in an oven at 100° for 5–6 hours. Brush the contents of the beaker into a mortar, and thoroughly grind together. Wrap in an 18.5 cm No. 2 Whatman filter paper, and transfer to the stainless steel container of a "Goldfisch" fat extraction apparatus. Extract with anhydrous ether overnight at low heat. The next morning, evaporate off the ether and dry the fat one-half to three-quarters of an hour at 100:, cool and weigh.

Method 10-7. *Fat—Method III (Field Test).*

APPARATUS

Hobart Model F-100 Fat Percentage Measuring Kit, obtainable from The Hobart Manufacturing Co., Troy, Ohio (Figure 10-7). This consists of an infrared heater, 2 funnels, 3 test tubes, 2 screens, a reading device, cleaning brushes and a cord and plug, mounted in a carrying case.

DETERMINATION

A 2 oz sample of ground beef is weighed onto a screen lying on top of one of the funnels; the tip of the funnel projects into one of the test tubes. The heater (which is fastened above the funnel) is turned on; after 15 minutes it automatically switches off, and a red light comes on and a bell rings to signal the end of the heating period. The fat which has tried out into the test tube during this period is then read on a scale calibrated directly in percentage. The apparatus is designed

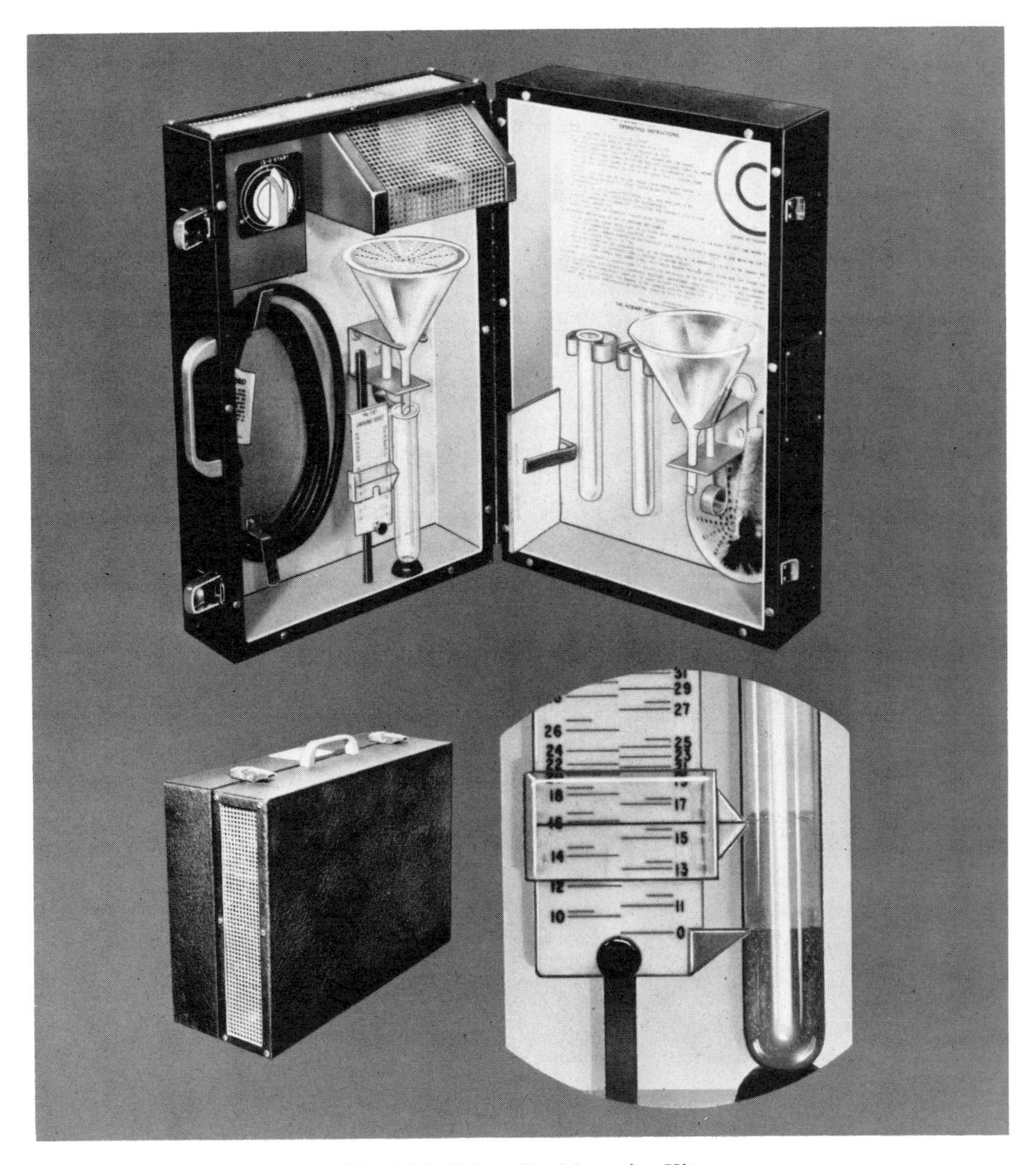

FIG. 10-7. Hobart Fat Measuring Kit.

to determine fat percentages between 10 and 40.

When the machine is operated in a room in which the temperature is below 65°F, the case must be closed during the heating period.

NOTES

1. This kit is intended to be used for determining fat in ground beef only. Under the best conditions results are accurate to one percentage point.

2. A similar apparatus (the "dunn-right Analyzer") is manufactured by dunn-right Inc., P.O. Box 26, Mansfield, Mass.

Method 10-8 *Protein.*[12]

Weigh a piece of vegetable parchment paper approximately 2″ square, and record the weight as a tare. With a spatula place 1-1.5 g of the prepared sample on the paper and weigh immediately.

Obtain the weight of sample taken by subtracting the tare.

Roll the paper around the meat and drop it into a Kjeldahl flask. Add catalyst, sodium or potassium sulfate, and sulfuric acid, and proceed with the determination of protein in the usual manner (*see* Chapter 1.) Make appropriate variations in the volumes of standard acid to receive the distillates from fat, lean, and mixed fat and lean samples, respectively. Multiply the percent of nitrogen determined by 6.25 to convert it to protein.

Method 10-9. *Added Water.*

Federal and some state regulations state with regard to sausage products: "For the purpose of facilitating grinding, chopping and mixing, not more than 3% of water or ice may be added to sausage which is not cooked and to luncheon meat; sausage of the type which is cooked, such as Frankfort style, Vienna style and Bologna style, may contain not more than 10% of added water or moisture to make the product palatable." The standard fomula for calculating added water for the purpose of enforcing this regulation assumes that meat naturally contains four times as much water as protein, and that therefore, if W = percent water and P = percent meat protein, in the sample, percent added water = $W - 4P$. It must be emphasized that for the purpose of making this calculation the total protein must be corrected for any protein contributed by additives such as dry skim milk and soy flour.

A special Federal regulation covering smoked hams and similar pork products (Agricultural Research Service, USDA, *CFR Amendment* 62–44, June 22, 1962) permits the addition of up to 10% of added water if the product is labeled "Water Added". The determination of the percent of added water is based as in the case of sausages on the ratio of water to protein in the fresh meat, but there is not complete agreement on what this ratio is.*

*Analyses in 1959–1966 by Perrin and Long[15a] of 25 authentic samples of boned and skinned raw ham showed an average moisture/protein ratio of 3.515.

Lactose

Federal and some state regulations permit the presence of a total of not over 3.5% of certain optional added ingredients, including "cereal, vegetable starch, starchy vegetable flour, soya flour, dried milk or dried skim milk". In addition, up to 2% of dextrose or corn syrup solids are allowed. The only satisfactory way of estimating dry skim milk in sausage is by determining the percent of lactose and multiplying by the appropriate factor (1.94–2.00). For distinguishing between lactose and other sugars present, the AOAC Yeast Adsorption Method **23.029** using washed yeast suspension **23.027(b)** suffices in many cases, but gives high results when maltose-containing ingredients such as corn syrup solids are present. The following method[16] has proved satisfactory in our hands both in the presence and absence of maltose:

Method 10-10. *Lactose.*

Reagents

a. *Yeast suspension.* Suspend three packages of commercial dried yeast (or equivalent in moist yeast cake) in about 100 ml of water. Collect the cells by centrifugation and wash twice with water. After final centrifugation, suspend the cells in 150 ml of water. Freeze in 20 ml portions. The cells may be kept frozen for long periods of time.

b. *Acetate buffer pH* 5.6. Prepare solutions containing 2.4 parts of 1*M* acetic acid (57.75 ml glacial acetic acid/L) to 22.6 parts 1*M* sodium acetate (82 g anhydrous sodium acetate/L).

c. *Antifoam.* Any water-soluble antifoam known not to interfere with the test may be used. (Dow-Corning Silicone AF antifoam has been satisfactory.)

d. *Enzymes.* Glucose oxidase, 15,000 oxidase units/g, and maltase, standardized at 600 *p*-nitrophenyl glucoside units. (Both may be obtained from Nutritional Biochemicals Corp., Cleveland, Ohio.)

e. *Enzyme suspension.* Triturate together in water enough glucose oxidase and maltase for samples to be run on the same day. Use 25 mg of glucose oxidase and 50 mg of maltase for not more than 50 mg of expected corn syrup solids in a 50 ml aliquot of meat solution.

A convenient set-up for 5 samples is to triturate 125 mg of glucose oxidase and 250 mg of maltase in water, transfer to a 25 ml volumetric flask, add a trace of antifoam, dilute to volume, and shake well. Use 5 ml of this suspension for each 50 ml of meat solution.

Determination

Preparation of meat sample. Place a 20 g sample in a 200 ml volumetric sugar flask, add a small amount of water, and break up by agitation. Add about 100 ml of water and warm on a steam bath 30 minutes, shaking frequently. Cool to room temperature. Add 4 ml of HCl and dilute to volume, using the bottom of the fat layer as the meniscus. Add 10 ml of 20% phosphotungstic acid solution, mix well, and let stand for a few minutes. Filter through a moist filter paper. (Centrifugation before filtration is recommended but not indispensable.) Pipet 160 ml of the filtrate into 200 ml volumetric flask and neutralize to pH4.8–5.0 (chlorophenol red indicator). Dilute to volume and mix.

Preliminary yeast treatment. Centrifuge 20 ml of thawed yeast suspension in a 100 ml centrifuge tube, and decant the water. To the packed cells, add about 60 ml of meat solution, and mix thoroughly. Incubate at 30° for 30 minutes, with air bubbling through the mixture. Centrifuge to remove cells. Use 50 ml of the supernatant solution for enzyme treatment. (This preliminary yeast treatment may be omitted if sucrose is known to be absent from the sample.)

Enzyme treatment. Add the following to a 200 ml volumetric flask: 4 ml of pH 5.6 acetate buffer, 50 ml of yeast-treated meat solution (or 50 ml of original meat solution), 1–2 ml of 1% water suspension of antifoam, and 5 ml of enzyme suspension containing 25 mg of glucose oxidase and 50 mg of maltase. (Total volume should not be more than 70 ml nor less than 50 ml in order to keep the buffer concentration within the optimum range for enzymes.)

Place the flask in a water bath so controlled that the contents are kept at 39 ± 2°. Bubble air rapidly through the mixture for the entire 3-hour incubation. If foaming is excessive, add additional antifoam.

After incubation, heat the contents of flask just to boiling, remove from the heat, and add 95% alcohol *slowly and carefully* with mixing, bringing nearly to volume. Cool to room temperature and adjust the volume with water. Shake the flask well, and filter the contents through a Whatman No. 2 filter paper. Transfer 100 ml of the filtrate (or any desired aliquot) to a beaker and evaporate on a steam bath to about 20 ml or until the alcohol aroma disappears. Do not permit to go to dryness. After evaporation, adjust the volume to 50 ml.

Reducing sugar determination. Determine the reducing sugars on the evaporated material (diluted to 50 ml) by the Munson-Walker method (Method **16-10**), or any other applicable method.

Method 10-11. *Lactose and Dextrose in Frankfurters free of Maltose.*[17]

Before the above enzyme method for lactose was devised by Hankin and Wickroski, a method depending on the adsorption of dextrose by yeast was employed by the Connecticut Agricultural Experiment Station for the separation and determination of lactose in frankfurters. As noted previously, this method does not distinguish between lactose and maltose, and consequently cannot be used to estimate dry skim milk in the presence of corn syrup solids. It does, however, serve to determine the total dextrose present and it also affords a check on the lactose content of those frankfurters to which no maltose-containing products have been added:

Determination

Weigh 12.5 g of the sample into a 250 ml beaker, and add 100 ml of water; mix thoroughly and boil for 5 minutes. Pour off the extract through a paper pulp mat in a Büchner funnel, using suction. Again boil the residue of the meat in a beaker with 50 ml of water and pour off as before. Wash the residue twice with approximately 40 ml portions of boiling water. Combine the extracts in a 250 ml flask, add 5 ml of 20% phosphotungstic acid and cool. Add 2 ml of HCl, dilute to the mark, mix, and filter through a dry paper. Neutralize a 200 ml aliquot of the filtrate with NaOH solution and dilute to 250 ml (Solution A). Use a 50 ml aliquot (2 g of meat) for the determination of total reducing sugars (Method **16-7**).

To remove *dextrose*, place 10 ml of a 25%

suspension of washed Fleischmann's yeast in a 100 ml tube and centrifuge. Pour off the water and dry the walls of the tube with filter paper. Add about 60 ml of Solution A to the yeast in the tube, mix thoroughly, let stand for 15 minutes, stirring frequently enough to keep the yeast in suspension. (Our determinations stood 1 hour but it appears that 15 minutes is sufficient.) Again centrifuge, pour off the supernatant liquid through a small, dry filter and determine copper reduction on a 50 ml aliquot.

The difference between the two reductions is due to adsorbable sugar (dextrose); the reduction after yeast treatment is due to lactose.

Where this method indicates the presence of only a fraction of a percent of lactose, it is well to confirm the presence of lactose (and/or maltose) in the sausage by the following qualitative test:

Method 10-12. *Qualitative Test for Lactose.* (*AOAC Method* ***23.026***).

To 10 g of comminuted sample in a small beaker add 20 ml of hot (70–90°) water. Mix thoroughly and filter. Transfer 4 ml of the filtrate to a test-tube, add 3–4 drops of 5% methylamine hydrochloride solution, and boil 30 seconds. Remove from the flame, add 3–5 drops of 20% NaOH solution and shake 10 seconds. If lactose is present, the solution turns yellow immediately, and then slowly changes to carmine.

Method 10-13. *Starch—Method I.*[18]

Treat in a 250 ml beaker 10 g of finely divided sample with 75 ml of 8 percent solution of KOH in alcohol and heat on a steam bath until the meat is all dissolved (30–45 minutes). Add an equal volume of alcohol, cool, and allow to stand for at least one hour. Filter by suction through a thin layer of asbestos in a Gooch crucible. Wash twice with a warm 4 percent solution of KOH in 50 percent (by volume) alcohol, and then twice with warm 50 percent alcohol. Discard the washings. Retain as much of the precipitate as possible in the beaker until the last washing. Place the crucible with its contents in the original beaker and add 40 ml of water and 25 ml of H_2SO_4. Stir during addition of the acid and make sure that it comes into contact with all of the precipitate. Allow to stand about 5 minutes, add 40 ml of water, and heat just to boiling, stirring constantly. Transfer the solution to a 250 ml volumetric flask, add 2 ml of 20 percent phosphotungstic acid, let cool to room tempeature, and make to the mark with water. Filter through starch-free paper, pipet 100 ml of filtrate into a 200 ml volumetric flask, neutralize with NaOH (1+1), make to volume, and determine the dextrose present in a 50 ml portion of the filtrate by Method **16-8(a)**. Weight of dextrose × 0.9 = weight of starch.

Method 10-14. *Starch—Method II.*[19]

REAGENTS

a. *Anthrone—sulfuric acid solution.* Dissolve 0.5 g of anthrone in 250 ml of H_2SO_4. (This reagent is good for 3–4 days if kept near 0°. If kept longer, a high blank results.)

b. *Dextrose stock solution*—Dissolve 0.100 g of anhydrous dextrose in 100 ml of water.

c. *Perchloric acid*—52%. Add 270 ml of 71% $HClO_4$ to 100 ml of water, and store in a glass-stoppered bottle.

PREPARATION OF SAMPLE

a. *Sugar and fat extraction*—Weigh 2.0 g of comminuted meat into a 100 ml conical centrifuge tube. Add 25.0 ml of alcohol-petroleum ether (1+3), stopper, shake vigorously, and centrifuge 5 minutes at 2500 r.p.m. Decant and discard the solution. Add 10 ml of hot (80°) alcohol, shake, and centrifuge 5 minutes at 2500 r.p.m. Discard this solution and repeat the extraction with alcohol at 80°.

b. *Starch extraction*—Add 5.0 ml of water to the residue, and stir. Add 6.5 ml of 52% $HClO_4$; stir or shake 5 minutes, and let stand 15 minutes. Add 20.0 ml of water and centrifuge 5 minutes. Pour the starch solution into a 100 ml volumetric flask. Add 5.0 ml of water to the residue in the tube, stir, add 6.5 ml of 52% $HClO_4$, and let stand 30 minutes. Stir and wash the entire contents of the tube into the volumetric flask. Make to volume and filter through a Whatman No. 12 filter paper.

DETERMINATION

Pipet 5.0 ml of the filtered starch solution into a 200 ml volumetric flask, and dilute to volume with water. Pipet 5 ml of the diluted solution into a cuvette tube, cool in a water-bath, and add 10 ml of the anthrone reagent, (**a**). Mix thoroughly, and heat 7.5 minutes at 100°. Remove the tube from the bath, cool rapidly to 25° in a water bath, and read the absorbance at 630 mμ. (The color is stable 30 minutes.) Calculate the micrograms of dextrose from the standard curve.

Since the 5 ml of the final dilution reacted with anthrone corresponds to 2,500 μg of sample, if $A = \mu g \times$ dextrose found from the curve,

$$\text{percent dextrose} = \frac{100\,A}{2{,}500} = 0.04\,A.$$

Then percent starch = percent dextrose × 1.06*, and percent cereal = starch × 1.25 = dextrose × 1.32.

Preparation of standard curve—Add 1.0, 2.0 and 3.0 ml of dextrose stock solution, (**b**), to separate 200 ml volumetric flasks and make to volume with water. Also dilute 2.0 ml of the stock solution to volume in a 100 ml volumetric flask. Treat 5 ml of each solution (equivalent to 25, 50, 75 and 100 micrograms of dextrose) as under "Determination" and plot a curve from the four readings.

Method 10-15. Nitrate and Nitrite.

Regulations permit a maximum of 200 parts per million of sodium nitrite in finished sausage products. The following method[20] for nitrate and nitrite appears to be the most reliable:

REAGENTS AND APPARATUS

a. *Cadmium column*—Place zinc rods in a 20% solution of $CdSO_4$. Remove the deposit of Cd after 3–4 hours and place in a flask. Cover with dilute HCl and disintegrate 1–2 seconds in a homogenizer. Wash with water. To the column (Figure 10-8), add a 1″ plug of glass wool, a 3.5 cm layer of washed sea sand, and a 3.5 cm layer of the spongy Cd. Wash the column with 0.1N HCl, water, and dilute buffer solution [reagent (**c**) diluted with 9 volumes of water].

*This factor is an experimental one derived by adding a known quantity of soluble starch to frankfurters (letter of Willie Glover, December 22, 1966). Since the theoretical factor is 0.90, and it is uncertain whether the starch employed was anhydrous, it would be advisable for persons using this method to check this recovery factor on mixtures containing accurately known percentages of starch.

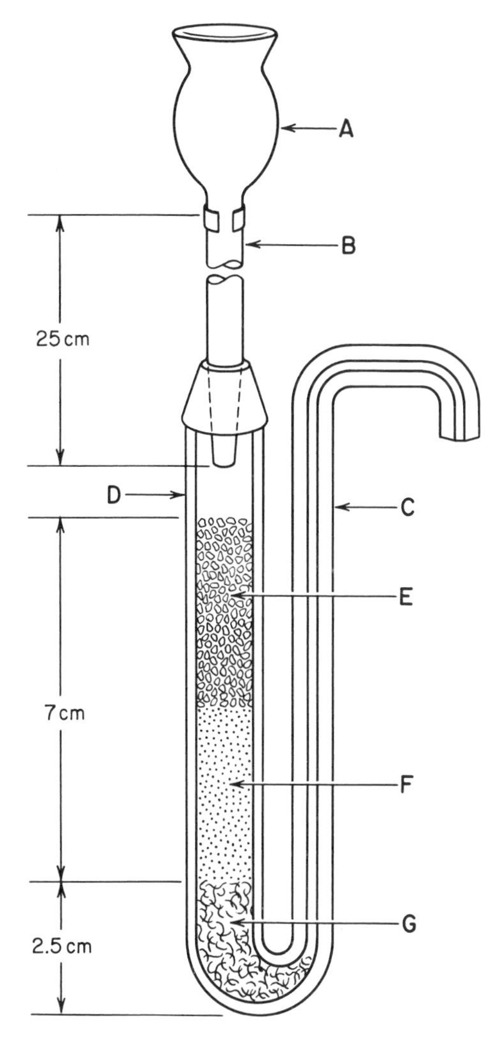

FIG. 10-8. Apparatus for nitrate reduction. A) 25 ml reservoir; B) Inlet capillary, 0.4 mm i.d., 5–6 mm o.d.; C) Outlet capillary, 2 mm i.d., 7–8 mm o.d.; D) Column, 1.2 cm i.d., 1.4–1.5 cm o.d.; E) Spongy cadmium, depth 3.5 cm; F) Sea sand, depth 3.5 cm; and G) Glasswool plug, depth1″.

b. *Saturated solution of potassium aluminum sulfate.*

c. *Buffer solution, pH* 9.6–9.7—To 500 ml of water add 20 ml HCl, mix, add 50 ml of NH_4OH (sp. g. 0.880), and make to one liter.

d. *Orange I reagent*—Warm 360 ml of water and 50 ml of acetic acid to 50° and pour into a 600 ml dark glass reagent bottle containing 0.25 g of powdered sulfanilic acid. Shake until dissolved, then add 0.20 g of 1-naphthol and dissolve by shaking. Cool to room temperature, and add 90 ml of dilute ammonia (10% NH_3). The pH of this reagent should be 4.0 ± 0.05.

e. *Standard sodium nitrite solution*—Prepare a solution containing 0.25 g of $NaNO_2$ in 500 ml.

Preparation of Standard Nitrite Curve

Take suitable aliquots (1–40 ml) of the standard $NaNO_2$ solution, (**e**), place in a series of 200 ml volumetric flasks with 5 ml of reagent (**c**), and make to the mark. Dilute 20 ml of each solution to 100 ml with water. To 5 ml of each dilute solution in a 20 ml test tube, add 10 ml of reagent (**d**) and 5 ml of water. Place in a water-bath at 25–30° for 30 minutes in the light to develop the color. Measure the absorbance in a 1 cm cuvette at 474 mμ. Plot the values against the concentrations of nitrite. These concentrations, covering the range 2.5–100μg of $NaNO_2$ in the final solution obtained for colorimetric measurement, can be conveniently expressed in terms of the corresponding concentrations, 50–2,000 p.p.m. of $NaNO_2$, in a 10 g sample of meat.

Preparation of Meat Extract

Macerate 10 g of meat with 70 ml of hot (70°) water and 5 ml of reagent (**c**) in a Waring Blendor or similar homogenizer. Add Celite 545, filter the mixture into a flask, and wash the residue with hot water. Raise the temperature of the filtrate to 80–90°, and adjust the pH to 5.5–6.5 with reagent (**b**), using a pH meter. Cool, filter into a 200 ml volumetric flask, make to volume with water, and filter through a No. 42 Whatman paper. The final pH should be between 5.5 and 6.5.

Determinations

a. *Determination of nitrite*—To 20 ml of the protein-free filtrate in a 100 ml volumetric flask add 5 ml of reagent (**c**), and dilute to the mark with water. Pipet 5 ml of the diluted solution into a 20 ml test tube, and add 10 ml of reagent (**d**) and 5 ml of water. Develop the color as under "Preparation of standard curve", and determine the parts per million of $NaNO_2$ in the meat from the curve.

b. *Determination of nitrate*—To 20 ml of the protein-free filtrate add 5 ml of reagent (**c**), pour the mixture into the reservoir of the Cd column, and allow to pass through the column, rejecting the first 10 ml. of eluate. Wash the column with water, collect 100 ml of eluate in a graduated cylinder, and shake to mix. Transfer a 5 ml aliquot of the mixture to a 20 ml test tube, add 10 ml of reagent (**d**) and 5 ml of water, and proceed as under "Determination of nitrite" for development of the Orange 1 color. Read off total nitrite from the standard curve. After subtraction of the value for nitrite obtained in the separate nitrite determination, the factor 1.231 is used to calculate p.p.m. of $NaNO_3$ in the meat, while the factor 1.465 gives p.p.m. of KNO_3.

c. *The meat blank*—The blank contributed by meat and reagents is insignificant in determining sodium nitrite, but meat contains substances which after reduction of the extract in the Cd column contribute color equivalent to up to 30 p.p.m. of $NaNO_2$. A correction for this blank may be made as follows:

To 30 ml of the protein-free meat extract add two drops of HCl and pass through a column of Deacidite FF (that has been treated with saturated NaCl and washed free of chloride with water). Wash the column with water and collect 30 ml of eluate. To this add two drops of NH_4OH, and shake well. Treat 20 ml of this solution with 5 ml of reagent (**c**), pass through the Cd column, (**a**), and proceed as under "Determination of nitrate."

Method 10-16. Sulfite.

The use of sulfites for the preservation of meat products is forbidden because, while the sulfites are not particularly good as true preservatives, they do redden meat and make it look fresh, and the odor of SO_2 covers that of putrefaction; the sulfites also have the undesirable property of destroying thiamine. Because no sulfite at all is permitted, it is usually unnecessary to run a quantitative determination on meat products; if such a determination is desired, reference should be made to Chapter 14.

The following qualitative test is simple and suitable for field use:

To a half-teaspoonful of hamburg add 8 drops of 0.02% malachite green solution, and mix vigorously 1–2 minutes with a spatula, Meat containing sulfite decolorizes the dye; normal meat turns bluish-green.

(This test has been said to work even with putrid meat, but it is nevertheless inadvisable to rely on it as a test for sulfite in hamburg that is obviously unfit for consumption.)

Method 10-17. *Benzoic and Sorbic Acids.*[21]

Transfer 50–200 g of ground and mixed sample to a 500 ml volumetric flask; add enough water to make a total volume of about 400 ml, and shake. Add 2–5 g of $CaCl_2$ and shake again. Make distinctly alkaline to litmus paper with 10% NaOH solution, dilute to the mark with water, shake thoroughly, let stand at least 2 hours, shaking frequently, and filter.

To 50 ml of filtrate add 5 ml of HCl (1+3) and extract with ether. If the mixture emulsifies, add 10–15 ml of petroleum ether (b.p. below 60°) and shake. If this treatment fails to break the emulsion, centrifuge. Wash the ether layer with two 5 ml portions of water, evaporate the greater portion of the ether in a porcelain dish on the steam bath, transfer the remainder to a small porcelain crucible, and allow it to evaporate spontaneously. Cut a hole in an asbestos board large enough to admit two-thirds of the crucible, cover the crucible with a small round-bottomed flask filled with cold water, and heat over a small flame until any benzoic and/or sorbic acid present sublimes and condenses upon the bottom of the flask. Run an infrared curve on the sublimate and compare with curves of authentic benzoic and sorbic acids.

Antioxidants

See Chapter 14

Ascorbic and Erythorbic Acids

As was noted previously on page 185, ascorbic acid is employed in smoked sausage to assure maximum color development in a minimum time. Federal regulations permit the use of 3/4 oz of the acid or 7/8 oz of the sodium salt per 100 pounds for this purpose. Its derivative d-*iso*ascorbic acid ("erythorbic acid") and the sodium salt of this acid are frequently substituted for ascorbic acid (and its sodium salt) because they are cheaper and serve the same color-developing and preserving functions; they have no vitamin properties, but this is of no concern in sausage manufacture. These compounds are sometimes added to fresh sausage to promote longer shelf-life, but this practice is frowned on by authorities because regulations forbid the addition of preservatives to fresh meat. Further, the addition of ascorbic acid alone does not appear to be effective in prolonging shelf-life beyond a day or so. However, Zipser *et al.*[22] have recently shown that when twice the legal limit of sodium ascorbate (0.108%) is combined with 0.5% of sodium tripolyphosphate ($Na_5P_3O_{10}$), frozen cured cooked pork is protected from oxidative rancidity for several months. The authors have no knowledge that this combination has been used commercially on pork sausage.

The method of Schmall *et al.*, *Anal. Chem.* **25**: 1486 (1953), **26**: 1521 (1954), is recommended for the determination of either ascorbic or erythorbic acid in meat, because this method is reasonably rapid and not subject to the interferences encountered with other methods. It does not distinguish between the two substances, but this is not usually necessary in meat analysis. Paper chromatographic procedures for the separation of ascorbic and erythorbic acids are available.[23]

Other Vitamins

See Chapter 20

Color

See Chapter 18

Soy Flour

The detection of soy flour in sausages and similar meat products, and its determination when present, are necessary both to ensure that its proportion plus that of other additives does not exceed 3.5%, and to make the correction in the total protein that is required before added water can be calculated. The following qualitative test[24] is specific when the "hour-glass" cells can be identified, but if the formula for calculating added water appears to yield anomalous results it may be desirable to run the quantitative method even when these cells were not detected.

Method 10-18. *Qualitative Test. AOAC Method* **23.021.**

Mix 10 g of finely-divided sample in a 250 ml beaker with 75 ml of 8% alcohol KOH solution, and heat on a steam bath until all the meat is dissolved (30–45 minutes). Transfer the liquid and residue to a 100 ml graduated sedimentation tube, dilute to 100 ml with alcohol, and let settle. Decant the supernatant liquid as completely as possible, and cover the residue with about 50 ml of warm water. Stopper the tube and shake vigorously; let stand a few minutes until foam subsides; then transfer to a 50 ml centrifuge tube, and centrifuge. Pour off and discard the supernatant liquid, and add 10 ml of HCl to the centrifuge tube. Stopper and shake, or mix contents thoroughly with a glass rod. Add about 15 ml of 25% alcohol, mix, and centrifuge. Pour off the superantant liquid, and examine the residue under a microscope for the characteristic "hour-glass" or I-shaped cells (sometimes called "bearer cells"), preferably with polarized light.

NOTE: It was reported to us that soy products intended for addition to sausage were required to contain a TiO_2 marker. If this were always true, identification of the marker by emission spectrography or otherwise would be relatively simple. However, samples submitted recently to the Connecticut Agricultural Experiment Station by inspectors, that apparently contained a soy product, were found to be free of titanium.

Method 10-19. *Quantitative Method.*[25]

Weigh 10 g of the finely chopped meat into a 100 ml centrifuge tube and add gradually, with constant stirring, 60 ml of hot water. Suspend the tube in a water bath, kept just below the boiling point, for 15–20 minutes with occasional stirring. Wash off the rod with 2–3 ml of hot water. Centrifuge the tube 5 minutes, return to the bath and add a slurry of 1 g of filtercel in 5 ml of hot water. Stir thoroughly without disturbing the meat and again centrifuge 10–15 minutes. Pour off and discard the supernatant liquid, which will probably be cloudy, especially in the presence of dried skim milk. Add 50 ml of 8% KOH in 95% alcohol to the tube and return to the water bath, stirring until the alcohol boils. Reduce the temperature of the bath to about 80° and continue heating with occasional stirring for 10 minutes after the particles of meat have disappeared (about 20 minutes in the water bath is necessary). Filter the alkaline solution through a Gooch crucible having a thin pad of asbestos covered with filtercel, using gentle suction and keeping the mixture well stirred during filtration. Wash the tube and crucible with four 25 ml portions of 95% alcohol. Remove the excess alcohol with strong suction. Invert the crucible over a large-stem funnel in a 200 ml Erlenmeyer flask and tap gently until the cake falls out. The cake usually comes out in one piece, although sometimes it will be necessary to remove the asbestos with a glass rod. After moistening with a few ml of 2.5% HCl, break up the cake and wash it into the flask with a stream of 2.5% HCl from a wash bottle. Wash the centrifuge tube, crucible and rod with the HCl (using a rubber-tipped rod), and add the wash solution to the Erlenmeyer flask. The total volume of the acid solution used should approximate 60 ml. Through a rubber stopper attach an air-reflex condenser to the flask. Suspend in a boiling saturated salt solution for 3 hours (boiling temperature about 105°). Remove from the salt bath, cool and add 1 ml of 20% phosphotungstic acid. Neutralize to pH 6.5–7.0, first adding 6 ml of 25% NaOH and then completing neutralization with 10% NaOH from a dropper, Make to volume in a 100 ml flask and filter about 75 ml, Centrifuge, in a 100 ml tube, 10 ml of a 25% yeast suspension (Fleischmann or similar preparation). Discard the liquid and dry the sides of the tube. Add about 75 ml of the neutralized filtrate, shake and allow to stand 1 hour with occasional mixing. Centrifuge until clear. Remove a 50 ml aliquot and determine reducing sugars according to Methods **16-8A** and **16-6B**, *Copper by titration with sodium thiosulfate*. Subtract a blank of 2.3 mg of Cu_2O before computing dextrose. If no cereal is present in the sausage, the percent non-fermentable sugars as dextrose × 14.4 = percent soybean flour in the sample. If cereal is present, 0.33% for each percent of cereal must be subtracted from this total.

NOTE: Fredholm[26] has recently proposed a method for the determination of pentoses and pentosans as an indication of the presence of soy meal in meat products. He found 0.045% pentoses in straight beef as against 0.167% in beef to which 5% of soy meal had been added. Pork yielded similar results. Untreated liver paste showed 0.26% pentoses, while the same paste containing 5% soy meal gave a figure of 0.50% pentoses.

Method 10-20. *Added Blood in Hamburg.*

Cases have been encountered where butchers have attempted to conceal the presence of high proportions of fat in their hamburg by adding blood, whose red color tends to mask the white fat particles. The following method[27] serves to determine the presence of one percent or more of added blood:

REAGENTS

a. *Catalase solution*—Prepare a solution containing 5 mg/ml of crude catalase (Nutritional Biochemicals Corporation).

b. *Stock iron solution*—Dissolve 200 mg of clean Fe wire in 20 ml of H_2SO_4 (1+5) by heating, and dilute to 1 liter with water. (This solution contains 200 μg Fe/ml.)

c. *Standard iron solution*—Dilute 2.5 ml of the stock Fe solution, (**b**), to 100 ml with water. (Solution contains 5.0 μg Fe/ml).

d. *Antifoam*—Dow-Corning Antifoam AF is satisfactory.

PREPARATION OF SAMPLE

Place 10 g sample in a 250 ml beaker, add 50 ml of water, mix and break the sample up with a flat-tipped glass rod. Let stand 5–10 minutes, stirring once or twice.

Decant the liquid and filter through a wet Whatman No. 1 filter paper. (Small amounts of meat falling onto the filter are not detrimental, but large quantities slow the filtration rate.) After all of the liquid has passed through the filter, add 10 ml of water to the sample, mix, and decant the liquid onto the filter. *Collect only the first* 50 *ml of filtrate*, and mix. (In rare cases it may be necessary to wash the meat with further amounts of water to collect the desired 50 ml of filtrate).

DETERMINATION

Add 6 ml of 30% H_2O_2 to 25 ml of filtrate, mix well, and let stand 30 minutes with occasional mixing, adding antifoam, (**d**), as needed. Then add about 1 ml of catalase solution, (**a**), and let stand until O_2 evolution ceases, adding antifoam as needed. [Reaction is complete when no O_2 is released on further addition of catalase. Completion may be checked by mixing one drop of the solution with a drop of a 1% solution of V_2O_5 in dilute H_2SO_4 (6+94); if no pink or red color develops, all H_2O_2 has been decomposed.] Add antifoam to dissipate all foam, and transfer the contents of the beaker to a 50 ml volumetric flask (reserving enough space for addition of the 5 ml of other reagents). Add 1 ml of 12.5% H_2SO_4 and 4 ml of 10% Fe-free $Na_2WO_4.2H_2O$ solution, make to volume with water, and mix. Let stand 15 minutes, and filter through a Whatman No.1 filter paper.

To separate test tubes add 6.0, 7.0. 8.0, 9.0 and 10.0 ml of the filtrate, and dilute the contents of each tube with water to 10 ml.

Prepare a standard iron curve as follows: To separate test tubes add 0–5.0 ml of the standard iron solution, (**c**), in 0.5 ml increments (0–25 mg of Fe), and make to 10 ml with water.

Using a hood, to each tube of the standard and sample solutions add 2 ml of NH_4OH, mix, add 0.1 ml of thioglycolic acid, and mix well. Transfer each solution to a cuvette and read its absorbance at 540 mμ against the 0 level standard solution as a blank. (If the tubes have stood for a long time before reading, the colors of the solutions may be restored by shaking well.)

CALCULATIONS

Plot a standard curve of absorbance vs. micrograms Fe from the readings on the standard solutions, and determine μg Fe/ml solution at each level of the assay solution from this curve. Then: Average μg Fe/ml solution × 50 = μg Fe/10 gm sample = A (if 25 ml original filtrate from sample were used).

μg. Fe/10 gm fat-free sample = $100A$ /(100% − % fat in sample) = Y. Percent added whole blood (X) = $Y-101.6/31.5$.

Tenderizers

Papain is the usual proteolytic enzyme employed in tenderizing meat. While there are published methods for determining the potency of papain preparations (*see* Ref. 14, Method **17.016–17.019**), the authors were unable to locate any reference to experiments on the chemical detection of added tenderizers in meat. Reference 4, pages 37–44* does however contain a discussion, illustrated with photomicrographs, of the effects of papain, bromelin and ficin on muscle fibers, and it is probable that a competent microscopist would be able to identify tenderized meat from observation of physical changes in a sample similar to those described in that article.

A practice has developed to some extent in recent years of injecting proteolytic enzymes into live animals just before slaughter. So far as we know, no chemical method for the detection of meat resulting from this inhumane process has been published.

Horse Meat

Because horse meat is considerably less expensive than beef, there has always been a temptation for unscrupulous producers to substitute it for beef or pork in sausages or even to market the fresh meat as beef. Because horse fat is yellow, the presence of deep yellow fat in meat samples raises a suspicion of the presence of horse meat, but since, as previously noted, the fat of some types of beef may also be quite yellow, fat color is not a certain indication of horse meat. For uncooked meat, the most reliable tests are biological ones (precipitin or complement fixation tests[28]) or the disc electrophoresis method.[29] None of these will work (because of the denaturation of protein by heat) on cooked products such as frankfurters. In such cases it has proved necessary to rely on distinctive properties of horse fat, such as its high content of linolenic acid. The following method[30] has been employed in the laboratory of the Connecticut Agricultural Experiment Station with satisfactory results in the identification of horse meat, but it must be pointed out that while the method may indicate the percentage of horse fat in the total fat, there is no accurate way of translating this figure into an estimate of the percentage of horse *meat* in the sample:

See also C. E. Weir *et al.*: *Food Research* **23**:411, 423 (1958).

Method 10-21. *Chemical Test for Horse Fat.*

Extraction of Fat

Pass the meat sample twice through a meat grinder, place in a beaker and warm to about 60° in a drying oven. Remove the beaker from the oven and add sufficient petroleum ether to completely cover the meat. After the ether has ceased to boil, cover the meat with more ether and allow to stand overnight at room temperature. Filter the fat-ether mixture through a dry filter and allow the ether to evaporate on a steam bath in the presence of a stream of nitrogen or carbon dioxide. Store the ether-free fat in a refrigerator until used.

Saponification of Fat

Reflux 10 g of the extracted fat with 100 ml of 0.5 *N* alcoholic KOH for 30 minutes. Transfer the hot liquid to a distilling flask and remove approximately 80 ml of the alcohol by distillation. Add 250 ml of distilled water to the flask and transfer the solution to a large separatory funnel. While the liquid is still warm, add 15 ml of 5 *N* H_2SO_4, 250 ml of saturated salt solution, and 50 ml of dry ethyl ether. Shake the funnel vigorously and allow the layers to separate. Withdraw the ether layer and wash with three 15 ml portions of saturated salt solution. Filter the washed ether through a dry filter into a 50 ml volumetric flask and cool to refrigerator temperature (5°–10°). Make to volume with dry ether at this same temperature.

Precipitation of Hexabromide

Place 10 ml (2 g) of the ether extract containing the fatty acids into a tared centrifuge tube (6″ × 1″).

Add 15 ml of anhydrous hexabromide-saturated ether plus 2 ml of glacial acetic acid. (The anhydrous ethyl ether is saturated at refrigerator temperature with hexabromide formed from the bromination of fatty acids present in horse fat.) Cool the tube for 15 minutes in an ice-salt bath, the temperature of which is $-5°$ to $-10°$. To this cooled solution add from a buret, dropwise, 2 ml of the brominating solution (2 ml of bromine plus 8 ml of glacial acetic acid). Keep the tube in the ice-salt bath during the bromination and for 15 minutes thereafter. Stopper the tube and place in a refrigerator held at 5°–10° overnight. Remove the tube from the refrigerator and place it in an ice bath for 15 minutes. Remove the stopper and centrifuge for 2–4 minutes at a speed of 900–1,000 rpm (12 inch head). Decant all the ether from the tube. (There should be no precipitate carried away with the decanted ether provided the centrifuge is run at the suggested speed.) To the tube add 10 ml of previously cooled anhydrous hexabromide-saturated ethyl ether (prepare this wash ether by adding 2 ml of glacial acetic acid to 25 ml of the anhydrous hexabromide-saturated ether). Stir the precipitate vigorously and cool the tube in an ice bath for 15 minutes. Centrifuge the tube for 2–4 minutes and decant the ether. (If this procedure is carefully followed, the precipitate remaining in the tube after the last washing should be white.) Should the precipitate still hold some occluded bromine, repeat the washings until all color of bromine has disappeared. Avoid excessive washing. Dry the tube at 105°–110° for 30 minutes. Record the weight of hexabromide in mg/g of fat.

Crowell found the following average values for beef, pork, and horse fat:

Type of fat	mg hexabromostearic acid/g. fat
Beef	2.0
Pork	7.5
Horse	56.8

Fat Constants

Methods for the determination of fat "constants" such as index of refraction, melting-point, iodine and saponification numbers, etc., are the same for animal as for vegetable fats, and are described in Chapter 13. These "constants" are, however, of very little value in distinguishing between the fats of different animals. (One possible exception is horse fat, which has an unusually high iodine number.)

Amino Acid Separations

The determination of the proportions of the various amino acids present in meat and other proteins is done nowadays mostly with the help of automatic chromatographic apparatus.[31] Such separations are made in connection with nutrition studies, and are not employed for control purposes.

Poultry Meat

While smoked turkey is an article of commerce for the gourmet trade, by and large poultry meat comes on the market as fresh meat or, at most, with no treatment beyond freezing or cooking. It forms an ingredient of chicken or turkey soups or pies, but there are no products analogous to the many sausages made from beef, veal or pork, and the only poultry organs sold separately are chicken livers. For discussion of the preparation of poultry for market the reader is referred to Text Reference 3, Vol. II, pp. 974–994.

Poultry Grades

In market parlance, "chickens" are birds of the hatch of the current year, not sexually mature; hens which have laid eggs are "fowls", and mature roosters are "cocks". Chickens weighing less than three pounds are "broilers"; those weighing 3 lb. are "fryers"; and chickens over 3 lb. are "roasters". Dressed poultry of each of these categories is divided into three market grades designated as A, B and C.

Proximate Analysis of Poultry

Table 10-6[32] gives the proximate analyses and calorie contents of several of the food birds:

More extended analyses will be found in Selected Reference D, from which the following analysis of domestic ducks is taken:

	Total edible portion	Flesh only
Water, %	54.3	68.8
Protein, %	16.0	21.4
Fat, %	28.6	8.2
Total carbohydrate, %	0-0	0.0
Ash, %	1.0	1.2
Calories/100 g	326	165

Instances have been encountered of the "plumping" of poultry by injection of water into the tissues. Detection by analysis of such watering is based on the same assumption as is the case of frankfurters and ham, namely, that the moisture/protein ratio is constant. Table 10-6 should be consulted in this connection.

Methods of Analysis

Preparation of Sample and Proximate Analysis. Proceed as for beef, pork, etc.

TABLE 10-6

Composition of Poultry

Cut		Number of analyses	Refuse	Water	Protein N× 6.25	Protein By difference	Fat	Ash	Fuel value per pound, cals.
Chicken	edible portion	3	—	74.8	21.5	21.6	2.5	1.1	505
	as purchased	3	41.6	43.7	12.8	12.6	1.4	0.7	295
Fowl	edible portion	26	—	63.7	19.3	19.0	16.3	1.0	1045
	as purchased	26	25.9	47.1	13.7	14.0	12.3	0.7	775
Goose	edible portion	1	—	46.7	16.3	16.3	36.2	0.8	1830
	as purchased	1	17.6	38.5	13.4	13.4	29.8	0.7	1505
Turkey	edible portion	3	—	55.5	21.1	20.6	22.9	1.0	1360
	as purchased	3	22.7	42.4	16.1	15.7	18.4	0.8	1075
Quail	as purchased	1	—	66.9	21.8	—	8.0	1.7	775

Sea Foods

Chemical Composition

Between 1958 and 1961, Thurston and co-workers reported proximate analyses, plus sodium and potassium determinations, on several samples each of 46 species of fresh and salt-water fish;[33] their results are summarized in Table 10-7. During the same period, Teeri *et al.*[34] published figures on the moisture, protein and B-vitamin contents of 17 types of sea food on sale in the New England market; these are given in Table 10-8.

TABLE 10-7
Proximate Composition and Alkali Metal Content of Fish Fillets[33]

	Fresh Water Fish	Ocean Fish
MOISTURE, PERCENT:		
Maximum	84.1	90.3
Minimum	29.8	63.9
Average	77.7	79.7
OIL, PERCENT:		
Maximum	63.5	22.6
Minimum	0.7	0.2
Average	4.4	1.9
PROTEIN, PERCENT:		
Maximum	22.1	23.2
Minimum	5.9	8.4
Average	17.6	18.1
ASH, PERCENT:		
Maximum	1.48	1.53
Minimum	0.41	0.96
Average	1.08	1.16
SODIUM, MG/100 G:		
Maximum	95.	195.
Minimum	46.	30.
Average	58.	71.
POTASSIUM, MG/100 G:		
Maximum	418.	508.
Minimum	223.	210.
Average	327.	395.

Analyses by Proctor *et al.*[35] indicate the following average composition of the drained fish in Maine canned sardines: Moisture, 61.9, protein, 22.9, fat, 12.1, ash, 3.87, chloride, 1.05, phosphorus, 0.40, calcium, 0.46, and cholesterol, 0.14, percent; iron, 2.99, fluorine, 4.50, niacin, 5.43, pyridoxine, 0.24, riboflavin, 0.14, and thiamine, 0.026, mg/100 g. Other analyses may be found in Text Reference 3, Vol. II, Chapter XXII, and Selected References F and G.

"Whole Fish Protein Concentrate"[36]

As far back as September 15, 1961, an application was filed with the U.S. Food and Drug Administration for approval of a definition and standard of identity for a food supplement to be called "fish protein concentrate, whole fish flour", that was to be prepared by solvent extraction of whole fish of any wholesome species. The Administration objected to the fact that the proposed product contained the heads, tails, fins, viscera and intestinal contents of the fish, which are not normally used for food. However, political interest in this product on the part of states with large fish industries remained strong, and to this was added an equally strong interest on the part of those agencies desiring to make a cheap source of high-quality protein available to the populations of economically-disadvantaged countries. As a result, the question was referred to the Advisory Committee on Marine Protein Resource Development, Food and Nutrition Board, National Academy of Sciences-National Research Council, and finally, on January 30, 1967, the Commis-

TABLE 10-8

Moisture, Protein and B Vitamin Contents of Sea Food on the New England Market

Species	Moisture	Protein	Biotin	Folic acid	Niacin	Pantothenic acid	Riboflavin	Vitamin B12
	%	%	μg/ 100 g	μg/ 100 g	μg/ 100 g	μg/ 100 g	μg/ 100 g	μg/ 100 g
FRESH FISH								
Cod	81.8	16.8	0.19	1.75	1,980	122	31	0.45
Cusk	80.7	17.4	0.12	1.98	2,200	115	32	0.30
Haddock	81.2	17.7	0.29	0.81	3,020	140	53	0.53
Haddock Roe	70.5	24.6	12.28	63.70	1,350	3,620	690	5.10
Halibut	79.7	18.6	1.89	1.95	6,210	299	51	9.71
Mackerel	71.4	20.1	1.50	1.24	7,620	464	251	4.85
Salmon	75.0	20.0	0.85	2.15	7,190	745	99	2.89
Smelts	80.2	15.6	3.03	3.66	1,450	638	120	3.44
Swordfish	76.7	19.3	0.16	2.04	7,390	203	50	0.57
MOLLUSCS								
Clams	87.9	8.5	2.34	2.65	1,260	622	188	62.30
Oysters	87.5	6.8	0.72	3.70	1,310	184	166	14.60
Scallops	79.7	16.4	0.32	0.51	1,150	143	65	1.34
CRUSTACEANS								
Lobster (claw)	78.5	17.3	5.18	0.55	1,240	1,990	46	0.46
Lobster (tail)	76.4	20.3	4.79	0.56	1,670	1,270	50	0.49
Shrimp	78.4	19.9	1.01	7.38	3,120	372	27	0.91
CANNED FISH								
Crab	80.1	16.3	4.60	0.30	1,370	490	84	0.46
Salmon	66.2	22.4	1.48	0.67	5,630	570	198	2.07
Tuna	69.6	28.3	0.45	3.25	14,100	200	105	1.48

sioner of Food and Drugs issued Food Additive Regulation 121.1202 covering "whole fish protein concentrate" which contained, among others, the following provisions:

a). The additive is derived from whole, wholesome hake and hakelike species of fish handled expeditiously and under sanitary conditions.

b). The additive consists essentially of a dried fish protein processed from the whole fish without removal of heads, fins, tails, viscera or intestinal contents. It is prepared by solvent extraction of fat and moisture with isopropyl alcohol or with ethylene dichloride followed by isopropyl alcohol; solvent residues are reduced by conventional heat drying and/or microwave radiation; and there is a partial removal of bone.

c). The protein content is not less than 75 percent, and the quality of this protein rates at least 100 by AOAC Method **39.133–39.137.**

d). Maximum percentages of moisture and fat are respectively 10 and 0.5 percent.

e). Solvent residues do not exceed: Isopropyl alcohol, 250 p.p.m.; ethylene dichloride, 5 p.p.m.

f). The fluoride content does not exceed 100 p.p.m.

g). The product is free of *Escherichia coli* and pathogenic organisms (including *Salmonella*), and the total plate count is not over 10,000/g.

h). There are no more than a faint characteristic fish odor and taste.

i). When consumed regularly by children up to 8 years old, the amount of this additive in the total diet must not exceed 20 g per day.

Degrees of Freshness

Fish that is not preserved by refrigeration or other means decomposes and becomes unpalatable much more rapidly than the flesh of land animals, e.g., beef, pork, poultry, meat, etc). The contrast is even more striking between game animals which are deliberately allowed to become "high" in flavor, and fish which are at their best when cooked the minute they leave the water. Fish of this freshness can only be savored by those who catch them, but as a practical matter in judging when to condemn fish as unfit to eat, it has been found desirable to divide market fish into five categories of freshness on the basis of their organoleptic properties. These classes have been described as follows:[37]

Class	Description
0	No perceptible odor in flesh.
1	An odor which, while slightly "off", is only superficial in character, not deep-seated and not repugnant (often discribed as "fishy").
2	A barely perceptible but deep-seated odor that is somewhat repugnant and connotes decay (often described as "stale").
3	An odor similar to that of Class 2, but having enough intensity to be readily noticeable; distinctly more repugnant than Class 2 (often described as "taint").
4	An odor of great intensity, and decidedly more repugnant than Class 3 (often described as "putrid").

When employed by a perceptive and experienced person, the nose is at least as reliable a means of judging the freshness of fish as is any method of chemical analysis. Nevertheless, because organoleptic methods are subjective, they are susceptible to court attack. For this reason a number of confirmatory chemical indices of decomposition have been worked out. Hillig *et al.*[38] found that the volatile acid number, formic and acetic acid contents, volatile bases, volatile amines and trimethylamine nitrogen showed the highest degree of correlation with organoleptic judgment in the case of frozen fish; of these tests they considered those for volatile acid number, formic acid and volatile bases to be the most useful, and employed the other tests chiefly to obtain confirmatory evidence.

Hillig *et al.*[38] and Campbell[39] list the following values of the various tests as indicating decomposition:

Species	Primary Indices			Secondary Indices		
	Volatile acid No. (ml 0.01 N NaOH/100g)	Formic acid (mg/100 g)	Volatile bases (ml 0.01 N acid/100 g)	Acetic acid (mg/100 g)	Volatile amines (ml 0.05 N $KMnO_4$/100 g)	Trimethyl-amine (mg N/100g)
Cod	40	any amount	150	40	350	30
Haddock	40	any amount	150	40	350	30
Perch	30	any amount	150	30	350	25
Shellfish	75	trace	367	74	513	42

The texts of these methods are given below.

In addition to the above tests, other indices have been employed for certain types of shellfish. Duggan and Strasburger[40] found that absolutely fresh shrimp were free of indole, while this substance averaged 19 micrograms per 100 grams for Class 2 shrimp, 140 for Class 3, and 684 for Class 4. Burnett[41] suggested ammonia as an indicator of decomposition in crabmeat; he found an average of 70 micrograms of NH_3 per gram of Class 2 crabmeat, as against 0.4 μg/g in Class 0 and 14 μg/g in Class 1.

Methods of Analysis

Method 10-22. *Preliminary Treatment and Preparation of Sample.* (*AOAC Methods **18.002*** and ***18.004***).

To prevent loss of moisture during preparation and subsequent handling, use as large samples as practicable. Keep ground material in a container with an air-tight cover. Begin all determinations as soon as practicable. If any delay occurs, chill the sample to inhibit decomposition. Prepare samples for analysis as follows:

a. *Fresh fish.*—Clean, scale, and eviscerate large fish in the usual way. In the case of small fish (6″ long or less), use 5–10 whole fish, including heads if desired. For large fish, cut from each of at least three fish, three transverse slices, one-inch thick: one slice from just back of the pectoral fins, one slice halfway between first slice and vent, and one slice just back of vent. Skin and bones may be separated if desired. For fat determinations, include the skin because many fish store large quantities of fat directly beneath the skin.

Pass the sample rapidly through a meat chopper three times. Remove unground material from the chopper after each grinding and mix throughly with the gound material. The meat chopper should have holes as small as practicable (1/16–1/8″ dia.) and should not leak around the handle end. As an alternative procedure for soft fish, a high-speed blender may be used. Blend several minutes, stopping the blender frequently to scrape down the sides of the cup.

b. *Canned fish and other canned marine products.*—Blend the entire contents of the container in a blender or pass through a meat chopper three times as in (**a**).

c. *Canned marine products in oil.*—Drain 2 minutes on a No. 8 sieve, and prepare the solid portion as in (**b**). The oil and brine may be analyzed separately if desired, or may be reincorporated with the solids.

d. *Fish packed in salt or brine.*—Drain the brine and rinse off adhering salt crystals with saturated NaCl solution. Drain again two minutes, and proceed as in (**a**).

e. *Dried smoked or dried salt fish.*—Cut large samples into small pieces, mix, and quarter down to about 1/4 pound. Cut, shred, grind, or otherwise comminute the 1/4 lb sample as finely as possible, so that reasonably representative samples may be weighed for analysis after thorough mixing. (Duplicate or triplicate determinations may be necessary to establish uniformity of sample.)

f. *Shellfish other than oysters, clams and scallops.*—If received in the shell, wash the sample as in (**g**), and separate the edible portions in the usual way. Prepare the edible material for analysis as in (**b**).

g. *Shell oysters, shell clams and scallops.*—Wash the shells with water to remove all loose salt and dirt, and drain well. Shuck into a clean dry container enough oysters or clams to yield at least one pint of drained meats. Transfer the meats to a skimmer (a flat bottomed metal pan or tray with 2 inch sides, of not less that 300 sq. in. in area, with perforations 0.25″ in dia. arranged 1.25″ apart in a square pattern), pick out pieces of shell, drain 2 minutes on the skimmer, and proceed as in (**h**) or (**i**).

h. *Shucked clams or scallops.*—Prepare as in (**b**).

i. *Shucked oysters.*—Blend the meats (including liquid) 1–2 minutes in a high speed blender.

Method 10-23. *Total Solids.* (*AOAC Methods* ***18.006*** *and* ***18.007***).

a. *For all marine products except raw oysters.*—Cut into short lengths about 2 g of asbestos fibers (Gooch type). Place the cut fibers and a glass stirring-rod about 8 cm long with a flattened end into a flat-bottomed covered metal weighing dish (about 9 cm in diameter). Dry the dish, asbestos and rod in an oven 1 hour at 100°, cool and weigh. Weigh quickly into this, to the nearest milligram, 9.5–10.5 g of prepared sample. Add 20 ml of water and mix the sample thoroughly with asbestos. Support the end of the rod on the edge of the dish, and evaporate just to dryness on the steam-bath, stirring once while still moist. Drop the rod into the dish, and heat at 100° 4 hours in a gravity oven, or 1 hour in a preheated forced-draft oven set for full draft. Cover the dish, cool in a desiccator, and weigh promptly.

b. *For raw oysters only.*—Weigh quickly, to the nearest milligram, 9.5–10.5 g of prepared sample into a covered weighed flat-bottomed metal dish about 9 cm in diameter and 2 cm high. Spread the sample evenly over the bottom of the dish. Then either:

(1) Evaporate just to dryness on the steam bath and dry 3 hours in an oven at 100°; or

(2) Insert directly into a preheated forced-draft oven set at full draft, and dry 1.5 hours at 100°.

Cover, cool in a desiccator, and weigh promptly.

Method 10-24. *Ash.* (*AOAC Method* ***18.008***).

Dry a sample representing about 2 g of dry material, and proceed as in Method 16–4, using a temperature not exceeding 550°. If the material is high in fat, make a preliminary ashing at a low enough temperature to permit smoking off of the fat without burning.

Method 10-25. *Salt.* (*AOAC Method* ***18.010***).

Weigh 10 g of shellfish meats, liquid, or mixed meats and liquid, or a suitably sized sample (depending on NaCl content) of other fish products, into a 250 ml Erlenmeyer flask or a beaker. Add a known volume of standard 0.1*N* $AgNO_3$, more than sufficient to precipitate all the Cl as AgCl, and 20 ml of HNO_3. Boil gently on a hot plate or sand bath until all solids except AgCl have dissolved (usually 15 minutes). Cool, add 50 ml of water and 5 ml of saturated $FeNH_4(SO_4)_2.12H_2O$ solution, and titrate with 0.1*N* NH_4SCN until the solution becomes a permanent light brown. Subtract ml of 0.1*N* NH_4SCN from ml of 0.1*N* $AgNO_3$ added, and calculate the difference to NaCl. For a 10 g sample, each ml of 0.1*N* $AgNO_3$ = 0.058% NaCl.

Method 10-26. *Protein.*

Proceed as in Chapter 1. The general factor 6.25 has usually been employed to calculate crude protein in fish, but the percentage of true protein is probably much less than is indicated by this formula.

Method 10-27. *Fat.* (*AOAC Method* ***18.013–18.014***).

A. *By acid hydrolysis.*— Weigh into a 50 ml beaker 8 g of well-mixed sample, and add 2 ml of HCl. Using a stirring rod with an extra-large flat end, break up coagulated lumps until the mixture is homogeneous. Add an additional 6 ml of HCl, mix, stirring occasionally with the rod. Cool the solution and transfer to a Mojonnier fat extraction tube. Rinse the beaker and rod with 7 ml of alcohol, add to the extraction tube, and mix. Rinse with 25 ml of ether, added in three portions; add the rinsings to the extraction tube, stopper the tube with a cork, and shake vigorously for 1 minute. Add 25 ml of petroleum ether (b.-pt. 30–60°) to the extraction tube, and repeat the vigorous shaking. Centrifuge the tube 20 minutes at about 600 r.p.m.

Draw off as much as possible of the ether-fat solution through a filter consisting of a pledget of cotton (packed just firmly enough in the stem of a funnel to permit the ether to pass freely) into a weighed 125 ml beaker-flask containing porcelain chips or broken glass. (This flask must previously have been dried with a counterpoise at 100°, let stand in air to constant weight, and then weighed against the counterpoise.)

Re-extract the liquid remaining in the tube twice, each time using only 15 ml of each ether. Shake well on addition of each ether. Draw off the clear ether solutions through the filter into the same flask as before, and wash the tip of spigot, funnel and end of funnel stem with a few ml of an equal-volume mixture of the two ethers that is free of suspended moisture. Evaporate the combined extracts slowly on the steam bath, then dry the fat in a 100° oven to constant weight (about 30 minutes) and weigh. (Owing to the size of the flask and the nature of the material. there is less error in cooling in air than in cooling in a desiccator.) Correct this weight by a blank determination on the reagents. Report as percent fat by acid hydrolysis.

Drying to constant weight takes only about 40 minutes for fish fat. Long heating periods may increase the weight of fat because of oxidation. If a centrifuge is not available, extraction can usually be made by letting the Mojonnier flask stand until the upper layer is practically clear, then swirling the flask and again letting stand until clear. If a troublesome emulsion forms, pour off from the Mojonnier flask as much of the ether-fat solution as possible after letting the flask stand, then add 1–2 ml of alcohol to the flask, swirl, and again let the mixture separate.

B. *By the rapid modified Babcock method.*—Weigh 9.0 ground and weighed sample into a Paley-type Babcock cheese bottle (Kimble No. 508, 20% size), stopper, and add about 30 ml of a reagent prepared by mixing equal volumes of glacial acetic acid and 70–72% perchloric acid. Place in a water-bath (a two-liter stainless steel beaker is satisfactory) maintained at $92 \pm 2°$, and swirl occasionally until no lumps remain (usually about 20 minutes). Remove from the bath, add reagent until the fat is well up in the calibrated neck of the bottle, centrifuge 2 minutes at about 600 rpm, and read the fat column to the bottom of the top meniscus, using dividers. This reading is directly in percent fat when a 9.0 g sample is used.

With very fat fish it may be necessary to take a sample of less than 9 g. In such case, multiply the reading by the factor 9/g sample to obtain percent fat.

Species Identification by Disc Electrophoresis

The separation of ionic components of a mixture by passing an electric current through the solution is an old technique that formed the basis of determining transference numbers of inorganic ions. Addition of a gel to such a solution can increase the sharpness of separation of two ions because the frictional properties of gels aid separation by sieving at the molecular level. These facts plus the utilization of a synthetic gel of adjustable pore size form the basis of the disc electrophoresis method for the separation of serum proteins, which has the advantage over other methods of producing high resolution in very brief runs. A more extended account of the theory of this method is given by Ornstein.[42]

Because the proteins of no two species of animals are identical, the method as outlined below produces a series of rings whose numbers and relative positions and sizes serve as "fingerprints" to identify the species of land or sea animal from which a sample of meat or fish came:

Method 10-28. *Disc Electrophoresis of Meat and Fish.*[29]

APPARATUS

a. *Electrophoresis apparatus.*—(*See* Fig. 10-9) Canalco Model 6 or 12, obtainable from Canal Industrial Corporation, 5635 Fisher Lane, Rockville, Md. 20852. These models handle 6 and 12 samples respectively at a time.

b. *Power source.*—(*See* Fig. 10-10) The Model 200 source of the above company is satisfactory for the Model 6 apparatus, while their Model 1400 (or equivalent) is needed for the Model 12 apparatus.

REAGENTS

Stock solutions.—Because it is more convenient to use the ready-prepared solutions obtainable from the manufacturer of the electrophoresis

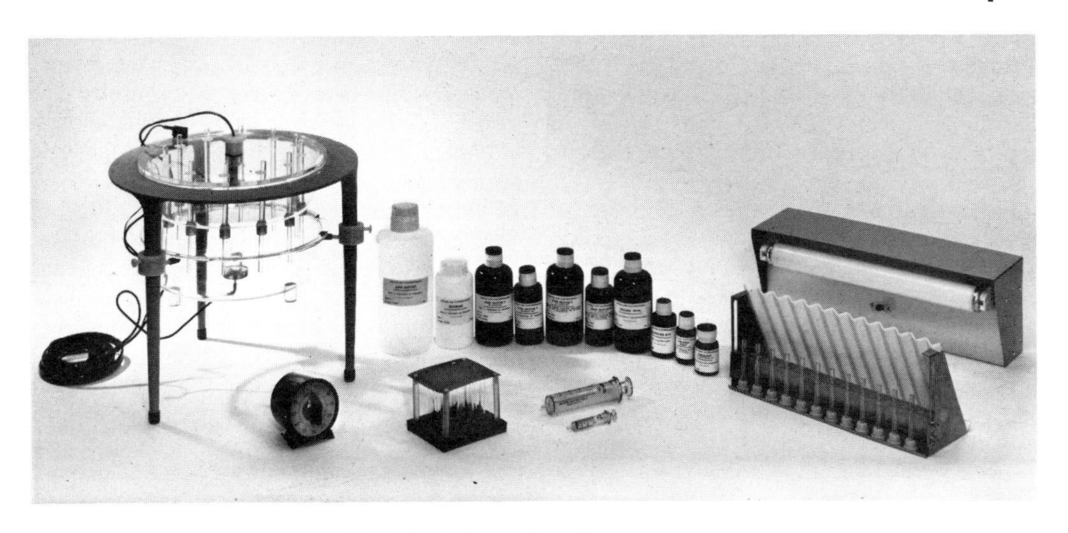

FIG. 10-9, Electrophoresis apparatus.

FIG. 10-10. Power source.

apparatus (see (a) above), methods of preparation of these solutions are not given here. Those desiring to prepare their own reagents can obtain the directions from the manufacturer.

PREPARATION OF SAMPLE

Place a known weight of sample (25–50 g) in a Blender cup, add a quantity of water twice the weight of the sample, and blend 15–20 seconds at high speed. Pour the mixture into centrifuge tubes and centrifuge 10 minutes at 1800 rpm. Decant the supernatant liquid and filter to obtain the sample extract.

Because identification of the sample species is dependent on comparison of the bands shown by the sample with those given by known species under the same conditions, extracts obtained from authentic samples of the suspected species under consideration should preferably be run side-by-side with the sample extract.

DETERMINATION

Remove the stock solutions from the refrigerator and allow then to warm to room temperature. The following instructions for gel formation apply only to materials at or near room temperature (21°):

Prepare the following working solutions in Erlenmeyer flasks, mixing by swirling gently with a circular motion to avoid formation of air bubbles. (Solutions may be measured out and transferred with pipettes, or with a syringe of suitable size fitted with capillary tubing and an adapter. The latter technique is more convenient and affords excellent reproducibility. *All working solutions must be used the same day they are prepared.*)

Separating gel solutions (12 *ml required for* 12 *samples*).—Mix equal parts of Stock Solutions A and C.

Stacking gel solutions (16 *ml required for* 12 *samples*).—Mix one part each of water and Stock Solutions B, D, and E with four parts of Stock Solution F. (Caution: The stacking gel is photosensitive and must be protected from light. Red Erlenmeyer flasks give adequate protection. Alternatively, flasks may be painted on the outside or covered with metal foil.)

Rinse solution (16 *ml required for* 12 *samples*).—Mix one part of Stock Solutions B and E with six parts of water.

Place base caps on the sample columns and stand in the loading rack. Mix in separate small beakers equal volumes of each prepared sample extract and stacking gel solution, and transfer each mixture to one of the columns. Add a small quantity of water to each column to form a layer over the sample gel mixture, as follows: Remove the plunger from the barrel of a 10 ml syringe and add 1.5–2 ml of water to the syringe. Form a drop of water at the needle tip by applying slight pressure to the end of the applicator with the index finger. Rest the needle tip against the top column wall. Water will now flow down the inner wall of the column and form a layer over the sample gel mixture. A layer 3–4 mm deep is sufficient. (This procedure eliminates the meniscus which would otherwise cause the bands of the separated components to be curved.)

Polymerize the sample gel by placing a photopolymerizing light source as close as possible to the gels. Allow 20–40 minutes for polymerization, which will be accompanied by a slight opalescence. When polymerization is complete, remove the supernatant water, either by drawing off with a syringe fitted to a needle or by blotting up with absorbent tissue. Add a 1 cm layer of the rinse solution, and repeat.

With the columns upright, now add 0.20 ml of stacking gel solution to each column, and layer with water as described above. Place the columns about 30 cm from the photopolymerizing light source. Adjust the distance of the columns from the light source so that polymerization (as evidenced by the appearance of opalescence) starts 10–15 minutes after exposure. When polymerization begins, move the light source closer to the columns for an additional 15 minutes to ensure total polymerization.

Drain the water layer from the column. Combine equal parts of the separating gel solution and Stock Solution G in a flask and mix by swirling, taking care not to entrap air bubbles in the mixture. Rinse the inner surface of each column twice with this mixture, and then fill each column to the top, adding enough excess gel to form a rounded bead or inverted meniscus.

Seal each column at the top by gently laying a patch of Saran wrap (approx. 10 × 10mm) on it so as to form a flat surface. (The total time from the addition of Stock Solution G to sealing the column should not exceed 10 minutes.) Protect the gel columns from strong light, and allow to stand undisturbed for 30 minutes while chemical polymerization occurs. For good results, start electrophoresis within 30 minutes after polymerization has taken place (1 hour after mixing the separating gel—catalyst solution).

Remove the Saran wrap and the base cap from each column. This must be done exceedingly carefully to avoid tearing or deforming the gel. Holding the column securely, squeeze the sides of the cap to draw one side away from the glass and allow air to flow freely to the gel surface as the cap is pulled away completely.

Prepare 2 liters of 10X Buffer. Insert the upper bath stoppers in the column adapters from the top, and pour the required amount of 10X Buffer into each bath. Add 5 ml of the tracking dye to the upper bath only.

Moisten the sample end of each sample with a little water and insert into a column adapter, pressing it all the way through and at the same time withdrawing the stopper from the top. The upper ends of the columns should fit evenly with the top surfaces of the adapters.

With a syringe fitted with a needle, place a drop of buffer on the bottom surface of each column to prevent entrapment of air bubbles when the lower bath is lifted into place. Use the syringe to displace any air bubbles on the top surfaces of the columns. Raise the lower bath into position so that the lower ends of the columns extend about 1 cm below the surface of the buffer, and connect the lower electrode. Turn on the power supply switch. Allow at least 1 minute warm-up time before beginning electrophoresis.

Adjust the power supply for the desired current, immediately after beginning electrophoresis. For most proteins a current of from 3 to 5 ma per gel column is ideal. Currents over 5 ma should be avoided because they cause excessive ohmic heating within the gel, resulting in band distortion. Use of a current-regulated power supply for disc electrophoresis is desirable because the gel resistance increases during the course of separation. With a constant-flow voltage power supply, this would result in a gradual decrease in the current flow, with a consequent increase in time required for a run. Runs should be kept as short as possible to limit diffusion of bands and loss of resolution.

Within several minutes after starting a flow of current, a thin disc of tracking dye may be observed migrating through the stacking gel. As this disc passes through the inner face between stacking gel and separating gel, it divides into two distinct discs. The leading disc is free Bromophenol Blue dye; the trailing disc is a dye sample complex. Let electrophoresis continue until the tracking dye band has migrated about $1\frac{1}{4}''$ into the separating gel. This will take about 30 minutes at 5 ma per column. During the run, separation of many protein components can be observed by the formation of refracting lines in the separating gel.

At the conclusion of the run, turn off the power, and remove the lower bath. Drain the buffer from the upper bath by placing a bottle or beaker under one of the column adapters. Buffer from the lower bath can be poured into the upper bath, and then drained into a container in the same way. However, *do not mix buffer from the upper bath with that from the lower bath.*

(Store buffer separately under refrigeration. Buffer can be re-used for approximately 72 samples so long as the upper buffer does not become contaminated with a sample or drop in pH below 8.2.)

Fill a 20 ml syringe with cold tap water, and insert a gel-removing (dissecting) needle at the sample end of the column, between glass and gel, so that the needle tip reaches the separating gel. Keep the needle flat against the glass surface to avoid scratching the gel, and rotate it completely around the circumference of the gel. Remove the needle and insert it from the other end to a depth of about 1 cm, holding it against the surface of the glass column and at the same time forcing a stream of water through the needle. Carefully remove the gel.

Remove all gels from their columns, immediately place in separate tubes of stain, and leave there at least one hour.

Insert destain tubes (fire-polished end down) into *dry* destain tube caps, and add approx. $\frac{1}{2}$ ml of stacking gel to each tube to form a gel plug. Place the tubes near the polymerizing light for 30 minutes to polymerize the gel.

When staining is complete, decant the stain from each tube, leaving the gel in the tube. Rinse once with 7% acetic acid, then transfer each gel from its test tube to a prepared destain tube so that the sample end is upward, and fill each destain tube with 7% acetic acid.

Remove the destain tube cap and insert the open end into a column adapter of the upper bath. (To avoid undue stress upon the gel plugs, the upper bath stopper should be removed before insertion of the destain tube. Use upper bath stoppers to plug any adapters that are not to be used.)

Pour 1 liter of 7% acetic acid into each of the two baths. With a syringe and needle, inject acetic acid as needed at the top of each destain tube to displace any air bubbles that may be present.

Turn on the power supply and allow it to warm up for 1 minute. Set it for 12.5 ma per gel. Destaining will require from 45 to 60 minutes. When the sections of gel containing no separated fractions are clear of stain, turn off the power supply. Remove the destain tubes, transfer the gel columns to test tubes filled with 7% acetic acid, and stopper for viewing and storage.

Clean and rinse all equipment thoroughly.

Method 10-29. *Volatile Acid Number.* (*AOAC Methods* ***18.016*** (**a**) *and* ***18.018–18.020***).

APPARATUS

Steam distillation assembly.—Fig. 10-11. The assembly consists of: (1), a 3 liter boiling flask giving steam at a constant rate so as to produce a constant rate of distillation; (2), a distillation flask; (3), a condenser, and (4), 200 ml volumetric flasks as receivers. A standard distillation flask with side arm (about 9 mm o.d.) attached near the center of the neck, and with a steam inlet tube (about 10 mm o.d.) is satisfactory. The heating coil of the steam generator is made by winding 5 feet of 28 gauge Chromel wire (or equivalent) around a hollow pipe of about 0.25″ diameter, and beating red hot to detemper the wire. Leads into the boiler flask are brass, copper, or other nonferrous metal about $\frac{3}{32}$″ in diameter.

Any similar distillation assembly may be used if it is of capacity to handle the volumes specified in the method and yields a $57 \pm 2\%$ recovery of acetic acid on distillation.

STANDARDIZATION OF THE DISTILLATION APPARATUS

Place the apparatus, so that it is free from drafts and sudden changes in temperature. Make a mark on the boiler flask at the 1.5 liter level, fill to this mark with water, heat to boiling, and boil several minutes before starting distillation. Transfer 150 ml of water to the distillation flask, add one drop of H_2SO_4 (1+1), connect the condenser, insert the steam inlet tube into the distillation flask, and bring the contents of the flask to incipient boiling with a burner. Connect the steam inlet tube to the steam supply from the boiler, and steam distil. Regulate the rate of evolution of steam and height of the small flame

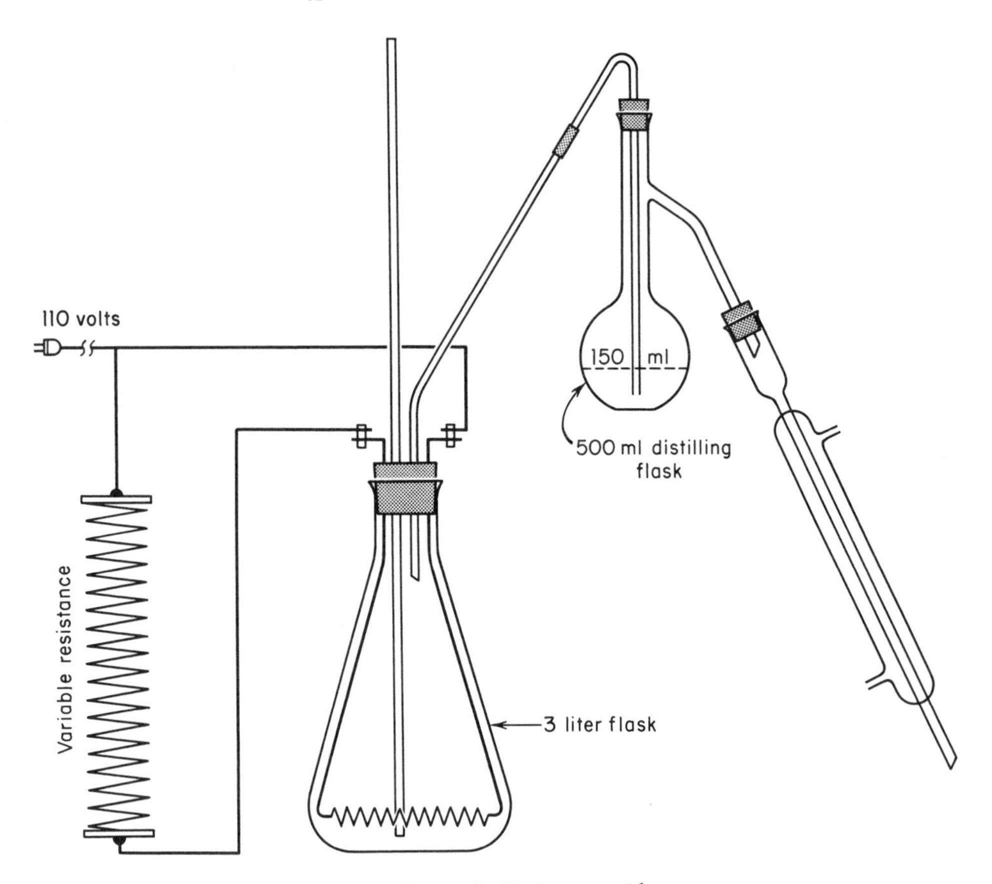

FIG. 10-11. Steam distillation assembly.

of the burner under the distillation flask so that the volume of liquid in the distillation flask is kept constant at 150 ml and the distillate collects at a rate of 200 ml/hour. (The period of collection may vary 5 minutes for 200 ml of distillate. The 150 ml in the distillation flask should remain constant within ± 10 ml. Boiling may be stopped to permit constancy of the 150 ml volume by momentarily interrupting the steam supply. A few trials will show the conditions necessary to maintain a constant volume in the distillation flask and a constant rate of distillation.) Determine the blank on two successive 200 ml portions of distillate by titrating with 0.01*N* alkali (phenolphthalein indicator) in a CO_2-free atmosphere.

Transfer 50 ml of about 0.1*N* acetic acid (concentration must be accurately known) to the distillation flask; add 1 drop of H_2SO_4 (1 + 1) (avoid contact with the neck of the flask) and 100 ml of water. Collect 200 ml of distillate and titrate with 0.1*N* alkali. Correct for titration blanks and compute the percent of the acid that was distilled. The distillation technique and apparatus are satisfactory when recovery is 57 ± 2%. Apparatus so adjusted gives recoveries (±2%) of formic, propionic and butyric acids of 40.5, 81 and 92%, respectively [*J. Assoc. Offic. Agr. Chemists*, **21**, 684, 688 (1938)], on 200 ml of distillate.

Preparation of Solution

Comminute the sample (include the entire contents of canned products) by passing three times through a food chopper, mixing after each grinding. Weigh 50 g of comminuted material into a tared 500 ml wide-mouthed Erlenmeyer flask, add about 150 ml of water, stopper the flask, and shake vigorously for 1 minute to effect thorough suspension of the material. Add 25 ml of 1*N* H_2SO_4, mix, precipitate the proteins with 20% phosphotungstic acid solution (40 ml is usually enough), make to 300 g with water, shake vigorously for about 1 minute, and filter through a 24 cm rapid folded paper.

Distillation and Computation

Pipet 150 ml of the prepared solution into the distillation flask, and make acid to Congo red paper with H_2SO_4 (1 + 1). Steam distil as under "standardization". Collect 200 ml of distillate, titrate with 0.01*N* NaOH to a phenolphthalein endpoint, and designate as *A*. Collect a second 200 ml of distillate, titrate with 0.01*N* $Ba(OH_2)$ to a phenolphthalein endpoint, and designate as *B*.

Volatile acid number = 4 (*A* − blank).

Method 10-30. Formic Acid. (*AOAC Method 18.021*).

Add 2 drops of saturated $Ba(OH)_2$ solution to distillate *B*, and evaporate to dryness on the steam bath. Add about 8 ml of water to the residue, and 1 ml more of 1*N* HCl than necessary to liberate the volatile acids. Filter through a small filter-paper into a 125 ml Erlenmeyer flask with a $ joint, and wash the paper with water in such a manner that the total filtrate amounts to 30–40 ml. Add 10 ml of a 500 ml aqueous solution of 12 g of NaCl and 25 g of sodium acetate trihydrate, and 10 ml of a 5% $HgCl_2$ solution. Connect the flask with a $ air condenser, and heat 2.5 hours on the steam bath.

With suction through a glass siphon attached to the funnel by a rubber stopper, transfer the precipitate of Hg_2Cl_2 to a previously-weighed microfunnel (Büchner type, 2 ml capacity with coarse fitted disk—Corning No. 3 6060) provided with a 2 mm asbestos mat. Rinse the flask with water followed by alcohol. Dry 30 minutes at 100°, cool, and weigh. Use a similarly treated funnel of the same type as a counterpoise.

Wt. Hg_2Cl_2 (mg) × 0.0975 =
mg formic acid in distillate.

To calculate the formic acid originally present in the aliquot of the sample taken for distillation divide mg formic acid found by 0.24 (= fraction of the formic acid distilled into the second 200 ml distillate) and multiply by 4, to obtain the formic acid in 100 g of the sample being analyzed.

Method 10-31. Volatile Bases.[43]

Apparatus

Steam distillation assembly—Use the assembly described under Method 10-29, "Volatile Acid Number", modified by another 500 ml flask inserted between the steam boiler and the distillation flask. The additional flask has the side arm bent upwards to make connection with the inlet tube of the distillation flask. In this assembly the new flask is made the distillation flask and the other flask (which was the distillation flask of the volatile acid assembly) contains 150 ml of a 15 g/ 1,000 ml suspension of $Ca(OH)_2$ to trap volatile acids. Methyl orange indicator in this flask will show when the $Ca(OH)_2$ has been consumed.

Preparation of Solution

Comminute the sample by passing three times through a food chopper, mixing after each grinding. Weigh 20 g into a 250 ml centrifuge bottle and pipet 100 ml of water into the bottle. Shake vigorously about 2 minutes, centrifuge and decant.

Determination

Pipet 25 ml of the decanted material into the distilling flask of the steam distillation assembly and add 125 ml of water. To prevent foaming add about $\frac{1}{4}$ ml of white beeswax U.S.P. Employ a 200 ml Erlenmeyer flask containing about 10 ml of water as a receiver. Fit the condenser with an adapter to conduct the distillate below the surface of the water in the receiver. Heat the $Ca(OH)_2$ suspension to incipient boiling, and the sample in the distilling flask to full boiling before connecting the inlet tube. At this point remove the flame from the distilling flask until steam is passing freely through the sample. (These precautions are necessary to prevent the material from being sucked back into the steam boiler.)
Steam distil about 100 ml of distillate, and titrate this distillate with 0.01*N* HCl. Transfer the titrated material to a 200 ml volumetric flask, dilute to the mark, shake, and designate as Solution *A*.

Ml 0.01*N* HCl/4.17 × 100 = ml 0.01*N* volatile bases/100 g fish. (Volatile acids will be retained in the flask containing the $Ca(OH)_2$ suspension, while aldehydes, ketones, alcohols and volatile bases, including amines, will be in the distillate.)

Method 10-32. *Acetic Acid.*

If a separate determination of acetic acid is desired, reference should be made to the chromatographic method for the separation of C_2 to C_4 saturated fatty acids in *Official Methods of Analysis of the Association of Official Agricultural Chemists*, 10th Ed. (1965), Methods **18.022—18.024**. This method employs the titrated distillate *A* of the Volatile Acid Number determination, Method 10-29, as the sample aliquot.

Method 10-33. *Volatile Amines.*[43]

Distillation

Pipet a 25 ml portion of the decanted material, Method **10–31**, "Preparation of Solution," into the distilling flask, add 125 ml of water, acidify with H_2SO_4 (1 + 1) to phenol red paper, add $\frac{1}{4}$ ml of beeswax, and steam distil as for volatile bases. Collect the distillate in a 100–110 ml volumetric flask containing 5–10 ml of water until the total volume is 90–100 ml, dilute to 110 ml, and shake (Solution *B*). (Volatile bases, including amines, will be retained in the distilling flask; volatile acids will be trapped in the flask containing the $Ca(OH)_2$ suspension; aldehydes, ketones and alcohols, if present, will be in the distillate.)

Oxidation of Solution *A*

Transfer 25–100 ml (depending on the probable concentration of reducing substances) of Solution *A*, Method **10–31**, "Determination," to a 300 ml Erlenmeyer flask, dilute to about 100 ml with water, and then add 10 ml of 10% NaOH solution and 25 ml of standard 0.05*N* $KMnO_4$. Place a short-stemmed funnel in the neck of the flask, heat to about 60°, and place the flask in a boiling water bath for 20 minutes. Cool, and add 10 ml of 6*N* H_2SO_4 and 10 ml of 20% KI solution. Titrate with standard 0.05*N* $Na_2S_2O_3$, adding starch indicator when most of the iodine has been consumed. Subtract the titer from 25.0 ml to determine ml 0.05*N* $KMnO_4$ used. If the oxidation mixture turns green during the heating period, or if more than 6 ml of $KMnO_4$ solution is used in the oxidation, repeat with a smaller aliquot of Solution *A*.

Oxidation of Solution *B*

Proceed as above.

Calculations

1. 100 × ml. 0.05*N* $KMnO_4$ used to titrate Solution *A* aliquot /0.208 × ml Solution *A* taken = *X*.
2. 100 × ml 0.05*N* $KMnO_4$ used to titrate Solution *B* aliquot /0.038 × ml Solution *B* taken = *Y*.
3. Ml. 0.05*N* $KMnO_4$ consumed in oxidation of volatile amines per 100 g of fish = *X* – *Y*.

Method 10-34. Trimethylamine Nitrogen.[44]

REAGENTS

a. *Toluene*—Shake with 1 *N* H_2SO_4, distil, and dry over anhydrous Na_2SO_4.

b. *Picric acid stock solution*—Dissolve 2 g of dry picric acid in 100 ml of dry toluene.

c. *Picric acid working solution*—Dilute 1 ml of the stock solution, (b), to 100 ml with H_2O—free toluene.

d. *Potassium carbonate solution*—Dissolve 10 g of K_2CO_3 in 100 g of water.

e. *Formaldehyde solution*—Shake commercial 40% formaldehyde with $MgCO_3$, and filter. Dilute 10 ml of the filtrate to 100 ml with water.

f. *Trimethylamine stock solution*—To 0.682 g of trimethylamine hydrochloride add 1 ml of HCl and dilute to 100 ml with water.

g. *Trimethylamine standard solution*—Check the basic nitrogen content of the trimethylamine stock solution, (f), by adding alkali to 5 ml aliquots and distilling into standard acid in a micro-Kjeldahl distillation apparatus. To prepare the standard solution, add 1 ml of the stock solution to 1 ml of HCl and dilute to 100 ml with water.

EXTRACTION

Weigh 100 g of minced or chopped, well-mixed fish. Add 200 ml of 7.5% trichloroacetic acid, and blend. If more convenient, the mixture may be shaken occasionally for several hours. (The mixture is stable and may be used without filtration.)

DETERMINATION

Pipet an aliquot (preferably containing 0.01–0.03 mg of trimethylamine N) into a 20 × 150 mm Pyrex test tube. For trimethylamine values (mg N/100 g) in the range 1–5 take 1 ml of extract and dilute to 4.0 ml with water. Add 1 ml of the formaldehyde reagent, (e), 10 ml of toluene (measured with an automatic pipet), and 3 ml of K_2CO_3 solution, (d). Stopper the tube with a polyethylene stopper and shake vigorously by hand about 40 times. Pipet 5 ml of the toluene layer into a small test tube containing about 0.3 g of anhydrous granular Na_2SO_4. (Avoid removing droplets of the aqueous layer.) Stopper the tube with a polyethylene stopper and shake gently a few times to dry the toluene. Decant into a dry colorimeter tube, add 5 ml of the picric acid working solution, (c), and mix by swirling gently. Determine the absorbance at 410 mμ, and calculate mg trimethylamine N/100 g fish from a curve prepared by carrying aliquots of the standard trimethylamine solution, (g), through the same procedure.

NOTE: Do not use stopcock grease; a mixture of sugar and glycerol ground together may be employed if necessary. Do not wash the tubes with soap or other detergent. Rinse them with water and occasionally clean them thoroughly with nitric acid.

Method 10-35. *Indole in Shrimp.* (*AOAC Method* ***18.036–18.038***).

APPARATUS AND REAGENTS

a. *Distillation apparatus*—Use a separate steam generator for each unit. The steam generator may be made from a 1-liter Erlenmeyer flask and connected to an all-glass steam distillation apparatus with minimum use of rubber tubing. The distillation flask (capacity not less than 500 ml) is connected to a straight bore condenser through a spray trap. A 500 ml Erlenmeyer flask is an effective receiver. Foil-wrapped rubber stoppers may be used in the absence of an all-glass apparatus (unprotected natural or synthetic rubber connections and stoppers cause variable distillation blanks).

Guard against traces of chlorine in the water, as they may partly or entirely inhibit development of the indole color.

b. *Color reagent*—Dissolve 0.4 g of p-dimethylaminobenzaldehyde in 5 ml of acetic acid, and mix with 92 ml of H_3PO_4 and 3 ml of HCl. Because the purity of this reagent exerts a strong influence upon the intensity of the reagent blank, purify the yellow commercial compound as follows: Dissolve 100 g in 600 ml of HCl (1 + 6). Add 300 ml of water, and precipitate the aldehyde by slow addition of 10% NaOH solution with vigorous stirring. As soon as the precipitate appears white, stop the addition of NaOH solution, filter, and discard the precipitate. Continue the neutralization until nearly all of the aldehyde has been precipitated, but do not carry to completion because the last 4–5 grams may be colored. Filter, and wash the precipitate with water until the washings are no longer acid. Dry the purified product (which should be practically white) in a desiccator.

c. *Acetic acid, purified*—If the acid on hand turns pink with the color reagent, purify as follows: Add 25 g of $KMnO_4$ to 500 ml of acetic acid in a $ one-liter round-bottomed flask, add 20 ml of H_2SO_4, and distil 400 ml in an all-glass still.

d. *Dilute hydrochloric acid*—Dilute 5 ml of HCl to 100 ml with water.

e. *Indole standard solution*—Accurately weigh 20 mg of indole into a 200 ml volumetric flask and dilute to the mark with alcohol. Keep refrigerated and discard after two weeks.

PREPARATION OF SAMPLE

Weigh 25 or 50 g (depending on the quantity of indole expected) of peeled raw or cooked shrimp. Transfer to a high speed blender, add 80 ml of alcohol, and mix several minutes until homogeneous. Transfer the mixture quantitatively to the distillation flask, and rinse the mixing chamber with a minimum quantity of alcohol.

DETERMINATION

Connect the flask for steam distillation and gently apply steam until distillation is well started, using care not to pass in steam so vigorously as to cause excessive foaming. Apply enough heat to the distillation flask to maintain a volume of 80–90 ml. Collect 450 ml of distillate in about 45 minutes. Wash the condenser with a small quantity of alcohol and allow to drain into the receiving flask.

Transfer the distillate to a 500 ml separatory funnel and add 5 ml of the dilute HCl and 5 ml of saturated Na_2SO_4 solution. Extract successively with 25, 20 and 15 ml portions of $CHCl_3$, shaking vigorously at least 1 minute each time. Combine the 25 and 20 ml extracts in a 500 ml separatory funnel and wash with 400 ml of water, 5 ml of saturated Na_2SO_4 solution, and 5 ml of the HCl. Save the wash water. Filter the combined extracts through a cotton plug into a dry 125 ml separatory funnel. Wash the 15 ml portion of $CHCl_3$, using the same wash water, and combine with the other portions in the same 125 ml separatory funnel.

Add 10 ml of the color reagent to the combined extracts, shake vigorously exactly 2 minutes, and let the acid layer separate as completely as possible. Transfer 9.0 ml of the acid layer to a 50 ml volumetric flask, dilute to the mark with acetic acid, mix well, and read the absorbance at 560 mμ.

Prepare a standard curve by steam distilling and otherwise treating as above a series of freshly prepared dilutions of the standard indole solution, (**e**). Also run a distillation blank omitting the addition of indole.

Method 10-36. *Ammonia in Crabmeat.*[41]

(Use NH_3-free water throughout; ordinary distilled water is suitable.)

REAGENTS

a. *Bromine solution*—Dissolve 0.5 ml of bromine in 100 ml of water. Add 35 ml of 2*N* NaOH and mix. (Solution is stable for 3 days when stored in a dark bottle.)

b. *Thymol solution*—Dissolve 2.0 g of thymol in 10 ml of 2*N* NaOH, and dilute to 100 ml with water. (Solution is stable for 3 days when stored in dark bottle.)

c. *Ammonia standard solution*—Dissolve 0.314 g of NH_4Cl, previously dried 1 hour at 100°, in water and dilute to 100 ml. Transfer 10.0 ml to a 200 ml volumetric flask, and dilute to volume with water. 1 ml = 50μg NH_3.

d. *Phosphotungstic acid*—10% aqueous solution (w/v).

PREPARATION OF SAMPLE

Remove the meat from the shell (if in shell), and grind three times through a food chopper, mixing after each grinding.

DETERMINATION

Place 50 g of prepared sample in a blender. Add 100 ml of 10% phosphotungstic acid solution, (**d**), and blend about 2 minutes at low speed. Filter through a fluted paper (S & S No. 588, or equivalent) into a 250 ml Erlenmeyer flask, collecting the first 20 ml in a separate container and returning to the filter paper (first 20 ml may be cloudy).

Pipet 25.0 ml of the filtrate (representing 9.0 g crabmeat of 78% moisture) into a 125 ml Erlenmeyer flask, and neutralize to litmus paper by adding 10% NaOH solution dropwise plus 2 drops in excess. Add 10.0 ml of thymol solution, (**b**), and mix. (Ignore any white precipitate which may form.) Add 5.0 ml of bromine solution, (**a**), mix, and let stand 20 minutes. (A blue solution indicates decomposition.)

Transfer the solution to a 125 ml separatory funnel. Rinse the flask with 2–5 ml portions of water, adding the rinsings to the funnel. Add 30 ml of n-butyl alcohol (reagent grade), and shake for about 30 seconds. Discard the lower aqueous layer and drain the alcohol layer through a 1″ column (about 28 mm diameter) of anhydrous Na_2SO_4 into a glass-stoppered flask.

Pipet 5.0 ml into a suitable container, add an equal volume of n-butyl alcohol, and mix. Determine the absorbance at 475 and 682 mμ in a 1 cm cell, using n-butyl alcohol in the reference cell. The final color solution is stable.

Net Absorbance = Absorbance_{682} − Absorbance_{475}.

STANDARD REFERENCE CURVE

Transfer 0, 2.0, 4.0, 6.0, 8.0, 10.0, 12.0, and 16.0 ml of standard ammonia solution to 125 ml Erlenmeyer flasks, and dilute each to 25 ml with water. These standards represent 0, 100, 200, 300, 400, 500, 600, and 800 μg NH_3, respectively.

Add 2 drops of 10% NaOH solution to each flask, and proceed as in the determination, beginning: "Add 10.0 ml of thymol solution, (**b**), . . .", except do *not* dilute the final color solution with an equal volume of n-butyl alcohol.

Determine the net absorbance and plot μgNH_3 versus net absorbance for the standard curve.

CALCULATIONS

From the net absorbance of the sample, determine μg NH_3 from the standard curve.

μg NH_3/g = μg NH_3 from curve/4.5.

Chlortetracycline

Chlortetracycline (aureomycin®) has been added by commercial fishermen to the ice in which ocean fish are packed, to reduce bacterial decomposition during the time elapsing between catching the fish and their delivery at the piers. The following microbiological method is designed to detect and determine this antibiotic in such fish (as well as in cooked fish):

Method 10-37. *Chlortetracycline.* (Cylinder Plate Method).[45]

The test organism employed is *B. cereus var. mycoides* (ATCC 9634 PCI—213).

Make slant subcultures every 2 weeks on Difco Pen Assay base agar (B-270), incubate 18 hours at 30° and then refrigerate for future use.

Prepare spore suspensions in Roux flasks using the slant medium supplemented with $MnSO_4 . 4H_2O$ to obtain a final concentration of 10 μg Mn^{++} per ml. Inoculate the Roux flask (150–200 ml medium) using the growth from one slant suspended in 5 ml of sterile physiological saline solution. Incubate 5–7 days at 30° and determine the degree of sporulation by direct smear. When 50% or more of the cells are in the spore state, wash off the Roux flask growth with 25–50 ml sterile saline and centrifuge the suspension. Wash the cells two or three times with sterile saline (25 ml), centrifuging between each washing. Finally resuspend the cells in 100 ml of sterile saline and heat shock in a 70° water bath for 30 minutes and then store in the refrigerator. Determine by standard curve tests the optimum percentage inoculum for suitable zone clarity and dose sensitivity.

The culture medium is Difco Pen Assay base agar (B-270) autoclaved 15 minutes at 121°; adjust to pH 5.7 after sterilization using 10–15 ml of 0.1 *N* HCl per liter of agar.

In preparing the assay plates, inoculate the cooled (50°) assay agar, mix thoroughly and distribute 5 ml medium into each flat-bottomed Petri dish (Pyrex No. 3162) equiped with an unglazed ceramic cover. After solidifying (10–20 minutes at room temperature), place 4 or 6 cups on the agar surface and allow cups to seal and seat themselves (5–10 minutes). Stainless steel assay cups (8 mm O.D. × 10 mm high) may be applied to the agar surface using a lucite template with 4 or 6 holes. A cylinder dropping device may be more convenient for a large number of determinations.

The diluent employed for the chlortetracycline standard curve and the samples is sterile pH 4.5 potassium phosphate buffer (0.1*M* KH_2PO_4 in distilled water).

Construction of the chlortetracycline hydrochloride standard curve involves the preparation of a 100 μg/ml stock solution:

Dissolve an appropriate weight of powder in distilled water, and then acidify with HCl to approximately 0.01N HCl, i.e., dissolve the powder in sufficient water, add 10 ml of 0.1N HCl, and dilute to 100 ml with water. Prepare further dilutions in pH 4.5 phosphate buffer to obtain 0.4, 0.2, 0.1, 0.05, 0.025, 0.0125, 0.00625, 0.00312 μg/ml. The stock solution (100 μg/ml) may be used for one week when held refrigerated.

To prepare the samples add 20 ml of pH 4.5 buffer to each 5-gram aliquot of tissue and homogenize in a Waring Blendor using a stainless steel micro-blender cup; check the final volume and dilute to 25 ml if necessary. When high potency tissues or liquid specimens are under test, dilute with buffer to an estimated 0.05 μg/ml.

Place test samples and standard solutions in the respective cups using a drawn capillary pipette. Each test plate contains four cups with the unknown and two cups filled with 0.05 μg/ml standard. Test each sample on at least two such plates. Prepare at least 12 cups for each CTC concentration on the standard curve. Incubate all test plates overnight (16–18 hours) at 30°. Measure the resulting inhibitory zone diameters on a Quebec colony counter equipped with an etched glass mm scale. Obtain the average zone size for each CTC standard concentration and plot on three cycle semi-log paper (log dose vs. mm zone size). Average the zones obtained with the sample under test, adjusting to fit the standard curve as determined by the 0.05 μg/ml standard and read off the value in μg/ml from the curve. Multiply the curve value by the appropriate dilution factor to obtain the potency in the original sample.

Struvite Crystals

Not infrequently crystals of magnesium ammonium phosphate ("Struvite") are found in cans of fish—particularly tuna fish and shrimp—and mistaken by the consumer for fragments of glass. No doubt these crystals result from liberation of ammonia from the flesh that is so slow as to favor the formation of well-characterized and fairly large crystals of the relatively insoluble double salt. The facts that these crystals are relatively hard, sharp and transparent lend credence to the suspicions of the chemically naïve that such fish are contaminated with broken glass.

Distinction in the laboratory between these crystals and glass fragments (which have been found in fish, although rarely) is simple. Microscopic examination will usually readily reveal whether or not the material is crystalline (which glass, of course, is not). The crystals dissolve in dilute acid. On ignition they form a white ash, while glass melts. There is little need to make specific tests for magnesium, phosphorus or ammonia.

TEXT REFERENCES

1. U.S. Dept. of Agriculture, Agricultural Research Service, Meat Inspection Division, *Regulations Governing the Meat Inspection of the U.S. Dept. of Agriculture*, Definition **1.1 (u)** (1960).
2. Leach, A. E., and Winton, A. L.: *Food Inspection and Analysis*, 4th ed. New York: John Wiley & Sons (1920), 208–211.
3. Jacobs, M. B. (ed.): *The Chemistry and Technology of Food and Food Products*. New York: Interscience Publishers, Inc. (1951), II, 903.
4. American Meat Institute Foundation: *The Science of Meat and Meat Products*. Chicago: W. H. Freeman & Co. (1960), 187–189.
5. Fisher, H. J., *Conn. Agr. Expt. Sta. Bull.*, **617**: 59 (1958).
6. Ref. 4, p. 209.
7. *Ibid.*, p. 205.
8. Assaf, S. A., and Bratzler, L. J.: *J. Agr. Food Chem.*, **14**: 488 (1960).
9. Ref. 4, p. 190.
10. Whitaker, J. R.: *Advan. in Food Res.* New York, London: Academic Press (1959), IX, 1–47.
11. Ref. 4, pp. 216–221.
12. Benne, E. J., Van Hall, N. H. and Pearson, A. M.: *J. Assoc. Offic. Agr. Chem.*, **39**: 937 (1956).
13. Fisher, H. J.: *Conn. Agr. Expt. Sta. Bull.*, **596**: 32 (1955); **602**: 36 (1956); **609**: 43 (1957).
14. *Official Methods of Analysis of the Association of Official Agricultural Chemists*, 10th ed. (1965), Method **23.007**.
15. Fisher, H. J.: *Conn. Agr. Expt. Sta. Bull.*, **609**: 43 (1957).

15a. Perrin, C. H., and Ferguson, P. A.: *J. Assoc. Offic. Anal. Chemists*, **51**: 971 (1968).
16. Hankin, L., and Wickroski, A.: *J. Assoc. Offic. Agr. Chem.*, **47**: 696 (1964).
17. Bailey, E. M.: *Conn. Agr. Expt. Sta. Bull.*, **401**: 869 (1936); **415**: 695 (1937); **426**: 14 (1938).
18. *Official Methods of Analysis of the Association of Official Agricultural Chemists*, 5th ed. (1945), Method **28.19**.
19. Glover, W., Kirschenbaum, H. and Caldwell, A.: *J. Assoc. Offic. Agr. Chem.*, **49**: 308 (1966).
20. Follett, M. J., and Ratcliff, P. W.: *J. Sci. Food Agr.*, **14**: 138 (1963); Landmann, W. A.: *J. Assoc. Offic. Agr. Chem.*, **49**: 875 (1966).
21. Modification of Ref. 14, Method **27.072(c)–27.073**.
22. Zipser, M. W., Kwon, T., and Watts, B. M.: *J. Agr. Food Chem.*, **12**: 109 (1964); Zipser, M. W., and Watts, B. M.: Ibid., **15**: 80 (1967).
23. Weeks, C. E., and Deutsch, M. J.: *J. Assoc. Offic. Agr. Chemists*, **50**: 793 (1967).
24. Ref. 14, Method **23.021**.
25. Bennett, O. L.: *J. Assoc. Offic. Agr. Chemists.*, **31**: 515 (Method 1) (1948).
26. Fredholm, H.: *Food Technol.*, **21**: 198 (1967).
27. Hankin, L.: *J. Assoc. Offic. Agr. Chem.*, **48**: 1122 (1965).
28. Ref. 2, pp. 236–237.
29. Letter of Claudy, N., Canal Industrial Corp., January 20, 1967. This method is essentially the same as the method in *J. Assoc. Offic. Anal. Chem.*, **50**: 205, 282 (1967), and a method submitted by Gershman, L. L., Boston District, U.S. Food & Drug Administration, on November 18 and December 1, 1966. *See also* Payne, W. R., Jr., *J. Assoc. Offic. Agr. Chem.*, **46**: 1003 (1963); Womack, M. S., and Grimmott, J. A., private communications, Food & Drug Administration, U.S. Dept. Health, Education and Welfare, respectively September, 1964 and November, 1965.
30. Crowell, G. K.: *J. Assoc. Offic. Agr. Chemists.*, **27**: 448 (1944).
31. Moore, S., and Stein, W. H.: *Ann. N. Y. Acad. Sci.*, **49**: 265 (1947); *J. Biol. Chem.*, **176**: 237 (1948); **178**: 79 (1949); Spackman, D. H., Stein, W. H., and Moore, S.: *Anal. Chem.*, **30**: 1190 (1958); Piez, K. A., and Morris, L.: *Anal. Biochem.*, **1**: 187 (1960).
32. Ref. 2, p. 211.
33. Thurston, C. E., *et al.: Food Res.*, **23**: 621 (1958); **24**: 497 (1959); **26**: 40, 497 (1961); *J. Agr. Food Chem.*, **9**: 314 (1961).
34. Teeri, A. E., *et al.: Food Res.*, **22**: 148 (1957); **25**: 480 (1960).
35. Proctor, B. E., *et al.: Food Res.*, **26**: 284 (1961).
36. Goddard, J. L.: *Federal Register*, February 2, 1967.
37. Hillig, F.: *J. Assoc. Offic. Agr. Chemists.*, **39**: 773 (1956).
38. Hillig, F., Shelton, L. R., Jr., and Loughrey, J. H.: Ibid., **45**: 724 (1962).
39. Campbell, C.: Ibid., **45**: 731 (1962).
40. Duggan, R. E., and Strasburger, L. W., Jr.: Ibid., **29**: 177 (1946).
41. Burnett, J. L.: Ibid., **48**: 624 (1965).
42. Ornstein, L.: *Ann. N. Y. Acad. Sci.*, **121**: 321 (1964).
43. Hillig, F., *et al.: J. Assoc. Offic. Agr. Chem.*, **41**: 785 (1958).
44. Dyer, W. J.: *Ibid.*, **42**: 292 (1959); Bethea, S., and Hillig, F.: Private communications, Food & Drug Administration, U.S. Dept. of Health, Education and Welfare, November 1964; *J. Assoc. Offic. Agr. Chemists*, **48**: 731 (1965).
45. Ma, R. M., and Morris, M. P.: Private communication, Food & Drug Administration, U.S. Dept. of Health, Education and Welfare, November 1964; *J. Assoc. Offic. Agr. Chemists.*, **48**: 731 (1965).

SELECTED REFERENCES

A. American Meat Institute Foundation: *The Science of Meat and Meat Products*. Chicago: W. H. Freeman & Co. (1960).

B. Bailey, B. E., Carter, N. M., and Swain, L. A.: *Marine Oils With Particular Reference to Those of Canada, Bull.* **89**. Ottawa: Fishery Research Board of Canada (1952).

C. Chatfield, C., and Adams, G.: *Proximate Composition of American Food Materials*, U.S. Dep. Agr. Circular **549** (1940).

D. Leverton, R. M., and Odell, G. V.: *The Nutritive Value of Cooked Meat*, Oklahoma Agr. Exp. Sta. Misc. Pub. MP–49 (1958).

E. Tressler, D. K., and Lemon, J. McW.: *Marine Products of Commerce*. New York: Reinhold Publishing Co. (1951).

F. Watt, B. K., and Merrill, A. L.: *Composition of Foods*, Agricultural Research Service, Consumer & Food Economic Division, U.S.D.A., Agriculture Handbook 8 Revised (December 1963).

G. Winton, A. L., and Winton, K. B.: *The Structure and Composition of Foods*. New York: John Wiley and Sons (1937), III.

CHAPTER 11

Fruits and Fruit Products

Section I. Fresh, canned and frozen fruits; fruit juices and concentrated juices; dried fruit

The distinction between "fruit" and "vegetable" is completely arbitrary. Botanically, a fruit is that portion of a seed plant containing the reproductive bodies (seeds) together with any adhering or supporting structures. In other words, it is the ripened ovary of the plant. Popularly, a fruit is the edible, more or less sweet, pulpy portion of a plant, and the term is so used in this text. Under this definition, rhubarb (the petiole or leaf stem) is considered a fruit, while squash and tomato, technically fruits, are considered vegetables.

The term "vegetable" is not used technically. In common language, it is the edible portion of a plant usually consumed during the principal part of a meal.

Sampling

In many cases the size and method of sampling is decided by someone other than the analyst. His only decision may be whether to use the entire sample for tests or reserve a portion for check analyses, exhibits or other reasons. Often the inspector or another person submitting the sample will divide it into two portions—one for analysis and one for reserve.

Fresh Fruit: The purpose of the investigation largely dictates the size and method of sampling. In general, individual fruits are drawn at random from the lot sampled. As many as can be conveniently handled should be drawn in order to reduce to a minimum the naturally occuring variation in composition among individual units.[1a,1b,13]

Canned and Consumer Size Frozen Fruit, and Dried Fruit: These usually bear an identifying mark on the container or carton such as a batch number, code number, processing date, or another symbol which serves to identify a particular batch or lot. In inspection of purchased merchandise or in regulatory analysis, each so identified lot should be sampled and analyzed separately. Inspection manuals[2] and such texts as *Official Methods of Analysis*, AOAC, usually include methods of sampling. A good general rule is to sample the square root of the number of units in the lot.[3]

Frozen Fruit, Large Containers: Sample barrels and large boxes of frozen fruit by drilling with a stainless steel auger bearing a serrated cutting edge.[4a,4b] Remove cores by using a wooden ram.

Dried Fruit, Large Boxes: Remove the top

and one end of the box and withdraw 1/8 of its contents as follows: Visually divide the top surface into quarters. With a sharp knife make two vertical cuts along the inner boundary of one quarter, extending *halfway* to the bottom and meeting at the center of the block. Remove all fruit in the angle between the two cuts, working rapidly to avoid loss of moisture; break up all lumps, mix thoroughly and withdraw enough sample to fill a quart Mason jar, replacing the remainder in the box. Seal the jar, and keep in the refrigerator until analysis begins. This is the official AOAC method of sampling.[4b] Prepare sample for analysis, after removing pits from stone fruit and determining their proportion in the sample, by passing it through a food chopper three times, mixing quickly and thoroughly after each grind.

Fruit Juices: Mix sample by shaking and, unless otherwise directed, filter through absorbent cotton or a rapid filter.

Methods of Analysis

Method 11-1. *Moisture in Dried Fruit.*

Spread a 5–10 g ground sample evenly over the bottom of a previously dried and tared metal dish about 8.5 cm in diameter, provided with a tight-fitting cover. Proceed as in Method 1-2 (Moisture, Vacuum Oven Method) by drying 6 hours at 70° and 100 mm of mercury. After 6 hours, replace the cover tightly, remove to a desiccator, cool and weigh.

For raisins and other fruit high in sugar, weigh a 5 g sample into a tared dish containing about 5 g of finely shredded asbestos. Moisten the contents of the dish with a little hot water, mix thoroughly, and evaporate barely to dryness on a steam bath. Place the dish in the oven and continue drying under vacuum for 6 hours as directed above.

Sulfur dioxide in Dried Fruit. See Method 14-1.

Method 11-2. *Determination of Head-Space, Net Weight, and Drained Weight of Canned Fruit.*[5,6]

NOTE: These determinations should be made on each of a representative number of units, according to batch numbers or other identifying marks.

DETERMINATION

In the case of an hermetically sealed container with lid attached by a double seam (the conventional canned food can), weigh the dry, unopened can on a suitable heavy-duty balance having a sensitivity of 0.01 or 0.02 oz and record as "gross weight." Cut out the lid without removing or altering the height of the double seam, and measure the vertical distance from the top level of the container to the top level of the food (a depth gage is convenient for this purpose). Record as "head-space."

Pour the contents of the can evenly over the surface of a #8 U.S. Standard woven-wire sieve (size of opening 2.38 mm) held over a pan to collect the drainage. Use an 8″ sieve for a container holding less than 3 pounds net and a 12″ sieve for larger containers. Carefully invert all fruit (such as peach halves) having cups or cavities that fall cup-side up. No other handling of the fruit during draining is permissible. Incline the sieve, without shifting the contents, to facilitate drainage, and allow to drain for two minutes from the time the contents were poured on the sieve. Immediately transfer the drained fruit to a clean, dry, tared pan in one rapid motion. Record this weight as "drained weight".

Wash, dry and weigh the empty container and lid; subtract this weight from the "gross weight". Record the difference as "net weight".

Method 11-3. *Water Capacity of Can and Fill of Container.*[5]

Fill the container, opened as in Method 11-2, with distilled water at 68° F (20°) to 3/16″ vertical distance from the top level of the container, and weigh. Subtract the weight of the empty can and record as "water capacity of the container". Keeping the water at 68°F., drain off the water to the level of the food originally in the can, which was determined as "head-space". Determine the weight of the remaining water in the container. Divide this weight by the weight recorded as "water capacity of container" and multiply by 100. Record this figure as "percent of fill". In general, this figure should not be less than 90 percent.

For glass jars and other containers not provided with a lid attached by a double seam, the "water capacity of the container" should be determined by filling it with water to the top of the container, and the "percent of fill" should be calculated on this basis.

Comments on Methods 11-2 and 11-3

FDA Standards of Fill of Container at times set limits on the amount of free liquid present (drained weight) and the percent of fill of container. Grade standards issued by the Canadian Department of Agriculture[7] include requirements for net weight, drained weight and head-space for each can size of listed canned fruits and vegetables. Many Federal and Military purchasing specifications issued by U.S. Government agencies include requirements for certain fill-of-container factors. A recommended fill of container is not incorporated in United States Standards for grades of canned foods issued by the Department of Agriculture, since fill of container, as such, is not a factor of quality for the purpose of these grades. Many of these standards, however, contain the statement: "It is recommended that each container be filled as full as practicable and that the product occupy not less than 90 percent of the volume of the container."

Fill of Container of Frozen Fruits

The AOAC has adopted as official a method devised by Wallace and Osborn[8] for fill of container of frozen fruits. The method determines the volume of the frozen fruit by displacement of light mineral oil from a specially devised over-flow can, as described in *Methods of Analysis*, AOAC, 10th ed. (1965), Method **20.004.** Since the method is of interest to relatively few chemists, it is given here by reference only.

Method 11-4. *Drained Weight and Approximate Fruit Content of Frozen Fruit/Sugar Mixtures.* (*AOAC Method* ***20.006***).[9, 10, 11]

Determine the gross weight. Immerse the frozen package in water agitated and maintained at a temperature of 20° ± 1°. (If the package is not airtight, place in a suitable plastic bag, remove excess air by use of vacuum, and tie off.) Avoid agitation of the package during thawing by use of a clamp or weight. When the center of the package reaches bath temperature (2 to 3 hours for a 10 to 16 oz package—this should be verified by a previous test), remove from the bath, blot off adhering water, and open the package with minimum disturbance. Tare a #8 U.S. Standard sieve with a light-weight drip-pan (use an 8″ sieve for a package whose net weight is under 3 pounds, a 12″ sieve for a larger package). With the screen tilted and supported for drainage into a suitable receptacle, distribute the entire contents of the package evenly over the screen in one sweeping motion. After 2 minutes from the time drainage begins, place the sieve with the fruit on the drip-pan and weigh. This weight, minus the weight of the sieve plus drip-pan, is the drained weight. Clean and dry the emptied package and subtract its weight from the gross weight to determine the net weight. Calculate drained weight as percent of net weight.

Recombine the fruit on the screen, any drippings in the drip-pan, and the liquor drained through the sieve, and mix thoroughly in a high-speed blender (Waring or equivalent) for 2 to 5 minutes. When the sample has become homogeneous, filter a small portion of the blended sample through a strong lens paper or other suitable medium. Reject the first few drops, then determine the refractive index of the filtrate at 20° by Method **11-15**. Report as percent soluble solids (sucrose), obtained from Table 23-3. Calculate the percent of fruit from the equation given below.

Calculation of Approximate Fruit Content

The percent of drained weight is a rough approximation of the put-in weight of fruit in the package. Soft fruits, such as strawberries and red raspberries, give somewhat lower figures than the true put-in weights. According to Falscheer and Osborn,[9] about 2/3 of the put-in weight of strawberries, and 3/4 or more of put-in weight of red raspberries, remain on the sieve. They found that 85 to 90 percent of the put-in weight of other fruit remained on the sieve.

A more accurate ratio of fruit to sugar can be obtained from the percent of soluble solids, as measured by the refractometer, using the equation:

$$\% \text{ of Fruit} = \frac{(100\text{-M})100}{100\text{-F}},$$

in which M is the percent soluble solids of the fruit and sugar mix as determined and F is the soluble solids of the fruit as obtained from reports of analyses of authentic fruit. (See Section III on "Authentic Fruit Analyses," this chapter.)

FRESH AND CANNED FRUITS, CHEMICAL ANALYSIS

Certain determinations such as total solids and water insoluble solids are made on the original sample, pulped through a food chopper, Hobart mixer or similar device. Others are, or may be, made on a "prepared solution" as described in Method 11-14. If the sample represents fresh fruit, or unsweetened canned fruit, weigh out 150 g of fruit, and add 150 g of sugar, instead of 300 g as directed in Method 11-14.

It may suffice, in certain examinations of canned fruit, to analyze the syrup drained from the fruit, since soluble constituents have reached equilibrium between the fruit and syrup. In this case separate the syrup by draining through a screen, and treat as a juice or jelly.

Juices and Juice Concentrates

Total solids are usually determined by means of a Brix spindle or a refractometer and reported either as degrees Brix, or as percent sugars (sucrose) as taken from refractive index-sugar Tables 23-3 or 23-5. If the percent of acid is more than 1 percent as anhydrous citric acid, as determined by titration to phenolphthalein, a positive correction for citric acid content of citrus juices and concentrates must be added to the Brix, or solids by refractometer, according to Fig. 11-1.

Most other constituents are determined by methods given in Section II of this chapter.

Pineapple Juice

The FDA standard for quality of canned pineapple juice is:

Soluble solids by Brix spindle at 20°, not less than 10.5° Brix.

Acidity, not more than 1.35 g anhydrous citric acid per 100 ml.

Ratio Brix/acidity not less than 12 to 1.

Finely divided "insoluble solids" (Method 11-6), not less than 5% nor more than 30%.

Pineapple juice may be sweetened if declared on the label as "sugar added". It may be artificially sweetened with saccharin, sodium saccharin, calcium cyclamate, sodium cyclamate or combinations thereof.* The label must bear the statement "artificially sweetened with . .".

The USDA standard for grades of canned pineapple juice have adopted the above quality characteristics for their Grade C juice. USDA Grade A juice requirements are:

Soluble solids by Brix spindle at 20°, not less than 12.0° Brix.

Acidity, not more than 1.10 g anhydrous citric acid per 100 ml.

Ratio Brix/acidity, not less than 12:1.

Finely divided "insoluble solids" not less than 5% nor more than 26%.

Grape Juice

There is no FDA Standard of Identity or Quality for grape juice. USDA Standards for Grades for Canned Grape Juice have established Grade A and Grade B requirements for both unsweetened and sweetened juice. They recognize three types: Type I, from Concord (slip-skin) grapes; Type II, from single-variety grapes other than Concord; and Type III, a combination of 2 or more varieties. The grade defines canned grape juice as juice from which tartrate crystals have been removed. USDA grade requirements are:

* The use of cyclamates was banned in 1970. *See* Chapter 14, p. 324.

Grade A

	UNSWEETENED	SWEETENED
Brix, at 20°	Not less than 15.1°	Not less than 17.1°
Acidity, calculated as tartaric	Not less than 0.60 g nor more than 1.20 g per 100 ml	Not less than 0.60 g nor more than 1.20 g per 100 ml
Brix/acid ratio	Not less than 14:1 nor more than 28.1	Not less than 14:1 nor more than 28:1

Grade B

	UNSWEETENED	SWEETENED
Brix, at 20°	Not less than 14.8°	Not less than 16.0°
Acidity, calculated as tartaric	Not less than 0.45 g nor more than 1.40 g per 100 ml	Not less than 0.45 g nor more than 1.40 g per 100 ml
Brix/acid ratio	Not less than 11.5:1	Not less than 11.5:1

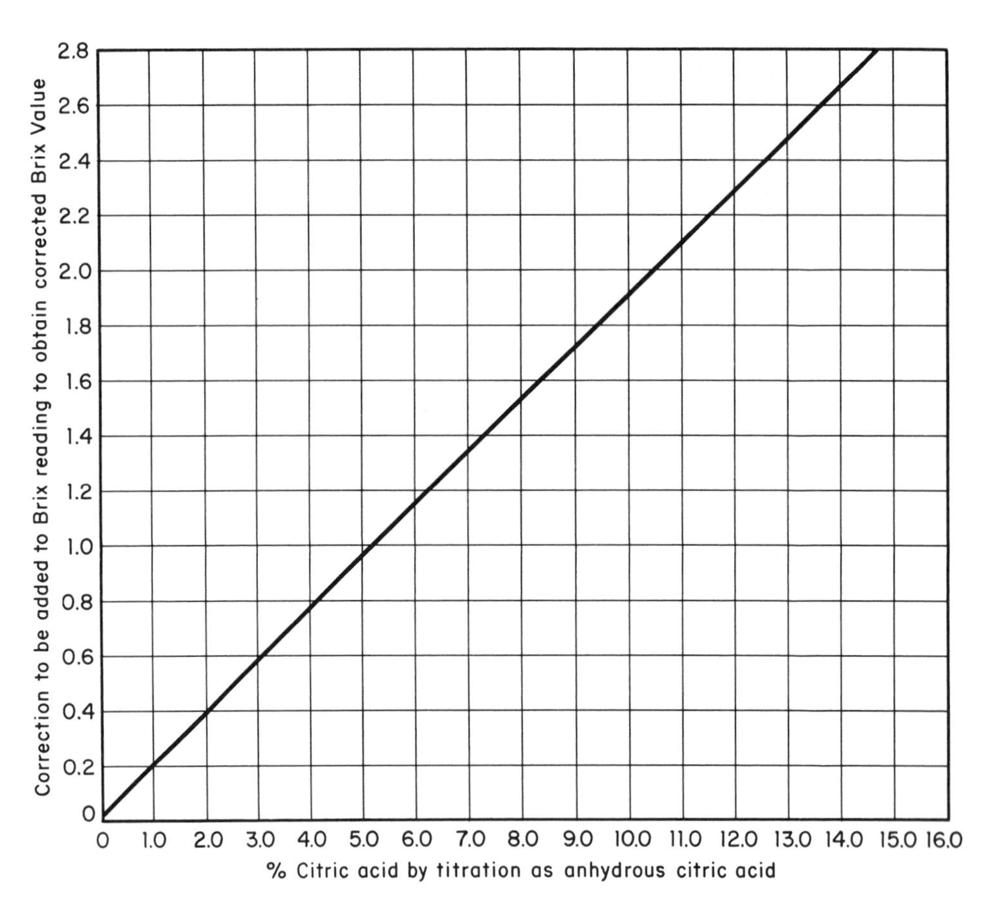

FIG. 11-1. Acidity corrections for Brix values. Plotted from a paper by L. W. Stevens and W. E. Baier, *Anal. Chem.*, **11**: 447 (1939).

The Standards for Grades state that titration for acidity may be made using phenolphthalein "or any other satisfactory indicator". This would appear to sanction an electrometric titration.

Citrus Juices and Concentrates

As of January 1, 1966 the U.S. Food and Drug Administration has recognized 7 different "orange juices" and 4 different "concentrates". Their definitions and analytical requirements are listed in Table 11-1.

The FDA has also promulgated "Definitions and Standards for Frozen Concentrate for Lemonade" and "Frozen Concentrate for Colored Lemonade". The definitions for these frozen concentrates are:

Shall be made from lemon juice and/or frozen lemon juice or concentrated lemon juice and/or frozen concentrated lemon juice, plus sugar and/or invert sugar syrup and water. The soluble solids shall not be less than 48% (48° Brix) corrected for acidity. The lemon oil content may be adjusted in accord with good manufacturing practice, and lemon pulp may be removed or added to make an ade containing no more pulp than if made from lemon juice from which pulp has not been separated.

The colored product is identical except for the addition of natural or artificial color suitable for use in food. Artificial color must be declared.

The ade made from this product according to labeled directions shall have an acidity of not less than 0.70 g per 100 ml expressed as anhydrous citric acid, and its soluble solids shall not be less than 10.5% by weight.

TABLE 11-1

FDA Definitions of Identity: Orange Products

Product	FDA* Section No.	Definition
Orange Juice	27.105	Unfermented juice from *Citrus sinensis* with seeds and excess pulp removed. May be chilled, but not frozen.
Frozen Orange Juice	27.106	Same as "Orange Juice", but frozen.
Pasteurized Orange Juice or Chilled pasteurized orange juice. Added sugars must be declared on the label.	27.107	"Orange Juice" to which unfermented juice from *Citrus reticulata* may be added up to 10%, and heat-treated to reduce pectinesterase activity and viable organisms. Solids may be adjusted by addition of concentrated juice, and Brix/acid ratio by addition of sweetening ingredients to normal ranges. Shall contain not less than 10.5% orange juice soluble solids, and Brix/acid ratio shall not be less than 10:1.
Canned Orange Juice. Added sugars must be declared.	27.108	"Orange Juice" or "Frozen Orange Juice" to which unfermented juice from *Citrus reticulata* may be added up to 10%, and orange oil and orange pulp content may be adjusted, conforming to good manufacturing practice, and sweetening ingredients to adjust Brix/acid ratio to that normal to orange juice, and sealed and processed by heat. Shall test not less than 10° Brix and Brix/acid ratio shall not be less than 9:1.
Frozen Concentrated Orange Juice (or Orange Juice Concentrate). Added sugars must be declared on label. If the dilution ratio is more than 3:1 the dilution ratio must be declared on the label, or may be declared as Brix°, if over 1 pint.	27.109	"Orange Juice" to which juice from *Citrus reticulata* up to 10%, and/or juice from *Citrus aurantium* up to 5% may be added, water removed, and the product frozen. Orange oil, orange essence from juice, orange pulp and sweeteners may be added to adjust to normal composition. The dilution ratio to produce orange juice shall not be less than 3+1. The orange juice soluble solids of the reconstituted juice shall not be less than 11.8%.

TABLE 11-1 (continued)

Product	FDA Section No.	Definition
Canned Concentrated Orange Juice (or Orange Juice Concentrate). The dilution ratio in containers over 1 pint may be declared as Brix°.	27.110	Complies with requirements of "Frozen Concentrated Orange Juice", except for being sealed and processed by heat.
Orange Juice from Concentrate (or Reconcentrated Orange Juice). Added sugars must be declared on the label.	27.111	Made by diluting "Frozen Concentrated Orange Juice" and/or "Concentrated Orange Juice for Manufacturing", to which may be added orange juice, or frozen, and/or pasteurized orange juice. Orange oil, orange pulp and/or sweeteners may be added to adjust to normal composition. The finished orange juice shall contain not less than 11.8% orange juice soluble solids.
Orange Juice for Manufacturing.	27.112	Prepared from "orange juice" except that the juice may be below standards for maturity, in Brix and Brix/acid ratios for such oranges. Juice from *Citrus reticulata* may be added up to 10%. Pulp and orange oil may be adjusted to good manufacturing practice.
Orange Juice with Preservative. Shall be labeled "—— added as a preservative".	27.113	Complies with "Orange juice for Manufacturing" except that sodium benzoate or sorbic acid may be added up to 0.2%.
Concentrated Orange Juice for Manufacturing. Must be labeled to show soluble solids as degrees Brix.	27.114	Complies with requirements for composition and labeling of "Frozen Concentrated Orange Juice" except that it may not be frozen, or may be prepared from oranges below standards for maturity in Brix and Brix/acid ratio. Concentration of orange juice solids shall not be less than 20° Brix.
Concentrated Orange Juice with Preservative. Must be declared "—— added as a preservative".	27.115	Shall comply with "Concentrated Orange Juice for Manufacturing", except that sodium benzoate or sorbic acid up to 0.2% may be added.

* FDC Regulations, part 27: Canned Fruits and Fruit Juices, Definitions and Standards. See also reference 5, part 27, sec. 27.105 *et seq.*

The U.S. Department of Agriculture, through its Agricultural Marketing Service (now called Consumer and Marketing Service), has issued Standards for grades of various citrus products, for voluntary use. Colors of citrus juices and concentrates are rated by comparing with USDA color standards. These standards may be obtained from Process and Marketing Service, U.S.D.A., Washington, D.C. The requirements for these grades are listed in Table 11-2. Most of these products do not have a requirement for acidity as such in their Standard for Grade. A few products do have specific acidity limits, reported as grams of anhydrous citric acid per 100 ml. These are:

	Grade A	Grade B (C)
Canned blended grapefruit and orange juice.	0.8 to 1.70	0.65 to 1.80
Canned lemon juice.	5.0 to 7.0	4.5 to 7.5
Frozen concentrate for lemonade, reconstituted	0.7 min	0.7 min.
Frozen concentrate for limeade, reconstituted.	0.7 min.	0.7 min.

The Food and Drug Directorate of Canada[12] has issued regulations defining certain citrus and other juices which have the force and effect of law. These definitions are:

Grapefruit juice shall be the juice obtained from grapefruit and shall contain in 100 ml juice at 20°, not less than 9.5 g soluble solids, and between 1.0 and 2.2 g of acid expressed as anhydrous citric acid.

Lemon juice shall be the juice obtained from lemons and shall contain in 100 ml juice at 20°, not less than 8.0 g soluble solids, and 5.0 g of acid expressed as anhydrous citric acid.

Lime juice shall be the juice obtained from limes and shall have a specific gravity (20°/20°) of not less than 1.030 nor more than 1.040, and shall contain in 100 ml juice at 20°, not less than 8.0 g of soluble solids and 5.5 g of acid expressed as anhydrous citric acid. Its optical rotation at 20° in a 200 mm tube shall lie between plus 5 and minus 1.5 degrees Ventzke.

Orange juice shall be the juice obtained from oranges and shall contain in 100 ml at 20°, not less than 10 g of soluble solids, not less than 0.5 and not more than 1.9 g of acid expressed as anhydrous citric acid, and shall have a soluble solids/acid ratio of not less than 8:1.

Concentrated fruit juice shall be fruit juice which has been concentrated to at least one-half its original volume by the removal of water, with or without the addition of vitamin C, food color, or stannous chloride.

An estimate of the fruit juice content of juices, concentrates, nectars, ades, etc., can be made by comparing the analyses of the products in question with analyses of authentic juices given in Section III of this chapter. A review of the heading "Interpretation of Analytical Results" in Section II, Page 241 will be helpful to the interpretation. As mentioned there, allowance must be made for additions of ingredients not derived from fruit that affects the analysis. For example, the presence of sodium benzoate or sodium sorbate increases the ash content, and added sugar increases the soluble solids content.

Citrus juice concentrates should be diluted to the juice equivalent, following directions on the label, before analysis. Consideration must be given also to the geographical origin of the sample.

There is considerable variation in the composition of single fruits on a tree. Sites and Reitz[13] analyzed approximately 1800 individual oranges from one Valencia orange tree in Florida, all picked within a 6-day period. They concluded that variation in composition was related to the position on the tree, fruits receiving more light having the highest solids content. These workers reported variations as follows:

Range in Composition of Oranges on One Tree

	Min.	Max.	Average
Soluble solids by refractometer, percent	5.90	13.50	10.24
Acidity, as citric, percent	0.50	1.39	0.89
Brix/acid ratio	4.80	21.00	11.56
Ascorbic acids mg/100 ml	18.20	59.60	37.10
Juice per orange, percent	32.70	65.80	49.20

Essential oils may be determined by the "water-insoluble volatile oil method" given for spices, using a Clevenger lighter-than-water trap described in Method 15-5. This method is official in United States Standards for grades of certain citrus juices and citrus products, under the designation "recoverable oil".

Method 11-5. *Volotile Oils (Recoverable Oils) in Citrus Juices and Concentrates.*

APPARATUS:

3L standard taper round-bottomed flask, to fit a Clevenger lighter-than-water oil trap and a West condenser. (A cold-finger reflux condenser may be used.)

DETERMINATION

	Size of Sample
Frozen concentrate for lemonade	1L, plus 1L water
Frozen concentrated grapefruit juice	400 g, plus 1.5L water
Frozen concentrated orange juice	2L of reconstituted juice
Canned grapefruit juice	2L
Dehydrated orange juice	2L of reconstituted juice
Canned blended grapefruit and orange juices	2L
Frozen concentrated blended grapefruit and orange juices	400 g plus 1.5L water
Frozen concentrated limeade	1L of concentrate plus 1L water

TABLE 11-2

USDA Standards for Grades of Citrus Products—Analytical Limits

Product / Grade[b]	Brix °–Min[a]		Brix/Acid Ratio[d]		Recoverable Oil, ml/100 ml Max.		Free and suspended pulp, ml/100 ml Max.	
	A	B(C)	A	B(C)	A	B(C)	A	B(C)
Grapefruit Juice:								
Single Strength–Unsweetened	9.0[c]	10.0[c]	8:1 to 14:1	7:1 min.	0.020	0.025	10	15
Single Strength–Sweetened	11.5[c]	11.5[c]	9:1 to 14:1	9:1 min.	0.020	0.025	10	15
Reconstituted–Unsweetened	10.0[c]	10.0[c]	8:1 to 14:1	7:1 min.	0.020	0.025	10	15
Reconstituted–Sweetened	11.5[c]	11.5[c]	9:1 to 14:1	9:1 min.	0.020	0.025	10	15
Canned Blended Grapefruit and Orange Juice:								
Unsweetened	10.0[c]	9.5[c]	9:1 to 17:1[e,l]	7.5:1 min.	0.035	0.055	12	18
Sweetened	11.5[c]	11.5[c]	10:1 to 17:1[f,l]	10:1 min.[f]	0.035	0.055	12	18
Frozen Conc. Grapefruit Juice:								
Unsweetened	38 to	42	9:1 to 14:1	7:1 to 16:1	0.027 to 0.067[h]	.067[h]	10[j]	10[j]
Sweetened	38 to	48[g]	10:1 to 13:1	8:1 to 13:1	0.027 to 0.067[h]	.067[h]	10[j]	10[j]
Conc. Grapefruit Juice for Manf.	[k]		6:1 min.	5.5:1 min.			10[j]	12[j]
Frozen Conc. Blended Grapefruit and Orange Juice:								
Unsweetened	40 to	44	10:1 to 16:1	8:1 to 18:1	0.097[h]	0.113[h]	12[j]	18[j]
Sweetened	40 to	48[i]	11:1 to 13:1	9:1 to 13:1	0.097[h]	0.113[h]	12[j]	18[j]
Frozen Concentrate for Lemonade[m]	10.5[n]	10.5[n]	20:1[l,n]	20:1[n]	0.025[n]	0.035[n]		
Frozen Concentrate for Limeade	10.5[n]	10.5[n]	18:1 max.[n,l]	18:1 max.[n]	0.008 to 0.025[n]	0.008 to 0.035[n]		
Canned Concentrated Orange Juice:								
Unsweetened	41.8[o]	41.8[o]	11.5:1 to 18:1	9.5:1 to 20:1	0.0028[h]	0.0034[h]		
Sweetened	42.0[o]	42.0[o]	12:1 to 18:1	10:1 to 20:1	0.0028[h]	0.0034[h]		

Frozen Conc. Orange Juice:						
Unsweetened	41.8°	42.0°	11.5:1 to 19.5:1[p]	10:1 min.	0.035 [j]	0.040 [j]
Sweetened	42.0°	42.0°	12:1 to 19.5:1[p]	10:1 min.	0.035 [j]	0.040 [j]
Dehydrated Orange Juice[q]			12:1 to 18:1[n]	10.5:1 to 19:1[n]	0.011 to 0.017[n]	0.009 to 0.025[n]
Dehydrated Grapefruit Juice:						
Unsweetened[q]			8:1 to 14:1[n]	7:1 to 14:1[n]	0.010 to 0.017[n]	0.009 to 0.025[n]
Sweetened[q]			11:1 to 14:1[n]	11:1 to 14:1[n]	0.010 to 0.017[n]	0.009 to 0.025[n]

[a] Brix° by refractometer corrected for citric acid—See Fig. 11-11.
[b] Some grades listed as A and B, others as A and C.
[c] Uncorrected for citric acid.
[d] Acid reported as percent anhydrous citric for Brix/acid ratio.
[e] If Brix is 11.5° or more, ratio may be not less than 8:1.
[f] If Brix is 15° or more, ratio may be less than 10:1.
[g] Shall contain not less than 3.47 lbs. soluble grapefruit solids per gallon.
[h] ml/100 g of concentrate.
[i] Brix of concentrate without sweeteners is 38 minimum.
[j] As ml/100 ml reconstituted product.
[k] 37.5° to 63.2°, depending on dilution factor. See U.S. Standard of 9/21/68.
[l] Acidity reported as grams anhydrous citric acid per 100 ml.
[m] Soluble solids corrected for acidity—48.0%.
[n] Basis of reconstituted ade or juice.
[o] Reconstituted juice shall contain 11.8% min. soluble orange solids.
[p] For California and Arizona juice. For others ratios are 12.5:1 to 19.5:1.
[q] Shall contain 3% moisture and 250 p.p.m. SO_2 maximum.

Transfer the sample to the 3L flask, attach the condenser and trap and proceed as in Method 15-5, *Volatile Oils, Clevenger trap method.* Report as ml oil/100 g.

Method 11-6. *Free and Suspended Pulp in Citrus Juices and Pineapple Juice.*[14, 15] (At times called "fine centrifugal pulp" or "finely divided insoluble solids".)

These are terms used in United States Standards for grades of various citrus juices and juice concentrates, and the FDA Definition and Standard for Canned Pineapple Juice to designate particles of membrane, core and peel and similar extraneous material that settles out upon centrifuging under prescribed conditions.

PREPARATION OF SAMPLE

Frozen Concentrated Grapefruit Juice—Skim floating fruit cells and pulp from the reconstituted juice.

Concentrated Lemon Juice for Manufacturing, and Canned Lemon Juice—Pour the lemon juice, or the concentrated lemon juice after reconstituting, through a 20 mesh sieve.

Canned Grapefruit Juice, Canned Blended Grapefruit Juice and Orange Juice, and Canned Pineapple Juice—Stir the juice thoroughly.

DETERMINATION

Measure 50 ml of the prepared sample into each of two 50 ml long-cone, graduated centrifuge tubes approximately 4 $\frac{3}{16}''$ over-all length. Place the tubes in a suitable centrifuge, the approximate speed of which is related to the diameter of swing in accord with the table immediately below. The word "diameter" means the over-all, tip-to-tip distance of opposing centrifuge tubes in operating condition.

Diameter inches	Approximate revolutions per minute	Diameter inches	Approximate revolutions per minute
10	1,609	15½	1,292
10½	1,570	16	1,271
11	1,534	16½	1,252
11½	1,500	17	1,234
12	1,468	17½	1,216
12½	1,438	18	1,199
13	1,410	18½	1,182
13½	1,384	19	1,167
14	1,359	19½	1,152
14½	1,336	20	1,137
15	1,313		

Centrifuge exactly 10 minutes (in the case of canned pineapple juice, 3 minutes). Multiply the reading to the top of the layer of pulp by 2 to obtain the percentage of "free and suspended pulp".

NOTE: The FDA Definition and Standard for canned pineapple juice requires not less than 5 nor more than 30% of "finely divided insoluble solids".

Method 11-7. *Titratable Acidity of Fruit Juices.*

Determine by titration with standard sodium hydroxide solution, using phenolphthalein as indicator, and 10 to 25 ml of unfiltered juice as sample, following the procedure of Method 11-19. Report as ml $N/10$ NaOH per 100 ml or as g per 100 ml for the predominating acid. Citrus juices and tomato juice are usually reported as grams of anhydrous citric acid, grape juice as grams of tartaric acid, and apple juice as grams of malic acid, per 100 ml of juice (or reconstituted juice in the case of concentrates). 1 ml $N/10$ NaOH is equivalent to 0.0064 g of anhydrous citric acid, 0.0075 g of tartaric acid and 0.0067 g of malic acid.

The FDA Standard and the USDA Standards for Grades for canned pineapple juice direct that 10 ml of unfiltered juice be added to 25 ml of freshly boiled, distilled water and the resulting mixture titrated with 0.1 N sodium hydroxide solution, using 0.3 ml of 1% phenolphthalein solution as indicator.

Fruit Acids

Occasionally an analyst finds it necessary to separate, identify and determine quantitatively the individual fruit acids in a fruit or a fruit product. For example: Fruit juices, such as apple or cherry, may contain citric acid added as an acidulant, or to conceal watering of the juice. These fruits naturally contain only very small quantities of citric acid. In such cases, comparison must be made with analyses of authentic juices, and methods must be available for the determination of these small quantities in the presence of larger quantities of other organic acids.

The AOAC has worked on this problem since 1940, as have laboratories such as

USDA research units and others, and methods are now available for the major fruit acids.

Method 11-8. *Citric and Isocitric Acids. Chromatographic Method.* (*AOAC Method **20.056–20.059***).

REAGENTS

a. *tertiary-Amyl alcohol in chloroform*, 30%—Wash USP $CHCl_3$ three times with about 0.5 volume of water to remove the alcohol. Dilute 300 ml of *tert*-amyl or *n*-butyl alcohol to 1 liter with the washed $CHCl_3$, and shake well with about 50 ml of water. Let the liquids separate, and discard the water. To the *tert*-amyl alcohol-$CHCl_3$ layer add an excess of anhydrous powdered Na_2SO_4. Shake well and filter through a dry paper.

b. *tert.-Amyl alcohol in chloroform*, 40%—Prepare as in **a**, using 400 ml of *tert*-amyl or *n*-butyl alcohol.

c. *Silicic acid suitable for chromatography*—Reagent grade "100-mesh" powder suitable for chromatography (Mallinckrodt Chemical Co. No. 2847 or equivalent).

d. *Lead acetate solution*—Dissolve 75 g of normal $(CH_3COO)_2Pb.3H_2O$ in water, add 1 ml of acetic acid, and dilute to 250 ml.

e. *Metaphosphoric acid*, 20%—Store in a refrigerator.

f. *Sodium sulfide solution*—Dissolve 4 g of $Na_2S.9H_2O$ in water and dilute to 100 ml. Store in the refrigerator.

g. *Sodium thiosulfate standard solution*—0.01N (2.482 g/liter). Standardize against 0.01N KIO_3 (0.3567 g/liter) as follows: To 5 ml of the KIO_3 solution add 1 ml of 2M H_3PO_4 and 1 ml of 10% KI solution, and titrate with the $Na_2S_2O_3$ solution, using starch indicator at the end point.

h. *Filter paper—Cl-free 9 cm.* Wash well with hot water and dry.

i. *Sodium hydroxide standard solution*—0.01N. Protect from CO_2.

j. *Potassium chloride solution*—0.9319 g of dried KCl/liter of 0.085M H_3PO_4.

k. *Silver iodate*—Protect from light.

l. *Orthophosphoric acid*—0.085M.

m. *Thymol blue indicator*—0.1%. Dissolve 0.1 g of thymol blue in water, add 0.1N NaOH until the solution turns blue, and dilute to 100 ml.

APPARATUS

a. *Chromatographic tube*—Approximately 13 mm i.d. and 400 ml long, with a 200 ml reservoir at the top. Plug the end of the tube with cotton.

b. *Piston*—To fit the tube for packing silicic acid.

c. *Centrifuge tube*—Approx. 3 × 11 cm; 60 ml capacity.

d. *Device for titrating in a CO_2-free atmosphere*—Equip a 125 ml pear-shaped separatory funnel with a rubber stopper containing 5 holes. Insert in these holes: 1) a tube with a drawn-out tip extending to the stopcock, for admitting CO_2-free air; 2) the standard acid buret tip; 3) the standard alkali buret tip; 4) a funnel for transferring the eluate; and 5) a tube for exhaust vapors.

e. *CO_2-free air*—Pass air twice through 20% NaOH solution, and then through water containing phenolphthalein and enough 0.1N NaOH to produce a pink solution.

STANDARDIZATION OF THE SILICIC ACID COLUMN

Mix thoroughly, in a mortar, 6 g of the silicic acid and an amount of 0.5N H_2SO_4 determined as follows: Ignite about 1 g of silicic acid, accurately weighed, in a small crucible at red heat about 15 minutes (a gas burner is satisfactory). Cool in an efficient desiccator and weigh. Calculate ml 0.5N H_2SO_4 required, V, from the formula: $V = W(1.9A-1)$; where W = g silicic acid used for the column, and A = the ratio of anhydrous to hydrated silicic acid (as determined by ignition).

Add chloroform a little at a time, and mix, making a uniform slurry that pours readily. With a $CHCl_3$ wash bottle, transfer all of the slurry to the chromatographic tube, pouring it down a thin rod and stirring until all air bubbles are removed. Cut a circle of coarse filter paper (Whatman No. 4 or equivalent) with a cork borer to fit snugly in the tube. Saturate with $CHCl_3$ and push down with the piston until the silicic acid is packed in a firm column. Remove the piston, letting the paper remain at the top of the column. Just before transferring the sample to the column, pour off the excess $CHCl_3$ and place an empty graduate under the tube.

Prepare 5 ml of an aqueous solution of citric and isocitric acids having a total acidity equivalent to 12 ml of 0.01N acids (about 4 mg of each acid).

(If both laevo and inactive malic acids, as well as tartaric acid, are also included, the total acidity should be contributed by 4 mg of each acid—that is, the equivalent of 30 ml of 0.1*N* acids.) Transfer the solution to the centrifuge tube, and add 1*N* NaOH until alkaline to phenolphthalein, then 2 drops excess. Heat in boiling water 15 minutes, cool to about 20°, and add 5.5 volumes of alcohol, 0.5 ml of 0.1*N* acetic acid, and 0.5 ml of the Pb acetate solution. Mix for 5 minutes (either at intervals or continuously), centrifuge, and decant the clear supernatant liquid. Test the liquid with a drop of the Pb acetate solution; if a precipitate forms within 1 minute, add it to the precipitate in the centrifuge tube. Stir or mix the precipitate with 20 ml of acetone, centrifuge, and decant and discard the acetone. Lay the tube on its side until the acetone evaporates, or remove the acetone with a very gentle current of air at room temperature. When the precipitate is dry, add 0.5 ml of 2*N* H_2SO_4 and mix with a rod to a smooth slurry. Add 1 g of silicic acid and mix until the powder does not adhere to the sides of the tube, adding a little more silicic acid if necessary.

Transfer through a funnel to the prepared column, rinse the centrifuge tube with about 5 ml of the 30% *tert*-amyl alcohol in chloroform, and pour the rinsings through the funnel. With a long, thin rod stir the powder and solvent until all air bubbles are removed. Apply pressure to the column until the solvent just sinks into the gel. Wipe the centrifuge tube, funnel and rod with cotton, and place the cotton in the chromatographic tube; rinse the centrifuge tube, funnel and rod with 2 ml of the solvent, pour onto the cotton, and push the cotton down to the top of the gel. Let the solvent sink into the gel. Add 200 ml of the solvent to the reservoir and apply pressure until the solvent elutes at a rate of 1–1.5 ml/minute.

Transfer the eluate in 10 ml portions promptly (*see* Note under "Determination, **a**") to the titrating apparatus. Rinse the graduate with 10 ml of freshly-washed neutral $CHCl_3$ and then with 10 ml of CO_2-free water. Add thymol blue indicator and 0.01*N* NaOH until, after thorough mixing by forcing CO_2-free air through the apparatus, the lower layer is colorless and the upper aqueous layer is blue. Back-titrate with standard acid and alkali until one drop of the alkali produces the characteristic blue of the indicator.

From the titration values determine the threshold volume and the volume required to elute each acid for the particular apparatus and reagents used. The acids elute in the following order: Acetic acid is eluted in the second and third 10 ml portions; both active and laevo-malic acids appear in the 100-160 ml fractions. When the malic acid is all removed (about 170 ml), pour off the remaining solvent, add the 40% *tert*-amyl alcohol in chloroform, and continue the elution. Both citric and isocitric acids appear in the 180-300 ml fractions. Continue the elution until tartaric acid is eluted (about 330-440 ml). Titrate immediately the eluate containing the citric and isocitric acids.

Determination

Take a quantity of juice or prepared solution whose titratable acidity is equivalent to about 30 ml of 0.01*N* acid, and whose solids content does not exceed 2 g. Transfer to a centrifuge tube, adjust to a volume of 5 ml by evaporation or addition of water, and proceed as under "Standardization of the Silicic Acid Column," third paragraph, beginning "add 1 *N* NaOH until alkaline to phenolphthalein." Continue up to "Back-titrate with standard acid and alkali" in the next-to-last paragraph.

a. *Total citric and isocitric acids*—After the malic acid is eluted, change to the 40% *tert*-amyl alcohol in chloroform, elute, and titrate 10 ml aliquots promptly as in the standardization. 1 ml of 0.01*N* NaOH = 0.64 mg of anhydrous citric and isocitric acids. Correct the titration for the blank. After each titration collect the lower layer and aqueous layer in separate containers. After the citric and isocitric acids are eluted, wash the combined lower layers with small quantities of water and alkali, separate, and add the aqueous portion to the titrated combined citric and isocitric acid solutions. Save this solution for the determination of normal citric acid.

Note: If the eluted free acid is allowed to stay in contact with the eluate, esters may be formed causing low results. After titration the aqueous solutions may be held until convenient for the determination of normal citric acid.

b. *Determination of the reagent blank*—Prepare

a silicic acid column as under "Standardization", add to it 1 g of silicic acid and 0.5 ml of 2 N H_2SO_4, and proceed with the elution and titration as under "Standardization", except to omit the addition of the solution of organic acids.

c. *Normal citric acid* [*Anal. Chem.* **23**: 467 (1951)]—Adjust the solution containing the citric and isocitric acids to a convenient volume (50 ml or less), and take an aliquot containing not over 4 mg of citric acid (as estimated from the titration). Add 2 ml of H_2SO_4 and cool. While holding below 22°, add 1 ml of 20% metaphosphoric acid, dilute to about 35 ml, and add 2 ml of 12% KBr and 5 ml of 4% $KMnO_4$. Mix and let stand 10 minutes without stirring. Cool to below 10° and add cold 3% H_2O_2 drop by drop, while stirring, until the solution is colorless.

Transfer to a 125 ml separatory funnel and rinse the container with about 25 ml of petroleum ether, adding the rinsings to the separator. Shake well, separate, and discard the aqueous portion. Wash the petroleum ether 4 times with 3 ml portions of water, draining and discarding the aqueous layer each time. (Halides must be completely removed from the petroleum ether and the tip of the funnel.) Add 3 ml of the Na_2S solution to the petroleum ether, shake well, and drain the aqueous layer into a 25 ml volumetric flask. Extract with another 3 ml Na_2S solution, and then wash with 2 ml portions of water until all color is removed, draining both the Na_2S extracts and the rinsings into the flask. Discard the petroleum ether. Add 2.0 ml of 2 M H_3PO_4 to the contents of the flask, mix, and then add a very small quartz or porcelain chip to facilitate smooth boiling. Boil 5 minutes, cool, and add *exactly* 5.00 ml of the KCl solution (**j**). Dilute to the mark and transfer the solution (without rinsing) to a 50 ml Erlenmeyer flask containing 0.25 g of dry $AgIO_3$. Shake vigorously 5 minutes and filter immediately through a dry Cl-free paper.

To 5 ml of the filtrate add 1 ml of 10% KI solution and two drops of 0.085 M H_3PO_4, and titrate at once with 0.01N $Na_2S_2O_3$, using starch indicator. Correct the titration for a blank determination on 5 ml of water and 6 ml of the Na_2S solution in a 25 ml volumetric flask. (The blank titration includes the values for both the KCl and any halide in the reagents.) Ml 0.01N $Na_2S_2O_3 \times 0.064$ = mg anhydrous normal citric acid in the filtrate aliquot.

d. *Isocitric acid*—Subtract the normal citric acid from the total acids to obtain isocitric acid.

Method 11-9. Tartaric Acid, Titration Method. (*AOAC Method* 20.045–20.047.)

Apparatus

Device for filtering at 0°.—Knorr extraction filter tube, about 20 mm i.d., with body about 11 cm long and stem 6–8 mm o.d. and about 10 cm long, provided with a removable, close-fitting nickel, monel metal, glass or porcelain disk at the bottom of the body of the tube (Allihn fritted glass filter, Ace Glass Co. No. 8571 with 10 cm stem, is satisfactory).

Remove the stem of an 11 cm. 60° glass funnel at the apex, and enlarge the opening to about 1 cm by grinding off the glass. Cut about 1 cm from the small end of a 1-hole rubber stopper that fits snugly in the funnel outlet. Pass the stem of the filter through the funnel stopper and then through a filtration bell jar of sufficient size to accommodate a 300 ml Erlenmeyer flask, connected to a source of vaccum by a two-way stopper.

Reagent

Lead acetate solution—Dissolve 75 g of $Pb(C_2H_3O_2)_2.3H_2O$ in water, add 1 ml of acetic acid, and dilute to 250 ml.

Removal of Pectin

Take a sample of juice of "prepared solution", Method 11–14, with a titratable acidity of about 3 ml of 1N acid and a solids content not exceeding 20 g. Designate as A, ml 1N alkali required to neutralize the sample. Adjust the volume of the sample to about 35 ml and heat to 50°. Transfer the adjusted sample to a 250 ml volumetric flask, rinse with 10 ml of hot water and finally with alcohol; cool, dilute to the mark with alcohol, shake, and let stand until the precipated pectin separates, leaving a clear liquid. (Let stand overnight if necessary.) Transfer to a centrifuge bottle, add 0.2 g of filter-aid, shake vigorously, centrifuge, and decant through a retentive paper, covering the funnel with a watch-glass. Pipet 200 ml of filtrate into another centrifuge bottle.

If the sample contains alcohol, esters of organic acids may be present, and saponification is necessary. In this case adjust the volume to 35 ml, add $A+3$ ml of $1N$ KOH, heat to about 60°, and let stand over night. Add $A+6$ ml of $1N$ H_2SO_4, transfer to a 250 ml volumetric flask, and proceed as above.)

DETERMINATION

To the solution in the centrifuge bottle add a volume of the Pb acetate solution equal to $A+3$ ml, or, if saponification was made, $A+6$ ml. Add 0.2 g of filter-aid, shake vigorously for 2 minutes and centrifuge. Test the supernatant liquid with a few drops of the Pb acetate solution; if a precipitate forms, add more Pb acetate solution; shake and centrifuge. Decant and let drain thoroughly by inverting the bottle for several minutes. To the material in the centrifuge bottle add 50 ml of 80% alcohol, shake vigorously to disperse the precipitate, add 150 ml more 80% alcohol, shake, centrifuge, decant, and drain.

To the Pb salts in the centrifuge bottle add about 150 ml of water, shake thoroughly, and pass in H_2S to saturation. (Unsaturation is indicated by noticing a partial vacuum when the bottle is stoppered and shaken, and the stopper carefully removed.) Transfer to a 250 ml volumetric flask, dilute to the mark with water, and filter through a folded paper. Transfer 100 ml of the clear filtrate to a 250 ml iodine flask, tared with 2 or 3 glass beads. (A Harvard trip balance sensitive to 0.1 g is convenient.) Evaporate on gauze over a flame to about 30 ml, remove from the flame, add a second 100 ml aliquot, and evaporate to 19 ± 0.5 g. Neutralize with 30% KOH solution, one drop at a time, using phenolphthalein, and add *one* drop of the alkali in excess. Add 2 ml of acetic acid, 0.2 g of filter-aid (Celite 545 is satisfactory), and slowly, with agitation, add 80 ml of 95% alcohol. Cool in cracked ice-salt mixture, shake vigorously two minutes, place in a refrigerator, and hold overnight at 0°.

Cover the filtering disk of the Knorr tube with a thin layer of asbestos, and place over it a thin layer of filter-aid. Place cracked ice in the outer funnel, wash the filter mat with ice-cold alcohol, and let stand a few minutes to thoroughly cool the filter. Swirl the flask to suspend the filter-aid and precipitate, and filter at 0°, sucking the mat dry. Wash the stopper with about 15 ml of ice-cold 80% alcohol, letting the wash liquid run into the precipitation flask. Stopper and shake to wash the flask well. (A stirring-rod bent at a 45° angle 1 inch from the end helps in washing the inside of the filter tube.) Conduct the wash liquid completely around the inside of the filter tube and suck dry. Wash the flask and filter tube with two 15 ml portions of ice-cold 80% alcohol. While filtering, keep the flask cold with cracked ice. Remove the ice from the outer funnel and transfer the precipitate and pad to the precipitation flask with boiling CO_2-free water. Heat almost to boiling and titrate with $0.1N$ alkali, using phenolphthalein. One ml $0.1N$ alkali = 0.015 g tartaric acid. Tartaric acid found /0.64 = tartaric acid in sample taken.

Method 11-9A. *Tartaric and other Fixed Acids (Gas Chromatographic Method).*

Proceed as in Method 2-13, except to take the following quantity of sample in place of the 15 ml of wine:

For a fruit juice, 1 ml.

For a fruit beverage, if X = estimated percent juice, ml sample $= \frac{100}{X}$.

Method 11-10. *Malic and Lactic Acids.*[18, 19]

This method, developed by the USDA Western Regional Research Laboratories, depends on removal of cations on a cation exchange resin column, then adsorption of acids on an anion exchange resin column. The weakly held acids, including lactic, are eluted by ammonium carbonate solution, then the strength of the eluting solution is increased to remove malic acid.

REAGENTS

a. *Amberlite IRA*-400, carbonate form, 60–80 mesh (Rohm & Haas Chemical Co.). The resin, in the chloride form, is ground and sieved wet, regenerated with 5% sodium hydroxide solution, washed with 1 N ammonium carbonate and finally with distilled water. The resin can be stored for some time under water.

b. *Dowex 50*, hydrogen form, 60–100 mesh, cross-linkage 12% (Dow Chemical Co.). This should be stored in air-tight containers.

c. *Ammonium carbonate*, 0.25 N (14.25 g/L).

d. *Ammonium carbonate*, 1 N (57.0 g/L).

e. *Sulfuric acid*, 96%, analytical grade, nitrate free.

f. *2,7-Naphthalenediol*, 1 g/100 ml of 96% sulfuric acid.

NOTE: If a limited number of determinations are to be made, this reagent can be incorporated in the sulfuric acid, reagent **e**, at the rate of 176 mg/L. It is stable for several months refrigerated storage. A color may develop, but this disappears after heating.

g. *p-Hydroxydiphenyl*, 1% in 0.1 *N* sodium hydroxide.

h. *Standard lactate solution*, lithium or zinc lactate.

i. *Sodium carbonate*, 1 *N*.

j. *Copper sulfate*, $CuSO_4.5H_2O$, 4%.

DETERMINATION

Prepare column 1 by inserting a No. 00 one-hole stopper carrying a short piece of glass tubing into one end of a 35 cm chromatographic glass tube of 14 mm O.D. Place a small circle of nylon bolting cloth at the bottom of the tube to retain the resin. Be sure no nylon threads project below the end of the column. No stopcock or clamp is necessary. Load the column with a water slurry of the prepared IRA-400 resin (reagent **a**) so that the height after settling is 10 cm.

Load column 2, a chromatographic tube 14 O.D. and 20 cm long similarly equipped with a rubber stopper and nylon bolting cloth, with a water slurry of Dowex 50 (reagent **b**), so that the height of the resin after settling is 10 cm. Support this column above column 1 so that its effluent drips directly into the anion column 1.

Add carefully to the upper column an aliquot of the solution to be analyzed containing not more than 3 meq. of total acids, nor less than 0.008 meq. (0.54 mg) of malic acid and 0.2 mg of lactic acid, and allow it to run freely through both columns. Remove the lower column and wash the upper column with three 10 ml portions of distilled water, allowing each portion to drain down to the top of the resin bed before the next is added. Now wash the lower column with three 10 ml portions of distilled water. (This second washing may be omitted if lactic acid is not to be determined.)

Elute the lower column with 5 successive 10 ml portions of 0.25 *N* ammonium carbonate (reagent **c**), collecting the eluate in a 100 ml volumetric flask. Dilute to the mark and reserve for the determination of lactic acid.

Place a 50 ml volumetric flask under column 2 and wash with 10, 10, 10, 10 and 9 ml portions of 1 *N* ammonium carbonate (reagent **d**). Dilute to the mark. This second eluate contains the malic acid.

ESTIMATION OF LACTIC ACID

Transfer 1 ml of the lactic acid eluate to a 25 × 200 mm test tube, add 0.05 ml of 4% copper sulfate solution, and add exactly 6.0 ml of 96% sulfuric acid from a burette, allowing the acid to run slowly down the side. Mix during the addition. Place the test tube in a boiling water bath for 5 minutes, remove, and cool below 20°. After cooling, add exactly 0.1 ml of p-hydroxydiphenyl solution (reagent g), mix well and place in a warm water bath held at 30° for 30 minutes, shaking occasionally. Now place the test tube in a boiling water bath for 90 seconds, remove and cool to room temperature. Transfer the colored solution to a spectrophotometer cell and measure percent transmittance at 560 mμ.

Make a curve of spectrophotometer readings against lactic acid through the range of 0 to 10 μg of lactic acid, using a solution of pure lithium or zinc lactate for the standardization. Determine the quantity of lactic acid present in the 1 ml aliquot of the lactic acid eluate from the standard curve and report as percent lactic acid (or mg/100 ml).

ESTIMATION OF MALIC ACID

Transfer 1 ml of the malic acid eluate, containing from 5 to 80 μg of malic acid, to a 25 × 200 mm test tube. Carefully add exactly 6.0 ml of 96% sulfuric acid drop by drop from a burette, running down the side of the test tube, swirling the liquid in the test tube during the addition. Mix well, then add 0.1 ml of 2,7 naphthalenediol solution (reagent **f**), and heat in a boiling water bath for 20 minutes. Cool, and transfer a portion to a spectrophotometer cell. Read the percent absorbance at 390 mμ. The curve of malic acid concentration vs. percent absorbance is a straight line from 5 to 80μg. Run a reagent blank with 0.1 ml color reagent, 1.0 ml water and 6.0 ml sulfuric acid. This should be colorless after heating. Report as percent malic acid or g/100 ml.

Method 11-11. *Methyl Anthranilate in Grape Juice and Grape Beverages.*[20]

This is AOAC Method **8.024-8.028** for non-alcoholic beverages. Applicable to samples containing less than 500 mg/L.

Methyl anthranilate (or other anthranilic acid esters) is a characteristic flavor constituent of certain slip-skin grapes, particularly Concord and Niagara grapes, but has not been found in *Vitis vinifera*, or European types of wine grapes. Artificial grape flavors, used in the manufacture of grape beverages and beverage bases, commonly contain anthranilic acid esters.

REAGENTS

a. *Dilute HCl*, 83 ml HCl to 1L with water.

b. *Sodium nitrite solution*, 2 g of $NaNO_2$ in 100 ml water.

c. *Hydrazine sulfate solution*, 3 g of $N_2H_4 \cdot H_2SO_4$ in 100 ml water.

d. *Potassium or sodiumalpha-naphthol-2-sulfonate solution*, 5.3 g of potassium salt or 5.0 g of sodium salt in 100 ml water.

e. *Sodium carbonate solution*, 25 g of Na_2CO_3 in 75 g water.

f. *Methyl anthranilate standard solution*, 0.25 g dissolved in 60 ml of ethanol and diluted with water to 250 ml.

APPARATUS

a. *Steam generator filled with water.* A gallon oil can with a spout serves the purpose.

b. *750 ml Kjeldahl flask with shortened neck*, about 10″ overall height, used as a distillation flask.

c. *Spray tube* with a perforated bulb at the end, passing through a rubber stopper and extending to the bottom of the distillation flask.

d. *Kjeldahl connecting bulb* with a bent connecting tube.

e. *Spiral glass condenser* with an extended outlet tube reaching to the bottom of a 500 ml receiving flask.

f. *Stop watch with second hand.*

DETERMINATION

Size of sample: 200 ml of grape juice, 50–100 ml of a grape beverage, 10–50 ml of a beverage concentrate. Add water, if necessary, to make 200 ml.

Place just enough water in the receiving flask to seal the end of the condenser outlet tube. Transfer the accurately measured sample to the distillation flask, and dilute with water, if necessary, to make 200 ml. Insert the stopper-carrying spray tube and connecting bulb into the distilling flask and connect with the condenser. Immerse the distillation flask in a water bath to the level of the contents.

When the contents of the distillation flask reach the temperature of the nearly boiling bath, connect with the steam generator and pass steam rapidly through the sample until about 300 ml of distillate have been distilled.

Disconnect the assembly and wash out the condenser with a little water. Add 25 ml of the dilute HCl (**a**) and 2 ml of 2% $NaNO_2$ solution (**b**) to the distillate and mix. Let stand exactly 2 minutes. Add 6 ml of the hydrazine sulfate solution (**c**) and mix well 1 minute, so that the liquid comes in contact with all parts of the flask that the solution may have touched when it contained free nitrous acid. Keep the liquid in the flask in rapid motion while quickly adding 5 ml of the sulfonate reagent (**d**), then immediately add 15 ml of the sodium carbonate solution (**e**). Transfer the colored solution to a 500 ml volumetric flask, rinse out the receiving flask with water, transfer the rinsings to the volumetric flask, dilute to the mark and mix.

Compare the colors of aliquots with the colors of a set of standards prepared as nearly as possible at the same time. Calculate the results as mg methyl anthranilate/L.

INTERPRETATION

Methyl anthranilate has been found in Eastern, slip-skin grapes, *Vitis labrusca*, but not in the European wine grape, *V. vinifera*, grown extensively in California.[20,23] Limited sampling of scuppernong and muscadine grapes, *V. rotundifolia*, revealed no anthranilic esters in these grapes.

Practically all of the grape juice processed in the United States is made from Concord grapes. Sale and Wilson[20] found up to 3.80 mg/kilo in Concord juice, with an average value close to 2 mg/kilo. They also reported that the methyl anthranilate content decreased somewhat during storage, and that the skins contained a substantial amount of this ester, up to 19.5 mg/kilo in the pressed skins. Power and Chestnut[23] reported 2 mg/L in authentic and commercial samples of Concord juice.

Method 11-12. *Determination of Tin in Fruit and Vegetable Juices.*[22]

Transfer 200 ml of juice to an 800 ml Kjeldahl flask, add 25–50 ml of HNO_3, then add cautiously 100 ml of H_2SO_4. Place the flask on an asbestos mat with a 2″ hole. Warm slightly, and discontinue heating if foaming becomes excessive. When the reaction has quieted, heat cautiously, rotating the flask at intervals to prevent caking. Whenever the mixture darkens, cautiously add a few ml of HNO_3. An oxidizing atmosphere should be maintained in the flask at all times. Continue digestion until all organic matter is destroyed and copious fumes of SO_3 are evolved. The final solution should be practically colorless.

Cool somewhat, and add 75 ml of water and 25 ml of saturated ammonium oxalate solution to expel any remaining oxides of nitrogen. Evaporate again until fumes of SO_3 appear in the neck of the flask.

Cool the contents of the flask and quantitatively transfer to a 250 ml Erlenmeyer flask. Add 15 ml of HCl and 100 mg of clean iron wire. Stopper with a 1-hole rubber stopper fitted with a Bunsen valve and boil gently until the wire is dissolved. Remove from the hot plate and stopper immediately.

Cool to 15°, add 5 ml of starch indicator (1 g of soluble starch in 200 ml of water) and 1 g of sodium bicarbonate. Titrate immediately with 0.01 *N* iodine.

$$1 \text{ ml } 0.01\ N \text{ iodine} \mathrel{\widehat{=}} 0.5935 \text{ mg Sn.}$$

A blank should be run on all reagents to correct for traces of tin.

NOTE: Sheet aluminum, free from tin, may be substituted for iron wire, if desired.

Chemical Indices of Decomposition of Fruit

Metabolic by-products of microbiological organisms are at times of diagnostic aid in evaluating the quality of fruit (and vegetable) products. Among these are ethyl and other alcohols, diacetyl, and acetylmethylcarbinol (AMC), as indications of rot in the fruit or of fermentation of fruit sugars by yeasts. Holck and Fields[24] found that AMC and ethyl alcohol in apple juice did not decrease substantially during storage in glass up to 219 days at temperatures up to 71°F, and no substantial loss in diacetyl occurred during this time. 20 percent loss occurred in apple juice stored in tin for 228 days, even at 91°F.

Fields[25] found that the dominating organisms in fermenting apple juice and fermenting residues on dirty plant equipment were Saccharomyces. These produced both AMC and diacetyl. Most organisms producing rot in apples formed AMC but only a very few formed diacetyl as a metabolite in the processed juice.

Method 11-13. *Determination of Diacetyl and/or AMC in Fruit Juice.* (AMC is oxidized to diacetyl, so that AMC and/or diacetyl is determined).

Reagents

Should be of analytical grade, of known purity. Eastman reagent numbers T170, 951 and 1591 of alpha naphthol, creatine and diacetyl respectively, are satisfactory.

a. *Alpha naphthol solution.* 5g/100 ml ethanol.

b. *Alkaline creatine solution.* 0.3 g creatine in a solution of 40 g of potassium hydroxide diluted to 100 ml with water.

c. *Diacetyl (2.3-butanedione) stock solution.* 1 g diluted to 1 L in a volumetric flask. 1 ml = 1 mg.

Standard Curve

Prepare a standard curve from various aqueous dilutions of the acetyl stock solution ranging from 1 μg to 10 μg/ml, developing the colors, and measuring the absorbances at 545 mμ.

NOTE: In some instruments the standard curve is linear only up to about 5 μg/ml. It is also desirable to run an absorption spectrum. In some instruments the peak may not be just at 545 mμ.

Determination

Transfer 300 ml of juice to a distilling flask attached to an efficient condenser, using all-glass connections. Distill 25 ml into a graduated cylinder.

Pipette 10 ml of the distillate into a test tube, add 5 ml of 5% alcoholic alphanaphthol solution and 2 ml of the alkaline creatine solution. Stopper, shake vigorously for 15 seconds, let stand for 10 minutes, then shake again for 15 seconds and immediately read the absorbance at 545 mμ against a reference cell containing a reagent blank treated in the same way, except for using water instead of 10 ml of distillate.

Report as the diacetyl equivalent of diacetyl and/or AMC in parts per million in the 25 ml of distillate from the 300 ml of juice. To convert diacetyl to AMC, multiply by 1.02.

INTERPRETATION

Field[26] suggested in 1962 that apple juice with an AMC content of 1 p.p.m. or less be judged acceptable, and between 1.1 and 1.6 p.p.m. be listed as questionable, while values above 1.6 p.p.m. be presumptive evidence of poor raw material and/or of poor sanitary conditions. Holck and Fields confirmed this conclusion in their 1965 report, and concluded further that *any* diacetyl present indicated poor sanitary conditions because of the association of diacetyl with yeast. These authors[27] later raised the figure for AMC content indicating undesirable quality from 1.6 to 1.8 p.p.m., and extended this criterion to apple jelly.

Hill and Wenzel[28] have applied the diacetyl test as an index of poor quality of concentrated orange juice. They showed that concentrates containing over 1.1 ppm of diacetyl frequently developed an off-flavor described as "buttermilk flavor".

Section II. Fruit products: Preserves, jellies, marmalades, butters, purées

United States and/or Canadian Definitions and Standards, USDA Standards of Grade, or Federal and Military purchasing specifications, have been issued for most of the products discussed in this chapter.

Sampling

Review the section on sampling in Section I, p. 220. The same principles apply to fruit products. Pulp a representative portion of the sample by passing through a food chopper, Waring blendor or Hobart mixer, after removing any pits from stone fruit. Weigh these separately and determine the percentage of seed.

Certain analyses, such as total solids and water-insoluble solids, are made directly from this pulped sample. Others are, or may be, made on a "prepared solution".[29]

Method 11-14. *Prepared Solution.* (*AOAC Method **20.003***).

a. *Preserves, marmalades and similar products*: Weigh 300 g of the pulped sample into a 1.5L or 2L beaker, add about 800 ml of water, and boil gently for 1 hour, covering the beaker with a watch glass. Occasionally replace any water lost by evaporation. Transfer the contents of the beaker to a 2L volumetric flask, cool and dilute to the mark. Filter the required volumes through absorbent cotton or a rapid filter paper for subsequent analysis.

b. *Jellies and syrups*: Mix the sample thoroughly to insure uniformity, and weigh 300 g directly into a 2L volumetric flask, using a wide-mouthed transfer funnel and warm water. Dissolve the jelly, heating on a steam bath if necessary. Apply as little heat as possible to minimize inversion of sucrose. Cool, dilute to the mark, and shake well. If insoluble matter separates, filter before removing aliquots for analysis.

Method 11-15. *Soluble Solids* (*Refractometric*).

This is usually determined on the pulped sample of a fruit product, or on the drained syrup, by an Abbé refractometer at 20°. If preliminary pulping is necessary, filter a small portion of the pulped sample through a pledget of absorbent cotton in a small funnel. Discard the first few drops, then place 2 or 3 drops (just sufficient to cover) on the refractometer prism. Determine the direct refractometer reading at 20°. Ascertain the equivalent percent of sucrose from Table 23-5.

NOTE: FDA Standards of Identity for Preserves and Fruit Butters[30] make no correction for water insoluble solids. For other products correct for insoluble solids by the formula:

$$X = \frac{A(100-b)}{100},$$

where X = the percent of soluble solids, corrected,
A = the percent of soluble solids as determined by the refractometer, and
b = the percent of water insoluble solids.

***Method 11-16.** Soluble Solids (Brix Spindle).*

A reasonably close approximation of the solids content may be made with Brix spindles, calibrated to read percentage of sucrose by weight in pure sugar solutions, for routine control work. A set of 12″ spindles graduated in 0.1° Brix at 20°, each in a limited range (0° to 6°, 6° to 12°, up to 66° to 72° Brix), is available from most laboratory supply houses.

Transfer a portion of the sample into a clear glass cylinder, wide enough to permit the spindle to come to rest without touching the side. The liquid should be poured down the side to avoid entrapment of air bubbles. The temperature of the solution should be as close as possible to the temperature at which the Brix spindle was calibrated (usually 20°).

Read the temperature of the solution from a thermometer inserted just after the spindle reading is noted. If more than 1° from the temperature of calibration, a correction should be applied. A convenient correction table is given in *Official Methods of Analysis*, **43.005**, 10th ed. (1965). The correction is subtracted for temperatures below the temperature of calibration, and added if above this temperature. This correction ranges from 0.04° to 0.09° Brix per degree centigrade, depending on the amount of sugar in the solution tested.

If the sample is too dense to determine density directly, combine equal weights of the sample and distilled water, mix thoroughly, allow the mixture to attain the temperature of calibration, and take the Brix reading. This spindle reading multiplied by 2 gives the percentage of total solids in the sample.

Water-Insoluble Solids

The AOAC recognizes two methods for the determination of water-insoluble solids in fruit, preserves and marmalades: "official" and "rapid". We, along with others, have found the rapid method to give comparable results with the longer official method, and it has the distinct advantage that drying can be completed within 10 to 15 minutes.[31] This is the method we choose to give here:

***Method 11-17.** Water-Insoluble Solids, Rapid Method (AOAC Method **20.013–20.014**).*

APPARATUS

a. *Waring blendor or similar comminuting device.*

b. *Aluminum or tinned iron weighing dishes,* about $5\frac{1}{4}''$ wide, with tight-fitting cover (16 mm film holders obtainable from camera stores).

c. *Dietert Moisture Teller* (H. W. Dietert Co., Detroit, Mich.) or a forced-draft oven set at 100°.

DETERMINATION

Fit a 15 cm Whatman No. 4 or 41-H filter paper or equivalent into a 12.5 cm Büchner funnel, add half of a 7 cm paper (used later to wipe any insoluble solids from the funnel), wash with boiling water, apply suction and dry, using the Moisture Teller and pan or the forced draft oven. Transfer to a weighing dish, cool for 1 hour in a desiccator, and weigh on a balance sensitive to 1 mg, using a weighing dish and paper as tare.

Weigh 25 or 50 g of well-mixed sample from the blendor to the nearest 10 mg, transfer with hot water to a 400 ml beaker, adjust to about 200 ml with hot water, stir and boil gently for a few minutes. Place the prepared filter paper on the Büchner funnel, and attach to a suction flask, but do not turn on the suction. Pour 50–100 ml of boiling water on to the filter, and when a steady flow of water passes through the filter, transfer the sample to the filter, portionally if necessary. Wash the insoluble solids with boiling water, collecting 850–900 ml of water. Keep the solids from matting on the filter paper during washing by portional addition of boiling water. After washing is completed, apply suction and aspirate thoroughly. Transfer the filter paper and contents to the weighing dish, using the extra half filter paper to swab off the sides of the beaker and Büchner funnel, and dry at 102° ±3° for about 15 minutes. Cool in a desiccator and weigh. Report as percentage of water-insoluble solids.

The "Official Method of Analysis" used by the Canada Food and Drug Directorate differs in certain details from the above method. The Canadian method uses a folded 15 cm Whatman No. 54 filter paper in a funnel, and dries in an air-oven overnight. Three 25 g replicates are weighed into 400 ml beakers, 200 ml of water are added, and the beakers are covered with a watch glass and boiled gently for 30 minutes. The solutions are filtered, and the filter papers are washed with eight 100 ml portions of hot (90°) water, and dried overnight at 110°.

Method 11-18. *Alcohol-Insoluble Solids (Alcohol Precipitate).*

This is a measure of the amount of pectin present. It is of minor diagnostic value as an estimate of the amount of fruit used in the manufacture of jams and jellies, since added pectin is permitted in these products. It is of more value in the analysis of fruits and fruit juices. The results obtained by this procedure are sometimes called "crude pectin". The official AOAC method, in use since 1924, is given below:

DETERMINATION

Transfer 100 ml of "prepared solution", Method 11-14, to a 400 ml beaker (if the product is unsweetened add 1 or 2 lumps of cube sugar, 4 to 8 g). Reduce to a volume of 20-25 ml by evaporation. If water-insoluble matter separates during evaporation, add more sugar. Cool to room temperature and, slowly and with constant stirring, add 200 ml of alcohol. Let stand at least 1 hour, then filter through a 15 cm qualitative filter paper and wash the precipitate with alcohol. Do not allow the precipitate to dry before transferring it from the paper.

Wash the precipitate back into the original beaker with hot water, rinsing the paper thoroughly. Evaporate to about 20 ml, and add 5 ml hydrochloric acid (1 + 2.5). If water insoluble matter separates, stir well, and, if necessary, warm to dissolve. Again precipitate with 200 ml of alcohol, let stand 1 hour and filter as before. Wash the precipitate and paper thoroughly to remove all hydrochloric acid. Rinse the precipitate from the paper into a platinum dish with hot water, evaporate to dryness on a steam bath, and dry in an oven at 100° to constant weight. Ignite and reweigh. Calculate the loss in weight to percent of alcohol precipitate.

NOTE: In many cases the precipitate is colorless and almost invisible, so take great care that none is lost during the dissolving and transferring operations. If the precipitate is very pure, and small in quantity, it may not be visible at first. In this case, add a few crystals of sodium chloride to the alcohol to flocculate the precipitate.

Method 11-19. *Titratable Acidity.*

Pipette 10 g of pulped fruit or 25 ml of "prepared solution" of jelly, preserves or fruit into about 200 ml of recently boiled or neutralized distilled water containing 0.3 ml of 1% alcoholic phenolphthalein for each 100 ml of solution being titrated. Titrate with 0.1*N* sodium hydroxide to a faint but distinct pink color. An illuminated white background facilities recognition of the end-point.

NOTE: If the solution being titrated is highly colored, the AOAC recommends the following procedure: Stop titration with alkali just before the end-point is reached, and transfer a measured quantity (2 or 3 ml) into a small beaker containing about 20 ml of neutral distilled water. If the pink end-point has not been reached, pour the extra diluted portion back into the original solution, add more standard alkali, and continue the titration in this manner until a permanent faint pink color is reached. Report as ml/0.1*N* alkali per 100 g of fruit, jelly or preserves. Samples of citrus or tomato products may be reported as citric acid. 1 ml 0.1*N* alkali is equivalent to 0.0064 g of anhydrous citric acid.

Titration for total acidity may be made by a glass electrode pH meter or electrometer, following the method accompanying the instrument used. Before use, the instrument should be checked against appropriate standard buffer solutions. Take a size of sample or sample solution that will titrate between 10 ml and 50 ml of 0.1*N* alkali, and dilute to a volume of 100–200 ml with water before titration.

Method 11-20. Ash. (*Review Method 1-9, Total Ash*).

Weigh 25 g of a well-mixed sample of fresh or canned fruit, preserves or marmalade, or 25 ml of fruit juice, into a 100 ml flat-bottomed platinum dish. If ash on the water-soluble portion of the product is desired, transfer 100 ml of a "prepared sample" into the platinum dish. Proceed as in Method 1-9, ashing at 525°. If later determination of ash constituents is desired, sufficient replicates may be ashed simultaneously for this purpose.

Ash Constituents.

The ash constituents of most value in estimating the amount of fruit or fruit juice present are potassium (expressed as K_2O) and phosphorus (expressed as P_2O_5), since these elements are present in comparatively large amounts in fruit ash. They may be determined by Methods 1-17, 1-18 or 1-19 or AOAC Methods **20.022**, **20.032** or **20.033**.

Saccharine Constituents.

Method 11-21. Sucrose.

a. *By polarization*—Treat a portion of the prepared solution, Method 11-14, equivalent to 52 g of sample, by Method 16-6, "Determination" (**a**), using clarifying agent (**c**).

b. *By copper reduction* (*AOAC Method* **20.069**)—Transfer a portion of the prepared solution, Method 11-14, representing (if possible) about 2.5 g of total sugars, to a 200 ml volumetric flask; dilute to about 100 ml and add an excess of saturated neutral Pb acetate solution, Method 16-6, clarifying agent (**c**) (2 ml is usually enough). Mix, adjust to 200 ml, and filter, discarding the first few ml of filtrate. Add dry K or Na oxalate to the filtrate to precipitate the Pb, mix and filter, discarding the first few ml of filtrate.

Transfer a 50 ml aliquot of this Pb-free filtrate to a 100 ml volumetric flask, add 10 ml of HCl (1 + 1), and let stand 24 hours at room temperature. Then exactly neutralize with concentrated NaOH solution (using phenolphthalein), and dilute to 100 ml.

Treat aliquots of this solution and the uninverted solution (25 ml, or smaller if over 200 mg of reducing sugars are present) by Method 16-6, "Determination (**b**)", "Precipitation of cuprous oxide", and "Calculations".

Method 11-22. Invert Sugar. See Method 16-7.

Interpretation of Analytical Results

The basic question to be solved by analysis of jams and jellies is: do they comply with FDA or other official definitions and standards, e.g., are they made from 45 parts of fruit constituents to 55 parts, by weight, of saccharine ingredient, and concentrated to a legal total solids content?

In factory quality control, this is done most expeditiously by comparison of certain analytical results as determined on the finished jam or jelly with the same analyses on the lot of fruit from which the product was made. In other cases, where the fruit is not available for analysis, as in market samples, comparison must be made with analyses of authentic fruit as obtained from the literature. (See Section III of this chapter.)

The analyses that have been found to be of most diagnostic value are: total, or water-soluble, solids, water-insoluble solids, ash (either total ash, or ash of the water-soluble portion) and ash constituents, such as potassium and phosphorus. Most of the analyses of authentic fruits and preserves reported by chemists of the FDA were made on the "prepared solution". This has the decided advantage of using a much larger sample (150 or 300 g), thus minimizing the effect of sample deviations incurred in the use of smaller samples. Other quoted analyses were made on the pulped sample.

Actually, there is so little difference in the results of determinations of ash and ash constituents made on the pulped fruit or preserve, from those made on a "prepared solution" aliquot, that the values may be used interchangeably for the estimation of fruit content of a preserve, jelly or fruit juice.

The FDA, through the last 35 years or

more, has collected and analyzed a large number of representative samples of fruit and fruit juices from the main production areas of the United States. Precautions were taken to insure the authenticity of each sample, and that no foreign constituents were present. FDA employees then supervised the manufacture of commercial batches of jams and jellies from this fruit or juice. Samples were drawn during the manufacturing process of all raw materials used (fruit constituent, sugar or other saccharine ingredient, pectin, fruit acids, etc.) as well as of the finished product. All these were carefully analyzed. Much of this work has been reported by Sale[32] and Osborn.[33] *See* Section III of this chapter for analyses of these authentics.

Sale then calculated the amount of fruit that was processed with each 55 lbs of sugar to make the finished product. For example: 100 lbs of an analyzed jam contained 0.2 lb of ash, and 100 lbs of the fruit used contained 0.5 lb of ash. Then the amount of fruit (X) in 100 lbs of jam is:

$$\frac{100 \times 0.2}{0.5}, \text{ or } 40 \text{ lbs.}$$

By analysis 100 lbs of fruit were found to contain 8.0 lbs of invert sugar, so the 40 lbs of fruit found to be present in 100 lbs of the jam contributed 3.2 lbs of invert sugar. Therefore 70.0—3.2, or 66.8 lbs of invert sugar were contributed by the sugar used in making the finished jam. These 66.8 lbs of invert sugar are equivalent to 66.8 × 0.95, or 63.5, lbs of sucrose used to make 100 lbs of the jam, or, in terms of 55 lbs of sugar, as used in the FDA definition of jam,

$$\frac{40 \times 55}{63.5},$$

or 34.65, lbs of fruit were used with 55 lbs of sugar.

Similar calculations may be made from the other indices of fruit content mentioned earlier. If, as is usually the case, a sample of the fruit used is not available, a close approximation of the fruit content may be calculated by using averages of these indices in analyses of authentic samples of the fruit in question, as given in Section III.

It has been argued that minimum instead of average content of these indices in authentic fruits should be used in calculating the quantity of fruit present. Scrutiny of the raw data from which the analyses of authentic fruit were summarized shows that it is very seldom that any one sample of fruit in a series of samples will show minimum quantities of all constituents analyzed. A truer picture of the fruit content of a preserve or jelly will be obtained by the use of averages of these constituents.

The steps in the calculation of the amount of fruit used per 55 lbs of fruit, as made by the Connecticut Agricultural Experiment Station laboratories, are:

1. Calculate separately the percentage of fruit from the ratios of percentage of ash, K_2O, P_2O_5 (and water-insoluble solids in the case of jam) to the percentages of these same constituents in authentic fruit, and take the average of these fruit percentages (A).

2. Subtract from the total percent of sugar of the jam or jelly as determined by analysis, the estimated percent of sugar corresponding to the percentage of fruit calculated as A. The difference is the added sugar (B).

3. Then $X = \frac{55A}{B}$, where X = the pounds of fruit per 55 lbs of sugar.

U.S. Food and Drug laboratories calculate the quantity of fruit used for each 55 lbs of sugar by an equation developed by Bonney.[34] The figure "5790" in this equation is the amount of invert sugar equivalent to 55 lbs of sucrose, times 100

$$\frac{55 \times 100}{0.95}.$$

$$X = \frac{5790B}{AD - BC}, \text{ where:}$$

X = number of pounds of fruit to 55 lbs of sugar in the original batch;

B = percent ash (or other fruit constituent) in the jam sample;
A = percent sugar, as invert (or percent soluble solids) in the sample;
D = average percent ash (or other fruit constituent) in authentic samples of the fruit involved; and
C = average percent sugar, as invert (or percent soluble solids) in authentic samples of the fruit involved.

Values for X should be calculated, using each of the fruit constituents determined by analysis, and their *average* used as an estimate of the quantity of fruit present.

These indices of fruit content must be used with caution, since addition during manufacture of certain ingredients that are foreign to fruit may give an abnormally high figure for one or more of these indices. For example pectin, if added in a quantity beyond that necessary to correct for normal deficiency, may contribute more or less ash, depending on the purity of the pectin. Fruit acids, added in a quantity beyond that necessary to correct for normal deficiency, will vitiate the use of titratable acidity as an index of fruit content.

Additional information may be obtained by the analyst by comparing certain ratios of these indices with the same ratios of average values taken from analyses of authentic samples of the fruit in question. Thus:

$$\text{high}\ \frac{\text{alcohol prec.}}{\text{ash}}\quad \text{or}\quad \text{low}\ \frac{\text{insoluble solids}}{\text{alcohol prec.}}$$

indicates added pectin;

$$\text{high}\ \frac{\text{ash}}{\text{acidity}}\quad \text{or}\quad \text{low}\ \frac{\text{insoluble solids}}{\text{acidity}}$$

indicates added acid.

Other ratio comparisons will occur to the analyst as being of diagnostic value.

NOTE: Grape juice loses argol upon storage, thus lowering both potassium (K_2O) and ash content. Section III gives analyses of both fresh and stored juices. Grape jelly is usually made (and Federal Specification Z-J-191 requires this) from stored juice. The choice of authentic analyses depends upon the analyst's knowledge of the product being analyzed. The amount of phosphorus (P_2O_5) in grape juice is not affected by storage.

The authors of the *Federal Purchasing Specification for Jams (Preserves), Fruit, Z-J-96a*, in a *Deviation List* for this specification dated October 7, 1963, have circumvented the difficulty of interpretation of analyses by requiring that "the chemical values of the finished product on a lot average basis shall not be less than those listed below":

Kind of Fruit:	Insoluble solids %	Ash %	K_2O mg/100 g	P_2O_5 mg/100 g
Apricot (solid pack)	0.74	0.308	178.2	21.0
Apricot (other than solid pack)	0.58	0.305	176.3	21.1
Blackberry (w/seeds)	2.49	0.169	82.3	13.8
Cherry	0.48	0.170	93.6	16.8
Plum	0.57	0.209	112.6	9.0
Grape (Concord)	0.27	0.188	98.2	10.7
Loganberry	2.35	0.177	84.9	11.8
Peaches (solid pack)	0.43	0.167	88.8	16.4
Peaches (other than solid pack)	0.43	0.171	94.4	16.2
Pineapple	0.62	0.151	74.9	5.8
Raspberry (Red)	2.48	0.167	82.0	14.5
Raspberry (Black)	3.76	0.186	94.6	11.9
Strawberry	1.09	0.154	79.2	10.5

Jams manufactured from combinations of more than one fruit shall contain amounts of the above constituents proportionate to the quantities of the various fruits present in the mixture.

TABLE 11-3
FDA Soluble Solids and Fruit/Sugar Ratio
Requirements for Standardized Jellies, Preserves and Fruit Butters

Product	Soluble Solids by Refractometer, not Corrected for Water-Insoluble Solids		Parts by Weight Fruit	to Sweetener
	GROUP I	GROUP II		
	68% Minimum	65% Minimum		
Preserves and Jams	Blackberry	Apricot		
	Black Raspberry	Cranberry		
	Blueberry	Damson, Damson Plum		
	Boysenberry	Fig		
	Cherry	Gooseberry		
	Crabapple	Greengage, Greengage Plum		
	Dewberry	Guava	45 Minimum	55 Maximum
	Elderberry	Nectarine		
	Grape	Peach		
	Grapefruit	Pear		
	Huckleberry	Plum		
	Loganberry	Quince		
	Orange	Red Currant, Currant (Other than Black Currant)		
	Pineapple			
	Raspberry			
	Red Raspberry			
	Rhubarb			
	Strawberry			
	Tangarine			
	Tomato			
	Yellow Tomato			
	Youngberry			
Fruit Butters	43% Minimum		5 Minimum	2 Maximum
Fruit Jellies	65% Minimum		45 Minimum	55 Maximum

The regulations established under Canada Food and Drugs Act[35] define two types of jam:

1. *Jam* shall contain not less than 45 percent of the named fruit, except for strawberry, which shall contain not less than 52 percent fruit, and not less than 66 percent water-soluble solids as determined by the refractometer.

2. *Jam with added pectin* shall contain not less than 27 percent of the named fruit, except for strawberry, which shall contain not less than 32 percent fruit, and not less than 66 percent water-soluble solids as determined by the refractometer.

Section III. Analysis of Authentic Fruits

This section consists entirely of analyses of fruits and fruit products, as selectively chosen from the literature and other sources. Much of this information has not been published previously.

The analyses of fruit products in this section include only products whose manufacture has been continuously watched by a qualified observer to negate any possibility of sophistication, or products made by the investigator in his own laboratory or pilot plant under simulated factory conditions.

Analyses of Authentic Samples of Citrus Juices

Oranges

California Oranges: 1949 *season*, representing 27 samples from 16 packing houses in the 4 Southern California counties producing oranges:[36]

Valencia Oranges. Sunkist Growers and Wisconsin Alumni Research Foundation collaborated in a 2-year program to determine the composition of California Valencia Oranges.[38, 39] Samples consisted of fresh-picked fruit from Sunkist packing houses in all of the major orange areas of California. The fruit were graded for size and only #88 oranges (approximate diameter 2.84″) were included. These analyses are given in Table 11-4. Note also the other orange analyses on p. 247.

Lemons

The cooperative study of citrus fruit also included a similar study of California Eureka lemons.[38,39] In this study a sample of 100 fruits was obtained from each of the 8 producing areas. All fruits included in each sample were size #150 (approximate diameter 2.2″). The analyses are given in Table 11-5 on p. 249. *See also* the other analyses on p. 248.

	Soluble Solids, % (Corr.)	Ash, %	K_2O, mg/100 ml	P_2O_5, mg/100 ml	Acidity, as % anhyd. citric	Vitamin C, mg/100 ml
Highest	13.96	0.46	261	47.6	1.62	59.6
Two lowest	8.82	0.33	181	31.6	1.02	44.9
	10.50	0.36	185	33.0	1.03	46.2
Average	12.39	0.41	224	38.7	1.33	51.0

1950 *season*, 21 samples from Southern and Central California packing houses:[37]

	Soluble Solids, % (Corr.)	Ash, %	K_2O, mg/100 ml	P_2O_5, mg/100 ml	Acidity, as % anhyd. citric	Vitamin C, mg/100 ml
Highest	15.18	0.48	268	58.7	1.3	53
Two lowest	10.80	0.33	181	34.8	0.8	35
	11.66	0.34	195	36.8	0.9	35
Average	13.03	0.42	231	44.8	1.0	44

Sugars, as invert, ranged from 3.45% to 10.23% (ave. 6.98%) before inversion and 7.92% to 12.50% (ave. 10.36%) after inversion.

Grape Juice

[See also authentic data in Table 11-11 (Osborn)]

Lipscomb[46] collected authentic samples of Concord grape juice, both stored and unstored, during the 1961–1962 season from the principal Concord grape areas of the United States. This investigation was motivated by changes that have taken place in recent years in commercial production of Concord grape juice. The enzymes pectinase and diastase are used; this affects the pectin and starch constituents; storage conditions for precipitation of argols are closely regulated at 30–32°F; and extraction methods have changed. The results are summarized in Table 11-6 on p. 250.

Mills and Petree[47] analyzed 11 samples of Washington State Concord type grapes representing the 1940 season.

Willard B. Robinson *et al.*[48] analyzed over 150 samples of New York Concord grapes,

(continued on p. 250)

TABLE 11-4
Valencia Oranges

	Peel			Edible Portion			Juice		
	1957	1958	Ave.	1957	1958	Ave.	1957	1958	Ave.
	G/100 G	G/100 G	G/100 G	G/100 G	G/100 G	G/100 G	G/100 G	G/100 G	G/100 G
Protein (Kjeldahl N × 6.26)	1.75	1.31	1.53	1.31	0.95	1.13	1.13	0.88	1.00
Amino nitrogen (Van Slyke)	0.25	0.30	0.28	0.06	0.06	0.06	0.06	0.06	0.06
Fat (ether extract)	0.16	0.30	0.23	0.43	0.16	0.30	0.40	0.18	0.29
Soluble solids, total	14.40	16.98	15.69	13.33	12.78	13.06	12.92	12.59	12.59
Sugar									
Total	7.22	7.87	7.55	8.70	9.49	9.10	9.90	9.54	9.72
Sucrose	2.04	1.93	1.99	4.14	4.68	4.41	4.91	4.56	4.73
Reducing	5.18	5.94	5.56	4.56	4.81	4.69	4.99	4.98	4.99
Acid as anhydrous citric	0.34	0.24	0.29	0.68	0.82	0.75	1.02	1.02	1.02
Moisture	71.77	73.26	72.52	85.05	85.41	85.23	86.99	87.27	87.11
Ash	0.82	0.74	0.78	0.51	0.45	0.48	0.31	0.37	0.34
	MG/100 G	MG/100 G	MG/100 G	MG/100 G	MG/100 G	MG/100 G	MG/100 G	MG/100 G	MG/100 G
Calcium	157.0	164.0	161.0	36.1	37.3	36.7	8.0	11.0	9.5
Magnesium	20.3	24.0	22.2	11.3	11.6	11.5	10.1	11.4	11.3
Iron	0.79	1.00	0.80	0.68	0.86	0.77	0.35	0.31	0.33
Phosphorus	21.8	19.8	20.8	22.8	21.3	21.8	20.0	18.9	19.5
Potassium	240	183	212	182	163	173	164	162	163
Sodium	3.6	2.4	3.0	2.0	0.6	1.3	0.8	0.6	0.7
Sulphur	21.0	21.0	21.0	11.0	12.0	11.5	8.0	9.0	8.5
Ascorbic acid									
Reduced	133.0	140.0	136.5	38.2	40.8	39.5	42.8	44.1	43.5
Total	—	—	—	41.6	43.0	42.3	—	—	—
Biotin	0.00472	0.00548	0.00510	0.00122	0.00113	0.00118	0.00077	0.00081	0.00079
Carotenoids									
Beta-carotene	—	0.25	0.25	—	0.15	0.15	—	0.17	0.17
Total	6.6	13.2	9.9	3.2	3.6	3.4	2.5	3.2	2.8
Choline	25.0	21.0	23.0	8.7	14.5	11.6	7.2	8.8	8.0
Folic acid	0.01410	0.00950	0.01180	0.00495	0.00336	0.00416	0.00260	0.00320	0.00290
Inositol	267	246	257	205	203	204	182	137	159
Niacin	0.927	0.849	0.888	0.467	0.515	0.491	0.355	0.396	0.376
Pantothenic acid	0.593	0.386	0.490	0.311	0.240	0.276	0.245	0.168	0.207
Pyridoxine	0.221	0.130	0.176	0.066	0.063	0.065	0.066	0.047	0.057
Riboflavin	0.104	0.078	0.091	0.033	0.033	0.033	0.028	0.025	0.027
Thiamine	0.12	0.12	0.12	0.13	0.12	0.13	0.10	0.09	0.10

Analyses of oranges from other areas, as reported by FDA analysts, are given below:
Florida Oranges, 1936–1937 season, 28 samples:[40]

	Soluble Solids, % (Corr.)	Ash, %	K_2O, mg/100 ml	P_2O_5, mg/100 ml	Acidity, as % anhyd. citric	Vitamin C, mg/100 ml
Highest	13.9	0.58	303	42.6	1.31	61
Two lowest	9.9	0.35	169	22.9	0.70	33
	10.0	0.38	184	23.6	0.77	36
Average	11.7	0.44	237	34.7	1.00	50
Std. deviation	1.1	0.047	24.7	5.8	0.19	8.2

1950 season, 15 samples:[41]

	Soluble Solids, % (Corr.)	Ash, %	K_2O, mg/100 ml	P_2O_5, mg/100 ml	Acidity, as % anhyd. citric	Vitamin C, mg/100 ml
Highest	13.7	0.49	282	45.3	1.18	59.5
Two lowest	10.5	0.42	212	21.6	0.74	29.1
	11.0	0.43	222	30.7	0.75	30.3
Average	12.0	0.46	249	36.6	0.89	45.0
Std. deviation	0.87	0.025	23.6	6.0	0.14	10.4

Texas Oranges: 93 samples of oranges from Rio Grande Valley during 1960–61 season. (3 samples of Jaffa variety and 1 each of Temple and Pineapple varieties analyzed are not included.)[42]

	Soluble Solids, % (Corr.)	Ash, %	K_2O, mg/100 ml	P_2O_5, mg/100 ml	Acidity, as % anhyd. citric	Vitamin C, mg/100 ml
			Marrs Variety—13 samples			
Highest	11.38	0.401	235	38.6	1.24	56.2
Two lowest	8.38	0.293	180	27.6	0.55	35.2
	8.97	0.310	183	27.7	0.56	35.6
Average	10.11	0.355	207	31.0	0.72	45.0
			Navel Variety—11 samples			
Highest	11.93	0.363	220	39.9	0.63	42.8
Two lowest	9.17	0.314	176	31.8	0.42	33.8
	9.82	0.326	186	33.5	0.42	34.0
Average	10.68	0.342	201	36.0	0.47	37.8
			Hamlin Variety—31 samples			
Highest	12.86	0.403	230	43.0	0.85	60.2
Two lowest	8.91	0.289	167	25.2	0.54	31.8
	9.13	0.294	170	25.4	0.57	37.4
Average	10.57	0.338	192	31.2	0.69	47.7
			Valencia Variety—38 samples			
Highest	13.27	0.420	241	42.8	1.18	60.4
Two lowest	10.38	0.329	173	26.7	0.69	37.2
	10.70	0.333	181	27.0	0.73	39.8
Average	11.75	0.366	211	32.5	0.96	47.1

Lemons

Summary, 1928–1947 Seasons.[43] (Nos. of samples are in parentheses.)

	Soluble Solids, % (Corr.)	Ash, %	P_2O_5, mg/100 ml	K_2O, mg/100 ml	Acidity, as % anhydrous citric	Vitamin C, mg/100 ml	Sugars, as invert	
							before inversion	after inversion
Highest	10.80	0.38	28.8	213	7.84	63	not determined	
Lowest	7.05	0.24	17.0	120	5.18	32		
Average	8.14(34)	0.29	22.6(38)	155(14)	5.99(38)	42(20)		

1940 Season. Canned California lemon juices, 20 samples.[43]

	Soluble Solids, % (Corr.)	Ash, %	P_2O_5, mg/100 ml	K_2O, mg/100 ml	Acidity, as % anhydrous citric	Vitamin C, mg/100 ml	Sugars, before inversion	Sugars, after inversion
Highest	9.94	0.34	31.1		7.09	63	2.44	2.49
Two lowest	7.78	0.25	12.5*	not determined	5.71	32	1.09	1.18
	7.90	0.25	12.6*		5.78	33	1.19	1.23
Average	8.69	0.28	25.4		6.12	42	1.68	1.70
Std. deviation	0.56	0.03	4.9		0.34	0.07	0.43	0.44

*These 2 cans were the only ones not lacquered; this may explain the low results.

Grapefruit

Florida Grapefruit, 13 samples, 1935–1936 season:[44]

	Soluble Solids, % (Vacuum Oven, 70°)	Soluble Solids, % (Refractometer, 20°)	Ash, %	P_2O_5, mg/100 g	K_2O, mg/100 g	Acidity, % Anhydrous Citric	Sugars, as invert	
							Before Inversion	After Inversion
Highest	10.86	10.90	.464	28.9		1.60	6.78	8.37
Lowest	7.95	7.80	.320	17.9	not	1.15	3.66	4.66
Average	9.17	9.13	.403	20.2	determined	1.29	5.17	6.40

Texas Grapefruit, 7 samples, 1939–1940 season:[44]

	Soluble Solids, % (Vacuum Oven, 70°)	Soluble Solids, % (Refractometer, 20°)	Ash, %	P_2O_5, mg/100 g	K_2O, mg/100 g	Acidity, % Anhydrous Citric	Sugars, Before Inversion	Sugars, After Inversion
Highest	12.20	12.36	0.44	33.7	220.8	1.70	9.05	9.48
Lowest	9.55	9.66	0.32	24.8	167.5	1.09	6.98	7.21
Average	11.04	11.18	0.40	28.1	192.8	1.36	7.93	8.40

Limes

California Lime Juice, 15 samples from 1933, 1940, 1944 and 1947 seasons:[45]

	Soluble Solids, % (Corr. for citric acid.)	Ash, %	K_2O, mg/100 g	P_2O_5, mg/100 g	Acidity as citric, %	Vitamin C, mg/100 ml	Total as invert, %	Reducing, as invert, %
Highest	10.20	0.34	150.6	30.3	7.64	34	4.01	1.93
Lowest	6.93	0.18	92.0	14.2	5.67	19	0.52	0.54
Average	8.87	0.24	114.2	23.7	6.73	28	1.49	1.32

TABLE 11-5

California Eureka Lemons

	Peel			Edible Portion			Juice		
	1957	1958	Average	1957	1958	Average	1957	1958	Average
	G/100 G	G/100 G	G/100 G		G/100 G		G/100 G	G/100 G	G/100 G
Protein (Kjeldahl N × 6.25)	1.44	1.50	1.47		1.13		0.44	0.50	0.47
Amino nitrogen (Van Slyke)	0.18	0.22	0.20		0.05		0.03	0.04	0.04
Fat (ether extract)	0.25	0.34	0.30		0.32		0.16	0.22	0.19
Soluble solids, total	8.90	10.47	9.69		8.82		7.97	8.63	8.30
Sugar									
Total	3.21	3.81	3.51		1.51		1.19	1.15	1.17
Sucrose	0.03	0.15	0.09		0.18		0.11	0.06	0.09
Reducing	3.18	3.66	3.42		1.33		1.08	1.09	1.09
Acid as anhydrous citric	0.41	0.38	0.40		4.93		5.91	6.05	5.98
Moisture	81.87	81.36	81.62		90.08		92.84	91.88	92.36
Ash	0.61	0.61	0.61		0.29		0.25	0.24	0.25
	MG/100 G	MG/100 G	MG/100 G		MG/100 G		MG/100 G	MG/100 G	MG/100 G
Calcium	130.0	137.0	133.5		22.3		5.6	6.5	6.1
Magnesium	13.9	16.0	15.0		5.8		5.8	6.3	6.1
Iron	0.59	1.05	0.82		0.65		0.21	0.21	0.21
Phosphorus	12.3	12.7	12.5		15.8		9.6	10.9	10.3
Potassium	176	143	160		108		101	102	102
Sodium	6.6	6.4	6.5		2.5		1.0	1.0	1.0
Sulphur	11.0	12.7	11.9		5.3		2.0	3.3	2.7
Ascorbic acid									
Reduced	138.0	120.0	129.0		52.7		44.8	42.7	43.8
Total	—	—	—		54.7		—	—	—
Biotin	0.00262	0.00234	0.00248		0.00058		0.00028	0.00022	0.00025
Carotenoids									
Beta-carotene	0.03	0.03	0.03		<0.01		<0.01	<0.01	<0.01
Total	0.21	0.39	0.30		0.06		0.05	0.04	0.04
Choline	10.0	12.0	11.0		10.0		5.9	5.0	5.5
Folic acid	0.00548	0.00440	0.00494		0.00182		0.00094	0.00082	0.00089
Inositol	201	231	216		109		76	56	66
Niacin	0.358	0.354	0.356		0.129		0.069	0.072	0.071
Pantothenic acid	0.390	0.247	0.319		0.194		0.106	0.101	0.104
Pyridoxine	0.081	0.263	0.172		0.080		0.030	0.072	0.051
Riboflavin	0.057	0.100	0.079		0.021		0.013	0.011	0.012
Thiamine	0.067	0.048	0.058		0.042		0.029	0.032	0.031

TABLE 11-6
Average Analyses of Concord Grape Juice
1961–1962 Season

Producing Area	Michigan	Arkansas	Washington	New York-Pennsylvania	Overall
Number of Samples	16	14	12	18	60
% Soluble Solids*	15.6	15.7	17.4	15.9	16.1
	(0.07)	(0.13)	(0.08)	(0.11)	(0.10)
% Ash	0.31	0.33	0.36	0.31	0.32
	(0.024)	(0.021)	(0.039)	(0.057)	(0.046)
P_2O_5, mg/100 g	29.0	26.9	38.1	29.6	30.5
	(0.87)	(1.02)	(0.72)	(2.60)	(1.61)
% acidity, as tartaric	1.03	0.69	0.76	1.12	0.92
	(0.009)	(0.004)	(0.008)	(0.006)	(0.007)
% protein (N × 6.25)	0.354	0.263	0.480	0.426	0.380
	(0.004)	(0.002)	(0.004)	(0.006)	(0.004)
% sugar before inversion	12.74	13.19	14.79	12.99	13.33
	(0.108)	(0.110)	(0.096)	(0.096)	(0.103)
% sugar after inversion	12.65	12.85	14.69	12.84	13.16
	(0.111)	(0.254)	(0.132)	(0.138)	(0.156)
K_2O, mg/100 g	185.3	193.9	202.8	173.0	187.1
	(4.79)	(3.74)	(2.77)	(4.90)	(4.26)

* Figures in ( ) represent the standard deviation within a sample.

Authentic Samples of Washington State Concord Type Grapes, 1940 Season
(7 Island Belle[a], 4 Concord Grapes)

	Solids, % (Vac. Oven 70°)	Sugars as Invert, % Before Inversion	After Inversion	Alcohol Prec., %[b]	Ash, %	P_2O_5, mg/100 g	Acid, % as Tartaric
Maximum	20.83	18.22	18.31	0.25	0.42	59.1	1.55
2 minima	13.83	11.29	11.38	0.16	0.34	27.0	0.45
	14.91	11.77	11.62	0.17	0.36	27.8	0.47
Average	16.38	14.03	14.09	0.19	0.39	38.1	1.07

[a] Island Belle grapes are Concord type, but less solid than Concords.
[b] Determination made on 5 Island Belle grapes only.

collected from processing plants in the 3 major grape-producing areas of the state, the Hudson Valley, the Chautauqua area and the Finger Lakes area. These workers separated each sample into two portions. One consisted of hot-pressed grape juice, while the other portion was seedless pureed grapes, made by removing the seeds by hand and homogenizing in a Waring blendor. All samples were frozen immediately after preparation and held at 10° F until analyzed. Their results are given in Table 11-7 on pp. 252 and 253.

Pineapple Juice

[See also authentic data in Table 11-11 (Osborn)]

McRoberts[49] and co-workers made an extensive survey of Hawaiian pineapple juice production, covering the entire industry (9 plants) on Oahu, Maui and Kauai during the July to October season of 1940.

Most of the commercial pineapple juice today comes from Hawaii. It is rarely pressed from the entire edible part of the fruit, but is a manufactured product, blended from juices obtained at various steps in fruit canning operations. It is thus closely geared to the canning line. The various forms of the canned fruit--slices, dice, crushed, etc.—change from day to day, with corresponding changes in the source of the juice. There are, however, three main sources:

1. *Core*—The cylindrical center removed from fruit before slicing.
2. *End-of-Table Material*—broken, excessively trimmed or blemished slices not meeting the Grade A standard. This may be crushed and canned as such, or pressed for juice.
3. *Eradicator Meat*—fruit meat removed from the inside of the skin after separation for slicing.

These components are blended by each packer to conform to his own standards for sweetness, acidity and pulp. These are judged by tests for solids (Brix), titratable acidity, and suspended pulp, respectively. In general, cores and eradicator meat are mixed together previous to extraction of the juice. Other minor sources of juice are that flowing from the machine peelers and corers, and juice drained from "crushed pineapple" before canning.

See Table 11-8 (p. 254) for analyses of authentic juices.

Apples

Lopez[52] analyzed 7 processing and 3 table varieties of apples, collecting the samples from lots for processing plants and cold storages in the main producing areas of Virginia. The maturity and condition of the fruit were representative of fruit normally received at packing and processing plants. Of the varieties analyzed, Delicious, Winesap and Lowry are table fruit; the other 7 are important processing varieties. Ash and pectin were determined on the apple flesh. All other determinations were made on the expressed juice. Pectin was determined by the Carre-Haynes method. All other determinations were made by *AOAC Methods*, 8th ed. The summary of these analyses is given in Table 11-9 on p. 255.

A considerable amount of literature has been published on analytical studies of apples in addition to the analyses given in this text. Caldwell[53] reported analyses of a large number of varieties of Virginia apples in 1928. Todhunter[54] and Fellers[55] made similar studies of commercial varieties of apples in 1937. Another study of Massachusetts apples was reported by W. B. Esselen and others[56] in 1947. An early study by Van Slyke[57] reported analyses of 75 varieties.

The Connecticut Agricultural Experiment Station[58] investigated the changes in composition of apples with increased maturity. The percentage of malic acid decreased 40 percent, while the total sugars increased 53 percent during eleven weeks before final harvest. The results are given in Table 11-10 on p. 256.

This Experiment Station[59] also reported analyses of 24 experimental samples of apple juice made in 1951 (see p. 257, top).

The total sugar content of fruit increases slightly for a short time after harvest, then gradually declines during storage due to loss of sugar during respiration. A Canadian study on apples[60] showed a drop in total sugar from about 12.5 percent to 10 percent during 6 months storage at 2°.

Cherries, Red Sour Pitted

Fifteen authentic samples of Montmorency red sour cherries (pitted) from the Michigan

(continued on p. 257)

TABLE 11-7

The Chemical Composition of New York State Grapes, 1947

Section of State	Grape juice						Seeded grapes					
	No. of samples	Mean ± S.E.	Minimum	Maximum	S.D. of single determination	Coefficient of variability	No. of samples	Mean ± S.E.	Minimum	Maximum	S.D. of single determination	Coefficient of variability
					Ash, %							
Hudson Valley	26	0.435±0.0063	0.36	0.50	±0.032	7.4	25	0.460±0.0120	0.32	0.58	±0.060	13.0
Finger Lakes	50	0.456±0.0083	0.36	0.62	±0.054	13.0	50	0.463±0.0093	0.33	0.67	±0.066	14.3
Chautauqua	86	0.420±0.0109	0.25	0.64	±0.101	24.0	86	0.465±0.0088	0.25	0.64	±0.082	17.6
All	162	0.434±0.0065	0.25	0.64	±0.083	19.1	161	0.464±0.0058	0.25	0.67	±0.074	15.9
					Phosphorus, mg P_2O_5/100 g							
Hudson Valley	25	32.7 ±1.47	22.0	52.8	±7.33	22.4	25	38.7 ±1.62	27.6	56.8	±8.08	20.9
Finger Lakes	44	31.4 ±0.92	22.8	45.2	±6.08	19.4	47	36.2 ±1.21	24.4	53.2	±8.28	22.9
Chautauqua	83	26.7 ±0.45	14.0	44.0	±4.06	15.2	82	32.0 ±0.59	16.1	66.0	±5.39	16.8
All	152	29.0 ±0.48	14.0	52.8	±5.94	20.5	154	34.4 ±0.59	16.1	66.0	±7.35	21.4
					Potassium, mg K_2O/100 g							
Hudson Valley	26	237 ±8.0	139	308	±41	17.3	25	253 ±8.0	168	320	±40	15.8
Finger Lakes	44	237 ±6.2	123	335	±41	17.3	42	259 ±7.9	173	391	±51	19.7
Chautauqua	82	215 ±4.8	105	311	±44	20.5	81	241 ±4.8	158	358	±43	17.8
All	152	225 ±3.6	105	335	±44	19.6	148	248 ±3.8	158	391	±46	18.5
					Soluble Solids, Brix Reading							
Hudson Valley	26	15.12±0.24	12.9	17.8	±1.20	7.9	—	—	—	—	—	—
Finger Lakes	48	16.37±0.21	13.1	19.5	±1.48	9.0	—	—	—	—	—	—
Chautauqua	82	16.74±0.18	11.7	20.0	±1.65	9.8	—	—	—	—	—	—
All	156	16.36±0.12	11.7	20.0	±1.54	9.4	—	—	—	—	—	—

					Acidity, pH							
Hudson Valley	26	2.81±0.035	2.52	3.11	±0.18	6.4	—	—	—	—	—	—
Finger Lakes	46	2.95±0.018	2.62	3.17	±0.12	4.1	—	—	—	—	—	—
Chautauqua	84	2.93±0.020	2.51	3.32	±0.18	6.1	—	—	—	—	—	—
All	156	2.92±0.014	2.51	3.32	±0.17	5.8	—	—	—	—	—	—
					Total Acid, % Tartaric Acid							
Hudson Valley	26	1.24±0.039	0.97	2.02	±0.20	16.1	—	—	—	—	—	—
Finger Lakes	46	1.36±0.024	0.94	1.75	±0.16	11.8	—	—	—	—	—	—
Chautauqua	83	1.26±0.012	1.04	1.65	±0.12	9.5	—	—	—	—	—	—
All	155	1.29±0.012	0.94	2.02	±0.16	12.4	—	—	—	—	—	—
					Insoluble Solids, %							
Hudson Valley	—	—	—	—	—	—	25	1.63±0.030	1.41	1.94	±0.15	9.2
Finger Lakes	—	—	—	—	—	—	49	1.70±0.053	1.13	3.98	±0.37	21.8
Chautauqua	—	—	—	—	—	—	86	1.65±0.017	1.17	2.01	±0.16	9.7
All	—	—	—	—	—	—	160	1.66±0.019	1.13	3.98	±0.24	14.5

TABLE 11-8

Analyses of Authentic Samples of Pineapple Juice

Finished Juice 26 Samples	Water-Insol. Solids %	Soluble Solids Corr. for WIS %	Sugar (invert)		Vitamin C mg/100 g	Pulp % by vol.	Ash		K_2O		P_2O_5		Total Acidity ml $N/10$ per 100 g
			Before inv. %	After inv. %			Whole Product %	Prepared Soln. %	Whole Product mg/100 g	Prepared Soln. mg/100 g	Whole Product mg/100 g	Prepared Soln. mg/100 g	
Maximum	0.31	15.30	11.59	13.36	12.8	17.2	0.465	0.464	244.2	245.0	30.3	28.7	146.7
Minimum	0.14	13.14	6.51	10.88	6.5	6.9	0.226	0.227	104.4	106.8	13.8	13.8	71.5
Average	0.22	14.02	8.00	12.24	9.3	12.1	0.360	0.359	181.8	183.8	21.9	20.0	108.1
Component Juices—Averages													
Core Juice 4 Samples	0.33	13.81	8.31	12.43	11.1	14.0	0.280	0.281	140.0	140.9	9.9	7.3	82.7
End of Table Juice 12 Samples	0.16	14.94	9.29	13.05	7.1	8.2	0.370	0.368	180.9	181.8	16.2	14.5	115.6
Eradicator Meat Juice 16 Samples	0.17	13.87	9.41	11.97	9.4	7.2	0.394	0.394	198.0	197.6	24.4	23.0	114.3

TABLE 11-9

A Statistical Analysis on Values Obtained for Several Chemical and Firmness Determinations Made on Ten Varieties of Virginia Processing and Table Apples

	Firmness[a]	pH	Acidity as malic	Soluble solids	Soluble solids-acid ratio	Reducing sugars	Sucrose	Total sugars	Ash	Sp. Gr. of juice	Tannin	Pectin as Ca Pectate
	LBS		%	%		%	%	%	%		%	%
Albemarle Pippin (6)[b]												
Mean value	16.5	3.36	0.71	15.61	22.12	7.90	4.50	12.40	0.25	1.0600	0.03	0.57
Standard deviation	2.1	0.08	0.08	2.28	2.59	1.13	1.16	2.03	0.03	0.01	0.01	0.06
Delicious (14)												
Mean value	13.0	3.99	0.25	15.46	62.45	9.59	3.58	13.17	0.26	1.0641	0.03	0.41
Standard Deviation	1.4	0.10	0.04	1.83	11.40	1.39	0.59	1.53	0.08	0.01	0.01	0.06
Golden Delicious (9)												
Mean value	12.9	3.62	0.40	14.60	37.19	8.43	3.66	12.08	0.25	1.0565	0.03	0.64
Standard Deviation	2.0	0.08	0.08	1.32	5.37	0.65	0.60	0.89	0.07	0.00[c]	0.00[c]	0.15
Grimes Golden (5)												
Mean value	11.6	3.50	0.50	15.26	31.13	6.97	4.57	11.54	0.23	1.0557	0.04	0.62
Standard Deviation	1.8	0.05	0.08	1.08	5.16	1.20	1.65	1.28	0.03	0.00[c]	0.00[c]	0.05
Rome Beauty (10)												
Mean value	13.4	3.50	0.36	12.87	36.46	7.16	3.49	10.65	0.18	1.0455	0.04	0.56
Standard deviation	2.9	0.07	0.07	1.27	4.63	0.77	0.55	0.80	0.06	0.00[c]	0.01	0.11
Stayman (11)												
Mean value	12.7	3.40	0.60	14.46	24.33	7.11	4.87	11.98	0.28	1.0558	0.03	0.61
Standard deviation	1.7	0.04	0.08	1.99	3.35	1.15	1.48	2.03	0.09	0.01	0.00[c]	0.14
Winesap (11)												
Mean value	15.1	3.44	0.59	14.49	24.66	8.37	3.44	11.81	0.28	1.0549	0.04	0.76
Standard deviation	1.5	0.08	0.07	1.80	3.33	0.93	0.83	1.08	0.04	0.01	0.01	0.07
York Imperial (9)												
Mean value	20.8	3.39	0.72	15.23	23.02	8.07	4.31	12.38	0.24	1.0566	0.03	0.53
Standard deviation	3.6	0.13	0.25	2.45	7.42	1.04	1.53	1.92	0.06	0.01	0.01	0.05
Jonathan (3)												
Mean value	9.4	3.48	0.53	13.71	26.03	8.68	2.23	10.91	0.23	1.0528	0.03	0.57
Standard deviation	0.38	0.03	0.05	1.43	0.55	1.06	0.78	1.78	0.02	0.81	0.03	0.06
Lowry (3)												
Mean value	13.0	4.25	0.22	17.46	78.84	9.64	3.59	13.23	0.25	1.0647	0.03	0.32
Standard deviation	0.49	0.09	0.02	0.80	4.54	0.03	0.64	0.83	0.03	0.01	0.01	0.03

[a] Indicates resistance of tissue against applied pressure. [b] Number of samples per variety. [c] Value smaller than 0.005.

TABLE 11-10
Change in Composition of McIntosh Apples with Maturity
1953 Season (116 Samples)

	July 15 Picking			August 14 Picking			September 28 Picking		
	Max.	Min.	Ave.	Max.	Min.	Ave.	Max.	Min.	Ave.
pH	3.30	3.15	3.21	3.40	3.12	3.28	3.40	3.18	3.30
Acidity (as malic acid), percent	1.37	0.93	1.03	1.06	0.69	0.80	0.79	0.51	0.63
Ash, percent	0.47	0.29	0.37	0.38	0.22	0.30	0.34	0.20	0.26
Invert sugar, percent	6.12	5.17	5.64	7.38	6.36	6.82	8.97	7.71	8.41
Sucrose, percent	1.17	0.57	0.78	1.44	0.95	1.15	2.27	1.11	1.44
Total sugars, percent	6.86	5.79	6.42	8.79	7.46	7.97	11.12	9.23	9.85
Potassium, p.p.m.	2,230.	960.	1,399.	1,640.	870.	1,147.	1,510.	690.	1,068.
Calcium, p.p.m.	220.	105.	166.	183.	48.	102.	105.	25.	66.
Magnesium, p.p.m.	150.	93.	113.	105.	58.	82.	88.	58.	72.
Phosphorus, p.p.m.	200.	85.	147.	192.	115.	148.	213.	95.	139.
Manganese, p.p.m.	1.2	0.5	0.7	1.0	0.3	0.6	1.0	0.3	0.4
Iron, p.p.m.	23.0	4.2	13.2	43.0	3.0	12.0	12.0	3.0	5.3
Aluminum, p.p.m.	10.0	1.0	2.5	2.8	1.0	1.7	2.0	0.8	1.3
Zinc, p.p.m.	15.7	2.0	4.5	52.0	3.0	14.5	79.0	2.0	9.7
Copper, p.p.m.	13.0	4.1	7.4	15.0	4.1	8.1	15.0	5.3	9.7
Boron, p.p.m.	4.5	2.1	2.9	5.1	1.5	2.9	6.9	1.3	2.9

	Ash g/100 ml	Acid as Malic g/100 ml	Ascorbic Acid g/100 ml	Invert Sugar g/100 ml	Sucrose g/100 ml	K	Ca	Mg	Fe	Cu	Zn	B
						Parts per million						
Average	0.21	0.44	0.47	7.86	2.37	915	35	40	10	1.0	2.1	1.7
Maximum	0.26	0.58	1.26	8.94	2.90	1060	46	45	14	1.4	3.6	3.2
2 Minima	0.16	0.27	0.20	7.02	1.81	730	28	36	7	0.6	1.4	0.8
	0.16	0.25	0.20	7.14	1.94	790	28	37	7	0.8	1.6	1.2

1962 season were used and reported by Boland and Bloomquist.[61] *AOAC Methods*, 9th ed., were used with amino acids being determined by formol titration. Their results are in good agreement with those of Osborn mentioned earlier:

Composition of Michigan R,S,P, Cherries

	Average	Maximum	2 Minimum	Std. Deviation
Insoluble solids, percent	1.06	1.10	0.87 0.90	0.26
Soluble solids, refract., percent	15.2	16.0	12.7 12.8	2.32
Ash, percent	0.382	405	343 357	0.026
K_2O, mg/100 g	197	207	178 182	11.40
P_2O_5, mg/100 g	36.7	42.3	31.6 32.5	3.70
Sugars as invert, percent	9.79	10.87	8.33 8.70	1.26
Acidity as malic, percent	1.01	1.18	1.00 1.10	0.17
Protein, percent	0.65	0.82	0.49 0.50	0.11
Amino acids, mg/100 g	2.88	3.61	2.07 2.42	0.39

Minor Fruit

Biale[62] has reviewed the literature on the composition and biochemistry of tropical and subtropical fruit. Some of these are grown to some extent in southern areas of the United States (avocado, cherimoya, date, guava, mango, papaya, papaw, passion fruit, sapote). There are also data on such fruit as banana, persimmon, pineapple and citrus.

Money[63] has published analytical data on British fruit, as has Hulme[64] on the composition of British apples and pears.

The Osborn Data

Osborn,[65] of the Food and Drug Administration, reviewed FDA files for many years back, and summarized previously unpublished data on the analysis of authentic fruit and fruit juices, and of jellies and jams made therefrom under observation of FDA inspectors and chemists.

As originally published, the paper included maximum, first and second minimum and average results and standard deviations, on all fruits and fruit products analyzed. We have condensed these data in Table 11-11 (pp. 258-

(Continued on p. 268)

TABLE 11-11

Authentic Data on Composition of Jams, Jellies, Fruit Juices and Other Mixtures Containing Fruit

	Soluble Solids, %	Sugars as Invert, %	Ash, %	K_2O, mg/100 g	P_2O_5, mg/100 g	Ash		Acidity, ml 0.1*N*/100 g
						K_2O, %	P_2O_5, %	
Apple as Original Fruit—20 Samples (3/9/43)								
Range	12.7–16.8	7.5–12.8	.203–.462	104–222	8.1–24.4			37.0–95.3
Average	14.3	10.95	.309	158	19.0	51.5	6.3	71.7
Apple—Fresh Juice, Series I—5 Samples (3/9/43)								
Range	Adjusted to	11.10–11.44	.239–.322	114–176	11–17.5	35.0–59.7	4.2–6.2	32.4–81.0
Average	13.3	11.23	.277	151	14.7	54.7	5.3	69.0
Apple—Fresh Juice, Series II—6 Samples (10/5/43)								
Range	10.3–13.4		.206–.347	95–179	11.8–22.3	46.1–59.4	4.2–7.2	
Average	12.1		.294	162	17.4	55.0	5.9	
Apple—Concentrated Jelly Juice, Series I—12 Samples (3/9/43)								
Range	Adjusted to	10.90–12.10	.234–.363	123–191	9.9–26.8	48.3–61.0	4.2–7.8	51.1–85.6
Average	13.3	11.47	.311	169	19.8	54.5	6.3	66.2
Apple—Concentrated Jelly Juice, Series II—5 Samples (10/5/43)								
Range	Adjusted to		.245–.342	151–182	10.8–21.9	53.1–61.7	4.4–7.2	
Average	13.3		.301	174	17.7	58.3	5.8	
Apple-Jelly, Adjusted—19 Samples (10/5/43)								
Range	Adjusted to		.102–.164	49–95	4.6–12.5	47.0–68.5	4.1–8.9	
Average	65.0		.128	73	8.4	56.5	6.5	
Apricot—Original Fruit, solid pack—34 Samples (7/14/43)								
Range	11.9–17.0	7.9–12.1	.596–.996	330–573	36.9–70.8	55.0–61.8	5.1–8.3	
Average	15.0	9.9	.765	446	52.2	58.3	6.9	

	Soluble Solids, %	Sugars as Invert, %	Ash, %	K_2O, mg/100 g	P_2O_5, mg/100 g	Ash		Acidity, ml 0.1*N*/100 g
						K_2O, %	P_2O_5, %	
Apricot—Jam, solid pack—34 Samples (7/14/43)								
Range			.290–.449	157–269	16.9–32.7	53.2–63.2	5.1–8.3	
Average			.356	208	24.2	58.3	6.8	
Apricot—Original Fruit, Other than solid pack—34 Samples (7/14/43)								
Range	11.1–18.7	7.1–14.2	.542–.852	309–512	30.7–67.2	56.0–60.8	5.7–9.9	
Average	14.7	10.2	.756	441	55.7	58.3	7.5	
Apricot—Jam, other than solid pack—19 Samples (7/14/43)								
Range			.259–.411	134–243.7	13.0–32.4	55.2–61.0	5.4–9.2	
Average			.341	198.5	25.3	58.2	7.4	
Apricot—Fresh Juice—Adjusted—1 Sample (10/5/43)								
	13.1		.763	413	63	54.4	8.8	
Black Raspberry—Original Fruit, Series I—5 Samples (1/21/43)								
Range	8.6–17.0	4.30–12.10	.459–.505	236–285	26.4–54.8	48.0–56.4	5.8–11.1	101.0–144.7
Average	12.2	7.91	.480	251	39.0	53.2	8.1	119
Black Raspberry—Jelly Juice, Adjusted—6 Samples (1/21/43)								
Range	Adjusted to	7.30–8.40	.324–.656	181–349	17.1–43.8	50.0–56.0	4.5–9.1	74.3–182.0
Average	11.1	8.05	.461	247	30.0	53.7	6.5	133.3
Black Raspberry—Jelly, Adjusted—6 Samples (1/21/43)								
Range	Adjusted to		.149–.311	88–162	8.2–23.4	49.6–60.0	4.7–9.3	
Average	65.0		.220	119	14.9	54.9	6.8	
Black Raspberry—Original Fruit, Series II—23 Samples (9/2/42)								
Range	7,10–15.8	3.92–9.90	.297–.598	132.0–312.0	20.0–61.8	44.4–53.1	5.1–14.3	
Average	12.8	7.76	.469	236.1	39.1	50.2	8.4	

TABLE 11-11 (continued)

	Soluble Solids, %	Sugars as Invert, %	Ash, %	K_2O, mg/100 g	P_2O_5, mg/100 g	Ash		Acidity, ml 0.1*N*/100 g
						K_2O, %	P_2O_5, %	
Black Raspberry—Jam, Adjusted—23 Samples (9/2/42)								
Range	Adjusted to		.143–.271	62.9–136	9.4–25.4	43.4–55.0	4.7–13.2	
Average	65.0		.218	112	16.4	51.1	7.6	
Black Raspberry—Fresh Juice—1 Sample (10/5/43)								
	10.8		.432	204	36.1	47.1	8.3	
Black Raspberry—Stored Juice—1 Sample (10/5/43)								
	10.9		.417	211	36.7	50.7	8.8	
Blackberry—Original Fruit, Series I—11 Samples (1/21/43)								
Range	8.6–15.8	5.10–12.22	.196–.478	82–244	23.4–49.6	41.8–51.5	6.5–19.2	67.4–227.2
Average	11.2	7.80	.355	173	35.6	48.5	10.4	158.2
Blackberry—Jelly Juice, Adjusted—11 Samples (1/21/43)								
	Adjusted to	6.80–7.94	.229–.420	93–202	18.5–34.6	41.0–54.7	5.3–13.5	77.2–208.6
	10.0	7.33	.323	159	28.3	49.1	9.0	146.7
Blackberry—Jelly, Adjusted—11 Samples (1/21/43)								
Range	Adjusted to		.107–.187	45–95	8.9–16.5	42.0–58.5	5.4–14.7	
Average	65.0		.149	75	13.4	50.3	9.3	
Blackberry—Original Fruit, Series II—35 Samples (9/1/42)								
Range	8.20–19.8	5.11–12.3	.319–.550	156–267	23.8–69.5	44.9–52.7	5.3–12.9	
Average	12.23	8.08	.403	197	39.3	48.8	9.7	
Blackberry—Jam, Adjusted—35 Samples (9/1/42)								
Range	Adjusted to		.152–.259	79.4–128	10.1–31.2	45.5–56.2	4.9–12.0	
Average	68.0		.195	96.2	18.0	49.3	9.2	

	Soluble Solids, %	Sugars as Invert, %	Ash, %	K_2O, mg/100 g	P_2O_5, mg/100 g	Ash		Acidity, ml 0.1*N*/100 g
						K_2O, %	P_2O_5, %	
Blackberry—Fresh Juice—2 Samples (10/5/43)								
Average	9.9		.351	178	30.5	50.8	8.6	
Blackberry—Stored Juice—2 Samples (10/5/43)								
Average	9.8		.346	177	28.8	51.1	8.4	
Cherry—Original Fruit—41 Samples (9/2/42)								
Range	12.9–22.0	8.40–14.6	.326–.583	172–330	33.8–57.0	49.3–58.7	7.7–12.6	
Average	15.6	10.1	.406	219	40.1	54.0	9.9	
Cherry—Jam, Adjusted—41 Samples (9/2/42)								
Range	Adjusted to		.144–.278	79.3–154.4	15.8–26.7	49.1–60.7	7.6–13.3	
Average	68.0		.192	106.6	19.1	55.4	10.0	
Crabapple—Jelly Juice, Adjusted—9 Samples (1/30/43)								
Range	Adjusted to	10.80–11.80	.274–.467	167–263	17.1–38.9	50.0–63.0	4.8–9.5	98.7–171.6
Average	15.4	11.18	.370	210	26.1	51.4	6.2	123.0
Crabapple—Jelly, Adjusted—9 Samples (1/30/43)								
Range		41.6–66.9	.128–.211	78–125	8.5–18.8	51.6–68.2	5.1–9.4	
Average		60.8	.169	98	12.1	58.5	7.2	
Crabapple—Fresh Juice—3 Samples (10/5/43)								
Average	11.2		.388	212	30.2	54.6	7.8	
Crabapple—Stored Juice—1 Sample (10/5/43)								
	13.4		.364	200	29.6	55.1	8.1	
Cranberries—Original Fruit—24 Samples (11/52)								
Range	9.6–11.5		.17–.23	78.5–116	6.0–14.8	46.3–55.9	2.6–7.7	
Average	10.7		.19	96.8	9.5	50.2	5.0	

TABLE 11-11 (continued)

	Soluble Solids, %	Sugars as Invert, %	Ash, %	K_2O, mg/100 g	P_2O_5, mg/100 g	Ash		Acidity, ml 0.1N/100 g
						K_2O, %	P_2O_5, %	
Currant (Red)—Original Fruit—17 Samples (1/30/43)								
Range	9.1–15.2	4.90–9.76	.431–.643	155–342	19.4–75.7	40.8–56.2	3.8–18.2	278.7–337.7
Average	11.3	6.82	.512	267	43.0	51.9	6.6	314.6
Currant (Red)—Jelly Juice, Adjusted—17 Samples (1/30/43)								
Range	Adjusted to	5.41–7.50	.368–.736	142–418	15.7–49.6	38.0–57.9	3.4–13.4	236.4–380.3
Average	10.5	6.54	.468	248	26.4	52.6	5.8	309.8
Currant (Red)—Jelly, Adjusted—17 Samples (1/30/43)								
Range			.178–.352	67–194	5.6–23.4	37.5–62.6	3.1–13.1	
Average			215	116	12.3	54.2	5.9	
Currant (Red)—Fresh Juice—2 Samples (10/5/43)								
Average	10.9		.433	223	23.6	51.5	5.5	
Currant (Red)—Stored Juice—2 Samples (10/5/43)								
Average	11.2		.438	216	22.8	49.2	5.2	
Damson Plum—Original Fruit—5 Samples (2/10/43)								
Range	12.3–17.6	5.20–8.27	.370–.710	191–404	21.6–71.0	51.6–59.8	4.8–10.0	286.0–369.4
Average	14.7	6.99	.556	308	38.5	54.9	6.7	329.5
Damson Plum—Jelly Juice, Adjusted—5 Samples (2/10/43)								
Range	Adjusted to	6.30–7.70	.418–.594	230–330	21.7–58.8	52.3–57.0	4.2–10.0	252–410
Average	14.3	7.02	.524	285	32.8	54.6	6.2	340.1
Damson Plum—Jelly, Adjusted—5 Samples (2/10/43)								
Range	Adjusted to		.185–.272	102–149	10.1–26.4	52.8–56.8	4.4–10.0	
Average	65.0		.242	132	15.7	55.0	6.4	

	Soluble Solids, %	Sugars as Invert, %	Ash, %	K_2O, mg/100 g	P_2O_5, mg/100 g	Ash		Acidity, ml 0.1*N*/100 g
						K_2O, %	P_2O_5, %	
Fig, Kadota—Original Fruit—3 Samples (10/12/42)								
Range	15.4–21.8	13.1–18.2	.346–.574	154–298	26.8–45.4	44.4–51.9	7.7–9.0	
Average	18.9	15.9	.422	208.7	34.4	48.9	8.2	
Fig, Kadota—Jam, Adjusted—5 Samples (11/12/42)								
Range	Adjusted to		.155–.257	72.2–.257	11.2–20.2	46.6–52.8	7.1–7.9	
Average.	65.0		.198	98.7	14.8	49.6	7.4	
Grape, Concord—Original Fruit—27 Samples (5/5/43)								
Range	12.0–19.0	9.20–15.80	.341–.597	178–338	14.4–56.8	49.0–60.0	3.0–10.1	107.7–175.5
Average	16.9	13.21	.474	264	30.2	55.8	6.3	140.8
Grape, Concord—Fresh or Stored Jelly Juice (with Argols) Series I—22 Samples (5/5/43)								
Range	Adjusted to	10.60–12.70	.287–.594	165–324	15.3–45.6	43.6–60.8	3.7–9.2	97.8–172.0
Average	14.3	11.73	.424	235	25.9	55.8	6.1	126
Grape, Concord—Stored Jelly Juice (without Argols) Series I—19 Samples (5/5/43)								
Range	Adjusted to	10.70–14.24	.126–.300	51–154	14.6–38.0	38.8–55.8	6.8–17.5	56.9–134.0
Average	14.3	12.53	.210	103	21.3	48.3	10.5	91.8
Grape, Concord—Jelly, Adjusted—37 Samples (5/5/43)								
Range	Adjusted to		.054–.277	29–153	7.1–20.4	43.0–64.5	4.1–19.4	
Average	65		.145	79	10.8	53.5	8.4	
Grape, Concord—Fresh Juice, Series II—2 Samples (10/5/43)								
Average	17.5		.351	170	19.6	49.2	6.0	
Grape, Concord—Stored Juice, Series II—2 Samples (10/5/43)								
Average	17.7		.257	142	19.6	40.2	8.4	

TABLE 11-11 (continued)

	Soluble Solids, %	Sugars as Invert, %	Ash, %	K_2O, mg/100 g	P_2O_5, mg/100 g	Ash		Acidity, ml 0.1*N*/100 g
						K_2O, %	P_2O_5, %	
Loganberry—Original Fruit, Series I—5 Samples (2/13/43)								
Range	10.3–12.3	5.70–7.80	.360–.489	153–254	25.1–41.8	42.5–53.4	5.6–11.6	248–364
Average	11.3	6.84	.437	218	31.3	49.6	7.0	311.8
Loganberry—Jelly Juice, Adjusted—5 Samples (2/13/43)								
Range	Adjusted to	6.0–7.46	.324–.508	141–246	17.8–26.3	43.0–50.8	3.9–8.0	230–345
Average	10.5	6.63	.417	203	22.3	48.5	5.5	307.6
Loganberry—Jelly, Adjusted—5 Samples (2/13/43)								
Range	Adjusted to	57.3–65.9	.157–.229	69–118	8.1–13.6	44.1–51.9	4.3–8.7	
Average	65.0	62.8	.191	95	10.9	49.8	5.9	
Loganberry—Original Fruit, Series II—11 Samples (8/28/42)								
Range	8.70–12.2	4.10–7.47	.347–.495	166–248	24.2–46.1	45.0–50.5	6.2–9.8	
Average	10.98	6.47	.419	204.8	33.9	48.8	8.1	
Loganberry—Jam—11 Samples								
Range			.169–.231	80.4–109.0	10.5–21.2	46.0–51.0	5.8–9.8	
Average			.197	95.1	15.3	48.2	7.7	
Peach—Original Fruit—Solid Pack—36 Samples (6/29/43)								
Range	8.50–14.2	6.50–11.5	.302–.495	155–279	29.8–53.3	50.4–58.0	6.8–13.0	
Average	11.4	8.76	.385	208	39.0	54.0	10.2	
Peach—Jam (solid pack fruit)—36 Samples (6/29/43)								
Range			.158–.228	81.4–134.0	15.0–24.6	44.0–63.2	6.9–13.8	
Average			.184	99.4	18.7	54.1	10.2	

	Soluble Solids, %	Sugars as Invert, %	Ash, %	K_2O, mg/100 g	P_2O_5, mg/100 g	Ash		Acidity, ml 0.1*N*/100 g
						K_2O, %	P_2O_5, %	
Peach—Original Fruit (Other than solid pack)—15 to 27 Samples (6/29/43)								
Range	8.70–13.8	6.80–10.5	.301–.481	196–281	34.9–58.6	51.0–59.1	7.9–12.8	
Average	11.2	8.58	.399	231	41.6	55.9	10.1	
Peach—Jam (Other than solid pack fruit)—13 Samples (6/29/43)								
Range			.152–.243	86.4–143.4	15.3–26.1	51.5–59.6	7.2–12.2	
Average			.194	108.7	18.8	56.1	9.8	
Pineapple—Original Fruit—27 Samples (1938–1949)								
Range	12.3–16.8	10.1–14.5	.243–.452	114–247	9.2–26.7	46.5–55.4	2.6–6.9	
Average	15.2	12.9	.364	188.0	15.7	51.4	4.4	
Pineapple—Jam—27 Samples (1938–1940)								
Range			.121–.212	58.9–129	5.1–12.8	44.3–60.7	2.7–7.7	
Average			.175	93.0	7.8	53.0	4.5	
Pineapple, Hawaiian—Jelly Juice—5 Samples (2/13/43)								
Range	Adjusted to	12.20–12.80	.293–.412	140–214	15.4–25.8	47.1–55.4	3.7–6.9	68.6–118.3
Average	14.3	12.48	.349	178	19.4	50.7	5.7	95.7
Pineapple, Hawaiian—Jelly—5 Samples (2/13/43)								
Range		62.6–65.5	.130–.184	65–98	6.8–12.2	47.8–53.9	3.7–7.2	
Average		63.7	.162	84	9.0	51.6	5.7	
Plum, Italian—Original Fruit—5 Samples (11/12/42)								
Range	13.0–23.7	8.07–13.7	.352–.571	178–323	24.3–40.2	47.1–56.7	6.3–9.1	
Average	16.5	9.94	.438	234.8	33.6	53.2	7.7	
Plum, Italian—Jam—5 Samples (11/12/42)								
Range			.150–.243	80.6–142.3	11.6–16.9	52.2–58.6	6.9–8.8	
Average			.189	104.7	14.8	55.1	7.9	

TABLE 11-11 (continued)

	Soluble Solids, %	Sugars as Invert, %	Ash, %	K_2O, mg/100 g	P_2O_5, mg/100 g	Ash		Acidity, ml 0.1N/100 g
						K_2O, %	P_2O_5, %	
Quince—Original Fruit, Series I—4 Samples (8/29/42)								
Range	11.3–16.3	6.50–9.96	.352–.441	180.9–244.0	17.3–44.9	50.0–55.3	4.9–10.2	
Average	13.6	8.10	.388	204.5	30.6	52.6	7.7	
Quince—Jam—4 Samples (8/29/42)								
Range			.161–.204	89.1–111.4	8.3–21.5	49.0–55.8	5.2–10.5	
Average			.179	94.9	14.3	53.0	7.8	
Quince—Original Fruit, Series II—5 Samples (2/13/43)								
Range	11.2–14.6	6.64–7.81	.318–.453	159–248	21.4–40.7	50.0–56.4	5.7–9.8	67.1–176.1
Average	12.9	7.45	.384	209	31.2	54.4	8.1	112.7
Quince—Jelly Juice Adjusted,—5 Samples (2/13/43)								
Range	Adjusted to	6.96–8.90	.386–.418	187–227	20.2–35.8	48.0–57.0	4.8–9.3	73.0–167.7
Average	13.3	7.97	.401	212	30.9	52.4	7.7	116.6
Quince—Jelly, Adjusted—5 Samples (10/5/43)								
Range			.178–.203	85–115	9.4–18.1	47.9–56.8	5.1–9.2	
Average			.188	102	14.7	53.9	7.8	
Red Raspberry—Original Fruit, Series I—54 Samples (10/23/42)								
Range	7.20–14.4	3.66–10.2	.321–.535	149.4–286.0	22.2–70.1	43.7–53.4	6.8–14.6	
Average	10.8	6.76	.397	194.3	43.6	48.9	10.9	
Red Raspberry—Jam—51 Samples (10/23/42)								
Range			.152–.279	73.9–148	10.9–33.3	42.1–53.8	6.7–14.5	
Average			.192	95.5	20.3	49.8	10.5	
Red Raspberry—Original Fruit, Series II—12 Samples (2/18/43)								
Range	9.0–15.0	4.70–10.85	.326–.455	165–240	23.2–51.2	47.9–52.8	5.1–12.5	158.8–285.1
Average	11.8	7.52	.395	200	41.5	50.6	10.5	193.2

	Soluble Solids, %	Sugars as Invert, %	Ash, %	K_2O, mg/100 g	P_2O_5, mg/100 g	Ash		Acidity, ml 0.1N/100 g
						K_2O, %	P_2O_5, %	
Red Raspberry—Jelly Juice, Adjusted—12 Samples (2/18/43)								
Range	Adjusted to	6.60–7.90	.305–.528	155–276	28.5–47.3	48.0–52.0	7.2–13.4	136.1–319.3
Average	12.5	7.28	.387	196	39.2	50.7	10.3	191.0
Red Raspberry—Jelly, Adjusted—12 Samples (2/18/43)								
Range	Adjusted to		.134–.234	76–138	13.7–22.3	47.4–59.0	8.4–13.0	
Average	65.0		.178	94	18.7	52.7	10.6	
Strawberry—Original Fruit, Series I—63 Samples (8/28/42)								
Range	5.20–13.0	3.20–9.81	.239–.612	117–268	19.9–72.3	43.8–54.6	6.0–12.5	
Average	8.24	5.59	.383	191	36.9	50.0	9.6	
Strawberry—Jam—63 Samples (8/28/42)								
Range			.105–.262	53.1–133	7.0–30.3	46.1–56.7	5.1–13.2	
Average			.186	96.5	15.4	51.9	8.2	
Strawberry—Original Fruit, Series II—14 Samples (4/10/43)								
Range	5.4–10.4	3.60–7.94	.305–.555	143–256	25.5–50.8	45.6–54.7	5.7–11.4	124.1–257.2
Average	7.7	4.97	.392	200	35.2	51.0	9.0	163.4
Strawberry—Jelly Juice, Adjusted—14 Samples (4/10/43)								
Range	Adjusted to	4.40–6.12	.268–.500	138–262	14.7–45.0	46.0–54.1	4.1–10.3	94.6–262.1
Average	8.0	5.25	.407	207	30.4	50.8	7.4	170.0
Strawberry—Jelly, Adjusted—14 Samples (4/10/43)								
Range	Adjusted to		.129–.228	68–129	7.3–20.9	46.2–57.2	4.1–10.6	
Average	65.0		.189	99	14.4	52.4	7.6	

267). In so doing, analyses of a few minor fruit such as elderberry, guava and prickly pear have been omitted.

These analyses represent fruit from the principal production centers of the United States. Each fruit sample was collected by a Food and Drug inspector, who observed the entire processing into jam or jelly, following the established manufacturing process used by each plant. These samples are, then, "authentic" in the accepted definition of that word.

Where fruit juices were "adjusted", the soluble solids of the fruit and jelly conform with the requirements of the FDA's Definitions and Standards for the respective fruit jellies. *See* Text Reference 5, *Regulation* 29.2.

TEXT REFERENCES

1-a. Denny, F. E.: *Botan. Gaz.*, **73**:41 (1922).
-b. Schultz, E. F. and Schneider, G. W.: *Proc. Am. Soc. Hort. Sci.*, **66**: 36 (1955).
2. U. S. Department of Defense: *Sampling Procedures and Tables for Inspection of Attributes, MIL-STD*-105 *D*. U. S. Government Printing Office (1963).
3. Munch, J. C. and Bidwell, G. L.: *J. Assoc. Offic. Agr. Chemists,* **11**: 220 (1928).
4-a. Mills, P. A.: *J. Assoc. Offic. Agr. Chemists,* **30**: 274 (1947).
-b. *Official Methods of Analysis.* AOAC, 10th ed. **20.001** (1965).
5. Anon.: *Code of Federal Regulations, Title* 21—Food and Drugs *Parts* 1 *to* 129, *Sec.* 10.6. Washington: U. S. Government Printing Office (January 1, 1966).
6. Ibid. Sec. 27.51(b)(1).
7. Anon.: *Office Consolidation of the Canada Agricultural Products Standards Act and the Processed Fruit and Vegetable Regulations, Schedule D.* Ottawa: Queen's Printer (1966).
8. Wallace, W. W. and Osborn, R. A.: *J. Assoc. Offic. Agr. Chemists,* **36**: 860 (1953).
9. Fallscher, H. O. and Osborn, R. A.: *J. Assoc. Offic. Agr. Chemists,* **36**: 270 (1953).
10. Osborn, R. A. and Hatmaker, C. G.: *J. Assoc. Offic. Agr. Chemists,* **37**: 309 (1954).
11. Fallscher, H. O.: *J. Assoc. Offic. Agr. Chemists,* **38**: 611 (1955).
12. Anon.: *The Food and Drugs Act and Regulations.* Department of National Health and Welfare. Ottawa: Queen's Printer (1966).
13. Sites, J. W. and Reitz, H. J.: *Proc. Am. Soc. Hort. Sci.,* **54**: 1 (1949); **55**: 73 (1950); **56**: 103 (1950).
14. Anon.: *United States Standards for Grades of* (1) *Canned Grapefruit Juice,* (2) *Canned Blended Grapefruit Juice and Orange Juice,* (3) *Canned Lemon Juice,* (4) *Frozen Concentrated Grapefruit Juice.* Washington.: USDA Consumer and Marketing Service.
15. Reference 5, Sec. 27.55.
16. Ferris, L. W.: *J. Assoc. Offic. Agr. Chemists,* **40**: 333 (1957).
17. Ferris, L. W.: *J. Assoc. Offic. Agr. Chemists,* **36**: 266 (1953).
18. Goodban, A. E. and Stark, J. B.: *Anal. Chem.,* **29**: 283 (1957).
19. Barker, S. B. and Summerson, W. H.: *J. Biol. Chem.,* **138**: 535 (1941).
20. Sale, J. W. and Wilson, J. B.: *J. Agri. Res.,* **33**: 301 (1926).
21. Luh, B. S. and Pinochet, M. F.: *Food Res.,* **24**: 423 (1959).
22. Major, A., Jr. and Hall, M. J. (Food and Drug Adm., Dept. of Health, Education & Welfare): *Private communication* (1964).
23. Power, F. B. and Chestnut, V. K.: *J. Am. Chem. Soc.,* **43**: 1741 (1921).
24. Holck, A. A. and Fields, M. L.: *J. Food Sci.,* **30**: 604 (1965).
25. Fields, M. L.: *Food Technol.,* **18**: 1224 (1964).
26. Fields, M. L.: *Food Technol.* 16, No. **8**: 98 (1962).
27. Holck, A. A. and Fields, M. L.: *Food Technol.,* **19**: 1734 (1965).
28. Hill, E. C. and Wenzel, F. W.: *Food Technol.* **11**: 240 (1957).
29. *Official Methods of Analysis.* AOAC, 10th ed. (1965), **20.003(b)(c)**.
30. Reference 5. Sec. 29.1(a)(5) and 29.3(a)(c).
31. Osborn, R. A.: *J. Assoc. Offic. Agr. Chemists,* **32**: 177 (1949); **33**: 349 (1950).
32. Sale, J. W.: *J. Assoc. Offic. Agr. Chemists,* **21**: 502 (1938).
33. Osborn, R. A.: *J. Assoc. Offic. Agr. Chemists,* **47**: 1068 (1964).
34. Bonney, V. B. (U. S. Food and Drug Administration): Private communication (1930).

35. Reference 12, regulations B11.201.
36. Gnagy, M. J. and Armstrong, J. F. (Food and Drug Administration). Private communication (1950).
37. Sale, J. W. (Food and Drug Administration): Private communication (1951).
38. Joseph, G. H. *et al.*: *J. Am. Dietet. Assoc.* **38**: 552 (1961).
39. Birdsall, J. J. *et al.*: *J. Am. Dietet. Assoc.*, **38**: 555 (1961).
40. Osborn, R. A. (Food and Drug Administration): Private communication (1938).
41. Ibid. (1953).
42. Hart, S. M.: *J. Assoc. Offic. Agr. Chemists*, **44**: 633 (1961).
43. Osborn, R. A. (Food and Drug Administration): Private communication (1953).
44. Woodfin, C. A. (Food and Drug Administration): Private communication (1936).
45. Osborn, R. A. (Food and Drug Administration): Private communication (1963).
46. Lipscomb, G. Q. (Food and Drug Administration): Private communication (1965).
47. Mills, P. and Petree, L. G. (Food and Drug Administration): Private communications (1940).
48. Robinson, W. B. *et al.*: *N. Y. State Agr. Exp. Sta., Bull.* **285** (1949).
49. McRoberts, L. H. (Food and Drug Administration): Private communication (1946).
50. Gunderson, E. (Food and Drug Administration): Private communication (1966).
51. Loughrey, J. H. (Food and Drug Administration): Private communication (1961).
52. Lopez, A. *et al.*: *J. Food Res.*, **23**: 492 (1958).
53. Caldwell, J. S.: *J. Agri. Res.*, **36**: 289, 391, 407 (1928).
54. Todhunter, E. N.: *Wash. State Univ. Agr. Exp. Sta., Pop. Bull.* **152** (1937).
55. Fellers, C. R.: *Mass. Agr. Exp. Sta. Tech. Bull.* **15** (1928).
56. Esselen, W. B. *et al.*: *Mass. Exp. Sta. Bull.* **44** (1947).
57. Van Slyke, L. L.: *N. Y. State Agr. Exp. Sta. Bull.* **258** (1904).
58. Fisher, H. J.: *Conn. Agr. Exp. Sta., Bull.* **596** (1955).
59. Fisher, H. J.: Ibid., *Bull.* **574** (1953).
60. Krothkow, K. and Helsen, V.: *Can. J. Res.*, **24**:126 (1946).
61. Boland, F. E. and Bloomquist, V.: *J. Assoc. Offic. Agr. Chemists*, **48**: 523 (1965).
62. Biale, J. B.: *Advan. Food Res.*, **10**: 293. New York: Academic Press (1960).
63. Money, R. W. and Christian, W. A.: *J. Science and Agriculture*, **1**: 8 (1950).
64. Hulme, A. C.: *Advan. Food Res.*, **8**: 297. New York: Academic Press (1958).
65. Osborn, R. A.: *J. Assoc. Offic. Agr. Chemists*, **47**: 1068 (1964).

SELECTED REFERENCES

A. Anon.: "*Vegetables and Fruits, Facts and Pointers*" (a series of fact sheets on history, botanical name, grades and standards, nutrient value and composition). Washington: United Fruit & Vegetable Association (1952).
B. Bartholomew, E. and Sinclair, W. B.: *The Lemon Fruit*. Berkeley: Univ. of Calif. Press (1951).
C. Braverman, J. B. S.: *Citrus Products*. New York: Interscience Publications (1949).
D. Chase, E. M.: *Composition of California Lemons. U.S.D.A. Bulletin*, **993** (1921).
E. Cruess, W. V.: *Commercial Fruit and Vegetable Products*. 4th ed. New York: McGraw-Hill (1958).
F. Eastman, E.: "Measurement of Color Changes in Foods". *Advances in Chemistry Series*, **3**: 3-12. Washington: American Chemical Society (1950).
G. Harding, P. L. and Fisher, D. F.: "Seasonal Changes in Florida Grapefruit". *U.S. Dep. Agr., Tech. Bull.* **886** (1945).
H. Joslyn, M. A.: "The Chemistry of Protopectins (A Review Through 1961)". *Advan. Food Res.*, **11**: 1 New York: Academic Press (1962).
I. Kefford, J. F.: "Chemical Constituents of Citrus Fruit." *Advan. Food. Res.*, **9**: 225. New York: Academic Press (1960).
J. Kertesz, Z.: *The Pectic Substances*. New York: Interscience Publishers (1951).
K. Kirschner, J. G.: "Chemistry of Fruit and Vegetable Flavors." *Advan. Food Res.*, **2**: 259. New York: Academic Press (1941).
L. Lipsky, H. T.: "Development of Military Specifications and Standards, Production Tests for New Products". *Food Technol.*, **19**: 1075 (1965).
M. Macara, T.: "Chemical Composition of Fruits Grown in England". *Analyst*, **56**: 35 (1931).
N. Marshall, R. E.: *Cherries and Cherry Products*. New York: Interscience Publishers (1955).
O. Owen, H. S.: "Methods at WRRL Laboratories

for Extraction and Analysis of Pectic Materials". *Publication AIC*–340, USDA Albany: Western Regional Research Laboratories (1952).

P. Rygg, G. L. and Getty, M. R.: "Seasonal Changes in Arizona and California Grapefruit". *U.S. Dep. Agr., Tech. Bull.* **1130** (1955).

Q. Sinclair, W. B.: *The Orange*. Berkeley: University of California Press (1961).

R. Smock, R. M. and Newbert, A. M.: *Apples and Apple Products*. New York: Interscience Publishers (1950).

S. Tressler, D. K. and Joslyn, M. A.: *Fruit and Vegetable Juice Processing Technology*. Westport: Avi Publishing Co. (1961).

T. Von Loesecke, H. J.: *Bananas*. New York: Interscience Publishers (1949).

CHAPTER 12

Nuts and Nut Products

According to Webster's 3rd New International Dictionary, nuts are hard-shelled, dry fruit or seed, having a more or less distinct rind or shell and an interior kernel. Popular nomenclature includes the peanut, which is in reality a legume whose pods grow underground. In general, they are characterized by a high oil content. An exception is the chestnut, which is a starchy fruit with less than 2 percent fat. Cashew nuts contain both fat and starch in appreciable quantities.

As prepared for food, nuts may be:

1) eaten as is,
2) roasted (some nuts, such as cashews and peanuts, contain a bitter substance that must be removed by roasting),
3) ground to a fine paste, such as peanut butter, almond paste (Marzipan), or
4) converted to a meal, after partial removal of oil by pressing, as in peanut meal or peanut flour.

Peanut Butter

The Food and Drug Administration first issued a proposed Definition and Standard of Identity for Peanut Butter in 1959.[1] It proposed that the product contain not less than 95 percent ground peanuts. Considerable controversy followed until the final proposal of December 6, 1967 was adopted as a regulation by publication in the *Federal Register* (November 7, 1968). Former proposals considered two standards: one for peanut butter, and the other for a lesser quantity of ground peanuts, to be designated "peanut spread". The December 1967 proposal provided only for a standard for peanut butter, saying that there was no persuasive evidence for a separate standard for "peanut spread".

The regulation defines peanut butter as the food "prepared by grinding . . . shelled and roasted peanuts, to which may be added safe and suitable seasoning and stabilizing ingredients that do not in the aggregate exceed 10% . . . The fat content . . . shall not exceed 55%." Artificial color, flavor, sweeteners, and chemical preservatives, including antioxidants and vitamins, are not included in the definition.

An interim Federal purchasing specification Z-P-00196B, was issued April 20, 1962, and is still in force. This limits total additives, including hydrogenated or partially hydrogenated peanut oil, to a total of 10 percent. Sweetening ingredients are limited to sucrose and/or dextrose, equivalent to 1.5 percent by

TABLE 12-1
Proximate Composition of Nut Kernels*

	Moisture %	Protein %	Crude Fiber %	Fat (ether ext.) %	Carbohydrate (by diff.) %	Ash %
Almond, dried	4.7	18.6[b]	2.6	54.2	16.9	3.0
Almond Meal, from press cake	7.2	39.5[b]	2.3	18.3	26.6	6.1
Brazil Nut	4.6	14.3[a]	3.1	66.9	7.8	3.3
Cashew, roasted	5.2	17.2[c]	1.4	45.7	27.9	2.6
Chestnut, dried	8.4	6.7[c]	2.5	4.1	76.1	2.2
Coconut, dried	3.5	7.2[c]	3.9	64.9	19.1	1.4
Filbert (hazel-nut)	5.8	12.6[c]	3.0	62.4	13.7	2.5
Macadamia	3.0	7.8[d]	2.5	71.6	13.4	1.7
Peanut—Spanish	Moisture	26.9[a]		50.8		
Virginia	free	26.05[a]		48.4		
Runner	basis	26.41[a]		50.3		
Piñon (pine-nut)	3.0	13.0[c]	1.2	60.5	19.4	2.9
Pistachio	5.3	19.3[c]	1.9	53.7	17.1	2.7
Walnut—English	3.5	14.8[c]	2.1	64.0	13.7	1.9
Black	3.1	20.5[c]	13.1	59.3	1.7	2.3

* These figures are from *USDA Agriculture Handbook No. 8, Composition of Foods,* except for peanuts, which are from Publication AIC-61, U.S.D.A. Agricultural Research Administration.

[a] Nitrogen/protein factor 5.46
[b] Nitrogen/Protein factor 5.18
[c] Nitrogen/protein factor 5.30
[d] Nitrogen/protein factor 6.25

weight of sucrose. (The degree of sweetness for sucrose is 1 and for dextrose is 0.5.) It shall contain not more than 1.4 percent of salt.

The U.S. Department of Agriculture has established standards for A and C Grade peanut butter. The only difference in composition between the two grades is that Grade A shall not contain more than 8 mg of water-insoluble inorganic residue (WIIR) per 100 grams of peanut butter and Grade C not more than 20 mg.

Coconut

General Services Administration has issued a Federal Specification, Z-C-571 d, for the use of all United States agencies in purchasing sweetened, prepared coconut. This defines three styles of shreds: shredded long-thread, fancy-shred and short-shred, depending on shred length. There are also five styles of flakes listed according to particle size. Two types are defined: Type 1, containing not less than 60 percent coconut, not more than 10 percent moisture and not less than 2 percent of propylene glycol; Type 2, containing not less than 50 percent coconut and not more than 18 percent moisture. This type shall be pasteurized.

The product shall be made from fresh or desiccated sweet coconuts without removing any oil, and shall be processed with sucrose, starch (as an anticaking agent in powdered sugar), glycerine, propylene glycol and sorbitol, or combinations of any two or more, and shall be flavored with salt.

Methods of Analysis

Method 12-1. *Preparation of Sample.* (These are the methods prescribed by AOAC).

Nuts in Shell: Remove meat or kernels from a representative sample, and separate all shell and dividing wall fragments. Include skin or spermoderm in all nuts, including peanuts and coconuts unless the purpose of the analysis dictates otherwise. Grind the sample as described under nut meats.

Nut Meats: Nut meats, shredded coconut, or small pieces: Grind not less than 250 g through an Enterprise No. 5 food chopper equipped with a revolving knife blade and a plate with $\frac{1}{8}''$ holes. (Other types of food choppers, graters or comminuting devices that produce a smooth homogeneous paste without loss of oil may be used.) Mix the sample well and store in air-tight glass containers.

Nut Butters: Mix thoroughly with a stiff-bladed knife or spatula, warming if necessary. An electric stirrer may be used if the sample is of proper consistency. Store in an air-tight jar.

Method 12-2. *Moisture. See* Method 1-2.

Dry a 2 g sample, prepared by Method 12-1, to constant weight (about 5 hours) at 95–100° and not over 100 mm of mercury. Distillation with toluene, Method 1-4, may also be used. These methods are not applicable to products containing glycerol or propylene glycol. In such cases dry in a vacuum oven at 60° and 100 mm of mercury. If an ether extract (fat) determination is to be made, weigh the sample for moisture by oven-drying into an aluminium foil dish that can be cut up and transferred to an extraction thimble after drying.

NOTE: *Federal Specification Z-C-*571*d*, prepared coconut, states: "Total moisture will be expressed as percent total volatiles, minus the percent of propylene glycol."

Method 12-3. *Ether Extract (crude fat). See* Method 1-12.

Use 2 g sample prepared by Method 12-1, or the residue from a moisture determination by drying. Extraction time, using a Soxhlet extractor, is about 16 hours. If large quantities of soluble carbohydrates are present, extract first with water, washing the aqueous extract with ether and adding the washings to the ether extract.

Method 12-4. *Crude Fiber.*

Proceed as in Method 1-11.

Method 12-5. *Nitrogen (crude protein).*

Remove most of the fat from the weighed prepared sample by extraction with petroleum ether, then proceed as in Method 1-7 or 1-8. The conventional factor for nitrogen to protein is 6.25. Agriculture Handbook No. 8 uses 5.18 for almonds, 5.46 for peanuts and Brazil nuts and 5.30 for other nuts.

Sugars.

This determination, which is seldom required, may be made by AOAC methods **25.008** and **25.009**, after extracting the weighed sample, prepared by Method 12-1, with petroleum ether.

Method 12-6. *Ash.*

Weigh a 5-10 gram sample, prepared by Method 12-1, into a platinum dish, ash first at low temperature to let fat smoke off without burning, then continue ashing at 525° as directed in Method 1-9.

Method 12-7. *Sodium Chloride. (AOAC Method* ***25.010****).*

Weigh a 2 g sample of ground nut meal or nut butter into a platinum evaporating dish and thoroughly incorporate 10 ml of 10% calcium acetate solution. (For nut butters and nut paste, disperse the sample in 10 ml of acetone before adding the calcium acetate solution, Remove the acetone at room temperature in a current of air.) Dry on the steam bath and ash at 525-550°. Complete ashing is not necessary.

Dissolve the ash in 25 ml of nitric acid (1 + 3), add a known excess of 0.1 *N* silver nitrate solution, heat to boiling, and cool. Add 5 ml of iron indicator (saturated solution of ferric ammonium sulfate, $FeNH_4(SO_4)_2.12H_2O$, chloride free), and titrate the excess of silver nitrate with 0.1 *N* thiocyanate solution to a light brown end point. Run a blank chloride determination on the calcium acetate, and correct the titration of the sample if necessary.

1 ml 0.1 *N* silver nitrate = 5.85 mg NaCl.

Woodroof, in his recent book "*Peanuts*",

includes an electrometric titration for salt, using a pH meter in which the connection of the electrodes is reversed.[2]

Method 12-8. *Water-Insoluble Inorganic Residue in Peanut Butter. (AOAC Method* ***36.021****).*

This determination measures the amount of sand and dirt (dirty-faced nuts) in peanuts used in the manufacture of peanut butter. In the trade it is generally referred to as WIIR.

DETERMINATION

Transfer 100 g of the well-mixed sample to a 250 ml beaker, add about 150 ml of petroleum ether in 10–25 ml portions, mixing thoroughly between each portion. Cover and let settle about ½ hour; decant and discard the solvent layer. Repeat this extraction with about 125 ml of petroleum ether; again decant and discard the solvent. Repeat this treatment with 125 ml solvent, this time washing down the sides of the beaker with the solvent, and again decant and discard the solvent.

Evaporate remainder of the solvent from the sample with the aid of gentle heat, add 150 ml of chloroform, mix thoroughly, and let settle for about 20 minutes, stirring the top layer occasionally. Decant and discard the chloroform layer and floating peanut tissues, being careful not to disturb any sediment in the bottom of the beaker. Repeat the extraction with chloroform, rinsing all particles from the sides of the beaker. (If many skin tissues are present, add just sufficient carbon tetrachloride to the chloroform to float these tissues away from the heavy residue of salt, sand, etc.) Dry the residue in air.

Add 50 ml of hydrochloric acid (1 + 35) to the residue, and than add 90 ml of boiling water. Let stand one half hour with occasional stirring to dissolve any difficultly soluble salts (phosphate, carbonate or calcium sulfate). Decant through a quantitative filter and transfer any residue onto the filter with a stream of hot water from a water bottle. Wash the residue several times with hot water and test the final washings for sulfates with barium chloride, using 3–5 ml of a saturated solution.

If the test for sulfates is positive, test the residue on filter by treating it with 25 ml of hydrochloric acid (1 + 35), added a little at a time. Test the filtrate with 20 drops of the saturated barium chloride solution, allowing 5 minutes for a precipitate to form. Wash the residue on the filter until a silver nitrate test shows all chlorides have been removed.

Transfer the residue on the filter to a platinum crucible and ignite in a muffle furnace at 500°, cool, and weigh to nearest milligram. If the test above shows sulfates have not been completely removed, make a quantitative determination of either calcium or sulfate, calculate to $CaSO_4$, and correct the weight in the crucible for the $CaSO_4$ present. Report the corrected weight as "WIIR".

Aflatoxins

In 1961–1962 reports began to appear of a highly toxic substance, later named *aflatoxin*, in peanut meal from Africa, Brazil and other areas. Following this, British workers traced this toxin to metabolites of a mold, *Aspergillus flavus*. Since then, a voluminous literature on the subject has been published, summarized in part by Trager *et al.*,[3] and interdepartmental working parties from various government laboratories here and abroad have worked on the subject. The AOAC established an Aflatoxin Methodology Working Group, which made reports[3,4,5,6] that culminated in the adoption of an official method by AOAC in 1966.

The U.S. specification for peanuts, shelled, MIL-P-831c, directs that tests for aflatoxin on deliveries to government agencies shall be made by the Processed Products Standardization and Inspection Branch, Fruit and Vegetable Division, Consumer and Marketing Service, USDA. This is the method given here, which is essentially the same as the AOAC method, published in *J. Offic. Assoc. Agr. Chemists*, **49**: 229 (1966).

Method 12-9. *Aflatoxin in Peanuts, Peanut Meal and Peanut Butter.* (Aflatoxin Chemical Assay Procedure of the Aflatoxin Methodology Working Group, June 3, 1965, Fruit and Vegetable Division, Consumer and Marketing Service, U.S. Department of Agriculture).

This is an official method of the U.S. Department of Agriculture Consumer and Marketing Service, and is reproduced here as written, with approval of the Department. It represents the best information available at this time. However, it should be realized that there still are several unknown factors in the analytical methodology for aflatoxin. Thus, in certain sections, several alternatives are presented, and it is left to the discretion of the individual chemist to decide which approach best fits his needs and still gives valid results.

Reagents

The reagents should be carried through the procedure to assure that they are free of interfering fluorescing impurities. Occasionally it has been observed that some organic solvents when stored in metal containers will develop fluorescent impurities which interfere with the final TLC determination. Therefore where possible it is recommended that the solvents be stored in glass bottles.

a. *Benzene, ACS.*
b. *Chloroform, ACS.*
c. *Ethanol, USP* 95%.
d. *n-Hexane, ACS.*
e. *Methanol, ACS.*
f. *Diatomaceous earth, acid washed* (Johns-Manville Celite 545, acid washed).
g. *Aflatoxin standards.* Can be obtained from Arthur D. Little, Inc., Cambridge, Mass. Aflatoxin has been shown to be an extremely potent toxic material to many animals. Neither the effect of aflatoxin on man nor the possible routes of entry are presently known, but this material should be handled as a potentially very toxic substance. Manipulations should be carried out under hoods whenever possible, and particular precautions should be taken when the toxin is in a dry form because of its electrostatic nature and resulting tendency to be dispersed in working areas.
h. *Silica Gel G-HR.* (Brinkmann Instruments, Inc. Westbury, N.Y.)
i. *Cotton.* Absorbent, washed and dried. Place 50 g of absorbent cotton in a beaker and wash with 1 L of $CHCl_3$. Remove the $CHCl_3$ using a Büchner funnel. Repeat two more times, using 750 ml of $CHCl_3$ for each wash. Remove the $CHCl_3$ by evaporation and store in a closed container. (An alternative method is to wash with $CHCl_3$ in a continuous extractor).
j. *Nitrogen gas.* Not required, but recommended.
k. *Carborundum boiling chips.*

Apparatus

a. *Hollow polyethylene stoppers*, such as Nalgene size 00.
b. *Explosion-proof Waring Blendor*, with 1 quart jar and lid with a small perforation (drill $\frac{1}{8}''$ hole approximately $\frac{1}{2}''$ from the center of the lid).
c. *Centrifuge*, 500 ml capacity.
d. *Chromatographic column*, 45 mm × 500 mm or 600 mm, fitted with a stopcock.
e. *Vials*, 4 dram, screw cap with foil liners (Kimble 60910-L).
f. *Heating block*, alluminum or brass, drilled to accommodate the vials (not required, but convenient).

Apparatus for Thin-Layer Chromatography

a. *Glass plates*, 20 × 20 cm (8″ × 8″).
b. *DESAGA-Brinkmann applicator*, or equivalent.
c. *Mounting board.*
d. *Spotting template.*
e. *Microsyringe*, 10 μl capacity.
f. *Desiccating storage cabinet*, Fisher 8-645-5 or equivalent.
g. *Storage rack*, 200-3. (Research Specialties Co., Richmond, Calif. or equivalent.)
h. *Thomas-Mitchell tank*, 9″ × 9″ × 3.5″ stainless steel trough and glass cover. (Arthur H. Thomas Co., or equivalent).
i. *Long wave UV lamp* (15 *watt*) *or Chromatovue cabinet equipped with* 1 *or* 2 15 *watt lamps.*

Extraction

Transfer 50 g of peanut butter quantitatively to a Waring Blendor using 100 ml of hexane and 250 ml 55:45 (V/V) methanol : water. Blend $3\frac{1}{2}$ minutes at full speed. Centrifuge in 500 ml bottle at 1800–2000 rpm for 30 minutes.

For defatted peanut meal use a 100 g sample, omit the hexane, and blend with 500 ml of the aqueous methanol. For peanuts, take a 100 g sample and blend with 500 ml of the aqueous methanol. Immediately after blending, and before any separation occurs, take a portion of the slurry for centrifuging. Discard the remainder. Centrifuge at 1800–2000 rpm for 30 minutes. The aqueous methanol (middle layer) is sometimes cloudy, but this can be disregarded. Proceed with the partition chromatography without delay.

Partition Column Chromatography

Transfer 50 ml from the aqueous methanol layer to a 600 to 1000 ml beaker, add 5 ml of H_2O, and swirl. Add 55 g of the Celite and mix thoroughly with a spoon or spatula until the mixture appears uniform when pressed against the bottom or side of the beaker. Put a cotton plug (about the size of a golf ball) loosely in the bottom of a 45 × 500 or 600 mm chromatographic column. Transfer the Celite-sample mixture to the column in portions of about $\frac{1}{3}$, packing each addition firmly with a tamping rod to make a smooth column. Wash the beaker with hexane and elute the column at a flow-rate of 20–60 ml/min with washings and additional hexane to make a total of 500 ml. When the hexane is about to disappear into the Celite, change the receiver, wash the Celite-sample mixing beaker with two portions of 50:50 (V/V) $CHCl_3$: hexane and add to the column. Continue adding this solvent mixture until 600 ml is collected in the receiver. This fraction contains the aflatoxins eluted from the column. (The column should not run dry any time during the elution procedure, and the total elution time should be no more than 1 hour.)

Add a few carborundum chips to the $CHCl_3$: hexane eluate, evaporate to near dryness on the steam bath, and transfer the residue quantitatively to a 25 or 50 ml flask with $CHCl_3$. Add 2–3 boiling chips and remove the $CHCl_3$ by evaporation, preferably under a gentle stream of nitrogen. Seal with a hollow polyethylene stopper and cap. Save for thin-layer chromatography.

Thin-Layer Chromatography

The TLC procedures described in this method are based on the principle of matching spot intensities. Many laboratories are satisfactorily using the principle of extinction for determining aflatoxin concentrations. For extinction, the minimum volume of sample solution which still allows detection of a fluorescent spot is determined via the TLC procedure below. The quantity of aflatoxin causing this fluorescence is obtained by comparison with the minimum amount of standard aflatoxin B_1 which is detectable on the same TLC plate.

Preparation of plates: Weigh 30 g of Silica Gel G-HR into a 300 ml glass-stoppered Erlenmeyer flask, add 60 ml of water, shake vigorously for not more than 1 minute, and pour into the applicator. Immediately coat five 20 × 20 cm glass plates with about a 0.25 mm thickness of the silica gel suspension, and allow the plates to rest undisturbed until gelled (about 10 minutes). Plate thicknesses from 0.2 to 0.5 mm have been used successfully. Dry the coated plates at 80° for at least 2 hours and store in a desiccating cabinet with desiccant until immediately before use. The drying time is related to plate thickness and drying temperature. Adequacy of drying can be checked by the ability of the analyst to separate all four aflatoxins on the TLC plate.

Preliminary TLC: (This step can be eliminated when the approximate amounts of aflatoxins are known.) Saturate the liner (blotter paper) of an insulated developing tank with 75 ml of $CHCl_3$ and fill the trough with 40–50 ml of 7% methanol in $CHCl_3$ (V/V). Cover the tank and allow to equilibrate for 30 minutes.

Uncap the vial containing the sample extract. Add 500 μl of $CHCl_3$ and reseal with the polyethylene stopper. Make a puncture in the polyethylene stopper just large enough to accommodate the needle of a 10μl syringe. In subdued light and as rapidly as possible spot 5 and 10 μl on a line 4 cm from the bottom edge of a TLC plate. (The quantities of extraneous material vary in different extracts; 10 μl of the sample may overload the plate and be unable to be used.)

Recap the vial to prevent evaporation and save for quantitative analysis. On the same plate spot 1, 3 and 5 μl portions of standard aflatoxins with a concentration of 1.0 μg/ml of B_1 and 0.7 μg/ml of G_1. (Ideally, the standard solution should be prepared from pure, recrystallized B_1, B_2, G_1 and G_2. Since pure aflatoxins are not generally available, an extract whose aflatoxin content is known may be used. Concentrations of 0.5 to 1.5 μg/ml of B_1 and 0.3 to 1.0 μg/ml of G_1 are satisfactory. The standard solution should not needlessly be exposed to light. It can be preserved when not in use by storing at freezer temperatures in a glass-stoppered volumetric flask placed in a closed jar containing a small amount of $CHCl_3$;

this should be replaced periodically. The $CHCl_3$ in the outside jar eliminates concentration of the standard by evaporation, and the jar prevents contamination with condensing water as the standard is warmed to room temperature each time it is used.)

Also spot 5 μl of a qualitative standard containing B_1, B_2, G_1 and G_2 to show that the four aflatoxins are properly resolved under the conditions used. Draw a line across the plate 2–3 cm from the top edge as a stop for the solvent front. Also draw a line about 0.5 cm in from each side edge. Immediately insert the plate into the equilibrated tank, and seal. (Aflatoxins may decompose on the plate after spotting if the plate is not developed immediately because they are sensitive to light, air, acids and bases.) Withdraw the plate from the tank when the solvent front has reached the stop line, 12–14 cm above the origin. After the solvent has evaporated, illuminate the plate from below by placing it flat, coated side up, on the long wave UV lamp in a darkened room. As an alternative the plate may be viewed in a Chromatovue cabinet or illuminated from above with the UV lamp, Observe the pattern of four fluorescent spots of the qualitative standard. In order of decreasing Rf's they are B_1, B_2, G_1, and G_2. Note the small color differences, the bluish fluorescence of the B contrasted with the slightly green G aflatoxins. Examine the patterns from the sample for fluorescent spots having Rf's identical with and appearances similar to those of the standards. From this preliminary plate estimate a suitable dilution for the quantitative TLC analysis. Take into account the quantity of extract used for the preliminary TLC in the final calculations.

Quantitative TLC: If, according to the preliminary plate, a new dilution of the sample extract is required, evaporate it to dryness on the steam bath and redissolve it in the estimated volume of $CHCl_3$. (Usually 0.5 to 3.0 ml.)

Spot successively 3.5, 5 and two 6.5 μl portions of the sample extract. All the spots should be about the same size and no larger than 0.5 cm. On the same plate spot 3.5, 5, and 6.5 μl of B_1 and G_1 standards. As an internal standard spot 5 μl of B_1 and G_1 on top of the two 6.5 μl portions of the sample. Also spot 5μl of B_1, B_2, G_1, and G_2 as qualitative standards. Proceed as above under *Preliminary TLC*.

INTERPRETATION OF THE CHROMATOGRAMS

Four clearly identifiable spots should be visible in the qualitative standard. If not, repeat the chromatography, correcting or adjusting the conditions to obtain the proper resolution.

Examine the pattern from the sample spot containing the internal standard for the aflatoxin B_1 and G_1 spots. If these spots cannot be identified, dilute the extract with $CHCl_3$ and rechromatograph. The Rf's of B_1 and G_1 used as the internal standard should be the same as, or should differ only very slightly from, those of the respective standard aflatoxin spots. Since the spots from the sample extract are compared directly with the standard aflatoxins on the same plate, the magnitudes of the Rf's are unimportant. These vary from plate to plate.

Compare the 5 μl sample pattern with that containing the internal standard. A fluorescent spot in the sample thought to be B_1 or G_1 must have a Rf identical to and color similar to those of the B_1 or G_1 used as internal standards. Identify the unknown spot of the sample as B_1 or G_1 only when the unknown spot and the internal standard spot are superimposed. The spot from the sample and the internal standard combined should be more intense than either sample or standard alone. Compare the sample pattern with qualitative standard to determine if B_2 and G_2 are present.

Compare the fluorescent intensities of the B_1 spots of the sample with those of the standard spots and determine which of the sample portions matches one of the standards. To aid in the determination, the UV light may be attenuated by moving the plate away from the lamp so that any particular pair of spots can be compared at extinction. Interpolate, if the sample spot intensity is found to be between those of two of the standard spots. If the spots of the smallest portion of sample are too intense to match the standards, dilute the sample and rechromatograph. Make a comparison of the G_1 spots in the same manner.

Assume that B_1 and B_2 have the same fluorescence intensity-to-weigh relationships*, and compare the B_2 spots of the sample with the B_1 standard spots to make the quantitative estimate of B_2. Likewise, assume G_2 has the same

* This assumption is probably in error and is currently under review.

fluorescence intensity to weight relationship as G_1, and compare the G_2 spot of the sample with the G_1 spots of the standard.

Calculate aflatoxin B_1 in μg/kg from the formula:

$$\mu g/kg = \frac{S \times Y \times V}{X \times 10^{**}},$$

where S = μl of aflatoxin in the B_1 standard equal to the unknown;
Y = concentration of aflatoxin B_1 standard, μg/ml;
V = volume in μl of the final dilution of the sample extract.
X = μl of sample extract spotted that provides a fluorescent intensity equal to S, the B_1 standard; and
** = The 50 ml of the aqueous methanol extract that is used for the Celite column. This is equivalent to 10 g of the sample. If the final extract dilution does not represent 10 g, calculate the correct sample weight and substitute.

Calculate aflatoxin G_1 in like manner.

Calculate B_2 and report as "μg B_2/kg based on fluorescence of B_1". Calculate G_2 and report as "μg G_2/kg based on fluorescence of G_1".

THIN-LAYER CHROMATOGRAPHIC CONFIRMATION OF AFLATOXIN G_1 OR G_2

(This confirmatory TLC procedure is optional for those laboratories with appropriate chromatographic equipment.) If aflatoxins G_1 and/or G_2 constitute more than 20 percent of the total aflatoxins, confirm the amounts and their identity by chromatography using the following solvent system:

Shake 46:35:19 (V/V) benzene:ethanol:water in a separatory funnel. Let the mixture sit overnight at room temperature. Carefully separate the upper and lower phases. Use the upper phase in the trough and the lower phase in the bottom of the chromatographic tank.

Respot the sample and standards on a silica gel plate as under *Quantitative TLC*. Put 50 ml of lower phase in the bottom of an insulated, unlined developing tank. Put 50 ml of the upper phase in the trough. Without equilibrating, insert the chromatoplate in the trough and seal. Let the solvent rise to the stop line 12–14 cm above the origin (30–50 minutes) and withdraw the plate. In order of decreasing Rf's, the qualitative standard gives B_1, B_2, G_1 and G_2 as before, but many extraneous fluorescent substances found in samples will have quite different Rf's relative to those of the aflatoxins in the two solvent systems. The G_1 and G_2 aflatoxins of the sample should have the same Rf's relative to those of the aflatoxins in the two solvent systems. The G_1 and G_2 aflatoxins of the sample should have the same Rf's as those of the respective standards. Make a quantitative estimate for G_1 and G_2 as described above.

Method 12-10. *Sorbitol in Coconut.* (Developed by Hause *et al.*[7], Merck and Co. Chemical Division. Reprinted by permission of the copyright owner, the American Chemical Society).

APPARATUS

a. *Gas Chromatograph*—The Barber-Colman Model 15 gas chromatograph equipped with a Lovelock ionization detector and containing a 56 microcurie Ra-226 foil source was employed in the original work on this method, but any similar instrument can be used.

b. *Columns*—Glass U-tubes, 6 ft. long, 6 mm. i.d.

REAGENTS

a. *Argon gas*—Obtainable from The Matheson Co. Inc., P.O. Box 85, East Rutherford, N.J. 07073.

b. *Potassium hydroxide solution in methanol*—0.5 *N*. Dissolve 33 g of KOH pellets (85% purity) in methanol and dilute to 1 liter.

c. *Reference standard sorbitol hexaäcetate*—Reflux 2–3.5 g of reagent-grade sorbitol 1 hr in a mixture of 30 ml each of pyridine and acetic anhydride, and remove the solvents by evaporation *in vacuo*. Dissolve the crystalline residue in $CHCl_3$, and wash well with water. Dry the $CHCl_3$ solution over Na_2SO_4, filter, and concentrate to a thick slurry *in vacuo*. Recrystallize to a constant melting point (99–100°) from ethyl acetate. (1.857 g $\approx$ 1 g sorbitol.)

d. *Reference standard mannitol hexaäcetate*—Treat 2–3.5 g of reagent-grade mannitol as under (c). The final melting point should be 124–125°. (1.857 g $\approx$ 1 g mannitol)

e. *Packing support*—Gas Chrom P, 100–140

mesh, obtained from Applied Science Laboratories, Inc., 140 North Barnard St., State College, Pa., is purified as follows:

Resize the material to 100–140 mesh, and place a 55 g portion in a beaker and overlay to a depth of 2–3″ with HCl. Swirl occasionally for 5 minutes, and remove the liquid with a sintered glass filter stick. Repeat this washing with HCl until the supernatant liquid is free of color (5–6 times). Then wash free of acid with water by decantation, removing the final wash by filtration through a sintered glass funnel. Wash the filter cake with three 200 ml portions of acetone, spread out on a filter paper, and air-dry at 60°.

Now gently stir the dried acid-washed material with 400 ml of the 0.5 *N* KOH in methanol for 15 minutes, filter, wash free of base with methanol, and air-dry on filter paper at 60°. Reclassify by flotation in methanol, decanting off the fines. Filter off the remaining portion, spread on filter paper and air-dry at 60°.

f. *Packing coating*—Fluoralkyl silicone polymer QF-1-0065 (10,000 centistokes) (also called FS 1265), obtainable from Applied Science Laboratories, Inc.

Preparation of Packing and Conditioning of Column

Dissolve 0.52 g of the QF-1, (**f**), in 100 ml of redistilled methyl ethyl ketone (Eastman No. 383 2-butanone) at 25°, and slowly add 16 g of the packing support (**e**), with swirling. After 2 minutes' swirling, pour the slurry rapidly onto a sintered glass filter, applying just enough vacuum to keep the funnel from overflowing, but not so much that the cake becomes dry before transfer is complete. Draw air through the cake until no more foam appears on the lower side of the sintered filter. Dry the damp powder in air at 60° to constant weight.

(The amount of liquid phase deposited on the support can be approximated by evaporating the filtrate in a tared flask and heating the flask at 100° *in vacuo* to constant weight. The concentration of the QF-1 solution can be varied to obtain the desired 1% coating.)

Transfer the coated material to a column, packing in the usual manner. Condition at 260° for 16 hours with an argon flow rate of 50 ml/minute. (Do not connect to the detector during this period.)

Preparation of Sample

Add 525 mg of mannitol to 50 g of shredded coconut sample, and extract the mixture with three 150 ml portions of water, Add sufficient pyridine to remove the water present, and distil at atmospheric pressure to a volume of about 30 ml. (If enough pyridine was added, the distillation temperature will have reached 115°.) Add 30 ml of acetic anhydride, and reflux 1 hour. Remove the bulk of the acetic anhydride and pyridine by vacuum distillation, and wash the residue into a liter volumetric flask with acetone. Make to volume with acetone, and mix thoroughly.

Determination

Connect the conditioned column to the detector, and adjust the chromatograph as follows: Column temperature, 217°; detector bath temperature, 270°; flash heater temperature, 325°; cell voltage, 1,000; relative gain, 10 (1×10^{-7} amp.); argon gas flow rate about 35 p.s.i.g. (100 ml/minute as measured by a flowmeter at the effluent of the column).

Inject a 0.5–2.0 μl aliquot of the acetone solution of the sample (size adjusted to give $\frac{1}{3}$–$\frac{2}{3}$ full scale deflection) with a 10 μl Hamilton syringe, and record the ratio of the peak heights of the mannitol hexaäcetate to that of the sorbitol hexaäcetate. At the same time run a series of eight acetone solutions of known mixtures of reference sorbitol and mannitol hexaäcetates, each containing mannitol hexaäcetate equivalent to 525 mg/liter of mannitol, and sorbitol hexaäcetate concentrations (in terms of sorbitol) of 0.525, 1.05, 1.575, 2.10, 2.625, 3.675, 4.725 and 5.25 g/liter respectively. Plot the manitol/sorbitol peak height ratios against the sorbitol concentrations for these standard solutions. Interpolate the peak height ratio of the sample solution from this curve to obtain the concentration X of sorbitol in this solution in g/liter. Then for a 50 g sample, percent sorbitol $= 2X$.

Notes

1. The authors found 0.50 percent of sorbitol normally present in shredded coconut.
2. Average recovery of sorbitol by this method is 94.1 percent.
3. Limited experiments by the authors of this method indicated that substitution of penta-

erythritol tetrapropionate for mannitol hexaäcetate as an internal standard would permit the simultaneous determination of mannitol and sorbitol.

4. Ready-packed columns can be obtained from Applied Science Laboratories.

5. *See also* Methods 16-27 and 16-28.

Water-Insoluble Solids in Prepared, Sweetened Coconut (Rapid Method).

Determine by Method 11-17.

Analyses of Authentic Samples of Peanuts

There are three varieties of peanuts used commercially in the United States. These are the large-seed Runner and Virginia peanuts and the small-seed Spanish variety. Runner peanuts contain a little more fat and a little less protein than do the other two. Peanut butter is usually made from a blend of at least two of these varieties and often all three.

Over half of the shelled peanuts produced in the United States are used for manufacture of peanut butter. According to statistics obtained from the Peanut Butter Manufacturers' Association, this amounted to over 495 million pounds in 1965. Another 20% each goes into salted peanuts and candy.

Smith and his co-workers at the Food and Drug Administration in 1962 studied the composition of roasted peanuts and of peanut butter made therefrom.[8] They also found that the degree of roast and the extent of grinding have less effect on the composition than does the variation from sample to sample within any one variety. Their analyses of 47 authentic samples of roasted peanuts are summarized in Table 12-2. They reported that all of the

TABLE 12-2

Composition of American-Grown Peanuts, Roasted Preliminary to manufacture of Peanut Butter[8]

	Moisture, %	Protein, % (N × 6.25)	Ether Extract, %	Sucrose, %[b]
Runner (13)[a]				
Average	1.56	31.5	51.6	4.78
Maximum	2.4	32.2	53.6	6.29
2 Minimum	0.7	30.5	48.2	4.26
	0.8	30.7	48.5	4.28
Spanish (16)				
Average	1.84	32.9	48.6	4.77
Maximum	4.9	34.2	51.8	7.08
2 Minimum	1.0	31.4	46.0	3.92
	1.2	31.8	46.8	3.97
Virginia (3)				
Sample 1	1.3	31.5	48.5	6.67
Sample 2	2.1	31.9	48.6	6.49
Sample 3	1.6	30.6	45.4	8.31
Mixed Varieties (15)				
Average	1.50	32.0	49.6	5.25
Maximum	2.5	33.0	52.4	6.37
2 Minimum	0.8	31.1	46.7	4.57
	0.9	31.3	47.1	4.63
Grand Average	1.6	32.1	49.7	5.07

[a] Figures in ( ) are numbers of samples analyzed.

[b] Corrected for volume of insoluble solids by multiplying values actually found by 0.97 (*Official Methods, AOAC*, 10th ed., **22·044**).

roasted peanuts contained reducing sugars after inversion, but none before inversion.

Smith and his co-workers also analyzed peanut butter made from each of the 47 lots of roasted peanuts summarized in Table 12-2. These were made under observation at 40 manufacturing establishments located throughout the country, following the usual manufacturing practice at each plant. In all, 49 batches of peanut butter were made from these 47 lots. They recommend that the protein content of peanut butter be used as the best index of peanut content, since this constituent varies least of the three major components. The average protein content of Smith's 49 peanut butters is 30.03 percent. Fat content is of lesser value as an index because of its inherent variability and because of the unknown amount of fat which may be added as a stabilizer.

Eheart *et al.* of Virginia Agricultural Experiment Station, made an earlier study of Virginia and Spanish peanuts grown in various locations in Virginia.[9] The study, ranging over several years, involved analysis of some 30 samples for each variety strain. Raw peanuts were analyzed, and reported on the moisture-free basis.

Their first study, covering harvests of five years, included 7 strains of Virginia peanuts and 5 strains of Spanish peanuts, with 30 samples of each strain being analyzed. Summarizing according to variety and strain, these workers found as follows:

TABLE 12-3
Chemical Composition of Virginia and Spanish Peanuts (Dry Basis)

	Protein (N × 6.25), %	Ether Extract, %
Virginia		
Average	29.5	45.4
Maximum	29.8	46.9
2 Minimum	29.0	44.5
	29.4	44.7
Spanish		
Average	32.1	47.1
Maximum	34.1	48.7
2 Minimum	29.8	45.7
	31.0	46.1

In a later study in 1952 on 10 strains of Virginia peanuts, these authors reported the analysis of 40 samples from 3 locations for each of these 10 strains (*see* p. 282).

Macadamia Nuts

Food technologists at the University of Hawaii have done considerable work on analysis, processing and storage of smooth-shell Macadamia nuts. The nuts were husked, air-dried, then roasted. Separate analyses were made of the air-dried and roasted nuts.[10,11] *See* Table 12-4, p. 282.

Coconut—Authentic Analyses

Coconut for food purposes (other than for fat extraction) is shipped from Florida, Central America and Mexico, following widespread destruction of coconut groves in the South Pacific during World War II. Desiccated coconut may arrive in this country from any area where the coconut palm is grown. *Agriculture Handbook No.* 8 (revised Dec. 1963) gives the average composition of coconut meat as shown on p. 282.

Rokita[12] in 1947 reported analyses of 8 samples of whole, fresh coconuts from

Chemical Composition of Virginia Type Peanuts (Dry Basis)

	Protein (N × 6.25), %	Ether Extract, %
Average	28.9	47.5
Maximum	30.0	48.5
2 Minimum	28.3	46.9
	28.6	47.0

TABLE 12-4
Analyses of authentic samples of Macadamia Nuts

Constituent	Air-dried	Air-dried, Roasted
Moisture, %	1.40	1.10
Ether extract, %	76.95	77.25
Total sugars, %	5.56	4.98
Reducing sugars, %	0.04	0.07
Free amino nitrogen, %	0.04	0.03
Constants of oil:		
Free fatty acid, mg KOH/g oil	0.44	0.38
Wijs iodine number	80.9	70.29

	Moisture %	Protein (N × 5.30) %	Fiber %	Fat %	Ash %	Carbohydrates (by difference) %
Fresh	50.9	3.5	4.0	35.3	0.9	5.4
Dried	3.5	7.2	3.9(?)	64.9	1.4	19.1

6 locations in Central America, Florida and Mexico, compositing from 4 to 6 nuts for each sample. She found an average of about 51 percent kernel, 24 percent coconut milk and 25 percent shell. The meat, or kernel, averaged 43.5 percent, and the "milk" 95.1 percent of moisture. Analysis of the meat alone, converted to the moisture-free basis, was reported as:

Ether extract (fat)	59.3%
Crude fiber	11.7%
Protein	7.5%

Sanders[13] made a more exhaustive analysis of coconut in 1944, examining authentic samples of Jamaica, Florida and Honduras coconuts:

Analysis of Whole Kernel (endosperm and spermoderm)

	Jamaica	Florida	Honduras
% of nut as kernel	51.8	53.5	49.9
Fat, moisture-free, %	67.4	58.3	65.7
Fiber, moisture-free, %	6.8	13.2	8.7
Protein (N × 6.25), moisture-free, %	11.1	8.5	7.7
*Sucrose, %	3.5	4.3	5.1
Ash, %	1.0	1.2	1.2
Moisture, %	33.7	36.2	30.4

* Sanders found only traces of invert sugar, located in the spermoderm.

One of us[14] reported average figures on analysis of 11 authentic desiccated coconuts:

Analysis of 11 samples of Philippine Desiccated Coconut

Moisture	2.77%
Fat	64.99%
Sucrose	5.48%
Invert sugar	0.19%
Salt	0.43%

Fisher[14] also reported the calculated composition of 12 samples of commercial "flaked coconut" collected during a market survey in 1960, labelled with such additional phrases as "Moist", "Extra Moist", "Sweetened":

Calculated Composition of Commercial "Flaked Coconut"

Desiccated coconut	27.3 to 53.6%
Sucrose	23.5 to 29.9%
Salt	0.1 to 1.0%
Added water	4.2 to 19.1%
*Sorbitol and/or propylene glycol	9.6 to 27.9%

* by difference from 100%

TEXT REFERENCES

1. *Code of Federal Regulations, Title* 21—*Food and Drugs, Parts* 1 *to* 129. *Section* 46.1. Washington: U.S. Govt. Printing Office (1966).
2. Woodroof, J. G.: *See* Selected Reference G., p. 283
3. Trager, W. T., Stoloff, L. and Campbell, A. D.: *J. Assoc. Offic. Agr. Chemists*, **47**: 993 (1964).
4. Salwin, H.: *J. Assoc. Offic. Agr. Chemists*, **49**: 63 (1966).
5. Campbell, A. D. and Funkhouser, J. T.: *J. Assoc. Offic. Agr. Chemists*, **49**: 730 (1966).
6. Anon.: *J. Assoc. Offic. Agr. Chemists*, **49**: 229 (1966).
7. Hause, J. A., Hubicki, J. A. and Hazen, G. C.: *Anal. Chem.*, **34**: 1567 (1962).
8. Smith, H., Horwitz, W. and Weiss, W.: *J. Assoc. Offic. Agr. Chemists*, **45**: 734 (1962).
9. Eheart, J., Young, R. W. and Allison, A. H.: *Food Res.*, **20**: 497 (1955).
10. Cavaletto, C., De la Cruz, A., Ross, E. and Yamamoto, H. Y.: *Food Technol.*, **20**: 1084 (1966).
11. De la Cruz, A., Cavaletto, C., Yamamoto, H. Y. and Ross, E.: *Food Technol.*, **20**: 1217 (1966).
12. Rokita, P. B. (U.S. Food and Drug Administration): Private communication (1947).
13. Sanders, J. W. (U.S. Food and Drug Administration): Private communication (1944).
14. Fisher, H. J.: *Conn. Agr. Exp. Sta. Bull.* **660**: 52 (1963).

SELECTED REFERENCES

A. Series of articles on peanuts and peanut butter by Freeman, A. F., Hall, R. H., Morris, N. J., O'Connor, R. T. and/or Willich, R. K. (Southern Regional Research Laboratory, New Orleans):
   a. "Roasting, cooling, blanching and picking peanuts". *Food Technology*, **6**: 71 (1952).
   b. "Effect of roasting and blanching on thiamine content". Ibid., **6**: 199 (1952).
   c. "Effect of roasting, blanching and sorting on oil and free fatty acid content". Ibid., **7**: 366 (1953).
   d. "Determination of color by reflectance". Ibid., **7**: 393 (1953).
   e. "Effect of processing and storage on oil stability". Ibid., **8**: 101 (1954).
   f. "Effect of roasting on palatability". Ibid., **8**: 377 (1954).
   g. "Effect of processing and storage on Vitamin A incorporated into peanut butter". Ibid., 381 (1954).
   h. "De-oiling of peanuts to yield a potentially useful food product". Ibid., **11**: 332 (1957).

B. "Peanut Butter". Southern Utilization Branch, U.S.D.A. *Report* AIC–370 (1954).

C. Banes, D.: "Mycotoxins Methodology". *Food Technol.*, **20**: 755 (1966).

D. Campbell, A. D.: "Report on Mycotoxins". *J. Assoc. Offic. Agr. Chemists*, **50**: 343–370 (1967).

E. Hamilton, R. A. and Fukunaga, E. T.: "Growing Macadamia Nuts in Hawaii". *Hawaii Agr. Exp. Sta. Bull.* **121** (1959).

F. Moltzau, R. H. and Ripperton, J. C.: "Processing of Macadamia Nuts". *Hawaii Agri. Exp. Sta. Bull.* **83** (1938).

G. Woodroof, J. G.: *Peanuts: Production, Processing, Products.* Westport: Avi Publishing Co. (1966).

H. Woodroof, J. G.: *Tree Nuts: Production, Processing, Products.* Westport: Avi Publishing Co. (1967).

CHAPTER 13

Oils and Fats

There is no sharp chemical or nutritional distinction between oils and fats. Most of these products go through some sort of refining process before they enter commerce as foodstuffs. (Olive oil is the only vegetable oil that is not refined before being consumed.) At this stage they consist largely of saponifiable triglycerides of straight-chain aliphatic saturated and unsaturated "fatty acids" insoluble in water, with a very small proportion, 3 percent or less, of other substances (phospholipids, sterols, tocopherols, carotenoids, vitamins and coloring matter). In common parlance, oils are those products that are liquid at ambient temperature, while fats are solids. They may be of animal or vegetable origin. In this text "oil" includes "fat" as a generic term except when the context indicates otherwise.

The fatty acid components of oils are mostly aliphatic acids containing from 6 to 24 carbon atoms, although fatty acids containing as many as 40 carbon atoms have been reported.

By some metabolic quirk of nature, only acids with even numbers of carbon atoms are combined with glycerol to form the natural oils and fats. (This is the consensus among oil chemists although there is some dissenting opinion.)

Government Standards

No Federal Definition or Standard of Identity or Quality have been promulgated for edible oils and fats by the Food and Drug Administration.

The Canadian Department of National Health and Welfare, under authority of the Canada Food and Drugs Act[1] has issued regulations defining certain edible oils and fats. Antioxidants and an antifoaming agent are permitted. The specified "constants" are:

Oil or Fat	Sp. G. 20°/20°	Ref. Index 20°	Saponification Value	Hanus Iodine Value	Acid Value
Olive Oil	0.912–0.918	1.468–1.470	185–195	77–94	7
Cotton Seed Oil	0.919–0.928	1.472–1.474	—	100–116	—
Cacao Butter	—	1.453–1.458 (40°)	188–200	32–41	5
Corn Oil	0.918–0.924	1.473–1.475	188–193	111–130	—
Peanut Oil	0.913–0.920	1.468–1.472	185–196	83–100	—
Soya Oil	0.921–0.925	1.472–1.476	—	—	—
Sunflower Oil	0.918–0.925	1.474–1.477	185–195	125–141	—

Laboratory Examination of Oils

Other than fundamental research on composition, the primary concern of the analyst is identification of the oil or oils through its chemical and physical attributes. The regulatory chemists will also look for substitution, in whole or part, of cheaper oils; the food processor's chemists will evaluate the oil against the processor's standards of quality and purchase specifications; and the oil processor's laboratory is primarily interested in the maintenance of quality standards. All chemists will concern themselves with evidence of decomposition, whether incipient or advanced to rancidity.

Many of the methods of analysis or tests included here have been used for several decades. Some have been incorporated in governmental manuals and pharmacopoeias and have thus attained a certain official status. This is particularly true of methods for the determination of so-called oil "constants" "numbers" or "values" such as iodine value, and identification tests for specific oils. Because of their established use they were adopted by such organizations as AOAC, AOCS, ASTM and others. These societies rewrote the directions more precisely, standardized the reagents and apparatus, but did not originate the methods.

Methods of Analysis

The American Oil Chemists' Society (AOCS) has granted permission to include certain *Official and Tentative Methods*, 1966 Revision, in this text. These methods are so identified in their title. Since the AOCS does change these methods from time to time, current revisions of "Official and Tentative Methods of the American Oil Chemists' Society" should be consulted.

Unless stated otherwise, mention of a specific oil in this text signifies that oil as refined for food use.

Method 13-1. *Moisture.*

PREPARATION OF SAMPLE

Liquid oils—Filter oils that are turbid or cloudy.

Solid fats—Melt solid fats, using no more heat than is required for complete melting. Filter the melted fat through paper, using a hot water funnel. Mix the sample thoroughly before each weighing.

DETERMINATION

Unless the purpose of the determination dictates otherwise, oils and fats should not be filtered before weighing out the sample. Fats may be softened with gentle heat, but not melted, since water tends to settle out. The AOCS directs mixing with an efficient mixer just before weighing.

Determine moisture and volatile matter on a 5 g sample by Method 1-1 air oven at 100° (but without filtering), or by Method 1-2, vacuum oven using the following table for permissible oven temperature: (This latter method is official for both AOAC and AOCS.)

Internal Oven Pressure mm of mercury	Permissible Oven Temperature Minimum	Permissible Oven Temperature Maximum
100	72°	77°
90	70°	75°
80	67°	72°
70	65°	70°
60	62°	68°
50	58°	63°
40	54°	59°

Moisture alone may be determined by Method 1-4, using 50-200 g samples and toluene as the immiscible solvent. The AOCS gives precise dimensions for the distillation apparatus used in AOCS tests.

Physical Measurements

The specific gravity, melting point and refractive index of oils and fats are of considerable value in identifying these products. Detailed instructions will not be given here since the techniques are familiar to every chemist.

Specific gravity is determined by a pycnometer, usually at 25°/25°. If the product is not liquid at 25°, determine specific gravity at 60°/25°. The AOCS gives a formula for calculating this constant at 60°/25°:

$$\text{Sp.G. at } 60°/25° = \frac{F}{W[1+(0.000025\times 35)]},$$

where F = the weight of the sample at 60°, and W = the weight of an equal volume of water at 25°.

The melting point is determined by use of a capillary tube, fused at one end, and a thermometer graduated to 0.2°.

The refractive index may be determined by an Abbé or butyro-refractometer, at a temperature controlled to ±0.1°. AOCS reports the refractive index at 40° for oils and at 60° for fats. AOAC reports this constant at 20 or 25° for oils and 40° for fats. A butyro-refractometer may also be used. In this case the instrument reading may be converted to refractive indices by use of Table 13-1, taken from *Methods of Analysis*, AOAC, 10th ed. (1965). In fact, this is the instrument of choice for testing glycerides, since it gives an additional significant decimal point for index of refraction over the Abbé. It might be mentioned here that a high refractive index usually accompanies a high iodine value.

TABLE 13-1

Butyro-refractometer readings and indices of refraction

Reading	Index of Refraction	Reading	Index of Refraction
40.0	1.4524	60.0	1.4659
40.5	1.4527	60.5	1.4662
41.0	1.4531	61.0	1.4665
41.5	1.4534	61.5	1.4668
42.0	1.4538	62.0	1.4672
42.5	1,4541	62.5	1.4675
43.0	1.4545	63.0	1.4678
43.5	1.4548	63.5	1.4681
44.0	1.4552	64.0	1.4685
44.5	1.4555	64.5	1.4688
45.0	1.4558	65.0	1.4691
45.5	1.4562	65.5	1.4694
46.0	1.4565	66.0	1.4697
46.5	1.4569	66.5	1.4700
47.0	1.4572	67.0	1.4704
47.5	1.4576	67.5	1.4707
48.0	1.4579	68.0	1.4710
48.5	1.4583	68.5	1.4713
49.0	1.4586	69.0	1.4717
49.5	1.4590	69.5	1.4720
50.0	1.4593	70.0	1.4723
50.5	1.4596	70.5	1.4726
51.0	1.4600	71.0	1.4729
51.5	1.4603	71.5	1.4732
52.0	1.4607	72.0	1.4735
52.5	1.4610	72.5	1.4738
53.0	1.4613	73.0	1.4741
53.5	1.4616	73.5	1.4744
54.0	1.4619	74.0	1.4747
54.5	1.4623	74.5	1.4750
55.0	1.4626	75.0	1.4753
55.5	1.4629	75.5	1.4756
56.0	1.4633	76.0	1.4759
56.5	1.4636	76.5	1.4762
57.0	1.4639	77.0	1.4765
57.5	1.4642	77.5	1.4768
58.0	1.4646	78.0	1.4771
58.5	1.4649	78.5	1.4774
59.0	1.4652	79.0	1.4777
59.5	1.4656	79.5	1.4780

Method 13-2. *Hanus Iodine Value. AOAC Method* ***26.020-26.021.*** (It goes back in principle to 1884[2]).

The iodine value is the quantity of iodine absorbed/gram of oil. It is an index of the degree of unsaturation (double bonds) of the oil. Two methods are in general use, the Hanus and Wijs. The AOAC gives both methods, while the AOCS has endorsed the Wijs method. We prefer the Hanus method because it is easier to use. The difference numerically between the two methods is negligible.

REAGENT

The *AOAC Official Methods* give a convenient procedure for preparing this reagent: Measure

825 ml of acetic acid (99.5%), and dissolve in it 13.615 g of I with the aid of heat. Cool, and titrate 25 ml of this solution with 0.1 N $Na_2S_2O_3$. Measure another 200 ml portion of acetic acid and add 3 ml of bromine. To 5 ml of this solution add 10 ml of 15% KI solution and titrate with the 0.1 N $Na_2S_2O_3$ solution. Calculate the quantity of bromine solution required to double the halogen content of the remaining 800 ml of I solution as follows:

$$A = B/C,$$

where A = ml bromine solution required; B = 800 × thiosulfate equivalent of 1 ml of I solution; and C = thiosulfate equivalent of 1 ml of bromine solution. If necessary, reduce the mixed I and bromine solutions by dilution with acetic acid to the proper concentration. Store in a cool, dark place.

DETERMINATION

Weigh about 0.5 g of filtered fat, 0.2500 g of filtered oil (0.1000 g if the oil is known to have a high absorbent power) into a 500 ml glass-stoppered flask and dissolve in 10 ml of $CHCl_3$. Add by means of a pipette 25 ml of Hanus reagent, draining the pipette a definite time, and let stand exactly 30 minutes. (There should be at least a 60 percent excess of I in the quantity added.)

Add 10 ml of 15% KI solution, shake thoroughly and add 100 ml of freshly boiled and cooled water, washing down any free I on the stopper. Titrate the iodine with 0.1N $Na_2S_2O_3$, adding it gradually, with constant shaking, until the yellow solution turns almost colorless. Add a few drops of 1% boiled and cooled soluble starch solution and continue titration until the blue color disappears. Towards the end of the titration, stopper the bottle and shake violently to remove any traces of I left in the $CHCl_3$.

Run 2 blank determinations along with that on the sample, draining the Hanus reagent pipette the same length of time for the blanks as for the sample.

$$\text{I Value} = \frac{[(B-S) \times N \times 12.69]}{\text{wt. of sample}},$$

where B = the titre of the blank, S that of the sample, and N = the normality of the thiosulfate solution.

NOTE: The oil must be filtered through a dry filter and all glassware must be absolutely clean and completely dry.

Method 13-3. *Thiocyanogen Value.* (*AOCS Official Method* ***Cd 2-38***).

This value is also a measure of unsaturation of oils and fats and is expressed in terms of the equivalent number of centigrams of I absorbed per gram of oil (percent iodine absorbed). Since thiocyanogen does not add to linoleic and linolenic acids in the same proportion as iodine, it is possible by determining both iodine and thiocyanogen values to calculate the fat or fatty acid composition.

APPARATUS

All glassware used must have been cleaned with acid-dichromate cleaning solution and completely dried in an air oven at 105° before use.

REAGENTS

Caution: All reagents must be completely dry.

a. *Anhydrous acetic acid.* Add 40 ml of acetic anhydride to 500 ml of acetic acid (99.5% or better) in a flask equipped with a $\mathfrak{T}$ air condenser. Boil gently for 3 hours. Attach a calcium chloride tube to the end of the condenser and cool to room temperature. Test the acetic acid and the acetic anhydride for reducing substances by diluting 2 ml with 10 ml of distilled water and adding 0.1 ml of 0.1 N $KMnO_4$ solution. The pink color must not be completely discharged at the end of 2 hours.

b. *Thiocyanogen solution.* AOCS and AOAC give directions for making the lead salt, $Pb(SCN)_2$. This is now unnecessary. It can be purchased from Eastman, No. P8183.

Preparation of 0.2 N thiocyanogen solution. Suspend 50 g of $Pb(SCN)_2$ in 500 ml of anhydrous acetic acid (a). Dissolve 5.1 ml of bromine in another 500 ml portion of anhydrous acetic acid. Add the bromine solution to the $Pb(SCN)_2$ suspension in small portions, shaking after each addition until the solution is completely decolorized. When all the bromine solution has been added, allow the precipitated $PbBr_2$ and excess $Pb(SCN)_2$ to settle out.

Filter by vacuum through an oven-dried qualitative paper into a 2L filter flask. When filtration is complete, transfer the Büchner funnel to a second 2L filter flask and pass the filtrate through the filter a second time. The filtrate must now be perfectly clear. Store in an amber glass-stoppered bottle in a cool place (65–70° F.). The solution will keep about a week, according to AOAC.

DETERMINATION

1. Weigh from 0.1000 to 0.7000 g of the oil into a 500 ml glass-stoppered bottle or flask. The weight of sample must be such that there will be an excess of 150–200 percent of $(SCN)_2$ reagent over the amount absorbed. For example: for an oil with an expected thiocyanogen value of 150 or higher, weigh about 0.12–0.14 g; for a cyanogen value of 100, weigh 0.21–0.25 g; for a value of 50, weigh about 0.4–0.5 g, etc.

2. Through a pipette add 25 ml of the $(SCN)_2$ solution, mix by swirling until oil dissolves, and store in a dark place for 24 hours at 65–70° F.

3. Prepare 3 blank determinations with each group of samples. To one add 1.66 g of KI. Swirl rapidly for 2 minutes, add 30 ml of water and titrate at once with 0.1 *N* $Na_2S_2O_3$. Add 1% boiled soluble starch solution toward the end of the titration and titrate to the disappearance of the blue color.

Treat the other 2 blanks exactly like the samples, as directed in (2) and (4).

4. At the end of 24 hours add 1.66 g of KI and 30 ml of water and titrate as directed in (3).

NOTE: If the difference in blank titrations before and after 24 hours exceeds 0.2 ml of 0.1 *N* thiosulfate/25 ml of $(SCN)_2$ solution the $(SCN)_2$ solution should be discarded.

CALCULATIONS

The thiocyanogen value in terms of its iodine equivalent =

$$\frac{(B-S)\times N\times 12.6}{\text{wt. of sample}},$$

where B and S = the titration of the blank and sample respectively, and N = the normality of the thiosulfate solution.

The equations for composition are slightly different, depending on whether the iodine values and thiocyanogen values were determined on the mixed fatty acids or the mixed triglycerides. The Hanus values and the Wijs values are close enough together numerically that they do not affect the equations.

The constants for fatty acids are:

Acid	Expressed in %	Iodine Value (I.V.)	Thiocyanogen Value (T.V.)
Linolenic	X	273.7	167.1
Linoleic	Y	181.1	96.7
Oleic	Z	89.9	89.3
Saturated and unsaponifiable	S	0*	0*

* These values are not actually 0, but are assumed so for calculation. When linolenic acid is present the saturated acid must be determined independently. When these values are determined on the mixed fatty acids, the calculated percentages of the hypothetically pure fatty acids are: $Y = 1.194$ I.V.–1.202 T.V.; $Z = 2.421$ T.V.–1.293 I.V.; $S = 100–(Y+Z)$.

The derivation of these formulae and those following is given in *Official and Tentative Standards, AOCS, Official Method* **Cd 2-38.**

When linolenic acid is present:

$X = 1.5902$ T.V. $- 0.1290$ I.V. $+ 1.3040\,S - 130.40$

$Y = 1.3565$ I.V. $- 3.2048$ T.V. $- 1.6423\,S + 164.23$

$Z = 1.6146$ T.V. $- 1.2275$ I.V. $- 0.6617\,S + 66.17$

When I.V. and T.V. are determined on the oils (triglycerides), the constants for the triglycerides are:

Glyceride	Expressed as %	Calculated from Fatty Acids: Iodine Value (I.V.)	Calculated from Fatty Acids: Thiocyanogen Value (T.V.)
Linolenin	X	261.8	159.8
Linolein	Y	173.3	92.5
Olein	Z	86.0	85.5
Saturated and Unsaponifiable	S	0*	0*

* These are not actually 0, but are assumed so for calculation. When no linolenin is present: $Y = 1.246$ I.V. $- 1.253$ T.V.; $Z = 2.525$ T.V. $- 1.348$ I.V.; $S = 100-(Y+Z)$. When linolenin is present: $X = 1.6610$ T.V. -0.1322 I.V. $-1.3056\,S-130.56$; $Y = 1.4137$ I.V. -3.3449 T.V. $-1.6441\,S+164.41$; $Z = 1.6839$ T.V. -1.2805 I.V. $-0.6615\,S+66.15$.

The sum of linolenin and linolein as so determined equals "poly-unsaturated fats".

The constants given above are based on the assumption that the T.V. of the triglycerides can be calculated stoichiometrically from the T.V. of the fatty acids.

Identification and Estimation of Individual Fatty Acids, or Fatty Acid Groups

These methods, being more useful to research, are given by reference only.

The *AOCS Official and Tentative Methods* include a method for the determination of polyunsaturated acids by measuring the U.V. absorption at specified wave-lengths of, first, the naturally occuring conjugated polyunsaturated constituents, and then the nonconjugated polyunsaturated constituents after alkali isomerization. The amounts of conjugated diene, triene, tetraene and pentaene acids are calculated from these data by simultaneous equations. Analysts interested in this method should refer to AOCS Method **Cd 7-58**, also adopted by AOAC as Method **26.041-26.047**.

Another AOCS method of interest here is that for *trans* isomers. The unsaturated constituents of most vegetable oils (and fats) contain only nonconjugated double bonds in *cis* configuration. Partial hydrogenation or oxidation may isomerize these to the *trans* form. Animal oils may naturally contain some *trans* isomers. These *trans* bonds show absorption in the infra-red region at about 10.3 μ (*cis* bonds and saturated compounds do not show this absorption). *See* AOCS Method **Cd 14-61** or AOAC Method **26.048–26.054**.

Hydrogenation of oils of course reduces the quantity of unsaturated oils present, i.e., oleates are converted to stearates. Isomerism also occurs: oleic acid is converted to iso-oleic acid, a solid.

AOCS Method **Cd 6-38** provides a means of separation of liquid and solid fatty acids, and an estimation of the percent of iso-oleic acid, through formation of their lead salts (the classical Twitchell reaction).

This method is of somewhat limited scope; it is not applicable to oils of the coconut group, to butter fat or to other oils containing low molecular weight saturated fatty acids, or to oils containing high molecular weight unsaturated fatty acids such as rape oil. It is therefore given here by reference only. Compare AOAC Method **26.040.**

AOCS Method **Ce 1-62**, included in AOAC Methods as **26.055–26.059**, separates methyl esters of oils having 8 to 24 carbon atoms by gas chromatography.

Method 13-4. *Saponification Value.* (Koettstorfer Value).

This method, dating back to 1879[3], has been official in the AOAC since *Bureau of Chemistry Bulls.* **107** and **137** (1909) forerunning the AOAC. It is also an official method of the AOCS.

The saponification value is defined as the number of milligrams of KOH necessary to completely saponify 1 g of oil. In other words, it is a measure of the average molecular weight of the mixed triglycerides constituting the oil.

REAGENT

Alcoholic potassium hydroxide. Reflux 1.2–1.5L of ethanol containing 5–10 g of KOH and 5 g of Al granules (or foil) under a reflux for 30–60 minutes. Distill 1 L of alcohol, after discarding the first 50–75 ml. Dissolve 40 g of KOH, low in carbonates, in 1 L of the alcohol distillate, keeping the temperature below 15°.

DETERMINATION

Weight about a 5 g sample to the nearest mg into a 250 ml Erlenmeyer flask. Pipette the alcoholic potash reagent into the flask, draining the pipette a definite number of seconds. Conduct a blank determination along with the sample, using the same pipette and draining for the same length of time.

Connect the sample flasks and the blank flask with air condensers, and boil gently but steadily until saponification is complete (AOAC says about 30 minutes, AOCS says 1 hour).

Clarity and homogeneity of the sample solution is a good index of completeness of saponification.

Cool and titrate the excess KOH with 0.5 *N* HCl, using phenolphthalein indicator.

$$\text{Saponification value, } SV, = \frac{28.05\ (\text{blank-sample titrations})}{\text{weight of sample}},$$

assuming exactly 0.5 N HCl was used in the titration.

The same method may be applied to a sample of fatty acids. If no unsaponified fat is present, this titration, called the Neutralization Value, is identical with the Saponification Value. The mean molecular weight (M) of the fatty acids may be calculated from the Neutralization Value (N) by the formula $M = \frac{56,100}{N}$, assuming the fatty acids are free from glycerides and other impurities.

Method 13.5, *Reichert-Meissl, Polenske and Kirschner Values.*

These are all AOCS methods. The AOAC has official methods for the first two values, but has superseded the Kirschner Value by a specific method for butyric acid (*Methods of Analysis*, 10th ed., **26.034**).

DEFINITIONS

Reichert-Meissl Value is the number of ml of 0.1 N alkali required to neutralize the soluble volatile fatty acids (chiefly butyric and caproic) of 5 g of oil.

Polenske Value is the number of ml of 0.1 N alkali required to neutralize the insoluble fatty acids (mostly caprylic, capric and lauric) of 5 g of oil.

Kirschner Value is the number of ml of 0.1 N alkali required for the volatile soluble fatty acids of 5 g of oil not precipitated by Ag_2SO_4.

These methods are particularly applicable to the identification of butter fat and the coconut group of oils, and are therefore useful in distinguishing butter from margarine and in identifying coconut and palm kernel oils. They all go back to early USDA Bureau of Chemistry bulletins predating AOAC Official Methods. The methods are highly empirical, hence the procedure and the dimensions of the distilling assembly must be followed exactly.

APPARATUS

An all-glass distillation assembly conforming to specifications given in *AOCS* Official Method **Cd 5-40** or *Methods of Analysis*, AOAC, 10th ed., Figure 26:2(B). This assembly can be obtained from most laboratory supply houses, made to these specifications. It is essential that the distilling flask be round-bottomed and that all dimensions be within the tolerances prescribed.

REAGENTS

a. *50% NaOH by weight.* This solution must be carbonate-free. Let the sediment settle, and use only the clear supernatant liquid.

b. *Glycerol-soda solution.* 20 ml of 50% NaOH solution **(a)** added to 180 ml of U.S.P. grade glycerol.

DETERMINATION OF R-M VALUE

Weigh accurately a 5 ± 0.1 g sample into the clean, dry 300 ml distilling flask. Add 20 ml of glycerol-soda solution **(b)** and heat with swirling over a flame or hot plate until completely saponified, as shown by the mixture becoming perfectly clear. Cool the contents to about 100° (5 minutes) and add 135 ± 1 ml of recently boiled, distilled water, drop by drop at first to avoid loss of water vapor. Now add 6 ml of dilute H_2SO_4 $(1+4)$ and a few pieces of pumice stone or silicon carbide. Connect the flask to the distillation aparatus, resting the flask on a piece of asbestos board with center hole 5 cm in diameter. Distill without previous melting of the fatty acids.

Regulate the flame so as to collect 110 ml distillate in 30 ± 2 minutes (measure the time from the passage of the first drop from the condenser into the receiver). The distillate must drip into the receiver at a temperature not higher than 20°.

When 110 ml have been collected, remove the flame, substitute a 25 ml cylinder for the receiving flask, and disconnect the distillation head from the condenser. Mix the contents of the receiver without violent shaking, and immerse the flask almost completely in water at 15° for 15 minutes. Filter through a 9 cm dry, moderately retentive filter paper (S and S White Ribbon is satisfactory) and titrate 100 ml with 0.1 N NaOH and phenolphthalein indicator. The pink end-point should persist for 2–3 minutes.

Prepare and conduct a blank determination, similar in all respects.

Reichert-Meissl Value $= 1.1 \times (S\text{–}B)$,

where S is the titration of the sample and B the titration of the blank, in ml of 0.1 N NaOH.

DETERMINATION OF POLENSKE VALUE

Remove the remainder of the water-soluble fatty acids from the insoluble acids on the filter by washing with three 15 ml portions of water at 15°, each portion previously having been passed through the condenser, the 25 ml cylinder and the 110 ml receiving flask. Discard the washings.

Dissolve the insoluble acids on the filter by passing three 15 ml portions of ethanol, previously neutralized to phenolphthalein, through the condenser, the 25 ml cylinder, the 110 ml receiver, and finally through the filter paper.

Combine the alcohol washings and titrate with 0.1 *N* NaOH and phenolphthalein to a pink end-point persisting 2-3 minutes. Carry the blank through the entire procedure.

Polenske Value = ml 0.1 *N* NaOH required, after correction for the blank.

DETERMINATION OF KIRSCHNER VALUE

Neutralize 100 ml of the Reichert-Meissl distillate, very accurately to a faint pink phenolphthalein color with 0.1 *N* $Ba(OH)_2$ solution. Conduct the titration in a closed vessel to prevent absorption of CO_2. Now add, without dilution, 0.3 g of finely powdered $AgSO_4$ and allow to stand 1 hour with frequent shaking. Filter.

Collect 100 ml of the filtrate, transfer to a 300 ml distilling flask, add a few pieces of pumice stone to prevent bumping, 35 ml of water and 10 ml of dilute H_2SO_4 (1+4). Attach to the Reichert-Meissl distillation apparatus and distill at the rate of 110 ml in about 20 minutes. After 110 ml have been collected, filter and titrate 100 ml of the filtrate with 0.1 *N* $Ba(OH)_2$ and phenolphthalein to a pink end point persisting for 2–3 minutes. Prepare and conduct a blank through the entire procedure.

$$\text{Kirschner Value} = \frac{A \times 121\ (100 + B)}{10{,}000},$$

where A = titration of sample, less the blank; and
B = ml of 0.1 *N* $Ba(OH)_2$ required to neutralize 100 ml of Reichert-Meissl distillate.

NOTE: There have been some fears expressed that benzoates or benzoic acid present as a preservative in oil-containing foods may elevate these values. Bolton[4], quoting work done by Grimaldi in 1912, says that only considerable quantities of benzoic acid have any effect on Reichert-Meissl or Kirschner values. The Polenske value is unchanged. 0.5 percent benzoic acid raises the Reichert-Meissl value of margarin from 5.57 to 6.45. Benzoic acid (or benzoates) are very rarely used as preservatives in excess of 0.1 percent.

Method 13-6. *Unsaponifiable Residue. (AOCS Tentative Method* ***Ca6b-53*** *and AOAC Method* ***26.071.****)*

DETERMINATION

Weigh accurately 2–2.5 g of the prepared sample into a 200 ml Erlenmeyer flask with a 24/40 outer joint. Add 25 ml of ethanol and 1.5 ml of KOH solution (50% by weight—prepared by dissolving 60 g of KOH, ACS grade, in 40 ml of water and cooling). Place on the steam bath, attach a reflux condenser and boil gently for 30 minutes with occasional swirling. Take care to avoid loss of alcohol during the saponification.

Transfer the saponified mixture while still warm to a 250 ml separatory funnel, using a total of 50 ml of water to rinse the flask. Cool to 20–25°. Wash the flask with 50 ml of ether, adding the ether washings to the separatory funnel. Shake vigorously, and allow the layers to separate and clarify. Decant or siphon off the upper ether layer as closely as possible into a second 250 ml separatory funnel containing 20 ml of water. Repeat the extraction 2 more times, each time decanting the ether layer as before into the 250 ml separatory funnel. Some oils high in unsaponifiable matter, e.g., marine oils, may require more than 3 extractions to completely remove unsaponifiable matter. Make another extraction with 50 ml of ether and evaporate separately. There should be less than 1 mg of residue if extractions were complete.

Rotate the combined ether extracts gently with the 20 ml of water (violent agitation will cause troublesome emulsions). Allow the layers to separate completely and draw off and discard the aqueous layer. Repeat the washing twice more with 20 ml of water, discarding the washings each time. Wash the ether solution 3 times with alternate 20 ml portions of 0.5 *N* KOH and water, shaking vigorously each time. If an emulsion forms, drain off as much of the aqueous layer as possible, leaving the emulsion behind in the other layer, and proceed with the next washing. After the third KOH washing, wash with 20 ml portions of water until the washings are no longer alkaline to phenolphthalein.

Transfer the ether solution to a 250 ml lipped

conical beaker, rinse the separatory funnel and its pouring edge with ether, and add the rinsings to the main ether solution. Evaporate to about 5 ml and transfer quantitatively to a tared 50 ml Erlenmeyer flask or a fat extraction flask, using several small portions of ether as rinses. Remove the ether by evaporation on a steam bath; when nearly all has been removed, add 2 or 3 ml of acetone, and, while heating on the steam bath, remove all solvent by passing a gentle current of clean dry air through the flask.

Dry to constant weight, preferably in a vacuum oven at 70–80° and not over 200 mm of mercury. Cool and weigh. After weighing, dissolve the residue in 2 ml of ether, add 10 ml of ethanol which has been neutralized to a phenolphthalein end-point. Titrate with 0.02 *N* NaOH to the same end-point. Correct the weight of the residue for its free acid content (1 ml 0.02 *N* NaOH = 0.0056 g oleic acid). Conduct a reagent blank (without the oil) and subtract the blank from the weight of the residue corrected for fatty acid. Calculate as percent unsaponifiable matter. Reserve the residue for a test for mineral oil, Method 13-16.

Method 13-7. *Acid Value.*

Weigh 50 g of prepared sample if the expected free fatty acid is 0.2 percent or under (or 25 g if the free fatty acid is 0.2–1.0 percent) into an Erlenmeyer flask. Add 50–100 ml of hot neutral ethanol that has been freshly distilled from ethanol held over KOH. Titrate with 0.1 *N* NaOH to a faint permanent pink persisting for 30 seconds.

Calculations:

$$\% \text{ Free fatty acids as oleic} = \frac{\text{ml } 0.1\, N \text{ NaOH} \times 28.2}{\text{weight of sample}}.$$

For coconut and palm kernel oils the free fatty acids are frequently expressed as lauric acid. For these oils the percent of free fatty acids as lauric is:

$$\frac{\text{ml } 0.1\, N \text{ NaOH} \times 20.0}{\text{weight of sample}}.$$

Acidity of oils is frequently expressed in terms of Acid Value or Acid Number. This is defined as the number of mg of KOH required to neutralize 1 gram of oil.

$$\text{Acid Value} = \% \text{ fatty acid as oleic} \times 1.99.$$

The Ester Value, which is a measure of the quantity of glycerin combined as glycerides, = saponification value – acid value.

Method 13-8. *Peroxide Value.* (*AOCS Method* ***Cd8-53.*** It has also been adopted as *AOAC Method* ***26.024.***)

This is an indication of the extent of oxidation suffered by an oil. It is roughly parallel to the depth of color obtained in the Kreis test.

REAGENTS

a. *Acetic-chloroform solvent.* Mix 3 volumes of acetic acid with 1 volume of $CHCl_3$.

b. *Freshly prepared saturated KI solution.* Test by adding 2 drops of 1% soluble starch solution. Discard if the solution turns blue, and requires more than 1 drop of 0.1 *N* $Na_2S_2O_3$ solution to discharge the color.

c. 0.1 *N and* 0.01 *N* $Na_2S_2O_3$ *solutions.* Prepare the latter by dilution of the 0.1 *N* solution with freshly boiled and cooled water just before use.

DETERMINATION

Weigh a 5.00 ± 50 mg sample into a 250 ml glass-stoppered Erlenmeyer flask. Add 30 ml of acetic $CHCl_3$ solvent (**a**) and swirl to dissolve. Add 0.5 ml saturated KI solution (**b**), using a Mohr pipette; let stand 1 minute with occasional shaking, then add about 30 ml of water. Titrate the liberated I with 0.1 *N* $Na_2S_2O_3$, adding the titrant drop by drop, with vigorous shaking, until the yellow I color almost disappears, then add about 0.5 ml of 1 percent soluble starch indicator and continue the titration, still shaking vigorously, until the blue color disappears. Run a blank on the reagents. This titration must not be more than 0.5 ml 0.1 *N* $Na_2S_2O_3$.

Peroxide value (meq. peroxide/kg): $S \times N \times 1000/\text{g}$. sample, where S = the corrected titration and N the normality of the $Na_2S_2O_3$ solution.

Acetyl value: This is a measure of the hydroxy acids present as glycerides in a fat. It is infrequently determined these days. This value is more variable than the other values listed herein, with a wide range of reported results. It is calculated from the saponification of the sample of oil before and after acetylation. The 10th edition of *Methods of Analysis* has starred this as a "surplus method" (**26.016–26.017**).

Interpretation of Oil Analyses

Most of the methods for oil "values" or "constants" are empirical, so any deviation from stated procedures may lead the analyst far astray. All of these tests should be supplemented by tests for specific oils when these are available. Acid, saponification and ester values of pure fatty acids or their esters bear definite relationships to the average molecular weight (MW):

$$\text{Fatty acids: MW} = \frac{56{,}100}{\text{Acid Value}}.$$

$$\text{Fatty acid triglycerides: MW} = \frac{3 \times 56{,}100}{\text{SV}}.$$

These equations stem from the definition of Saponification Equivalent, which is the number of grams of oil saponified by 56.1 g (the molar weight) of KOH.

The saponification values of the more common oils range from about 190 to 200. Low values may be associated with a high percentage of unsaponifiable matter, such as occurs in some of the marine oils, or may even indicate adulteration with mineral oil. High values indicate a low average molecular weight oil, such as butter fat or coconut oil. Hydrogenation does not affect the saponification value to any extent.

The iodine and thiocyanogen values have a fairly linear relationship with the refractive indices and specific gravities of oils. With a few exceptions, high I.V. and T.V. are associated with high SpG and IR. The Wijs method for iodine value tends to give a bit higher figure than does the Hanus method—in the case of iodine values over 100, as much as 2 % higher. Since they are measures of the number of double bonds in the fatty acid chain, most of the animal fats, butter fat, and the coconut and palm oils give low IV or TV values. Consult Method 13-3 for equations for calculation of the percentages of oleic, linolenic, linoleic, and total saturated acids in an oil. Hydrogenation, of course, greatly lowers both the IV and TV below those of the original oil. If the IV value is considerably higher than the TV, polyunsaturated fatty acids are indicated.

The Reichert-Meissl, Polenske and Kirschner values are of use in distinguishing butter fat from coconut and palm kernel oil. The Reichert-Meissl value is a measure of butyric and caproic acids, the Polenske value is a measure of caprylic plus capric acids, and the Kirschner value measures primarily the butyric acid content. Butterfat shows high Reichert-Meissl and Kirschner values, but low Polenske values, while coconut and palm kernel oils have high Polenske values. See Chapter 6 for a further discussion.

The identification of mixed oils requires the determination of as many physical measurements and "values" as possible, plus considerable skill in interpretation, including such factors as taste and odor. Once a roughly quantitative mixture of oils is postulated, a mixture of the oils assumed to be present should be analyzed and compared with that of the sample under scrutiny.

Tables of "Values" or "Constants" of oils and fats are available in Handbook of Chemistry and Physics and in many of the books on edible oils, some of which are included in Selected References at the end of this chapter. Because of this ready availability, we are including an abbreviated table, listing the oils and fats that may be encountered in this country. This was compiled, as far as available, from various sources published since 1950.

The tables mentioned above should be consulted for information on the more unusual oils.

TABLE 13-2

Physical and Chemical Characteristics of Selected Oils and Fats

Oil or fat	Solidifying or Melting point	Refractive Index—40°	Iodine Number	Thio-cyanogen Number	Saponi-fication Value	Reichert-Meissl Number	Polenske Number	Kirschner Number	Squalene mg/100g
Butterfat	28–36°	1.452–1.458	22–40		218–238	24–36	1.0–3.5	20–28	
Cocoa Butter	28–35	1.450–1.458	35–40	35–40	190–198	0.3–1.0	0.5		0
Coconut	23–28	1.448–1.450	7–10	7.5–10.5	248–265	6.0–8.0	12–18	0.5–2.0	2
Corn	−10 to −20[s]	1.470–1.473[a]	105–130	72–80	187–193	<0.5	<0.5		16–42 (*av.* 28)
Cottonseed	−2 to +4[s]	1.463–1.470[a]	100–114	61–68	190–198	<1.0	0.7–0.9		3–15 (*av.* 8)
Cottonseed Stearin	16–22		80–103						
Lard	30–48	1.458–1.461	52–74		193–202	<1.0	<1.0		
Lard Oil	−2 to +4[s]	1.460–1.462	67–82		193–198	<0.5	<1.0		
Olive	−6 to +7	1.466–1.468[a]	78–90	75–83	186–196	0.5–2.0	1.0–2.0		136–708 (*av.* 383)
Palm	27–46	1.450–1.464	44–55	45–48	195–205	<0.5	<1.0	0.0–1.0	
African Palm Kernel	22–30	1.449–1.452	14–23	13–18	240–255	4.0–8.0	8–12	1–2	
S. American Palm Kernel	24–28		25–31		220–232				
Peanut	−1 to +3	1.462–1.465	84–100	60–68	188–194	<0.5	<0.5		8–49 (*av.* 27)
Poppy Seed	−15 to −19[s]	1.474–1.476	130–150	77–79	186–196	<0.5	<1.0		
Rape Seed	−9	1.467–1.470	93–105		168–180	<1.0	<0.5		24–28
Rice Bran	2[s]	1.465–1.468	90–108	66–75	180–188	<0.5	<0.5		330
Safflower Seed	−12 to −18[s]	1.468–1.470	130–145	80–88	186–194				
Sesame Seed	−4 to 0	1.464–1.468	103–115	74–78	188–195	<0.5	<0.5		3–9
Soya Bean	−10 to +18[s]	1.470–1.472	118–140	72–83	189–196	<1.0	<0.5		
Sunflower Seed	−16 to −19[s]	1.468–1.470	120–138	78–81	188–194	<0.5	<0.5		8–19 (*av.* 12)
Tea Seed	−5 to −12[s]	1.460–1.463	81–91	75–78	188–196	<1.0			8–16

s = solidifying point   a = 25°

Tests for Specific Oils

Method 13-9. Cottonseed Oil.

This test, known as the Halphen test, dates back to at least 1897. The test is invalid for oils heated over 250°. Lard from pigs fed on cottonseed meal will often give a positive reaction. It is said to show a positive test on as low as 1 percent cottonseed oil. Kapok oil is the only oil known to interfere.

REAGENT

Halphen reagent. Mix CS_2 containing 1 percent of S with an equal volume of amyl alcohol.

TEST

Mix equal volumes (10 ml) of Halphen reagent and the oil sample in a test tube and heat in a water bath at about 75° until foaming ceases and CS_2 is expelled. Then almost immerse in an oil bath or saturated brine bath at 110–115° for 2 hours.

If appreciable quantities of cottonseed oil are present, a red color will appear in 1 hour or less. For small amounts, around 1 percent, 2 hours are necessary. This test may be made roughly quantitative by running the test simultaneously with a series of test tubes containing known dilutions of cottonseed in another vegetable oil, *e.g.* peanut oil.

Method 13-10. Sesame Oil. (This is the Villavecchia test).

TEST

Add 2 ml of furfural, reagent grade, to 100 ml of ethanol. Mix 0.1 ml of this reagent with 10 ml of the oil sample plus 10 ml of HCl. Shake well for 15 seconds and let stand until the emulsion breaks (not more than 10 minutes). If no pink or crimson color appears in the lower layer, the test is negative. If any color is observed, add 10 ml of water, shake again, and observe the color as soon as the layers separate. If the red color persists, sesame oil is present. If it disappears, the test is recorded as negative. As little as 0.5 percent of sesame oil can be detected.

This test, like the Halphen test, may be made roughly quantitative by comparing with known quantities of sesame oil run simultaneously.

Method 13-11. Teaseed Oil.

This is the Fitelson test.[5] It is applicable only to mixtures of teaseed and olive oils.

APPARATUS AND REAGENT

a. Set of *matched, clear glass test tubes* (18 × 150 mm is a convenient size) or *AOCS color tubes.*

b. *Mohr pipettes* to deliver exactly 0.2, 0.8 and 1.5 ml.

c. *Anhydrous ether,* distilled over sodium.

TEST

Place in a test tube exactly 0.8 ml of acetic anhydride, 1.5 ml of $CHCl_3$ and 0.2 ml of H_2SO_4. Mix, and cool in an ice bath to 5°. Add 7 drops of the oil under test. If cloudiness appears, add acetic anhydride dropwise to clear. Hold at 5° for 5 minutes. Add 5 ml of cold (5°) ether, stopper with a cork and mix immediately by inverting. Replace in the ice bath and observe the color. The depth of red color formed is proportional to the amount of teaseed oil present. The test will give reliable results with 10 percent or more of teaseed oil in olive oil. An estimate of the proportion of teaseed oil present may be made by comparing the test sample with known mixtures tested simultaneously with the sample.

Method 13-12. Peanut Oil.

This test has been in use in various modifications since it was proposed by Bellier in 1899. It is AOAC Method **26.078–26.079**. This test depends on the relative insolubility of arachidic acid, a component of peanut oil, in acidified ethanol. According to the AOAC, it is applicable only in the presence of olive, corn, cottonseed and soybean oils.

REAGENTS

a. *Alcoholic KOH solution,* 1.5 *N.* Purify ethanol by refluxing and distilling as directed for reagent (**a**), Method 13–4. Dissolve 10 g of KOH in 50–75 ml of this purified alcohol, cool and dilute to 100 ml.

b. *Hydrochloric acid* Sp.g.1.16). Dilute 83 ml concentrated HCl (sp.g.1.19) to 100 ml with water. Check the sp.g. by means of a specific gravity spindle.

c. 70% *ethanol by volume.* Dilute 700 ml ethanol to 950 ml with water. Check by sp.g. or refractive index and adjust to 70% if required.

TEST

Weigh a 0.92 g sample into a 125 ml Erlenmeyer flask fitted with a ₮ air condenser. Add 5 ml of alcoholic KOH reagent (**a**), insert the air condenser and heat 5 minutes on a steam bath. Swirl the contents once or twice during saponification. Add 50 ml of 70% alcohol (**c**) and 0.8 ml of HCl, reagent (**b**), using a burette or Mohr pipette to dispense the acid. Warm to dissolve any precipitate that may form. Using a thermometer as a stirring rod, cool with continuous agitation so that the temperature falls at the rate of about 1°/minute. (Cooling may be accomplished in air or by occasionally dipping the flask in a water bath with a temperature not more than 5° below that of the solution. Do not immerse the flask below the level of its contents, and agitate continuously while cooling.) Note the temperature at which a definite turbidity occurs, by looking through the solution toward a good light, or toward a dark background with good light coming from one side.

The presence of peanut oil is indicated if turbidity appears before the temperature reaches 9° (in the case of olive oil) or 13° (in the case of corn, cottonseed or the soybean oil).

If confirmation is desired, apply the Renard test as given in AOAC Method **26.077**. This test gives a quantitative approximation of the amount of peanut oil present.

Rape Oil

This oil is not commonly encountered in this hemisphere. It, or related mustard oil, has been identified as an adulterant of European olive oil. Winton and Winton in their book have included tests for erucic acid, the characteristic fatty acid of cruciferous oils, which include rape, mustard and charlock.[6] Since this is not a common adulterant, these tests are given by reference only.

According to the Wintons, low saponification value and high iodine value point to the possibility of the presence of rape oil in olive oil.

Method 13-13. *Fish and Marine Oils.*[7] (*AOAC Method* ***26.084.***)

Dissolve a 6 g sample in 12 ml of equal parts of acetic acid and $CHCl_3$ in a test tube. Add Br dropwise until a slight excess is indicated by a change in color. Keep the solution at about 20° during this addition. Let the mixture stand for 15 minutes or more, then place the test tube in boiling water. If marine oils are present, the solution will remain cloudy due to ether-insoluble brominated glycerides. Vegetable oils alone remain perfectly clear.

Method 13-14. *Squalene.*[8] (*AOAC Method* ***26.072–26.074.***)

Squalene, an unsaturated hydrocarbon, is present in olive oil in considerably greater quantity than in other common edible oils. Fitelson used it as an index of adulteration of olive oil with other oils. This is the oil that is *most subject* to adulteration.

REAGENTS:

a. *Concentrated potassium hydroxide solution* —60 g of KOH in 40 of ml water.

b. *Dilute potassium hydroxide solution*—28 g of KOH in 1L of water.

c. *Petroleum ether*—Skellysolve B(b.p. 63–70°) or equivalent.

d. *Aluminium oxide absorbent*, 80–200 mesh absorption alumina for chromatographic analysis. Keep in a tightly closed container, away from moisture.

e. *Pyridine sulfate bromide solution*, 0.1 *N*. Dissolve 8 g of Br in 20 ml of acetic acid (99.5%). Prepare another solution by adding, gradually with cooling, 5.45 ml of H_2SO_4 to a mixture of 20 g of acetic acid and 8.15 ml of pyridine. Mix the 2 solutions, cool, and dilute to 1L with acetic acid.

f. *Sodium thiosulfate solution*, 0.05 *N*. Prepare just before use by diluting the 0.1 *N* solution. See reagent (**a**), Method 2-10.

APPARATUS

Adsorption column, prepared fresh for each determination just before use. Place a small wad of cotton in the constricted end of a glass tube 30 cm long and 8 mm i.d. (for convenience the column may have a Teflon stopcock in the stem and a top reservoir of not less than 40 ml capacity). Add Al_2O_3 absorbent (**d**) in about 10 small increments until the column is about 10 cm high. Apply gentle suction and tamp down each portion of the absorbent with a flattened glass rod. Place a small wad of cotton on top and tamp down. Wash with about 15 ml of petroleum ether (**c**), and turn off the suction. Leave a shallow layer of petroleum ether on top until ready for use.

DETERMINATION

Weigh accurately (± 20 mg) a 5 g sample into a 125 ml Erlenmeyer flask furnished with an upright glass-stoppered air condenser. Add 3 ml of concentrated KOH (**a**) and 20 ml of ethanol. Reflux for 20 minutes, swirling occasionally. Cool somewhat, and while still warm add 50 ml of petroleum ether (**c**). Mix and transfer to a separatory funnel. Rinse the flask with 20 ml of ethanol, then 40 ml of water, adding the rinsings to the separatory funnel. Shake vigorously, let the layers separate completely, and slowly drain the soap solution into a second separatory funnel containing 50 ml of petroleum ether.

Pour the petroleum ether extract from the top of the separatory funnel into another separatory funnel containing 20 ml of H_2O.

Reextract the soap solution with another 50 ml of petroleum ether and combine the ether extracts. Gently rotate the separatory funnel to avoid emulsification, then draw off and discard the water layer. Repeat the washing by shaking the petroleum ether solution with 20 ml of water, and discard the lower layer.

Wash the petroleum ether solution once with dilute KOH solution (**b**), then vigorously with successive 20 ml portions of water until alkali-free as shown by a phenolphthalein test. After the last washings, swirl the contents of the separatory funnel and drain off the last drops of water.

Pour the petroleum ether solution through the top of the separatory funnel into a lipped conical beaker. Rinse the separatory funnel with petroleum ether and add the rinsings to the beaker. Add a few broken porcelain bits, or silicon carbide, and evaporate almost all of the solvent on a steam bath. Remove the last traces of solvent with a stream of CO_2 or N while warming the beaker. Do not expose the contents to the air while the beaker is still warm.

Dissolve the unsaponifiable residue in 5 ml of petroleum ether and transfer to the absorption tube. The filtrate, which is caught in a 250 ml glass-stoppered flask, should emerge dropwise at about, but not exceeding, 1 ml/minute. Gentle pressure may be used to force the solution down the column. When the solution has been nearly drawn into the column, add about 5 ml of petroleum ether, previously used to rinse the beaker. Continue adding solvent in 5-10 ml increments previously used to rinse the beaker, until 50 ml have passed through the absorption tube. Always keep the top of the column covered with liquid. (if a column with Teflon stopcock and reservoir is used, proceed as above through the 5 ml rinse, then rinse the beaker with 40 ml of petroleum ether, add it to the reservoir, and pass it through the column.)

Add a few pieces of broken porcelain and evaporate most of the solvent on a steam bath. Finally pass a current of CO_2 or N through the heated flask until the last traces of solvent are expelled. (All traces of solvent must be removed at this point.) Cool to room temperature in an inert atmosphere.

Dissolve the unadsorbed residue in 5 ml of $CHCl_3$ and enough of the pyridine reagent (**e**) to provide at least a 50 percent excess (10 ml is usually adequate). Let the mixture remain in the dark for 5 minutes, then add 5 ml of 10% KI solution and 40 ml of water. Mix thoroughly, wash down any free iodine on the stopper, and titrate with 0.05 *N* $Na_2S_2O_3$ solution. Toward the end, add 1% soluble starch solution as indicator, shake the flask vigorously, and continue titrating until the blue color disappears.

Conduct a blank determination on the pyridine reagent (**e**) and calculate ml of 0.05 *N* $Na_2S_2O_3$ equivalent to the absorbed halogen. Conduct a blank determination on all reagents. This should show practically no halogen consumption.

$$1 \text{ ml } 0.05\ N\ Na_2S_2O_3 = 1.71 \text{ mg squalene.}$$

Report results as mg squalene/100 g.

Interpretation: The average squalene content of olive oil is close to 400 mg/100 g. Fitelson[9] reports squalene contents as follows:

Squalene Contents of Edible Vegetable Oils, mg/100 g

Oil	Olive	Cotton-Seed	Peanut	Corn	Soy	Sun-Flower	Tea-seed	Sesame	Rape
No. samples	103	41	37	31	48	5	3	3	2
Max.	708	15	49	42	22	19	16	9	28
Min.	136	3	8	16	5	8	8	3	24
Average	383	8	27	28	12	12	12	5	26

Method 13-15. *Mineral Oil (Qualitative).* (*AOAC Test **26.085.***)

TEST

Transfer 1 ml of oil to an Erlenmeyer flask, add 1 ml of concentrated KOH solution (reagent (**a**), Method 13-13) and 25 ml of ethanol. Boil under a reflux air condenser, shaking occasionally, until saponification is complete (about 5 minutes). Add 25 ml of water and mix; turbidity indicates mineral oil. This test is sensitive to about 0.5 % mineral oil.

Method 13-16. *Mineral Oil (Quantitative).* (*AOAC Method **26.086.***)

Dissolve the unsaponifiable residue obtained in Method 13-6 in a small quantity of petroleum ether, and filter through a small cotton plug into a Babcock 18g milk-test bottle. Add a few bits of broken porcelain and remove the solvent by heating on a steam-bath while passing a current of air through the bottle. Cool, add 5 ml of H_2SO_4, mix and keep the bottle in a boiling water bath 30 minutes, shaking occasionally.

Remove the bottle from the bath, cool, and fill with water until the surface rises well into the graduated neck. Centrifuge for 5 minutes at about 1200 rpm, and read the volume of free oil on top. If enough oil is available, determine its density, using a small Sprengel tube. The weight of mineral oil equals its volume times 0.88. Its refractive index should be below 1.500 at 20°.

Identification of Oils and Fats of Animal Origin by Gas Chromatography

The question at times arises whether a food is of animal or vegetable origin, *i.e.*, whether it complies with *Kosher* requirements. It may be answered through identification of its sterol as phytosterol or cholesterol. Presence of the latter is of course decisive proof of animal origin of a food:

Method 13-17. *Cholesterol and Phystosterols in Mixtures of Animal Fat with Vegetable Fats and Oils.*[11a]

APPARATUS

a. *Column chromatographic tubes*—2.5 cm dia. × 30 cm long, with Teflon stopcocks.

b. *Gas chromatograph*—Barber-Colman Series 5,000 or an equivalent, with hydrogen flame detector and recorder of at least 0–50 mv capacity.

c. *Glass U-tubes for gas chromatography*—6′ × ¼″ i.d. (Obtainable from Kontes Glass Co., Vineland, N.J., or gas chromatographic supply houses.)

REAGENTS

a. *Florisil*—PR (pesticide residue) grade, 60–100 mesh (obtainable from Floridin Co., Pittsburgh, Pa). Dry at least 150 g at 260° for 2 hours. Let cool to room temperature in a wide- mouthed jar with a screw cap. Weigh 100 g into a liter round-bottomed flask and add 10.0 ml of water, swirling the mixture so that water does not collect in one area. Shake vigorously for 10–15 minutes until the flask feels cool (exothermic reaction stops). If necessary, break up any small clumps which have formed, using a stirring rod, and allow to equilibrate overnight before using. Keep stoppered, and do not store longer than a week. Determine its activity by analyzing a mixture of 10 g each of cholesterol, hexacosanol (or tetracosanol) and lanosterol. All of the hexacosanol should elute in the third fraction (as determined by gas chromatography), and more than 90% of the cholesterol should elute in the fourth fraction. If a lot of Florisil fails to meet these performance standards, vary the water content in increments of 0.5 ml/100 g and recheck.

b. *Hexane*—Either redistill the 95% grade (b. pt. 68–70°) after drying over Na, or use "distilled-in-glass" hexane (b. pt. 68–69°), obtainable from Burdick & Jackson Laboratories, Inc., Muskegon, Mich., or Mallinckrodt Chemical Works, 2nd and Mallinckrodt Sts., St. Louis, Mo.

c. *Anhydrons ethyl ether*—Alcohol-free (0.01% max.) Fisher No. E-138 or equivalent.

d. *Cholesterol*—A grade, obtainable from Calbiochem, Box 54282, Los Angeles, Calif., as their No. 2281, or from Applied Science Laboratories, Inc., P.O. Box 440, State College, Pa., as their No. 19508.

e. 1-*Hexacosanol* (*or* 1-*tetracosanol*)—Obtainable from Lachat Chemicals, Inc., Chicago, Ill.

f. *J × R silicone rubber*—Obtainable from Applied Science Laboratories, Inc.

g. *Gas Chrom-Q*—100–120 mesh. Obtainable from Applied Science Laboratories, Inc.

h. *Dimethyldichlorosilane* (*DMCS*)—Obtainable from Applied Science Laboratories, Inc.

i. *Lanosterol*—Obtainable from Calbiochem and Applied Science Laboratories, Inc.

j. *β-Sitosterol*—Obtainable from Calbiochem as their No. 567141, or from Applied Science Laboratories, Inc., as their No. 19517.

k. *Stigmasterol*—Obtainable from Calbiochem as their No. 5699, or from Applied Science Laboratories, Inc., as their No. 19518.

l. *Cholesterol—sitosterol standard solution*—Prepare solutions in chloroform of cholesterol, (**d**), and β-sitosterol containing respectively 4 mg/ml and 8 mg/ml, and mix equal volumes of each.

m. *Cholesterol standards for area plot*—Prepare at least three solutions of USP cholesterol in chloroform in the range of 0.05–0.50 mg/ml for an area plot in the range of the peak area of the sample.

n. *Cholesteryl acetate*—Eastman No. 2391.

Saponification

Weigh a 5.00 ±0.01 g sample into a saponification flask (a 200 ml Erlenmeyer with a ₸ 24/40 outer joint is recommended). Add 50 ml of alcohol and 3 ml of KOH solution (3+2). Saponify by boiling, with occasional swirling, on a steam bath for 30 minutes under a reufix condenser. (No loss of alcohol should occur during saponification.) Transfer the alcoholic soap solution while still warm to a 500 ml separatory funnel, using a total of 100 ml of water. Rinse the saponification flask with 100 ml of ether, and add the ether to the separatory funnel. Shake vigorously and let the layers separate and clarify. Drain the lower layer and pour the ether layer through the top into a second separatory funnel containing 40 ml of water. Rinse the pouring edge with ether, adding the rinsings to the second separatory funnel. Make two more extractions of the soap solution with 100 ml portions of ether in the same manner.

Rotate the combined ether extracts gently with the 40 ml of water (violent shaking at this stage may cause troublesome emulsions). Let the layers separate, and drain off the aqueous layer. Wash with two additional 40 ml portions of water, shaking vigorously. Then wash the ether solution three times with alternate 40 ml portions of approximately 0.5 *N* (33 g/liter) KOH and water, shaking vigorously each time. If an emulsion forms during the washing, drain as much of the aqueous layer as possible, leaving the emulsion in the separatory funnel with the ether layer, and proceed with the next washing. After the third KOH treatment, wash the ether successively with 40 ml portions of water until the washings are no longer alkaline to phenolphthalein.

Evaporate the washed ether extract on the steam bath to about 5 ml, cool, and transfer to a 50 ml fat extraction flask through a powder funnel containing a pledget of cotton and 10–15 g of anhydrous Na_2SO_4, and evaporate to dryness under nitrogen. Dissolve the unsaponifiable matter in 2–3 ml of $CHCl_3$ and transfer to a 2 dram vial, using a small quantity of $CHCl_3$ to rinse the flask. Evaporate to dryness under nitrogen, and store under nitrogen at −15° until ready for column chromatography.

Column Chromatography

Fill a column (2.5 cm dia. × 30 cm long) three-fourths full with redistilled hexane. Add 30 g of the prepared Florisil in small portions, tapping the tube between additions to ensure even packing. After all of the Florisil has been added, drain the excess hexane through the stopcock until its level is about 1½" above the surface of the packing. Add about ½" of anhydrous Na_2SO_4, wash down the sides with hexane, and drain off the solvent until it touches the top of the Na_2SO_4.

Transfer the unsaponifiable matter to a 4 dram vial (using ethyl ether), and evaporate the ether on the steam bath. Add 0.5 ml of $CHCl_3$, dissolve the sample, and transfer to the top of the column. Elute dropwise into a 125 ml Erlenmeyer flask until the level of the liquid reaches the top of the Na_2SO_4. Rinse the vial with two 5 ml portions of hexane, and elute with 40 ml of hexane (Fraction 1) at a flow rate of 6 ml/minute. Collect three additional fractions (flow rate, 6 ml/minute) in separate 300 ml Erlenmeyer flasks, eluting with 120 ml of 5% anhydrous ethyl ether in hexane (Fraction 2), 120 ml of 15% anhydrous ethyl ether in hexane (Fraction 3), and 150 ml of 30% anhydrous ethyl ether in hexane (Fraction 4, sterol fraction). Do not allow the levels of the eluting solvents to fall below the top of the Na_2SO_4 layer. Evaporate the sterol fraction (Fraction 4) on a steam bath to a volume of about 5 ml, using boiling chips. Transfer to a tared 50 ml fat extraction flask containing three to five boiling chips, and evaporate to dryness under nitrogen. Dissolve the sterols in $CHCl_3$ and transfer with 2-3 ml of $CHCl_3$ to a 2 dram vial. Evaporate to dryness under nitrogen and add about 0.2 ml of $CHCl_3$. Store at $-15°$ under nitrogen until ready for gas chromatography.

GAS CHROMATOGRAPHY

Preparation of Column. Coat the Gas Chrom-Q support with substrate as follows: Weigh 1.2 g of J × R silicone rubber, and dissolve in 200 ml of toluene or 1:1 methylene chloride-toluene, heating to dissolve. (The polysiloxane dissolves very slowly in the methylene chloride-toluene solvent.) Add this solution to 40 g of Gas Chrom-Q, and let stand 10 minutes with occasional gentle stirring. Dry either in a rotary evaporator held in a 50° bath, or by heating on the steam bath with occasional stirring, followed by removal of the residual solvent in a vacuum oven at 50°. Treat a U-tube [Apparatus, **(c)**] with DMCS by filling the tube with a 5% solution of DMCS in chloroform, removing the DMCS solution, and rinsing the tube with methanol until the rinsings are neutral to indicator paper. Pack the coated material into the tube by adding small portions while vibrating the tube at the packing level with a "Vibro-Graver". Fill to within 1″ on the exit side and to 3″ on the entrance side. Fill the remaining space with silanized glass wool prepared by immersing glass wool in a 5% solution of DMCS in $CHCl_3$ and rinsing with methanol until the rinsings are neutral to indicator paper. Condition the column by slowly heating the packed column to 280° (stepwise over two days) while 2 psi of nitrogen is flowing through the column, then holding over night at 280° with no nitrogen flow, and finally reducing the temperature to 250° under a N_2 pressure of 30 psi and holding at that temperature for 5-7 days.

Gas Chromatograph Settings. For hydrogen flame detection, using a Barber-Colman Model 5000 gas chromatograph or the equivalent, adjustments should be as follows: Column temperature, 220–250°; nitrogen pressure, 25 psi; hydrogen pressure, 30 psi; air pressure, 40 psi; electrometer range, 5×10^{-8} to 1×10^{-9} amperes.

Determination. Chromatograph in duplicate 2 μl of a solution of the sample (sterol fraction) in 1.00 ml of chloroform. Use the cholesterol standards, **(m)** (a minimum of three) to obtain a suitable standard curve, alternating injection of sample and cholesterol standard. Carefully compare the retention time of the cholesterol standard with sample peaks of similar retention time. Calculate the area of the cholesterol peak by triangulation (peak height × width at half height). Use the average of two injections per standard to prepare an area plot (mg cholesterol *vs* peak area). Determine mg cholesterol/100 g sample from the area plot as follows: mg cholesterol/100 g sample = μg cholesterol (from area plot) × 10.

Confirmatory Test. The presence of cholesterol can be confirmed by gas chromatography of the sterol acetates, prepared as follows: Transfer the sterol fraction (Florisil Fraction 4) to a 4 dram vial (having a screw cap with an aluminum or a Teflon liner), using ethyl ether. Evaporate the ether to dryness on the steam bath under a gentle stream of nitrogen. Cool, add 3 ml of pyridine and 1 ml of acetic anhydride. Cap the vial and swirl the contents in the steam bath until the sterols dissolve, and continue heating on the steam bath for ½ hour. Evaporate, using a stream of nitrogen, until no odor of pyridine can be detected. Chromatograph the sterol acetates, together with a sample of cholesteryl acetate, and compare the retention times of the peaks.

NOTES

1. Ready-silanized glass wool may be obtained from Applied Science Laboratories, Inc., or

Analabs, Inc., P.O. Drawer 5397, Hamden, Conn. 06518. This latter company also supplies DMCS and pure cholesterol, lanosterol, β-sitosterol and stigmasterol, as well as hexacosanol and tetracosanol. Custom-prepared columns of 3% J × R silicone rubber on Gas Chrom-Q may be obtained from either source.

2. The above-outlined determination is designed primarily for the detection of cholesterol as an indication of the presence of animal fat. If the quantitative distribution of the various phytosterols in a sample is also desired, standard curves should be prepared by chromatographing aliquots of solutions of the sterols such as Reagent (**I**).

3. AOAC Method **26.065–26.070** is similar to the above method, but was designed primarily for detecting adulteration of butter with vegetable oil, whereas Method 13-17 is planned to handle the reverse of this—namely, the contamination of vegetable oils with animal fats. This latter problem is of particular importance in examining *Kosher* foods.

4. Method 13-17 also offers a chemical means of distinguishing rice bran oil from olive oil. Fitelson had established that these two oils were distinguishable from other vegetable oils by their high squalene values (average about 330 mg squalene/100 g), but could not be differentiated on this basis.[14] Eisner and Firestone's gas chromatographic analyses have shown that the sterols of the two oils differ in that olive oil is free of stigmasterol, while an average of 5.8 percent of the total sterols in rice bran oil is stigmasterol.[13] It should therefore be possible to distinguish between straight rice bran and olive oils by Method 13-17, but reasonably certain identification of mixtures of the two oils by this means could call for more information on the range of the stigmasterol contents of rice bran oil sterols. (Eisner and Firestone found between 4.1 and 7.1 percent in six samples.)

Tests for Oxidative Rancidity and Keeping Quality of Oils

The Peroxide Value, Method 13-8, is a measure of the extent of oxidation undergone by an oil. It is expressed as meq. H_2O_2/1000 g of oil. Some laboratories consider that peroxide values of 20 meq. in animal oils, and 70 meq. for vegetable oils, as the rancidity point.[15] Jacobs[16] confirms the figure "20" for animal oils but prefers to set the rancidity point at 75 for hydrogenated vegetable oils and 125 meq. for natural cottonseed and similar oils.

The classical Kreis phloroglucinol test, as modified by Kerr, is given in many texts on food analysis. This is a qualitative test, depending on the formation of epihydrinaldehyde. The depth of the resultant pink color is roughly proportional to the degree of rancidity:

Method 13-18. *Kreis Test for Rancidity.*

Pipette 10 ml of the sample into a test tube, add 10 ml of HCl, stopper, and shake vigorously for half a minute. Add 10 ml of a 0.1% solution of phloroglucinol in ether, stopper, again shake vigorously, and note the color. A pink or red color in the acid layer indicates rancidity.

Pool and Prater[17] quantitized this test by dissolving 5 ml of the oil under test in 5 ml of $CHCl_3$, adding 10 ml of a 30% solution of trichloroacetic acid in glacial acetic acid, then adding 1 ml of a 1% solution of phloroglucinol in glacial acetic acid. After stirring with a stream of air they immersed the test tube in a waterbath at 45° for exactly 15 minutes, removed the test tube, added 4 ml of ethanol and immediately measured transmittance at 545 mμ. They concluded that transmittance over 70 percent was negative, below 60 percent denoted incipient rancidity, and low figures (around 10 percent) showed strong rancidity.

Mention should be made here of the methods used in the food industry for measuring the potential oxidative stability of oils and fats. The most widely used methods are: The Active Oxygen Method (Swift stability test), the Schaal Oven method and the ASTM Bomb Method. Eastman Chemical Products researchers have compared these three methods. Their papers should be consulted for details.[18]

Glycerides

Natural oils and fats are mixtures of triglycerides. The mono- and diglycerides do not occur naturally to any great extent unless the oil is partially hydrolyzed. Recent analyses by chromatographic separation show less than 0.3 percent of the mono- and di- compounds. These are made commercially, either by glycerolysis (which results in products containing 60 percent or less of the monoglyceride) or, in much higher concentration, by distillation.

The alpha monoglyceride may be determined by reaction with periodic acid to form aldehydes. The beta modification does not so react. This is the basis of the AOCS method:

Method 13-19. *Alpha-monoglycerides.* (*AOCS Method* ***Cd 11-57***).

INTERFERENCES: Carbohydrates with adjacent hydroxyl groups, or a carbonyl group adjacent to another carboxyl or hydroxyl group or to an amino group. Glycerin must be separated before the oxidation with periodic acid.

REAGENT

Periodic acid, reagent grade (may be obtained from G. Frederick Smith Chemical Co., Columbus, Ohio), 5.4 g in 100 ml of water. Dilute to 1L with glacial acetic acid.

NOTE: To test the purity of the periodic acid add 50 ml of the periodic acid solution above to 0.5 g of C.P. glycerin dissolved in 50 ml of water. Prepare a blank using only 50 ml of water. Titrate with 0.1% thiosulfate solution as directed under "Determination". The titer of the glycerin solution divided by that of the blank will be between 0.75 and 0.76 if the periodic acid is of sufficient purity.

DETERMINATION

Weigh a solid sample in flake form *as is*. For other products, prepare as directed in Method 13-1, and weigh the quantity of sample indicated in the table (top of next column) into a small beaker.

Dissolve the sample in reagent grade $CHCl_3$, transfer to a 100 ml glass-stoppered volumetric flask with the aid of $CHCl_3$, and bring to the mark with $CHCl_3$. Pour the entire contents of the flask into a 500 ml glass stoppered-Erlenmeyer

Monoglyceride %	Approximate size of Sample g	Weighing Accuracy g
100	0.30	±0.0002
75	0.40	±0.0002
50	0.60	±0.0003
40	0.70	±0.0005
30	1.00	±0.001
20	1.50	±0.001
10	3.00	±0.002
5	6.00	±0.004
3 or less	10.00	±0.01

flask, add 100 ml of water (pipette) and shake vigorously. Let stand until the layers separate and the $CHCl_3$ layer is, at the most, slightly cloudy. This usually takes from 1 to 3 hours. (If emulsions form, causing a poor separation, use 100 ml of 5% acetic acid instead of water.) Carry along 2 blanks of 50 ml of $CHCl_3$ and 1 of 50 ml of water, proceding as for the sample.

Pipette 50 ml of the periodic acid solution into a 400 ml beaker and add a 50 ml aliquot of the $CHCl_3$ sample solution. Swirl to mix, cover with a watch glass and let stand 30 minutes. Do not allow the temperature of the solution or the blanks to exceed 35°. Add 20 ml of 15% KI solution, swirl to mix, and allow to stand 1 minute but not more than 5 minutes. Avoid strong sunlight.

Add 100 ml of water, and titrate with 0.1 *N* $Na_2S_2O_3$ solution, using an electric stirrer to keep the solution well mixed. Titrate until the brown color disappears from the aqueous layer, then add about 2 ml of 1% soluble starch solution and continue the titration until the blue color disappears. Vigorous agitation is necessary to remove the last traces of iodine from the $CHCl_3$. Read the burette to 0.01 ml.

The sample titration should be less than 0.8 that of the blank. If not, use a smaller sample. If the difference in titrations of the blank and sample is less than 4, repeat, using double the sample, but not more than 10 g. Calculations, assuming the monoglyceride is monostearin, are as follows:

$$\% \text{ alpha monoglyceride} = \frac{(B-S) \times N \times 17.927}{W},$$

where B = the titration of the 50 ml $CHCl_3$ blank, S = the titration of the sample, N = the normality of the $Na_2S_2O_3$ solution, and W = the sample weight. (17.927 is the molecular weight of monostearin divided by 20.) If calculation to other monoglycerides is desired, substitute $\frac{MW \text{ of desired monoglyceride}}{20}$ for the value 17.927.

Several attempts have been made to quantitatively separate the component glycerides of an oil (or fat). O'Connor and co-workers proposed an I-R spectrophotometric method[19] for detecting and determining mono-, di- and triglycerides. Adequate use of this method awaits assignment of specific absorption bands to characterize the individual compounds.

AOAC gives a method[20] for separating these glycerides by column chromatography. The percentage of mono-, di- and triglycerides are calculated from data obtained by examination of various fractions from the column.

COMMERCIAL FOOD USES OF MONOGLYCERIDES[21]

Bread	0.25–0.30% of the end-product
Cakes	3–7%
Pastries	0.5–4.0%
Cake Mixes	1.0–1.5%
Chewing gum	0.5–1.0%
Ice cream	0.10–0.15%
Household shortenings	2%
High-ratio baker's shortening	up to 15%
Convenience foods with a pasta base	1.0%
Margarine	up to 0.5%

Bartlett, of the Food and Drug Directorate of Canada, and Iverson, of U.S. Food and Drug Administration[22], collaborated in a statistical study of the estimation of the fatty acid composition of oils by isothermal gas chromatography. Working with the methyl esters, they demonstrated that with only two simple measurements: peak height, and retention time for each peak, the fatty acid composition of oils could be determined.

Antioxidants

See Chapter 14, Preservatives

Margarine

There is no food, except perhaps alcoholic beverages, more subject to government regulation than margarine (sometimes called oleomargarine).

Margarine was invented by Hippolyte Mège-Mouries at the behest of Napoleon III, under a French patent granted in 1869. He digested beef tallow with gastric enzymes, slowly crystallized the purified fat, and then pressed it. He recovered 80% of a soft yellow fraction which he named "oleomargarine", sometimes called "oleo oil", and a hard white fat which he named oleostearin. He churned this "oleomargarine" with skim milk and a little sodium bicarbonate. The result was the product originally known as margarine. It was made in France in 1870, soon spread to other European countries, and in the United States through a patent granted in 1873. Now margarine has been accepted in most countries as a butter substitute, under proper labeling.

Hydrogenated vegetable oils appeared on the market early in the 20th century, and were soon adopted by the margarine industry as a better source of fat than those of animal origin. According to a Bureau of Census report in 1967,[23] out of 1,708 million pounds of oil used for margarine in 1966, well over 92 percent consisted of soybean, corn and cottonseed oils in that order, with unreported amounts of peanut and safflower oils. Probably 98 percent or more of present day margarines are of vegetable origin.

The plasticity and other physical properties of margarine are largely controlled by partial hydrogenation to a predetermined iodine value. Most margarines are now made by a continuous process. These contain from 17 to 18.5 percent skim milk.

Definition and Standard of Identity

A Federal definition and standard has been promulgated for oleomargarine (also named margarine).[24] It is the plastic food made by mixing a fat or oil with a pasteurized milk ingredient and finely ground soybeans and water, or water alone. The fat or oil may be of animal and/or vegetable origin, or a stearin derived therefrom. All or part of the fat or oil may be hydrogenated. Artificial color, flavor or preservatives are permitted, along with not over 0.5 percent of certain emulsifiers. It may be enriched with Vitamins A and D. The finished product shall contain not less than 80 percent fat. Vitamin A, up to 15000 U.S.P. units/pound, with or without Vitamin D, may be added.

The problem of the sale of colored margarine has vexed many states. Before Federal law permitted it, some states prohibited colored butter substitutes entirely; others permitted it, but taxed it heavily, and one state allowed the sale of colored margarine only if colored pink! Probably all the states in this country have laws of one kind or another regulating traffic in margarine.

Permitted preservatives and their maximum amounts are: benzoic acid or sodium benzonate, 0.1%; potassium sorbate, 0.1%; citric acid, isopropyl citrate, 0.02%; steryl citrate, 0.15%; EDTA, 75 ppm. Permitted emulsifiers or surfactants are: lecithin, mono- and/or diglycerides or sodium sulfo-acetates thereof, up to a total of 0.5%. Any safe and suitable artificial flavor that imparts a flavor in semblance of butter may be used.

Liquid oleomargarine (or liquid margarine) conforms to the definition and standard of identity for oleomargarine, except that it is in liquid rather than plastic form.

The Canadian Dominion government has not promulgated a standard for margarine, leaving this function to the provincial governments. One such standard is that of the province of Ontario. This Act[27] requires quantitative listing of each ingredient, including individual oils. It specifies a maximum permissible yellow color, expressed as Lovibond tintometer readings. Margarine shall contain not less than 80% fat, not more than 16% water, and not more than 1.4% milk solids.

Methods of Analysis

K. A. Williams[25] observed that the interpretation of butter and margarine analyses presents the most difficult problems which the food analyst has to face. This is due to the variation in composition, and complexity, of margarine, and the wide natural variations in butter.

The methods for examination of butter given in Chapter 6, Dairy Products, are, by and large, applicable to margarine. Some countries require addition of an "indicator ingredient" to facilitate differentiation from butter, *e.g.*, sesame oil, detected by Method 13-10, 0.3% starch, detected by the iodine test, or cottonseed oil, detected by the Halphen test, Method 13-9.

Unsaturated Fats: These may be calculated from the iodine value and thiocyanogen value of the separated oil, Method 13-3. See also "Identification and Estimation of Individual Fatty Acids", p. 289.

Differentiation Between Butter and Margarine

These tests are made on the clarified, filtered fat. See Method 6-30 for preparation of sample. Gross sophistication may be shown by physical tests or be determined by the conventional analytical values. When margarine is added to butter in small amounts, however, the Reichert-Meissl value, Method 13-5, is the only reliable guide. Margarine containing coconut oil or palm-nut fat may complicate the interpretation, since these fats may have Reichert-Meissls value up to 7 or even 8. However, they have high Polenske values and low Kirschner values, in contrast to butter fat.

Positive evidence of the presence of phytosterol, AOAC Method **26.065-26.070**, is conclusive proof of the use of vegetable oil margarine.

Method 13-20. *Critical Temperature of Dissolution (CTD). AOAC Method* ***15.137.***

This test clearly distinguishes all present-day margarines from butter. It was developed by Felman and Lepper[26] from the well-known Valenta test as a rapid field method and was later adopted as an official test by AOAC.

REAGENT:

Alcohol isoamyl alcohol mixture.

Determine the specific gravity of 95% ethyl alcohol and adjust to 95.0% by volume. Redistill *iso*amyl alcohol, collecting the distillate between 128–132°. Mix 2 volumes of the 95% ethyl alcohol with 1 volume of redistilled *iso*amyl alcohol (both measured with pipettes or volumetric flasks). Keep tightly stoppered.

APPARATUS

a. *Pyrex test tubes* (18 × 150 mm), calibrated with water from a burette at 2 ml and 4 ml.

b. *Micro-burner.*

c. *Glass tube of 2-3 ml capacity*, drawn to a free-flowing tip.

d. *Thermometer*, 0°–100°, graduated in degrees.

DETERMINATION

Using the drawn out glass tube, fill the test tube to the 2 ml mark with the clear filtered oil prepared as directed under "Butter", Chapter 6. Immediately add the mixed alcohol reagent to the 4 ml mark (or add 2 ml using a pipette). Using the thermometer as a stirring rod, mix, and heat in the flame of the micro-burner. Stir and heat until the mixture becomes homogeneous and clear. *Do not boil.* Remove from the heat and continue stirring until a definite turbidity appears in the *mixture proper*. Record the temperature at which the first discernible turbidity occurs (opalescence will immediately follow throughout the entire mixture with a further drop in temperature).

NOTE: Preparation of the mixed alcohol reagent is the critical part of this determination. The ethyl alcohol must be 95% by volume, the *iso*amyl alcohol must fall within the 128–132° distillation range, and the alcohols must be measured by volumetric apparatus to make the 2+1 mixture.

One of us has used this method in Connecticut as a field test, using a 2+1 mixture of ordinary 95% ethanol and reagent grade *iso*amyl alcohol, without further precautions. A flat-bottomed glass tube etched at the 2 ml level is used instead of the calibrated test tube (**a**), and an automatic pipette to measure the mixed alcohol reagent. A Reichert-Meissl determination was later made in the laboratory on all suspicious samples.

INTERPRETATION

FDA field chemists tested over 150 samples of all grades of market butter and 85 samples of all known brands of margarines collected from all parts of the country in 1949.[26] The over-all CTD range for the butter samples was 42°–53°, and for the margarine samples 66°–75°, with 90 percent of the margarines falling between 68° and 73° inclusive. Several samples of "process butter" were also tested by this method. The CTD of these samples were all within the 42°–53° range found for butter.

Shortenings

These are mixtures of various oils and fats, usually solid, with a melting point around 30°. They may be either for general use or tailored to fit a specific bakery need. *Federal Purchasing Specification EE-S-321b*, Shortening Compound, issued in 1963, recognizes General Purpose types, and Bakery types, as well as one for deep fry cookery. This last type may be either plasticized or fluid. These specifications are typical of modern industrial shortenings.

Type III or bakery shortenings are distinguished by the requirement that they contain 7.0–9.0 percent monoglycerides, if designed for use in yeast raised baked goods, and 2.0–3.5 percent if the product is to be used for cakes and icings. This is determined by Method 13-19.

All shortenings listed in this specification have limits for free fatty acids, peroxide value, stability by the accelerated oxidation method, melting point and smoke point, and moisture. AOCS methods are required to be used. The specification should be consulted for details.

TEXT REFERENCES

1. *Office Consolidation of the Food and Drug Act and the Food and Drug Regulations:* regulations B-09.001–B-09.009. Ottawa: Queen's Printer (1970).
2. Hübl: *Dingler's Polytech. J.*, **25**: 251 (1884).
3. Koettsdorfer: *Z. anal. Chemie*, **18**: 109 (1879).
4. Bolton: Selected Reference F., 52.
5. Fitelson, J.: *J. Assoc. Offic. Agr. Chemists*, **19**: 496 (1936), **20**: 418 (1937).
6. Selected Reference X, 535–537.
7. Ibid., 538.
8. Fitelson, J.: *J. Assoc. Offic. Agr. Chemists*, **26**: 499 (1943), **28**: 282 (1945), **29**: 247 (1946).
9. Ibid.: **28**: 282 (1945).
10. Cannon, H. and Firestone, D.: *J. Assoc. Offic. Agr. Chemists*, **43**: 577 (1964).
11. Eisner, J. H., Wong, N. P., Firestone, D. and Bond, J.: *J. Assoc. Offic. Agr. Chemists*, **45**: 337 (1962).

11a. Firestone, D. (Food and Drug Administration): Private communication (December 28, 1967).

12. Cannon, J. H.: *J. Assoc. Offic. Agr. Chemists*, **38**: 338 (1955).
13. Eisner, J. and Firestone, D.: *J. Assoc. Offic. Agr. Chemists*, **46**: 542 (1963).
14. Fitelson, J.: *J. Assoc. Offic. Agr. Chemists*, **26**: 509 (1943).
15. Gearhart, W. N., Stuckey, B. N., Austin, J. J.: *J. Am. Oil Chemists' Soc.*, **34**: 427 (1957).
16. Jacobs, M. B.: *Chemical Analysis of Foods*, 3rd ed. New York: D. Van Nostrand Co. (1958), 397.
17. Pool, M. F. and Prater, A. N.: *Oil and Soap*, **22**: 215 (1945).
18. a. Gearhart, W. N., Stuckey, B. N. and Austin, J. J.: *J. Am. Oil Chemists' Soc.*, **34**: 427 (1957).
    b. Stuckey, B. N., Sherwin, E. R. and Hannah, F. D., Jr.: *J. Am. Oil Chemists' Soc.* **35**: 581 (1958).
19. O'Connor, R. T., Dupre, E. E. and Fengl, R. O.: *J. Am. Oil Chemists' Soc.* **32**: 88 (1955).
20. *Methods of Analysis*, AOAC, 10th ed., **26.097–26.102** (1965).
21. Selected Reference P.
22. Bartlett, J. C. and Iverson, J. L.: *J. Assoc. Offic. Agr. Chemists*, **49**: 21 (1966).
23. National Assoc. of Margarine Manufacturers: Private communication (1967).
24. a. *31 Fed. Register*, 5433 (4/6/66); *32Fed. Register* 2893 (2/15/67).
    b. *Code of Fed. Regulation, Title 21, Part 1–129, Sec. 45.1, 45.2*, Office of the Federal Register, Washington, (1966).
25. Selected Reference W, 211.
26. Felman, H. A. and Lepper, H. A.: *J. Assoc. Offic. Agr. Chemists*, **33**: 492 (1950).
27. *Ontario Oleomargarine Act.* Chapter 268, Reg. 458. Toronto: Queen's Printer (1963).

SELECTED REFERENCES

A. Anderson, A. J. C.: *Margarine.* New York: Academic Press (1954).

B. *British Standards Institutions, B.S. 628–632, 65–658, 684.* Methods of Analysis, Fats and Oils. Available through American Standards Asso.

C. "Committee on Food Research, Conference on Deterioration of Fats and Oils". *Quartermaster Corps Manual.* Washington: Office of the Quartermaster General (1945).

D. Bailey, A. E.: *Melting and Solidification of Fats.* New York: Interscience Publishing Co. (1950).

E. Boekenoogen, H. A.: *Analysis and Characterization of Oils.* New York: Interscience Publishing Co. (1964), I.

F. Bolton, E. R.: *Oils, Fats and Fatty Foods.* London: J. and A. Churchill (1928).

G. Deuel, H. J.: *The Lipids, Their Chemistry and Biochemistry.* New York: Interscience Publishing Co. (1951–1957), I–III.

H. Eckey, E. W.: *Vegetable Fats and Oils* (ACS Monograph). New York: Reinhold Publishing Co. (1954).

I. Gearhart, W., Stuckey, B. N. and Austin, J. J.: "Comparison of Methods for Testing Stability of Fats and Oils, and Foods Containing Them". *J. Am. Oil Chemists' Soc.*, **34**: 427 (1957).

J. Hilditch, T. P.: *The Chemical Constitution of Natural Fats*. 3rd ed. London: Chapman and Hall Ltd. (1956).

K. Lea, C. H.: *Rancidity in Edible Fats*. New York: Chemical Publishing Co. (1939).

L. Lewkowitsch, J.: *Chemistry and Technology of Oils, Fats and Waxes*. 6th ed. London: MacMillan Co. Ltd. (1921–1923), I–III.

M. Lundberg, W. D. (ed.): *Autooxidation and Anti-oxidants*. New York: Interscience Publishing Co. (1961), I–II.

N. Markley, K. S. (ed.): *Fatty Acids, Their Chemistry, Properties, Production and Uses*. 2nd ed. New York: John Wiley & Sons, Inc. (1960–1968). Parts 1–5.

O. Mehlenbacher, V. C.: *Analysis of Fats and Oils*. Champaign: Garrard Press (1960).

P. Miller, M. E. and Dwoskin, P. B.: *U.S. Dep. Agr. Marketing Research Report No. 659*. Washington: U.S. Dep. Agr. Marketing Research Service (1964).

Q. Ralston, A. W.: *Fatty Acids and Their Derivatives*. New York: John Wiley and Sons (1948).

R. Salle, E. M. (ed.): *Official and Tentative Methods of the American Oil Chemists Society*. (rev.). Chicago: AOCS (1966).

S. Schwitzer, M. K.: *Margarine and Other Food Fats*. New York: Interscience Publishing Co. (1956).

T. Stuckey, B. N. and Osborne, C. E.: "Review of Analytical Techniques". *J. Am. Oil Chemists' Soc.* **42**: 228 (1965).

U. Swern, D. (ed.): *Bailey's Industrial Oil and Fat Products*. New York: Interscience Publishing Co. (1964).

V. Waterman, H. J.: *Hydrogenation of Fatty Oils*. New York: American Elsevier Publishing Co. (1951).

W. Williams, K. A.: *Oils, Fats and Fatty Foods*, 4th ed. New York: American Elsevier Publishing Co. (1966).

X. Winton, A. L. and K. B.: *Analysis of Foods*. New York: John Wiley and Sons Inc. (1945), Part 2-B.

Y. Witcoff: *The Phosphatides*. New York: Reinhold Publishing Co. (1951).

CHAPTER 14

Preservatives and Artificial Sweeteners

Preservatives

Foods processed in the United States and Canada must, by law, declare the names of any chemical preservatives used. Some preservatives are considered "food additives" under the U.S. Food, Drug and Cosmetic Act. Food additive regulations promulgated under this act may set certain limits, either as to the amount of permitted preservative, or the foods in which it may be used. Antioxidants are considered preservatives under these regulations.

Qualitative Tests for Chemical Preservatives

Method 14-1A. *Sulfur Dioxide (Sulfites).* (*See also* Method 10-16).

H_2S is evolved by addition of metallic zinc and an acid to the aqueous solution or suspension of the food, and is recognized by the black or brown color formed with lead acetate.

Add sulfur-free zinc and several ml of hydrochloric acid to about a 25 g sample in an Erlenmeyer flask, adding water if necessary. H_2S is detected by holding a lead acetate test paper in the neck of the flask.

Method 14-1B. *Benzoates and Salicylates.*

Place a 50 ml sample, or an equivalent quantity of aqueous solution, in a separatory funnel, add 5 ml of HCl (1+3) and extract twice with 50 ml of ether. If the mixture emulsifies, break the emulsion by addition of 10–15 ml of petroleum ether, and shake. If the emulsion still fails to break, centrifuge or allow the mixture to stand over night until most of the aqueous layer separates.

Wash the ether layer twice with 5 ml portions of water, drain off the water, and remove most of the ether by evaporation on a steam bath. Let the remainder evaporate spontaneously and add one drop of 0.5% neutral ferric chloride solution. (Neutralize ferric chloride solution by adding dilute ammonia to incipient precipitation of ferric hydroxide, and filtering.)

A salmon-colored precipitate indicates the presence of benzoates; a violet color indicates salicylates.

If coloring matter interferes with this test, take up the residue from the ether extract with 25 ml of ether, transfer to a separatory funnel, add several drops of 10% NH_4OH and 25 ml of water. Shake, let separate, and filter the aqueous layer through a wet filter into an evaporating dish. Evaporate the filtrate almost to dryness and test with neutral ferric chloride solution as above.

Method 14-1C. *Boric Acid and Borates.*

A characteristic test for borax and boric acid is the reaction with turmeric. Feigl in his classic text on spot tests says this reaction is sensitive to a dilution of 1:2,500,000.

Acidify the solution under test with HCl and

immerse a strip of turmeric paper. (Freshly prepared paper is more sensitive.) Remove the test-paper and let it dry spontaneously. If borax or boric acid is present, the paper will turn a red-brown, changing to a dark blue-green when held over an open bottle of ammonia.

Oxidizing substances hinder the reaction. If these are present, make the solution under test decidedly alkaline with lime water, evaporate to dryness, and ignite under low red heat until organic matter is thoroughly charred. Cool, digest with 10–15 ml of HCl (1+4). Test to be sure the solution is distinctly acid. If so, test it with turmeric paper as above.

Ferric iron also turns turmeric a dark red-brown, but the color so formed does not change to blue-green in ammonia fumes.

Method 14-1D. *Formaldehyde.*

This preservative is rarely used in modern food processing, although it is occasionally employed in milk, and has been reported in fish. Some investigators have claimed that formaldehyde is formed by the slow oxidation of trimethylamine, a decomposition product of fish. It is also one of the compounds found in wood-smoke.

DETECTION

For semisolid or solid foods, macerate 100 g with an equal quantity of water. Transfer the slurry to a 500 or 800 ml Kjeldahl flask (short-necked preferred), and make it distinctly acid with phosphoric acid. Attach the flask to an upright condenser through a Kjeldahl trap and slowly distill 40 to 50 ml. For milk and other liquid foods, use 100 to 200 ml, acidify with phosphoric acid and distill as above.

The distillate may be tested for HCHO in several ways. AOAC has made the chromotropic acid test official. The reagent is a saturated solution (about 500 mg/100 ml) of chromotropic acid in 72% H_2SO_4 (3 parts of concentrated acid to 2 parts of water by volume).

To 5 ml of the chromotropic acid reagent in a test tube add, with mixing, 1 ml of distillate. Place the tube in a boiling water bath for 15 minutes. Presence of HCHO indicated by a light-to-deep violet color.

Feigl has simplified the test by mixing a drop of distillate in a test tube with 2 ml of 72% H_2SO_4, then adding a little chromotropic acid to obtain the violet color.

The Hehner-Fulton test is also official in AOAC Methods:

Oxidizing Reagent. Slowly add concentrated H_2SO_4 to an equal volume of saturated bromine water. To 6 ml of concentrated H_2SO_4 slowly add, with cooling, 5 ml of distillate. Place 5 ml of this mixture in a test tube, slowly add with cooling 1 ml of aldehyde-free milk, then 0.5 ml of the oxidizing reagent, and mix. A purplish-pink color indicates the presence of HCHO.

To test for HCHO in milk, dilute 3 ml of concentrated H_2SO_4 with 0.5 ml of water, add 0.5 ml of the oxidizing reagent (or a crystal of NaBr) and, without mixing, immediately overlay with 1 ml of the milk to be tested. A violet ring develops at the interface.

Method 14-1E. *Formic Acid.*

Formic acid has been occasionally recommended as a preservative for fruit juices, and has been approved in the past by some European countries. No incident of the use of this preservative has come to our attention. It can be detected by reduction by nascent hydrogen to formaldehyde, and subsequent test with chromotropic acid as directed above.

Two or 3 ml of the distillate obtained as described under "formaldehyde" above, is mixed with an equal volume of HCl (1+3). Magnesium ribbon is added until evolution of gas ceases. Three ml of concentrated H_2SO_4 and a few crystals of chromotropic acid are then added. In the absence of HCHO, a violet color indicates the presence of formic acid.

Method 14-1F. *Hydrogen Peroxide.*

This preservative is permitted in certain foreign countries as a preservative, but is of limited value because of its instability. It is occasionally used in milk in this hemisphere. A simple vanadium oxide test will show its presence.

REAGENT

Dissolve 1 g of vanadium pentoxide, V_2O_5, in 100 ml of 6% by volume sulfuric acid. To test for the presence of peroxides in milk add 10–20 drops of test reagent to about 10 ml of milk and mix. A pink or red color indicates the presence of hydrogen peroxide.

Method 14-1G. *Fluorides, Soluble.*

Harrigan[1] developed a modification of the aluminum oxine test that increased the sensitivity. This was incorporated in the AOAC qualitative quenching test for fluorine **(27.026)**:

REAGENTS

a. *Aluminum solution*—2.22 g $AlNH_4(SO_4)_2 \cdot 12H_2O$ plus 3 drops of conc. HCl in 250 ml of water.

b. *Oxine reagent*—Dissolve 5 g of 8-hydroxyquinoline (oxine) in 100 ml of 2*N* acetic acid [1 ml is equivalent to about 5 ml of aluminum solution (**a**).]

c. *Ammonium acetate solution*—77 g in 500 ml of water.

d. *Aluminum oxine reagent*—Warm 250 ml of aluminum solution (**a**) to 50–60° and add excess of oxine reagent (**b**). Add ammonium acetate solution (**c**) slowly until a permanent precipitate forms, then 20–25 ml additional. Let the precipitate settle, filter through a fritted glass crucible, wash with 8 to 10 30-ml portions of cold water, and dry at 120–140°. Store in a desiccator. For test use, dissolve 5 mg in 10 ml of $CHCl_3$. (This solution should be prepared daily.)

Transfer about 150 ml of a liquid food, or the equivalent of an aqueous extract of a solid or semi-solid food, into a beaker. Add 5 ml of 5% K_2SO_4 solution, 3 ml of acetic acid and 10 ml of 10% barium acetate solution, and boil. Transfer the mixture to a centrifuge bottle, centrifuge to compact the precipitate, decant and discard the supernatant liquid, and wash the precipitate onto a small filter. Ignite the filter paper in a platinum dish. Transfer the ignited residue to a 4 or 5 ml porcelain crucible.

Wet a piece of filter paper with the $CHCl_3$ solution of aluminum oxine reagent (**d**) over a spot larger in diameter than the crucible. Air-dry the filter paper. Add fluorine-free conc. H_2SO_4 to cover the ash in the crucible, crimp the paper over the crucible edge so that the spot extends over the rim, and set a small beaker of water on the crucible to weight down the paper. Heat the covered crucible for 5 minutes at 50–60°. Observe the aluminum oxine spot under UV light. In the presence of fluorine, the fluorescence of the aluminum oxine spot is quenched in the area that was over the crucible. Sensitivity is about 0.05 mg of fluorine. Conduct a blank on the sulfuric acid.

Quaternary Ammonium Compounds and Monochloroacetic Acid

These compounds were used for a short while as preservatives, but were banned by FDA and, to our knowledge, are not now used in this country in foods. They are occasionally used as sanitizing agents in such places as restaurants and diet kitchens. Traces of these compounds may therefore show up in foods prepared in such establishments. What little interest there may be in these compounds is largely limited to their detection. Chapter 27, *Official Methods of Analysis*, **AOAC**, 10th ed., includes qualitative tests for these preservatives.

Sulfur Dioxide and Sulfites

These substances are used in foods for various purposes: inhibiting enzymatic activity and preventing oxidative discoloration in dehydrated fruits and vegetables, and as a fumigant, bleaching agent, or preservative. While the U.S. Food, Drug and Cosmetic Act lists sulfur dioxide and salts of sulfurous acid as "generally recognized as safe" (except in meats and foods recognized as sources of Vitamin B1), many government and commercial purchase specifications prescribe limits for SO_2 in certain foods. See Table 17-1.

Canadian limits are prescribed by regulation. For example, wines are limited to 70 ppm of SO_2 in the free state, or 350 ppm in the combined state. For dehydrated fruits and vegetables, Canadian law sets a limit of 2500 ppm, and in fruit juices, jams and jellies, pickles and relishes, syrups and molasses, and comminuted tomato products, 500 ppm.

The classical Monier-Williams method for determination of sulfurous acid was devised in 1927.[2] Thompson and Toy[3] modified it in 1945, using nitrogen instead of carbon dioxide to sweep out the evolved SO_2, which they titrated electrometrically. The AOAC adopted the Monier-Williams method in 1929. Improvements in chemical glassware and reagents led to the modification in the 10th edition (1965) of AOAC *Methods of Analysis.*

The Monier-Williams method determines total (free plus combined) SO_2. Government regulations and Official Methods, both in this country and in Canada, prescribe the Monier-Williams method or a modification, as do most Federal and Military specifications. Tolerances and limits established in these regulations and specifications are in terms of total SO_2.

Method 14-2. *Total Sulfurous Acid, Monier-Williams Distillation Method.*[2, 4]

This method is rewritten from the official AOAC method **27.078–.080** and is not applicable to dehydrated vegetables containing volatile S-compounds, such as onion, garlic, leek, cabbage, and cauliflower.

REAGENTS

a. H_2O_2—*3% solution.* Test 30% H_2O_2, ACS or AR grade, to insure compliance with its sulfate specification. Determine the H_2O_2 content by titration with $KMnO_4$ standard solution, dilute to about 6% H_2O_2 content, neutralize to a methyl red endpoint and dilute to the calculated volume to give 3.0% H_2O_2.

b. *Methyl red indicator*—0.25% in ethanol. Adjust to transition color.

APPARATUS

See Fig. 14-1. Connect 1 outside neck of a

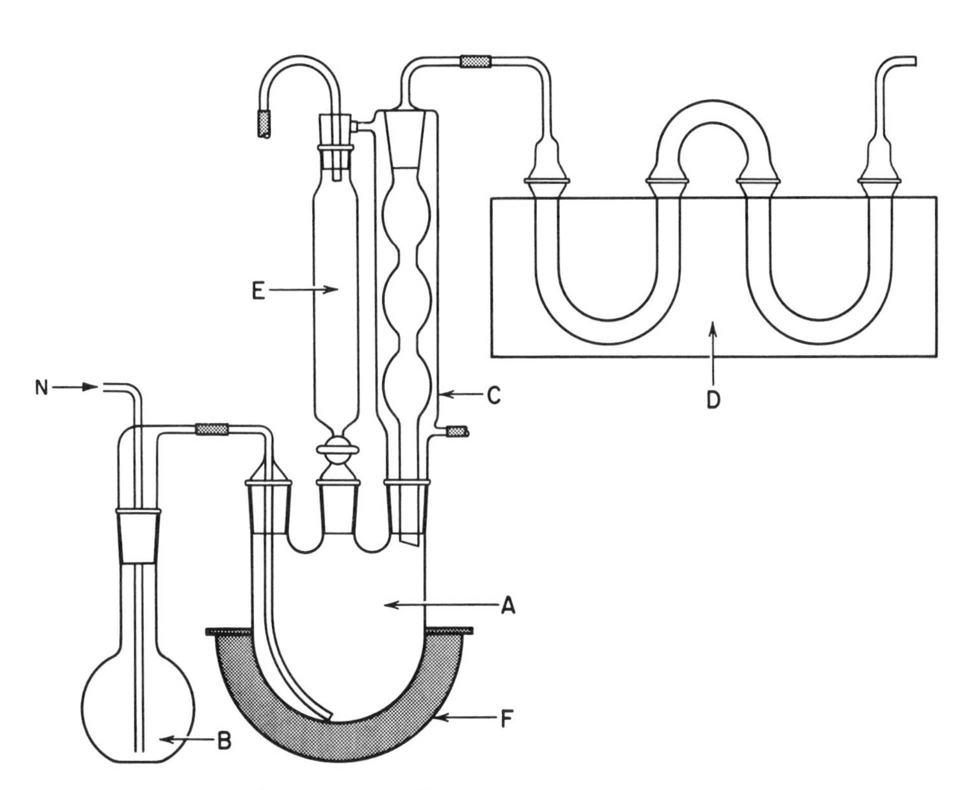

FIG. 14-1. Apparatus for modified Monier-Williams method for sulfur dioxide. A) Distilling flask; B) Gas washing bottle; C) Allihn condenser; D) Balljoint U-tubes; E) Straight-side separatory funnel; F) Heating mantle.

₮ 24/40 1 L, 3 neck distilling flask (A), through a ₮ 24/40 gas inlet tube with curved end reaching nearly to the bottom of the flask, to a 250 ml gas washing bottle (B), leading to a source of nitrogen. Connect a vertical 30 cm Allihn condenser (Scientific Glass Co. JC-5450 or equivalent) (C) to the other outside neck of the distilling flask. Connect this condenser through a 6 inch silicone tubing to two balljoint ₮ 35/20 U-tubes (55 $\pm$ 5 mm long, made from 20 mm diameter tubing) (D), connected by a crossover tube, ₮ 35/20 balljoints, 55 ± 5 mm between centers. To the center neck attach a 125 ml ₮ 24/40 straight-sided separatory funnel (E).

Add two 25 mm pieces of solid glass rod and 10 ml of 3 mm glass beads at the exit side, and add 10 ml of 3% H_2O_2 containing a drop of methyl orange indicator to each U-tube.

Grind 4.5 g of pyrogallol with 5 ml of water and transfer the slurry into gas washing bottle (B). Repeat the grinding and transfer with 2 more 5 ml portions of water. Pass water-pumped nitrogen from a tank through a 2-stage regulator into the gas washing bottle to flush out air. Add to the gas washing bottle, through a long-stemmed funnel, a freshly prepared, cooled solution of 65 g of KOH in 85 ml of water. Turn off the nitrogen and connect the exit tube of the gas washing bottle and the gas inlet tube with $\frac{1}{4} \times 6''$ silicone tubing, Clamp off both ends of (B).

Attach a piece or rubber tubing to a short glass U-tube inserted through a rubber stopper in the neck of the separatory funnel (E). Blow into rubber tubing and close separator stopcock. Let stand a few minutes to check for leaks as shown by the liquid levels in U-tubes.

Place distilling flask (A) in heating mantle (F) controlled by a variable transformer.

NOTES

1. The 6 inch lengths of silicone tubing connecting distilling flask (A) with gas washing bottle (B), and Allihn condenser (C) with the U-tubes (D), should be pre-boiled in HCl $(1+20)$ and rinsed with water before use.

2. Thrasher[5] simplified the assembly shown in Fig. 14-1 by substituting the alternative trapping system shown in Fig. 14-2.

DETERMINATION

Place a sample containing at least 45 mg of SO_2 in distilling flask (A), using water for transfer if necessary. Dilute to about 400 ml with water, add 90 ml of HCl $(1+2)$ to separatory funnel (E) and force the acid into the flask with gentle pressure. Start the nitrogen flow at a slow, steady stream of bubbles. Heat the flask to cause refluxing in 20-25 minutes (about 80 volts on a 7 amp transformer). When refluxing is steady, apply line voltage and reflux $1\frac{3}{4}$ hours. Turn off the water in the condenser and continue heating until the inlet joint of the first U-tube (D) shows condensation and slight warming. Remove the separatory funnel and turn off the heat. When the joint at the top of the condenser cools, remove the connecting assembly and rinse into the second U-tube (D). Attach the cross-over tube to the exit joint of first U-tube, rotate until the open ends touch, add a drop of methyl red indicator, and titrate with 0.1 N NaOH. Titrate the contents of the second U-tube similarly (1 ml of 0.1 N NaOH $=$ 3.203 ml of SO_2).

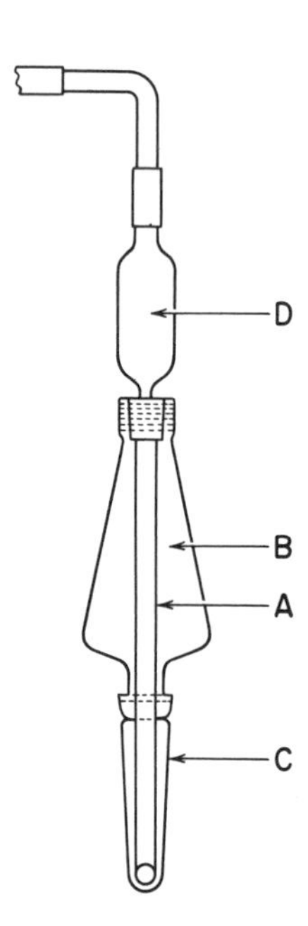

FIG. 14-2. Alternative trapping apparatus for sulfur dioxide. A) Gas dispersion tube, size 2, porosity C (Scientific Glass Apparatus Co. No. G-5420, or equivalent); B) Kuderna-Danish Concentrator Flask, ₮ 24/25 (Kontes Glass Co. No. K-57000B); C) Concentrator Tube, ₮ 24/25 (Kontes Glass Co., No. 57000C); D) Trap: a 50 ml transfer pipette, cut 11 cm below bulb.

A confirmatory gravimetric determination may be made after titration by rinsing the contents of the U-tubes into a 400 ml beaker, acidifying with about 4 drops of 1 *N* HCl, and adding an excess of filtered 10% $BaCl_2$ solution. After standing overnight, the mixture is filtered by decanting through a tared sintered glass crucible, and washing the precipitate by decantation with hot water 3 times. The precipitate is then transferred to the crucible, washed with 20 ml of ethanol and 20 ml of ether, and dried to constant weight at 105–110°:

$$\frac{\text{mg } BaSO_4 \times 274.46}{\text{weight of sample}} = \text{ppm } SO_2.$$

Determine a blank on the reagents, both by titration and gravimetrically, and correct the results accordingly.

Method 14-3. *Total Sulfurous Acid—Colorimetric Method.* (Applicable to dried fruit. *AOAC Method,* ***27.081.***[6, 7]

This method is based on the well-known Schiff reaction between *p*-rosaniline, formaldehyde and sulfur dioxide, forming a purple color. It was developed by the USDA Western Regional Laboratory in 1959[5], and determines both free and combined SO_2.

REAGENTS

a. *Formaldehyde solution,* 0.015%—Prepare from 40% by volume HCHO (USP) by diluting in 2 steps: 10:1000 and 75:2000.

b. *Acid-bleached p-rosaniline hydrochloride* (National Aniline Div., Allied Chemical and Dye Corp.)—Transfer 100 mg of *p*-rosaniline HCl to a 1 L volumetric flask. Add 200 ml of water and 160 ml of HCl (1 + 1), dissolve the salt, and dilute to the mark. Let stand 12 hours before use.

c. *Sodium tetrachloromercurate*—Transfer 23.4 g NaCl and 54.3 g $HgCl_2$ to a 2 L volumetric flask. Dissolve in about 1900 ml of water and dilute to the mark.

d. *Sulfur dioxide standard solution*—Dissolve about 170 g of sodium bisulfite, $NaHSO_3$, in water and dilute to 1 L. Standardize with 0.01 *N* iodine solution; 1 ml is equivalent to about 100 μg SO_2.

STANDARD CURVE

Add 5 ml of mercurate reagent (**c**) to a series of 100 ml volumetric flasks, then add 0, 1.0, 2.0, 3.0, etc., ml of standard SO_2 solution (**d**). Dilute to volume and mix. Transfer 5.0 ml portions of each to 200 mm test tubes containing 5 ml of rosaniline reagent (**b**). Add 10 ml of HCHO solution (**a**) to each, mix and hold at 22° for 30 minutes. Read the absorbance at 550 mμ against the zero standard and plot to obtain a standard curve.

DETERMINATION

Weigh 25 ± 0.02 g of ground, dried fruit onto an 11 cm Whatman No. 1 filter paper. Transfer the sample and paper to a 500 ml blendor jar containing 475 ml of 0.1 *N* NaOH. Cover, and blend 2 minutes. Let stand about 5 minutes (a clear layer will form at the bottom of the jar). With a pipette transfer 10 ml of the clear lower layer to a 500 ml volumetric flask containing 15 ml of 0.1 *N* HCl. Mix gently, add 25 ml of mercurate reagent (**c**) and dilute to the mark. Run a blank on all reagents, omitting the 10 ml of fruit extract.

Transfer 5 ml of the sample solution to a 200 mm test tube containing 5 ml of rosaniline reagent (**b**). Add 10 ml of HCHO solution (**a**), mix, and hold for 30 minutes at 72°. Read the absorbance at 550 mμ against the blank. Refer to the standard curve and report results as ppm of SO_2.

[If the same colorimeter tube or cell is used for successive samples, clean between use with HCl (1 + 1) and water.]

Results obtained by the authors of this method indicate that its precision equals that of the Monier-Williams gravimetric method, but is less than that of the Monier-Williams volumetric method.

Benzoate and Hydroxybenzoate Compounds

Methyl-, ethyl- and propyl-*p*-hydroxybenzoic compounds have become popular in recent years as preservatives, although more so in Europe than in the United States. These compounds have been given the generic term "parabens" by some writers. Considerable difficulty has been encountered in separating these compounds from each other and from sorbic, benzoic and salicylic acids.

Schwartzman, AOAC Referee on Preservatives, and his co-workers, have made a promising approach by first separating these acids (which they call "preserving acids") from the food by steam distillation at a controlled rate, extracting the acids, and separating the individual acids by thin layer chromatography. The compounds are then identified by their behaviour under UV or IR light. These authors are now studying a TLC quantitative procedure.[8,9,10] Their reports will be published in subsequent issues of *J. Assoc. Offic. Anal. Chemists.*

The methods as now published will undoubtedly be revised and then be subjected to collaborative study by the AOAC before final adoption by that Association. We prefer therefore to give these methods by reference only, rather than include them in a transitional form.

Boric Acid and Borates

Method 14-4. *Borates and Boric Acid, Quantitative.*[11]

This method is based on the fact that boric acid reacts with polyvalent alcohols to form stronger complex acids which give a sharp end-point to phenolphthalein indicator. Thompson, the originator of the method, recommended neutral glycerol. Later mannitol was found to be better adapted to the purpose. Fructose and invert sugar have also been recommended.[12]

Determination

Transfer a 10–100 g sample, depending on the material and the quantity of borates present, to a platinum dish. Make the sample distinctly alkaline with NaOH solution. Evaporate to dryness, then ignite at 400–500° until organic matter is thoroughly charred. Cool, digest with about 20 ml of hot water, and make distinctly acid with dilute HCl (phenolphthalein indicator). Filter into a 100 ml volumetric flask and wash with hot water, catching the washings in the volumetric flask, until the volume in the flask amounts to 50–60 ml.

Return the filter paper to the platinum dish, wet it thoroughly with lime-water, dry on the steam bath, and ignite to a white ash. Dissolve the ash in a few ml of HCl (1+3), and filter into the 100 ml volumetric flask, washing out the dish with a few ml of water and add the washings to the 100 ml flask. Add 0.5 g of $CaCl_2$, and 10% HaOH solution until a light pink phenolphthalein color forms. Dilute to the mark with lime-water, mix, and filter through a dry filter.

To 50 ml of the filtrate add 1 *N* H_2SO_4 drop by drop until the pink color disappears. Now add methyl orange indicator and continue the addition of 1 *N* H_2SO_4 until the yellow color changes to pink. Boil a minute or two to expel CO_2, cool, and add 0.2 *N* NaOH drop by drop from a burette until the liquid assumes a yellow tint. Avoid excess alkali. The boric acid is now free H_3BO_3 with no uncombined H_2SO_4 present. Record the burette reading.

Add 1 to 2 g of neutral mannitol and a few drops of phenolphthalein indicator. Titrate again with the 0.2 *N* NaOH to reappearance of the pink color. Repeat the alternate addition of mannitol and standard NaOH solution until a permanent pink end-point is reached.

One ml of 0.2 *N* NaOH is equivalent to 0.0124 g of boric acid, H_3BO_3.

Sorbic Acid and Sorbates

These preservatives have been recommended to retard mold and yeast growth in pickle brine, sweet pickle syrup and soft drinks. They have been approved by FDA as a mold-inhibiting ingredient of consumer-sized packages of certain firm cheeses such as cheddar, Swiss, Monterey jack, provolone and romano, *e.g.*, the standard of identity of cheddar cheese[13] provides for up to 0.3% maximum of sorbic acid, while in process cheeses, cheese spreads and club cheese up to 0.2% is permitted. Sorbic acid is also permitted, in the form of the potassium salt, in margarine.[14]

Sorbic acid and its calcium, sodium and potassium salts are exempted from tolerance requirements as food additives under the *Federal Food, Drug and Cosmetic Act* as being "generally recognized as safe (GRAS)."[15]

Sorbic acid and sorbates are permitted under certain restrictions in Canada. They may be used in unstandardized foods except meat, poultry and fish, and in fruit juices, jams, jelly or marmalade with pectin, pickles and bread up to 1000 ppm, and in certain hard cheeses up to 3000 ppm.[16]

Several methods for the detection of sorbates may be found in chemical literature. These include spectrophotometric absorption in the visible, UV or IR, partition and gas chromatography, and by conversion to colored compounds. A review of these methods has been made by Ciaccio[17] in his report on a method for the determination of sorbates in low-calorie salad dressing. We have chosen to include here a method developed by Food and Drug Administration chemists as being applicable to many varieties of food.

Method 14-5. *Infrared Determination of Sorbic Acid in Foods.*[10]

APPARATUS

a. *Steam distillation apparatus*—Scientific Glass Apparatus Co., Bloomfield, N.J., No. JD-3170 or equivalent.

b. *Recording infrared spectrophotometer*—recording between 9.0 and 10.5μ.

REAGENTS

A reagent grade sorbic acid for preparing standard solutions "puriss grade" may be purchased from Aldrich Chemical Co., Inc., Milwaukee, Wis. 53210.

PREPARATION OF SAMPLE

Solids and semi-solids—Blend 50–100 g with about 100 ml of water and transfer to the distilling flask.

Liquids—Pipette 50–100 ml into the distilling flask.

DETERMINATION

Add 200 g magnesium sulfate, $MgSO_4.7H_2O$, 100 ml water and 5 ml H_2SO_4 to the sample in the distilling flask. Add 10 ml HCl and 30 ml $CHCl_3$ to the receiving flask. Steam distill, collecting about 450–500 ml distillate. Transfer entire contents of the receiving flask into a 1 L separatory funnel, add 10 ml HCl, shake 1 minute, let separate, and drain $CHCl_3$ layer into a 250 ml beaker. Extract with 6 more 30 ml portions of $CHCl_3$, rinsing the receiving flask with the $CHCl_3$ before each extraction.

Evaporate the combined $CHCl_3$ extracts on the steam bath under an air current to about 5 ml. Remove from steam bath and continue the evaporation to dryness under an air current at room temperature. Dissolve the residue in $CHCl_3$, transfer quantitatively to a 50 ml flask and dilute to volume with $CHCl_3$. Scan the IR spectrum in a 1 mm cell, using $CHCl_3$ as reference, between 9.0 and 10.5μ. Draw a baseline from 9.90 to 10.5μ and measure the peak absorbance at 10.0μ.

Prepare a standard solution of sorbic acid in $CHCl_3$ containing 2 mg/ml. Scan its IR spectrum between 9.0 and 10.5μ, using $CHCl_3$ as reference, and determine its absorbance.

Calculate percentage of sorbic acid present in the sample by the following formula:

$$\%\ \text{sorbic acid} = \frac{[(A_s/A_{sa})\ (\text{g of std. sorbic acid/ml})\ (\text{ml final volume})]\ 100}{\text{wt. of sample in grams}}$$

where A_s = absorbance of sample, and A_{sa} = absorbance of standard sorbic acid solution.

Confirmation: Evaporate to dryness a sample aliquot containing about 2 mg sorbic acid. Add 200 mg KBr and make a disk. Scan its IR spectrum and compare with spectrum of standard sorbic acid in KBr.

NOTE: Sorbic acid is used in most foods at the 0.05–0.22% level. The authors of this method report that other common preservatives do not interfere.

Dehydroacetic Acid (DHA)

Dehydroacetic acid is permitted as a preservative for cut or peeled squash by FDA.[19] A tolerance of 65 ppm is prescribed.

It has been used occasionally on fresh or dried fruits.[18]

Chlorohydrin and Epoxide Residues in Dehydrated Plant Material Treated with Ethylene (or Propylene) Oxide

Ethylene oxide and propylene oxide are used extensively for sterilization of dehydrated vegetables and spices. These epoxides are effective in killing, or greatly reducing, viable micro-organisms, with no adverse effect on odor and flavor. Residual amounts of these epoxides are said to be dissipated shortly after application, leaving only traces of the corresponding glycol as the reaction product.

Recently British workers[20] for Crosse and Blackwell, Ltd., London, found that these glycols further reacted in the presence of chlorides to form chlorohydrins. These are considerably more toxic than the glycols, and Wesley and his co-workers found as high as 980 ppm of chlorohydrins in samples of spices fumigated at the point of origin. These compounds were identified as 2-chloroethanol (from ethylene oxide), and 1-chloro-2-propanol and 2-chloro-1-propanol (from propylene oxide). They proposed a method for the volumetric determination of these chlorohydrins said to be sensitive to about 10 ppm.

Ragelis and co-workers at the U.S. Food and Drug Administration[21] later developed a gas chromatographic method sensitive to 1 ppm or less. Both methods are given.

Method 14-6. *Chlorohydrin Residues in Dehydrated Plant Material—Volumetric Determination for Amounts Above* 10 *ppm.*[20, 22]

DETERMINATION

Weigh 20 to 100 g of ground sample into the boiling flask of a steam distillation assembly. Add water sufficient to form a slurry. Steam-distill 150 ml of distillate into a 500 ml Erlenmeyer flask. Make the distillate alkaline with 5 ml of 20% NaOH solution.

Heat the alkaline distillate for 30 minutes on a steam bath to hydrolyze the chlorohydrins. Acidify with a small excess of dilute HNO_3, add an excess of 0.05 N standard $AgNO_3$ solution, and back titrate with 0.05 N KCNS solution to a permanent light brown color, using 5 ml of saturated ferric alum (ferric ammonium sulfate) solution as indicator.

One ml 0.05 N $AgNO_3$ = 0.000402 g ethylene chlorohydrin or 0.000473 g propylene chlorohydrin.

NOTE: This method is not specific for chlorohydrins, since it merely determines any steam-volatile organo-halogen compound yielding inorganic halides upon hydrolysis with alkali. The presence of chlorohydrins should be confirmed by gas chromatography to establish their presence.[22]

Method 14-7. *Chlorohydrin Residues in Dehydrated Plant Material—Gas Chromatographic Method for Amounts in the* 1 *ppm Range.*[21]

APPARATUS* (*See* Fig. 14-3.)

a. *Chromatographic column*—400 mm long × 20 mm i.d. (K-42055); equipped with a coarse porosity fritted disk, a Teflon stopcock, and a female 24/40 ₮ fitting, with glass hooks at the opposite end (Fig. 14-3, C).

b. *Chromatographic reservoir*—500 ml capacity (quotation no. 25042); equipped with a 24/40 ₮ male fitting at the bottom and a 24/40 ₮ female fitting at the top. Both the male and female joints should be equipped with glass hooks. (Fig. 14-3, B.)

c. *Adapter*—Gas outlet type (K-18300); with ₮ 24/40 male joint at bottom and male fitting equipped with glass hooks. (Fig. 14.3, A)

d. *Evaporative concentrator*—Kuderna-Danish type (K-57000); consists of (1) flask (Fig. 14-3, F), 1 L capacity with a 24/40 ₮ female fitting at top and a 24/25 ₮ male fitting at bottom, (2) Snyder distilling column (K-50300) with 24/40 ₮ male and female joints, and (3) receiving flask (Fig. 14-3, G) (K-62140); volumetric, 10 ml capacity, size 10-F, with 24/25 ₮ female fitting. The male

*All apparatus, except the gas chromatograph, (f), may be purchased from Kontes Glass Co., Vineland, N.J.; numbers in parentheses refer to catalogue numbers for ordering from Kontes. Equivalent materials may be substituted for any of the above apparatus.

joint of the flask and the female joint of the receiving flask should be equipped with glass hooks.

e. *Flask—Erlenmeyer* (*K-61725*); 500 ml capacity, all glass with 24/40 ₸ female fitting and pouring lip.

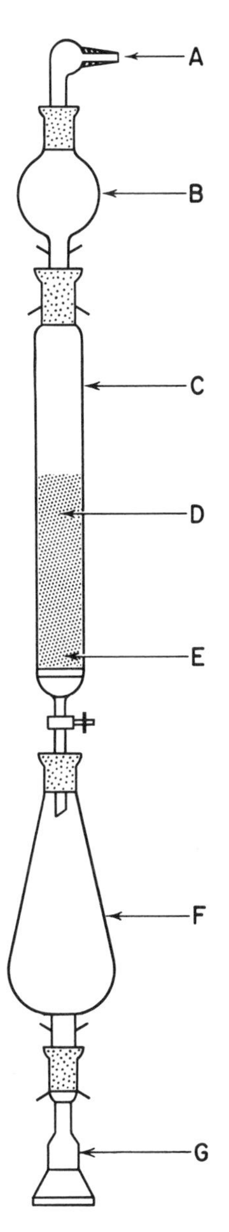

FIG. 14-3. Apparatus for extraction of chlorohydrins A) Gas outlet adapter; B) Chromatographic reservoir; C) Chromatographic column; D) Sample; E) Celite; F) Flask, 1 L; G) Receiving flask, 10 ml.

f. *Gas chromatograph*—Beckman GC-4; with dual flame ionization detector and stainless steel column, 3 mm i.d. × 10 ft. long, packed with 20% by weight Carbowax 20M on 60/80 mesh acid-washed Chromosorb W. Temperatures should be as follows: column 112°; detector 230°; and injection port 188°. Use a helium carrier gas flow rate of 100 ml/min.

REAGENTS

a. *Ether*—Reagent grade, anhydrous; Fisher Scientific Co., or equivalent.

b. *Florisil*—60 to 80 mesh, activated by the manufacturer for 3 hours at 1250° F; Floridin Co., Hancock, West Va.

c. *Celite 545*—Analytical filter aid; Johns-Manville Co.

d. *Chlorohydrins*—(1) 2-chloroethanol (No. 131, White Label), and (2) 1-chloro-2-propanol (No. P-1325), containing approximately 25 percent 2-chloro-1-propanol. Eastman Kodak Co., Rochester, N.Y. (The composition of these materials was verified by gas chromatography, nuclear magnetic resonance, and infrared spectrophotometry.)

e. *Standard solutions*—Dissolve 10 mg of 2-chloroethanol or 1-chloro-2-propanol isomer mixture in a 25 ml volumetric flask in anhydrous ether and adjust to volume.

f. *Nitrogen*—Water-pumped or equivalent purity nitrogen in a cylinder equipped with regulator and valve to control flow rate at 5 psig.

DETERMINATION

Sample should be ground to pass through a 20-mesh, and preferably a 30-mesh, screen. Pour 1 g of Celite 545 into the chromatographic column (Fig. 14-3, E) and tap gently to level the adsorbent layer. Wash with 10 ml of anhydrous ether. Add 50 g of the sample (Fig. 14-3, D) on top of the Celite, and tap the column to level the contents. Place a 500 ml Erlenmeyer flask under the column, open the stopcock and elute with 50 ml of anhydrous ether, using a nitrogen source at 2.5 psi attached at adapter (Fig. 14-3, A) to force it through the column. Just before the solvent reaches the sample level add a second 50 ml of ether and continue the percolation. Repeat this operation with 2 additional 50 ml portions of ether, drawing the final 50 ml portion completely through the column.

Add 30 g of Florisil to a second chromatographic column and tap gently to settle the contents. Place the 1 L Kuderna-Danish flask with attached 10 ml volumetric receiving flask (Fig. 14-3, F and G) under the column. Insert 3 carborundum boiling stones into the 1 L flask and attach to the column. The tip of the column should extend as far as possible below the standard taper fitting of the 1 L flask without hindering movement of the stopcock.

Pass the ether eluate that was collected in the 500 ml flask from the first (Celite) column through the Florisil column. Wash the 500 ml flask with 2 separate portions of anhydrous ether, passing each in sequence through the column.

Fit the Snyder distilling column into the 1 L flask containing the ether extracts. Set the assembly in a hood over a vigorously boiling water bath and concentrate the extract to about 5 ml. Remove the concentrate from the steam bath, allow the residual solvent in the distilling column to drain into the 10 ml receiving flask, and cool to room temperature. Remove the receiving flask from the assembly, adjust to volume with ether, stopper, and mix.

Gas chromatographic analysis.—Inject 2 μl of sample solution from the volumetric receiving flask into the gas chromatograph. Determine the area under the chromatographic peak and from the area of a previously injected 2 μl standard solution, calculate the corresponding mg of chlorohydrin extracted from a 50 g sample as follows:

mg chlorohydrin extracted

$$= (2.5 \times \text{area of sample/area of standard}) \times \text{mg chlorohydrin standard}.$$

ppm chlorohydrin

$$= (\text{mg chlorohydrin extracted/g sample}) \times 10^3.$$

Results

Recoveries by Ragelis and his co-workers of 2-chloroethanol or 1-chloro-2-propanol and 2-chloro-1-propanol added to 50 g of flour or ground black pepper at the 5–10 ppm levels have ranged from 76 to 86%.

Method 14-8. Epoxide Residues in Dehydrated Plant Material Resulting From Ethylene (or Propylene) Fumigation.[23, 24]

This is a modification of the Lubatti method used by Griffith Laboratories, Chicago, Ill. It is applicable to a wide variety of products. Recovery of the epoxide is said to be better than 90% within a reproducibility of ±0.2 ppm.

Apparatus (*See* Fig. 14-4).

a. *Two 250-ml gas washing bottles.*

b. *3 L round bottomed reaction flask* with 3 standard taper necks.

c. *Glas-Col mantle* for 3 L flask.

d. 12″ Liebig standard taper condenser.

e. *Safety tube.*

f. *Flow-meter*, capable of measuring 1.5 L air/minute.

g. *Three absorption tubes* (25 × 250 mm or 29 × 300 mm test tubes).

h. *Three gas dispersion tubes*, extra coarse porosity.

i. *Filter pump* or aspirator.

Reagents

a. *Magnesium bromide solution*, 450 g of $MgBr_2$, reagent grade, in 1 L of approximately 0.02 *N* H_2SO_4.

b. 0.02 *N NaOH solution.*

c. *Bromo-cresol green indicator*, 0.07% in 50% ethanol.

Apparatus Assembly

Follow Fig. 14-4. Half fill the first gas washing bottle with 40% NaOH solution. Insert pledgets of absorption cotton in the second. A standard taper gas distributor tube extends to the bottom of the 3 L reaction flask through one neck, the middle neck is closed with a stopper, and a Liebig condenser is attached by the third neck. The top of the condenser is connected through an empty safety tube and flow meter to a series of 3 absorption tubes and finally a source of vacuum.

DETERMINATION

Transfer from 200 to 400 g of sample (which may be fine grind or coarsely chopped) into the 3 L flask, attach the Glas-Col mantle, the gas distributor tube and the condenser. Pipette exactly 20.0 ml of the magnesium bromide solution into each of the 3 absorption tubes, attach the gas dispersion tubes, and insert in the aeration train. Pass air through the train at the rate of 1.0 to 1.5 L/minute for 1 hour.

Stop the aeration, remove the flask stopper, and thoroughly wet the sample with 200 to 300 ml of cold distilled water. Let stand 10 to 15 minutes, then add 2 L of freshly boiled, hot distilled water and stopper the flask. Mix the slurry with a heavy stirring rod, while swirling the contents of the flask. Insert the flask into the aeration train, adjust the heat control of the mantle so as to maintain the slurry at boiling temperature, and pass air continuously through the aeration train for an additional 1½ hours, at the same rate as before.

Turn off the airflow after 1½ hours and disconnect the equipment. Transfer the contents of the 3 absorption tubes to a 250 ml Erlenmeyer flask. Rinse the 3 gas dispersion tubes with a minimum amount of distilled water, catching the rinsings in the respective absorption tubes. Add the rinsings to the contents of the Erlenmeyer flask, again rinsing out the emptied absorption tubes with a minimum amount of distilled water and adding the rinsings to the flask.

Add 5 or 6 drops of bromocresol green indicator to the contents of the flask and titrate with 0.02 *N* NaOH solution. When the color changes to green, titrate drop by drop with continuous stirring to the first appearance of a definite blue end point. (A magnetic stirrer is a convenient means of continuous stirring.) When the definite blue color appears, stopper the flask with a clean stopper and shake. If the color does not change consider this the end-point. If it becomes less blue, add the 0.02 *N* NaOH drop by drop to reappearance of the blue tinge.

Run a reagent blank by pipetting 3 × 20.0 ml of the magnesium bromide solution into an Erlenmeyer flask, adding 10–15 ml of distilled water and titrating as before.

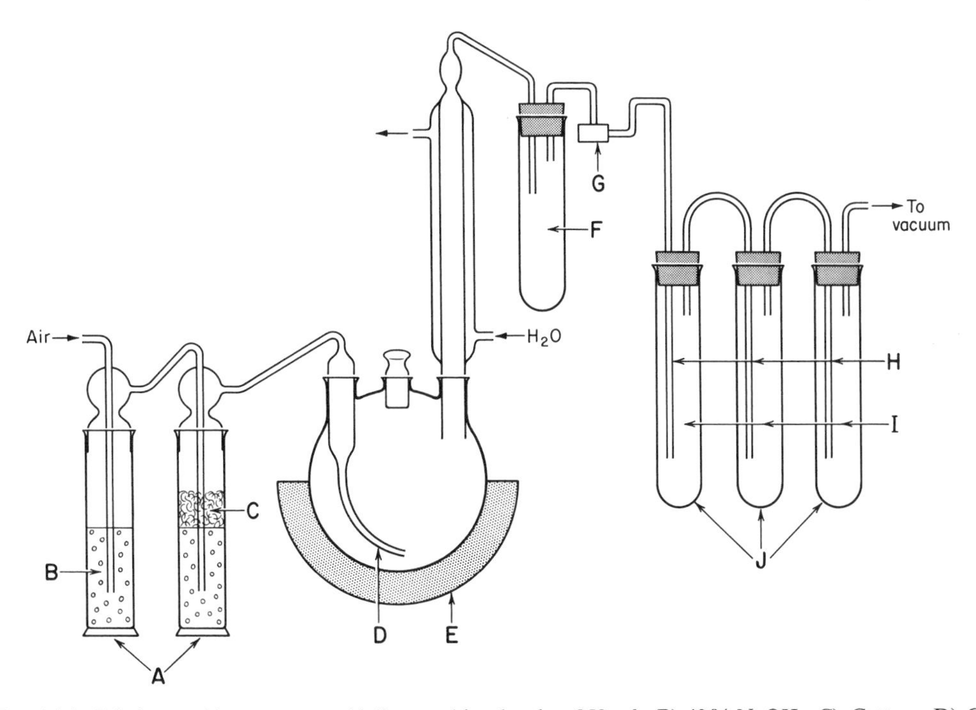

FIG. 14-4. Ethylene oxide apparatus. A) Gas washing bottles, 250 ml; B) 40% NaOH; C) Cotton; D) Gas distributor tube, S 29/42 (Ace, No. 5295-D); E) Glass-col mantle; F) Safety tube (empty); G) Flow meter; H) Gas dispersion tubes, extra coarse porosity; I) $MgBr_2$ solution; J) Absorption tubes (test tubes).

CALCULATIONS

$$\text{ppm ethylene oxide} = \frac{(\text{blank-sample titration}) \times \text{normality NaOH} \times 44{,}000}{\text{sample weight in grams}}.$$

$$\text{ppm propylene oxide} = \frac{(\text{blank-sample titration}) \times \text{normality NaOH} \times 58{,}000}{\text{sample weight in grams}}.$$

Biphenyl (Diphenyl) in Citrus Fruits

Biphenyl may be used as a post-harvest mold inhibitor for citrus fruits. FDA has established a tolerance of 110 ppm in or on citrus fruit.[25] The usual mode of application is by tissue paper wraps impregnated with biphenyl, either by vapor application or by a coating of biphenyl dissolved in melted paraffin wax. Packing boxes and box linings or pads may also be treated in this manner.[26]

The method for determination of biphenyl given here was supplied by Monsanto Chemical Company and the Institute of Paper Chemistry, who developed the method jointly.[26,27]

Method 14-9. *Determination of Biphenyl in Citrus Fruit and Fruit Wraps.*

APPARATUS AND REAGENTS

a. *Spectrophotometer*—Capable of measuring absorption at 245 mμ.

b. *Liquid—Liquid Clevenger extractor* (*see Fig. 14-5*).

c. *Acetic-permanganate solution*—Equal parts by volume of 6% aqueous $KMnO_4$ solution and glacial acetic acid.

d. *Purified n-heptane*—Suitable for solvent in gas chromatography. May be obtained from Mallinkrodt, 2nd and Mallinkrodt Sts. St. Louis, Mo., or from Burdich and Jackson Labs, Inc., Muskegon, Mich. 49442.

A. EXTRACTION OF BIPHENYL FROM THE FRUIT

1. *Juice:* Weigh 12 fruit, halve the fruit transversely, and ream 6 stem-end halves and six blossom-end halves on a mechanical juicer. Measure the volume of juice and transfer it or a 500-ml aliquot to a 1 L, flat-bottomed flask with sufficient water (if required) to bring the final volume to approximately 500 ml. Add boiling chips, and connect a modified Clevenger liquid extractor and condenser. Dow Corning Fluid No. 550 may be added as an anti-foam agent.

Add 20 ml of purified heptane through the condenser, and boil the contents of the flask for two hours over a 300-watt electric hot plate. Allow the extractor to cool to room temperature,

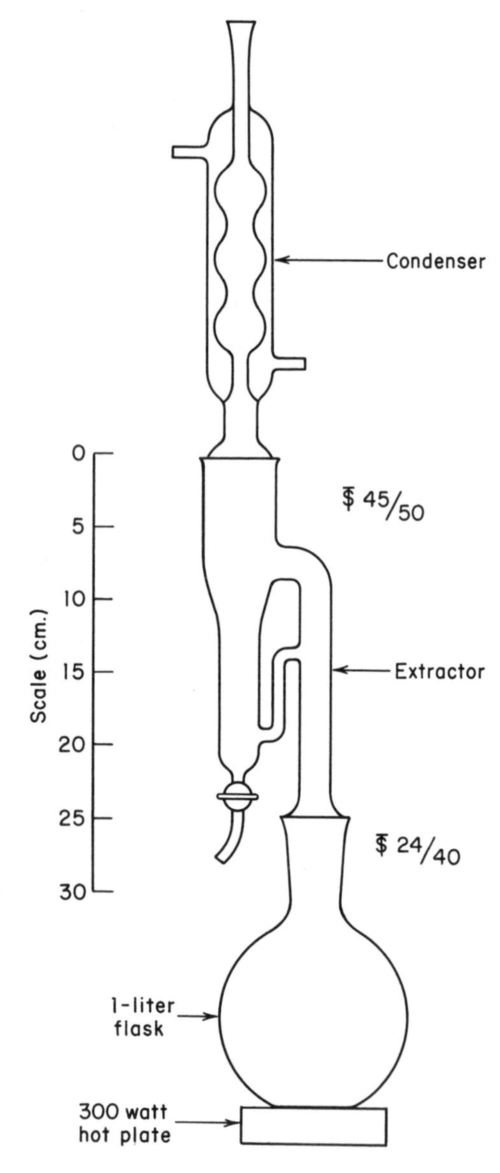

FIG. 14-5. Clevenger extractor.

and withdraw the heptane layer through a loose cotton plug (held in a small funnel), into a 50 ml volumetric flask. Wash the extractor twice with heptane and dilute to the mark with heptane.

2. *Whole Fruit:* Weigh the 6 stem-end halves combined with the 6 blossom-end halves remaining from (1). Cut each half into 4 uniform sections by one transverse and one longitudinal cut, and combine opposite sections to yield a sample of 24 sections (eighths), or the equivalent of 3 whole fruit, and weigh. Grind the fruit portionwise in a Waring Blendor with a total of 800–1000 ml of water to form a slurry or paste. Transfer one-third of the slurry quantitatively to a 1 L extraction flask and complete the extractions as in (1).

3. *Peel*: Cut the reamed peel (1) into four sections by one transverse and one longitudinal cut, following the directions for whole fruit. Manually remove and discard the pulp or "rag", and weigh a sample composed of opposite sections as directed for whole fruit. Grind the peel and perform the extraction as described for whole fruit.

4. *Carton*: With a circle-cutter obtain one 5.0 square inch circle from each treated surface of the carton. Pulp the circles with water in a Waring Blendor. Transfer the pulp slurry quantitatively to the extractor and complete the extraction as directed for juice.

5. *Pad or Tissue Wrap*: Weigh the whole pad or tissues, and determine the area in square inches or square feet. Fold the pad or combined tissues lengthwise along the middle and then lengthwise again. Make $\frac{3}{4}$ inch cuts with scissors $\frac{1}{8}$ inch to $\frac{1}{4}$ inch apart along the fold-line to form a "fringe", then make three or four cuts across the "fringe" to yield a sample of $\frac{1}{8}$ to $\frac{1}{4}$ inch squares. Weigh a sample of 1.0 g into the boiling flask and complete the extraction as directed for juice.

NOTE: Handle pads or tissues as little as possible.

REMOVAL OF INTERFERING SUBSTANCES BY OXIDATION WITH POTASSIUM PERMANGANATE

Place 25 ml of the juice extract in heptane, or 1–5 ml diluted to 25 ml in heptane of the fruit, peel, carton, or pad extracts, in a 250 ml, glass-stoppered bottle. (The amount of dilution depends on the level of biphenyl present, which may vary by several orders of magnitude.) Add 50 ml of acetic-permanganate solution, tie the stopper in place, and shake the mixture vigorously on a mechanical shaker for thirty minutes. Transfer the mixture to a 250 ml separatory funnel and withdraw the acid layer. Wash the sides of the funnel with water and withdraw the washings. Shake the heptane layer with 25 ml of 5% sodium carbonate solution, and separate the heptane layer.

Determine the biphenyl spectrophotometrically at 248 mμ against solvent which was treated in the same way. Compare the readings against a standard curve plotted from readings obtained from known amounts from 2 to 25 ppm of biphenyl. Refined biphenyl may be obtained from Aldrich Chemical Co., Inc., 2371 N. 30th St., Milwaukee, Wisc. 53210. Express the biphenyl content in appropriate units.

Monsanto has also developed an infra-red spectrophotometric method in which the diphenyl is extracted with cyclohexane and the absorbance of the final extract is measured at 13.75μ and at 14.29μ. See *Monsanto Information Bulletin* No. 2.[26]

Antioxidants

There are 4 compounds generally used to retard oxidation, and thus inhibit development of rancidity in oils and fats and fatty foods. These are: butylated hydroxyanisole (BHA), butylated hydroxytoluene (BHT), propyl gallate (PG), and nordihydroguaiaretic acid (NDGA). These are all characterized by the U.S. Food and Drug Administration as "generally recognized as safe", in amounts not exceeding 0.02% of the fat or oil present, including the volatile oil content of the food.[28]* They are also used widely in Canada, being permitted in lard and shortenings. Combinations of PG and NDGA are not permitted.[29]

Chemists of the Kansas City District of the Food and Drug Administration have done considerable work on the identification and estimation of the 4 antioxidants mentioned above. Earlier work was also reported by Mahon, Anglin, Chapman, and Sahasrabudhe

*NDGA was deleted from this list by FDA. See Fed. Reg. April 11, 1968.

of the Canadian Food and Drug Directorate [30,31,32,33,34,35,36].

Qualitative Tests for PG, NDGA, BHA and BHT in Oils, Fats and Waxes.[32] (These are AOAC Methods **26.105-26.107.**)

REAGENTS

a. 1% *Barium hydroxide.* 1 g of $Ba(OH)_2$ in 100 ml of freshly boiled distilled water. Keep in a tightly stoppered bottle.

b. *Ehrlich reagent (diazobenzene sulfonic acid).* Mix 1 part of 0.5% aqueous solution of $NaNO_2$ with 100 parts of 0.5% solution of sulfanilic acid in HCl (1 + 19). Prepare the $NaNO_2$ solution fresh every 3 weeks. Keep both solutions refrigerated. Mix the $NaNO_2$ and sulfanilic acid solutions freshly each day in the proportion of 1 to 100.

c. *Dianisidine solution.* Dissolve 250 mg of dianisidine (3,3′-dimethoxybenzidine) in 50 ml of anhydrous methanol. Add 100 mg of activated charcoal, shake 5 minutes and filter. Mix 40 ml of the clear filtrate with 60 ml of 1 *N* HCl. Prepare daily and protect from light.

d. *Activated Florisil adsorbent*—60—100 mesh. Activated by the manufacturer at 260° or 650° (available from Floridin Co., 2 Gateway Center, Pittsburgh, Pa. 15222).

Test the Florisil for BHT retention as follows: Add 0.2 mg of BHT in 25 ml petroleum ether to the prepared column below, elute with 150 ml of petroleum ether, and apply the BHT test, after evaporating the ether just to dryness, as in Method 14-10D. If BHT is not eluted, activate the remaining Florisil by heating 2 hours at 650°, cool, add 6.5% of H_2O by weight, and homogenize by shaking 1 hour in a closed container.

PREPARATION OF FLORISIL COLUMN FOR CLEANUP OF BHT EXTRACT

Insert a small glass wool plug into a chromatographic column 25 cm long × 20 mm in diameter with a Teflon stopcock, and add about 12 g of Florisil with gentle tapping. Wash with two 15 ml portions of petroleum ether, adding the second portion when liquid level drops to just above the top of the Florisil. When the level of the second portion is about 1 cm above the Florisil, close the stopcock. Do not let column become dry.

Tests

Method 14-10A. *Propyl gallate (PG).*

Weigh about 30 g of fat (melted by gentle warming) or oil, dissolve in about 60 ml of petroleum ether, and transfer to a 250 ml separatory funnel. Add 15 ml of H_2O and shake gently 1 minute. Let separate and drain the aqueous phase into a 125 ml separatory funnel, leaving any emulsion in the organic phase. Repeat the extraction of petroleum ether with 2 additional 15 ml portions of water and reserve the petroleum ether solution for further extraction with acetonitrile (**14-10B**). Add 15 ml of ether to the combined aqueous extractions and shake 1 minute. Discard the aqueous phase and evaporate the ether just to dryness in a small beaker. Add 4 ml of 50% alcohol to the residue, swirl, and add 1 ml of NH_4OH. If the solution turns rose, propyl gallate is present. (The color is unstable and fades after a few minutes.)

Method 14-10B. *Nordihydroguaiaretic acid (NDGA).*

Extract the petroleum ether solution from Method 14-10 A by shaking 2 minutes with 20 ml of acetonitrile. Let the layers separate and drain the acetonitrile into a 1 L separator. Repeat the extraction with 2 additional 30 ml portions of acetonitrile and discard the petroleum ether. Dilute the combined acetonitrile portions with 400 ml of water, add 2–3 g of NaCl, and shake 2 minutes with 20 ml of petroleum ether. Let the layers separate and drain the diluted acetonitrile into a second 1 L separator. Extract the dilute acetonitrile with 2 additional 20 ml portions of petroleum ether and reserve the diluted acetonitrile solution for further extraction. Combine the petroleum ether portions in a 100 ml beaker and set aside for BHA and BHT tests.

Add 50 ml of ether-petroleum ether (1 + 1) to the diluted acetonitrile and shake 2 minutes. (*Caution:* Vent the separator.) Let the layers separate, discard the acetonitrile, and evaporate the ether just to dryness in a small beaker. Add 4 ml of 50% alcohol, swirl, and then add 1 ml of 1% $Ba(OH)_2$ solution and mix. If NDGA is present, the solution turns blue and fades rapidly.

Method 14-10C. Butylated hydroxyanisole (BHA).

Take one-third of the combined ether solutions reserved for the BHA-BHT tests (*see* Method 14-10 B) and evaporate just to dryness in a small beaker, using gentle heat, under an air current. Add 2.5 ml of alcohol to dissolve the residue and dilute with 2.5 ml of water. Swirl, add 1 ml Ehrlich reagent, immediately add 1 ml of 1 *N* NaOH, and swirl again. If the solution turns red-purple, BHA is present.

Method 14-10D. Butylated hydroxytoluene (BHT).

Pass the remaining two-thirds of the combined ether solutions through the Florisil column and elute with 150 ml of petroleum ether. Collect the eluate in a 200 ml beaker and evaporate just to dryness. Add 2.5 ml of alcohol, swirl, and dilute with 2.5 ml of water. Add 2 ml of the dianisidine solution and mix. Add 0.8 ml of 0.3% $NaNO_2$ solution. Mix, and let stand 5 minutes; then transfer to a small separatory funnel. Add 0.5 ml of $CHCl_3$, shake vigorously 30 seconds, and let separate. If the $CHCl_3$ turns pink to red, BHT is present. Confirm BHT by comparing the spectrophotometric curve of the colored $CHCl_3$ extraction with a control prepared from reference standard BHT as follows: Dissolve about 15 mg of BHT in 5 ml of aqueous alcohol (1 + 1), add 2 ml of dianisidine solution, and proceed as above.

Quantitative Determination of Antioxidants

Some work has been done in recent years on quantifying antioxidants in foods (text references 30 to 36, inclusive). Scheidt and Conroy[34] detected amounts as low as 1μg of BHT, BHA, PG and NDGA, while Sahasrabudhe[36], using TLC technique, quantitatively separated these 4 antioxidants, along with the 2 isomers of BHA, from one aliquot of fat.[38]

Takahashi, using argon ionization gas chromatography, estimated BHA and BHT in cereals.[38,39] Anderson and Nelson also developed methods for this estimation.[37] McCauley and co-workers[41] have developed a method for multiple determination of BHT, BHA, PG, Ionox-100, thiodipropionic acid (TDPA) and 2,4,5-trihydroxybutyrophenone (THBD), using vacuum sublimation and UV, IR and mass spectrometry. Mitchell[40] published a paper on chromatographic identification of BHT, BHA, PG and NDGA, alone or in combination.

Most, if not all, of the methods mentioned in this section are, or will be, subjected to further collaborative study. Such study will most likely result in some modifications. We are not giving the details of these methods in their present form, since they may be obsolete before this text is published. Those interested should consult the original papers and watch for further reports as published in forthcoming issues of scientific journals.

The Meat Inspection Division of USDA[42] uses the following control method for the determination of BHA in lard, adapted from Mahon and Chapman's procedure:[43]

Method 14-11. Butylated Hydroxyanisole in Lard and Shortening.

Reagents

a. 2,6-*Dichloroquinonechlorimide*, 0.01% solution in absolute ethyl alcohol. This reagent must be freshly prepared for optimum results.

b. *Borax buffer*, 2.0% acqueous solution of $Na_2B_4O_7.10H_2O$.

c. *Alcohol*, 72% ethyl alcohol by volume.

d. *Butylated hydroxyanisole standard solution*, containing 10 μg in 72% alcohol.

Determination

Place 10 g of the fat in a 300 ml separatory funnel and dissolve in 50 ml of petroleum ether. Extract the petroleum ether solution of the fat by shaking the contents of the funnel with three separate 25 ml aliquots of 72% ethyl alcohol (**c**) for 3 minutes per extraction. Make a fourth extraction, using 60 ml of 72% ethyl alcohol and shaking for 1 minute. Combine the 4 extracts, dilute to a volume of 200 ml with 72% ethyl alcohol, and filter through a folded filter. The clear alcohol extract contains the butylated hydroxyanisole. Place two suitable aliquots of the extract (5 ml for 0.01% butylated hydroxyanisole) in separate 18 mm colorimeter tubes and dilute to

10 ml with 72% ethyl alcohol. To each tube add 2 ml of freshly prepared 0.01% 2,6-dichloroquinonechlorimide reagent (**a**). Mix the contents of the tubes, add 2 ml of 2% aqueous borax solution (**b**), and mix the contents again. Prepare a blank containing 10 ml of 72% ethyl alcohol plus the reagents. After 15 minutes measure the absorbancy relative to the blank in a spectrophotometer at 620 mμ. Calculate the amount of butylated hydroxyanisole by reference to a calibration curve prepared over the range of 10 to 50 μg of butylated hydroxyanisole (**d**) per tube, using the same butylated hydroxyanisole preparation that was added to the fat. (It is necessary to prepare a new calibration curve for each type or batch of antioxidant preparation used.)

Method 14-12. *Determination of Propyl Gallate in Lard.*

The rapid control method used by USDA Meat Inspection Division for propyl gallate was adapted from a spectrophotometric method developed by Kahn[44], and later adopted by AOAC as Method **26.108–110:**

REAGENTS

a. 2% *Ammonium acetate in distilled water.*

b. 0.04% *Ferrous sulfate* ($FeSO_4.7H_2O$) *in distilled water.* Prepare just before use.

c. *Ethyl alcohol.* Purify by adding approximately 0.1% potassium hydroxide and 0.1% potassium permanganate to absolute alcohol and distilling in all glass apparatus.

d. 1% *sodium chloride solution in 25% (vol.) ethyl alcohol* (use purified alcohol).

e. *Standard solution of propyl gallate*—50 μg/ml of Eastman No. P6265 or equivalent propyl gallate in 25% alcoholic salt solution (**d**).

DETERMINATION

Add 20 g of fat to a 250 ml separatory funnel and put in solution with 40 ml of carbon tetrachloride, warming if necessary. Add quantitatively 100 ml of the 1% sodium chloride solution in 25% alcohol (**d**). Shake gently for five minutes. Let stand until separation is complete. Discard the CCl_4 and transfer most of the alcohol portion to a centrifuge tube and spin until completely clear. Pipette an aliquot (should contain 60–600 μg of propyl gallate) of the clear solution into a 25 ml volumetric flask and make to approximately 20 ml with 1% salt solution in 25% alcohol (**d**). Add 2 ml of 2% ammonium acetate solution (**a**) and 2 ml of 0.04% ferrous sulfate solution (**b**), mix and make to volume with alcohol-salt solution. Develop the color for 10 minutes, transfer to a cuvette and read the optical density at 515 mμ wave length. Zero the spectrophotometer by using a reagent blank of 21 ml of alcohol-salt solution, 2 ml of 2% ammonium acetate solution and 2 ml of 0.04% ferrous sulfate solution. Calculate the percentage of propyl gallate from a curve made by carrying known amounts of propyl gallate (**e**) through the procedure.

PREPARATION OF STANDARD CURVE

Add 0, 2, 4, 6, 8, 10 and 12 ml aliquots of the standard propyl gallate solution to 25 ml volumetric flasks and carry through the above procedure, beginning where the aliquot is pipetted from the centrifuge tube. Read the optical densities and plot the curve. A curve prepared in this manner is identical to one prepared by adding known amounts of propyl gallate to lard and running through the procedure.

Artificial Sweeteners

Until recently two artificial sweeteners and their derivatives were permitted in foods in the United States and Canada, provided such foods met label requirements: the cyclamates (salts of cyclohexylsulfamic acid) and saccharin (o-benzoicsulfimide) and its sodium, calcium and ammonium salts. Dulcin (4-ethoxyphenylurea) and P-4000 (5-nitro-2-propoxyaniline) have not been permitted for many years. Cyclamates are generally considered to be 30 times as sweet as sucrose and saccharin 300 times as sweet, while dulcin and P-4000 have sweetness factors of 250 and 4000 respectively.

In 1969, after certain laboratory tests on animals had suggested that cyclamates or their metabolite cyclohexylamine could be carcinogenic if ingested at high levels, the U.S. Food and Drug Administration limited the proportions of cyclamates in foods to amounts that would yield total daily consumptions not exceeding 3500 milligrams for adults and 1200 milligrams for children. Later regulations provided that these substances be totally excluded from foods as of the following dates in 1970: January 1, beverages and beverage mixes; February 1, all foods not otherwise excepted; April 1, jams, jellies, desserts, ice cream, meal substitutes; July 1, dietary supplements (vitamin-mineral combinations), drugs (for non-therapeutic use); September 1, canned fruits and vegetables, fruit and vegetable juices, lemonade concentrate, non-carbonated fruit drinks, ice tea mixes.[45]

Qualitative Tests for Artificial Sweeteners

Method 14-13A. *Cyclamates.*

To a 100 ml sample of a beverage, or a 100-140 g/liter aqueous dilution of the decanted or filtered liquid from a canned fruit, add 2 g of 10% $BaCl_2$ solution. Let stand 5 minutes and filter. Add 10 ml of HCl and 0.2 g of $NaNO_2$ to the clear filtrate. A white precipitate of $BaSO_4$ indicates the presence of cyclo-hexylsulfamates (sucaryl).

Method 14-13B. *Dulcin.*

Extract a 100 ml sample, prepared as in Method 14-13 A, and made alkaline with 10% NaOH solution if necessary, with 2 or 3 50 ml portions of ethyl ether. Divide the ether extracts equally between 2 porcelain evaporating dishes and let the ether evaporate at room temperature. Dry the residues for 10–15 minutes in an oven at 105–110° C.

Moisten residue in the one dish with HNO_3 and add 1 drop of water. Presence of dulcin is indicated by a brick-red or orange precipitate.

For a confirmatory test, expose the dry residue in the other evaporating dish to HCl fumes for 5 minutes and add 1 drop of anisaldehyde. An orange-red to deep red color indicates presence of dulcin. This test will detect 25 mg/liter.

Method 14-13C. *P*-4000.

Add 10% NaOH solution to about 200 ml of a liquid food, or to the aqueous extract of about 100–200 g of a solid or semi-solid food, until alkaline (pH 7.5–8). Extract with three 50 ml portions of petroleum ether. Wash the combined extracts with water. Transfer the washed extracts to a beaker, add 4 ml of HCl (1 + 1), and evaporate the solvent. Add a small piece of mossy tin and let stand on the steam bath for 5 minutes. Decant the liquid into a test tube and add bromine water, drop by drop. A rose-red to a deep burgundy color, disappearing with excess bromine water, indicates the presence of P-4000.

Method 14-13D. *Saccharin.*

Acidify the sample, or aqueous extract with HCl and extract with 2 or 3 portions of ethyl ether, wash the extracts once with 5 ml water and evaporate the combined ether extracts to dryness spontaneously at room temperature. (If the residue tastes sweet, saccharin is indicated.) Dissolve the residue in a little hot water and test a small portion for salicylates as in the confirmation $FeCl_3$ test, Method 14-1B.

Dilute the remainder of the solution to about 10 ml, add 2 ml of H_2SO_4 (1+3), and heat to boiling. Add a slight excess of 5% $KMnO_4$ solution dropwise, partly cool, and dissolve about 1 g of NaOH pellets in it. Filter into a silica dish. Evaporate to dryness and heat for 20 minutes at 210–215° C. Dissolve the residue in water, acidify with HCl and test the ether extract for salicylates as in Method 14–1B. This treatment eliminates any salicylates naturally present.

NOTE: Mitchell[46] has worked out a rapid paper chromatographic method for the simultaneous detection of these artificial sweeteners.

Cyclohexylsulfamic Acid and Salts (Cyclamates)

Alkali, calcium and magnesium cyclohexylsulfamates were exempted from the requirement of tolerances as additives as being "generally recognized as safe" (GRAS), but see above. These compounds were used as artificial sweeteners and flavor-enhancers. Their use had been proposed in canned fruit for which a Federal standard of identity was issued.[47,48] These were sold under the trade name of Sucaryl.

Method 14-14. Determination of Cyclamates in Foods.

(This is AOAC Method **27.016** for clear carbonated beverages. It was adapted by Wilson[49] from a method furnished by Abbott Laboratories, based on conversion of cyclamates to sulfates by nitrous acid.)

Cyclamates were generally used in the range of 0.20–0.30% in beverages, and up to 0.50% in canned fruits. The latter is about equivalent to 15 percent of sucrose. They were permitted for use in dry dessert mixes up to 2–5%, along with Saccharin.

PREPARATION OF SAMPLE

Fruits. Blend a representative sample in a Waring blendor, and dilute so that 100 ml represents 10–300 mg of sodium or calcium cyclamate. Centrifuge or filter, and dilute to 500 ml or 1000 ml in a volumetric flask.

Beverages. If clear, use as is; if cloudy, filter until clear.

DETERMINATION

To a 100 ml clear aliquot containing 10–300 ml of the cyclamate salt, add 10 ml of HCl and 10 ml of 10% $BaCl_2$ solution. Stir, and let stand 30 minutes. Filter clear if a precipitate forms, and wash the precipitate with water. Add 10 ml of 10% $NaNO_2$ solution to the filtrate. Stir, cover with a watch glass, and heat on a steam bath for at least 2 hours with occasional stirring. Then leave in a warm place overnight (on top of a 100° oven is suitable).

Filter through a tared crucible, wash, then place the crucible on an asbestos mat over a Bunsen flame for at least 10 minutes, ignite, cool in a desiccator and weigh.

$BaSO_4 \times 0.8621$ = sodium cyclohexylsulfamate,

$BaSO_4 \times 0.9266$ = calcium cyclohexylsulfamate.

Dulcin and P-4000[50]

Since dulcin and P-4000 are non-permitted sweeteners, identification by qualitative tests (Methods 14-13B and 14-13C) is usually sufficient. If quantitative determination is required, AOAC Method **27.023** may be used for dulcin and AOAC Method **27.045** for P-4000.

Method 14-15. Saccharin.

This method, involving separation of saccharin from the food by ether extraction and evaporation, fusion of the residue with alkali carbonates to form sulfates, and precipitation of the sulfates with $BaCl_2$, has long been used. It has also been adopted by AOAC as an official method.

PREPARATION OF SAMPLE

a. *Beverages and beverage bases, fruit juices and syrups*: Transfer a 100–200 ml sample to a 250 ml volumetric flask. (Pour carbonated beverages back and forth several times in beakers to remove CO_2.) For alcoholic liquids, dealcoholize the measured sample by evaporation on the steam bath to about half volume; in the case of heavy alcoholic syrups, dilute with an equal volume of water before beginning the evaporation. Transfer to a 250 ml volumetric flask. Add 5 ml of acetic acid, mix, then add a slight excess of 20% neutral lead acetate solution, mix thoroughly, dilute to the mark, remix and filter.

b. *Solid and semi-solid samples:* Transfer 50–75 g to a 250 ml volumetric flask with the aid of a little hot water and dilute with boiling water to about 200 ml. Let stand 2 hours, shaking occasionally, and add 5 ml of acetic acid and a slight excess of 20% neutral lead acetate solution. Dilute to the mark, let stand 20 minutes and filter.

DETERMINATION

Transfer a 150 ml aliquot of the filtrate to a separatory funnel, add 15 ml of HCl and extract with three 80 ml portions of ethyl ether, shaking for 2 minutes each time. Wash the combined extracts once with 5 ml of water, evaporate the ether, and transfer the residue by the aid of a little ether to a platinum crucible. If difficultly soluble substances are present, use small portions of water and ether alternately. Remove the ether by evaporation on a steam bath and add sufficient 10% Na_2CO_3 solution (usually 2–3 ml) to make the solution strongly alkaline. Rotate so that all the residue comes in contact with the liquid, and evaporate to dryness on a steam bath.

Add a weighed 4 g amount of a mixture of equal parts by weight of Na_2CO_3 and K_2CO_3. Heat gently at first and then to complete fusion. This should take about 30 minutes. (A convenient method of fusion is to insert the crucible into a closely-fitting hole cut into a square of heavy asbestos board, so that about one third of the crucible projects above the board, and heat the lower part of the crucible with a Bunsen or Meker burner).

Cool, dissolve the melt in water, add about 5 ml of bromine water, acidify with HCl, filter, wash the filter paper with a little water, dilute the filtrate to about 200 ml, heat to boiling, and slowly add an excess of 10% $BaCl_2$ solution. Let the mixture stand overnight, filter off the precipitated $BaSO_4$, wash until free from chlorides, dry, ignite in a tared crucible, cool and weigh. Run a blank on 4 g of the fusion mixture used and correct the weight of $BaSO_4$ accordingly.

Corrected weight of $BaSO_4 \times 0.7848$
= weight of saccharin.

Saccharin × 1.317
= weight of saccharin sodium.

TEXT REFERENCES

1. Harrigan, M. C.: *J. Assoc. Offic. Agr. Chemists*, **41**: 584 (1958).
2. Monier-Williams, G. W.: *Reports on Public Health and Medicine, Subject No.* 43. London: Ministry of Health (1927).
3. Thompson, J. B. and Toy, E.: *Ind. Eng. Chem., Anal. Ed.*, **17**: 612 (1945).
4. Thrasher, J. J.: *J. Assoc Offic. Agr. Chemists*, **45**: 905 (1962).
5. Thrasher, J. J.: Ibid., **49**: 834 (1966).
6. Nury, F. S., Taylor, D. H. and Brekke, J. E.: *Agr. Food Chem.*, **7**: 351 (1958).
7. Nury, F. S. and Bolin, H. R.: *J. Assoc. Offic. Agr. Chemists*, **48**: 796 (1965).
8. Schwartzman, G.: *J. Assoc. Offic. Agr. Chemists*, **49**: 79 (1966).
9. Pinella, S., Fallo, A. D. and Schwartzman, G.: *J. Assoc. Offic. Agr. Chemists*, **49**: 829 (1966).
10. Calloway, G. F. and Schwartzman, G.: *J. Assoc. Offic. Agr. Chemists*, **48**: 794 (1965).
11. Thompson, J. B.: *J. Soc. Chem. Ind.*, **12**: 432 (1893).
12. Woodman, A. G.: *Food Analysis*, 4th ed. New York-London: McGraw-Hill Book Co. (1941) 110.
13. *Code of Federal Regulations, Title* 21—*Food and Drugs.* Parts 1 to 129, Sec. 19.500(d). Washington: Government Printing Office (as of Jan. 1, 1966).
14. Ibid. Sec. 45.1(c)(xi).
15. Ibid. Sec. 121.101(d)(2).
16. *Office Consolidation of the Food & Drugs Act and Regulations:* Table XI, Parts 2, 3. Ottawa: Queen's Printer (1966).
17. Ciaccio, L. L.: *Food Technol.*, **20**: 73 (1966).
18. Ma, R. M. and Morris, M. P. Food and Drug Adm., Dept. of Health Education and Welfare: Private communication (1965).
19. Ref. 13. Sec. 121.1089.
20. Wesley, F., Rourke, B. and Darbishire, O.: *J. Food Sci.*, **30**: 1037 (1965).
21. Ragelis, E. P., Fisher, B. S. and Klimeck, R. A.: *J. Assoc. Offic. Agr. Chemists*, **49**: 963 (1966).
22. Wesley, F.: Private communication (1967).
23. Sair, L. Vice President, Griffith Laboratories, Chicago, Ill.: Private communication (1967).
24. Lubatti, O. F.: *J. Soc. Chem. Ind.*, **54**: 421-T (1935).
25. Ref. 13. Sec. 120.141.
26. "Biphenyl for Citrus Fruit Preservation". *Organic Chemicals Intermediates, Information Bulletin No.* 2. St. Louis: Monsanto Company, Organic Chemicals Div. (undated).
27. Dickey, E. E. and Green, J. W. Institute of Paper Technology: Private communication (1967).
28. Ref. 13, Sec. 121.101(d)(2).
29. Ref. 16. Reg. B. 16.100, Table XI, Part 4 and Reg. B.16.005.
30. Heidrick, P. L. and Conroy, H. W.: *J. Assoc. Offic. Agr. Chemists*, **45**: 244 (1962).
31. Schwein, Wm. G. and Conroy, H. W.: *J. Assoc. Offfc. Agr. Chemists*, **48**: 489 (1965).
32. Scheidt, S. A. and Conroy, H. W.: *J. Assoc. Offic. Agr. Chemists*, **49**: 807 (1966).
33. Schwein, Wm. G., Miller, B. J. and Conroy, H. W.: *J. Assoc. Offic. Agr. Chemists*, **49**: 809 (1966).
34. Mahon, J. H., Chapman, R. A.: *Anal Chem.*, **23**: 1106, 1120 (1951), **24**: 534 (1952).
35. Anglin, C., Mahon, J. H. and Chapman, R. A.: *J. Agr. Food Chem.*, **4**: 1018 (1952).
36. Sahasrabudhe, M. R.: *J. Assoc. Offic. Agr. Chemists*, **47**: 888 (1964).
37. Anderson, R. H. and Nelson, J. P.: *Food Technol.*, **17**: 915 (1963).
38. Takahashi, D. M.: *J. Assoc. Offic. Agr. Chemists*, **47**: 367 (1964), **48**: 694 (1965).

39. Takahashi, D. M.: Ibid., **49**: 704 (1966).
40. Mitchell, L. C.: *J. Assoc. Offic. Agr. Chemists*, **40**: 909 (1957).
41. McCauley, D. F., Fazio, T., Howard, J. W., Di Ciurcio, M. and Ives, J.: *J. Assoc. Offic. Agr. Chemists*, **50**: 243 (1967).
42. Philbeck, R. H. Consumer and Marketing Service, U.S.D.A.: Private communication (1967).
43. Mahon, J. P. and Chapman, R. A.: *Anal. Chem.*, **23**: 1120 (1951).
44. Kahn, S.: *J. Assoc. Offic. Agr. Chemists*, **35**: 186 (1952).
45. Fowler, G. W.: Paper given at summer meeting of New England Food and Drug Officials, Bethel, Maine, June 4-5, 1970.
46. Mitchell, L. C.: *J. Assoc. Offic. Agr. Chemists*, **38**: 943 (1955).
47. 32 *Federal Register* (4/19/67) 6144.
48. Reference 13: Section 121.104 (d) (4).
49. Wilson, J. B.: *J. Assoc. Offic. Agr. Chemists*, **38**: 559 (1955).
50. Cox, W. F.: *J. Assoc. Offic. Agr. Chemists*, **40**: 785 (1957).

SELECTED REFERENCES

A. Bosund, I.: "Action of Benzoic and Salicylic Acids on the Metabolism of Micro-organisms". In: *Advan. in Food Res.*, **11**: 331. New York: Academic Press (1962).
B. Braverman, J. B. S.: "Sulfur Dioxide in Pretreatment of Fruits and Vegetables, Chemistry and Technology". In: *Advan. in Food Res.*, **5**: 97. New York: Academic Press (1954).
C. Desrosier, N. W.: *Technology of Food Preservation.* Westport: Avi Publishing Co. (1963).
D. Desrosier, N. W.: *Radiation Technology in Food, Agriculture, Biology.* Westport: Avi Publishing Co. (1960).
E. Dunn, C. G.: "Quaternary Ammonium Compounds". In: *Advan. in Food Res.*, **2**: 117. New York: Academic Press (1949).
F. Furia, T. E.: *CRC Handbook of Food Additives, Chapters 4, 5 and 14.* Cleveland: Chemical Rubber Co. (1968).
G. Jacobs, M. B.: *Synthetic Food Adjuncts.* New York: D. Van Nostrand Co. Inc. (1947).
H. Troller, J. A. and Olsen, R. A.: *J. Food Sci.*, **32**: 228 (1967).

C H A P T E R 15

Spices and Condiments

Introduction

Traffic in spices existed very early in the history of commerce. In some countries spices were even used as a medium of exchange. These spices were gathered from wild plants, air dried, and shipped throughout the then known world. Some spices lent themselves to cultivation and, slowly at first, to genetic improvement in species and varieties.

Early history of the traffic in spices reveals extensive contamination and sophistication in both whole and ground spices, the latter, of course, offering more leeway in concealment of the adulterant. During the twelfth century the Pepperers' Guild, a voluntary mercantile association, was formed. One of this guild's functions was to maintain the quality of the spices in which their members dealt. In England this later became the Grocers' Company or Guild and was given official status as the custodian of the weights standards; certain members of the guild were appointed as "garblers" or public food inspectors.

It was not until the science of analytical chemistry was developed, and the microscope became a practical tool for the identification of foreign substances, that consistent detection of adulteration became possible and the laws could be enforced. Modern food laws have greatly reduced the incidence of gross adulteration of ground spices with such things as sawdust, nut shells, roasted cereals, and foreign leaves, although such practices are occasionally encountered. Emphasis is now placed more on quality or grade and flavor profile of spices, with occasional spot checks for foreign material and evidences of contamination by insects, rodents, animal excrement, etc. Microscope examination by an analyst familiar with the histology of spices is recommended on shipments from new suppliers, with spot checks on occasional shipments from approved suppliers. In times of scarcity or disruption of ocean traffic, adulteration may again become more prevalent.

Specifications and Standards

There is no official standard under the current United States Food, Drug and Cosmetic Act of 1938. The Department of Agriculture, and later the Food and Drug Administration[1], published definitions and standards which were only advisory and had no legal status. These standards were largely for total ash, acid-insoluble ash and crude fiber, with occasional references to other constituents such as starch, non-volatile ether extract and foreign plant material.

Canada has issued official standards under its Food and Drugs Act[2] which do have the force and effect of law. These are given in Table 15-1. They are somewhat similar to, but not identical with, the U.S. advisory standards. Since the Canadian standards have been revised as late as 1961, they probably represent more nearly the composition of spices in the spice trade of today.

A good picture of the composition of present day spices is given in the *U.S. Federal Specification EE-S-631e*, dated December 27, 1970.[3] This defines the botanical species, gives quantitative requirements, methods of test, and specifies the sieving requirements for ground spices. The quantitative requirements

TABLE 15-1
Canadian Food and Drugs Act Standards for Spices

	Ash, % Max.	Acid–Insol. Ash, % Max.	Crude Fiber, % Max.	Other Constituents
Allspice	6	0.4	25	
Caraway	8	1.5	—	
Cardamom	8	3	—	
Cayenne	8	1.25	28	Starch, 1.5% max.; non-volatile ether extract, 15% min.
Celery Seed	10	2	—	
Cinnamon	5	2	—	
Cloves	8	0.5	10	Volatile ether extract, 15% min.; clove stems, 5% max.
Coriander	7	1.5	—	
Dill Seed	10	3	—	
Ginger	7.5	2	9	Calcium as CaO, 1% max.; ginger starch, 45% min.; cold water extract, 13.3% min.; water soluble ash, 2% min. (all on dry basis). Moisture, 10% max.
Ginger, Jamaica	7.5	2	9	Same as "Ginger", except cold water extract, 16.6% min.
Ginger, Limed	11	2	9	Calcium as CaO, 2% max.
Mace	3	0.5	7	Non-volatile petroleum and ethyl ether extracts, 33% max.
Marjoram	16	4.5	—	Stems and foreign material, 10% max.
Mustard Seed	8[a]	1.5[a]	—	
Mustard, Ground	8[a]	—	—	Starch, 1.5% max.; volatile oil, 0.4% min.
Nutmeg	5	0.5	—	Non-volatile ether extract, 25% min.
Paprika	8.5	1	23	Non-volatile ether extract, 18% max.; Hanus I., No. 125–136.
Pepper, Black	6	0.9	—	Non-volatile ether extract, 6% min.; pepper starch, 30% min.
Pepper, White	2.2	0.3	5	Non-volatile ether extract, 7% min.; pepper starch 52% min.
Thyme	12	4	—	
Celery Salt	—	—	—	Salt, 75% max., with or without anti-caking agent (2% max.).
Garlic Salt	—	—	—	*See* "Celery Salt".
Onion Salt	—	—	—	*See* "Celery Salt".

[a] Oil free basis.

TABLE 15-2
Requirements of Federal Specification EE–S-00631e (12/27/67)

Spice	Ash, % Max.	Acid–Insol. Ash, % Max.	Crude Fiber, % Max.	Moisture, % Max.	Vol. Oil, mg/100 g Min.	Other Requirements
Allspice	5.0	0.3	25.0	8.0	3.0	
Bay Leaf	4.5	0.5	30.0	7.0	1.0	Stem limit: same as sweet basil
Caraway	8.0	1.0	18.0	9.0	2.0	
Cardamom	5.0	3.0	30.0	9.0	3.0	
Cassia, Batavia	5.0	2.0	20.0	10.0	2.0	
Cassia, Saigon	6.0	2.0	20.0	10.0	3.0	
Celery Seed	11.0	1.5	—	10.0	2.0	Non-volatile ether extract, 12.0% min.
Cinnamon	See "Cassia"					
Cloves	6.0	0.5	10.0	8.0	15.0	Whole cloves, 5% stems max.
Coriander	7.0	1.0	25.0	9.0	0.3	
Cumin	9.5	1.0	7.0	9.0	2.5	Harmless vegetable matter, 5% max.
Fennel	9.0	1.0	14.0	7.5	2.0	
Fenugreek	—	—	—	6.0	—	
Ginger	6.0	1.0	8.0	12.0	2.5	Lime (CaO), 1% max.; starch, 42% max.
Ginger, Limed	10.0	2.0	8.0	12.0	2.5	Lime (CaO), 1% max.; starch, 42% max.
Mace	3.0	0.5	4.0	6.0	12.0	Non-volatile ether extract, 20–35.0%.
Marjoram	13.0	4.0	22.0	10.0	1.0	Whole spice, 10% stems max.
Mustard, White	5.0	1.0	22.0	—	—	
Mustard, Black	5.0	1.5	—	—	0.6	Vol. oil for mixture white & black, 0.4–0.7%; ash, 8.0% max.
Nutmeg	3.0	0.5	5.0	8.0	6.5	Non-volatile ether extract, 25% min.
Oregano	8.5	1.0	—	10.0	3.0	
Paprika, Hungary	6.5	0.5	23.0	11.0	—	Extractable color, 120–130 ASTA units
Paprika, Spain	8.5	1.0	21.0	11.0	—	Extractable color, 120–130 ASTA units
Paprika, U.S.	8.5	1.0	21.0	11.0	—	Extractable color, 120–130 ASTA units
Pepper, Black	7.0	1.0	12.5	12.0	2.0	Starch, 30% max.; N.V. methylene chloride extract, 7.0% min.
Pepper, White	2.0	0.3	5.0	15.0	1.5	Starch, 52% max.; N.V. ether extract 7% min.
Pepper, Red	8.0	1.0	28.0	7.0	—	Pungency, 12,000–15,000 Scoville units
Pepper, Cayenne	8.0	1.0	—	10.0	—	Pungency, 25,000–30,000 Scoville units
Rosemary	12.0	4.0	—	6.0	2.0	
Sage	10.0	1.0	—	10.0	1.5	Whole spice, 10% stems max., excluding petioles
Sweet Basil	15.0	1.0	—	8.0	0.4	Whole spice, 3% stems max., excluding petioles
Thyme	8.0	2.0	—	9.0	1.5	
Turmeric, Jamaica	7.0	0.5	6.0	9.0	4.0	Curcumin, 4.1–5.2%
Turmeric, Aleppy	7.0	0.5	6.0	9.0	4.0	Curcumin, 5.3–6.0%

TABLE 15-3
Botanical Characteristics of Spices

Spice	Botanical Name	Botanical Structure	Total Volatile Oil, ml/100 g	Principal Flavor Constituents
			RANGE	
Allspice	*Pimenta officinalis, Lindl*	Unripe berry	3.0–4.5	Eugenol & methyl eugenol
Anise	*Pimpinella anisum L.*	Fruit (seed)	1.5–3.5	Anethole
Bay Leaf (Laurel)	*Laurus nobilis L.*	Leaf	2	Cineole
Capsicums: Cayenne, Chili Pepper, Paprika	*Capsicum annuum L.* or *Capsicum frutescens L.*	Fruit	—	Capsaicin (non-volatile) carotenoids & oleoresin
Caraway	*Carum carvi L.*	Fruit (seed)	3.0–7.5	d-Carvonene, d-Limonene
Cardamom	*Elettaria cardamomum Maton*	Seed	2–8	Limonene, cineole, borneol
Cassia:				
China Cassia	*Cinnamomum cassia Bl.*	Bark	—	
Batavia Cassia	*Cinnamomum burmanni Bl.*	Bark	2–5	Cinnamic aldehyde,
Saigon Cassia	*Cinnamomum ioureirri Nees*	Bark	0.5–1.0	cinnamyl acetate,
Ceylon Cinnamon	*Cinnamomum zeylanicum Nees*	Bark	—	cinnamic aldehyde, eugenol
Celery Seed	*Apium graveolens L.*	Fruit (seed)	2–3	d-Limonene & selinene
Cinnamon, Ceylon	See "Cassia"			
Cloves	*Caryophyllus aromaticus L.*	Bud	12–20	Eugenol, eugenol acetate
Coriander	*Coriandrum sativum L.*	Fruit (seed)	0.1–1	d-Linalool, pinenes
Cumin	*Cuminum cyminum L.*	Fruit (seed)	2.5–4.5	Cumaldehyde, cymene
Dill	*Anethum graveolens L.*	Seed	2–4	Carvone, d-limonene, phellandrene
Fennel	*Foeniculum vulgare Miller*	Fruit (seed)	3–4	Anethole, fenchone
Fenugreek	*Trigonella foenum-graecum L.*	Seed	.02	—
Ginger	*Zingiber officinale Rosc.*	Rhizome	1–3	Zingiberene, phellandrene, d-camphene
Mace	*Myristica fragrans Houtt*	Seed coat of nutmeg	7–14	See nutmeg
Marjoram	*Majorana hortensis M.*	Leaf	0.7–3.5	Terpinene, terpineol
Mustard, White	*Sinapis alba L.*	Seed	0	Sinalbin (glucoside)
Mustard, Black	*Brassica nigra Koch*	Seed	0.5–1.0	Allyl isothiocyanate, sinigrin (glucoside)
Nutmeg	*Myristica fragrans Houtt*	Seed	7.15	Pinenes, d-camphene, dipentene, safrole
Oregano, Mexican	*Oreganum vulgare L.*	Leaf	—	Thymol, carvacrol, terpenes
Pepper, Black	*Piper nigrum L.*	Berry	1–3	Pinenes, 1-α-phellandrene, piperonal
Pepper, White	*Piper nigrum L.*	Mulled berry	—	
Peppermint	*Mentha piperita L.*	Leaf	0.4–1.0	l-Menthol, l-menthone, bipinene, pinene
Rosemary	*Rosmarinus officinalis L.*	Leaf	0.3–2.0	Borneol, pinene, camphene
Saffron	*Crocus sativus L.*	Stigma	0.5–1	—
Sage	*Salvia officinalis L.*	Leaf	About 2.5	Thujone, borneol
Savory	*Satureia hortensis L.*	Leaf	About 1	Carvacrol
Spearmint	*Mentha spicata L.*	Leaf	—	l-Carvone, limonene
Star Anise	*Illicium verum Hook*	Fruit	About 3	Anethole
Sweet Basil	*Ocimum basilicum L.*	Leaf	—	Methyl chavicol, cineole
Thyme	*Thymus vulgaris L.*	Leaf	2.5	Thymol, carvacrol
Turmeric	*Curcuma longa L.*	Rhizome	1.3–5.5	

of this specification are summarized in Table 15-2. The botanical and flavor characteristics of spices are given in Table 15-3. The information on "Principal Flavor Constituents" is abstracted from Parry's texts.*

General Methods of Analysis of Spices

All samples for analysis, if not already ground, should be ground to pass through a 20-mesh woven wire sieve, or one with 1 mm circular openings. The *Official Methods of Analysis* of the American Spice Trade Association (ASTA) recommend use of a Wiley mill. The ground sample should be mixed thoroughly by stirring or by rotating the container immediately before removing smaller portions for analysis (which should be dipped from the center of the material), since ground spices have a pronounced tendency to stratify. Prepared samples should be stored in tightly stoppered containers and kept in a refrigerator unless analysis is to be started immediately.

ASTA requires that nutmeg and celery seed be ground with a hand-operated pulverizing mill, that cassia be ground to pass through a U.S. #30 sieve, and that mace be blended with a mortar and pestle to effect an even distribution of oil if a Wiley mill is used for grinding. The AOAC further recommends that samples of spices used for the determination of starch by the diastase method be ground to an impalpable powder.

Moisture. See Method 1-4, Solvent Distillation, using toluene as the solvent. (ASTA directs the use of benzene for substances such as capsicum spices, chili powder, and other spices which contain larger amounts of sugars and other substances which may decompose at 100° to liberate water.)

Size of sample: Take an amount sufficient to yield 2 to 5 ml of water, usually about 50 g. For pepper, ginger, paprika, chili powder and chili pepper, use 30–35 g.

Karl Fischer Titration Method. See Method 1-6. Size of sample: 0.2–0.3 g.

This is a convenient, rapid method often used for routine quality control. See also the near-infrared spectrophotometric method 17-13 for dehydrated vegetables.

Total Ash. See Method 1-9. Size of sample: 2 g.

This is the AOAC method, following the proposed revision of Klemm.[4] The ASTA method is essentially the same, although less precise in its directions. The proposed AOAC method directs ashing at 550°, or 550–600°, depending on the spice being analyzed. ASTA directs ignition at 600° ± 20°.

Acid Insoluble Ash. See Method 1-10.

The ASTA method is practically identical, except for ashing at 600° instead of 625–650°.

Determination of total ash, and more particularly acid-insoluble ash, is of value in detecting abnormal amounts of sand and soil, as well as such refuse material as clove stems and pepper siftings. It should be part of a routine examination.

Protein (N X 6.25). *See Method 1-7. Size of Sample*: 2–4 g (for black or white pepper, and capsicum spices use 1 g).

Except for pepper, where it serves as a rough indication of the amount of piperine present, the determination of total nitrogen (crude protein) is of minor diagnostic value in the detection of adulterants of spices, Knowledge of protein content is of some importance in agronomy studies and other horticultural investigations. The seed spices, pepper and capsicums, are comparatively high in protein.

Method 15-1. *Crude fiber. Size of sample*: 2 g.

Follow the procedure for the determination of crude fiber, Method 1-11, through the final washing of the crude fiber mat with 1.25% H_2SO_4,

*See selected references at the end of this chapter.

water and alcohol. Wash 3 times with 15 ml of ether. Proceed as in Method 1-11, beginning "remove mat and residue and transfer to the ashing dish . . ."

The official ASTA method is essentially the same as the AOCS–AOAC Crude Fiber Liaison Committee method given in this text as Method 1-11, except that it does not provide for the correction due to loss of weight by ignition of the asbestos mat. This should be negligible (about 1 mg). The digestion and drying times and temperatures vary slightly in the two methods.

Crude fiber determination is of value in detecting refuse material such as pepper shell, mustard hulls, and clove stems, as well as ground fibrous adulterants such as olive pits, nut shells and sawdust.

Starch

While there is no method for the determination of starch as such in spices, there are two generally accepted methods, both depending on hydrolysis of starch to dextrose, that are in general use. Certain other carbohydrates, if present, are hydrolyzed also. The method depending on direct hydrolysis by dilute acid is comparatively simple, but may give results many times higher than does the diastase method, in spices such as cloves, cayenne and mustard that are very low in starch but high in fiber, tannins and pentosans (gums). These are also hydrolyzed by dilute acid into substances that reduce copper solutions to cuprous oxide. The acid hydrolysis method gives results comparable with the longer diastase method on spices that are high in starch and low in fiber, such as black and white pepper, ginger, and nutmeg.

Method 15-2. *Starch, Acid Hydrolysis Method.*

Weigh a 4 g sample into a fine porosity alundum crucible, or a funnel fitted with a Whatman No. 2 filter paper. Extract with five 10 ml portions of ethyl ether. Let the ether evaporate spontaneously, then extract with 150 ml of 10% ethyl alcohol, followed by 15–20 ml of absolute ethyl alcohol. Carefully wash the residue by means of a wash bottle into a 500 ml flask using 200 ml of water. Add 20 ml hydrochloric acid (sp.g 1.125. Reagent (**a**), Method 15-3) and hydrolyze; follow by reduction with Fehling solution as directed under Method 15-3, "Determination", third paragraph, beginning "connect with a reflux condenser."

Method 15-3. *Starch, Diastase Hydrolysis.*

This, the official ASTA method, determines as starch those carbohydrate materials which are hydrolyzed to dextrose by diastase followed by acid hydrolysis. The result is sometimes incorrectly referred to as "true starch". Except for minor procedural details, it is the AOAC diastase method for starch in spices, Method **28.012.**

APPARATUS

a. *Gooch crucible*—Poor sufficient acid-washed asbestos pulp into a Gooch crucible attached to a vacuum filter flask to make a mat of such thickness that the holes are just barely visible when looking through the crucible at a light source. Wash first with water to remove fine particles, then with ethanol and finally with ether. Dry at 110° for 30 minutes, cool in a desiccator and weigh.

b. *Water bath*—maintained at 55°.

REAGENTS

a. *Hydrochloric acid, sp.g.* 1.125—dilute 680 ml of HCl to 1 L with water.

b. *Ethyl ether*—anhydrous.

c. *Ethyl alcohol*—absolute.

d. *Sodium carbonate, about* 0.5 *N*—26.5 g of Na_2CO_3 diluted to 1 L with water.

e. *Sodium acetate solution*—dissolve 574 g of $NaC_2H_3O_2.3H_2O$ in water and dilute to 1 L.

f. *Malt diastase solution*—saturated solution of commercial malt diastase. (If preferred, this may be prepared from fresh barley malt as directed by AOAC Method **22.046**. The commercial malt diastase has proven satisfactory in our laboratories.)

g. *Soxhlet modification of Fehling's solution*:

1. *Copper sulfate solution*—34.639 g of $CuSO_4.\ 5H_2O$ in distilled water, diluted to 500 ml. Filter through asbestos.

2. *Alkaline tartrate solution*—dissolve 173 g of $NaKC_4H_4O_6.4H_2O$ (Rochelle salt) and 50 g of NaOH in water and dilute to 500 ml. Let stand for 2 days and filter through asbestos.

h. *Potassium iodide solution*—dissolve 42 g

of KI in water, and dilute to 100 ml. Make very slightly alkaline with Na_2CO_3 to avoid formation and oxidation of HI.

i. *Standard thiosulfate solution*—dissolve about 25 g $Na_2S_2O_3.5H_2O$ and 0.1 g of Na_2CO_3 in water. Dilute to 1 L. To standardize, accurately weigh 0.20–0.23 g of N.B.S. Standard Sample $K_2Cr_2O_7$ (dried for 2 hours at 110°) into a glass-stoppered flask, add 2 g of KI and dissolve in 80 g of Cl-free water. Add with swirling 20 ml of 0.1 *N* HCl and place in the dark for 10 minutes. Titrate with the standard $Na_2S_2O_3$ solution, adding starch indicator after most of the I has been consumed, as shown by the pale yellow color.

$$\text{Normality} = \frac{\text{wt.}K_2Cr_2O_7(\text{g})}{\text{ml } Na_2S_2O_3 \times 0.04904}.$$

DETERMINATION

Weigh 4 g of finely ground sample into a funnel containing a Whatman No. 2 or equivalent filter paper. Extract with five successive 10 ml portions of anhydrous ethyl ether. Discard the ether extracts. Allow the ether to evaporate from the residue on the filter, wash it with 150 ml of 10% alcohol, then with 15–20 ml of absolute alcohol.

Transfer the insoluble residue on the filter to a beaker, using about 50 ml of water. Gently rub the paper with a rubber policeman to aid the quantitative transfer. Place the beaker in boiling water and stir constantly for about 15 minutes, or until all starch is gelatinized. Cool to 55°, add 20 ml of malt diastase reagent (**f**), and hold at 55° for 1 hour. Make a spot test for starch with iodine solution. If a blue color results, repeat the malt diastase treatment.

Cool, transfer to a 250 ml volumetric flask and dilute to the mark with water. Filter, transfer 200 ml of the filtrate to a 500 ml glass-stoppered flask, add 20 ml of HCl reagent (**a**), connect with a reflux condenser, and heat in a boiling water bath for 2½ hours. Cool, nearly neutralize with 10% NaOH solution, then complete the neutralization with Na_2CO_3 reagent (**d**), using indicator paper. Transfer to a 500 ml volumetric flask, mix and dilute to the mark.

Filter the hydrolysate through a dry paper, discarding the first 10 ml portion of filtrate. Pipette 25 ml each of Fehling solutions (1) and (2) [reagent (**g**)] into a 400 ml beaker; add by pipette 50 ml of the filtrate (or 35 ml + 15 ml of water if the expected starch content is over 50%), and determine the reducing sugar by Munson and Walker copper-reduction method either by direct weighing of the cuprous oxide or by thiosulfate titration as described in Chapter 16, Method 16-6, "Determination (**b**)"

CALCULATIONS

Gravimetric Method: Find the mg of dextrose corresponding to the weight of Cu_2O from AOAC Table **43.012**, absence of sucrose.

Volumetric Method: Refer to Table 23-6. to find mg of dextrose corresponding to the weight of copper (Cu) determined by titration.

$$\text{Dextrose, } \% = \frac{\text{dextrose (mg)} \times 250 \times 500 \times 0.1}{\text{wt. of sample (g)} \times 200 \times \text{aliquot (ml)}}.$$

$$\text{Starch} = \%\ \text{dextrose} \times 0.90.$$

Method 15-4. *Volatile and non-volatile oils by ether extraction.*

This ether extraction method is not suitable for the determination of volatile ether extract in spices high in volatile oils, such as cloves, mace or nutmeg. It is valuable in determining non-volatile, or fixed, oils in such spices as celery seed, mustard and capsicums. The volatile oil portion of this method is included for those analysts who wish to compare their results with data on authentic spices given by Winton, Richardson, Leach, and other earlier workers (see the section on authentic spices in this chapter), and with earlier official standards which were established through use of this method. In general, one of the steam distillation methods described below (Methods 15-5 and 15-6) is preferable for determination of volatile oil in spices.

This method was developed in the U.S. Division of Chemistry, USDA, by Richardson in 1887 (Selected Reference H). It is still included in the *Official Methods* of the AOAC. The non-volatile oils portion of the method has been adopted by ASTA as an *Official Analytical Method.*

DETERMINATION

Weigh a 2 g sample of the ground spice into a paper extraction thimble or alundum crucible of medium porosity. Place the thimble in a Soxhlet or similar extraction tube, and extract with anhydrous ether (see Method 1-13 for preparation) for 20 hours, using a tared extraction flask. Remove the ether by placing the extraction flask in a forced draft hood at approximately room temperature until no appreciable ether odor is apparent. If total ether extract is to be determined, store the flask in a desiccator over freshly-boiled concentrated sulfuric acid for 18 hours, and weigh as total ether extract. Heat the extract gradually, finally in a drying oven at 110° (±2°), to minimum weight. (The difference between two final weighings a half hour apart should not exceed 1 mg.) This residue is the non-volatile ether extract, and the difference between this and the total ether extract is the volatile ether extract.

The extracted residue remaining in the thimble may be used for the determination of crude fiber.

Oils Volatile with Steam

The steam distillation method for volatile oils (for strict accuracy it should be designated as the method for water-insoluble steam volatile oils) was developed from distillation traps devised by J. F. Clevenger[5] in 1928. The oil may be separated from the trap and examined for its chemical and physical characteristics. The method was first adopted by the AOAC in 1934 after accumulation of data on authentic spices by Clevenger[6] and tests by collaborators[7]. Between 1934 and 1942 Clevenger collected authentic data on 20 spices, and supervised collaborative studies on his method. These are all reported in *J. Assoc. Offic. Agr. Chemists* for those years.

N. Aubrey Carson, as Associate Referee, improved the original Clevenger apparatus in several ways. Carson's collaborative studies were reported in *J. Assoc. Offic. Agr. Chemists* during the period 1950–1959. They form the basis of the present AOAC method and one of the ASTA methods, given here as Method 15-5.

Lee and Ogg[8] simplified the two Clevenger traps by making the graduated portions interchangeable with the adapter. The Lee and Ogg method has been adopted as an alternate Official Method by ASTA. It is given here as Method 15-6.

Method 15-5. *Water-Insoluble Steam Volatile Oils.* (*Official AOAC* and *ASTA Clevenger Trap Method.*)

APPARATUS*

All glassware should have glass joints.

a. 1 L *round bottomed, short necked flask*, with heavy-duty magnetic stirrer and Teflon coated stirring bar, preferably of ovoid shape.

b. *Heat source*, electric heating mantle, preferably without aluminum housing, connected to a variable voltage transformer.

c. *Clevenger type volatile oil traps:* Type A—lighter-than-water trap for oils with densities near or less than that of water. Type B—heavier-than-water trap for oils with densities greater than that of water. Scientific Glass apparatus Co. Nos. JP-5300 and JP-4800.

d. *West condenser*, 400 mm in length, with water-cooled drip tip.

DETERMINATION

Transfer enough prepared sample to yield 2–5 ml of oil, ground to pass through a U.S. #20 sieve and weighed to nearest centigram, into the round-bottomed flask. Add about 500 ml of water and a pea-sized pellet of an antifoam agent such as Dow Corning Antifoam A, or a light spray of Dow Corning Antifoam Spray. Insert stirring bar, enclose the flask in the electric heating mantle, attach the condenser and proper trap (filled with distilled water) and place on a magnetic stirrer. Start stirrer and apply heat through the mantle, with the transformer set at 90 volts (not over 3 amp). Distill at the rate of 1 to 2 drops a second until two consecutive readings at 1 hour intervals show no change in oil volume in the trap. (Not less than 6 hours distillation.) If the oil in the trap does not separate satisfactorily, or clings to the wall of the condenser, agitate the liquid by inserting a copper wire with a small loop at the end, through the condenser into the trap. If this fails, add a few drops of a saturated aqueous detergent solution through the top of

*Can be purchased from Scientific Glass Apparatus Co., Bloomfield, N.J.

the condenser and continue the distillation for 10 minutes, to wash the detergent from the trap.

If the density of the oil is nearly 1.000, as in cassia, or if the oil separates into two fractions of different density, as in allspice and nutmeg, add 1 ml of xylene, accurately measured to 0.01 ml, to the lighter-than-water trap. (Preferably this is done before distillation is started.) Cool, and read the volume of the collected oil to the nearest 0.02 ml, correcting for the amount of xylene, if any, added.

Calculation: % Volatile Oil (V/W)

$$= 100 \frac{\text{(Vol. of oil (ml)} - \text{Added xylene (ml) if any)}}{\text{Weight sample (g)}}.$$

Method 15-6. *Water-Insoluble Steam Volatile Oils. Alternate Official Lee and Ogg Trap Method.*

APPARATUS*

All glassware must have ₸ joints.

Lee and Ogg Type volatile oil traps: Type A—lighter-than-water trap for oil with density near, or less than, that of water. Type B—heavier-than-water trap for oils with density greater than that of water. (Scientific Glass Apparatus Co. No. JP-4800). Remaining apparatus same as in Method 15-5.

DETERMINATION

Transfer enough prepared sample, ground to pass through a U.S. #20 sieve, to yield 0.5–1.5 ml of oil (preferably over 1 ml), weighed to an accuracy of 5 mg, into the round-bottomed flask. Continue as in Method 15-5, beginning "add about 500 ml water . . ." Read the volume of oil in the trap to the nearest 0.005 ml.

NOTES ON STEAM DISTILLATION METHODS

1. A 2 L round-bottomed flask may be necessary for bulky leafy spices low in oil, such as sage or rosemary, if the Clevenger trap is used.

2. The trap and condenser should be thoroughly cleaned with dichromate-sulfuric acid cleaning solution, and rinsed with distilled water just before use to prevent adherence of oil droplets.

3. If the separated oil is required for further chemical or physical examination, drain off or pipette the oil from the trap into a glass-stoppered tube, let stand until clear or dry with a minute quantity of anhydrous sodium sulfate, and let settle before pipetting off portions for further test. Keep the separated oil in the refrigerator.

4. As a correction for solubility of xylene in water, and for small losses in distilling, ASTA directs that if xylene is added to the trap, 0.07 ml be added to the corrected volume of oil when the Clevenger trap is used, and 0.04 ml when the Lee and Ogg trap is used.

Method 15-7. *Non-Volatile Oils—Methylene Chloride Extract.* (This is the method designated by *Federal Specification EE-S-*631*d* for non-volatile oil in black pepper.)

DETERMINATION

Follow the procedure for volatile and non-volatile oils, Method 15-4, using reagent grade methylene chloride (dichloromethane), B.P. 40°, instead of anhydrous ethyl ether. Omit drying the extract in a desiccator over sulfuric acid. Instead, remove the solvent by placing the extraction flask on a steam bath until no solvent odor is apparent. Dry in a drying oven at 110° to minimum weight. Report the results as "percent non-volatile methylene chloride extract".

NOTE: The weighed residue may be reserved for determination of nitrogen by Method 1-7 as a measure of the crude piperine present.

Methods for Specific Spices

Capsicum Spices

Method 15-8. *Hanus iodine number of extracted oil of paprika.*

DETERMINATION

Transfer the 10 g weighed sample into a dry 200 ml glass-stoppered flask. Add 100 ml chloroform, measured by a pipette, into the flask, swirling the contents of the flask during its addition. Let stand 1 hour, shaking at 15 minute intervals, allow the spice to settle, and filter about 50 ml through a 12.5 cm folded filter paper, rejecting the first 10 ml. Pipette one 10 ml aliquot into a tared evaporating dish, remove the solvent by evaporation on a

* See footnote, p. 336.

steam bath, dry in an oven at 100° for 1 hour, cool and weigh. Use this weight as the sample weight in calculating the iodine number.

Transfer a second 10 ml aliquot of the cholroform extract (using the same 10 ml pipette) into a 500 ml glass-stoppered flask or bottle (the so-called "iodine flask" is convenient), add 25 ml of Hanus iodine solution and proceed as in Method 13-2, Iodine Absorption Number. Calculate the Hanus iodine number of the extracted oil. *Methods of Analysis*, AOAC 10th ed., states this should not be less than 130.

Method 15-9. *Extractable Color in Paprika.* (This is an official ASTA method).

APPARATUS

Spectrophotometer, capable of measuring accurately the absorbance at 450 mμ, with 1 cm matched cells or cuvettes.

REAGENTS

a. *Isopropanol*—at least 99% pure.

b. *Standard color solution*—0.500 mg/ml of reagent grade $K_2Cr_2O_7$ in 1.8*M* H_2SO_4.

c. 1.8 *M* H_2SO_4.

DETERMINATION

Weigh 0.100 g of ground paprika into a 250 ml Erlenmeyer flask (with unground capsicum, grind to pass through a U.S. No. 40 standard sieve). Pipette 100 ml of isopropanol (**a**) into the flask, stopper, swirl, and let stand at least 3 hours at 70°, or overnight (16 hours) at room temperature. Cool to room temperature, then filter through a Whatman #12 or equivalent paper, discarding the first 10 ml. Pipette 25 ml of the filtrate into a 50 ml volumetric flask, dilute to the mark with isopropanol, and mix. Determine the absorbance at 450 mμ, using 1 cm cells, against isopropanol as a blank.

Determine the absorbance of the standard color solution (**b**) at 450 mμ, against 1.8 *M* H_2SO_4 (**c**) as a blank.

CALCULATION

Absorbtivity (**a**) of standard color solution

$$= \frac{\text{absorbance of standard color solution at 450 m}\mu}{\text{cell length (cm)} \times \text{concentration (mg/ml)}}.$$

Extractable color, ASTA units

$$= \frac{\text{absorbance of extract at 450 m}\mu \times 200}{\text{cell length (cm)} \times \mathbf{a} \times \text{concentration of sample (mg/ml)}}.$$

NOTE: Other cell lengths may be used, but 1 cm cells are preferred.

Capsaicin in Capsicum Spices

Despite considerable research for an accurate, specific chemical method, no method has been developed, in our opinion, that meets the criteria for inclusion laid down in the preface. Most food chemists will not find need for such a method. In lieu of this, references are given (Text References 9, 10, 11, 12, 13, 14, 15, 16) for those chemists interested in this determination. These references contain extensive bibliographies which will be of aid in further literature search. Capsaicin content in capsicums ranges from 0.06 to 0.85 percent.[13]

The taste threshold method that follows is used by most spice houses, the U.S. Quartermaster Corps, British Pharmaceutical Society, and others:

Method 15-10. *Flavor pungency of hot red peppers.*

This is a taste threshold test-panel method, adapted from a method first published by Scoville in 1912[17], which is used in the spice trade and by *Federal Specification EE-S*-631*e*, *Spices*, as a measure of pungency for cayenne and other hot peppers. The dilution at which a sensation of pungency is first experienced in the mouth and throat is known as the Scoville value. It is of little value for determining the pungency of paprika and chili pepper, which contain relatively small amounts of the pungent principle. A review of this method may be found in *J. Am. Pharm. Assoc.* in 1924.[18]

DETERMINATION

Weigh 1.000 g of ground sample into a 100 ml glass-stoppered flask or bottle. Add 50 ml ethyl alcohol, measured by means of a pipette, and allow to stand 24 hours at room temperature or

3 hours in a water bath adjusted to 70°, with occasional shaking. Filter through a fast filter paper (S & S sharkskin or equivalent), catching the filtrate in a glass-stoppered flask.

Pipette 5 ml of the filtrate into a 100 ml volumetric flask, fill to the mark with 5% sucrose solution, and mix. This makes a 1 to 1,000 dilution.

PRELIMINARY TEST

Into a series of seven 100 ml graduated cylinders transfer successively 20, 10, 5, 2.5, 2, 1.25 and 1 ml portions of the 1:1000 dilution by means of a Mohr pipette or burette. Dilute each to 100 ml with 5% sucrose solution and mix. This will make a series of dilutions equivalent to Scoville values of 5,000, 10,000, 20,000, 40,000, 50,000, 80,000, and 100,000 Taste 5 ml of each in order, beginning with the most dilute solution, until the first definite sensation of pungency (burning in the throat) is noticed. Record the dilution as the Scoville value. Use 5 ml of 5% sucrose solution as a control. In all this tasting, the 5 ml portion being tasted should be tossed into the back of the throat and swallowed at once. This preliminary test, which can be done by one analyst, gives the approximate pungency range of the sample. If this information is sufficient for the purpose, the test should be repeated by a panel of five trained tasters, and three at least should obtain a definite sensation of pungency. Calculate the average, discarding any results obviously out of range.

If a more accurate value is desired, knowing the approximate Scoville rating, make a new series of appropriate dilutions, and repeat the tasting by the panel.

SUGGESTED DILUTIONS

Make a 1:5000 dilution by diluting 1 ml of the filtrate to 100 ml with 5% sucrose solution. If the preliminary Scoville value was 50,000 or more, transfer 5 ml of the 1:5000 dilution into each of a series of 100 ml graduated cylinders. Add 5% sucrose solution to each, the first diluted to 40 ml, the second to 45 ml, and so on in 5 ml increments, the last being diluted to 100 ml. Mix the contents of each cylinder thoroughly. This will give a series of Scoville values of 40,000, 45,000, etc., to 100,000.

For lower Scoville values, say 10,000 to 50,000, 10 ml of the 1:5000 dilution can be similarly diluted to volumes of 20 ml to 100 ml. For still lower Scoville ranges, say 2,000 to 20,000, a 1:1000 dilution (5 ml filtrate to 100 ml) can be further diluted to the proper range, *e.g.* 5 ml of 1:1000 dilution further diluted to 100 ml is equivalent to 20,000 Scoville, and if diluted to 50 ml is 10,000 and so on.

It is necessary to begin all taste tests with the most dilute solution, as an inadvertent tasting of a very pungent sample will dull the taste mechanism so that no more testing can be done for some time. Pure capsaicin has a Scoville value of about 15,000,000.

Seeds and Stems in Ground Capsicums

The U.S. Food and Drug Administration has adopted the general principle that seeds and stems in ground or comminuted vegetable products should not exceed the quantity of these tissues normal to the product. Chili pepper is made by grinding the whole chili pods. Paprika is sold as either "whole pod" or "seedless".

The following method* is based on the observation that acid phloroglucinol colors woody fiber pink or red, leaving other tissues unaffected. Since this method estimates the amount of woody tissues by *count* of the stained particles rather than by *weight*, the analyst should accumulate data on the counts found in samples of normal whole pods, and of seeded and stemmed pods, following the same procedure. The percent by weight of separated seeds and stems should also be recorded.

Method 15-11. *Stems and Seed in Ground Capsicums.*

REAGENT

0.5 g *phloroglucinol*, dissolved in a mixture of 50 ml of water and 50 ml of concentrated hydrochloric acid.

APPARATUS

a. *Low-form Petri dish*, with portion of rim broken off for ease in manipulation.

b. *Greenough, or wide-angle microscope.*

c. *Hand tally bacteria colony counter.*

*F. L. Hart: unpublished method.

DETERMINATION

Grind the sample so that practically all passes through a 40 mesh screen. Transfer about 50 g to a 400 ml beaker and extract by decantation at least 4 times with petroleum ether to remove oil and coloring matter. After the final extraction allow the product to dry by standing in air, screen through a 40 and 60 mesh screen combination, retaining only the portion that passes the 40 mesh and is retained on the 60 mesh screen. It is important that this screening be thorough, in order to eliminate the fines.

Rule $\frac{3}{8}$ inch squares in the centre of a 9 cm hardened filter paper, place the paper on a flat glass plate or Petri dish, and spread out evenly a very thin layer of the extracted and dried sample so that between 200 and 300 particles are in an area $\frac{3}{4}$ inch square (four of the $\frac{3}{8}$ inch squares). Using a magnification of about 7×, count the total number of particles in this area.

After counting the total number of particles, insert a few drops of phloroglucinol under the filter paper by inserting a medicine dropper through the broken portion of the rim, using just sufficient to completely wet the paper. Allow the paper to stand 3 to 5 minutes to develop the color. Both seed and stem particles are stained a characteristic pink color. The seed particles may be distinguished from stem by their rippled and wavy appearance under the microscope.

Count the seed particles and stem particles separately, using 20× to 30× magnification. Report the percent by count of each.

Normal California chili pods and paprikas contain from about 20 to 30% seeds, and 6 to 8% stems, depending on the size of the pods and the washing and cleaning procedures used commercially by the spice grinder. Small foreign paprika pods may contain a slightly higher amount of stems.

Pepper, Black or White

Method 15-12. *Crude Piperine.*

DETERMINATION

Transfer quantitatively into a 300–500 ml Kjeldahl flask the weighed methylene chloride residue obtained in Method 15-7, using warm chloroform. Remove the chloroform by evaporation on a steam bath and determine nitrogen by the Kjeldahl-Gunning-Arnold Method 1-7, using 0.1 *N* acid and alkali standard solutions for titration of the distilled ammonia. 1 ml of standard 0.1 *N* acid ≎ 0.0014 g of nitrogen or 0.02853 g of piperine.

The ASTA Official method is essentially the same as this, except that ASTA directs distillation of the ammonia into water containing 10 ml of saturated boric acid solution, and titration of the ammonia with standard 0.02 *N* hydrochloric acid, using a mixed indicator of methyl red and bromocresol green.

The AOAC Method[19] determines the nitrogen content of the non-volatile ether extract. This method is based on work done by Winton *et al.* in 1898 (Selected Reference L).

Piperine by Chemical Methods of Assay

Several methods have been developed for the determination of the pungent alkaloid of the pepper berry, piperine. These embrace the techniques of IR and UV spectrophotometry, column and thin-layer chromatography, colorimetry using cyclic aldehydes plus thiourea (the Komarosky reaction), and others. Graham of the University of Puerto Rico has reviewed these past investigations and proposed two methods.[20] One of these is included herein as Method 15-13:

Method 15-13. *Piperine in Black or White Pepper.*[20]

APPARATUS AND REAGENTS

a. *Beckman model DU spectrophotometer*, or equivalent, capable of measuring absorption at 635 mμ.

b. 200 × 16 mm *Pyrex test tubes*, permanently etched at the 25 ml mark.

c. *Fritted glass funnels*, medium porosity, or glass filtration columns, 150 mm × 5-20 mm (broken burettes will do).

d. *Piperine*, m.p. 129°–130°, m.w. 285.3 (supplied by K & K Laboratories, Inc., 121 Express Street, Engineers' Hill, Plainview, N.J), recrystallized three times from absolute methanol.

CONSTRUCTION OF STANDARD CURVE

Prepare a standard solution of recrystallized piperine in ethanol to contain 7 micromoles/ml.

Place 0, 0.1, 0.2, 0.3 to 1.0 ml in each of a series of 25 ml volumetric flasks. Remove the alcohol by placing the flasks in a boiling water bath for 10 minutes. Cool the flasks and fill to the mark with 85% phosphoric acid, mix well, and allow to stand at 28°–30° for 30 minutes, shaking every 5 minutes. Heat the flasks in a boiling water bath at 100° ± 1°. Cool, and allow to stand in a water bath at 28°–30° for 30 minutes. Measure the absorbance of each at 635 mμ. Plot absorbance (Y-axis) vs. concentration of piperine per ml of 85% phosphoric acid (X-axis) to form the standard curve.

DETERMINATION

Grind pepper to pass through a U.S. #60 to #80 standard sieve. Transfer 25 mg of the ground sample into each of 2 dry 25 ml volumetric flasks. Dilute to the mark with 85% phosphoric acid, shake well and allow to stand for 30 minutes at 28°–30°, shaking every 5 minutes. Filter through a medium porosity fritted glass funnel (or through a glass filtration column packed with glass wool).

Collect 10 ml of the filtrate, heat for 8 minutes in a boiling water bath at 100°, cool and allow to stand in a water bath held at 28°–30° for 30 minutes. Read the absorbance at 635 mμ against a reagent blank similarly treated. The blank should show less than 1% absorbance.

Standard Curve Equation:

$X = 0.4950\,Y - 0.0047$, where

X = millimols piperine per ml phosphoric acid, and

Y = absorbance at 635 mμ.

$$\%\ \text{Piperine} = \frac{X(0.28533)\ (\text{dilution factor}) \times 100}{\text{sample weight in mg}}.$$

Genest of the Canadian Food and Drug Directorate[21] separated piperine from its chief analog piperettine by UV spectrophotometry, and identified the analogs by paper and thin-layer chromatography.

We have included one other method, the ASTA method, because of its official status:

Method 15-14. *Piperine in Black or White Pepper. (ASTA Official Method).*

APPARATUS

Spectrophotometer capable of measuring absorbance at 580 mμ.

REAGENT

Chromotropic acid solution—Dissolve 20 g of 4,5-dihydroxy-2,7-naphthalene disulfonic acid in 200 ml of slightly warm water. If a black precipitate forms, remove it by filtration through a sintered glass filter or by centrifuging.

DETERMINATION

Weigh accurately about 0.5 g of the prepared ground sample into a 250 ml volumetric flask and add 100 ml of ethanol and boil 3 hours on a steam bath with occasional stirring. Cool and dilute to the mark with ethanol. Filter about 10 ml of the extract (see Note 1).

Transfer 1 ml of the filtrate into each of 2 25 ml glass-stoppered volumetric flasks. Add 0.5 ml of the chromotropic acid solution to one flask. Mark this "A_1" and mark the other flask "A_2". For a blank determination, add to a third 25 ml volumetric flask 1 ml of ethanol and 0.5 ml of chromotropic acid reagent. Add 10 ml of concentrated H_2SO_4 to all three flasks, rotating the flasks during addition to wash down the sides. Mix well and heat in a boiling water bath for 30 minutes. Seat the stoppers loosely for the first minute, then tightly for the remainder of the time. Cool to room temperature, and mix thoroughly, then dilute to the 25 ml mark with concentrated H_2SO_4 and mix.

Measure the absorbance at 580 mμ of solution A_1 against the blank, and of A_2 against a blank of concentrated sulfuric acid. Calculate the percent of piperine from the equation:

$$\%\ \text{Piperine} = 10.81\,\frac{A_1 - A_2}{\text{wt. sample (g)}}.$$

NOTES

1. If the extract is not clear after filtration, refilter the 10 ml aliquot until clear.
2. When transferring solutions to the absorbance cells, avoid excess contact with humid air to prevent striations. If striations do occur, wait 1 or 2 minutes before measuring the absorbance.
3. The factor 10.81 in the equation is based on an absorbance of 0.660 for the chromotropic acid complex derived from 1 micromole of piperine

in 25 ml of sulfuric acid solution:

$$\text{Constant} = \frac{\text{mol.wt.} \times 10^{-6} \times \text{dilution factor} \times 100}{B},$$

in which B = the absorbance of 1 cm of a solution containing 1 micromole of piperine in 25 ml. This value may vary slightly from laboratory to laboratory. Standardize with diethoxymethane or pure piperine (obtainable from Amend Drug Company, New York, N.Y.). Reference: *Anal. Chem.* **28**: 1621 (1956).

Mustard

There is some confusion as to the botanical classification of mustard. Linnaeus divided the mustards into two genera, *Brassica* and *Sinapis*. Botanists have disagreed over whether these are identical. According to a bulletin issued by ASTA[22], mustard is the seed of two varieties, *Brassica juncea* (black or brown) and *Sinapis alba* (white or yellow). *Federal Specification EE-S*-631*d* also includes *Brassica nigra* (black).

The seeds of *B. nigra* and *B. juncea* are so alike that both are accepted in commerce as "black mustard". The volatile oils derived from these two spices are practically identical, consisting almost entirely of allyl isothiocyanate. This does not occur as such in the seed, but in the form of a glucoside, sinigrin, which, in the presence of water and an enzyme myrosin, is hydrolyzed to the volatile allyl isothiocyanate, dextrose and potassium bisulfate.

Mustard occurs commercially in several forms—whole seed, unbolted ground seed (meal), mustard cake (ground mustard from which a portion of the fixed oil has been expressed), and mustard flour, which is ground mustard or mustard cake with the hulls largely removed. There is also "prepared mustard", which is defined by *Federal Specification EE-M*-821*d* (1963) as being made from the ground dried ripe seeds of white and/or black mustard, with or without mustard flour or mustard cake, together with vinegar and with or without addition of sweetening agents, salt, and spices. It shall contain no added starch-bearing materials other than those of turmeric, ginger, and black pepper, and shall be free from excess of hulls. It shall contain not less than 16.5% total solids (by drying at 100° to minimum weight), not more than 4% acid expressed as acetic, not more than 7% crude fiber on a moisture free basis, and not less than 3.8% nitrogen on a moisture free basis.

The seeds of *S. alba*, white or yellow mustard, contain a different glucoside, sinalbin, which upon hydrolysis yields a compound, p-hydroxybenzyl isothiocyanate, which is only sparingly volatile with steam. A bisulfate, sinapin acid sulfate, is liberated by this reaction also.

Chemical literature contains contradictory statements on the hydrolysis of black mustard. Some claim that black mustard does not contain sufficient myrosin to split all of the glucoside present, so that it is necessary to add ground, authentic white (or yellow) mustard (*S. alba*) to complete the hydrolysis. Others say this is not necessary. For a discussion, *see* Selected Reference D, Vol. VI, p. 57, and Roe[23]. The AOAC method does not add white mustard; otherwise it is quite similar to the official ASTA method. We prefer the AOAC method because of the collaborative study it has had.

Method 15-15. *Volatile Oil in Mustard Seed.* (*AOAC Method* ***28.023***).

Place 5.00 g of mustard flour or ground seed in a 200 ml flask, add 100 ml of water, stopper and macerate for 2 hours at about 37°. Add 20 ml of alcohol (the ASTA method adds 4–5 drops of anti-foam emulsion also) and distill about 60 ml into a 100 ml volumetric flask containing 10 ml of NH_4OH (1+2), taking care the tip of the condenser is below the surface. Add 20 ml of 0.1 *N* $AgNO_3$ to the distillate, let stand overnight, then heat to boiling on a water bath to agglomerate the Ag_2S formed. Cool, dilute to the mark with water and filter. Acidify a 50 ml aliquot of the filtrate with about 5 ml of HNO_3 and titrate with 0.1 *N* NH_4CNS, using 5 ml of 10% ferric alum [$FeNH_4(SO_4)_2.12H_2O$] as indicator.

1 ml of 0.1 *N* $AgNO_3$
= 0.004954 g of allyl isothiocyanate.

Terry and Curran[24] have devised a method which determines, by benzidine precipitation, the bisulfates that are split from the mustard glucosides by hydrolysis.

Method 15-16. *Crude Fiber in Mustard and Prepared Mustard.*

DETERMINATION

Ground Mustard, Mustard Flour: Proceed as in General Spice Method 15-1.

Prepared Mustard: Transfer a 10 g sample, weighed to 0.01 g to a 250 ml beaker, add 50 ml of alcohol and stir vigorously, then add 40 ml of ethyl ether, and stir. Let the mixture stand about 5 minutes, and decant through a hardened 11 cm filter paper in a Büchner funnel. Wash the sediment onto the filter paper with several washings of ether, removing any adhering particles on the wall of the beaker by means of a rubber policeman. Wash the material on the filter paper into a 500 ml Erlenmeyer flask with 200 ml of boiling 1.25% sulfuric acid (Method 1-11) and proceed as in Method 15-1, General Spice.

Method 15-17. *Acidity of Prepared Mustard.*

DETERMINATION

Weigh a 10 g sample into a 200 ml volumetric flask, dilute to the mark with water, and mix. Let the sediment settle, then filter through a dry filter paper, rejecting the first few ml. Titrate a 50 to 100 ml aliquot with 0.1 *N* sodium hydroxide solution, using phenolphthalein as indicator. 1 ml 0.1 *N* NaOH ≎ 0.0060 g acetic acid.

Method 15-18. *Starch in Prepared Mustard.*

REAGENTS

a. *Calcium chloride solution,* 30 g/100 ml water, adjusted to 0.01 *N* alkalinity.

b. *Alcoholic sodium hydroxide solution,* 70 ml alcohol + 30 ml 0.1 *N* NaOH.

c. *Iodine—KI solution,* 2g I + 6g KI in 100 ml water.

DETERMINATION

Place 5 g of prepared mustard in a 500 ml Erlenmeyer flask, and pipette in 100 ml of the calcium chloride solution, swirling gently until all lumps are broken. Add the calculated amount of 0.1 *N* NaOH solution to neutralize the acidity in the 5 g sample as calculated from Method 15-17.

Add glass beads. Connect to reflux condenser, first wetting the inside of the condenser and stopper with water and draining one minute. Heat gently (on an asbestos board with a center hole) to avoid initial foaming, and boil 15 minutes.

Leaving the condenser connected, cool the flask to room temperature in a pan of cold water. Remove the flask, stopper, and shake vigorously. Pour the contents into a centrifuge bottle and centrifuge at 1500 rpm 5 minutes. Withdraw as much as possible of the partially clarified middle layer (ca 75 ml) and filter through an 11 cm circle of absorbent cotton about 5 cm thick placed in a 60° funnel. Pipet 50 ml of the filtrate into a second centrifuge bottle containing 150 ml of alcohol, stopper, and shake vigorously several minutes. Centrifuge at 1500 rpm until clear (about 5 min).

Decant the liquid, using suction, through an asbestos pad in a Caldwell crucible without transferring the starch to the crucible. Transfer the pad to the same centrifuge bottle and rinse all particles adhering to the crucible into the bottle with water. Add several glass beads and water to about 100 ml. Stopper, and shake the bottle vigorously until the precipitate is as finely dispersed as possible. Add a slight excess of the I-KI solution (2–3 ml) and 30 ml of saturated $(NH_4)_2SO_4$ solution. Stopper, and shake the bottle. Rinse the particles adhering to the stopper into the bottle, and centrifuge until clear.

Decant the supernatant liquid with suction through an asbestos pad in a Caldwell crucible. Add 50 ml of the alcoholic NaOH solution to the precipitate in the centrifuge bottle. Stopper and shake vigorously. Wash the stopper with 70% alcohol. Centrifuge and decant the supernatant liquid through the same pad as before. Repeat the treatment with the NaOH solution until practically all blue disappears (usually 2–3 treatments). Without centrifuging, transfer the contents of the bottle to the Caldwell crucible, using 70% alcohol. Aspirate until the pad is dry; then transfer the pad to a 500 ml Kjeldahl flask. Rinse the bottle and crucible with 10 ml of HCl (sp. g. 1.1029) followed by five 10 ml portions of water, carefully removing all adhering particles. Attach the Kjeldahl flask to a reflux condenser, first adding glass beads to lessen bumping. Place on an asbestos board with center hole and boil 1 hour. Cool, neutralize with NaOH (1+1) (Me orange), and filter into a 200 ml volumetric flask; rinse the flask and filter thoroughly, and dilute to volume with water. Mix well, and determine dextrose in a 50 ml aliquot by Method 16-8 (**a**). (Blank on Fehling solution should not be more than 0.3 mg.)

$$\% \text{ starch} = \frac{[\text{g dextrose} \times 0.9(100 + A + B) \times 8]}{\text{weight of sample}},$$

where A = ml $1N$ NaOH used to neutralize acidity, Method 15-17, and B = g H_2O in sample as determined by loss of weight after evaporation of a 5 g sample, following the procedure in Method 6-32. Two or three ml of water may be added to the drying dish before drying to distribute the sample evenly.

Ginger

Commercially, ginger is extracted with alcohol to make Extract of Ginger, or with water to make Ginger Ale. The exhausted ginger resulting from either process constitutes the most common adulterant of commercial ground ginger. In such cases, attempts have been made to restore pungency by addition of cayenne or mustard. Color has been restored by addition of turmeric.

Dried ginger occurs in commerce as peeled or unpeeled rhizomes, at times bleached with sulfur dioxide or hypochlorite. It is often coated with lime to prevent insect attack. In the latter case, the lime, as CaO, should not exceed 1% (*Federal Specification EE-S-631d*). Other analytical limits are listed in Tables 15-1 and 15-2.

Method 15-19. *Alcohol Extract of Ginger.*

DETERMINATION

Weigh 2 g of ground ginger into a 100 ml volumetric flask and make to the mark with ethyl alcohol. Stopper, shake, and allow to stand for 24 hours, shaking at half hour intervals during the first 8 hours. Filter through a dry paper, rejecting the first few ml of filtrate. Evaporate a 50 ml aliquot (equivalent to 1 g) of the filtrate to dryness on the steam bath, using a flat-bottomed dish, then dry in an oven to constant weight at 105–110°.

Method 15-20. *Cold Water Extract of Ginger.*

DETERMINATION

Weigh 4 g of ground ginger into a 200 ml volumetric flask, and bring to the mark with water at room temperature. Stopper, shake, and allow to stand for 24 hours, shaking at half-hour intervals during the first 8 hours. Filter through a dry filter, rejecting the first few ml. Evaporate a 50 ml aliquot (equivalent to 1 g) of the filtrate to dryness on the steam bath using a flat-bottomed dish, then dry in an oven to constant weight at 100°.

NOTE: These 2 empirical methods were developed by Andrew L. Winton and co-workers at Conn. Agricultural Experiment Station in 1898. (Selected References G and L). According to these workers, less than 3.6% alcohol extract or less than 10.8% water extract denotes exhausted ginger. The normal range for authentic gingers is given as 3.63–6.58% (av. 5.78%) for alcohol extract and 10.2–17.55% (av. 13.42%) for cold water extract.

Method 15-21. *Volatile Oil and Oleoresin in Ginger.*

This is the AOAC Method.[25] Weigh 50 g of ground ginger into a Soxhlet extractor and extract completely with ethyl ether. (About 4 hours is usually sufficient.) Transfer the extract quantitatively to a 300 ml flask with stopper fitting a Clevenger or Lee and Ogg volatile oil trap (*see* Methods 15-5 and 15-6). Evaporate the ether on a steam bath until solvent odor is no longer detected. Add 50 ml of water, attach the flask to a Lee and Ogg lighter-than-oil trap, and distil the volatile oil.

Transfer the residue in the flask quantitatively to a separatory funnel and extract with ether. Transfer the ether extracts to a tared beaker, remove the ether by evaporation on a steam bath, and finally dry to constant weight in a vacuum desiccator. Report as ginger oleoresins.

Examination of Volatile Oils from Spices

The most convenient method of sampling is to pipette off the volatile oils recovered in the Clevenger or Lee and Ogg volatile oil traps. Oil from duplicate determinations may be combined. (Methods 15-5 and 15-6). The distilled oil from these traps may be pipetted, or drained off through the stopcock of the trap, into a glass-stoppered tube or graduate, avoiding transfer of water. Traces of water should be removed by addition of a very small amount of anhydrous sodium sulfate. Allow to settle and store in the refrigerator. This oil may be used for determination of such physical constants as refractive index and specific gravity, and for certain chemical determinations.

Method 15-22. *Phenols in Volatile Spice Oils.*

This method may be used for determination of eugenol in clove and allspice oils, and thymol plus carvacrol in thyme oil. It is essentially the method used by the U.S. and British Pharmacopoeias, except for the smaller volume of oil used:

DETERMINATION

Transfer 2 ml of distilled oil, using a pipette, into a Babcock milk bottle. Add 20 ml of 3% KOH solution, shake vigorously for 5 minutes, heat 10 minutes in a boiling water bath, remove and cool. After the liquids have separated, bring the oil within the graduated neck of the Babcock flask by addition of 3% KOH solution, and note the volume of residual oil. The difference between the volume of the original sample and that of the residual oil is the phenol content. Calculate the percentage by volume from this difference.

If sufficient sample is available, say 5–10 ml, the separation of phenols may be made in a cassia flask instead of a Babcock bottle.

Method 15-23. *Aldehydes and Ketones in Volatile Spice Oils.*

This method, based on the formation of a water-soluble addition product of aldehydes and ketones with alkali bisulfites, is the basis of U.S. and British Pharmacopoeia assays of volatile spice oils. It may be used for determination of cinnamic aldehyde in oil of cinnamon or cassia, carvone in spearmint, caraway or dill oil, and pulegone in pennyroyal oil.

REAGENTS

Saturated solution of sodium sulfite, neutralized to phenolphthalein indicator by drop-by-drop addition of 30% sodium bisulfate solution or, preferably, with 50% (by volume) acetic acid.

DETERMINATION

Transfer 10 ml of oil (if pennyroyal oil, 5 ml) to a 100 or 150 ml cassia flask, add 50 ml of saturated sodium sulfite solution and heat in a boiling water bath, shaking repeatedly. As the pink phenolphthalein color returns, neutralize with 30% sodium bisulfate or 50% acetic acid solution. When the pink color does not reappear after addition of 2–3 drops of indicator and heating for 15 minutes in the boiling water bath, cool to room temperature. When the liquids have separated completely, add sufficient sodium sulfite solution to raise the oil within the graduated portion of the neck of the cassia flask, and note the volume of the residual oil. The difference between the volume of the original sample and that of the residual oil represents the volume of aldehydes and/or ketones present. Calculate percentage by volume from this difference.

Composition of Authentic Spices

The older compilations of analyses of authentic spices, done during the late 19th and early 20th centuries, are primarily, although not entirely, of only historical interest today. Some of the methods used were inexact or not reported, and the exact details were not always specified, *e.g.*, determinations of ash, total solids, and essential oils. Some of the analyses reported are of samples of questionable authenticity. Then, too, the trade practices in cultivation, harvesting and sanitation have materially improved since the publication of these papers. The limits given in Tables 15-1 and 15-2 more nearly represent current trade practices. These older compilations appear in many texts that are easily available and will not be reproduced here. The list of selected references at the end of this chapter includes this source material.

Capsicum Spices

Much of the earlier literature on composition of the capsicum spices is of questionable value. In addition to the reasons stated at the beginning of this section on authentic spice analyses, the botanical characteristics of the genus *capsicum* have changed materially during the past 30 years. The large acreages of chili pepper and paprika now in production in California have resulted in varietal changes in composition. California paprika is markedly different from paprika grown in the Mediterranean countries.

Composition of Authentic California Grown Capsicum Spices[a]

Values expressed in percent on moisture free basis

	Protein (N × 6.25)	Crude Fat	Crude Fiber	Ash	Reducing Sugars
Californian type					
Smooth-skin (ancho) pods	13.9–18.4	12.9–20.3	16.4–20.1	5.1–6.4	16.0–25.6
Mexican type					
Wrinkled skin (Pasilla) pods	12.0–19.0	11.4–16.9	—	5.4–6.9	26.9-31.0
California paprika					
Long, "heat-free" smooth pods	13.8–16.4	12.0–17.3	16.9–20.1	4.8–6.2	13.9-26.2

[a] Unpublished data from the laboratories of the Gentry Corporation, Glendale, California, representing commercially ground spice, 1959, 1960 and 1965 seasons.

G. L. Tandon *et al.*[26] have reported the proximate analysis and the capsaicin, carotene, ascorbic acid and mineral contents of hot chili peppers, *Capsicum annuum, var. acuminatum*, grown in India. They analyzed separately the whole chillies, pericarp, seeds and stems of whole pods collected in local Indian markets. They found that about 90 % of the capsaicin and 95 % of the carotenoid pigments are localized in the 40 % constituting the pericarp, or fleshy portion, of the fruit. The capsaicin content of the whole pods ranged from 150 to 193 ppm.

The following table gives the proximate analysis of whole hot Indian chillies reported

Proximate Analysis of Whole Indian Hot Chilies

Values expressed in percent on moisture free basis

	Average	Range
Moisture	8.94	8.82- 9.10
Volatile Ether Extract	0.36	0.06- 0.46
Non-volatile Ether Extract	21.71	19.41-24.16
Total Ash	6.27	6.13- 6.46
Acid-insoluble Ash	1.05	0.35- 1.85
Protein (N × 6.25)	12.83	12.52-13.00
Crude Fiber	31.85	28.05-43.50
Carbohydrates, by difference	26.99	12.53-33.83

by Tandon and his co-workers.[26] Five samples were analyzed, each representing 50 Kg lots purchased in the open markets. Since these pods were whole, it is a fair assumption that they were authentic.

A book published in Spanish by the University of Murcia (Spain) in 1962 includes a summary of analyses of many samples of Spanish paprika examined over a 12-year period. This 208 pp. treatise includes much valuable information on the botany, agronomy, and chemistry of Spanish capsicums and paprika. It is listed under "Selected References" at the end of this chapter.

Proximate Composition of Spanish Paprika

Values expressed in percent

	Average	Range
Moisture, by toluene distillation	8.50	6.00–14.50
Ether extract, Soxhlet extractor	11.10	6.20–15.30
Total ash, ignition at 525°	6.00	5.00– 7.00
Acid-insoluble ash	0.10	0.05– 0.50
Crude fiber	20.00	17.00–25.00
Cellulose	15.00	13.00–20.00
Protein (N × 6.25)	14.00	12.00–15.00
	Fresh, mature pods	**Dried pods**
Pulp and peel	67.87–69.75	42.32–43.55
Seed	24.33–28.91	42.01–45.21
Stems and placenta	3.28– 6.11	7.12–10.22

Table 15-4 gives unpublished data from Gentry Corporation of Glendale, California, on the mineral and vitamin contents of authentic California (smooth-skin) and Mexican (wrinkled-skin) chili pods, 1952 crop:

TABLE 15-4

Mineral and Vitamin content of California Chili Pepper, 1952 crop

Reported as mg/100 g on moisture free basis

	Ca	Fe	P	Ascorbic Acid[a]	β-Carotene[a]	Thiamine	Riboflavin	Niacin
Mexican type (wrinkled-skin)								
Sample 1	193	17.3	377	81	22.9	0.329	1.61	6.6
Sample 2	164	9.3	341	75	51.1	0.419	1.49	6.7
Sample 3	110	7.8	286	101	51.4	0.421	1.93	6.9
Sample 4	106	8.3	303	332	33.6	0.351	0.78	6.6
Californian type (smooth-skin)								
Sample 1	186	13.9	392	45	25.0	0.323	1.37	9.1
Sample 2	163	11.2	318	86	37.2	0.421	1.39	5.8
Sample 3	190	8.8	312	69	32.6	0.422	1.50	6.9

[a] The wide variations in ascorbic acid and β-carotene are attributed to varying conditions under which the pods were dehydrated.

Pepper

Smith and his co-workers[27] reported the composition of authentic pepper, pepper shells and pepper siftings. Shells and siftings are occasionally used as adulterants of ground pepper. As of that date, these authors said that "in the long history of food adulteration pepper has probably received more attention from food analysts than any other product". They obtained 38 samples of pepper and white pepper berries representing four varieties from inspectors at ports of entry. After evaluating the results obtained from determination of 16 different constituents of pepper (reported in the original article), they concluded that the most valuable criteria for detecting addition of pepper shells were: crude fiber, dextrose, magnesium, the ratio of MgO to dextrose, and the product of MgO times crude fiber. These are reported in Table 15-5. Standard AOAC methods (1925 ed.) were used for most of these determinations made by these investigators.

TABLE 15-5
Results of analysis of authentic samples of whole pepper, pepper shells, and pepper siftings computed to a dry basis

	On dry basis			Ratios	
	Crude Fiber %	d-Glucose %	MgO %	MgO × CF	MgO × 1000 / d-Glucose
Lampong (18 samples)					
Maximum	15.3	57.2	0.44	6.4	8.9
Minimum	13.1	49.5	0.41	5.2	7.4
Average	13.9	52.5	0.42	5.8	8.1
Alleppi (9 samples)					
Maximum	14.5	57.9	0.43	6.2	7.9
Minimum	12.1	52.4	0.38	4.6	6.7
Average	13.0	56.1	0.41	5.3	7.3
Tellicherry (5 samples)					
Maximum	15.8	59.2	0.41	6.2	7.2
Minimum	12.9	56.6	0.39	5.3	6.8
Average	13.0	57.5	0.40	5.8	7.0
Singapore (2 samples)					
Sample 1	15.6	55.2	0.43	6.7	7.8
Sample 2	15.6	55.1	0.43	6.7	7.8
White and decorticated (4 samples)					
Maximum	5.5	85.4	0.24	1.2	3.3
Minimum	1.2	73.2	0.11	0.1	1.4
Average	4.1	77.9	0.16	0.7	2.1
Shells (4 samples)[a]					
Sample 1	31.6	21.2	0.80	25.3	37.8
Sample 2	24.4	29.3	0.64	15.6	21.9
Sample 3	25.3	29.7	0.64	16.2	21.5
Sample 4	30.0	21.1	0.79	23.7	37.4
Siftings (3 samples)					
Sample 1	18.3	12.8	—	—	—
Sample 2	20.8	14.8	—	—	—
Sample 3	24.6	22.5	0.61	15.0	27.1

[a] Samples 2 and 3 were found to contain a considerable amount of endosperm. Samples 1 and 4 are considered more nearly representative of true pepper shells.

Horseradish

Prepared horseradish is at times adulterated by admixture with ground parsnip or turnips, then fortified by addition of allyl isothiocyanate (synthetic mustard oil) to produce the desired pungency. Carol and Ramsey[28] found that the infrared spectra of volatile oil of horseradish and allyl isothiocyanate were practically identical, and differed distinctly from the spectrum of volatile oil of parsnip. The latter spectrum has a strong band at 6.65 μ which is absent in horseradish oil and synthetic oil of mustard. These latter two spectra have a strong absorption band at 4.75 μ, which is absent in oil of parsnip.

Unreported work by the Food and Drug Administration (1952) showed that horseradish contained from 0.22 to 0.86% allylisothiocyanate on a moisture free basis, while parsnip and turnip contained practically none (less than 0.08%). Horseradish also contains considerable starch (average of 23% on 10 samples examined), while turnips contain less than 10%, and parsnips from 11 to 18%.

Determination of Geographical Origin of a Spice

It is important, at times, to be able to identify the geographical origin of a spice, since the quality, and hence the cost, of spices from different areas vary. For example, East Indian nutmeg and mace are considered of higher quality than those from the West Indies, yet half of these spices imported into the United States comes from the latter area. With very few exceptions, conventional methods of analysis are of little value in solving this problem.

The use of gas chromatographic and spectrophotometric techniques by various researchers has greatly augmented our knowledge of the composition of volatile spice oils beyond that obtained by conventional methods. While details of such research methods are considered beyond the scope of this book, the authors believe that reference to a few research papers will indicate the type of information that may be obtained by these advanced techniques.

By the use of gas chromatographic separations and infrared analysis, Jennings and Wrolstad[29] separated volatile black pepper oil into at least 23 volatile components. Researchers will be interested in comparing this work with that done on the composition of black pepper oil by Hasselstorm *et al.*,[30] using conventional chemical analytical methods. Datta and co-workers[31], in research sponsored in part by a grant from ASTA, using similar techniques, examined volatile oils of black pepper, cassia, nutmeg, and ginger. By this means they were able to distinguish between spices of different geographical origin, e.g., Lampong from Malabar pepper, East Indian from West Indian nutmeg, African from Jamaican and Cochin ginger, and Saigon from Batavian cassia.

ASTA (Selected Reference A; Text Reference 32) has adopted a spectrophotometric method for distinguishing between East Indian and West Indian nutmeg and mace, based on the coupling with a diazonium salt of the phenols present in the volatile oils:

Method 15-24. *Phenols in Nutmeg and Mace. (ASTA Official Method).*

Purpose: To distinguish between East and West Indies Nutmeg or Mace.

APPARATUS

Spectrophotometer, capable of measuring absorbance at 400 mμ.

REAGENTS

a. *p-Nitroaniline*, recrystallized.

b. H_2SO_4, 24 *N*, 660 ml of H_2SO_4 in water diluted to 1 L.

c. *Sodium nitrite*, 0.2 *N*–13.8 g, $NaNO_2$/L. Use only freshly prepared solution.

d. *Diazonium salt solution*—dissolve 0.276 g of p-nitroaniline (**a**) in 5 ml of 24 *N* H_2SO_4 (**b**) in a 250 ml beaker and add 3 g of solid NaOH. Cool, add 20 g of ice (frozen from distilled water) and place the beaker in an ice-salt bath. Add 10 ml of 0.2 *N* $NaNO_2$ slowly (over 15 minutes) with continuous stirring. Stir for 15 minutes longer, then add 20 g of ice and 40 ml of ice-cold water. Keep the solution cool.

DETERMINATION

Pipette 1 ml of the steam volatile nutmeg or mace oil, as obtained from Method 15-6, into a 12 ml conical centrifuge tube, add 6 ml of $CHCl_3$ and 4 ml of 10% NaOH solution. Mix and centrifuge. Pipette 2 ml of the top layer into a 22 × 150 mm test tube, cool to 0° in ice-water bath, and keep cold. Add slowly, while shaking, using a cold pipette, 4 ml of the diazonium reagent (**d**). Keep the test tube cold all during this step to allow coupling of the phenol.

Pipette 1 ml of the coupled phenol solution into a 10 ml volumetric flask, and dilute to the mark with water. Measure the absorbance at 400 mμ against a blank derived from 2 ml of 10% NaOH solution carried through that portion of the "Determination" beginning "And slowly while shaking . . ." and ending "dilute to the mark with water".

INTERPRETATION

West Indian nutmeg and mace oils normally have absorbances between 0.000 and 0.020; East Indian oils normally have absorbances between 0.060 and 0.150.

Cinnamon and Cassia

Cinnamon and cassia are recognized in the spice trade as 2 distinct species, although this distinction does not appear to exist at the consumer level. Cinnamon, native to Ceylon, but also cultivated in other parts of tropical Asia, Sumatra and Java, is *Cinnamomum zeylanicum Nees*, while cassia is *C. cassia*, grown in China, Indochina and India. Cinnamon is the more expensive.

Dutta[33] has devised a method which is claimed to distinguish between these two species by differences in amount and ash content of the "mucilage" obtained from these two spices. By "mucilage" Dutta means the alcohol precipitate of the defatted, aqueous-acetic acid extract obtained from the ground spice. He reports the following distinctions as obtained by analysis of authentic samples of the ground spices:

	Cinnamon	Cassia
Mucilage, %		
Range	1.9–3.7	8.0–11.5
Average	2.6	9.5
Ash content of mucilage, %		
Range	17.2–24.6	6.1–7.8
Average	19.7	7.2

The original article should be consulted for details of the method.

Food Dressings

Mayonnaise was the first commercial dressing to gain wide distribution, appearing about the time of World War I. "Salad dressing", denoting the product made in part of cooked or partly-cooked starch and vegetable oils, did not appear on the market until the late 1920's.

Obligatory *Standards of Identity* have been promulgated by the U.S. Food and Drug Administration for three dressings.[34] These standards are (*see top p.* 351).

All these products contain an acidifying ingredient (a vinegar, lemon or lime juice, or citric acid—this latter to contribute only part of the acidity) and may contain disodium or calcium disodium EDTA not to exceed

	Vegetable Oil	Emulsifying Ingredient
Mayonnaise	65% minimum	egg-yolk-containing ingredient only
French dressing	35% minimum	egg-yolk solids and/or certain listed gums[a], totalling not more than 0.75% (optional)
Salad dressing	30% minimum	egg-yolk-containing ingredients equivalent to not less than 4% liquid yolks; certain listed gums totalling not more than 0.75% may also be present.

[a] Permitted gums are: acacia, carob bean, guar, karaya, extract of Irish moss, pectin, propylene glycol ester of alginic acid, sodium carboxymethylcellulose, methylcellulose USP, hydroxypropyl methylcellulose, and mixtures of two or more of these.

75 ppm, and dioctyl sodium sulfosuccinate (a wetting agent) up to 0.5% by weight of the gum ingredients used.

Salt, sugars, monosodium glutamate, spices (except that turmeric or saffron may not be used in mayonnaise and salad dressing) and harmless food seasonings and flavorings (except imitations) are also optional in all three dressings, providing they do not impart color simulating that of egg yolk. Tomato paste, tomato purée, catsup and sherry wine may be used in French dressing.

The Canadian requirements[35] for vegetable oil contents of mayonnaise and French dressing are the same as those of the United States. Canada, however, requires not less than 35% vegetable oil in salad dressing. Requirements for the other ingredients are essentially the same as those of U.S. law. Lactic and tartaric acids are permitted as acidulants, and the use of saffron and turmeric is banned in mayonnaise, but permitted in salad dressing and French dressing.

The *Federal Purchasing Specification for Mayonnaise and Salad Dressing* (*EE-M*-131*d*, *March* 1967) requires a somewhat higher oil content than the minima established by the FDA *Standards of Identity*:

Analytical Requirements for Mayonnaise and Salad Dressings Submitted for U.S. Government Purchase

	Mayonnaise	Salad Dressing	French Dressing
Fat, percent, not less than	80.0	40.0	35.0
Acidity, percent of moisture phase, not less than	2.5	2.25	2.75
Egg yolk, percent not less than	7.0	5.0	—
Total sugars, as invert, percent, not less than	—	8.0	12.0
Copper, ppm, not more than	2	2	2

Analysis of Food Dressings

Method 15-25. *Preparation of Sample.* (*AOAC Method* ***28.042***).

Emulsified dressings: Transfer the entire contents (1 quart or less) to a suitable container with a tight-fitting lid, such as a fruit jar, of larger capacity than the sample container, and mix with a spatula for a few minutes until homogeneous. Weigh out immediately successive portions for the various determinations. If the sample has stood for any appreciable time, it should be remixed before subsequent weightings.

Separable dressings: The AOAC adds a little powdered egg albumen to the dressing to prevent separation. This, of course, renders such a sample unsuitable for determination of nitrogen. For small containers, say up to 1 quart, the AOAC directs as follows: weigh the container, shake 1 minute and immediately empty the contents into a high-speed blender, such as a Waring Blendor, and let drain 1 minute. Weigh the emptied bottle and record the weight of the sample.

Add 0.20 g of powdered egg albumen for each 100 g of dressing, cover the blender, and stir for 5 minutes. Transfer the mixture to a container, of larger capacity than the sample bottle, with a tight-fitting lid. Shake the sample about 20 times, and stir with a spatula or large spoon about 20 times just before each portion is removed for analysis. Make all weighings immediately after sample preparation. Correct the results for the weight of added emulsifier.

Method 15-26. *Total Solids in Food Dressings.* (*AOAC Method* ***28.043***).

Quickly weigh about 2 to 5 g into a covered metal dish previously dried at 98–100°, cooled and weighed. Remove most of the moisture by heating the dish, without cover, on the steam bath. Replace the cover loosely and dry in a vacuum oven at 98–100° to constant weight (about 5 hours). Follow the precautions given in Method 1-2 and "Notes on Oven Moisture Determinations".

Admit dry air into the oven, tighten the cover of the dish and remove the dish to an efficient desiccator. Weigh soon after the dish has reached room temperature.

Method 15-27. *Total Fat by Acid Hydrolysis.*

[This method is official in AOAC Methods for several types of food (eggs, fish, noodles, etc.).]

Accurately weigh, by difference, about a 1 g sample of the well-mixed sample into a Mojonnier fat extraction tube. Add 10 ml of conc. HCl, shake, and set the tube into a water-bath heated to 70°. Bring to boiling and boil for 30 minutes, shaking the contents of the tube thoroughly every 5 minutes. Remove the tube from the bath, add water to fill the lower bulb (but not the neck) and cool to room temperature.

Add 25 ml of ether and shake vigorously for 1 minute. Continue as in Method 10-27, (a), beginning "Add 25 ml of petroleum ether (b.p. 30–60°) . . .," but drying the extract, after evaporation of the solvents, for 90 minutes instead of the 30 minute drying directed for fat in fish.

Method 15-28. *Total Nitrogen.* (*AOAC Method* ***28.048***).

Weigh into a 500 ml Kjeldahl flask by difference (to 10 mg) about 15 g of well-mixed sample. Avoid getting any of the sample on the inner side of the neck of the flask. Place the flask on a steam bath and heat until the egg is thoroughly cooked and the oil has separated. Cool, add 50 ml of petroleum ether, mix, and decant the solvent through a small filter paper. Repeat the petroleum ether extraction twice, wash the filter with petroleum ether, and add the filter paper to the sample in the flask. Place the flask on the steam bath for a few minutes to remove petroleum ether residues. Determine nitrogen as in Method 1-7, using 50 ml of sulfuric acid (more if necessary). Report as percent nitrogen. If percent protein is desired, multiply percent N by 6.25.

Method 15-29. *Total Phosphorus.* (*AOAC Method* ***28.049***).

Weigh by difference, to 10 mg, about 10 g of the well-mixed sample into a platinum dish. Add about 10 ml of 10% Na_2CO_3 solution and evaporate to dryness on a hot plate, or over-night in a 100° oven. Burn off the oil at a low temperature to allow the fat to smoke off without burning, then raise the temperature to 500–525° and continue ashing for about 1 hour at this temperature. Cool, add a few drops of water, and break up charred lumps with a flat-ended glass rod. Stir in 10 ml of HNO_3 (1+3), cover with a watch-glass, and place on a steam-bath for about 10 minutes. Wash off the watch-glass and stirring rod, and filter through a small filter paper into a 250–500 ml Erlenmeyer flask. Wash the charred material on the paper with water into the flask. Determine phosphorus by the gravimetric quinoline phosphomolybdate method, Method 1-18.

Calculation of Egg Content

The FDA *Standard of Identity* for egg yolks[36] requires not less than 43% total egg solids. *Federal Purchasing Specification for Mayonaise and Salad Dressings EE-M-131b* directs use of the following formula in calculating egg yolk content on a 43% total egg solids basis, regardless of the source of the egg constituent (liquid or frozen whole eggs, or liquid or frozen egg yolks):

$$\% \text{ yolk} = 94.26\text{ P} - 2.192\text{ N},$$
where $P = \% P_2O_5$ and $N = \%$ total N.

The Methods of Analysis, AOAC 6th ed (1945) had somewhat different formulae for calculating percent of egg in mayonnaise and salad dressings. These formulae are:

$$\% \text{ yolk} = 75.69\text{ P} - 1.802\text{ N};$$
$$\% \text{ white} = 60.80\text{ N} - 114.59\text{ P};$$
$$\% \text{ total egg} = \% \text{ yolk} + \% \text{ white};$$
$$\% \text{ white in egg component} = \frac{\% \text{ white}}{\% \text{ total egg}}.$$

Other formulae are:

$$\% \text{ vegetable oil} = \% \text{total fat (ether extract)} - (\text{yolk} \times 0.3188)$$

minor constituents (sugar, salt, spices, stabilizers)
$$= \text{total solids} - [(\text{yolk} \times 0.05047) + (\text{white} \times 0.1221) + \text{vegetable oil}].$$

Method 15-30. *Total Acidity.* (*AOAC Method* ***28.047***).

Weigh by difference to 10 mg about 15 g of the thoroughly mixed sample into a 500 ml Erlenmeyer flask. Add about 200 ml of distilled water and shake thoroughly. Titrate with 0.1 *N* NaOH solution, using phenolphthalein as indicator. Observe the first change of color to a pink tinge by comparison with a similarly diluted duplicate sample. Calculate as percent acidity as acetic (1 ml 0.1 *N* NaOH is equivalent to 6.0 mg acetic acid).

The Federal Specification for mayonnaise directs that acidity be reported as the amount of acid as acetic in the water phase of the dressing, according to the formula:

$$\% \text{ acid in water phase} = \frac{B \times 100}{A},$$

where $A = \%$ water in the sample, and
$B = \%$ acid as acetic in the sample.

Method 15-31. *Reducing Sugars Before and After Inversion.* (*AOAC Method* ***28.044***. This method is not applicable to starchy dressings).

Weigh by difference about 20 g of the well-mixed sample to 10 mg into a 250 ml centrifuge bottle or a 4 oz nursing bottle. Add about 75 ml of petroleum ether, shake and centrifuge. Remove as much of the solvent as possible (conveniently done by use of a pipette) and repeat the extraction with petroleum ether three or four times, until the solvent is colorless.

NOTE: The petroleum ether solution may be reserved for identification tests on the oil. Consult Chapter 13.

Remove the petroleum ether from the residue by means of a current of air, and add 5–10 ml of freshly prepared 5% metaphosphoric acid solution (prepared by first rinsing off any white coating on the lumps or sticks of acid with cold water before weighing them out). Transfer quantitatively to a 100 ml volumetric flask with water and dilute to the mark, mix and filter.

Transfer 80 ml, or as large an aliquot as possible of the filtrate into a 100 ml volumetric flask, neutralize with NaOH solution (1+1) to phenolphthalein, and dilute to the mark. Determine the reducing sugars before and after inversion according to Method 16-6B "Determination."

CALCULATION

Multiply the difference between percent reducing sugars after inversion and percent before inversion by 0.95 to give percent sucrose.

Sugars (sucrose, dextrose, honey, corn syrup and other similar sweeteners) are permitted ingredients in salad dressings. It is seldom that chemists are asked to determine them quantitatively. They may be determined as sugars before and after inversion by AOAC Method **28.044–28.046**.

Method 15-32. *Starch in Salad Dressings.* (*AOAC Method* ***28.058/28.059***).[37]

REAGENTS

a. *Calcium chloride solution*—30 g $CaCl_2.H_2O$/100 ml water adjusted to 0.01 *N* alkalinity.

b. *Alcoholic NaOH solution*—70 ml ethanol + 30 ml 0.1 *N* NaOH.

c. *I-KI solution*—2 g of I + 6g of KI in 100 ml of water.

Determine the total acidity of the prepared sample by Method 15-30. Transfer 4 to 5 g of the prepared sample, weighed to 10 mg, into a 500 ml Erlenmeyer flask, and add the calculated quantity of 0.1 *N* NaOH solution to neutralize the acidity. Add 100 ml of $CaCl_2$ solution (**a**), swirl gently to break up lumps of dressing, and proceed as in Method 15-18, beginning "add glass beads . . ." Determine dextrose in a 50 ml aliquot by Munson-Walker copper reduction method, Method 16-8 (**a**), weighing the CuO. Report as percent starch.

$$\% \text{ Starch} = \text{dextrose} \times 0.9.$$

Method 15-33. *Emulsion Stability.*

The *Federal Purchasing Specification for Mayonnaise and Salad Dressings EE-M-*131*b* directs the following test and standard:

Place about 100 g of the well-mixed product in a clean jar of approximately the same capacity, having a tight closure. Place in a drying oven at 100° F (38°) for 48 hours. For emulsified French dressings, the heating period should be 1 week. After that time note any separation of oil and water.

There should be no separation of mayonnaise or salad dressing after a 48 hour test, nor any separation of emulsified French dressing after 1 week's heat treatment.

Method 15-34. *Specific Weight.* (*Weight per Unit Volume*).[38]

This is a measure of the amount of air incorporated or entrapped in mayonnaise or emulsified salad dressings. It is of value as a measure of the degree of emulsification during manufacture.

Weight an unopened jar of dressing (1 quart size or less) in avoirdupois ounces, to the nearest 0.01 ounce. Remove the lid and measure the headspace to the nearest $\frac{1}{32}$ inch, using a head-space gauge. Remove the contents of the container, clean, dry and weigh the container, and record the net weight of the dressing.

Fill up the container with a measured amount of water to the head-space measured above, and report the volume in fluid ounces.

$$\text{Specific weight is } \frac{W}{V},$$

where W is the net weight of the dressing in avoirdupois ounces, and V is the volume occupied by the dressing in fluid ounces.

Gums and Similar Thickeners in Dressings

Considerable work has been done on the detection and estimation of thickeners, other than starch and pectin, in food dressings as well as other foods. This is inevitable because of the wide variance in chemical and physical characteristics of the many compounds, natural and synthetic, that have been used for this purpose, and the widely different physical characteristics of foods.

AOAC interest in this subject began with publication of a paper on detection of gums in foods by Hart[39] in 1937. Later he became Referee on Gums in Foods and published reports in 1939 and 1940.[39] Gnagy[40] became Referee in 1952 and continued for 10 years.

Detection of gums in food dressings was one of the early collaborative studies initiated under this general subject, first confined to work on non-starchy dressings and later extended by Redfern[41] to starchy dressings. The promulgation of *Standards of Identity for Mayonnaise, French Dressing and Salad Dressing*,[34] which prohibit gums or thickeners other than egg, augmented the need for reliable methods for their detection.

Method 15-35. *Detection of Gums in Mayonnaise and French Dressing.* (*AOAC Method* ***28.052.*** (Not applicable in the presence of starch).

REAGENTS

a. *Trichloroacetic acid*, 50 g diluted to 100 ml with water.

b. *Benedict's reagent* (*qualitative*), Dissolve 17.3 g of sodium citrate and 10 g of anhydrous Na_2CO_3 in about 80 ml of hot water; dissolve

1.73 g of $CuSO_4.5H_2O$ in 10 ml water. Filter the alkaline citrate solution, add the $CuSO_4$ solution slowly with constant stirring, and dilute to 100 ml with water.

DETERMINATION

Transfer a 100 g sample to a 250 ml beaker, add 35–40 ml of hot water and mix thoroughly. Heat to 65–70° in a water bath, add 10 ml of the 50% trichloroacetic acid soln, and maintain at 65–70° until the emulsion shows signs of breaking (in no case more than 10 min). Transfer the mixture to an 8 oz nursing bottle and insert a pipet guard [*J. Assoc. Offic. Agr. Chemists* **20**: 529 (1937)] (a wide-bore glass tube long enough to reach almost to the bottom of the centrifuge bottle, with lower end loosely stoppered; the tube is held in place by a slotted rubber stopper). Centrifuge 15–20 min. at about 1200 rpm. This should separate the mixture into a lower aqueous layer and an upper oily layer, with a layer of curd between. If separation does not occur, add 30–40 ml of toluene, mix, and repeat the centrifuging. Using a pipet inserted through the pipet guard, remove as much of the aqueous layer as possible and filter it into a 600 ml beaker. Add 5 vols of alcohol and let the mixture stand overnight to precipitate gums.

Decant or pipet off enough alcohol to leave not more than 225 ml, transfer the contents of the beaker to an 8 oz nursing bottle, centrifuge until the gum settles to the bottom, and decant the supernatant alcohol as completely as possible. Dissolve the residue in not more than 50 ml of hot water, add 1 or 2 ml of acetic acid and precipitate by adding alcohol to the 8 oz mark on the nursing bottle. Let stand overnight, or until the precipitate flocculates, centrifuge at 1200 rpm, and decant the alcohol. (A heavy flocculent precipitate at this point indicates a significant quantity of gum; a slight precipitate should not be considered a positive test for gums, as spices present in most mayonnaises and French dressings usually give such a precipitate.) Confirm the presence of gums as follows:

Add 35 ml of hot water to the precipitate in the nursing bottle, transfer to a small beaker, add 5 ml of HCl, and boil gently 2 minutes to hydrolyze gums to sugars. This solution may now be used for various qualitative tests for monosaccharide sugars as follows:

1. *Copper reduction test.* Transfer 1 ml of the hydrolyzed gum solution to a test tube, neutralize to litmus paper with approximately 2 *N* NaOH, and remove the paper. Add 5 ml of the Benedict qualitative solution and boil vigorously 1–2 min. Let cool spontaneously. A voluminous precipitate, which may be green, yellow, or red, indicates reducing sugars.

2. *Molisch test.* Transfer 5 ml of the hydrolyzed gum solution to a test tube, and add 2 drops of 15% alcoholic α-naphthol solution. Incline the tube and slowly pour down the inner side 3–5 ml of H_2SO_4 so that the 2 layers do not mix. A reddish-violet zone at the point of contact indicates carbohydrates. (5% Alcoholic thymol solution may be substituted for α-naphthol.)

Detection of Gums in Salad Dressings

See AOAC Method **28.053–28.055** and Text References 40 and 41. Gums are permitted in all dressings except mayonnaise, for which *see* Method 15–35.

Method 15-36. *Detection of Alginates in Food Dressings.* (*AOAC Method* ***28.056-28.057.*** Applicable to mayonnaise, French dressing and salad dressing.)

REAGENTS

a. *Acetone-alcohol mixture* (1 + 2).

b. *Dioxane.*

c. *Celite* 545.

d. *Ferric hydroxide-sulfuric acid reagent.*—Dissolve 10 g of $FeCl_3.6H_2O$ in about 100 ml of water in each of two centrifuge bottles, and precipitate the $Fe(OH)_3$ by adding excess NH_4OH (by odor). Wash the precipitate with about five successive portions of water, centrifuging and decanting until little odor of NH_3 remains. Break up the centrifuged precipitate each time before washing. Dry the precipitate on the steam bath or in an oven overnight, break up, and dry again. Mix with a spatula or grind in a mortar to obtain a moderately fine powder. Keep in a closed container. *See* AOAC Method **12.044.**

[Ferric hydroxide (moist) (code 1738, Baker & Adamson) may be used instead of precipitating $Fe(OH)_3$ as above. Transfer this product to centrifuge bottles, shake, centrifuge, decant, wash and dry as above.] Place 0.5 g of the dry powder in a 50 ml glass-stoppered graduated cylinder, add 50 ml of H_2SO_4, shake vigorously, and let settle until clear (usually 4–7 days). Some ferric sulfate appears to stick to the sides, but the reagent is ready for use after 7 days. Prepare fresh after 3 weeks. Check as follows before use:

Dissolve a small amount (1–5 mg) of commercial alginate in water containing 5 drops of 0.1 *N* NaOH, add 4 volumes of alcohol to precipitate the alginate, centrifuge, decant, and dry on the steam bath until no odor of alcohol remains, using an air current to remove the last traces of alcohol. Add 3 drops of 0.1 *N* NaOH, dissolve with the aid of glass rod, and add 2 ml of the Fe-H_2SO_4 reagent. The solution turns purple slowly, usually within 1 hour, depending on the amount of algin present, but the change may take longer. If the solution appears to be turning brown, add an additional 2 ml of the reagent, mix with a glass rod, and let stand.

Preparation of Sample

Weigh directly into a 250 ml centrifuge bottle 2 g, or enough sample to give 10–20 mg of alginate (about 2 g sample at the 0.5% level). Disperse in 50 ml of water and fill the bottle with acetone-alcohol (1+2). Shake vigorously and let stand until the precipitate begins to settle; then centrifuge at 1600–1700 rpm about 10 min. Decant, discard the liquid, disperse in 50 ml of water and reprecipitate two more times with alcohol. To the precipitate add 50 ml of dioxane, shake well, and filter through an asbestos-matted Gooch crucible with suction. Transfer the precipitate to the Gooch crucible, rinse the bottle and precipitate with several portions of dioxane, and suck dry. Return the precipitate and asbestos mat to the centrifuge bottle, add 50 ml of water, disperse the precipitate, and adjust to pH 8–9 with 3% NaOH solution. (With French dressing add 0.25 g of Celite 545, shake, and let stand 10 minutes, shaking several times. Centrifuge 10 minutes at 1600–1700 rpm. Decant the supernatant liquid, filtering if necessary to remove suspended particles.) To a 10 ml aliquot in a 50 ml centrifuge tube, add 40 ml of acetone-alcohol (1+2), stir vigorously, and let stand until a precipitate forms. If necessary, add 1 drop of saturated NaCl solution to start precipitation. Centrifuge, discard the liquid, and dry the precipitate on the steam bath.

Detection

(Start detection at the beginning of the day so the color changes can be observed.)

Moisten the precipitate with 3 drops of 0.1 *N* NaOH, rubbing thoroughly with a glass rod. Add 2 ml of Fe-H_2SO_4 reagent and let stand. Formation of a purple-red color indicates the presence of alginate. If a brown color forms, repeat the detection, using a smaller aliquot for final precipitation. Alginates are absent if no color develops on standing overnight.

Vinegars

Vinegars were made as far back as Old Testament days. Roman legions drank a diluted vinegar, made through a natural secondary fermentation of wine. Later methods were found by which vinegar could be made by a continuous primary and secondary fermentation of fruit juices.[48]

Vinegar is the acidic liquid developed by the alcoholic and subsequent acetous fermentation of starch or sugar-containing liquids. In the United States the term "vinegar" unmodified is restricted to cider vinegar. Other fruit vinegars, orange, pineapple, etc., are occasionally marketed.

No FDA Standard of Identity has been promulgated for vinegar. That Bureau issued an advisory definition and standard, without the force of law, in 1936.[43] This defined various vinegars, containing not less than 4 g/100 ml acetic acid, derived by alcoholic and subsequent acetous fermentation:

Cider Vinegar	From the juice of apples.
Wine Vinegar	From the juice of grapes.

Malt Vinegar	From an undistilled infusion of barley malt or cereals whose starch has been converted to malt.
Sugar Vinegar	From sugar, molasses or refiners syrup.
Glucose Vinegar	From a solution of glucose.
Spirit, Grain or Distilled Vinegar	From acetous fermentation of dilute distilled alcohol.

The Canadian definitions[44] are similar except that they have not defined "sugar vinegar" or "glucose vinegar". They further define vinegar as containing "not less than 4.1% or more than 12.3% acetic acid". They also permit addition of caramel, without label declaration, to all vinegars except spirit vinegar (also called "alcohol-, white- or grain vinegar").

The only additional analytical requirement in the Canadian regulations is for malt vinegar, which shall contain not less than 1.8 g of solids or 0.2 g of ash per 100 ml.

Adulteration and sophistication of vinegars was fairly common in the United States 25 or more years ago. This practice has declined materially, except for the substitution of distilled vinegar, or even dilute acetic acid, in part for fruit or malt vinegar, and the adulteration of distilled vinegar with dilute acetic acid.

Analysis of Vinegars

In general, the methods given in Chapter 11 for fruit juices and jellies, for ash and ash constituents, sugars, and titratable acidity, are applicable to vinegars. Results are usually reported as "per 100 ml" rather than on a weight basis.

Method 15-37. *Total Solids in Vinegar.* (*AOAC Method **28.062***).

Transfer 10 ml into a weighed 50 mm flat-bottom platinum dish, evaporate on a boiling water-bath for 30 minutes, and dry exactly 2½ hours in a water oven at the temperature of boiling water. Cool in a desiccator and weigh.

It is necessary to use a dish of the diameter and shape specified, and to dry for the exact time specified, in order to obtain concordant results. This is due to the fact that acetic acid is persistently retained in the dried solids and can be removed only by the repeated addition and evaporation of water to the dry residue. This latter procedure is not official in the AOAC method.

Method 15-38. *Total Acidity* as *Acetic Acid.* (*AOAC Method, **28.069***).

Dilute 10 ml of vinegar with recently boiled and cooled distilled water until it is only slightly colored, add phenolphthalein indicator solution, and titrate with 0.5 *N* NaOH to a slight pink color. 1 ml of 0.5 *N* NaOH is equivalent to 0.0300 g of acetic acid.

Method 15-39. *Non-Volatile Acids.* (*AOAC Method, **28.070***).

Transfer 10.0 ml to a porcelain or platinum dish, and evaporate on the steam bath just to dryness. Add 5–10 ml of water and repeat the evaporation at least 5 times until no acetic acid odor is detected. Take up with 150–200 ml of recently boiled and cooled distilled water and titrate with 0.1 *N* NaOH and phenolphthalein. This is conventionally reported as grams of malic (for cider vinegar) or tartaric (for wine vinegar) acid, although the non-volatile acid present is mostly lactic, the major portion of other fruit acids being converted to lactic acid during the fermentation. 1 ml of 0.1 *N* NaOH is equivalent to 6.7 mg of malic or 7.5 mg of tartaric acid, or 6.0 mg of acetic acid if volatile acidity is calculated by subtracting non-volatile acidity from total acidity. Total and volatile acidities are almost identical in distilled vinegar.

Method 15-40. *Permanganate Oxidation Number.* (*AOAC Method **28.084***).[45, 46]

REAGENTS

a. *$KMnO_4$ solution*, 31 g/L. Boil the solution 1 hour. Let stand over-night protected from dust, and filter through a Caldwell crucible and a 3 mm asbestos mat, using moderate suction. Make to volume. Standardization is unnecessary.

b. *Sodium thiosulfate*, 0.5 *N*, accurately standardized against 0.5000 g of NBS Standard Sample $K_2C_2O_7$, using 10 g of KI and 90 ml of water.

c. *KI solution*, 50 g/100 ml, filtered. Use only if colorless.

DETERMINATION

Adjust the sample to 4.00 g/100 ml acidity as acetic. Steam distill 50 ml of adjusted sample, using an all-glass distillation assembly, Fig. 10-11, and collect 50 ml of distillate. Regulate distillation so that about 45 ml remains in the distilling flask when 50 ml has distilled. Keep distillate and reagents at 25°.

Transfer the 50 ml of distillate to a 500 ml glass-stoppered Erlenmeyer flask, add 10 ml of H_2SO_4 (1+1) and 25 ml of the $KMnO_4$ solution (**a**), accurately measured, draining the pipette a definite time. Hold in a water bath at 25° for exactly 1 hour, then immediately add 20 ml of KI solution (**c**) and mix well. Titrate the liberated I with 0.5 *N* $Na_2S_2O_3$ (**b**).

Run a blank at the same time, using 50 ml of distilled water, 10 ml of H_2SO_4 (1+1), and 25 ml of $KMnO_4$ solution (**a**), draining the pipette the same length of time as before.

If the permanganate oxidation number, as calculated below, is greater than 15, repeat using 25 ml of adjusted sample + 25 ml of water. Calculate on the basis of 50 ml of the adjusted sample.

CALCULATION

Oxidation No. =

$$= \frac{(\text{ml } 0.5\ N\ Na_2S_2O_3 \text{ for blank} - \text{ml } 0.5\ N\ Na_2S_2O_3 \text{ for sample})}{2}.$$

Composition of Authentic Vinegars

Vinegars, the end result of a series of fermentations, contain certain constituents carried over from the unfermented raw material. This is particularly true of fruit vinegars. The determinations that are made to ascertain the genuineness of fruit juices may be used in part to determine the origin of the fruit vinegar. Consult Chapter 11.

Michael[47] described the manufacturing procedures current in the industry in 1949. There were three:

1. *Vat Method.* This is the oldest method, probably discovered accidentally through the spoilage of wine by subsequent acetous fermentation. In this method, the juice is allowed to ferment spontaneously in vats or barrels, the barrels being left partially filled with the bung open until the juice turns to vinegar.

2. *Upright Wooden Cask Generation Method.* Casks up to 15 feet high and 5–8 feet in diameter are filled with bundles of rattan placed upright, followed by a perforated section and another stack of rattan bundles. Inoculated hard cider or fermented fruit juice is introduced at the top of the generator and allowed to trickle down to an orifice at the bottom. This method is relatively inefficient, making cider vinegar to a maximum acidity of 6% acetic acid or less.

3. *The Modern Generator.* This is an upright cylindrical tank, up to 20 feet high and 10 to 14 feet in diameter. The generator is filled with corn cobs or rattan bundles, or, more frequently, with tightly coiled beechwood shavings. The tank is equipped with adjustable openings for the admission of air, since the rate of acetification is proportional to the amount of oxygen available to the organisms responsible for the acetification. Coke is sometimes used for the filler, particularly for manufacture of distilled vinegar. A pump draws off the fermenting liquid from the bottom and recirculates it through a rotating sprinkler at the top of the tank. The pump pipes are water-jacketed to maintain the juice at the optimum temperature of 30°. A normal charge in this generator is about 2750 gal, circulated at a rate of 750 gal/hour. This is the so-called "Frings process".

Vinegar is usually placed in filled tanks or barrels to develop higher alcohols, aldehydes and esters, similarly to the aging of wine. It is then "fined", either by filtration or by the addition of a clarifying agent such as casein, gelatin or bentonite clay, followed by filtration.

There are many tabulations of analyses of authentic vinegar in earlier literature (before 1920) that cannot now be used as references to establish the genuineness of vinegars, because the modern "quick-generator" methods make vinegar of different composition from that made by vat, or slow-generator, methods. Michael relies chiefly on the "permanganate oxidation" figure of the distillate, determined by Method 15-40. This is particularly valuable in determining adulteration of distilled vinegar with dilute commercial acetic acid.

McNeill and Henry reported the following oxidation numbers on authentic vinegars and acetic acid, adjusted to 4% acidity:[46]

Product	$KMnO_4$ Oxidation No. of Distillate
Destructively distilled acetic acid	0.1
Same, colored with caramel	0.1
Synthetic acetic acid	0.1
Same, colored with caramel	0.1
Distilled vinegar (white)	3.3–6.2
Cider vinegar	greater than 15
Molasses vinegar	greater than 15

They recommend "viewing with suspicion" oxidation numbers of distilled vinegar below 3.0.

Michael also reported the following analyses of six New Hampshire vinegars given in Table 15-6: (A private communication from Gilman K. Crowell, Chief of the New Hampshire Food and Chemistry Service, states that these were authentic cider vinegars, generated on beechwood shavings).

More recently a "submerged fermentation" method for manufacture of vinegar has been devised.[48] No packing or porous material is used. Finely dispersed air bubbles are circulated through the mash, thus facilitating the acetous fermentation. We have not found analyses of authentic vinegars produced by this method.

TABLE 15-6
Analyses of six New Hampshire vinegars

Sample	Acetic Acid, %	Total Solids, %	Alcohol % by wt.	$KMnO_4$ Oxidation No.
1	4.20	1.73	0.14	11.4
2	4.23	1.74	0.13	11.1
3	4.11	1.66	0.13	9.31
4	4.14	1.66	0.12	9.62
5	4.18	1.50	0.74	18.55
6	5.03	1.54	0.25	12.62

Analysis of Authentic Vinegars

The U.S. Food and Drug Administration collected and analyzed a series of 32 apple vinegars made in the apple regions of New York and Virginia in 1931–32.[49] FDA inspectors observed the entire processing operations and drew samples of the completed vinegars. The samples were later analyzed at the FDA laboratories in Washington. The results of these analyses are tabulated in Table 15-7 (*see p.* 360).

Fisher investigated the genuineness of 78 wine vinegars sold in Connecticut in 1946 and found gross adulteration.[50] He pursued the investigation in 1947, reporting the then known analyses of authentic wine vinegar.[51] Since these data may not represent current manufacturing practices they are given by reference only.

Webb and Galetto[52] of the University of California (Davis) examined 58 California red wine vinegar samples from 7 producers. These were all made by the modern submerged fermentation method. Of these, 34 were drawn at the end of the acetification stage before blending and filtering, and 22 were finished and bottled vinegars. Two made by the old style barrel fermentation were also examined for comparison. These authors place most reliance on the absorption in 1:60 dilution at about 280 mμ, and total tartrates as determined by the metavanadate method.[53]

TABLE 15-7

Authentic Vinegars from Virginia and New York—1930/1931 season

Results in grams/ml unless otherwise stated

Sample No.	Generator Packing	Solids	Ash	Alkalinity of Sol. Ash	Non-Sugar Solids	Sugar	Total Acid[b]	Non-Vol. Acid[b]
				ml N/10				
1	Coke	1.89	.33	43.8	1.75	.14	5.62	.01
2	Rattan	1.60	.33	39.4	1.51	.09	4.85	.04
3	Coke	2.23	.37	44.9	1.85	.38	6.22	.02
4	Coke	1.77	.33	39.8	1.65	.12	5.42	.02
5	Beechwood	1.52	.33	35.4	1.32	.20	4.69	.01
6	Beechwood	1.96	.40	47.9	1.78	.18	5.02	.05
7	Beechwood	2.64	.40	50.8	2.31	.33	6.18	.02
8	Corn Cob	1.75	.36	38.0	1.55	.20	5.08	.02
9	Pumice	1.58	.31	33.3	1.45	.14	5.24	.01
10	Pumice	1.50	.38	44.2	1.43	.07	6.07	.02
11	Pumice	1.33	.36	42.9	1.21	.12	5.65	.03
12	Pumice	1.43	.36	41.6	1.33	.09	6.29	.04
13	Pumice	1.53	.34	40.4	1.45	.08	5.38	.03
14	Pumice	1.52	.28	39.6	1.34	.18	5.60	.02
15	Pumice	1.52	.36	36.8	1.33	.19	5.70	.03
16	Rattan	1.47	.28	34.0	1.32	.15	5.08	.01
17	Rattan	1.40	.27	33.9	1.29	.11	5.03	.01
18	Rattan	1.40	.28	33.9	1.30	.10	5.27	.01
19	Beechwood[a]	1.75	.31	35.0	1.49	.26	5.73	.04
20	Beechwood[a]	1.54	.28	32.8	1.39	.15	5.90	.05
21	Coke & Cob	1.50	.28	30.6	1.36	.15	5.03	.01
22	Cob, Coke & Pumice	2.50	.35	38.1	1.97	.53	6.00	.05
23[c]		2.26	.29	29.8	1.91	.35	0.36	.14
24[c]		2.38	.37	36.4	2.13	.25	0.44	.14

[a] Rotating generators.
[b] Expressed as acetic acid.
[c] Stocks sampled from a concrete catch tank.

Sample No.	Alcohol	Protein N × 6.25	CaO	K	Al	$\frac{K}{Protein}$	SO_3
			mg/100 ml	mg/100 ml	mg/1000 ml		mg/100 ml
1	.16	.025	8.19	.163	1.4	6.5	2.8
2	.21	.018	8.63	.153	0.7	8.5	3.6
3	.20	.016	7.31	.164	3.8	10.0	3.5
4	.32	.022	7.75	.140	—	6.4	—
5	.39	.015	8.21	.150	1.2	10.0	2.9
6	.27	.015	8.99	.121	1.0	8.0	3.9
7	.17	.016	9.50	.163	—	10.2	4.0
8	.12	.016	10.44	.164	1.4	10.3	3.0
9	.40	.014	7.65	.134	—	9.5	—
10	.12	.030	10.86	.179	5.3	6.0	—
11	.20	.019	8.89	.166	3.7	8.7	—
12	.16	.019	9.60	.161	—	8.5	4.2

TABLE 15-7 (continued)

Sample No.	Alcohol	Protein N×6.25	CaO	K	Al	K/Protein	SO_3
			mg/100 ml	mg/100 ml	mg/1000 ml		mg/100 ml
13	.16	.018	8.30	.166	3.4	9.2	—
14	.09	.046	8.89	.161	—	3.5	2.9
15	.27	.092	9.72	.170	2.9	1.8	—
16	.30	.017	6.18	.136	0.4	8.0	2.4
17	.17	.015	6.03	.132	—	9.0	3.1
18	.24	.014	10.02	.138	—	9.8	2.9
19	.35	.015	8.74	.145	2.0	9.6	5.5
20	.22	.015	9.94	.138	—	9.2	1.6
21	.24	.028	8.48	.129	1.0	4.6	6.2
22	.47	.019	12.40	.162	3.7	8.5	2.9
23	4.26	—	15.70	—	—	—	4.1
24	5.17	—	—	—	—	—	7.3
8 additional samples, 1932 season, gave:			12.3				3.9
			11.4				3.9
			11.7				4.3
			9.9				4.1
			9.7				3.5
			9.8				4.1
			10.1				4.4
			9.4				3.4

They conclude: "Using total tartrates and absorbance analyses, together with analyses for total solids, ash, ash alkalinity, phosphorus, nitrogen, esters, acetoin, methylguanidine, acetic acid anhydride, amino acids and pigments . . . differentiation between red wine vinegars and other vinegars should be very nearly certain."

These authors did not analyze malt or cider vinegars. In a private communication, Dr. Webb stated he would not expect to find any tartrates and only low absorbances at 280 mμ in these vinegars. This check should be made by any analyst interested in using these methods to distinguish wine vinegars from other vinegars.

Analyses of Authentic Wine Vinegars[52]

Results expressed as grams/100 ml

	Volatile Acid as Acetic	Tartrates[a]	Absorbance, 280 mμ
34 unfinished vinegars	5.45–10.49 (mean 9.54 $\pm$ 0.63)	112–180 (mean 134 $\pm$ 13)	0.241–0.640 (mean 0.352 $\pm$ .083)
22 finished, bottled vinegars	4.25–6.96 (mean 5.44 $\pm$ 0.44)	52–225 (mean 120 $\pm$ 50)	0.028–0.338 (mean 0.185 $\pm$ .071)

[a] As mg/100 ml of tartaric acid.

The wide range in tartrates and absorbance in the finished vinegars is probably due to greater dilution with water of the higher acidity unfinished vinegars.

Synthetic Vinegars

The U.S. Department of Defense issued military specification MIL-V-35017 in September 1956 titled 'Vinegar, Dry, Synthetic''. The formula given is:

Lactose	500 lbs.
d-1 malic acid	500 lbs.
Glacial acetic acid	18 gal.
350-fold apple essence	3 gal.
citric acid	$\frac{1}{4}$ lb.
caramel coloring	$\frac{1}{8}$ lb.

The analytical requirements for the diluted, ready-to-use product (4 oz added to one quart of water) are:

sp.g.	1.04 to 1.05 @ 60° F.
ash	0.04 percent maximum
titratable acidity	5.5 to 6.0 percent as acetic acid.
pH	2.0 (with maximum deviation 0.2) by an electrometric pH meter at 25°.

This specification was issued by the Department of Defense for use by the Armed Forces for general issue.

Adulteration of Wine Vinegar with Cider Vinegar.

Sorbitol is a natural constituent of cider, and, since it is not fermentable, it is carried over into cider vinegar. Previous literature reports the sorbitol content of cider as 300–800 mg/100 ml. Grape juice is said to contain less than 6 mg/100 ml.

Minsker[54] reported an average sorbitol content of 210 mg/100 ml in 8 authentic cider vinegars, while wine vinegar gave no weighable amount by an *o*-chlorobenzaldehyde condensation method first studied by Litterscheid.[55] While the range of sorbitol in cider vinegar (185–250 mg/100 ml) found by Minsker is rather wide, this is of little moment, since adulteration of wine vinegar with cider vinegar must be of a rather gross nature to be profitable, and the method recommended by Minsker seems able to detect with certainty the addition of 10% of cider vinegar to wine vinegar.

***Method 15-41.** Sorbitol in Vinegar.*

REAGENTS

a. *Darco decolorizing charcoal*—Atlas Powder Co. Grade G60.

b. *o-Chlorobenzaldehyde*—Eastman Organic Chemicals No. 757, or equivalent.

DETERMINATION

Add 3 g of Darco decolorizing charcoal to 75 ml of vinegar in a 125 ml Erlenmeyer flask, cover with a watch glass, and heat on a steam bath for 15 minutes. Cool and filter through a folded filter. Pipette off 50 ml into a 150 ml beaker, evaporate to about 10 ml on an electric hot plate, and complete evaporation to dryness on a steam bath, using a current of air. Add 50 ml of absolute methanol and 0.5 g of the Darco charcoal to the residue and heat on a water bath at 60° for 10 minutes, breaking up the residue with a plastic or rubber policeman. Filter through a folded filter and wash the flask and residue with 20 and 10 ml portions of absolute methanol, collecting the filtrate and washings in a 125 ml glass-stoppered Erlenmeyer flask.

Evaporate the filtrate to about 5 ml on a steam bath, then to dryness at room temperature, using a current of air. Take up the residue in 5 ml of water, add 20 ml of HCl and 5 drops of *o*-chlorobenzaldehyde. Shake violently for a few seconds and let stand 30 minutes. Add 5 more drops of *o*-chlorobenzaldehyde and again shake. Let stand in a refrigerator at about 40° F for 35–40 hours.

Remove from the refrigerator, add 20 ml of water, and shake violently to break up the precipitate as finely as possible. Filter on a tared medium porosity sintered glass crucible, transferring with 50 ml of water followed by 50 ml of ethanol. Wash with water, then with alcohol, dry at 100° for 1 hour and weigh.

Divide the weight of the precipitate by 2.6 to obtain the weight of sorbitol obtained from 50 ml of vinegar. Using the average figure of 210 mg/100 ml of sorbitol found in cider vinegar, the approximate degree of adulteration with cider vinegar may be calculated.

TEXT REFERENCES

1. *Service and Regulatory Announcements. Food & Drug No.* 2 (*Revision of* 1936).
2. *Office Consolidation of the Food & Drugs Act and Food & Drugs Regulations.* Canada Dept. of National Health and Welfare. Ottawa: Queen's Printer (July 14, 1966).
3. *Federal Specification, EE-S-*631*e, Spices, Ground and Whole.* Washington: U.S. Government Printing Office (1967).
4. Klemm, G. C.: *J. Assoc. Offic. Agr. Chemists,* **47:**40 (1964).
5. Clevenger, J. F.: *J. Am. Pharm. Assoc.,* **17:**345 (1928).
6. Clevenger, J. F.: *Ibid.,* **21**:30 (1932).
7. Clevenger, J. F.: *J. Assoc. Offic. Agr. Chemists,* **16:**557 (1933), **17:**371 (1934).
8. Lee, L. L., and Ogg, C. L.: *J. Assoc. Offic. Agr. Chemists,* **39:**807 (1956).
9. Tice, L. F.: *Am. J. Pharm.,* **105:**320 (1933).
10. Joint Committee of the Pharmaceutical Society: *Analyst,* **84:**603 (1959), **89:**377 (1964).
11. Ting, S. W., and Barrons, K. C.: *Proc. Am. Soc. Hort. Sci.,* **40:**504 (1942).
12. Todd, P. H. Jr.: *Food Technol.,* **12:**468 (1958), **15:**270 (1961).
13. Suzuki, J. I., *et al.*: *Food Technol.,* **11:**100 (1957).
14. Hayden, A. and Jordan, C. B.: *J. Am. Pharm. Assoc.,* **30:**107 (1941).
15. Datta, P. R., and Susi, H.: *Anal. Chem.,* **33:**148 (1961).
16. Brauer, O., and Schoen, W. J.: *Angew. Botan.,* **36**(1):35 (1962).
17. Scoville, W. L.: *J. Am. Pharm. Assoc.,* **1:**453 (1912).
18. Wirth, E., and Gathercoal, E. N.: *J. Am. Pharm. Assoc.,* **13:**217 (1924).
19. *AOAC Methods of Analysis,* 10th ed., **28.008** (1965).
20. Graham, H. D.: *J. Food Sci.,* **30:**644, 651 (1965).
21. Genest, C., *et al.*: *J. Agr. Food Chem.,* **11**(6):508 (1963).
22. *Spices, What they are—Where They Come From.* New York: American Spice Trade Association (1961).
23. Roe, J. E.: *J. Assoc. Offic. Agr. Chemists,* **40:**781 (1957).
24. Terry, R. C. and Curran, J. M.: *Analyst,* **64:**164 (1939).
25. Ref. 19, **28.022.**
26. Tandon, G. L., Dravid, S. V. and Siddappa, G. S.: *J. Food Sci.,* **29:**1 (1964).
27. Smith, E. R.: *J. Assoc. Offic. Agr. Chemists,* **9:**333 (1926).
28. Carol, J., and Ramsey, L. L.: *J. Assoc. Offic. Agr. Chemists,* **36:**937 (1953).
29. Jennings, W. G. and Wrolstad, R. E.: *J. Food Sci.,* **26:**499 (1961).
30. Hasselstrom, T. E., Hewitt, E. J., Konigsbacher, K. S. and Ritter, J. J.: *J. Agr. Food Chem.,* **5:**53 (1957).
31. Datta, P. R., Susi, H., Higman, H. C. and Filipic, V. J.: *Food Technol.,* **16**(10):116 (1962).
32. Lee, L. L., Caruso, V. J.: *J. Assoc. Offic. Agr. Chemists,* **41:**446 (1958).
33. Dutta, A. B.: *J. Assoc. Offic. Agr. Chemists,* **44:**639 (1961).
34. *Code of Federal Regulations,* Title 21, *Food and Drugs,* parts 1-129. Reg. **25.1–25.3.** Washington: Office of the Federal Register (1966).
35. Ref. 2. Regulations **B.07.031–033.**
36. Ref. 34. Regulation **42.20.**
37. Fine, S. D.: *J. Assoc. Offic. Agr. Chemists,* **27:**263 (1944).
38. Hart, F. L.: Unpublished work (1934).
39. Hart, F. L.: *J. Assoc. Offic. Agr. Chemists,* **20:**532 (1937), **22:**605 (1939), **23:**597 (1940).
40. Gnagy, M. J.: *J. Assoc. Offic. Agr. Chemists,* **35:**358 (1952).
41. Redfern, S.: *J. Assoc. Offic. Agr. Chemists,* **29:**250 (1946).
42. Wiger, O. R.: *J. Assoc. Offic. Agr. Chemists,* **46:**623 (1963), **47:**389 (1964).
43. *Service and Regulatory Announcement, Food and Drug No.* 2, Rev. 5. U.S. Dept. Agr. (1936).
44. Ref. 2. Regulations **B.19.001–009.**
45. Michael, G. A.: *J. Assoc. Offic. Agr. Chemists,* **35:**229 (1952).
46. McNeill, R. E., and Henry, A. N.: *J. Assoc. Offic. Agr. Chemists,* **27:**263 (1944).
47. Michael, G. A.: *Assoc. Food & Drug Officials U.S., Quart. Bull.,* **13:**93 (1949).
48. Mayer, E.: *Food Technol.,* **17:**582 (1963).

49. Unpublished work, U.S. Food & Drug Administration.
50. Fisher, H. J.: *Conn. Agr. Exp. Sta. Bull.* **510** (1947).
51. Fisher, H. J.: *Conn. Agr. Exp. Sta. Bull.* **528** (1948).
52. Webb, A. D., and Galetto, W.: *Am. J. Enol. Viticult.*, **16**(2):79 (1965).
53. Amerine, M. A.: *Laboratory Procedures for Enology*. Davis: University of California (1960).
54. Minsker, F. J.: *J. Assoc. Offic. Agr. Chemists*, **45**:562 (1962).
55. Litterscheid, F. A.: *Z. Untersuch. Lebensm.*, **62**:653 (1931).

SELECTED REFERENCES

GENERAL

A. American Spice Trade Association: *Official Analytical Methods*. New York: American Spice Trade Association (1960).
B. *A Glossary of Species*. New York: American Spice Trade Association (1966).
C. Chipeault, J. R., Mizuno, G. R., Hawkins, J. M., and Lundberg, W. D.: "Antioxidant Properties of Spices". *Food Res.*, **17**:46 (1952), **20**:44 (1955); *Food Technol.*, **10**:209 (1958).
D. Guenther, E. E.: *The Essential Oils*, I-VI. New York: D. Van Nostrand Co. (1947-1952).
E. Parry, J. W.: *The Spice Handbook*. New York: Chemical Publishing Co. (1945).
F. Parry, J. W.: *Spices, their Morphology, Histology and Chemistry*. New York: D. Van Nostrand Co. (1962).

ANALYSIS OF AUTHENTIC SPICES

G. Winton, A. E., and Winton, R. B.: *Structure and Composition of Foods*, IV. New York: J. Wiley & Sons (1939).
H. Bureau of Chemistry: *Bulletin* **13:**11 (1887).
I. Bureau of Chemistry: *Bulletin* **152** (1912).
J. Canada Inland Revenue Dept.: *Bulletin* **73.**
K. Canada Inland Revenue Dept.: *Bulletin* **252.**
L. Conn. Agr. Exp. Sta., New Haven: *Annual Report* (1898).
M. Conn. Agr. Exp. Sta., New Haven: *Annual Report* (1899).
N. Conn. Agr. Exp. Sta., New Haven: *Annual Report* (1901).
O. Conn. Agr. Exp. Sta., New Haven: *Annual Report* (1903).
P. Conn. Agr. Exp. Sta., New Haven: *Annual Report* (1905).
Q. Jacobs, M. B.: *Chemical Analysis of Foods and Food Products*. New York: D. Van Nostrand Co. (1938).
R. Leach, A. E., and Winton A.: *Food Inspection and Analysis*, 4th ed. New York: J. Wiley & Sons (1920).
S. Michigan Dairy and Food Commission: *Bulletin* **94** (1903).
T. Sancho, J. -G.: *Pimientos y Pimienton, Estudia Quimica, Fisico*. University of Murcia (1962).
U. Woodman, A. G.: *Food Analysis*, 4th ed. New York: McGraw-Hill Book Co. (1941).

CONDIMENTS

V. Amerine, M. A.: "The Fermentation Industries after Pasteur." *Food Technol.*, **19**:757 (1965).
W. Balcolm, R. W.: "Composition of Vinegar." *U.S. Dep. Agr., Bur. Chem., Bull.* **132** (1910).
X. Bailey, E. M.: "Composition of Salad Dressi ngs". *Conn. Agr. Exp. Sta. Bull.* **329** (1931).

CHAPTER 16

Sugars and Sugar Products

General

Definitions

The compounds known as carbohydrates acquired that name at a time when the known members of the group had the empirical formula $C_x(H_2O)_y$ and were considered to be hydrates of carbon. By the time the structures of these compounds had been elucidated and it was discovered that the simple two-to-one hydrogen-oxygen relationship did not always hold, the name had become so well established that it was retained.

Carbohydrates may be defined simply as polyhydroxyaldehydes or ketones, and derivatives thereof. The term *sugar* is applied to the simpler carbohydrates (monosaccharides and oligosaccharides) which have a more or less sweet taste.

While other sugars will have to be taken into consideration in special cases, nearly all of the food analyst's concern will be with two hexoses, dextrose (glucose)* and levulose (fructose), and the three disaccharides sucrose, lactose and maltose. The trisaccharide raffinose is encountered as a minor ingredient of beet molasses, and melezitose occurs in certain honeys. Another term frequently met in sugar analysis is "invert sugar"; this refers to the equimolecular mixture of dextrose and levulose formed by the hydrolysis of sucrose.

The disaccharides are condensation products of the monosaccharides, and on hydrolysis each molecule of a disaccharide yields two hexose molecules which may or may not be different, and one molecule of water. Similarly three hexose and two water molecules are produced from each molecule of a trisaccharide (*see* top p. 366).

Milk is the only natural food product that contains lactose. Maltose is formed by the partial hydrolysis of starch.

*The terms usually employed in theoretical carbohydrate chemistry are "glucose" and "fructose", but (partly because of confusion caused by a former official standard defining "glucose" as "a thick, sirupy colorless product made by incompletely hydrolyzing starch") the names "dextrose" and "levulose" will be used in this book.

Relative Sweetness and Specific Rotations of the Sugars

Table 16-1[1] compares the relative sweetness (sucrose = 100) of several sugars, and their equilibrium specific rotations at 20° with sodium D light:

Disaccharide or trisaccharide	*Hexoses formed on complete hydrolysis*
Sucrose	Dextrose and levulose
Lactose	Dextrose and galactose
Maltose	Dextrose (two molecules)
Raffinose	Galactose, dextrose and levulose
Melezitose	Dextrose (2 molecules) and levulose

TABLE 16-1
Comparison of the Sweetness and Specific Rotations of Some Common Sugars

Sugar	Relative Sweetness	Specific Rotation (Circular Degrees)
Lactose	16	+ 52.5°
Raffinose	22	+104.0°
Galactose	32	+ 80.2°
Rhamnose	32	+ 8.3°
Maltose	32	+137.0°
Xylose	40	+ 18.8°
Dextrose	74	+ 52.5°
Sucrose	100	+ 66.5°
Invert sugar	130	—
Levulose	173	− 92.3°

Methods of Analysis

Because all of the monosaccharides and some of the disaccharides (lactose and maltose) contain a free aldehyde or ketone group, they act as reducing agents, and since solutions of nonreducing disaccharides such as sucrose yield monosaccharides on acid hydrolysis, such solutions also become reducing upon proper acid treatment. All of the older well-established methods for the chemical determination of hexose and disaccharide sugars are based on the fact that neutral solutions of these sugars (with or without prior acid hydrolysis) reduce alkaline solutions of salts of the heavy metals. The archetypal reaction is that between sugar solutions and Fehling's solution (a solution of copper sulfate, potassium sodium tartrate and sodium hydroxide), which produces on heating a precipitate of cuprous oxide proportional to the amount of sugar present. The official methods in their present forms vary in the compositions of the precipitating reagents used, but all of these reagents contain copper sulfate and an alkali. Rochelle salt (potassium sodium tartrate) is also present in most of them. By modifying the alkalinity of the copper solution and the reaction temperature, it is possible to devise conditions under which only one sugar is appreciably oxidized; an example is the Jackson-Mathews modification of the Nyns method[2], which is selective for levulose.

There are also variations in the techniques by which the amount of precipitated cuprous oxide is estimated. The simplest way is to dry and weigh the precipitate. If there is no interference from precipitation of substances other than Cu_2O, this method is capable of yielding results as accurate as any other, but in the case of certain plant materials that are relatively high in calcium and low in sugars, enough calcium can be precipitated by the tartrate to cause serious interference in a gravimetric estimation. Other methods that are more specific for copper depend on: 1), dissolving the Cu_2O in nitric acid and either electrolyzing for copper or titrating with sodium thiosulfate after addition of KI; or 2), dissolving in ferric sulfate solution and titrating with potassium permanganate.

Classical methods for the determination of sugars all depend on one or the other of two principles: the reduction of an alkaline copper solution or measurement of optical activity. There are many variations of the copper reduction method that have been designed for specific problems, and only a selection of those most often used has been included in the present volume. For information on others the reader is referred to Selected Reference D.

We have, however, included information on some of the more modern methods, such as that depending on the use of glucose oxidase, and gas and thin layer chromatographic techniques for the separation of individual sugars.

Method 16-1. *Preparation of Sample.* (*AOAC Method* ***29.001***).

A. *Solids* (*sugars, etc.*).—Grind, if necessary, and mix thoroughly to secure uniform samples. Raw sugars should be mixed thoroughly and in the shortest possible time, either on a glass plate with a spatula, reducing lumps (when present) with a glass or iron rolling pin, or in a large, clean, dry mortar, using a pestle to reduce the lumps.

B. *Semi-solids* (*massecuites, etc.*).—Weigh a 50 g sample, dissolve any sugar crystals in a minimum quantity of water, wash into a 250 ml volumetric flask, dilute to the mark, and mix thoroughly, or weigh a 50 g sample and dilute with water to 100 g. If insoluble material remains, mix uniformly by shaking before taking aliquots or weighed portions for various determinations.

C. *Liquids* (*molasses, syrups, etc.*).—Mix the materials thoroughly. If sugar crystals are present, dissolve them either by heating gently (avoiding loss of moisture by evaporation) or by weighing the whole mass, adding water, heating until completely dissolved, and, after cooling, reweighing. Calculate all results to weight of original substance.

Method 16-2. *Moisture.* (*AOAC Methods* ***29.006*** *and* ***29.008***).

A. *Vacuum drying* (*applicable to both cane and beet, raw and refined sugars.*—Dry a 2–5 g prepared sample in a flat dish (Ni, Pt or Al with tight-fit cover) for 2 hours at a temperature not over 70° (preferably 60°) and a pressure not exceeding 50 mm of mercury. Remove the dish from the oven, cover, cool in a desiccator, and weigh. Redry 1 hour, cool, and weigh. Repeat until the change in weights between successive dryings does not exceed 2 mg.

B. *Drying upon quartz sand* (*applicable to massecuites, molasses, and other liquid and semi-liquid products*).—Digest pure quartz sand that passes a No. 40 but not a No. 60 sieve with HCl, wash acid-free, dry, and ignite. Preserve in a stoppered bottle. Place 25–30 g of this prepared sand and a short stirring rod in a dish about 55 mm in diameter and 40 mm deep, fitted with a cover. Dry thoroughly, cover the dish, cool in a desiccator, and weigh immediately. Add enough of a diluted sample of known weight to yield about one gram of dry matter, and mix thoroughly with the sand. Heat on a steam-bath 15–20 minutes, stirring at 2–3 minute intervals or until the mass becomes too stiff to manipulate readily. Dry at under 70° (preferably 60°) and at a pressure not exceeding 50 mm Hg, making trial weighings at 2 hour intervals towards the end of the drying period (about 18 hours), until the change in weight is not over 2 mg.

NOTES:

1. Vacuum ovens should be bled with a current of dry air to ensure removal of water vapor.
2. Since dry sand, as well as the dried samples, absorb appreciable quantities of moisture on standing over most desiccating agents, all weighings should be made as quickly as possible after cooling samples in a desiccator.

Method 16-3. *Solids by Means of a Refractometer. AOAC Method* ***29.011*** *Modified.* (Applicable only to liquid samples containing no undissolved solids).

Determine the immersion refractometer reading of the solution at 20°, and obtain the corresponding percentage of dry substance from Table 23-5. Circulate water through the trough of the immersion instrument long enough for the temperatures of the prisms and sample to reach equilibrium, and continue the circulation during the observations, taking care to keep the temperature of the water constant.

If the determination is made at a temperature other than 20°, correct the readings to the standard temperature from Table 23-4.

Method 16-4. *Ash.* (*AOAC Method **29.012***).

Heat a sample of appropriate weight (usually 5–10 g) in a 50–100 ml Pt dish at 100° until moisture is expelled, add a few drops of pure olive oil and heat slowly over a flame or under an infrared lamp until swelling stops. Place the dish in a muffle at about 525° and leave until a white ash is obtained. Moisten the ash with water, dry on the steam bath and then on the hot plate, and re-ash in the muffle at 525° to constant weight.

Method 16-5. *Total Nitrogen.* (*AOAC Method **29.019***).

Determine nitrogen as in Chapter 1, Method 1-7, using a 5 g sample and, if necessary for complete digestion, a larger quantity of H_2SO_4.

Method 16-6. *Sucrose.*

PREPARATION AND USE OF CLARIFYING AGENTS

a. *Basic lead acetate solution.*[3]—Activate litharge (PbO) by heating 2.5–3 hours at 650–670° in a muffle (cooled product should have a lemon color). Boil 430 g of neutral lead acetate, 130 g of the freshly-activated litharge, and 1 liter of water, for 30 minutes. Let the mixture cool and settle; then dilute the supernatant liquid with recently boiled water to a specific gravity of 1.25.

NOTES:

1. Solid basic lead acetate may be substituted for the normal salt and litharge in preparing this solution, or may be used as the dry salt as a substitute therefor (about $\frac{1}{5}$ g for each 1 ml of the solution).

2. Because of the error caused by the volume of precipitate, basic lead acetate solution is not recommended for clarifying products of low purity. The dry salt, if not in excess, does not cause a volume error, but when molasses or any other substance producing a heavy precipitate is being clarified, some dry, coarse sand should be added to break up pellets of the salt and precipitate.

b. *Alumina cream.*[4]—Prepare a cold saturated solution of ammonia or potash alum in water. Add NH_4OH stirring constantly until the solution is alkaline to litmus. Let the precipitate settle, and wash by decantation with water until the wash water gives only a slight test for sulfates with $BaCl_2$ solution. Pour off the excess water and store the residual cream in a stoppered bottle.

Alumina cream is suitable for clarifying light-colored sugar products or as an adjunct to other agents when sugars are determined by polariscopic or reducing sugar methods.

c. *Neutral lead acetate solution.*[4a]—Prepare a saturated solution of neutral lead acetate and add to the sugar solution before diluting to volume.

This reagent may be used for clarifying light-colored sugar products when sugars are determined by polariscopic methods, and its use is imperative when reducing sugars are determined in the solution used for polarization.

To remove excess Pb used in clarification, add to the clarified filtrate anhydrous K or Na oxalate in small quantities until a test for Pb in the filtrate is negative; then refilter.

d. *Ion-exchange resins.*[5]—Amberlite IR-120 H, C.P.Medium Porosity, obtainable from Mallinckrodt Chemical Works, 2nd and Mallinckrodt Streets, St. Louis, Mo. 63160, as Catalogue No. 3405; and Duolite A-4(OH), obtainable from Chemical Process Co., 901 Spring Street, Redwood City, Calif.

A mixture of 2 g of the Amberlite resin and 3 g of the Duolite resin is used for final purification for sugar analysis of an alcoholic extract of a vegetable or other plant product. With some plant materials, clarification with basic lead acetate does not suffice to remove all non-sugar reducing substances. Directions for preparation of a vegetable sample and use of these resins for its clarification are as follows:

For fresh vegetables, thoroughly remove all foreign matter and rapidly grind or chop the material into fine pieces. Add a weighed sample to sufficient hot redistilled 95% alcohol that the final alcohol concentration, allowing for the moisture content of the sample, is about 80%; then add enough precipitated $CaCO_3$ to neutralize any acidity. Heat nearly to boiling on a steam or water bath for 30 minutes, stirring frequently. Pour the alcoholic extract through a filter paper or extraction thimble, catching the filtrate in a volumetric flask. Transfer the insoluble material to a beaker, cover with 80% alcohol, warm on a steam bath one hour, let cool, and pour the alcoholic solution through the same filter as before. If the second filtrate is highly colored, repeat the extraction. Transfer the residue to the filter, let drain, and dry. Grind the residue so that all particles will pass through a 1 mm sieve, transfer to an extraction thimble, and extract 12 hours in a Soxhlet apparatus with 80% alcohol. Dry the residue and save for a starch determination. (*See* Method 17-1.) Combine the alcoholic filtrates and dilute to volume at a definite temperature with 80% alcohol.

For dried materials, grind finely, mix well, weigh the sample into a beaker, and proceed as above, beginning with "cover with 80% alcohol."

In either case, place an aliquot of the alcoholic extract in a beaker and heat on a steam-bath to evaporate the alcohol. (Avoid evaporation to dryness by adding some water.) When the odor of alcohol disappears, add about 15–25 ml of water and heat to 80° to soften gummy precipitates and break up insoluble masses. Cool to room temperature. Prepare a thin mat of Celite Analytical Filter-Aid on filter paper in a Büchner or fritted-glass funnel, and wash with water until the filtrate comes clear. Filter the sample through this mat, wash the mat with water, dilute the filtrate and washings to an appropriate volume in a volumetric flask, and mix well.

Place a 50 ml aliquot in a 250 ml Erlenmeyer flask, add 2 g of the Amberlite resin and 3 g of the Duolite resin, and agitate about every 10 minutes for a period of 2 hours. Filter the solution through a fluted filter paper, discarding the first few ml. Take a 5 ml aliquot of the filtrate for sugar determination.

NOTES:

1. Isopropanol may be substituted for ethyl alcohol.[6]

2. If starch is not to be determined, the following shorter method of sugar extraction may be used:

For fresh vegetables, make the preliminary extraction as above, except to boil 1 hour on the steam bath. Then decant the alcoholic solution into a volumetric flask, and comminute the undissolved portion with 80% alcohol in a high-speed blender. Boil the blended material one-half hour on the steam bath, cool, transfer to the volumetric flask, dilute to the mark with 80% alcohol at room temperature, filter, and take an aliquot of the filtrate for evaporation of alcohol and resin treatment as above.

In the case of dry material, grind to pass a 20-mesh sieve, transfer a weighed sample to a volumetric flask, and add 80% alcohol and enough $CaCO_3$ to neutralize any acidity. Boil 1 hour on the steam bath, cool, adjust to volume at room temperature with 80% alcohol, filter, and take an aliquot of the filtrate for evaporation of alcohol and resin treatment.

e. *Zinc sulfate-barium hydroxide.*[7]—Heat an aqueous extract of about 1 g of plant material to just below the boiling point, stir and simultaneously add 10 ml of 10% $ZnSO_47H_2O$ and an equivalent quantity of saturated $Ba(OH)_2$ solution. (The "equivalent quantity" is determined by titrating 10 ml of the $ZnSO_4$ soltuion with the $Ba(OH)_2$ solution to a phenolphthalein endpoint.) Keep the solution warm for 3 minutes, cool rapidly, dilute to 100 ml and filter.

DETERMINATION

Both of the classical methods for determining sucrose call for hydrolyzing this disaccharide to an equal mixture of the two monosaccharides dextrose and levulose. Because the negative rotation of levulose is numerically higher than the positive rotation of dextrose, the hydrolysis of a sucrose solution results in reversing the sign of rotation of such a solution. It is for this reason that acid hydrolysis of a sucrose solution is called "inversion", and that the mixture of monosaccharides so produced is known as "invert sugar".

Inversion is an essential part of the polarimetric method of determining sucrose because the resulting change in rotation makes it possible to distinguish between this disaccharide and other optically active sugars likely to be present with it, particularly preformed invert sugar. It is an even more essential part, if possible, of the copper reduction method, because the sucrose molecule contains no free aldehyde or ketone group, and consequently will not act as a reducing agent until it is transformed into a mixture of dextrose and levulose, both of which are reducing substances owing to their respective possession of terminal aldehyde and ketone groups.

A. Polarimetric method (*AOAC Method* ***29.026***).

Polarimetric sugar determinations are usually made with a type of polarimeter called a "saccharimeter", whose scale is designed to read directly in percent of sucrose. A saccharimeter calibrated in the International Sugar Scale gives a reading of exactly 100 degrees when white light filtered through a potassium dichromate solution (of such concentration that % $K_2Cr_2O_7$ × length of solution in cm = 9) passes through a 200 mm tube filled with a "normal" solution of sucrose (26.000 g/100 ml) at 20°. Older saccharimeters graduated in the Herzfeld-Schönrock scale read 99.90° instead of 100° under these conditions, but such saccharimeters may be made to read directly in percent sucrose by taking a "normal" sample weight of 26.026 g.

If a polarimeter calibrated in circular degrees and using sodium light is employed for the method below, the readings obtained should be converted to the International Sugar Scale by multiplying by 2.889 before using the formulas for calculating sucrose [AOAC Method **29.020(c)**].

For directions for checking the accuracy of the calibration of the instrument, the reader is referred to Text Reference 3, Method **29.020(b).**

1. *Direct reading*—Dissolve a double normal weight of sample (52 g), or an even fraction thereof, in water in a 200 ml volumetric flask; add the necessary clarifying agent [(**a**), (**b**) or (**c**) above], avoiding any excess; shake, dilute to the mark with water, mix well, and filter, keeping the funnel covered with a watch-glass. Reject the first 25 ml of filtrate.

If a lead clarifying agent was used, remove excess Pb from the filtrate by adding anhydrous Na_2CO_3, a little at a time, avoiding any excess; mix well and filter again, rejecting the first 25 ml of filtrate.

Pipet 50 ml of the Pb-free filtrate into a 100 ml volumetric flask, and add 2.315 g of NaCl and 25 ml of water. Dilute to the mark with water at 20°, and polarize in a 200 mm tube at 20°. Multiply the reading by 2 to obtain the direct reading, *P*.

2. *Invert reading* (*inversion at* 60°)—Pipet 50 ml of the Pb-free filtrate into a 100 ml volumetric flask and add 20 ml of water. Add, little by little, while rotating the flask, 10 ml of HCl of sp.g 1.1029 at 20/4°. Place the flask in a water bath adjusted at 60°, agitate continuously about 3 minutes, and then leave the flask in the bath exactly 7 minutes longer. At once plunge the flask into water at 20°.

When the contents cool to about 35°, dilute almost to the mark. Leave the flask in the bath at 20° at least 30 minutes longer, and finally dilute to the mark. Mix well and polarize the solution in a 200 mm tube provided with a lateral branch and a water jacket, keeping the temperature at 20°. This reading must also be mutliplied by 2 to obtain the invert reading *I*. If it is necessary to work at a temperature other than 20°, which is permissible within narrow limits, the volumes must be completed, and both direct and invert polarizations must be made, at exactly the same temperature.

Calculate sucrose by the following formula:

$$S = \frac{100(P-I)}{132.56+0.0794(m-13)-0.53(t-20)},$$

where S = % sucrose; P = direct reading, normal solution; I = invert reading, normal solution; t = temperature at which readings are made; and m = g total solids in 100 ml inverted solution (total solids in 50 ml original solution).

For liquids, determine the total solids in the original sample as percent by weight by means of a refractometer, and take this fraction of the weight of original sample in 100 ml of invert solution as m.

3. *Invert reading* (*inversion at room temperature*) —Inversion may also be accomplished as follows: Pipet 50 ml of Pb-free filtrate into a 100 ml volumetric flask and add 20 ml of water and 10 ml of HCl of sp.g 1.1029 at 20/4°. Set aside at room temperature between 22° and 28° for 24 hours, or set aside 10 hours if the temperature exceeds 28°. Dilute to 100 ml at 20° and polarize as in (2). Under these conditions sucrose should be calculated by the following formula:

$$S = \frac{100(P-I)}{132.66+0.0794(m-13)-0.53(t-20)}.$$

NOTE (AOAC Method **29.031**): When the volume of combined insoluble matter and precipitate from clarifying agents exceeds 1 ml from 26 g, the following double dilution method should be used:

Weigh a 13 g sample and dilute the solution to 100 ml, using basic Pb acetate for dark-colored confectionery or molasses, and alumina cream for light-colored confectionery. Also weigh a 26 g sample and dilute this second solution with clarifier to 100 ml. Filter both solutions, and obtain the direct polariscopic readings. Invert each solution by one of the methods given above, and obtain the respective invert readings.

The true direct polarization of the sample = 4 times the direct polarization of the diluted solution minus the direct polarization of the undiluted solution. The true invert polarization = 4 times the invert polarization of the diluted solution minus the invert polarization of the undiluted solution. Calculate sucrose from the true polarizations thus obtained, using the formula appropriate to the manner of inversion that was used.

B. Copper reduction method (*Munson—Walker*) [*AOAC Methods* ***29.035*** (**a**) and (**b**) and ***29.038—29.045***].

REAGENTS

a. *Asbestos*—Digest asbestos, amphibole variety, with HCl (1+3) for 2–3 days. Wash acid-free, digest for a similar period with 10% NaOH solution, and then treat for a few hours with the hot alkaline tartrate solution, (**c**). Wash the asbestos alkali-free; digest several hours with HNO_3 (1+3); wash acid-free, and shake with water into a fine pulp. In preparing a Gooch crucible for a determination, make a film of the asbestos $\frac{1}{4}''$ thick and wash thoroughly to remove fine particles. If the precipitated Cu_2O is to be weighed as such, wash the crucible successively with 10 ml of alcohol and 10 ml of ether; dry 30 minutes at 100°, cool in a desiccator, and weigh.

b. *Copper sulfate solution*—Dissolve 34.639 g of $CuSO_4.5H_2O$ in water, dilute to 500 ml, and filter through glass wool or paper.

c. *Alkaline tartrate solution*—Dissolve 179 g of Rochelle salt (potassium sodium tartrate) and 50 g of NaOH in water, dilute to 500 ml, let stand two days, and filter through glass wool.

d. *Thiosulfate standard solution*—Prepare a solution containing 39 g of $Na_2S_2O_3.5H_2O$ per liter. Weigh accurately 0.2–0.4 g of pure electrolytic copper and transfer to a 250 ml Erlenmeyer flask roughly marked at 20 ml intervals. Dissolve the Cu in 5 ml of HNO_3 (1+1), dilute to 20 or 30 ml, boil to expel red fumes, add a slight excess of saturated Br_2 solution, and boil until the Br_2 is completely removed. Cool, and add 10 ml of sodium acetate solution (574 gm trihydrate/liter). Prepare a 42 g/100 ml KI solution, made very slightly alkaline to avoid formation and oxidation of HI. Add 10 ml of the KI solution and titrate with the $Na_2S_2O_3$ solution to a light yellow. Add enough starch indicator, (**e**), to produce a marked blue. As the end-point nears, add 2 g of KSCN and stir until completely dissolved. Continue titrating until the precipitate is perfectly white. One ml of the $Na_2S_2O_3$ solution should be equivalent to about 10 mg of Cu.

It is essential for the $Na_2S_2O_3$ titration that the concentration of KI in the solution be carefully regulated. If the solution contains more than 320 mg of Cu, at completion of the titration 4.2–5.0 g of KI should have been added for each 100 ml of total solution. If greater quantitites of Cu are present, add KI solution slowly from a buret with constant agitation in quantities proportionately greater.

e. *Starch indicator*—Mix about 2 g of finely powdered potato starch with cold water to a thin paste; add about 200 ml of boiling water, stirring constantly, and immediately discontinue heating. Add about 1 ml of mercury, shake, and let the solution stand over Hg.

f. *Potassium permanganate standard solution*—Approximately 0.1573 *N*. Dissolve 4.98 g of $KMnO_4$ in 1 liter of water. Boil the solution 1 hour and let stand overnight protected from dust. Thoroughly clean a 15 cm glass funnel, a perforated glass plate from a Caldwell crucible, and a glass-stoppered bottle (preferably of brown glass), with warm H_2SO_4-$K_2Cr_2O_7$ solution. Digest asbestos (for Gooch crucibles) on the steam bath for 1 hour with approximately 0.1 *N* $KMnO_4$ that has been acidified with a few drops of H_2SO_4 (1+3). Let settle, decant, and replace with water. To prepare the glass funnel, place the porcelain plate in its apex, make a pad of the prepared asbestos about 3 mm thick on the plate, and wash acid-free. (The pad should not be too tightly packed and only moderate suction should be applied.) Insert the stem of the funnel into the neck of the bottle and filter the $KMnO_4$ solution directly into the bottle without suction.

Transfer 0.35 g of dried (1 hour at 105°) NBS Standard Sample sodium oxalate to a 600 ml beaker. Add 250 ml of H_2SO_4 (5+95) that has been previously boiled 10–15 minutes and cooled to 27° ± 3°. Stir until the $Na_2C_2O_4$ dissolves. Then add 39–40 ml of the $KMnO_4$ solution at a rate of 25–35 ml/minute, stirring slowly. Let stand until the pink color disappears (about 45 seconds). Heat to 55–60°, and complete the titration by adding the $KMnO_4$ solution until a faint pink persists 30 seconds. Add the last 0.5–1 ml drop by drop with particular care to let each drop decolorize before adding the next.

Determine the excess of $KMnO_4$ solution required to turn the solution pink by matching with the color obtained by adding $KMnO_4$ solution to the same volume of the boiled and cooled dilute H_2SO_4 at 55–60°. This correction is usually 0.03–0.05 ml. From the net volume of $KMnO_4$ calculate the normality as follows:

$$N = \frac{\text{g } Na_2C_2O_4 \times 1{,}000}{\text{ml } KMnO_4 \times 66.999}.$$

One ml N $KMnO_4$ is equivalent to 0.06354 g Cu.

g. *Ferric ammonium sulfate solution*—Dissolve 135 g of $FeNH_4(SO_4)_2 \cdot 12H_2O$ in water, and dilute to 1 liter. Titrate 50 ml of this solution, acidified with 20 ml of 4 N H_2SO_4, with the $KMnO_4$ solution, (**f**), and use this titer as the zero-point correction.

h. *Ferrous phenanthroline indicator*—Dissolve 0.7425 g of *o*-phenanthroline in 25 ml of 0.025 M $FeSO_4$ solution (6.95 g $FeSO_4 \cdot 7H_2O$/liter).

Precipitation of Cuprous Oxide

Prepare a solution of the sample as in the polarimetric method, A,1, clarifying with neutral Pb acetate ["Preparation and Use of Clarifying Agents", (**c**)], or with ion exchange resins after alcohol extraction, (**d**), in the case of vegetables. (If Pb acetate was used, remove any excess of Pb by treatment with anhydrous $K_2C_2O_4$ and filtration.) Also prepare an invert solution according to the polarimetric method, (A, 2, or 3), except to bring the inverted solution to exact neutrality before dilution to volume.

Transfer to each of two 400 ml beakers 25 ml each of the $CuSO_4$ and alkaline tartrate solutions, (**b**) and (**c**). To one beaker add 50 ml of the clarified Pb-free sample solution that has not been inverted; to the other beaker add 50 ml of the neutralized invert solution. If the quantity of sugar in 50 ml of either solution is too great to be handled by 25 ml of the $CuSO_4$ solution (invert sugar exceeding 230 mg), use a smaller volume of sugar solution, but add sufficient water to make the total volume 100 ml.

Heat each beaker on an asbestos gauze over a Bunsen burner that has been so regulated that boiling begins in just 4 minutes, and boil exactly 2 minutes. (It is important that these time directions be strictly observed, but an electric heater that has been adjusted to cause 100 ml of a mixture of 25 ml of $CuSO_4$ solution, 25 ml of alkaline tartrate solution and 50 ml of water to come just to a boil in 4 minutes may be substituted for the Bunsen burner.) Keep the beaker covered with a watchglass during this heating.

Filter the hot solutions at once through asbestos mats in porcelain Gooch crucibles, using suction. Wash the precipitates of Cu_2O thoroughly with water at 60°. From this point determine Cu by one of the following methods. In any case, a blank determination should be run on 25 ml of the $CuSO_4$ solution, 25 ml of the alkaline tartrate solution, and 50 ml of water. If the weight of Cu_2O obtained exceeds 0.5 mg, correct the results of the sugar determinations accordingly. (The alkaline tartrate solution deteriorates on standing, and the quantity of Cu_2O obtained in the blank increases.)

Determination of Copper

By direct weighing. After washing thoroughly with hot water, wash the precipitated Cu_2O with 10 ml of alcohol and then with 10 ml of ether. Dry 30 minutes in an oven at 100°, cool, and weigh.

Note: If a series of determinations is to be made by the direct weighing method, it is possible, and indeed desirable, to leave the asbestos mat in each crucible and reuse it after washing with HNO_3, water, alcohol and ether, drying and re-weighing. The weight of the mat soon becomes very stable.

By titration with sodium thiosulfate. After washing the precipitate with water, cover the crucible with a watch-glass and dissolve the Cu_2O with 5 ml of HNO_3 (1+1) that is directed under the watch-glass with a pipet. Collect the filtrate in a 250 ml Erlenmeyer flask roughly marked in 20 ml intervals, and proceed as under "(**d**), Thiosulfate standard solution" (p. 371), beginning with "boil to expel red fumes".

By titration with potassium permanganate. After thoroughly washing the precipitated Cu_2O with water, transfer the asbestos pad and precipitate to a beaker with a glass rod. Add 50 ml of the $FeNH_4(SO_4)_2$ solution, (**g**), and stir vigorously until the Cu_2O is completely dissolved. (Examine for complete solution, holding the beaker above eye level. Cu_2O must be quantitatively transferred. If necessary, immerse the crucible in the solution and make sure adhering Cu_2O is dissolved. Remove the crucible with a glass rod and wash with water.) Add 20 ml of 4*N* H_2SO_4 and titrate with the standard $KMnO_4$ solution, (**f**). As the end point approaches, add one drop of the ferrous phenanthroline indicator, (**h**). At the end point, the brownish solution changes to green.

By electrolytic deposition. After thoroughly washing the precipitated Cu_2O with hot water, transfer the asbestos mat and precipitate to a beaker with a glass rod and rinse the crucible with 14 ml of HNO_3 (1+1), letting the rinsings flow into the beaker. After the Cu_2O dissolves, dilute to 100 ml with water, heat to boiling, and continue boiling about 5 minutes to remove oxides of N, cool, filter, transfer to a 250 ml beaker, and dilute to 200 ml. Add one drop of 0.1 *N* HCl and mix thoroughly.

For electrolysis use cylindrical Pt electrodes, about 1.5″ and 2″ respectively in diameter, and about 1.75″ high, that have been thoroughly cleaned, ignited, cooled in a desiccator and weighed. Insert these electrodes into the solution so that the surface of the cathode clears the anode by at least 5 mm and both electrodes almost touch the bottom of the beaker. Cover with a split watch-glass to avoid loss by spattering. Electrolyze with a current of 0.2–0.4 ampere until deposition is complete, usually overnight. (Test completeness by washing down the watch-glass and the sides of the beaker with water and seeing if there is any deposit of Cu on the new surface of the electrode so exposed.)

Without interrupting the current, slowly lower the beaker while washing the electrodes with a stream of water. Immediately immerse the electrodes in another beaker of water and break the current. (Washing may also be accomplished by siphoning.) Rinse the cathode with alcohol and dry a few minutes in an oven at 100°. Cool in a desiccator and weigh.

If the electrolyte is stirred by a rotating anode or mechanical stirrer, the current may be increased to 1–2 amperes; in this case the time required for complete deposition of Cu is reduced to about 1 hour.

If extreme care is taken to avoid spattering, the Cu_2O can be dissolved directly out of the crucible by letting the HNO_3 flow down the sides. When doing this, keep the crucible covered as much as possible with a small watch-glass, collect the filtrate in a 250 ml beaker, and wash the watch-glass and tip of the pipet with a jet of water. From this point proceed as above, beginning "dilute to 100 ml".

CALCULATIONS

By one of the above four methods, there will have been obtained the weight of Cu or Cu_2O produced by a known weight of sample, both before and after inversion. From Table 23–6 calculate the weights of invert sugar corresponding to these two amounts of Cu. For the invert solution use the column headed "Invert Sugar"; for the uninverted solution use the one of the three columns headed "Invert Sugar and Sucrose" that most closely corresponds to the quantity of total sugars in the aliquot.*

Deduct the percentage of invert sugar obtained before inversion from that obtained after inversion, and multiply the difference by 0.95 to obtain percent sucrose.

Method 16-7. Invert Sugar.

In the absence of other reducing sugars, *e.g.*, lactose, maltose and amounts of dextrose and levulose not arising from hydrolysis of sucrose,

*If direct weighing of Cu_2O has been employed, slightly more accurate results may be obtained by using the Munson and Walker Table **43.012** in *A.O.A.C. Methods of Analysis*, 10th Ed. (1965).

invert sugar can be determined by the above copper reduction method, using a solution of the sample that has not been inverted. Its determination is also a by-product of the determination of sucrose by this method.

When other reducing sugars are present, recourse must be had either to more nearly specific chemical methods for dextrose[8] and levulose[9] or to chromatographic methods of separation such as are discussed further in this chapter. In this connection it is scarcely necessary to note that "invert sugar" is not a specific sugar, but merely a name for the equimolecular mixture of dextrose and levulose that results from the hydrolysis of sucrose; consequently the specific methods do not determine invert sugar as such.

Method 16-8. *Dextrose.*

A. *Munson-Walker method*—Proceed as for invert sugar above, except to use the column headed "Dextrose" for calculating the weight of sugar corresponding to the weight of copper found.

B. *Zerban-Sattler modification of the Steinhoff Method* (*AOAC Method **29.184–29.185***)—Not applicable in the presence of other reducing sugars.

REAGENTS

a. *Soxhlet modification of Fehling copper solution—See* "Reagents, **b**" under the copper reduction method for sucrose.

b. *Sodium acetate solution*—Dissolve 500 g of sodium acetate trihydrate in about 800 ml of water, cool, and dilute to 1 liter.

c. *Potassium iodide-iodate solution*—Dissolve 5.4 g of KIO_3 and 60 g of KI in water, add 0.25 g of NaOH dissolved in a little water, and dilute to 1 liter.

d. *Sulfuric acid*—Approximately 2 *N*. Dilute 57 ml of H_2SO_4 to 1 liter.

e. *Saturated potassium oxalate solution*—Dissolve 165 g of $K_2C_2O_4.H_2O$ in 100 ml of hot water, and cool.

f. *Sodium thiosulfate standard solution*—0.1 *N*. Prepare as in Method 16-6, "Reagents, **d**", except to use 25 g of $Na_2S_2O_3.5H_2O$ per liter. 1 ml 0.1 *N* $Na_2S_2O_3$ = 6.354 mg Cu.

DETERMINATION

Prepare a solution of the sample by clarification with neutral Pb acetate and removal of excess Pb with $K_2C_2O_4$ as in the Munson-Walker method. Dilute a quantity of this solution containing 10 g of solids to 1 liter. Add 10 ml of this solution, 10 ml of the Soxhlet Cu solution (**a**), 20 ml of the Na acetate solution (**b**), and 10 ml of water to a 250 ml Erlenmeyer flask. Mix, close the flask with a rubber stopper fitted with a Bunsen valve, and immerse in a briskly boiling water bath exactly 20 minutes. Then cool in cold running water, venting the valve to prevent vacuum boiling. When cold, pipet 25 ml of the KI-KIO_3 solution (**c**) into the flask, and mix by gentle shaking. Add rapidly from graduated cylinders first 40 ml of the 2 *N* H_2SO_4 (**d**), and then 20 ml of the $K_2C_2O_4$ solution (**e**). Shake until the precipitate completely dissolves, and titrate the excess of I with the 0.1 *N* $Na_2S_2O_3$, **f**.

Run a blank, substituting 10 ml of water for the sample solution. The difference between the titer of the blank and that of the sample is a direct measure of the precipitated Cu_2O. From Table 16-2 obtain the amount of dextrose corresponding to the titer of 0.1 *N* $Na_2S_2O_3$.

Correction for reducing effect of maltose—If maltose is present, subtract from the observed titer of 0.1 *N* $Na_2S_2O_3$ the maltose correction obtained from the table by interpolation.

C. *Sichert-Bleyer modification* (*AOAC Method **29.187–29.189***)

REAGENTS

g. *Ferric ammonium sulfate solution*—Dissolve 120 g of $Fe_2(SO_4)_3(NH_4)_2SO_4.24H_2O$ and 100 ml of H_2SO_4 in water and dilute to 1 liter.

h. *Potassium permanganate solution*—0.1 *N*. Prepare as under Method 16-6, B, "Reagents, **f**", except to use 3.2 g of $KMnO_4$. Standardize as follows:

As under "Determination" analyze a 10 ml solution of 50 mg of pure dextrose. From Table 16-3, 50 mg of dextrose should give a titer of 15.38 ml. Therefore, to give the correction factor for the $KMnO_4$ solution, divide 15.38 by the titer found. Multiply all titers by this factor before referring to the table. Redetermine this factor every day analyses are made.

TABLE 16-2

Zerban-Sattler table for determination of dextrose with copper acetate reagent[a]

Titer	Dextrose (mg)										Maltose corrections (subtract from observed Titer) Maltose Present, mg		
	0.0	0.1	0.2	0.3	0.4	0.5	0.6	0.7	0.8	0.9	200	100	50
10	25.7	26.0	26.3	26.6	26.9	27.2	27.5	27.8	28.1	28.4	2.5	1.4	0.6
11	28.7	29.0	29.3	29.6	29.9	30.3	30.6	30.9	31.2	31.5	2.3	1.2	0.4
12	31.8	32.2	32.5	32.9	33.2	33.6	34.0	34.3	34.7	35.0	2.2	1.1	0.4
13	35.4	35.8	36.1	36.5	36.8	37.2	37.6	37.9	38.3	38.6	2.0	1.0	0.3
14	39.0	39.4	39.9	40.3	40.7	41.2	41.6	42.0	42.4	42.9	1.9	1.0	0.3
15	43.3	43.8	44.2	44.7	45.1	45.6	46.1	46.5	47.0	47.4	1.7	1.0	0.3
16	47.9	48.4	49.0	49.5	50.1	50.6	51.1	51.7	52.2	52.8	1.8	1.0	0.3
17	53.3	53.9	54.5	55.2	55.8	56.4	57.0	57.6	58.3	58.9	1.6	0.9	0.3
18	59.5	60.2	60.9	61.6	62.3	63.1	63.8	64.5	65.2	65.9	1.4	0.8	0.3
19	66.6	67.4	68.2	69.0	69.8	70.6	71.4	72.2	73.0	73.8	1.2	0.7	0.3
20	74.6	75.6	76.5	77.5	78.4	79.4	80.3	81.3	82.2	83.2	1.0	0.6	0.2
21	84.1	85.2	86.3	87.4	88.5	89.6	90.6	91.7	92.8	93.9	0.6	0.4	0.2
22	95.0	—	—	—	—	—	—	—	—	—	0.4	0.3	0.1

[a] Table may be interpolated for each .01 ml, but should *not* be extrapolated.

TABLE 16-3

Sichert-Bleyer table for determination of dextrose[a]

Titer 0.1 *N* permanganate	Dextrose (mg)									
	0	0.1	0.2	0.3	0.4	0.5	0.6	0.7	0.8	0.9
ML										
10	26.5	26.8	27.1	27.4	27.8	28.1	28.4	28.7	29.0	29.3
11	29.7	30.0	30.4	30.7	31.1	31.5	31.8	32.2	32.6	32.9
12	33.3	33.7	34.1	34.5	34.9	35.4	35.8	36.2	36.6	37.0
13	37.4	37.9	38.4	38.8	39.3	39.8	40.3	40.7	41.2	41.7
14	42.2	42.7	43.2	43.8	44.3	44.9	45.4	46.0	46.5	47.0
15	47.6	48.2	48.8	49.4	50.1	50.7	51.3	51.9	52.5	53.2
16	53.8	54.5	55.2	55.9	56.6	57.3	58.0	58.7	59.4	60.2
17	60.9	61.7	62.5	63.3	64.1	64.9	65.7	66.5	67.4	68.2
18	69.0	69.9	70.9	71.9	72.8	73.8	74.8	75.7	76.7	77.6
19	78.6	79.6	80.7	81.7	82.7	83.7	84.8	85.8	86.8	87.8
20	88.9	90.0	91.2	92.3	93.5	94.7	96.0	97.2	98.5	99.7

[a] Table may be interpolated for each .01 ml, but should *not* be extrapolated.

DETERMINATION

Prepare a solution of the sample containing 10 g solids/liter as in Method 16-8, **B**. Transfer to a 250 ml Erlenmeyer flask 10 ml of this solution, 10 ml of the Soxhlet Cu solution, (**a**), 20 ml of the Na acetate solution, (**b**), and 10 ml of water. Mix, and close the flask with a rubber stopper provided with a Bunsen valve. Immerse in a boiling water bath exactly 20 minutes. Filter the Cu_2O precipitate through a Gooch crucible prepared as for the Munson-Walker method, and wash the flask and crucible three times with hot water. It is not necessary to remove all of the precipitate from the flask.

Transfer the asbestos mat and crucible to a 150 ml beaker marked at 60 ml. Wash the flask with exactly 20 ml of the $FeNH_4(SO_4)_2$ solution, (**g**), in three portions, transferring quantitatively to the beaker containing the crucible. (The precipitate must be completely dissolved.) Finally wash the flask and crucible with hot water, and remove the crucible. Add hot water to the 60 ml mark. Heat the solution to boiling on a hot plate, let stand 3 minutes, and titrate with the 0.1 *N* $KMnO_4$, (**h**). The addition of 1 ml of H_3PO_4 toward the end of the titration facilitates reading the endpoint. The pink-gray end point persists about 20 seconds. Multiply the titer by the factor obtained above, and calculate mg of dextrose from Table 16-3.

D. *Glucose oxidase method*—In a mixture containing only dextrose and levulose, dextrose is equal to the difference between total reducing sugars and levulose as determined by the glucose oxidase method below. A variation on this method that employs a proprietary mixture of enzymes and chromogenic agent (*o*-dianisidine hydrochloride) is described in Text Reference 10.

Method 16-9*. Levulose.*

A. *Munson-Walker method**—Proceed as under *Invert Sugar*, except to use the column headed "Levulose" for calculating the weight of sugar corresponding to the weight of copper found.

B. *Jackson-Mathews Modification of the Nyns Selective Method.* (*AOAC Method* ***29.062–29.063***)

*Not applicable in the presence of other reducing sugars.

REAGENT

Ost solution—Dissolve 250 g of anhydrous K_2CO_3 in about 700 ml of hot water, add 100 g of pulverized $KHCO_3$, and agitate the mixture until completely dissolved. Cool, and add, agitating vigorously, a solution of 25.3 g of $CuSO_4.5H_2O$ in 100–150 ml of water. Dilute to 1 liter and filter.

DETERMINATION

Transfer 50 ml of the Ost solution to a 125 ml Erlenmeyer flask, and pipet in a volume of the sample solution (prepared as for the Munson-Walker method) containing not over 92 mg of levulose or its equivalent in reducing power of a levulose-dextrose mixture (dextrose has about one-twelfth the reducing power of levulose). Dilute with water to 70 ml. Immerse in a water bath regulated at 55° (preferably within 0.1°). Digest exactly 75 minutes, swirling at 10-15 minute intervals.

Filter the precipitated Cu_2O on a Gooch crucible with an asbestos mat (prepared as for the Munson-Walker method) and wash the flask and precipitate thoroughly, without attempting to transfer the precipitate quantitatively. Determine copper by one of the methods prescribed for the Munson-Walker method. (As it is usually difficult to transfer the Cu_2O precipitate quantitatively from the Erlenmeyer flask, preferably select a method involving solution of all of the Cu in HNO_3.)

See Table 16-4 for the levulose equivalent of the found Cu. If the sample contains dextrose in addition to levulose, the analytical result is not true but "apparent" levulose, because dextrose has an appreciable reducing action under the conditions of analysis. To determine the dextrose correction, analyze the sample also for total reducing sugars (invert sugar) by the Munson-Walker method, and compute true dextrose and levulose by a series of approximations. The difference between the percent of total reducing sugars and the percent of apparent levulose in the original sample is the "apparent" dextrose. Divide the apparent dextrose by the factor 12.4 and deduct the result from the apparent levulose to obtain a new approximation of the true levulose. Deduct the new levulose percentage from the percent of total reducing sugars to obtain a more nearly correct value for true dextrose, and again divide by 12.4. Deduct the quotient from the original value of apparent levulose. Continue to approximate in this manner until the percentage of levulose remains essentially unaltered by two successive approximations.

If the original sample contained sucrose, determine by the polarimetric method, and correct the Cu for the reducing action of sucrose before referring to Table 16-4. 1, 2, 3, 4 and 5 g of sucrose under the conditions of the present method give precipitates of 3.3, 5.7, 7.4, 8.5 and 9.0 mg respectively of Cu.

TABLE 16-4

Copper-levulose equivalents according to Jackson and Mathews modification of Nyns selective method for levulose

Expressed in mg. Linear interpolation yields accurate results

Cu	Levulose	Cu	Levulose
5	2.5	130	39.3
10	4.5	140	42.0
15	6.2	150	44.7
20	7.9	160	47.4
25	9.5	170	50.0
30	11.0	180	52.6
35	12.5	190	55.2
40	13.9	200	57.9
45	15.4	210	60.6
50	16.8	220	63.4
55	18.3	230	66.4
60	19.7	240	69.4
65	21.2	250	72.5
70	22.5	260	75.7
80	25.4	270	79.0
90	28.1	280	82.4
100	30.9	290	85.9
110	33.7	300	89.5
120	36.5	310	93.2

C. *Glucose oxidase method.* (*AOAC Method 6.078–6079.*)

REAGENTS

a. *Glucose oxidase preparation*—Add slowly, stirring constantly, 100 ml of water to 5 g of "Dee-O" (Takamine Laboratories, Clifton, N.J.). Stir about 1 minute, and centrifuge to obtain a clear solution. Add about 1 ml of $CHCl_3$ and refrigerate. This solution is stable at least one month.

b. *McIlvaine's citrate-phosphate buffer*—Dissolve 214.902 g of $Na_2HPO_4.12H_2O$ and 42.020 g of citric acid in water, and dilute to 1 liter.

DETERMINATION

Prepare an alcoholic solution of the sample and otherwise proceed as under Method 16-6, "Preparation and use of clarifying agents (**d**)". To a suitable aliquot of the clarified aqueous solution (containing 250 mg or less of levulose) add one-fourth its volume of the buffer (**b**), to give a pH of about 5.8. Add 30 percent as much of the glucose oxidase preparation as the estimated dextrose content (for 500 mg dextrose add 3 ml of the preparation) and a few drops of 30% H_2O_2. Let stand overnight at room temperature.

Determine levulose by the Munson-Walker method, using Table 16-5 below. Check the equivalents given in the table by analysis of solutions of pure levulose, and correct accordingly.

TABLE 16-5

Abbreviated Munson and Walker Table for Calculating Levulose

From Official and Tentative Methods of Analysis, AOAC, 5th Ed., 1940

Cuprous Oxide mg	Levulose mg	Cuprous Oxide mg	Levulose mg
10	4.5	300	148.6
50	23.5	350	174.9
100	47.7	400	201.8
150	72.2	450	229.2
200	97.2	490	253.9
250	122.7	—	—

NOTES

1. It was because the details of the above glucose oxidase method were worked out with its application to the determination of levulose in plants in mind[11] that alcoholic extraction and later clarification with ion exchange resins was specified for preparation of the sample solution. There is no apparent reason, however, why the method should not work with non-plant materials whose solutions have been clarified by other means, provided interfering reducing substances are absent.

2. Because glucose oxidase is specific for β-d-glucose, and the present method determines

as levulose all reducing sugars not destroyed by the enzyme, this method will give falsely high values for levulose if applied directly to aqueous solutions of solid samples high in dextrose that have not been allowed to stand at least two hours to complete mutarotation.

Method 16-10. *Lactoes and Maltose.*

In the absence of other reducing sugars or interfering substances, lactose may be determined by the Munson-Walker method in the same manner as dextrose or levulose, using the appropriate column in the Hammond table for calculating the weight of sugar corresponding to the weight of copper found. While the Hammond table does not contain a column for maltose, this sugar may be determined by the Munson-Walker method if reference is made to the maltose column in Table **43.012** of *Official Methods of Analysis of the Association of Official Agricultural Chemists*, 10th ed. (1965).

Special methods for these sugars in dairy products, flesh foods and alcoholic beverages are given in Chapters 6, 10 and 2 respectively.

Identification and determination of sugars by chromatographic methods

Method 16-11. *Gas Chromatography.*

Because in principle the gas chromatographic method of separation of two compounds depends on a difference in the rate of migration of the vapors of these compounds through a column (that either contains a liquid-coated, nonvolatile packing or is a coated capillary), this method can only work with substances that possess some volatility. Since the sugars are essentially non-volatile, they must be converted to volatile derivatives before this new and powerful technique can be applied to their separation and determination. The derivatives that were first tried were the O-methyl ethers[12] and acetates, but it was later discovered that the trimethylsilyl derivatives were superior in their ease of formation and readiness of separation. C. C. Sweeley *et al.*[31] showed that it was possible to separate on appropriate columns the trimethylsilyl derivatives of a large number of sugars, including the pentoses, galactose, dextrose, sucrose, lactose, maltose and melibiose. Alexander and Garbutt[14], whose primary interest was in the quantitative analysis of commercial corn sugar and corn syrups, modified the technique of Sweeley *et al.* by the use of sorbitol as an internal standard. Their method is given below. While it was written with the limited field of the determination of the dextrose isomers (α-D-glucose and β-D-glucose) in corn products in mind, it should be adaptable without essential modification to the separation and determination of other sugars in other products:

APPARATUS

a. *An Aerograph A-600-B chromatograph* equipped with a hydrogen flame ionization detector (obtainable from Wilkins Instrument and Research, Inc., Walnut Creek, Calif.), attached to a Leeds and Northrup Speedomax H recorder, was employed by the authors of this method, but instruments of other manufacturers can be substituted.

b. *Column*—A 6 ft × 0.25 in. o.d. coiled stainless steel column packed with 3% SE-52 on 100–120 mesh Chromosorb P. Obtainable from Applied Science Laboratories, State College, Pa.

REAGENTS

a. *Hexamethyldisilazane.* Obtainable from Peninsular Chem-Research, Gainsville, Fla.

b. *Trimethylchlorosilane.* Obtainable from General Electric Co., Silicone Products Div., Waterford, N.Y.

(Mixtures of these two compounds in the proper proportion may also be obtained from Applied Science Laboratories or Pierce Chemical Co., P.O. Box 117, Rockford, Ill. 61105)

c. *α-D-Glucose and β-D-Glucose.* Obtainable from Pfanstiehl Laboratories, Inc., Waukegan, Ill.

d. *Sorbitol.* Obtainable from Pfanstiehl.

e. *Anhydrous pyridine*. Store over KOH pellets.

PREPARATION OF SAMPLE

Anhydrous corn sugar—Weigh 1.000 g into a 100 ml volumetric flask, dissolve in the anhydrous pyridine, and make to volume. Pipet 2 ml of this solution into a 10 ml volumetric flask, and add 2.0 ml of a 10 mg/ml pyridine solution of sorbitol, and 6.0 ml of pyridine. One milliliter of this final solution contains 4 mg of sample.

Aqueous corn sugar—Determine the percentage

of solids (usually 20–25%) from the refractive index. Weigh a 4.0 ml sample into a 100 ml flask, dilute to volume with pyridine, and otherwise proceed as above.

Corn syrups—Such syrups normally contain 80% solids. Dilute with water to 20–25% solids, and determine the actual solid content from the refractive index. Weigh a 4.0 ml sample of this diluted solution into a 100 ml volumetric flask and make to volume with pyridine. For 54–72 D.E. syrups (25–50% dextrose), pipet 4.0 ml of this pyridine solution, 2.0 ml of the sorbitol solution, and 4.0 ml of pyridine e into a 10 ml volumetric flask. For syrups lower than 54 D.E., take 4.0 ml of the pyridine solution of the sample, 1.0 ml of the sorbitol solution, and 5.0 ml of the pyridine.

FORMATION OF THE SILANE DERIVATIVES[13]

To 1 ml of the pyridine solution of sample plus sorbitol (see "Preparation of sample") in a plastic-stoppered, 1-dram vial, add 0.2 ml of hexamethyldisilazane and 0.1 ml of trimethylchlorosilane. Shake vigorously about 30 seconds, and let stand 5 minutes or longer at room temperature. (The solution becomes cloudy from precipitation of ammonium chloride on addition of the trimethylchlorosilane, but this fine precipitate does not interfere with the subsequent gas chromatography. If the carbohydrate itself appears to remain persistently insoluble in the mixture, warm the vial 2–3 minutes at 75–85°).

DETERMINATION

Adjust the column temperature to 180° and the injector oven temperature to 230°. Use an argon or nitrogen flow rate of 25–30 ml/minute. For the Aerograph instrument the electometer settings are: Imput impedance, 10^9 ohms; output sensitivity, 1×; attenuator, 8–32.

With a 10.0 μl. Hamilton syringe, inject a 1.0 to 5.0 μl. portion of the silanized sample solution. Also run a reference standard composed of 25% α-D-glucose, 25% β-D-glucose and 50% D-sorbitol that has been treated in the same fashion as the sample. (Sample size should be adjusted to use $\frac{1}{3}$–$\frac{2}{3}$ of the recorder scale.) Total retention time will be about 35 minutes for argon and 29 minutes for nitrogen.

Measure the areas of all peaks with a Gelman planimeter.

CALCULATIONS

For anhydrous materials proceed as follows:

Divide the total dextrose area (α and β-D-glucose) of the sample by the sorbitol area.

This ratio divided by the corresponding ratio for the reference standard, and multiplied by 100, gives the percent dextrose.

For aqueous samples, divide the percent of dextrose, as calculated above, by the sample weight times the percent solids (expressed as a fraction), to obtain the percentage of dextrose in the sample. In the case of corn syrups, divide this value by 2 or 4 to correct for the 2:1 or 4:1 syrup/sorbitol ratios used.

NOTES

1. Alexander and Garbutt introduced the use of an internal standard because of difficulties in controlling sample size and because the peak areas produced by equal amounts of a sugar decreased considerably with the lapse of time after cleaning the flame detector (apparently due to deposition of an SiO_2 film).

2. Sweeley *et al.*[13] found that while SE-52 was the liquid phase of choice for a complete scan of the carbohydrates from the tetroses to the polysaccharides, separations of the aldohexoses were more complete on a column containing 15% polyethylene glycol succinate (EGS) on 80–100 mesh Chromosorb W (obtained from Applied Science Laboratories). These authors found that for the SE-52 columns they separated 20 components in 75 minutes.

Because of limited thermal stability, the polyester (EGS) column could not be operated above 200°; this limited analyses to those of sugars containing approximately six carbon atoms. The temperatures they used were 140, 150, and 170°, depending on the molecular weight range of the mixture being analyzed.

3. N,O-Bis(trimethylsilyl)acetamide has recently been proposed as a trimethylsilylating agent.[15]

4. Ready-prepared trimethylsilyl derivatives of the following sugars are obtainable from Pierce Chemical Co., either singly or in a kit, as 10% solutions in hexane: L-arabinose, β-D-fructose, D-galactose, α-D-glucose, β-D-glucose, lactose, maltose, D-mannose, D-ribose, L-sorbose, sucrose and D-xylose.

5. It should be noted that gas chromatographic methods identify a separated compound only to

the extent that they show that, with a particular column and a particular form of detection, the compound gives a peak on a chart at the same place as does the compound it is assumed to be. For complete identification, methods have been worked out to collect the separated substance and submit it to examination by an infrared spectrophotometer or a mass spectrometer. Because of the minute amounts of material involved, very special techniques of handling are required. A description of these is beyond the scope of this book, but those interested may consult the papers in Text Reference 16.

While the above methods of identification are the only ones that may be classed as absolute, they are beyond the resources of the average laboratory because of the special training and expensive equipment that are required. Recourse is therefore usually had to running the same sample by both gas chromatography and paper or thin layer chromatography. If the two methods agree as to the composition of a particular ingredient of the sample, the chances of misidentification become so slight that they may ordinarily be disregarded.

The above statements hold true with regard to the general examination by gas chromatography of materials whose constituents may be unknown. When, as is the case with the sugar products discussed in the present chapter, the possible ingredients are known, rigorous proofs of identity are unnecessary, and comparison with authentic samples of the various sugars, run at the same time by the same gas chromatographic method as the sample being analyzed, suffices.

Method 16-12. *Column Chromatography.*

The various types of column chromatography preceded the other chromatographic methods, and in fact Tswett coined the name "chromatography" for the process because of the series of different-colored bands that were obtained when solutions of plant pigments were filtered through columns of finely-divided adsorbents, and the columns were then washed with organic solvents.

One of the most-used adsorbing media for the separation of the sugars has been charcoal.[17] A method based on the use of this adsorbent to separate the constituents of honey (Method 16-42) is given in the "honey" section of the present chapter.

Method 16-13. *Paper Chromatography.*

Paper chromatography was first reported by Consden, Gordon and Martin[18]; it was probably suggested by the use of powdered cellulose as an adsorption medium in column chromatography. In its simplest form, a vertical sheet or strip of filter-paper serves as support for a stationary liquid phase (usually water) that undergoes partition with a mobile liquid phase (descending or ascending) that is immiscible with it.

Methods have been worked out for the paper chromatographic separation of the sugars, and are cited in Text Reference 19. Paper chromatography is, however, gradually being superseded by thin-layer chromatography because the latter technique is usually much shorter.

Method 16-14. *Thin-Layer Chromatography.*

The newer technique of thin layer chromatography on coated glass plates is not only much faster than column or paper chromatography, but it is more readily adaptable to the recovery of the separated substances than is paper chromatography. For information on the general details of this method (such as the coating of plates and their development) the reader is referred to the special texts on thin-layer chromatography.[20] Apparatus can be obtained from C. A. Brinkmann & Co., Inc., Cantiague Road, Westbury, L.I., N.Y.

Plates coated with layers of Silica Gel G Merck prepared with 0.1 N boric acid instead of water have been found particularly useful for the separation of sugars.[21] Amounts applied may reach 50 to 250 mg, depending on the amount of sugars in the mixture, but quantities of 5–50 mg are preferable. Solvent systems used are methyl ethyl ketone-acetic acid-methanol (60-20-20) and butanol-acetone-water (40-50-10). The first of these gives a good separation of dextrose from sucrose, and the second a good separation of levulose from sucrose and dextrose (the last two sugars are not separated by this solvent) in 60 minutes. The second solvent will also separate di- and trisaccharides; hR_f values found were: Raffinose 6, D-melibiose 17, lactose 26, melezitose 33 and maltose and sucrose 44-45. (The hR_f value $= 100 \dfrac{\text{distance of center of spot from starting point}}{\text{distance of solvent from starting point}}$)

To separate dextrose, levulose, sucrose and raffinose it was found necessary to employ a two-dimensional technique using ethyl acetate-isopropanol-water (65-23.5-11.5) and methyl ethyl ketone-acetic acid-methanol (60-20-20) consecutively.

There is a large number of spray reagents that may be used to visualize sugar spots, but the following have been particularly recommended because of sensitivity:

1. *Aniline phthalate*—Dissolve 0.93 g of aniline and 1.66 g of *o*-phthalic acid in 100 ml of water-saturated n-butanol. After spraying, heat the chromatograms 10 minutes at 105°. The yellow-brown spots fluoresce under ultraviolet light.

NOTE: Prepared aniline phthalate reagent in a spray can is available from Gelman Instrument Co., P.O. Box 1448, Ann Arbor, Mich.

2. *Naphthoresorcinol-sulfuric acid*[22]—Mix equal volumes of alcoholic 1,3-hydroxynaphthalene solution (0.2 g/100 ml ethanol) and aqueous 20% H_2SO_4. After spraying, heat 5-10 minutes at 100–105°.

Colors given by the naphthoresorcinol-H_2SO_4 reagent after development with methyl ethyl ketone-acetic acid-methanol are shades of green, blue, violet and brown, varying with the sugar present.

The method can be made quantitative by comparing the area of the spot given by a particular sugar in the sample with the areas given by known quantities of that sugar that have been processed similarly. It is also possible to scrape an untreated spot from the plate, extract the sugar therefrom, and analyze the extract by some chemical method, but because these spots contain only microgram quantities of the sugars, the method chosen must be extremely sensitive. Text Reference 19b contains on pages 258 and 259 references to colorimetric methods that have been employed in paper chromatography; these methods should be equally applicable to aqueous extracts of spots from thin layer chromatography.

Another method of attack mentioned in Text Reference 19b is that of Jaarma[23]; this author oxidized the sugars with silver nitrate containing a known proportion of radioactive silver, and measured the radioactivity of the precipitated silver; a precision of 3–40 μg was claimed for this method.

Confectionery

Introduction

The dictionary[24] defines "candy" as "a food made of a sugar paste or syrup often enriched and varied with coloring and flavoring (as chocolate) and filling (as fruit or nuts) and shaped into various attractive forms", and "confectionery" as "sweet edibles (as candy, cake, pastry, candied fruits, ice cream)". For the purpose of this chapter, and in the general understanding of the Food, Drug and Cosmetic Act, however, "confectionery" is considered to be synonymous with "candy" as above defined.

The origin of candy making is lost in antiquity, but no doubt the original sweetening agent was honey, and it is known that in classical times confections composed of mixtures of fruit and honey or flour paste and honey were served at banquets. In Europe, candy was first sold by apothecaries and spice stores, and later by cake shops. Prior to the nineteenth century all candy was made by hand, and it was not until the London Exposition of 1851 introduced candy-making machines that the development of the modern candy industry began.[25]

Many different materials are employed in the manufacture of candy; among them are sugar, cacao products, corn syrup, dairy products, peanuts, coconut and other nut meats, fruits, licorice, gum arabic, gelatin, cooking starch, molasses, essential oils, pectin and salt.[25] "Liqueur" chocolates containing alcoholic centers were once a popular confection, but the present law classes as adulterated any confectionery that "shall bear or contain any alcohol" (in excess of 0.5 percent).* (The product now being sold as "liqueur chocolate cherries" is misnamed, since these chocolates contain cherries and syrup but no alcohol.) The law also forbids the use in confectionery of any "non-nutritive article or substance", with the exception of "harmless coloring, harmless flavoring, harmless resinous glaze not in excess of four-tenths of one percent, harmless natural gum or pectin". There have been some disputes as to whether this section of the law does not outlaw completely those special dietary confections in which sugar is replaced by saccharin or a cyclamate. (It is generally conceded that this argument does not apply when mannitol or sorbitol is the sweetening agent, because these compounds are nutritive substances.)

Standards

There is no general Federal standard for confectionery, but the definitions and standards of identity for cacao products include various types of chocolate that occur in chocolate bars and the coatings of chocolates. In summary, the standards for these products are (*see p.* 383).

Methods of Analysis

Method 16-16. Preparation of Sample.

A. *General* (*AOAC Method* ***29.074***)—If the composition of the entire sample is desired, grind and mix thoroughly. If the sample is composed of layers or of distinctly different portions, and it is desired to examine these individually, separate with a knife or other mechanical means as completely as possible, and grind and mix each portion thoroughly.

B. *Chocolate* (*AOAC Method* ***12.001***)

1. Chill sweet or bitter chocolate until hard, and grate or shave to a fine granular condition. Mix thoroughly and preserve in a tighly stoppered bottle in a cool place.

2. Melt bitter, sweet, or milk chocolate by placing in a glass or metal container and partly immersing the container in a bath at about 50°. Stir frequently until the sample melts and reaches 45–50°. Remove from the bath, stir thoroughly, and (while still liquid) remove a portion for analysis, using a glass or metal tube of 4–10 mm diameter that is provided with a close-fitting plunger to expel the sample from the tube.

Moisture, Ash and Total Protein. See ***Methods 16-2, 16-4*** *and* ***16-5.***

Method 16-17. *Alkalinity of Soluble Ash.* (*AOAC Method* ***29.015–29.016***).

Add water to the ash in a platinum dish, heat nearly to boiling, filter through an ashless paper, and wash with hot water until the combined filtrate and washings measure about 60 ml. Cool, and titrate with 0.1 *N* HCl, using methyl orange indicator (0.05% in water). Express the alkalinity as ml *N* acid/100 g sample. This value multiplied by 0.06911 gives percent K_2CO_3.

NOTE: This method will find its chief use in confectionery in checking compliance with the Federal requirement that added alkalis in chocolate have a neutralizing value not in excess of that of 3% of K_2CO_3.

Method 16-18. *Milk Protein in Chocolate.* (*AOAC Method* ***12.038***).

Place 10 g of milk chocolate in a centrifuge bottle (250 ml or larger), and extract with two 100 ml portions of ether by shaking until uniform, centrifuging, and decanting the supernatant ether. Place in the neck of the bottle a perforated two-hole stopper carrying a short bent glass tube and a straight glass tube that extends into the bottle to about one-third of the way to the bottom. Place the bottle in a moderately warm but not hot place, and expel

*Shortly after the presence of alcohol in confectionery became illegal, a few brands of (mostly imported) cordial chocolates that contained *iso*-propanol in place of ethyl alcohol appeared on the market. Action against these on the ground that *iso*propanol was a deleterious substance soon caused their disappearance.

the ether by applying suction to the bent tube and drawing a moderate current of air through the bottle. When the ether is expelled, pipet 100 ml of water into the bottle, stopper, and shake vigorously for 4 minutes. Pipet in 100 ml of 1% $Na_2C_2O_4$ solution, stopper, and shake vigorously 3 minutes more. Let stand about 10 minutes, shake 1–2 minutes, and centrifuge about 15 minutes at about 1800 rpm.

Remove the bottle from the centrifuge and decant the supernatant liquid into a beaker. Pipet 100 ml of this solution into a dry 250 ml beaker and add 1 ml of acetic acid while stirring gently. Let stand a few minutes so the precipitate can partly separate and add, stirring, 4 ml of 10% tannic acid solution (not over 1 week old). Let the precipitate settle a few minutes; then filter with moderate suction on a Coors No. 1A Büchner funnel containing an S & S No. 589 filter paper that is overlaid with a layer of paper pulp made by shaking one 15 cm No. 1 Whatman filter, torn to bits, with water. (The filtrate should be clear.) Using a wash solution of 1% $Na_2C_2O_4$ that contains 1 ml of acetic acid and 2 ml of 10% tannic acid in each 100 ml, transfer all of the precipitate to the funnel with the aid of a policeman. Wash the filter once or twice. Then loosen the filter around its edge with a spatula, carefully roll up, and transfer to a Kjeldahl flask. Wipe off any particles of precipitate clinging to the funnel or spatula with small pieces of damp filter paper, and add to the Kjeldahl flask. Determine nitrogen

Official Name and Synonyms	Standard Requires
Chocolate (unqualified) (bitter chocolate, chocolate coating, bitter chocolate coating)	50–58% cacao fat (exclusive of other fats); added alkali (bicarbonate, carbonate or hydroxide of Na, K or NH_4, or $MgCO_3$ or MgO) not greater in neutralizing value than 3% K_2CO_3.
Sweet chocolate (sweet chocolate coating)	Not less than 15% chocolate; less than 12% milk solids
Bittersweet chocolate (semisweet chocolate, semisweet chocolate coating)	Not less than 35% chocolate; otherwise same as sweet chocolate
Milk chocolate (sweet milk chocolate, milk chocolate coating, sweet milk chocolate coating)	Not less than 3.66% milk fat, 12% milk solids and 10% chocolate; non-fat milk solids 1.20–2.43 times milk fat
Skim milk chocolate (sweet skim milk chocolate, skim milk chocolate coating, sweet skim milk chocolate coating)	Less than 3.66% milk fat; not less than 12% skim milk solids
Buttermilk chocolate (buttermilk chocolate coating)	Less than 3.66% milk fat; not less than 12% sweet cream buttermilk solids
Mixed dairy product chocolates (mixed dairy product chocolate coatings)	May contain less than 3.66% milk fat. Must contain not less than 12% milk constituent solids of components used. When components are same as for milk chocolate and skim milk chocolate, nonfat milk solids must be more than 2.43 times milk fat
Sweet chocolate and vegetable fat (other than cacao fat) coating	Not less than 15% chocolate
Sweet cocoa and vegetable fat (other than cacao fat) coating	Not less than 6.8% fat-free cacao

as in Method 1-7. Wt. N $\times 3 \times 6.38$ = total casein and albumin present in the 10 g sample taken for analysis. This value $\times 1.07$ = total milk protein.

Method 16-19. *Sucrose.*

A. *General—See* Method 16-6, "Determinations A" (with note) and "B".

B. *In chocolate* (*AOAC Method* ***12.040***)—Transfer 26 g of the prepared sample, 16-16 (**b**), to a 250 ml centrifuge bottle, add about 100 ml of petroleum ether, shake 5 minutes, and centrifuge. Carefully decant the clear solvent, and repeat the treatment with petroleum ether. Place the bottle containing the defatted residue in a warm place until the petroleum ether has evaporated. Add 100 ml of water and shake until most of the chocolate has become detached from the sides and bottom of the bottle. Loosen the stopper, and carefully immerse the bottle in an 85–90° water bath for 15 minutes, shaking occasionally to remove all chocolate from the sides. Remove from the bath, cool, and add sufficient basic Pb acetate solution [Method 16-6, "Preparation and use of clarifying agents, (**a**)"] to complete precipitation (5 ml is usually enough). Add water to make a total of 110 ml of added liquid. Mix thoroughly, centrifuge, and decant the supernatant liquid through a small filter. Precipitate the excess of Pb with powdered anhydrous $K_2C_2O_4$, and filter. Dilute 10 or 20 ml of the filtrate with an equal volume of water, mix, and polarize in a 200 mm tube at 20°. Obtain the invert reading as in Method 16-6, "Determination, (A, 2)". Multiply both readings by 2 to obtain the direct and invert polarizations "P" and "I". From the data obtained, calculate percent sucrose (S) from the following formulas:

$$S = \frac{(P-I)(110+X)}{143.0-t/2},$$

where $$X = \frac{0.2244\,(P-21d)}{1-0.00204\,(P-21d)},$$

when $$d = \frac{P-I}{143.0-t/2}.$$

Method 16-20. *Dextrose in Chocolate.* (*AOAC Method* ***12.041***).

Prepare a clarified and deleaded sample solution as in Method 16-19 (B), except to use only a 10 g sample. Proceed as in Method 16-8 (B or C).

Method 16-21. *Commercial Glucose.* (*AOAC Methods* ***29.033–29.034***).

A. *Substances containing little or no invert sugar*—Commercial glucose cannot be determined accurately, due to the fact that the proportions of dextrin, maltose and dextrose vary. (See, however, Method 16-53.) In products in which the quantity of invert sugar is too small to affect the result appreciably, commercial glucose may be estimated approximately by the following formula:

$$G = (a-S)\,100/211,$$

where G = % commercial glucose solids, a = direct polarization, normal solution, and S = % cane sugar.

Express the results in terms of commercial glucose solids polarizing at +211°S.

B. *Substances containing invert sugar*—Prepare an inverted half-normal solution of the sample as in Method 16-6, Determination, (A, 2), except to cool the solution after inversion, make neutral to phenolphthalein with NaOH solution, slightly acidify with HCl (1+5), and treat with 5–10 ml of alumina cream, Method 16-6, Preparation and use of clarifying agents, (**b**), before diluting to the mark. Filter and polarize at 87° in a 200 mm jacketed metal tube. Multiply the reading by 200 and divide by 196, to obtain the quantity of glucose solids polarizing at +211°S.

Method 16-22. *Lactose in Chocolate.* (*AOAC Method* ***12.039***).

Determine reducing sugars before inversion as in Method 16-6 (B) in an aliquot (usually 20 ml) of the Pb-free filtrate obtained in Method 16-19 (B). Determine reduced Cu as Cu_2O by the the volumetric thiosulfate method. Correct for Cu_2O due to sucrose as follows: Obtain the approximate percentage of lactose from the following formula, using the data obtained in Method 16-19 (B):

Approximate % lactose

$$= [P(1.1 + X/100 - S]/0.79.$$

From the calculated polarimetric sucrose/lactose ratio and total Cu_2O obtained as above, determine the quantity of Cu_2O to be subtracted from the total Cu_2O found, using the graph, Fig. 16–1. Convert the corrected Cu_2O to g lactose (L), using Table 23–6. Then obtain percent lactose from the following relationship:

$$\% \text{ lactose} = L(110 + X)/0.26\ C,$$

where X = value obtained in the polarimetric sucrose determination,

and C = the volume of solution (ml) used in the above lactose determination.

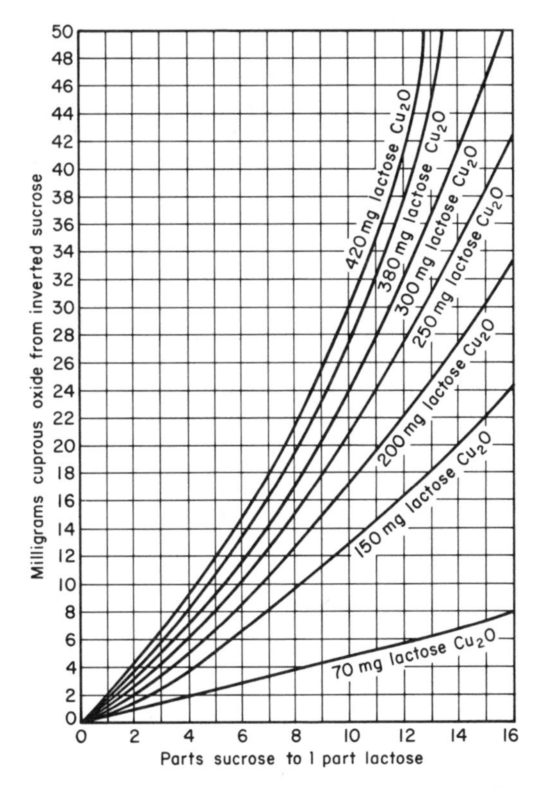

Fig. 16-1—Graph used in correcting cuprous oxide for effect of sucrose.

Method ***16-23A.*** *Starch — General.* (*AOAC Methods* ***29.085*** *and* ***22.046–22.047***).

REAGENT

Malt Extract—Use clean, new barley malt of known efficacy, and grind only as needed. Grind well, but not so fine that filtration is greatly retarded. Prepare an infusion of freshly ground malt just before use. For every 80 ml of malt extract required, digest 5 g of ground malt with 100 ml of water at room temperature. A 2 hour digestion period is sufficient; this may be reduced to 20 minutes if an electric mixer is used. Filter to obtain a clear extract, refiltering the first portions of the filtrate if necessary. Mix the infusion well.

DETERMINATION

Measure 25 ml of a suspension of a uniform mixture (representing 5 g of sample) into a 300 ml beaker, or add to the beaker 5 g of finely ground sample (previously extracted with ether if sample contains much fat); add enough water to make 100 ml; heat to about 60°, avoiding, if possible, gelatinization of the starch. Let stand about one hour, stirring frequently to secure complete solution of the sugars. Transfer to a wide-mouthed bottle, rinse the beaker with a little warm water, and cool. Add an equal volume of alcohol, mix, and let stand at least one hour.

Centrifuge until the precipitate is closely packed on the bottom of the bottle, and decant the supernatant liquid through a hardened filter. Wash the precipitate with successive 50 ml portions of 50% by vol. alcohol by centrifuging and decanting through the filter until the washings are sugar-free by the following test: Add to a test-tube a few drops of the washings, 3 or 4 drops of 20% alcoholic α-naphthol solution, and 2 ml of water. Shake well, tip the tube, let 2–5 ml of H_2SO_4 flow down its side, and then hold it upright. If sugar is present, the interface of the two liquids is colored faint to deep violet; on shaking, the whole solution turns blue-violet.

Transfer the residue from the bottle and hardened filter to a 250 ml volumetric flask. Add 50 ml of water, immerse the flask in boiling water, and stir constantly 15 minutes, or until all starch is gelatinized. Cool to 55°, add 20 ml of the malt extract, and hold at this temperature for one hour, or until the residue shows no blue tinge upon treatment with iodine solution and microscopic examination. Cool, dilute to 250 ml, and filter.

Place 200 ml of the filtrate in a 500 ml volumetric flask, add 20 ml of HCl (sp.g 1.125), attach a reflux condenser, and heat in a boiling water-bath for 2½ hours. Cool, nearly neutralize with 10% NaOH solution, finish the neutralization with Na_2CO_3 solution, and dilute to 500 ml. Mix the solution thoroughly, pour through a dry filter, and determine dextrose in an aliquot as in Method 16-8 (A.). Conduct a blank determination on the same volume of malt extract as was used with the sample, and correct the weight of dextrose accordingly. Wt. dextrose obtained × 0.90 = wt. starch.

*Method 16-23B. Starch in Chocolate (AOAC Methods **12.042–12.043** and **22.045**).*

1. *Direct acid hydrolysis method*—Weigh 4 g of sample if unsweetened, or 10 g if sweetened, into a small porcelain mortar; add 25 ml of ether and grind. After coarser material settles, decant the ether, together with fine suspended matter, onto an 11 cm paper of sufficiently fine texture to retain crude starch. Repeat this treatment until no more coarse material remains. After the ether has evaporated from the filter, transfer the fat-free residue to a mortar by means of a jet of cold water and rub to a smooth paste, filtering on the paper previously used. Repeat this process until all sugar is removed. (In case of sweetened products, the filtrate should measure at least 500 ml.)

Heat the insoluble residue 2½ hours with 200 ml of water and 20 ml of HCl (sp.g 1.125) in a flask provided with a reflux condenser. Cool and nearly neutralize with NaOH. Transfer to a 250 ml volumetric flask, dilute to the mark, filter, and determine dextrose in an aliquot of the filtrate by Method 16-8(A):

Wt. dextrose obtained × 0.90 = wt. starch.

2. *Diastase method*—Remove fat and sugar from a 4 g sample if unsweetened, or a 10 g sample if sweetened, as in (1). Carefully wash the wet residue into a beaker with 100 ml of water, heat to boiling over asbestos with constant stirring, and continue boiling and stirring for 30 minutes. Replace water lost by evaporation, and immerse the beaker in a water-bath kept at 55°–60°. When the liquid cools to bath temperature, add 20 ml of freshly-prepared malt extract, Method 16–23 A, and digest the mixture 2 hours with occasional stirring. Boil a second time for 30 minutes, dilute, cool, and digest as before with another 20 ml of the malt extract. Heat again to boiling, cool, and transfer to a 250 ml volumetric flask. Add 3 ml of alumina cream, 16-6, "Preparation and use of clarifying agents, (**b**)", dilute to the mark, and filter through dry paper. (The residue on the paper should show no sign of starch when examined microscopically.) Continue as in Method 16-23 A, "Determination", last paragraph.

*Method 16-24. Ether Extract. (AOAC Method **29.086–29.087**).*

A. *Continuous extraction method*—Measure 25 ml of a 20% mixture or solution into a very thin, readily frangible glass evaporating shell, or a thin Pb or Sn foil dish, containing 5–7 g of freshly ignited asbestos fiber. Alternatively, if possible to obtain a uniform sample, weigh 5 g of mixed finely divided sample into a dish and wash with water onto the asbestos in the evaporating shell, using if necessary a small portion of the asbestos fiber on a stirring rod to transfer the last traces of the sample from the dish to the shell.

Dry to constant weight at 100°, cool, wrap the glass shell loosely in smooth paper, crush between the fingers into rather small fragments, and carefully transfer the crushed mass (including the paper) to an extraction tube or fat extraction cartridge. If a metal dish is used, cut this into small pieces and place in the extraction tube. Extract with anhydrous ether or petroleum ether (b-pt. 45–60° and without weighable residue) in a continuous extraction apparatus for at least 24 hours. (In most cases it is advisable to remove the substance from the extractor after the first 12 hours, grind with sand to a fine powder, and re-extract for 13 hours more.) Transfer the extract to a weighed flask, evaporate the solvent, and dry to constant weight at 100°.

B. *Roese-Gottlieb method*—Introduce a 4 g sample, or a quantity of a sample solution equivalent to this weight of dry substance, into a Mojonnier or Röhrig fat extraction tube. Dilute to 10 ml with water, add 1.25 ml of NH_4OH, and mix thoroughly. Add 10 ml of alcohol and mix;

then add 25 ml of ether and shake vigorously about 30 seconds. Finally add 25 ml of petroleum ether (b.pt. below 60°) and shake again about 30 seconds. Let stand 20 minutes, or until separation is complete.

Draw off as much as possible of the ether-fat solution (usually 0.5–0.8 ml is left) through a small rapid filter into a weighed flask. (Weigh the flask using a similar flask as counterpoise.) Re-extract the liquid remaining in the tube with 15 ml portions of ether and petroleum ether, shaking vigorously 30 seconds and allowing to settle after the addition of each solvent. Proceed as above, washing the mouth of the tube and filter with a few ml of an equal mixture of the two solvents (previously freed from deposited water).

For a greater degree of accuracy, extract once more. If the previous solvent-fat solutions have been drawn off closely, the third extraction usually yields not over 1 mg of fat, or about 0.03 percent on a 4-gram charge. Evaporate the solvent slowly on a steam-bath and dry the fat in a boiling water oven to constant weight. Test the purity of the fat by dissolving in a little petroleum ether. If a residue remains, wash the fat out completely with petroleum ether, dry the residue, weigh, and deduct this weight.

Method 16-25A. *Fat in Chocolate.* (*AOAC Method* ***12.022***).

(Not applicable to chocolate containing milk ingredients or to products prepared by cooking with sugar and water and drying.)

APPARATUS

Knorr extraction tube, about 20 mm i.d., body about 11 cm long, stem 6–8 mm o.d. and about 10 cm long. Provided with removable, close-fitting perforated Ni, monel metal, glass or porcelain disk at bottom of larger tube.

DETERMINATION

Pack the extraction tube with a 6 mm tightly packed mat of purified asbestos that has been carefully freed of coarse pieces. (An Allihn type glass filter tube with a coarse fritted disk, such as Ace Glass, Inc. No. 8571, is also satisfactory.) Wash the filter with alcohol, ether, and a little petroleum ether. (All petroleum ether must be redistilled and boil below 60°.) Take care that no rubber particles adhere to the stem. Place a weighed Erlenmeyer flask at such a height that the tube stem passes through the neck into the body of the flask. Lengthen the stem if necessary. Fill the tube to about two-thirds capacity with the redistilled petroleum ether, and stir the sample thoroughly with a flat-ended rod, crushing all lumps. Let stand one minute and drain by suction. Regulate the suction so that the solvent collecting in the flask does not boil violently. Release the vacuum after each draining before adding more solvent. Add solvent from a wash-bottle while turning the tube between the thumb and finger so that the sides of the tube are washed down by each addition. Repeat the extractions, with stirring, until the fat is removed (usually 10 extractions). Remove the stopper and tube from the bell-jar, wash traces of fat from the end of the stem with petroleum ether, evaporate the solvent, and dry to constant weight at 100°.

Method 16-25B. *Fat in Chocolate.* (*AOAC Method* ***12.023***).

(Applicable to chocolate containing milk ingredients or to products prepared by cooking with sugar and water and drying.)

Weigh accurately into a 400 ml beaker 10–20 g of milk chocolate, and add 30 ml of water and 25 ml of HCl. Heat on the steam bath 30 minutes, stirring frequently; add 5 g of filter-aid (diatomaceous earth type) and 50 ml of ice water, and chill 30 minutes in ice water. Fit a heavy piece of butcher's linen, dress linen with about 45 threads/inch, or No. 40 filtering cloth (National Filter Media Corp., Hamden, Conn., or equivalent), into a Coors No. 1A Büchner funnel, and moisten with water; apply gentle suction, and completely overlay the cloth with filter-aid (3 g filter-aid suspended in 30 ml of water, poured over the funnel and allowed to drain).

Filter the hydrolyzed mixture by gentle suction, rinsing the beaker three times with ice-cold water (but do not suck the pad dry until transfer and washing are complete). Finally, wash three times with ice-cold water, tamping tightly with a flat-ended rod after the last washing. Suck dry. Strip the linen from the cake, and transfer the cake to the original beaker. With a small piece of filter paper, transfer to the beaker any material adhering to the funnel. Wash the funnel with petroleum ether, and add to the cake in the beaker. Evaporate the ether on the steam bath.

Break up the cake with a stirring-rod, and keep on the steam bath until the material pulverizes easily and appears dry. Place in an oven for 1 hour at 100°. Add 15 g of powdered anhydrous Na_2SO_4 and mix well. Transfer the dry mass to a large Knorr-type tube (about 175 ml capacity, fitted with a ½″ perforated metal disk overlaid with a dry asbestos pad about ⅜″ thick), or to a fritted 150 ml, 60–65 mm medium porosity Büchner funnel fitted with a layer of asbestos that is covered with a qualitative filter paper and overlaid with a 20 mesh wire screen. Wash the beaker with 50 ml of petroleum ether, pouring the rinsings into the tube or onto the Büchner funnel. Wash the material in the tube or on the funnel with six 50 ml portions of petroleum ether. Each time stir the sample thoroughly with the glass rod while crushing all lumps, let stand 2 minutes, and drain by gentle suction into a tared 250 ml flask. (Evaporate between extractions if necessary.) Evaporate the petroleum ether on the steam-bath, and dry the fat to constant weight.

Method 16-26. *Milk Fat in Milk Chocolate.* (*AOAC Method* ***12.029***.)

Estimate the quantity of milk fat in milk chocolate from the following formula:

$$C = \frac{AX+BY}{5},$$

where A = g butter fat in 5 g mixed fat,
$B = 5-A$ = g cacao fat in 5 g mixed fat,
C = Reichert-Meissl number of extracted fat,
X = Reichert-Meissl number of authentic butter fat, and
Y = Reichert-Meissl number of authentic cacao butter.

Then

wt butterfat, A, in 5 g mixed fat

$$= \frac{5(C-Y)}{(X-Y)},$$

and

$$\% \text{ butterfat} = \% \text{ total fat} \times \frac{(C-Y)}{(X-Y)}.$$

Roese-Gottlieb Number

—*See* Chapter 13.

Saccharin and the Cyclamates

—*See* Chapter 14.

Mannitol and Sorbitol

These sugar alcohols have found frequent use in "dietetic" confectionery both because of their physical properties, and because they are sweetening agents that are nutritive substances, rather than non-nutritive substances like saccharin and the cyclamates, and therefore bypass the legal prohibition against the use of non-nutritive substances in confectionery. Because they are less sweet than sugar and its other substitutes, they must be used in substantial amounts in products in which no other sweetener is present.

The following methods are slight modifications of methods designed by their authors for use on bakery products rather than confectionery. The first is a titration method that does not separate the two alcohols; it depends on thin-layer chromatography to identify the compound present. The second is a gas chromatographic method that does separate mannitol from sorbitol.

Method 16-27. *Mannitol and Sorbitol—Titrimetric.*[26]

APPARATUS AND REAGENTS

a. *TLC apparatus*—A suitable applicator (Desaga/Brinkmann or equivalent), 20 × 20 cm glass plates, and chromatographic tank and accessories (Arthur H. Thomas Co. No. 3106-FO5, or equivalent).

b. *Chromatographic columns*—12–14 mm × 45 cm with stopcock or adjustable pinch clamp. (Columns may be made from glass tubing.)

c. *NaOH solutions*—0.1 *N* and 5%.

d. *Acidic potassium periodate solution*—Dissolve 0.60 g of KIO_4 in 400 ml of water plus 20 ml of H_2SO_4, then dilute to 1 liter with water.

e. *Potassium iodide*—Powdered or granular; suitable for iodometry.

f. *Sodium thiosulfate*—0.02 *N*; prepare fresh before use by diluting a more concentrated solution and standardizing.

g. *Strongly basic ion exchange resin*—Amberlite IRA-400, 20–50 mesh or equivalent. Prepare the column as follows: Rinse enough resin into the chromatographic column with water to have 8″ of resin. Add an excess of water, close one end of the column, and invert it several times to remove all air bubbles and redistribute the resin according to particle size. After the resin settles, add a small glass wool plug and pass 150 ml .of 5% NaOH solution through the column at a rate of about 5 ml/minute. Wash the column with 200 ml. of water or until the eluate is neutral to pH paper. (*Note*: Do not let the column become dry, since air pockets will prevent good contact between the solution and the resin and limit the extent of ion exchange. Do not retain the column in the hydroxide form for more than a few hours).

h. *Silica gel R*—For thin layer chromatography (Brinkmann Instruments).

i. *Mobile solvent*—*Iso*propyl alcohol: acetone: water (4:2:1, v/v/v).

j. *Chromogenic agent*—0.5% $KMnO_4$ in 1 *N* NaOH.

k. *Standard solutions*—Prepare 1% solutions in water of sorbitol, sucrose, mannitol, and any other sugars or sugar alcohols which might be present.

l. *Hydrochloric acid*—0.1 *N*.

PREPARATION OF SAMPLE

Transfer 10 g of the sample, prepared as in Method 16-16 (A or B), to a 250 ml centrifuge bottle, add about 100 ml of petroleum ether, shake 5 minutes, and centrifuge. Carefully decant the clear solvent, and repeat the treatment with petroleum ether. Place the bottle containing the defatted residue in a warm place until the petroleum ether has evaporated. Air-dry and weigh the fat-free residue. Accurately weigh a 0.70–0.75 g portion of the dry residue into a 100 ml round-bottomed flask.

Extract with 50–75 ml of 80% alcohol under reflux for 30 minutes, using a water-cooled condenser. Cool, let the residue settle, and decant the supernatant liquid. Re-extract with 50–75 ml of 80% alcohol for 20 minutes. Combine the extracts, and evaporate to dryness with a flash evaporator at 50°. Wash the residue into a 200 ml volumetric flask with water, and make to volume.

DETERMINATION

Pipet 5.0 ml of the extract into a 250 ml ₸ flask fitted with a water reflux condenser. Add 10 ml of 0.1 *N* HCl and reflux 30 minutes. Without disconnecting the apparatus, add 25 ml of 0.1 *N* NaOH and reflux at a rapid boil for an additional 1.5 hours.

Transfer the solution quantitatively to a 100 ml volumetric flask, cool, and dilute to volume with water. Pipet 25.0 ml of this solution onto the ion exchange column and let it pass into the resin bed at about 3 ml/minute. When the solution has just passed into the column, rinse the sides of the column with about 10 ml of water and let this pass into the resin; then complete the elution with 200 ml of water at about 3 ml/minute, collecting the eluate in a 500 ml Erlenmeyer flask.

Add 50.0 ml of acidic potassium periodate reagent, (**d**), to the eluate, and heat on a steam bath for 15 minutes. Cool to room temperature and add 2 g of potassium iodide, (**e**). Let the solution stand for 5 minutes and titrate with 0.02 *N* sodium thiosulfate, using starch T.S. as the indicator. Perform a blank determination with potassium periodate and an amount of water equal to the eluate collected. [Each ml of 0.02 *N* sodium thiosulfate is equivalent to 0.3644 mg of *d*-sorbitol ($C_6H_{14}O_6$).]

CALCULATIONS

Ml $Na_2S_2O_3$ consumed × equivalence = mg sorbitol in aliquot; (ml blank − ml sample) × 0.3644 × 160 (dilution factor) = mg sorbitol in original sample taken.

NOTE: The titration is valid for not > 5.0 mg sorbitol. When the calculated result is 4.5 mg sorbitol or greater, take a smaller aliquot from the degraded solution, pass through the ion exchange column, and again titrate as indicated in the procedure.

THIN-LAYER CHROMATOGRAPHY

Prepare a slurry of 30 g of silica gel G and 60 ml of water, and coat five 20 × 20 cm plates (250 μ thickness). Dry the plates at 100° for 1 hour and store in a desiccator until used.

Spot a product equivalent to 20 μg sorbitol and about 2 μl each of the standard solutions at 15 mm distances. Heat the plates at 100° for a few minutes to dry the spots, and cool to room temperature.

Develop with mobile solvent (i) until the solvent front has ascended nearly the height of the plate. Remove the plate from the tank and dry. Spray the dried plates with a fine spray of the chromogenic agent (j). The carbohydrates are yellow spots on a purple background.

NOTE: Heating at 100° after spraying will facilitate the appearance of the spots.

Method 16-28. *Mannitol and Sorbitol—Gas Chromatographic.*[27]

REAGENTS AND APPARATUS

a. *Ethanol*—80%, v/v in water.

b. 1,4-*Sorbitan* (1,4-*anhydro-D-sorbitol*)—M.p. 115–116°. Recrystallize by refluxing in *iso*propyl alcohol; induce crystallization by adding *iso*-propyl ether.

c. *Gas chromatograph*—F & M model 700, with a flame ionization detector coupled to an F & M model 720 programming unit. Use helium, at a rate of 120 ml/minute, as the carrier gas. Maintain the detector and injection port at 310 and 300°, respectively.

Analyze the TMS ethers on 6′ × ⅛″ o.d. stainless steel columns packed with 10% SE-30 (silicone gum rubber) on 80–100 mesh Chromosorb W, programmed from 145 to 257° at 10°/min., Analyze the acetate esters on 3′ × ⅛″ o.d. stainless steel columns packed with a mixture of 7% QF-1 (trifluoropropylmethyl siloxane) and 1.7% BDS (butane-diol succinate polyester) on 100/120 mesh Chromosorb W, programmed from 160 to 217° at 4°/min.

PREPARATION OF SAMPLE

Proceed as in Method 16-27, "Preparation of sample", first paragraph. Then accurately weigh 150 ± 10 mg of 1,4-sorbitan, and thoroughly mix with the weighed-out sample portion in the flask. Continue as in Method 16-27, second paragraph, through "evaporate to dryness with a flash evaporator at 50°".

Dissolve the syrupy residue in 2–5 ml of pyridine for preparation of the derivatives.

PREPARATION OF DERIVATIVES

a. *TMS ethers*—Pipet about 0.5 ml of pyridine solution into a 5 ml vial and add 0.2 ml of hexamethyldisilazane and 0.1 ml of trimethylchlorosilane. Shake the mixture and let stand 1 minute.

b. *Acetate esters*—Pipet about 2 ml of pyridine solution into a vial, add 1–1.5 ml of acetic anhydride, and warm at 60° for 30 minutes. Add 5 ml of chloroform and wash three times with 50 ml of water. Dry the chloroform layer over anhydrous magnesium sulfate and evaporate to about 0.5 ml.

PREPARATION OF STANDARD MIXTURES

Weigh accurately 0.6 ± 0.1 g of hexitol and dissolve in 50 ml of pyridine. (It may be necessary to warm the solvent to completely dissolve the sample.) Similarly dissolve 0.5 ± 0.1 g of 1,4-sorbitan in 50 ml of pyridine. Prepare standard mixtures of the hexitols and 1,4-sorbitan by withdrawing and mixing different known volumes from each solution. (These solutions are stable and may be stored for 3 months without any apparent change in composition and concentration.) Prepare the trimethylsilyl ethers and acetate esters of the standard mixtures as described above. To obtain the standard hexitol-1,4-sorbitan curve, plot the observed molar ratios (obtained from the peak areas) *vs.* the calculated molar ratios from the weights in solution (*see* Fig. 16-2).

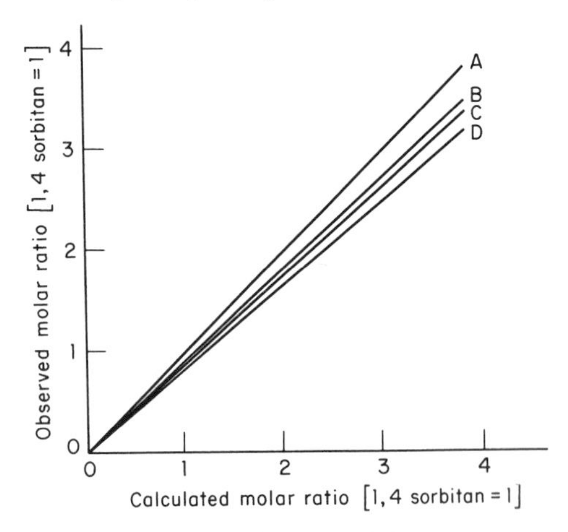

Fig. 16-2.—Standard curves of the acetate esters based on observed and theoretical molar ratios and response of hydrogen flame detector; theoretical response (A), xylitol (B), D-mannitol (C), and D-sorbitol (D).

DETERMINATION

Inject 0.5–1.5 μl of sample solution into a gas chromatograph with an appropriate column and conditions for the derivative being analyzed.

Calculate the quantity of hexitol present as follows:

% Hexitol = $R \times M \times W_1 \times 100/(1 + W_2)$, where R = corrected ratio of the peak areas (hexitol derivative/1,4-sorbitan derivative); M = ratio of the molecular weights of the derivatives (1,4-sorbitan derivative/hexitol derivative); W_1 = ratio of the weights used (1, 4-sorbitan/aliquot); and W_2 = ratio of the weight of fat extracted in *n*-hexane to the weight of the defatted sample.

Flavors—*See* Chapter 9.

Corn Products

Introduction

The term "corn sugar" as applied to dextrose is a misnomer, since the major sugar naturally present in corn (*i.e.*, maize) is sucrose. It is not generally realized that the stalks of certain varieties of corn contain as much sucrose as does sugar cane. The reason why corn has never become a commercial source of sucrose as have sugar cane and sugar beets is purely economic.

Commercial "corn syrups" and "corn sugar" are made by the acid hydrolysis of corn starch. This hydrolysis takes place, in steps, through carbohydrates of intermediate molecular weight ("dextrins") to maltose and finally dextrose. This is actually a complex process, both because the proportions of the various carbohydrates higher than maltose may vary and because recondensation of dextrose to higher polymers ("reversion") can take place. Commercial acid hydrolysis of starch aims at producing two types of products: syrups and crystallizable sugars. To produce "corn syrup", a starch slurry containing 40 to 43% solids is treated with sufficient hydrochloric acid to give a normality of 0.0120-0.014, with a corresponding pH of 1.7–1.9; this mixture is heated at 145° (45-lb. pressure) for 4-6 minutes to form a syrup of 42-55 dextrose equivalent*. Theoretically, the hydrolysis of 100 lbs of starch should yield 111 lbs of anhydrous dextrose, but in current commercial practice 100 lbs of dry substance starch produce only 95 lbs of dextrose and 14 pounds of other sugars.[28]

Besides corn syrups and crystalline dextrose, two other types of products are marketed for human food use: "corn syrup solids", made by dehydration of corn syrup, and "crude corn sugar", which is a solidified corn sugar "liquor" consisting essentially of dextrose.[29] In addition a product known as "hydrol", which is the final mother-liquor remaining from the refining of crystalline dextrose, is sold for use as an ensilage supplement and a stock feed ingredient.

In the production of corn syrup, besides the straight "acid process" of hydrolysis, an "acid-enzyme process" is sometimes used, in which acid hydrolysis is followed by treatment with a diastatic enzyme; this yields a syrup of high maltose content. In either process, when conversion has reached the desired point, the reaction product is neutralized with sodium carbonate to a pH of 4–5, fatty substances are removed by skimming or centrifuging, and the syrup is filtered. It is then evaporated to about 60% solids, and clarified and decolorized with carbon or ion-exchange resins. The clarified syrup is then vacuum-evaporated to the desired solids content.

*The term "dextrose equivalent" generally used by the industry to characterize its products is defined as the "percentage of reducing sugars, expressed as dextrose, calculated on a dry substance basis".

If "corn syrup solids" are desired, concentration is continued in spray or vacuum drum driers to produce an amorphous, powdery product whose moisture content is less than 3.5%.

"Crude corn sugar"is prepared by a more complete acid hydrolysis, followed by neutralization, filtration, clarification and concentration. Finally the concentrated liquor is seeded with dextrose crystals, cast in pans and allowed to crystallize. Crude corn sugar is sold in two grades numbered "70" and "80", whose dextrose equivalents are respectively about 83 and 91. It is shipped as slabs or pellets, or as bags of chips.

As noted above, in the manufacture of crystalline dextrose conditions of hydrolysis of the starch are adjusted to yield the maximum dextrose content. After purification and concentration as for corn syrup, the liquor is seeded with dextrose crystals and crystallization allowed to take place for about 100 hours. The dextrose monohydrate crystals so produced are then separated by centrifuging, washed, dried, and bagged. The mother liquor is re-converted, clarified, concentrated and passed through another crystallization cycle. The crystals so obtained are added to new starch hydrolysate and recrystallized to yield a second crop of dextrose monohydrate. The dark brown mother liquor from the second crystallization is the "hydrol" or "feeding corn sugar molasses" referred to above.

For the production of anhydrous dextrose, the monohydrate is redissolved in water to form a solution containing about 55% solids. This solution is treated with carbon, heated to 71° for 30 minutes, filtered, evaporated to crystallization, and the crystals separated by centrifuging. The crystals are then washed with a hot-water spray, dried, pulverized and packed.

Corn starch hydrolysates of any dextrose equivalent rating up to 100 may be produced, but the commercial products are divided into the following classifications:

Dextrose Equivalents	Common Name
Up to 13	Dextrins
13 to (but not including) 28	Malto-dextrins
28 to (,, ,, ,,) 38	Low conversion corn syrups
38 to (,, ,, ,,) 48	Regular conversion corn syrups
48 to (,, ,, ,,) 58	Intermediate ,, ,, ,,
58 to (,, ,, ,,) 68	High ,, ,, ,,
Greater than 68	Extra high ,, ,, ,,

The high-conversion syrups are further classified as acid-conversion and acid-enzyme (dual-conversion) syrups.

The relative percentages of dextrose, maltose and higher saccharides in various types of commercial corn syrups are indicated by the results of chromatographic analysis of typical samples shown in Table 16-6.[29]

Ough [30] identified in a 60 dextrose equivalent syrup by paper chromatography the following proportions of various sugars other than dextrose: Laminaribose, 0.2%; nigerose, 0.8%, maltose, 15.5%; cellobiose, 0.4%; kojibiose, 0.2%; β-β-trehalose, 0.2%; isomaltose, 1.6%; and gentiobiose, 0.3%.

Montgomery and Weakley[31] reported the following proximate analysis of a sample of hydrol: Moisture, 26.04%; ash, 8.80%; protein, 0.199%; reducing sugars, 55.14%; other carbohydrate (by difference), 9.61%; and 5-hydroxymethylfurfural, 0.31%. By column chromatography these authors found the composition of the carbohydrate fraction of another sample to be: Dextrose, 63%; brachiose, 15.2%; gentiobiose, 5.0%; maltose, 2.8%; α,α-trehalose, 1.7%; other disaccharides, 4.7%; and higher oligosaccharides, 29.4%.

Methods of Analysis

For sample preparation and most determinations the methods in the "General" section of this chapter may be used. Methods 16-8 (B) and (C) and the glucose oxidase method of Text Reference 10 were particularly designed for the determination of dextrose in corn products. Text References 30 and 32 cite procedures for the paper chromatographic separation and colorimetric determination by the anthrone method of the carbohydrates in corn syrups.

The following are standard methods of the Corn Industries Research Foundation, Inc.

Method 16-29. *Moisture by Azeotropic Distillation.*

APPARATUS—*See* Fig. 16-3.

DETERMINATION

In the wide-mouthed reaction flask thoroughly mix a sample of corn syrup or corn syrup solids (containing not over 5 ml of water) with sufficient Filter-Cel (analytical grade) to obtain maximum sample dispersion. (Water may be lost in this step, depending on the time consumed in manipu-

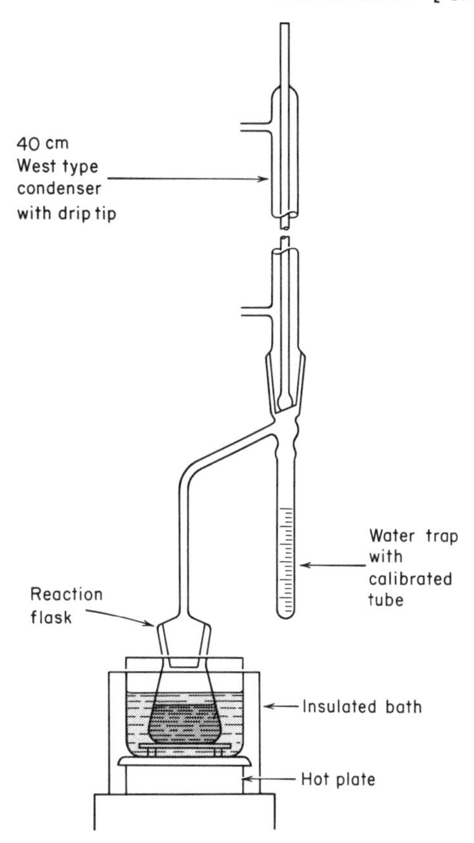

Fig. 16–3.—The bath rests on the hot plate and is surrounded by an insulated cylindrical shield and cover. The reaction flask rests on a one-quarter-inch mesh metal screen supported by metal legs one-half inch above bath bottom.

TABLE 16-6

Examples of Carbohydrate Composition of Commercially Available Corn Syrups

Type of Conversion	Dextrose Equivalent	Percent Saccharides: Mono-	Di-	Tri-	Tetra-	Penta-	Hexa-	Hepta-	Higher
Acid	30	10.4	9.3	8.6	8.2	7.2	6.0	5.2	45.1
Acid	42	18.5	13.9	11.6	9.9	8.4	6.6	5.7	25.4
Acid-enzyme[a]	43	5.5	46.2	12.3	3.2	1.8	1.5	—	29.5[b]
Acid	54	29.7	17.8	13.2	9.6	7.3	5.3	4.3	12.8
Acid	60	36.2	19.5	13.2	8.7	6.3	4.4	3.2	8.5
Acid-enzyme[a]	63	38.8	28.1	13.7	4.1	4.5	2.6	—	8.2[b]
Acid-enzyme[a]	71	43.7	36.7	3.7	3.2	0.8	4.3	—	7.6[b]

[a] The carbohydrate composition of acid-enzyme syrups will vary as a result of different processes used. The values given in the table are to be considered only as examples of ranges of values that are available commercially.

[b] Includes heptasaccharides.

lation; the loss may be estimated by another weighing after dispersion.) Add 100 ml of toluene, and stir until thoroughly mixed. Attach a clean and dry receiver and condenser, as illustrated in Fig. 16-3. Immerse the reaction flask in a bath, containing bath wax or oil, supported on an adjustable hot plate. Regulate the heat input so that the condensate falls from the condenser tip at a rate of about 2 drops a second; after one hour, increase the rate to about 4 drops/second.

Continue the distillation until the volume of water in the trap fails to increase (usually 8–12 hours). When finished, dismantle the assembly and immerse the calibrated portion of the trap in a water bath at 20°; hold there until the trap contents assume the bath temperature. Estimate the volume of water in the trap to the closest hundredth-ml. Run a blank determination on an accurately-weighed portion of water (about 3 g). Add to the sample water volume the difference between g of water employed and ml of water recovered.

$$\text{Percent moisture} = \frac{100\ (\text{corrected vol. } H_2O)}{\text{wt. sample}}.$$

(Because the correction for the density of water in the trap is less than the precision with which the volume can be read, this correction can be ignored.)

NOTES

1. In analyzing finely ground corn sugar by this method, since this material is more heat-labile than corn syrup and corn syrup solids, benzene must be substituted for toluene. This increases to 30–48 hours the time required for complete removal of water.

2. This method is similar in principle to Method 1-4.

Method 16-30. *Dextrose Equivalent.*

PRINCIPLE

Methods most commonly used for estimation of aldose-type sugars are based on their reducing action toward certain metallic salts. In the following modification of the Lane and Eynon procedure, dextrose and related sugars contained in the sample reduce copper sulfate in an alkaline tartrate system. Dextrose equivalent is defined as "reducing sugars expressed as dextrose and calculated as a percentage of the dry substance".

SCOPE

The method is applicable to crude and refined corn sugars (Note 1) and all starch hydrolyzates prepared by acid or enzyme conversion and combinations thereof.

REAGENTS

a. *Fehling Solution*

1. Dissolve 34.64 g of C.P. crystalline copper sulfate pentahydrate ($CuSO_4.5H_2O$) in distilled water and dilute to 500 ml.

2. Dissolve 173 g of C.P. potassium sodium tartrate tetrahydrate ($KNaC_4H_4O_6.4H_2O$) and 50 g of C.P. sodium hydroxide in distilled water and dilute to 500 ml.

Combine the solutions, adding (2) to (1), let stand 1 day at room temperature, and filter. Standardize as follows, immediately prior to use:

Dry a quantity of National Bureau of Standards dextrose in a vacuum oven for 2 hours at 100°. Dissolve 5.000 g in distilled water, dilute to 500 ml, and mix thoroughly. Pipet 25.0 ml of the Fehling solution into a boiling flask, bring to a boil and titrate with the standard dextrose solution as directed under "Determination". Adjust the concentration of the Fehling solution by dilution or addition of copper sulfate so that titration requires 12.02 ml of the standard dextrose solution.

b. *Methylene Blue Indicator*—1% aqueous solution.

DETERMINATION

Weigh accurately an amount of sample such that after dilution the solution contains about 1% reducing sugars (Note 2). Transfer the sample quantitatively to a 500 ml volumetric flask with the aid of hot water, cool to room temperature, dilute to volume and mix thoroughly.

Pipet 25.0 ml of the Fehling solution into a 250 ml boiling flask and bring to a boil over an open flame. By means of a buret, add the sample solution to within 0.5 ml of the anticipated end point (determined by a preliminary titration). Again heat the flask over the flame, with a motion to rotate the contents.

Boil the contents moderately for 2 minutes, then

quickly add 2 drops of methylene blue indicator (Note 3). Add immediately about 2 drops of the sample solution and again bring to a boil. Allow the cuprous oxide to settle slightly and observe the color of the supernatant liquid. Complete titration rapidly (Note 4) by adding sample solution dropwise and boiling after each addition, until the blue color disappears (Note 5).

Determine the dry substance concentration of the sample by an approved method.

CALCULATION

% Reducing Sugars (*as is*, calc. as dextrose)

$$= \frac{500 \text{ ml} \times 0.1202 \times 100}{\text{Sample Titer (ml)} \times \text{Sample Wt. (g)}}.$$

$$\text{Dextrose Equivalent} = \frac{\% \text{ Reducing Sugars} \times 100}{\% \text{ Dry Substance}}.$$

NOTES AND PRECAUTIONS

1. Crude and refined sugars include dextrose (anhydrous and hydrate), dextrose solutions, "70" and "80" sugars, first draw liquor, first greens (Corn Sugar Molasses) and hydrol (Feeding Corn Sugar Molasses).
2. In the analysis of refined sugar, use about 5 g of sample. For crude corn sugars, use approximately 7.5 g. For hydrol, use about 10 g. Concentration should be such that the sample titer falls in the range of 10 to 15 ml.
3. To prevent oxidation after addition of indicator, the hot flask should not be shaken or rotated outside the flame, or placed on a cold surface.
4. The entire titration should not consume more than 3 minutes.
5. A white background and a daylight lamp will assist in observing the end point.

Honey

Definitions

Honey has been officially defined as "The nectar and saccharine exudations of plants gathered, modified, and stored in the comb by honeybees (*Apis mellifica* and *A. dorsata*)". The definition further specifies that "Honey is levorotary and contains not more than 25% of water, not more than 0.25% of ash, and not more than 8% of sucrose."* The specification that honey be levorotary results in excluding from sale as "honey" the so-called "honeydew honey". Honeydew honey, while it is gathered by bees, has its source not in floral nectars but in the exudations of insects such as the sugar-cane leaf-hopper (*Perkinsiella saccharicida*) and the sugar-cane aphid (*Aphis sacchari*) and other aphids.[34]

Honey is sold in three forms: (1), *Comb honey*, which is honey contained in the cells of the comb; (2), *extracted honey*, which is honey that has been separated from the uncrushed comb by gravity or centrifugation; and (3), *strained honey*, which is honey that has been removed from the crushed comb by straining or other means.

*White *et al.*[33], as a result of their analyses of 490 authentic samples, concluded that the 25 percent moisture limit was too high, and the 0.25 percent ash limit too low (it was exceeded by 21 percent of their samples). They also found that only one of their samples would have exceeded a 7 percent sucrose limit.

Composition

Prior to 1956, the only extensive collection of analyses of authentic American honeys was that of Browne (*see* Selected Reference L). The methods employed in that study were largely empirical, and a critical study of methods of sugar analysis applicable to honey that was made by White *et al.* in 1962 [33] showed that the Browne methods for individual sugars were unreliable, chiefly because they did not take into account the occurrence of reducing disaccharides in honey.

The Browne report was superseded in 1962 by the appearance of *USDA Technical Bulletin* 1261, which details the results of analyses by White *et al.* by modern methods of 490 samples of honey and 14 of honeydew from the 1956 and 1957 crop years. These samples represented 83 single floral types, 93 blends of known composition, and four types of honeydew. Average values for the honey samples were found to be: Color, dark part of "White"; granulatory tendency, 1/8 to ½" layer; moisture, 17.2%; levulose, 38.19%; dextrose, 31.28%; sucrose, 1.31%; reducing disaccharides as maltose, 7.31%; higher sugars, 1.50%; pH, 3.91; free acidity, 22.03 milliequivalents/kg; lactone 7.11 meq/kg; total acidity, 29.12 meq/kg; lactone/free acid ratio, 0.335; ash 0.169%; nitrogen, 0.041%; diastase no. 20.8. The reducing disaccharides included—in addition to maltose—*iso*maltose, maltulose, turanose, nigerose and kojibiose.[36] Contrary to long-standing belief that the chief acid in honey was formic, Stinson *et. al.*[37] established that the major acid was gluconic. Honey also contains lactones, probably mostly gluconolactone.

Honeys are graded into seven color classifications: water white, extra white, white, extra light amber, light amber, amber, and dark amber.

As noted previously, honeydew is distinguished from normal floral honey by being dextrorotary. Dextrorotary honey can also be produced by feeding bees on cane sugar syrup or glucose, but honeydew is further distinguished by its molasses-like odor and (usually) very dark color; the ash is also higher than that of most honeys. Citrus honey is characterized by an appreciable content of methyl anthranilate.[38]

Methods of Analysis

Method 16-31. *Preparation of Sample.* (*AOAC Method* ***29.091***).

A. *Liquid or strained honey*—If the sample is free of granulation, mix thoroughly by stirring or shaking before weighing portions for determinations; if granulated, place the container (tightly stoppered) in a water-bath (without submerging) and heat 30 minutes at 60°; then if necessary heat at 65° until liquefied. (Occasional shaking is essential.) Mix thoroughly, cool rapidly as soon as sample liquefies, and weigh out portions for determinations to be made. If foreign matter, such as wax, sticks, bees, particles of comb, etc., is present, heat the sample to 40° in a water bath and strain through cheesecloth in a hot water funnel before weighing portions for analysis.

B. *Comb honey*—Cut across the top of the comb (if sealed), and separate the honey completely from the comb by straining through a No. 40 sieve. When portions of comb or wax pass through the sieve, heat the sample as in (A) and strain through cheesecloth. If the honey is granulated in the comb, heat until the wax is liquefied; stir, cool, and remove the wax.

Color

Honey is color-graded in a special comparator by comparing with a set of six glass standards. The necessary equipment and directions can be obtained from the Phoenix Instrument Co., 3803 North 5th Street, Philadelphia, Pa. 19140.

Method 16-32A. *Moisture by direct drying. See Method 16-1B.*

Method 16-32B. *Moisture by refractometer.* (*AOAC Method* ***29.095–29.096***).

Determine the refractometer reading at 20°, and obtain the corresponding moisture percentage from Table 16-7. If the reading is made at a temperature other than 20°, correct it to the standard temperature as directed in the footnote.

Method 16-33. *Ash.* (*AOAC Method* ***29.097***).

Weigh 5–10 g of honey into an ignited and weighed platinum dish. Place under a 375-watt infrared lamp with variable voltage input, and slowly increase the applied voltage until the sample becomes black and dry and there is no danger of loss by foaming. Heat in a muffle at 600° to constant weight (overnight). Cool and weigh.

TABLE 16-7

Chataway table showing relationship between refractive index, moisture content, and weight per gallon of honey

Refractive Index at 20°	Moisture	Wt Honey in lbs/gallon at 20°		Refractive Index at 20°	Moisture	Wt Honey in lbs/gall on at 20°	
	Percent	*lb*	*oz*		*Percent*	*lb*	*oz*
1.5041	13.0			1.4935	17.2		
35	.2	12	1	30	.4	11	13
30	.4			25	.6		
25	.6			20	.8		
20	.8	12	½	15	18.0	11	12½
15	14.0			10	.2		
10	.2	12	0	05	.4		
05	.4			00	.6	11	12
00	.6			1.4895	.8		
1.4995	.8	11	15½	90	19.0	11	11½
90	15.0			85	.2		
85	.2			80	.4		
80	.4	11	15	76	.6	11	11
75	.6			71	.8		
70	.8	11	14½	66	20.0		
65	16.0			62	.2	11	10½
60	.2			58	.4		
55	.4	11	14	53	.6	11	10
50	.6			49	.8		
45	.8			1.4844	21.0		
40	17.0	11	13½				

Temp. corrections: Refractive Index, 0.00023/°C or 0.00013/°F; lb/gallon, ½ oz/6°C or 11°F. If reading is made at temp. above 20°C (68° F) add the correction; if made below, subtract the correction. *Can. J. Research* **6**:532 (1932); **8**:435 (1933); *Canadian Bee J.* August 1935, p. 215; *J. Assoc. Offic. Agr. Chemists* **25**:99, 681 (1942).

Method 16-34. *Nitrogen.*[39]

REAGENTS

a. *Methyl red-methylene blue indicator*—Mix 2 parts of 0.2% alcoholic methyl red solution with 1 part of 0.2% alcoholic methylene blue solution.

b. *Sodium hydroxide-sodium thiosulfate solution*—Add 25 ml of 25% $Na_2S_2O_3.5H_2O$ solution to 100 ml of 50% NaOH solution.

c. *Boric acid solution*—Saturated solution.

d. *Hydrochloric acid*—0.01 *N*, diluted from standard 0.1 *N*.

APPARATUS

a. *Digestion rack*—Use a rack with electric heaters which will supply sufficient heat to a 30 ml flask to cause 15 ml of water at 25° to come to a rolling boil in not less than 2 nor more than 3 minutes.

b. *Distillation apparatus*—Use a one-piece micro-Kjeldahl distillation apparatus similar to Catalogue No. 7497 of Arthur H. Thomas Co., Philadelphia, Pa., or No. JM-4250 of Scientific Glass Apparatus Co., Inc., Bloomfield, N.J.

c. *Digestion flasks*—Use 30 ml regular Kjeldahl flasks.

DETERMINATION

Transfer 300 mg of honey (a sample which will require 3–10 ml 0.01 *N* HCl) to a 30 ml Kjeldahl flask. Add 1.9 ± 0.1 g of K_2SO_4, 40 ± 10 mg of HgO and 3.00. ± 1 ml of H_2SO_4. Add boiling chips which pass a No. 10 sieve, and digest for 1 hour after the acid comes to a true boil. Cool, add a minimum quantity of water to dissolve the solids, cool and place a thin film of petroleum jelly on the rim of the flask. Transfer the digest and boiling chips to the distillation apparatus, and check completeness of transfer by adding a drop of indicator to the final rinses. Place a 125 ml Phillips beaker or Erlenmeyer flask containing 2.5 ml of H_3BO_3 solution and 1-2 drops of indicator under the condenser, with the tip extending below the surface. Add 8–10 ml of the $NaOH\text{-}Na_2S_2O_3$ solution to the still, collect about 15 ml of distillate, and dilute to approximately 25 ml. Titrate to a gray end point or the first appearance of violet. Make a blank determination and calculate:

$$\% \mathrm{N} = \frac{(\text{ml HCl-blank}) \times \mathrm{N} \times 14.008 \times 100}{\text{wt. sample in mg}}.$$

Method ***16-35****. Direct Polarization.* [*AOAC Methods* ***29.101(b)*** *and* ***29.023***].

a. *Constant direct polarization*—Transfer 26 g of honey to a 100 ml volumetric flask with water, add 5 ml of alumina cream, Method 16-6, "Preparation and use of clarifying agents, (**b**)", dilute to the mark with water at 20°, and filter. To complete mutarotation, either let stand overnight or add dry Na_2CO_3 until just distinctly alkaline to litmus. (In the latter case do not allow to stand long enough for destruction of levulose to take place.) Polarize at 20° in a 200 mm tube.

b. *Direct polarization at* 87°—Polarize the solution obtained in (**a**) at 87° in a jacketed 200 mm metal tube.

NOTE: The work of White *et al.* reported in *USDA Technical Bulletin* 1261 has tended to make Methods 16-35 to 16-41 obsolescent, but because they still have value in detecting adulteration they have been included in this book.

Method ***16-36****. Invert Polarization.* (*AOAC Method* ***29.102***).

a. *At* 20°—Invert 50 ml of the solution, Method 16-35 (**a**), as in 16-6, Determination, (**A, 2** or **3**), and polarize at 20° in a 200 mm tube.

b. *At* 87°—Polarize the same solution at 87° in a jacketed 200 mm metal tube.

Method ***16-37****. Reducing Sugars.* (*AOAC Method* ***29.103***.)

Dilute 10 ml of the solution, Method 16-35 (**a**), to 250 ml, and determine reducing sugars in 25 ml of this solution by Method 16-6, "Determination," (B). Calculate to percent invert sugar.

Method ***16-38****. Sucrose.* (*AOAC Method* ***29.104***).

a. *By polarization*—Calculate from the data given by Methods 16-35 (**a**) and 16-36 (**a**), using the formula of Method 16-6, "Determination", (**A, 2** or **3**), whichever is applicable.

b. *By copper reduction*—Proceed as in Method 16-6 (**B**), "Calculations". To determine reducing sugars after inversion, dilute 10 ml of the solution, 16-36 (**a**), with a small quantity of water, neutralize with Na_2CO_3, dilute to 250 ml with water, and use 50 ml of this diluted solution for copper reduction.

Method ***16-39****. Levulose.* (*AOAC Method* ***29.105***).

Multiply the direct reading at 87°, Method 16-35 (**b**), by 1.0315, and from the product subtract algebraically the constant direct polarization at 20°, Method 16-35 (**a**). Divide the difference by 2.3919 to obtain g levulose in a normal weight (26 g) of honey. From this figure calculate the percent of levulose in the original sample.

Alternatively, determine levulose selectively by Method 16-9 (B).

Method ***16-40****. Dextrose.* (*AOAC Method* ***29.106***).

To obtain the approximate percentage of dextrose, subtract the percentage of levulose, Method 16-39, from the percentage of invert sugar, Method 16-37.

To determine dextrose more closely, multiply the percent of levulose, Method 16-39, by 0.915; this gives its dextrose equivalent in copper-reducing power. Subtract the figure so obtained from that of the reducing sugars, Method 16-37, also calculated as dextrose; the difference is the percent of dextrose in the sample. Because of the difference in reducing powers of different sugars, the sum of the dextrose so found and levulose, Method 16-39, will be greater than the quantity of invert sugar obtained by Method 16-37.

NOTE: This method was shown by White *et al.*[40] to overestimate the dextrose content because it fails to make allowance for the presence in honey of reducing disaccharides such as maltose. It is nevertheless included here because it still has value in detecting adulteration and is much less time-consuming than more accurate chromatographic methods such as Method 16-42. Both methods may eventually be superseded by gas chromatography, since this technique should, theoretically at least, be both specific and fast.

Method 16-41. *Dextrin.* (*AOAC Method **29.107***).

Using not over 4 ml of water, transfer 8 g of sample (4 g for dark-colored honeydew) to a 100 ml volumetric flask as follows: Let the sample drain from the weighing dish into the flask, dissolve the residue in 2 ml of water, add this solution to the flask, and rinse the weighing dish with two 1 ml portions of water, adding a few ml of absolute alcohol each time before decanting. Fill the flask to the mark with absolute alcohol, shaking constantly. Set the flask aside until the dextrin collects on the sides and bottom and the liquid is clear.

Decant the clear liquid through a filter paper and wash the residue in the flask with 10 ml of the absolute alcohol, pouring the washings through the same filter. Dissolve the dextrin in the flask with boiling water and filter through the paper already used, receiving the filtrate in a weighed dish prepared as in Method 16-2 (**b**). Rinse the flask and wash the filter a number of times with small portions of hot water, evaporate on the water bath, and dry to constant weight at 70° under a pressure not exceeding 50 mm of Hg.

After determining the weight of the alcohol precipitate, dissolve it in water and dilute to a definite volume (50 ml/0.5 g or part thereof of precipitate). Determine reducing sugars in this solution, both before and after inversion, as in Method 16-6, Determination, (B), expressing the results as invert sugar. Calculate sucrose from these results, and subtract the sum of the sucrose and reducing sugars before inversion from the weight of the total alcohol precipitate to obtain the weight of dextrin.

Method 16-42. *Sugars by Column Chromatography.*[41]

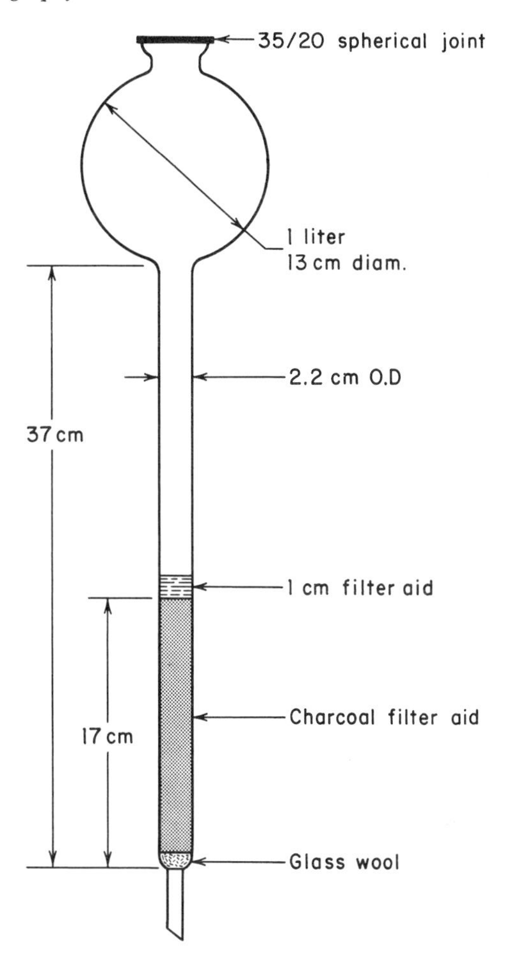

Fig. 16-4.—Analytical charcoal column used for honey analysis.

PREPARATION AND STANDARDIZATION OF ADSORPTION COLUMN

Column, shown in Fig. 16-4, is 22 mm outside diameter by 370 mm long, with 1 liter spherical section and 35/20 spherical ground joint at top. Adsorbent is 1+1 mixture of Darco G-60 charcoal and rapid filter-aid (Celite 545 or Dicalite 4200).

Insert a glass wool plug, wet from below, and add enough dry adsorbent to the dry tube

(23–26 cm) to compress to 17 cm when vacuum is applied with *gentle* tapping of the column. Remove excess charcoal from the walls of the column, and add a filter-aid layer at the top with *gentle* packing (1–1.5 cm). Wash the column with 500 ml of water and 250 ml of 50% alcohol and let stand overnight with 50 percent alcohol on it. The flow rate should be 5.5–8.0 ml/min. with water at 9 lb/sq. in. air pressure. Slower flow rates delay analyses excessively.

The following alternative wet packing procedure has been found to increase the column flow rate: Prepare a column with a glass wool plug and 10 mm of dry filter aid at the bottom. Then, with outlet open, add a suspension of 18 g of adsorbent mixture in 200 ml of water. After 5 minutes, apply 4 lb/sq. in. air pressure until the charcoal surface is stabilized. After application of 9 lb/sq. in. pressure, use suction to remove any excessive charcoal mixture beyond 17 cm depth, and place a layer of filter aid on the charcoal surface. Then continue washing as above.

The alcohol content of eluting solutions must be adjusted to the retentive power of the charcoal used. Wash the column alcohol-free with 250 ml of water, quantitatively add a 10 ml solution of 1.000 g of anhydrous dextrose to the top, and draw it into the column with suction; do not let dry. Add 300 ml of water to the top, break the suction, apply pressure (10 lb/sq. in. max.), and collect the eluate in five 50 ml portions in tared beakers. Include 10 ml from sample introduction in the first 50 ml fraction. Evaporate the fractions on the steam bath, dry in a vacuum oven at 89°–100°, and weigh.

Decant the remaining water from the top of the column, pass 50 ml of 5% alcohol and then 250 ml of water through the column, and repeat the chromatography, using 1.000 g of anhydrous dextrose in 10 ml of 1% alcohol, washing with 250 ml of 1% alcohol as above. Repeat the chromatography with 2% alcohol if necessary to select as solvent *A* that which removes dextrose in 150 ml.

Wash the column with 250 ml of water and then 20 ml of 5 percent alcohol. To the top, add 10 ml of 5% alcohol solution containing 100 mg of maltose and 100 mg of sucrose. Elute as above with 250 ml of 5% alcohol, weighing the evaporated 50 ml portions of the filtrate. Repeat, if necessary, with 7, 8, and 9% alcohol to find a solvent *B* that will elute at least 98% of the disaccharides in 200 ml. Solvent *A* previously selected must not elute disaccharides. Combinations found satisfactory with various charcoals are 1,7; 2,8; 2,9%. At conclusion, pass 100 ml of 50% alcohol through the column, and store under a layer of this solvent.

PREPARATION OF FRACTIONS

Wash the column with 250 ml of water and decant any supernatant. Pass 20 ml of solvent *A* through the column, and discard. Dissolve 1 g sample in 10 ml of solvent *A* in a 50 ml beaker. Transfer the sample (using a long-stem funnel) onto the column, and force into the column. Use 15 ml of solvent *A* to rinse the beaker and funnel, and add to the column. Collect all of the eluate, beginning with sample introduction, in a 250 ml volumetric flask. Add 250 ml of solvent *A*, and collect exactly 250 ml total (fraction A-monosaccharides). Decant excess solvent from the top, add 265–270 ml of solvent *B*, and collect 250 ml in a volumetric flask (fraction B-disaccharides). Decant the excess, add 110 ml of 50% alcohol (solvent *C*), and collect 100 ml in a volumetric flask (fraction C-higher sugars). Mix each fraction thoroughly. The column may be stored indefinitely, outlet closed, under 50% alcohol. Discard the packing after eight uses.

Determination of Levulose.

REAGENTS:

a. *Iodine solution,* 0.05 *N*—Dissolve 13.5 g of pure I in a solution of 24 g of KI in 200 ml of water, and dilute to 2 liters. Do not standardize.

b. *Sodium hydroxide solution,* 0.01 *N*—Dissolve 20 g of NaOH in water and dilute to 5 liters.

c. *Sodium hydroxide solution* 1 *N*—Dissolve 41 g of NaOH in water and dilute to 1 liter.

d. *Sulfuric acid solution,* 1 *N*—Add 56 ml of H_2SO_4 to water and dilute to 2 liters.

e. *Sulfuric acid solution, 2N* —Add 56 ml of H_2SO_4 to water and dilute to 1 liter.

f. *Sodium sulfite solution,* 1%—Dissolve 1 g of Na_2SO_3 in 100 ml of water. Make fresh daily.

g. *Starch solution,* 1%, freshly prepared.

h. *Bromocresol green solution*—Dissolve 150 mg of bromocresol green in 100 ml of water.

i. *Shaffer-Somogyi reagent.*—Dissolve 25 g each of anhydrous Na_2CO_3 and Rochelle salt in about 500 ml of water in 2-liter beaker. Add 75 ml of a solution of 100 g of $CuSO_4.5H_2O$ per liter, through a funnel with tip under surface, with stirring. Add 20 g of dry $NaHCO_3$, dissolve, and add 5 g of KI. Transfer the solution to a 1-liter volumetric flask, add 250 ml of 0.100 *N* KIO_3 (3.567 g dissolved and diluted to 1 liter), dilute to volume, and filter through fritted glass. Age overnight before use.

j. *Iodide-oxalate solution.*—Dissolve 2.5 g. of KI and 2.5 g. of K oxalate in 100 ml. of water. Make fresh weekly.

k. *Sodium thiosulfate standard solution*, 0.005 *N*—Prepare from standardized stock 0.1000 *N* solution. Make fresh daily.

DETERMINATION

Pipet 20 ml of fraction *A* into a 200 ml volumetric flask. Add 40 ml of 0.05 *N* I solution by pipet, then with vigorous mixing add 25 ml of 0.1 *N* NaOH over 30 seconds period, and immediately place the flask in an $18 \pm 0.1°$ water bath. Exactly 10 minutes after alkali addition, add 5 ml of 1 *N* H_2SO_4 and remove from the bath. Exactly reduce the I with Na_2SO_3 solution, using 2 drops of starch solution near the end point. Back-titrate with dilute I if necessary. Add 5 drops of bromocresol green solution and exactly neutralize the solution with 1 *N* NaOH; then make just acid to the indicator. Dilute to volume and determine the reducing value of 5 ml aliquots by the Shaffer-Somogyi method. Place 5 ml in a 25 by 200 mm test tube, add 5 ml of Shaffer-Somogyi reagent, and mix by swirling. Place in a boiling water bath and cap with a funnel or bulb. After 15 minutes, remove to a running water cooling bath with care, and cool 4 minutes. Carefully remove the cap, and add, down the side, 2 ml of iodide-oxalate solution and then 3 ml of 2 *N* H_2SO_4. (Do not agitate the solution while alkaline.) Mix thoroughly, seeing that all Cu_2O is dissolved. Return to the cold water and let stand 5 minutes, mixing twice in this period. Titrate in the tube with 0.005 *N* $Na_2S_2O_3$ and starch indicator. (A magnetic stirrer is most suitable for this purpose.) Make duplicate blanks and determinations. Deduct the titration from that of the blank and calculate levulose:

$$\% \text{levulose} = \frac{500\,[(\text{titer} \times 0.1150) + 0.0915] \times 100}{\text{mg sample}}.$$

Levulose correction for dextrose determination $= l.c. = [(\text{titer} \times 0.1150) + 0.0915] \times 40$. Bracketed quantity is mg levulose in 5 ml aliquot, valid between 0.5 and 1.75 mg levulose.

Determination of Dextrose.

REAGENT

Sodium thiosulfate solution, 0.05 *N*—Prepare from standardized stock 0.1000 *N* solution.

DETERMINATION

Pipet 20 ml aliquots of fraction *A* into duplicate 250 ml Erlenmeyers. Evaporate to dryness on a steam bath in an air current. Add 20 ml of water, pipet 20 ml of 0.05 *N* I and, as in the levulose determination, add 25 ml of 0.1 *N* NaOH slowly, and immediately place in an $18° \pm 0.1°$ water bath. Exactly 10 minutes from the end of alkali addition, add 5 ml of 2 *N* H_2SO_4, remove from the bath, and titrate with 0.05 *N* $Na_2S_2O_3$ using starch solution. Make duplicate blanks, using water. Subtract the titration value from that of the blank, and calculate dextrose:

$$\% \text{dextrose} = \frac{56.275\,[\text{titer} - (0.01215 \times l.c.)] \times 100}{\text{mg sample}},$$

where $l.c.$ = levulose correction from levulose determination. This equation is valid over the range 10–50 mg dextrose in 20 ml. In the presence of dextrose, 1 mg levulose requires 0.01215 ml 0.05 *N* $Na_2S_2O_3$, in the range 15–60 mg levulose.

Determination of reducing disaccharides as maltose.

Pipet duplicate 5 ml aliquots of fraction *B* into 25×200 mm test tubes, and add 5 ml of Shaffer-Somogyi reagent. Determine the reducing value as in the levulose determination, except to boil the tubes 30 minutes. The value for the 15-minute water blank may be used here. Calculate percent reducing disaccharides as maltose:

$$\% \text{“maltose”} = \frac{50\,[(\text{titer} \times 0.2264) + 0.075] \times 100}{\text{mg sample}}.$$

Maltose correction for sucrose determination = maltose titer × 0.92. Reducing value of maltose at 15 minutes is 92 percent of final value. The bracketed quantity is mg maltose in 5 ml aliquot, valid between 0.15 and 3.80 mg maltose.

Determination of Sucrose.

REAGENTS

a. *Hydrochloric acid solution,* 6 *N*—Add 250 ml of HCl to water and dilute to 500 ml.

b. *Sodium hydroxide solution,* 5 *N*—Dissolve 103 g of NaOH in water and dilute, after cooling, to 500 ml.

DETERMINATION

Pipet 25 ml of fraction *B* into a 50 ml volumetric flask. Add 5 ml of 6 *N* HCl and 5 ml of water. Mix, let stand in a 60° water bath 17 minutes, cool, and neutralize to bromocresol green with 5 *N* NaOH (a polyethylene squeeze bottle is excellent for holding and delivering alkali). Adjust to the acid color of the indicator, using 2 *N* H_2SO_4 to correct overrun. Dilute to volume and determine the reducing value of 5 ml aliquots by the Shaffer-Somogyi determination as for levulose. Subtract the titration from the blank, and calculate sucrose by reference to a curve constructed from the following table:

Sucrose in 5 ml aliquot oxidized, mg	0.005 *N* $Na_2S_2O_3$ required, ml
0.255	1.75
0.502	3.95
1.004	8.72
1.260	11.28

From the curve obtain S_1 = sucrose equivalent to maltose correction (see above for maltose) and S_2 = sucrose equivalent of sucrose titer.

$$\% \text{ sucrose} = \frac{50\ (2S_2 - S_1) \times 100}{\text{mg sample}}.$$

Method 16-43A. *Melezitose.*[42]

REAGENTS

a. *Yeast invertase*—1%. Dissolve 1 g of melibiase-free yeast invertase preparation in water and dilute to 100 ml.

b. *Buffer*—*M*/10 *acetate,* pH 4.5. Dissolve 6 g of glacial acetic acid in 500 ml of water, titrate with *N* NaOH to pH 4.5, and dilute to 1 liter.

DETERMINATION

To 25 ml of Method 16-42, fraction *B* in a 50 ml volumetric flask add 0.1 ml of the enzyme solution and 1.0 ml of buffer. Mix, let stand 1 hour at room temperature, make to volume, and determine the reducing value of a 5 ml aliquot by the Shaffer-Somogyi determination as for levulose. Subtract the titration value from the blank (with enzyme, buffer) and obtain the value for true sucrose from the table given under "sucrose." Calculate as for sucrose.

The difference between this value and that obtained as described under "sucrose" in Method 16-42 is considered due to melezitose. Multiply the difference, expressed as percent of honey sample, by 1.47 to obtain an estimation of the melezitose content of the honey.

NOTE: The amount of enzyme solution used will depend on the strength of the invertase solution.

Method 16-43B. *Higher Sugars, or "Dextrin".*

Pipet 25 ml aliquots of fraction *C*, Method 16-42, into 50 ml volumetric flasks. Add 5 ml of 6 *N* HCl and 5 ml of water, and heat in a boiling water bath 45 minutes. Cool, neutralize as for sucrose, Method 16-42, dilute to volume, and determine the reducing value by a Shaffer-Somogyi determination as for levulose. Subtract the titration value from the blank and obtain the dextrose equivalent from a curve constructed from the data below:

Dextrose, mg	Titer, ml
0.05	0.20
0.10	0.60
0.25	1.85
0.50	4.00
1.00	8.50
2.00	17.60

$$\% \text{ higher sugars} = \frac{40\ (\text{dextrose equiv.}) \times 100}{\text{mg sample}}.$$

NOTE: For most accurate work, Shaffer-Somogyi values must check within 0.04 ml.

Calibration of the entire procedure, including the column, using known synthetic mixtures of dextrose, levulose, sucrose, maltose, and raffinose (corrected for moisture), is recommended for critical work. The efficiency of the column separation may be checked by paper chromatography of fractions *A*, *B*, and *C*.

Method 16-44A. *Commercial Glucose—Qualitative Detection.* (*AOAC Method* ***29.118–29.120***).

REAGENT

Aniline-diphenylamine chromogenic reagent—Dissolve 500 mg of diphenylamine hydrochloride and 0.55 ml of redistilled aniline in 50 ml of acetone. Add 5 ml of 85% H_3PO_4. Prepare fresh daily.

PREPARATION OF SAMPLE

Dilute the sample with an equal volume of water. To 0.5 ml in a small centrifuge tube (11 × 100 mm test tube), add 4 ml of absolute alcohol, shake and centrifuge. Decant the clear or slightly cloudy supernatant liquid, dissolve the precipitate in 0.5 ml of water, reprecipitate with 4 ml of absolute alcohol, and centrifuge. Decant and dissolve the precipitate in 0.1 ml of water. Apply 2 μl to the origin of the paper chromatogram. Also apply control spots of authentic honey and/or honeydew, and of corn syrup, that have been treated as above.

PAPER CHROMATOGRAPHY

General details of the paper chromatographic procedure can be found in Mitchell, *J. Assoc. Offic. Agr. Chemists*, **40**:999 (1957), and Mitchell, Mills and Yarnell, *Ibid.*, **43**:3 (1960). Apparatus specifically designed for use with 8″ squares of Whatman No. 1 paper by the Mitchell techniques may be obtained from Arthur H. Thomas Co., P.O. Box 779, Philadelphia, Pa. 19105.

Either ascending or descending chromatography may be used. A suitable solvent for the latter is *n*-propanol-ethyl acetate-water 7:1:2. Equilibrate for 45 minutes and irrigate at least 40 hours, letting the solvent drip from the serrated lower edge of the paper. For shorter ascending use (about 6 hours), roll the paper into a cylinder, staple the edges, set in a cylindrical jar, and employ *iso*amyl-alcohol-pyridine-water 7:7:6 as the solvent. To obtain increased resolution, dry the paper and repeat the irrigation one or more times.

After irrigation, remove and dry the paper chromatogram. Dip it in the chromogenic reagent, let the acetone evaporate, and heat at 85–95° for 5–8 minutes until the control spots of corn syrup turn blue. A honey or honeydew sample containing 5% of commercial glucose shows a series of blue maltodextrin spots of low R_f converging to the origin. Honey and honeydew dextrin spots are distinctly brown or gray, not blue. If the paper is heated excessively, both honeydextrin and maltodextrin spots will approach the same shade of gray.

NOTE:[43] Avoid contamination of chromatographic papers by (a) rejecting first and last sheets in a package, (b) handling paper only by the rod or clips, (c) placing the paper only on scrupulously clean surfaces (glass, stainless steel, filter paper, etc.), and (d) using clean rods, clips, ruler, etc.

Use a hard pencil for ruling the starting line, dotting and identifying initial spots. In spotting, prevent the portion of the paper being spotted from making direct contact with the bench surface. Keep the diameter of initial spots 5 mm by spotting portionwise in the presence of a stream of air. It is advisable to gain familiarity with paper chromatographic procedures by chromatographing a series of standards before the sample extract is chromatographed.

Method 16-44B. *Commercial Glucose—Quantitative Estimation.* (*AOAC Method* ***29.121***).

An approximate determination can be made by Browne's formula as follows: Multiply the algebraic difference in the polarizations of the invert solution at 20° and 87°, 16-36A and B, by 77, and divide this product by the percentage of invert sugar found in the sample after inversion. Multiply the quotient by 100, and divide the product by 26.7 to obtain the percentage of honey in the sample. 100 − % honey = % glucose. NOTE: Indicated percentages of less than 10 by this formula should be disregarded unless confirmed by the qualitative test, Method 16-44A.

Method 16-45. *Commercial Invert Sugar. Qualitative.* (*AOAC Method* ***29.122–29.123***).

REAGENT

Dissolve 1 g of resublimed resorcinol in 100 ml of HCl (sp.g 1.18–1.19).

TEST

Dissolve 2 g of honey in 10 ml of water, and extract rapidly with washed ether for 30 minutes in a continuous extractor. [The extractor of Palkin, Murray and Watkin[43a] is highly satisfactory, but no longer available from supply houses. The similar Knapp extractor[43b] can be obtained from Scientific Glass Apparatus Co., Inc., Bloomfield, N.J., as their No. JE-2550; from Arthur H. Thomas Co., Philadelphia, Pa., as their No. 4985-G; and from Lab Glass, Inc., Vineland, N.J., as their No. LG-6990.] Concentrate the ether to about 5 ml, and transfer to a test tube. Add 2 ml of the freshly-prepared reagent, shake, and note the color. A cherry-red color appearing within 1 minute indicates the presence of commercial invert sugar. Yellow to salmon shades have no significance.

Method 16-46. *Commercial Invert Sugar. Quantitative Aniline Chloride Number.*[44]

APPARATUS—*Spectrophotometer.*

REAGENTS

a. *Aniline chloride reagent.*—Pipet 50 ml of aniline into a 125 ml Erlenmeyer flask, add 15 ml of hydrochloric acid containing 25% hydrogen chloride, and mix. Prepare the reagent fresh about every week or ten days. Redistill the aniline if its color is darker than a yellow or slight brownish yellow. Preserve aniline and the aniline chloride reagent in the refrigerator. (Caution: In using aniline or the reagent, take care to keep from contact with the skin. Wash off immediately if contact occurs.)

b. *Ethyl acetate (purified).*—Purify the ethyl acetate (absolute) by refluxing about 1000–1200 ml over 7 or 8 g of metaphenylenediamine hydrochloride for about 90 minutes, then distilling, rejecting the first 15 ml of distillate, and discontinuing distillation when the volume left in the distillation flask is about 100 ml.

c. *Isoamyl alcohol (purified).*—Reflux about 1 liter of reagent grade *iso*amyl alcohol over 35–40 g of KOH for 60–70 minutes. Distill in glass apparatus and reject the first 25 ml of distillate, leaving 125 ml of liquid remaining in the distillation flask.

DETERMINATION

On a balance accurate to 0.1 g, weigh 50 g of honey or syrup into a 100 ml volumetric flask (or beaker) and transfer with H_2O to the flask. Add water (about 50 ml), swirl to dissolve sample, wash down the neck of the flask, dilute to the mark, and mix again thoroughly. Measure exactly 25 ml of the clear solution into a 100–125 ml separatory funnel. Introduce 25 ml of ethyl acetate (purified) into the funnel with a volumetric pipet. Stopper the funnel and shake vigorously for 2 minutes. Allow the layers to separate and draw off the lower aqueous layer and any emulsion at the interface. Filter the ethyl acetate layer through a folded filter and pipet 10 ml of the clear ethyl acetate extract into a small Erlenmeyer flask (about 50 ml). Add an equal volume (10 ml measured by pipet) of the purified *iso*amyl alcohol and swirl to mix. Introduce into the mixed solution exactly 1 ml of aniline chloride reagent, stopper, mix quickly, and note the time. Set the flask aside in the dark at a temperature of about 20–25° (preferably in a bath) for just fifteen minutes after the addition of the aniline chloride reagent. Without delay determine the percent transmission at 520 mμ in a spectrophotometer, using a 1 cm cell and using as a comparison in the other or standard cell the clear solvent mixture, *i.e.*, equal volumes of ethyl acetate and amyl alcohol. From the transmission reading obtain the absorbance, *A*, at 520 mμ from the formula:

$$A = -\log = \frac{\text{transmission solution}}{\text{transmission solvent}}.$$

Multiply *A* by 100 to obtain the aniline chloride number.

NOTES

1. This method measures the hydroxmethylfurfural that is present in commercial invert sugar.

2. Seven authentic American honeys and one imported and probably unadulterated honey showed aniline chloride numbers ranging from 7 to 87 and averaging 28.9, while five adulterated honeys gave values ranging from 136 to 197 and averaging 161.[44]

3. Because there is evidence that old honeys may contain hydroxymethylfurfural, this test should be applied with caution to honeys of unknown age.

Method 16-47. *pH and Free, Lactone, and Total Acidity.* (*AOAC Method* **29.130**).

Dissolve 10 g of sample in 75 ml of CO_2-free water in a 250 ml beaker. Stir with a magnetic stirrer, immerse the electrodes of a pH meter in the solution, and record the pH. Titrate with 0.05 *N* NaOH at a rate of 5.0 ml/minute. Stop the addition of NaOH at a pH of 8.50. Immediately pipet in 10 ml of 0.05 *N* NaOH, and back-titrate at once with 0.05 *N* HCl (in a 10 ml buret) to a pH of 8.30. Calculate as milliequivalents/kg:

Free acidity
= (ml 0.05 *N* NaOH from buret − ml blank) × 50/g sample.

Lactones
= (10.00 − ml 0.05 *N* HCl from buret) × 50/g sample.

Total acidity = free acidity + lactones.

NOTE: Most collections of honey analyses appearing prior to publication of *USDA Technical Bulletin* 1261 gave figures for acidity calculated as percent of formic acid. Values for milliequiv./kg obtained by the present method can be recalculated to percent formic acid by multiplying by 0.004603. It should be noted, however, that: 1) It is now known that formic acid is *not* the major acid in honey; and 2), because determinations of endpoints were not precise by older methods, recalculations to percent formic acid are of limited value for comparing present-day samples with older authentic honeys.

Method 16-48. *Methyl Anthranilate.*[45]

APPARATUS

Parnas-Wagner all-glass micro-Kjeldahl distillation apparatus. May be obtained from Scientific Glass Apparatus Co., Inc., Bloomfield, N.J., as their Catalogue No. JM-4280; or from Arthur H. Thomas Co., Philadelphia, Pa., as their Catalogue No. 7492-N.

REAGENTS

a. *Potassium 1-naphthol-2-sulfonate solution*—Dissolve 1.25 g of the salt (practical grade) in 50 ml of hot water, add a little decolorizing charcoal, and filter after a few minutes. Store in an amber bottle. (Stable at room temperature for at least two weeks.)

b. *Sodium carbonate solution*—Dissolve 25 g of Na_2CO_3 in 75 ml of water.

DETERMINATION—Add 5 ml of water to about 10 g of honey that has been weighed to 1 mg. Mix, transfer to the distilling apparatus, using 5 ml more of water for washing. Distill (reducing the voltage from 115 to 85 when foaming begins), and collect 50 ± 1 ml. (The condenser water temperature should not exceed 15°; it is not necessary to submerge the tip of the condenser.) Transfer the distillate to a 100 ml volumetric flask; to another flask add 50 ml of water as a blank. To each flask add 2.5 ml of 1 *N* HCl and 0.4 ml of 1% $NaNO_2$ solution, and let stand 1 minute. Then add 0.6 ml of 3% hydrazine sulfate solution, wash down the necks of the flasks with a few ml of water, and let stand 1 minute. Finally add 1 ml of the naphthol sulfonate solution, followed immediately with 1.5 ml of the Na_2CO_3 solution. Mix, make to 100 ml, and determine the absorbance of the sample at 500 mμ in a 10 cm cuvette against the prepared blank. Calibrate against solutions of methyl anthranilate containing 0-50 μg in 50 ml.

NOTE: The presence of methyl anthranilate distinguishes citrus honeys from honeys of other floral sources. The predominant citrus honey is that of the orange; lesser amounts are derived from grapefruit and lemon. These honeys originate principally in Florida and California. White found 21 samples of citrus honey from Arizona, California and Florida to contain from 0.84 to 4.37 μg/g of methyl anthranilate, with an average of 2.87. By comparison, 12 non-citrus honeys averaged only 0.07 μg/g.

Pollen Analysis

Microscopic identification of the pollen grains present is the only *general* method of identifying the source (or sources) of a honey. This method may fail with strained honey if clarification has been complete enough to have filtered out all or nearly all of the pollen. The test is primarily one of identity; the use of low pollen counts to indicate sophistication with non-honey syrups is of doubtful value because of the wide variation in the amounts of pollen present in genuine honeys.

Persons desiring to make pollen tests should consult the following references:

1. Erdtman, G.: "*Investigations of Honey Pollen*". *Svensk Bot. Tidskr.*, **29** (1): 79 (1935).

2. Erdtman, G.: *Pollen Analysis*. Waltham: Chronica Botanica Co. (1943).

3. Griebel, C.: *Z. Lebensm. Untersuch. Forsch.* **59**: 63, 441 (1930).

4. Hyland, F., Graham, B. F. Jr., Steinmetz, F. H., and Vickers, M. A.: *Maine Air-Borne Pollen and Fungus Spore Survey*. Orono: University of Maine (1953).

5. Jamieson, C. A.: "*A Method of Distinguishing Between Pollens of Melilotus*". *Bee World* (Sept. 1939).

6. Niethammer, Anneliese: *Z. Lebensm. Untersuch.-Forsch.* **55**: 467 (1928).

7. Wodehouse, R. P.: *Hayfever Plants*. Waltham: Chronica Botanica Co. (1945).

8. Wodehouse, R. P.: "*Pollen Grains in the Identification and Classification of Plants, I-V.*" *Bull. Torrey Botan. Club*, **55**: 181, 449 (1928); **56**:123 (1929); **57**:21 (1930); *Am. J. Botany*, **16**:297 (1929).

9. Wodehouse, R. P.: *Pollen Grains, Their Structure, Identification and Significance in Science and Medicine*. New York: McGraw-Hill Book Co. (1935).

10. Young, W. J.: *U.S. Dept. Agr. Bur. Chem. Bull.* **110**:70 (1908).

Maple Products

Descriptive

The production of articles of food by condensation of the sap of the maple tree is exclusive to North America and originated before the coming of the white man. Maple syrup is the primary product, but considerable quantities of the maple sugar that results from nearly complete evaporation of the sap are also sold. Various other maple products are on the market, including one known as "maple butter" or "maple cream". This is a spread of butter-like consistency composed of microscopic sugar crystals interspersed with a thin coating of saturated syrup, which is produced by concentrating a syrup of 0.5-2% invert sugar content to supersaturation and then suddenly cooling it to room temperature.

Of the 13 species of maple native to the United States and Canada, only two, *Acer saccharum* (the sugar maple, hard maple, rock maple, sugar tree) and *Acer nigrum* (black sugar maple), are important in syrup production. In 1963 the states producing maple sugar ranked in this order: Vermont, New York, Ohio, Pennsylvania, Wisconsin, Michigan, New Hampshire, Massachusetts, Maine, Maryland and Minnesota. Canada's total maple crop is about the same as that of the United States.

Maple sap will flow from a wound in the sapwood, whether the wound is from a cut, a hole bored in the tree, or a broken twig. The sap flows any time from late fall after the trees have lost their leaves until well into the spring, each time a period of below-freezing weather is followed by a period of warm weather.

Primarily, however, the maple season comes in the early spring. The trees are tapped by boring a hole in them and inserting a wooden or metal spout, from which the sap flows into a bucket* or plastic bag. The sap in these is collected and conveyed to "evapora-

*At one time lead-coated ("terne plate") spouts and buckets were commonly employed, but these became obsolete when it was discovered that they contributed illegal amounts of lead to the syrup. Metal buckets now in use are made of galvanized iron or aluminum.

tor houses" where it is concentrated. Details of the equipment and processes used may be found in Selected Reference O.

Development of the desired maple flavor and color is the result of chemical reactions that occur while the sap is boiling in the evaporator, and the extent of these reactions is determined in part by the length of time the sap is boiled. Not all of the potential flavor is developed during the usual evaporation process, and to develop maximum flavor the syrup must be heated to a higher temperature for a longer time. Such high temperature heating has the disadvantage of favoring formation of an acrid "caramel" flavor, but a method has been developed to avoid this effect and produce a product known as "high-flavored maple syrup" whose flavor is four to five times as intense as that of ordinary syrup, yet is essentially free of caramel. This result is accomplished in several different ways, one of which (the atmospheric process) is to heat a high-grade syrup at atmospheric pressure to a boiling temperature of 250-255°F., hold at this temperature for $1\frac{1}{2}$-2 hours, and then cool and add sufficient water to bring the finished product to standard density. Details regarding this and the other processes may be found in Selected Reference O, pages 89-90.

Standards

Under the old U.S. Food and Drug Act, "maple syrup" was defined as a product "made by the evaporation of maple sap, or by the solution of maple sugar"; it was required to contain not more than 35% of water, and to weigh not less than 11 pounds to the gallon. Only syrup made by the direct evaporation of the sap was entitled to be called "maple sap syrup".

No standard for maple products has been adopted under the new Act, but grade standards for maple syrup have been set by the U.S. Department of Agriculture.[46] In summary these are as follows:

There are three grades: AA (Fancy), A and B. All must have a minimum density of 11 lb/gallon at 68°F. (65.46° Brix), must possess a characteristic maple flavor, and be clean, free from fermentation, and free from damage caused by scorching, "buddiness", and any other objectionable flavor or odor. The grades differ, however, in their minimum color and cloudiness requirements.

In addition to these Federal standards, the State of Vermont has adopted detailed regulations covering the sale of maple products manufactured in that state[47], and New York[48] and Wisconsin[49] have issued similar standards.

Authentic Analyses

Table 16-8[50] summarizes the results of analysis of many authentic samples of maple syrup:

TABLE 16-8
Summary of Analyses of Genuine Maple Syrups
All on Moisture-free Basis

	Number of analyses	Average	Max.	Min.	Range in % of average
Total ash	770	0.96	1.68	0.61	111
Soluble ash	770	0.59	1.23	0.30	158
Insoluble ash	770	0.37	1.01	0.12	241
Soluble ash ÷ insoluble ash	655	1.74	3.86	0.53	191
Alkalinity of soluble ash	655	0.74	1.22	0.41	109
Alkalinity of insoluble ash	655	1.00	2.08	0.41	167
Alkalinity soluble ash ÷ alkalinity of insoluble ash	655	0.80	1.83	0.21	203
Winton lead number	528	2.62	4.41	1.05	128
Malic acid value (calcium acetate)	1094	0.80	1.60	0.29	164
Conductivity value	174	148	230	110	81

The compilers of this table suggest the following scheme of tests as being sufficient for the rapid testing of syrup and enough to condemn most adulterated samples:

Determination	Limits of Value	
	Extreme	Ordinary
Conductivity value, 25°	110–230	113–205
Total ash, dry basis	0.61–1.68	0.69–1.47
Alkalinity of soluble ash, dry basis	41–122	48–109
Winton lead number	0.70–2.70	0.76–2.47

The refractometer reading must of course also indicate at least 65% solids.

The above figures should be applicable for checking adulteration of maple sugar and maple butter or cream if the determinations are run on syrups prepared by Method 16-49 (B, 2) below.

Methods of Analysis

Method 16-49. *Preparation of Sample.* (*AOAC Method* ***29.131***).

A. *Maple Syrup*

1. *For solids determination*—If the sample contains no sugar crystals or suspended matter, decant enough clear syrup for a determination. If sugar crystals are present, redissolve by heating at about 50°. If suspended matter is present, filter the sample through cotton wool.

2. *For other determinations*—If sugar crystals are present, redissolve by heating. If other sediment is present, distribute it evenly through the syrup by shaking. Transfer about 100 ml of syrup, with its suspended sediment, to a casserole or beaker, add one-fourth its volume of water, and evaporate over a flame. When the temperature of the boiling syrup approaches 104°, draw a small quantity into a thin-walled pipet of about 1 ml capacity, and cool to room temperature in running water. Wipe the outside of the pipet, let the syrup in the point (which may be diluted) escape, transfer some of the remaining syrup to a refractometer, and determine the solids content of the cooled syrup. Repeat this procedure from time to time until a reading is obtained corresponding to 64.5% solids (n_{20} = 1.4521), or to such other value as in the experience of the analyst will give a filtered syrup of 65.0% solids. Filter the syrup through a filter that will let 100 ml pass in 5 minutes, and adjust the filtrate to 65.0 ± 0.5% solids (refractometric) by thorough mixing with the appropriate quantity of water.

B. *Maple Sugar and Other Solid or Semisolid Products*

1. *For moisture and solids determination*—Grind in a mortar (if necessary) and mix thoroughly.

2. *For other determinations*—Prepare a syrup from the sample by dissolving about 100 g in 150 ml of hot water, boiling until the temperature approaches 104°, and proceeding thereafter as in (A, 2), beginning "draw a small quantity".

Method 16-50. *Moisture or Solids.* (*AOAC Method* ***29.134-29.135***).

A. *Maple sugar*—Proceed as in Method 16-2 (A), using sample prepared as in Method 16-49 (B, 1).

B. *Maple syrup, cream etc.*—Proceed as in Method 16-2 (A) or (B) or 16-3 (C).

Method 16-51. *Ash.* (*AOAC Method* ***29.136***).

Using 5 g of the prepared syrup, Method 16-49 (A, 2), proceed as in Method 16-4.

Method 16-52. *Soluble and Insoluble Ash.* (*AOAC Method* ***29.015***).

To the ash, Method 16-51, in a platinum dish add water, heat nearly to boiling, and filter through an ashless paper. Wash with hot water until the combined filtrate and washings measure about 60 ml. Return the paper and contents to the Pt dish, ignite carefully, cool, and weigh to obtain percent of water-insoluble ash. % total ash − % insoluble ash = % soluble ash.

Method 16-53. *Alkalinity of Soluble and Insoluble Ash.* (*AOAC Method* ***29.016-29.017***).

A. *Alkalinity of soluble ash*—Cool the filtrate from Method 16-52 and titrate with 0.1 *N* HCl, using 0.05% methyl orange indicator. Express the alkalinity in terms of ml 1 *N* acid/100 g sample.

B. *Alkalinity of insoluble ash*—Add an excess of 0.1 N HCl (usually 10–15 ml) to the ignited insoluble ash in the Pt dish, Method 16-52. Heat to incipient boiling on an asbestos plate, and cool. Transfer quantitatively to an Erlenmeyer flask, and titrate the excess of HCl with 0.1 N NaOH, using 0.05% methyl orange as indicator. Express the alkalinity in terms of ml 1 N acid/100 g sample.

Method 16-54. *Polarization.*

Determine direct and indirect polarizations at 20° and 87° as in Methods 16-35 and 16-36.

Method 16-55. *Sucrose.* (*AOAC Methods* ***29.143*** *and* ***29.144***).

A. *Polarimetric*—Proceed as in Method 16-6, Determination, (A), or calculate from the results of Method 16-54, using the appropriate formula from Method 16-6, Determination, (A, 2 or 3).

B. *Copper reduction method—See* Method 16-6, Determination, (B).

Method 16-56. *Reducing Sugars as Invert Sugar.* (*AOAC Method* ***29.145***).

A. *Before inversion*—Proceed as in Method 16-6, Determination, (B), using the aliquot of the solution used for direct polarization, Method 16-54. Only neutral Pb acetate may be used for clarification, and the excess of Pb must be removed with dry $Na_2C_2O_4$.

B. *After inversion*—Proceed as above, using the aliquot of the solution used for invert polarization. The same remarks apply with regard to clarification.

Method 16-57. *Commercial Glucose.*

A. *Products containing little or no invert sugar* (*AOAC Method* ***29.033***).—In syrups in which the quantity of invert sugar is too small to appreciably affect the result, commercial glucose may be estimated approximately by the following formula:

$G = (a-S)\ 100/211$, where G = % commercial glucose solids; a = direct polarization, normal solution; and S = % cane sugar.

Express the results in terms of commercial glucose solids polarizing at +211°S. (This result may be recalculated in terms of commercial glucose of any polarization desired.)

B. *Products containing invert sugar* (*AOAC Method* ***29.034***).—Prepare a half-normal solution as in Method 16-6, Determination, (A, 2), except to cool before inversion, neutralize to phenolphthalein with NaOH solution, slightly acidify with HCl (1+5), and treat with 5-10 ml of alumina cream, 16-6, Preparation and Use of Clarifying Agents, (**b**), before diluting to the mark. Filter and polarize at 87° in a 200 mm jacketed metal tube. Multiply the reading by 200 and divide by 196 to obtain the quantity of commercial glucose solids polarizing at +211°S. (The result may be recalculated in terms of commercial glucose of any polarization desired.)

Method 16-58. *Winton Lead Number.* (*AOAC Method* ***29.149–29.151***.)

REAGENT

Dilute basic lead acetate standard solution—Activate litharge by heating 2.5–3 hours to 650–670° in a muffle (the cooled product should be lemon yellow). In a 500 ml Erlenmeyer flask provided with a reflux condenser, boil 80 g of neutral Pb acetate trihydrate and 40 g of the freshly-activated litharge with 250 gm of water for 45 minutes. Cool, filter off any residue, and dilute the filtrate with recently boiled water to a density of 1.25 at 20°.

Prepare the dilute standard solution from this concentrated solution by adding 4 volumes of water.

DETERMINATION

Transfer 25 ml of the dilute standard basic Pb acetate solution to a 100 ml volumetric flask, add a few drops of acetic acid, and make to volume with water. Wash 25 g of sample into another 100 ml volumetric flask, add 25 ml of the dilute standard Pb acetate solution, shake, fill to the mark with water, and shake again. Let both flasks stand at least 3 hours, and then filter. Pipet 10 ml of the clear filtrate from each flask into a separate 250 ml beaker, add 40 ml of water and 1 ml of H_2SO_4, shake, and add 100 ml of alcohol. Let stand overnight, filter on a weighed Gooch crucible, wash with alcohol, dry at 100°, and ignite in a muffle at 550°. (A Bunsen burner may be used if the Gooch crucible is placed inside a larger crucible and heat is applied gradually until the outside of the Gooch crucible is barely a visible red.) Cool and weigh. Subtract the weight of the $PbSO_4$ from the sample from that of the $PbSO_4$ in the blank, and multiply by 27.33. This factor gives the lead number (g Pb/g sample) directly.

NOTES

1. Addition of the acetic acid to the blank solution is necessary to retain all of the Pb in solution when the reagent is diluted.

2. The chief constituent of the precipitate in this method may be lead malate, but lead salts of other organic acids, as well as lead sulphate and chloride and proteinaceous matter, are also present.

3. There are a number of pancake syrups on the market that are mostly cane sugar syrup but contain 10–15% of maple syrup (usually made by re-solution of maple sugar). Sometimes the percentage of maple syrup is declared on the label and it is desired to check the accuracy of the claim. Lead number determinations have been used for this purpose, but in 1923 Snell[51] offered data that appeared to indicate that the lead numbers of such mixtures underestimated the maple syrup content by about three percentage points. His tests were run on syrups that were from 50 to 80% maple. When Fisher[52] made similar tests on mixtures containing 0–20% of maple syrup he found the lead numbers to be higher rather than lower than the values calculated from the known dilution of the maple syrup. Because this question is still in dispute, and there is no generally accepted index of concentration other than the ash, for the present it is probably advisable not to take action against a sample for failure to meet its claimed maple content unless the ash figure unequivocally indicates such a deficiency. Unfortunately this method of calculation is based on the doubtful assumption that the cane sugar used in making the mixture is ash-free, and it is to be hoped that a more nearly specific method will soon be devised.

Method 16-59. Malic Acid Value.[53]

Transfer 6.7 g sample to a 200 ml beaker with 5 ml of water, add 2 ml of 10% Ca acetate solution, and stir. Stir in 100 ml of alcohol, and agitate the solution until the precipitate settles and the supernatant liquid is clear. Filter off the precipitate and wash with 75 ml of 85% alcohol. Dry the filter paper and ignite in a Pt dish. Add 10 ml of 0.1 *N* HCl to the residue, and warm gently until all of the $CaCO_3$ dissolves. Cool, and back titrate with 0.1 *N* NaOH (methyl orange indicator).

$$\text{Malic acid value} = \frac{\text{ml } 0.1\ N \text{ acid consumed}}{10}.$$

Correct this value by a blank determination.

NOTE: This test was based on the assumption that the calcium salt precipitated was essentially all calcium malate. A chromatographic method that is specific for malic acid is given in AOAC Method **29.155–29.159**, but has the disadvantage for routine work of being much more time-consuming.

Method 16-60. Conductivity Value.

APPARATUS

a. *Conductivity bridge*—Any commercially available conductivity bridge (these are usually self-contained instruments with two external connections, one to the power source and the other to the conductivity cell) may be used, but apparatus such as the Leeds and Northrup Bridge No. 4861 and Cell No. 4924 is convenient because means are provided for making adjustment for the temperature and the cell constant so that the conductivity at 20° can be read directly when the bridge is in the "SC" position.

Calibrate the scale (slide wire) of the bridge by means of a 1,000 ohm external fixed resistor with external leads. Attach this resistor to the connections for the conductivity cell, and adjust the slide wire to read 1,000 ohms. The bridge should give a zero reading.

b. *Conductivity cell*—Made of resistance glass with platinized electrodes firmly fixed and adequately protected from displacement. The cell may be of the dipping type for immersion in the test solution, or of a vessel type into which the solution is run and from which it is subsequently drained.

c. *Constant temperature bath*—This bath (for

controlling the temperature of the test solution and the cell) should maintain or supply water at 25 ± 0.1°.

DETERMINATION OF CELL CONSTANT

Dry 2–3 g of KCl to constant weight at 110°. Weigh two portions of the dried material, one weighing 0.3728 g and the other 0.7456 g (each ± 0.0002 g). Transfer to separate 500 ml volumetric flasks, and dilute to volume with water at 20°. These solutions will be 0.01 *M* and 0.02 *M* respectively. Transfer a portion of the 0.01 *M* solution to a beaker and adjust the temperature to 25 ± 0.1°.

If the bridge has a temperature compensating device, set this at 25°. Attach the cell leads to the bridge. If the cell is of the dipping type, place it in the 0.01 *M* KCl, taking care to immerse the electrodes completely; if it is a vessel type, fill it wi.h the solution. Adjust the bridge slide wire to give a null-point reading. Repeat until three successive concordant slide wire values (ohms resistance) have been obtained on separate portions of the KCl solution (taking care to shake adhering drops of liquid from the electrodes before immersing them in a fresh solution).

Repeat, using the 0.2 *M* KCl.

Calculate the cell constant by multiplying the observed resistances (scale readings in ohms) by 141.2 (the specific conductivity of 0.01 *M* KCl) and 276.1 (the specific conductivity of 0.02 *M* KCl) respectively. Average all results.

DETERMINATION OF CONDUCTIVITY VALUE OF SAMPLE

Add 70 ml of water to a 100 ml glass-stoppered graduated cylinder and fill to volume with the maple syrup to be tested. Stopper, mix thoroughly, adjust the temperature to 25 ± 0.1° and measure the resistance of the diluted syrup as above. Repeat until three concordant scale readings (ohms) are obtained.

$$\text{Conductivity value} = \frac{\text{cell constant} \times 1000}{\text{scale reading}}.$$

Flavor

The chief constituent of imitation maple flavors is an extract of fenugreek (*Trigonella foenum-graecum*, L.) seed. This seed contains trigonelline and choline, and tests for these betaines have been employed as indicators of the adulteration of maple-based pancake syrups with fenugreek. Recent information indicates, however, that such tests are not wholly reliable, and the only present reliable means of identification is tasting by an expert.

The substances responsible for the true maple flavor have proved difficult to identify, but three major components have been established to be vanillin, syringaldehyde and dihydroconiferyl alcohol.[55]

Molasses and Other Cane Syrups

Blackstrap molasses is the end-product of raw sugar manufacturing and refining. The composition varies with climate, sugar content, growth of the cane, methods of manufacture and other factors, but representative figures are: sucrose, 35%; invert sugar, 15%; ash, 10%; and moisture, 20–25%. The balance consists of organic nonsugars (proteins, organic acids, gums, etc.). Most of the 300 million gallons imported into the United States is utilized for fermentation into alcohol and as an ingredient of cattle feeds, but a small proportion is used for making vinegar, as an ingredient of chewing tobacco, etc.[56]

Cane syrup is generally understood to be the product resulting from the direct evaporation of cane juice without the removal of sugar. Regulations under the Federal Act of 1906 required it to contain "not more than 30% of water and not more than 2.5% of ash". These standards also recognized a "sugar syrup" defined as "made by dissolving sugar in water", which was required to contain not more than 35% of water.

Cane syrup may be clarified simply by skimming during evaporation, or the juice may be partially neutralized with lime and the precipitate allowed to settle. Most of it is sulfited.

Analyses of 10 authentic samples of cane syrup[57] were as follows:

	Maximum	Minimum	Average
Total solids, percent	76.9	72.6	74.7
Reducing sugars before inversion, percent	37.87	10.08	17.53
Sucrose, percent	59.09	32.50	52.64
Ash, percent	2.76	1.40	2.15
Winton lead no.	3.63	2.73	3.08

Edible molasses is a by-product of so-called "direct consumption sugars" made by the sulfitation process, *i.e.*, those white sugars, made in the cane factory directly from cane syrup without the intervening production of raw sugar, which use sulfur dioxide as a clarifying agent. If only one boiling is employed, the liquid separated from the crystallized sugar on centrifuging is known as "first molasses;" it is of a light brown color, has a density of about 80° Brix, and contains 50-55% sucrose. A second boiling and centrifugation yields a darker "second molasses".

Molasses was defined under the 1906 Act as "The product left after separating the sugar from massecuite, melada, mush sugar, or concrete", and was required to contain not more than 25% of water nor more than 5% of ash.

"High test molasses" is a name that has been given to a material produced by treating cane syrup with acid or invertase until about two-thirds of the sucrose has been inverted, and then further concentrating to 78 percent solids content. It is a clear, reddish-brown color and is composed largely of sucrose and invert sugar. The name "molasses" is a misnomer.[56]

Refinery invert syrups are inverted sucrose syrups, water-white to golden yellow in color, that contain about 75 percent solids, divided equally between sucrose and invert sugar; they are low in ash and other impurities.

"*Liquid sugars*" are sucrose solutions of about 68° Brix that have been highly purified by charcoal or ion-exchange resin treatment.

Methods of analysis—These products are analyzed by the methods given in the general section of this chapter for syrups. Cane syrup and molasses are often sulfited, and if a determination of SO_2 is desired, reference should be made to Chapter 14.

Note that the Karl Fischer Method 1-5 for moisture was designed specifically for molasses.

Other Sugar Products

While syrups and/or sugar are produced to a minor extent from a number of sources other than those previously mentioned (including the date palm *Phoenix sylvestris* in India and the Middle East), the only such product that has had any sale in the United States is the sorghum syrup made from the stalks of the sugar sorghum, *Andropogon sorghum*, L. This is produced in the southern states on a small farm basis. Animal- or steam-driven portable mills extract the juice, which is then concentrated by the "boil and skim" method, without the addition of lime or other chemicals. About 10 million gallons were produced in 1946.[44]

Analyses of seven authentic samples in 1949 were as follows:[57]

	Maximum	Minimum	Average
Total solids, percent	82.7	76.0	80.0
Reducing sugars before inversion, percent	63.79	31.64	44.13
Sucrose, percent	42.19	24.31	30.42
Ash, percent	3.28	1.89	2.51
Winton lead no.	3.86	3.33	3.60

Methods of analysis—These are the same as those for any other syrup.

TEXT REFERENCES

GENERAL

1. Oser, B. L. (ed.): *Hawk's Physiological Chemistry*, 14th ed., New York: McGraw-Hill Book Co. (1965). 62, 66.
2. Horowitz, W. (ed.): *Official Methods of Analysis of the Association of Official Agricultural Chemists*, 10th ed. (1965). Method **29.062–29.063**.
3. Ibid. Method **29.021 (a)**.
4. Ibid. Method **29.021 (b)**
4a. Ibid. Method **29.021 (d)**.
5. Ibid. Methods **6.074** and **6.076**. *See also* Williams, K. T., *et. al.*: *J. Assoc. Offic. Agr. Chemists*, **33:**986 (1950); 36:402 (1953).
6. Potter, E. F.: *J. Assoc. Offic. Agr. Chemists*, **48:**717 (1965).
7. Potter, E. F., and Long M. C.: *Ibid.*, **48:**728 (1965).
8. Ref. 2, Methods **29.184–29.189**.
9. *Ibid.* Methods **29.062–29.063** and **6.078–6.079**.
10. Rohmer, A. G., Henschel, E. R. and Engel, C. E.: *J. Assoc. Offic. Agr. Chemists*, **48:**844 (1965). *See also* "Glucostat" circular (revised July 1963) of Worthington Biochemical Corp., Freehold, N.J.
11. Williams, K. T., and Potter, E. F., *J. Assoc. Offic. Agr. Chemists*, **41**:307, 681 (1958); Potter, E. F., Wilson, J. R., and Williams, K. T.: Ibid., **42**:650 (1959).
12. Gee, M. and Walker, H. G., Jr.: *Anal. Chem.*, **34:**650 (1962).
13. Sweeley, C. C., Bentley, R., Makita, M. and Wells, W. W.: *J. Am. Chem. Soc.*, **85:**2497 (1963).
14. Alexander, R. J. and Garbutt, J. T.: *Anal. Chem.*; **37:**303 (1965).
15. Klebe, J. F., Finkbeiner, H. and White, D. M.: *J. Am. Chem. Soc.*, **88:**3390 (1966).
16. (Infrared spectrophotometry) Beroza, M.: *Gas Chromatog.*, **2:**330 (1964); Edwards, R. A. and Fagerson, I. S.: *Anal. Chem.*, **37:**1630 (1965). (Mass spectrometry) Gohlke, R. S.: *Anal. Chem.*, **31:**535 (1959), *Chem. Ind.*, 946 (1963).
    Widmark, K. and Widmark, G.: *Acta Chem. Scand.*, **16:**575 (1962); Muller, D. O.: *Anal. Chem.*, **35:**2033 (1963);
    Ryhage, R.: *Anal. & Chem.*, **36:**759 (1964);
    Ross, W. D., Moon, J. F. and Evers, R. L.: *J. Gas Chrom.*, **2:**340 (1964);
    Amy, J. W., Chait, E. W., Baitinger, W. E. and McLafferty, F. W.: *Anal. Chem.*, **37:**1265 (1965); Kemp, W., and Rogne, O.: *Chem. & Ind.*, 418 (1965).
17. Whistler, R. L. and Duraso, D. F.: *J. Am. Chem. Soc.*, **72:**677 (1950);
    Whistler, R. L.: *Science*, **120:**899 (1954);
    Wolfram, M. L., Thompson, A., O'Neill, A. N. and Galkowski, E. F.: *J. Am. Chem. Soc.*, **74:**1062 (1952).
18. Consden, R., Gordon, A. H. and Martin, A. J. P.: *Biochem. J.*, **38:**224 (1944).
19. (a) Hais, I. M. and Macek, K.: *Handbuch der Papier Chromatographie. Grundlagen und Technik*. Jena: Veb. Fischer Verlag (1958). I, 262–288.
    (b) Lederer, E. and Lederer, M.: *Chromatography Review of Principles and Applications*, 2nd ed. Amsterdam–New York–Princeton: Elsevier Publishing Co. 245–258. (1957).
    (c) Ough, L. D.; *Anal. Chem.*, **34:**660 (1962).
20. (a) Stahl, E. (ed.): *Thin-Layer Chromatography, a Laboratory Handbook*. Berlin–Heidelberg–New York: Springer-Verlag (1965).
    (b) Randerath, K.: *Thin-Layer Chromatography*, 2nd ed., revised. New York: Verlag-Chemie GmbH/Academic Press (1966).
    (c) Bobbitt, J. M.: *Thin-Layer Chromatography*. New York: Reinhold Publishing Co. (1963).
21. Ref. 20(a), 164.
    Prey, V., Borbalk, H. and Kausz, M.: *Mikrochim. Acta*, **6:**968 (1961).
22. Pastuska, G.: *Z. Anal. Chem.*, **179:**427 (1961).
23. Jaarma, M.: *Acta Chem. Scand.*, **8:**860 (1954).

CONFECTIONERY

24. *Webster's Third New International Dictionary*. Springfield: G. & C. Merriam Co. (1961).
25. Jacobs, M. B. (ed.): *The Chemistry and Technology of Food and Food Products*, 2nd ed. New York: Interscience Publishers, Inc. (1951). II, 1625–1626.
26. Hundley, H. H. and Hughes, D. D., Jr.: *J. Assoc. Offic. Anal. Chemists*, **49:**1180 (1966).
27. Jones, J. G., Smith, D. M. and Sasarabudhe, M.: *J. Assoc. Offic. Anal. Chemists*, **49:**1183 (1966).

CORN PRODUCTS

28. Petzer, W. R., Crosby, E. K., Engel, C. E. and Kirst, L. C.: *Ind. Eng. Chem.*, **45:**1075 (1953).
29. Selected Ref. K.
30. Ough, L. E.: *Anal. Chem.*, **34:**662 (1962).
31. Montgomery, E. M. and Weakley, F. B.: *J. Assoc. Offic. Agr. Chemists*, **36:**1096 (1953).
32. Dimler, R. J., Schaefer, W. C., Wise, C. S. and Rist, C. E.: *Anal. Chem.*, **24:**1411 (1952); (a)

Whistler, R. L., and Hickson, J. L.: *Anal. Chem.*, **27**:1514 (1955).

HONEY

33. Selected Ref. N, p. 38.
34. Selected Ref. F, p. 667.
35. White, J. W., Jr., Ricciuti, C. and Maher, J.: *J. Assoc. Offic. Agr. Chemists*, **35**:859 (1952).
36. White, J. W., Jr., and Roban, N.: *Arch. Biochem. Biophys.*, **80**:386 (1959); Watanabe, T. and Aso, K.: *Nature*, **193**:1740 (1959).
37. Stinson, E. E., Subers, M. H., Petty, J. and White, J. W., Jr.: *Arch. Biochem. Biophys.*: **89**:6 (1960).
38. White, J. W., Jr.: *J. Food Sci.*, **31**:102 (1966).
39. Selected Ref. N, p. 52.
40. White, J. W., Jr., Ricciuti, C. and Maher, J.: *J. Assoc. Offic. Agr. Chemists*, **35**:466, 478 (1954).
41. Selected Ref. N, pp. 46–50.
42. Ibid. pp. 50–51.
43. Ref. 2, Method **24.110.**

43a. Palkin, S., Murray, A. G. and Watkins, H. R.: *Ind. Eng. Chem.*, **17**:612 (1925).

43b. Knapp, I. E.: *Ind. Eng. Chem., Anal. Ed.*, **9**:315 (1937).

44. Winkler, W. O. (Food and Drug Administration, Dept. of Health, Education and Welfare). Private communication (1946).
45. White, J. W., Jr.: *J. Food Sci.*: **31**:102 (1966).

MAPLE PRODUCTS

46. Agricultural Marketing Service, U.S. Dep. Agr.: *United States Standards for Grades of Table Maple Syrup* (1940).
47. Vermont Dept. of Agriculture, Division of Markets: *Vermont Maple Syrup Grading and Marketing Law, Circular* **14** (1950).
48. *New York State Official Standards, Definitions, Rules and Regulations for Maple Products.* New York State Bureau of Markets (1956).
49. "Wisconsin Maple Products; Production and Marketing." *Wisconsin Dept. of Agriculture, Bull.* **335** (1956).
50. Snell, J. F., and Scott, J. M.: *Ind. Eng. Chem.*, **6**:219 (1914); also cited in Selected Reference I, p. 326.
51. Snell, J. F.: *J. Assoc. Offic. Agr. Chemists*, **16**:172 (1933).
52. Fisher, H. J.: *Conn. Agr. Expt. Sta. Bull.* **528**:33–35 (1949).
53. Selected Ref. I, p. 322.
54. Wendt, A. S., and Benjamin, E. J.: *J. Assoc. Offic. Anal. Chemists*, **49**:508 (1966).
55. Underwood, J. C. and Filipic, H. J.: *J. Assoc. Offic. Agr. Chemists*, **46**:334 (1963).

MOLASSES AND OTHER CANE SYRUPS

56. Selected Ref. P.
57. Deal, E. C., Food & Drug Administration: Private communication (September 1950).

SELECTED REFERENCES

GENERAL

A. Bates, F. J., *et al.*: *Polarimetry, Saccharimetry and the Sugars.* U.S. Dept. of Commerce National Bureau of Standards Circular C440 (1942).

B. Browne, C. E. and Zerban, F. W.: *Physical and Chemical Methods of Sugar Analysis.* 3rd ed. New York: John Wiley & Sons Inc. (1941).

C. Haworth, W. N.: *The Constitution of Sugars.* London: Edward Arnold (1929).

D. Horwitz, W., (ed.): *Official Methods of Analysis of the Association of Official Agricultural Chemists.* 10th ed. Washington: Association of Official Analytical Chemists (1965). Chapter 29.

E. Jacobs, M. B.: *The Chemical Analysis of Foods and Food Products.* 3rd ed. Princeton: D. Van Nostrand & Co. Inc. (1958). Chapter 10.

F. Leach, A. E. and Winton, A. L.: *Food Inspection and Analysis.* 4th ed. New York: John Wiley & Sons, Inc., (1920). Chapter 14.

G. Oser, B. L. (ed.): *Hawk's Physiological Chemistry.* 14th ed. New York: McGraw-Hill Book Co. (1965). Chapter 2.

H. Winton, A. L. and Winton, K. B.: *The Structure and Composition of Foods.* New York: John Wiley & Sons Inc. (1939). IV, 1–77.

I. Woodman, A. G.: *Food Analysis.* 4th ed. New York: McGraw-Hill Book Co. (1941). Chapter 6.

CONFECTIONERY

J. Jordan, S. and Langwill, C.: "Confectionery Analysis and Composition." In: *Manufacturing Confectioner* (1946).

CORN PRODUCTS

K. Technical Services Committee of the Corn Industries Research Foundation, Inc.: *Corn Syrups and Sugars*, 3rd ed. Washington: Corn Industries Research Foundation Inc. (1965).

HONEY

L. Browne, C. A.: "Chemical Analyses and Composition of American Honeys". In: *Bureau of Chemistry, U.S. Dept. of Agriculture, Bulletin* **110** (1908).

M. White, J. W., Jr.: *American Bee Journal* **101**:299 (1961).

N. White, J. W., Jr., Riethof, M. L., Subers, M. H. and Kushnir, I.: "Composition of American Honeys." Agricultural Research Service, *U.S. Dept. of Agriculture Technical Bulletin* **1261** (1962).

MAPLE PRODUCTS

O. Willits, C. O.: "Maple Sirup Producers Manual." U.S. Dept. of Agriculture, *Agriculture Handbook* **134** (Rev.) (1965).

MOLASSES AND OTHER CANE SYRUPS

P. Jacobs, M. B. (ed.): *The Chemistry and Technology of Food and Food Products*. 2nd ed. New York: Interscience Publishers Inc. (1951). III, 2128–2130.

CHAPTER 17

Vegetables and Vegetable Products

Introduction

The word "vegetable" is not used technically. Webster's Third New International Dictionary defines a vegetable as:

"1. A usually herbaceous plant that is cultivated for an edible part which is used as a table vegetable.
2. An edible part of a plant (as seeds, leaves or roots) that is used for human food and usually eaten cooked or raw as the main part of a meal rather than as a dessert—contrasted with fruit."

Thus tomato, egg plant, bell pepper, while botanically classified as fruit, are considered as vegetables under this definition.

Analytical procedures for the principal components of vegetables are mainly those given in Chapter 1, Proximate Analysis, and Chapter 11, Fruits. These chapters should be consulted.

Processors are interested in the total (or soluble) solids content of the raw vegetable, as well as visual examination of size, defective units, foreign material, etc., for costing purposes. For this purpose the soluble solids content may be determined by refractometer test on the expressed juice. While this may not be exact, such data are comparable from field to field and year to year.

Olericulturists and other horticulturists are interested in the quantity of vitamins or minerals, sugars, fiber, acidity, color, or other factors that influence the commercial or nutritional value of the vegetable. Methods for the determination of these constituents may, for the most part, be found in appropriate chapters of this text, *e.g.*, proximate analysis, sugars, vitamins, etc.

Authentic Analyses of Vegetables

Knowledge of the composition of authentic samples of vegetables, as published in the literature, will be of little value to an analyst of specific samples of vegetables, whether fresh or processed, for purposes of comparison. Vegetables are processed in all stages of maturity, from early succulent stages, as in peas, Chinese pea pods and baby lima beans, to the mature, or even stored, stage, as in dry beans, potatoes, onions and cabbage. These come from many varieties, some of which are from limited agricultural areas.

Most vegetable processors, and to a lesser extent handlers of fresh vegetables, accumulate data from year to year on the particular varieties they use. These are usually grown under uniform conditions of horticultural and harvesting practices, on contract farms where these conditions can be closely supervised. They cannot be compared even with the same varieties grown under different conditions. Then, too, there are some varieties or strains that are grown for the exclusive use of one processor.

It is for these reasons that analyses of authentic vegetables are not included in this chapter. Knowledge of the composition of vegetables grown in particular areas may often be found in publications of the Agricultural Extension Services of the several states, in Bulletins of the U.S. Department of Agriculture, and in the Selected References listed at the end of this chapter.

Measurement of Textural Properties of Vegetables and Fruit

Objective measurement of texture by mechanical devices constitutes a separate study beyond the scope of this book. These tests are empirical, being attempts to simulate conditions to which the food is subjected during harvesting, processing or consumption. Here we shall give a brief review of some of the instruments available and list some source material for those who wish to pursue the subject further.

The first attempt to measure texture mechanically appears to be that of Lehmann in Germany in 1907.[1] He devised two instruments to measure the tenderness of meat. One measured shearing force, the other breaking strength. Since then, a vast number of instruments has been developed. These have been reviewed by Kramer and Twigg as of 1959.[2]

Further reviews were made by Friedman[3] and Szczesniak[4] in 1963, and a general text on the subject was written by Matz in 1962.[5]

Some instruments, such as the MIT Denture Texturometer[6], are models of the human jaw and buccal cavity. These can be adapted to a large variety of foods. Other examples of this type are the General Foods Texturometer by Freidman[3], the Shear-press developed by Kramer[7], and the Instron Universal Testing Machine.[8]

Methods of Analysis

Method 17-1. *Determination of Starch in Vegetables.*

Reagents

a. *Dextrose standard solution*—Dissolve 0.100 g of anhydrous dextrose in 100 ml of water. This can be preserved by addition of 0.1 g of sodium benzoate.

b. *Anthrone-sulfuric acid reagent*—Dissolve 0.5 g of anthrone in 250 ml of 95% H_2SO_4. This is stable for 3–4 days if kept at 0°. (High blanks occur with old solutions.)

c. *Perchloric acid*, 52%—Add 270 ml of 71% $HClO_4$ to 100 ml of water. Store in a glass-stoppered bottle.

Preparation of Sample

Fresh starchy vegetables: Blend a convenient amount, sufficient to avoid sampling errors, in a Waring Blendor with an equal weight of water until homogeneous. If foaming occurs, add a few drops of amyl alcohol.

Canned starchy vegetables: Drain the contents of the can on an 8-mesh sieve. Transfer 100-200 g of the drained vegetable to a Waring Blendor and blend until homogeneous.

Dried starchy vegetables: Grind the sample to pass through a 60–80 mesh sieve.

Determination

For fresh or canned vegetables, weigh 5.00 g of the slurry (equivalent to 2.5 g original vegetable) into a 50 ml centrifuge tube.

For dried vegetables, weigh 0.200 g of ground sample into a 50 ml centrifuge tube, add a few drops of 80% ethanol to prevent clumping, then add 5 ml of water and stir thoroughly.

Extraction of sugars—Add 25–30 ml hot 80% ethyl alcohol to the contents of the centrifuge tube, stopper, shake vigorously, centrifuge at about 2500 rpm, decant, and discard the supernatant liquid. Repeat the washing, centrifuging and decanting three times, or until a qualitative test with the anthrone solution (**b**) shows no green color.

After the final extraction, add water to the

sugar-free residue to make 10 ml, and stir. Cool in ice-water and add, while stirring, 13 ml of 52% $HClO_4$. Stir for 5 minutes, and occasionally thereafter for 15 minutes while keeping the solution cold. Add 20 ml of water, centrifuge, and pour the supernatant solubilized starch solution into a 100 ml volumetric flask. Add 5 ml of water to the residue in the centrifuge tube, cool in ice-water, and slowly stir in 6.5 ml of 52% $HClO_4$. Solubilize as before for 30 minutes in the ice-water bath and wash the contents into the 100 ml volumetric flask. Dilute the combined extracts to the 100 ml mark with water and filter, discarding the first few ml.

Dilute 5–10 ml of the filtered solution to 500 ml, or to contain 5 to 20 μg of starch/ml.

Pipette 5 ml of this diluted solution into a cuvette, cool in a water bath and add 10 ml of anthrone reagent (**b**). Mix thoroughly and heat for $7\frac{1}{2}$ minutes in a boiling water bath. Cool rapidly to 25° in the water bath and read the absorbance at 630 mμ (the color is stable for 30 minutes). Calculate μg of dextrose from the standard curve.

Preparation of Standard Curve

Add 1 ml, 2 ml and 3 ml of the dextrose standard solution (**a**) to separate 200 ml volumetric flasks and make to volume with water. Treat 5 ml of each solution (equivalent to 25, 50 and 75 μg of dextrose) as under "Determination—Extraction of Sugars", beginning "After the final extraction", and plot a curve from these readings.

Sugars in Fresh and Dehydrated Vegetables. *See* Method 16-6. Use clarifying agent (**d**), ion-exchange resin, and the copper reduction Method 16-6B, "Copper by direct weighing."

Canned Vegetables

Definitions and Standards

A general definition for canned vegetables has not been promulgated by the U.S. Food and Drug Administration. Definitions of specific canned vegetables are included in the Standards of Identity. These name the vegetable and the packing medium, specify the degree of maturity, if appropriate, list mandatory and optional ingredients, and conclude the definition by the statement: "The food is sealed in a container and so processed by heat as to prevent spoilage."

As of now, definitions and standards have been individually promulgated for canned peas, green beans, corn, tomatoes and tomato products; and a blanket definition and standard has been established for 35 vegetables not specifically regulated. All of these standards are included in Parts 51 and 53 of *Code of Federal Regulations, Title* 21.[9]

Canada has promulgated definitions of canned vegetables under the Canada Food and Drugs Act.[10] These regulations are mandatory for products manufactured in Canada or imported commercially into Canada. They state that: "Canned vegetables shall be prepared by heat-processing properly prepared fresh vegetables, with or without (a) sugar or dextrose, (b) salt or (c) a conditioner, and shall be packed in hermetically sealed containers." A further regulation defines a conditioner as: "Calcium chloride, calcium citrate, calcium phosphate or calcium sulfate only, in an amount not greater than 0.026 percent, calculated as calcium, if the presence of such conditioner is declared on the label."

These definitions are very similar, but not identical, to the U.S. definitions and standards. Tomato pulp, purée, paste and catsup may contain up to 1000 ppm of benzoic or sorbic acid (expressed as the free acid), and sulfurous acid up to 500 ppm (as SO_2), if declared on the label.[10]

Grades and Standards for processed (canned, frozen, dehydrated) vegetables have also been issued under Canada's Agricultural Products Standards Act.[11] These go into much more detail than do their Food and Drug Standards. They are very similar to United States Standards for Grades issued by the U.S. Department of Agriculture.[12]

There is one important difference between U.S. and Canadian Standards issued by the respective Departments of Agriculture: the first are voluntary and the latter are mandatory. Under Canadian law, officials have the authority, under Section 3(1)(d) of the Agricultural Products Standards Act of 1955[11] to "prescribe the sizes, dimensions and other specifications of packages in which an agricultural product must be packed and the manner in which it must be packed and marked as a condition to application or use of the name of a grade so established".

Under this authority, the Canada Department of Agriculture has set up can sizes and volume designations, minimum net and drained weights, and headspace allowances for standard containers permitted for processed fruits and vegetables. These are given in Text Reference 11, Schedule C.

Method 17-2. *Fill of Container, Canned Vegetables.*

This is the general method prescribed in Standards of Fill of Container, U.S. Food and Drug Administration.[13]

Follow Method 11-2, using a No. 2 U.S. Standard sieve for tomatoes (size of opening 0.446″) and a No. 8 Standard sieve (size of opening 0.093″) for other vegetables. Drain for 2 minutes. (The *Official Methods* of Canada Food and Drug Directorate and Canada Department of Agriculture, Production and Marketing Branch, direct drainage of canned tomatoes for 30 seconds on a No. 2 Standard sieve.)

WATER CAPACITY OF CAN AND FILL OF CONTAINER. *See* Method 11-3.

Requirements for minimum fill and drained weights are not considered factors of quality for the purpose of establishing USDA Standards of Grades. These USDA Standards do, however, set "recommended fill of container" and "recommended minimum drained weights." The general rule for fill of container in the United States, unless expressed otherwise in a specific standard, is that the product and packing medium shall occupy not less than 90% of the total water capacity of the container in avoirdupois ounces. Drained weights are often listed in USDA Standards for Grades as minimum weights for each can size. Unless directed otherwise, they are determined by 2 minute drainage as directed above.

Certain canned vegetables do not lend themselves to the FDA General Method for Fill of Container.

Method 17-3. *Fill of Container, Canned Peas.*[14]

Weigh the container. If it is a can with lid attached by a double seam (the conventional tinned can), cut out the lid without removing or altering the height of the double seam. Transfer the contents to a suitable container, and pour back into the original container. The leveled peas, irrespective of the liquid, should completely fill the container when observed 15 seconds after they have been so returned.

A can with lid attached by double seam shall be considered completely filled if the headspace, measured from the top of the double seam to the leveled peas, is $\frac{3}{16}$″ or less below the top of the container.

The contents of the container may be reserved for such observations as defects, foreign material and tenderness as may be indicated.

In the United States, military specifications for canned foods issued by the Department of Defense, and Federal Specifications issued by General Services Administration, sometimes include specific fill of container requirements and tests. If the packing medium is a free-flowing liquid, as in canned tomatoes or beets, drained weights are determined by 2-minute drainage on a sieve as directed by Method 11-2.

If the packing medium is viscous, as in canned dried beans (*Federal Purchasing Specification JJJ-B-*101*a*), a requirement for "washed drained weight" may be a part of the specification.

Method 17-4. *Washed Drained Weight.*

Use an 8″, No. 8 sieve for can sizes smaller

than No. 2 (307 × 409),* and a 12″ No. 8 sieve for larger cans.

Pour the contents of the can evenly over the surface of the tared sieve. Immerse sieve and contents in water at 70°F., remove meat and fat particles if present, then swirl the sieve vigorously for 1 minute, taking care to break all chunks so that adhering sauce may be washed off. Withdraw sieve and contents from the water, then again immerse momentarily, twice in succession. Allow the product to drain 2 minutes, then weigh the sieve and contents. Report this weight less the weight of the sieve as "washed drained weight." Weighings are usually made in pounds and ounces.

The Canada Food and Drug Directorate's method for these products washes the sieve with a gentle stream of water.

Method 17-5. *Drained Solids of Beans with Pork, Beans or Vegetarian Beans (Canada Food and Drug Directorate).*

Weigh container and contents; open the container and pour the contents evenly over an 8-mesh sieve. Wash the contents of the sieve free of sauce with a gentle stream of cold water. Drain 5 minutes, then gently tap the sieve to release entrapped water. Transfer the washed contents to the dried, tared container or other suitable tared container, and weigh. Report as percent of net weight.

Method 17-6. *Net Weight and Drained Weight of Canned Fruits and Vegetables.* (Official Method, Canada Department of Agriculture.)

Measure the overall can height and diameter to the nearest $\frac{1}{16}$″. Weigh the container, remove the can top, measure the gross headspace in sixteenths of an inch from the top of the can to the product, and pour the contents evenly over an 8-mesh screen, fitted over a tared dish or pan. Use a sieve 8″ in diameter for a 28 fl. oz. or smaller can; for larger cans use a 12″ sieve. Halves of peaches, pears, etc. should be turned "cups down" to facilitate draining.

Allow the contents of the sieve to drain one-half minute. Weigh the drained liquid in the dish, and deduct the tare.

Wash, dry and weigh the empty can, and deduct its weight from the gross weight. Report this difference as "net weight". Deduct the weight of the drained liquid from the net weight and report the difference as "drained weight".

Refer to Text Reference 11, Schedule C, for minimum net and drained weight requirements.

Method 17-7. *Total Solids. AOAC Method* ***30.003****).*

This determination may be desired on either drained canned vegetables or the total contents of the container (vegetable plus packing medium). It is also applicable to puréed or comminuted vegetables, catsup and similar products.

Preparation of Sample

Canned Whole Vegetables.—Thoroughly grind in a food chopper the drained vegetables, or the entire contents of the container, as required by the purpose of the analysis. Thoroughly mix the portion used for analysis, and place the remainder in glass-stoppered containers. Store in a refrigerator. Unless analysis is to be completed in a reasonably short time, determine moisture in a portion of the prepared sample, and, to prevent decomposition, dry the remainder, mix thoroughly, and store in glass-stoppered containers. When samples of the dried vegetable are withdrawn for later analysis, again determine the moisture, and calculate the total solids results to the original moisture basis.[15]

Comminuted Products—Shake the unopened container thoroughly to incorporate any sediment. Transfer the entire contents to a large glass or porcelain dish, and mix thoroughly by stirring for at least one minute. Transfer the well-mixed sample to a glass-stoppered container, and shake thoroughly each time before removing portions for analysis.

Determination

Into a flat-bottomed metal dish with tight-fitting cover add diatomaceous earth filter-aid (Celite or equivalent) at the rate of about 15 mg per square centimeter of the area. Dry about 30 minutes at 110°, cool in a desiccator and weigh. Add to each dish a sample of such size that the dry solids

* Overall dimensions for cans are expressed in the manner used in the industry, with the diameter of the can in inches and sixteenths listed first, *e.g.*, "307 × 409" means 3 $\frac{7}{16}$″ diameter and 4 $\frac{9}{16}$″ height.

shall be between 9 and 30 mg/sq cm. Weigh as rapidly as possible to avoid loss of moisture. Mix with the filter-aid and distribute uniformly over the bottom of the dish. A little water may be used if necessary to facilitate distribution.

Bring samples to apparent dryness (remaining moisture content not more than about 50% of the dried solids) by one of the following methods:

1. Place samples on a boiling water bath, removing them when they reach apparent dryness.

2. Place samples in a forced-draft oven at 70°. The oven must have rapid air circulation and sufficient interchange of outside air to remove moisture rapidly. Examine the dishes at intervals of 30 minutes or less, and remove when the samples reach apparent dryness.

3. Place samples in a vacuum oven at 70° with release cock partially opened to allow rapid flow of air at not less than 310 mm mercury pressure. Examine the dishes at 30 minute intervals, and remove any that reach apparent dryness.

Place the partially dried samples directly on the metal shelf of a vacuum oven equipped with a thermometer in direct contact with the shelf. Temperature variation from one part of the shelf to another must not exceed 2°. Admit dry air by bubbling it through concentrated H_2SO_4 at a rate of 2-4 bubbles per second. Dry the samples for 2 hours at 69–71° (oven may be 65° at the start, but must reach 69–71° within the first hour) and a pressure of not more than 50 mm of mercury. As the dried sample will absorb appreciable amounts of moisture over most desiccants, cover the dishes quickly and weigh as soon as possible after they reach room temperature.

***Method 17-8.** Salt in Canned or Puréed Vegetables.*

Prepare the sample as in Method 17-7. Transfer 5 to 10 g of the prepared sample to a platinum dish, add 20 ml of 5% Na_2CO_3 solution, and continue as in Method 1-21, beginning "ignite as thoroughly as possible . . ."

NOTES

1. Verify complete retention of chloride in each type of material by trial, since loss can occur, especially with samples high in carbohydrates, from use of insufficient sodium carbonate during ignition, or, in any case, from ashing at a temperature above 500°.

2. The accuracy of this method is limited to ±0.2 mg Cl. The sample therefore should contain at least 20 mg of Cl or about 30 mg of NaCl.

***Method 17-9.** Salt in Canned or Puréed Vegetables. Rapid Control Test.*

REAGENTS

Ferric indicator, and 0.1 *N silver nitrate* and *thiocyanate solutions* as in Method 1-21.

Add a known volume of the 0.1 *N* silver nitrate solution in slight excess to a prepared sample of 5-10 g. Add 20 ml of nitric acid and boil gently until nitric oxide fumes cease to evolve (about 15–20 minutes), and cool. Add 50 ml of water, and 5 ml of the ferric indicator. Titrate with 0.1 *N* thiocyanate solution to a permanent light brown color.

CALCULATIONS—Same as Method 1-21.

Special Tests for Quality on Specific Canned Vegetables.

Federal Standards of Quality for certain canned vegetables list several criteria for judging the maturity of the vegetables as an index of quality. The methods prescribed are somewhat empirical, and must be followed exactly as prescribed in the regulations. For this reason, and because of their limited use, we advise interested chemists to consult the regulations cited.

Canned peas are judged, among other factors, by the percentage of alcohol-insoluble solids in the drained peas, and by the number of peas that will support a 2-lb weight.[16] Canned green beans are graded by the number of tough strings (those that support a suspended half-pound weight for 5 seconds) and by the percentage of fibrous material in the drained, deseeded beans.[18,19] (It might be noted that the apparatus prescribed for this test is no longer commercially available.[20]) Canned corn is judged by the percentage of alcohol insoluble solids present in the drained corn and by its consistency.[17]

Dehydrated Vegetables

The process of commercial dehydration of a vegetable, by whatever method employed, alters to some extent its composition. A certain proportion of volatiles is removed in addition to water. Carbohydrates may be hydrolyzed, proteins may be denatured, the quantity of certain vitamins may be reduced, and other changes may occur. Comparative analyses of the raw vegetables (trimmed and washed preparatory to dehydration) and the dehydrated vegetables will reveal the magnitude of these changes.

Proximate analysis of vegetables, whether raw or dehydrated, may be made by the methods given in Chapter 1 and those given earlier in this chapter, making allowance in the sample weight for the concentration accomplished by dehydration. Determination of certain constituents of dehydrated vegetables may require special techniques.

U.S. Government purchasing specifications for dehydrated vegetables prescribe limits for moisture and sulfur dioxide content, and specify methods that shall be used. These are set forth in Table 17-1.

The preparation of a representative sample of dehydrated vegetables presents certain problems, particularly on the larger pieces. The source of the sample, whether from selected stations during the production cycle, or from finished material in drums, cans or cartons, influences the size of the sample. Grab samples can be drawn from a production line station at preselected time intervals, composited, and reduced in size by quartering, riffling or other means. Drum or carton samples may be drawn by use of a suitable trier. In all cases the sampler should work as expeditiously as possible, since these products are extremely hygroscopic.

The final sample should consist of at least 200 g. This should be ground in a clean, dry Waring Blendor or a suitable mill for the minimum time necessary. A small portion should first be ground and discarded to clean out the mill. The ground material may be stored in small, tightly closed containers. It should be held at least a half hour before using for analysis, to equalize the moisture content. Samples used for moisture and fat (ether extract) should be ground to pass through a 20-mesh screen. Samples for the near-infrared method or the Karl Fischer moisture method should be ground to pass through a 30-mesh screen.

Moisture in Dehydrated Vegetables

Moisture may be determined by Method 1–2. If Federal or Military specifications are involved, refer to Table 17–1 for the required time and temperature. Note that the method prescribes 6 hours' drying at 70° for certain dehydrated vegetables and 16 hours at 60° for others.

The Karl Fischer procedure, Method 1–6, is used in many plant laboratories as a rapid quality control method. In such cases, it should be checked against the prescribed vacuum oven method and, if necessary, the appropriate correction applied. This may amount to 0.1 to 0.3% moisture over that obtained by the vacuum oven method.

The solvent distillation method, Method 1–4, is also applicable to plant control procedure. Benzene should be used as the solvent for dehydrated vegetables containing appreciable amounts of carbohydrates (beets, cabbage, carrot, onion, etc.) to avoid decomposition of sugars. This method should be checked against the prescribed vacuum oven method for necessary corrections.

Rader[21], an Associate Referee for the AOAC, has described a near-infrared spectrophotometric method applicable to dried vegetables and spices which is specific for water.

TABLE 17-1

Federal and/or Military Specifications for Dehydrated Vegetables

Analytical Requirements

Vegetables	Moisture max., %	SO_2 ppm	Method Specified: Moisture	Method Specified: SO_2	Processing requirements before dehydration
Beans, green[a]					
Uncooked	4.0	750 ± 250	Vac. oven, 6 hours at 70°	AOAC method **27.078** (Monier-Williams)[k]	Blanched, live steam
Cooked	4.0	”	”	”	Cooked
Beans, white or red[a]					
Precooked	4.0	—	”	—	Blanched, live steam
Quick-cooked	11.0	—	”	—	Cooked
Beans, lima (Fordhook)[a]					
Cooked	4.0	500 ± 200	”	AOAC method **27.078** (Monier-Williams)[k]	Cooked, frozen
Beets	5.0	—	”	—	Blanched, or precooked
Cabbage[b]					
Raw	5.0[c]	1500 ± 500	Vac. oven, 16 hours at 60°	AOAC method **27.078** (Monier-Williams)[k]	No heat treatment or blanch
Cooked	5.0[c]	”	”	”	Cooked
Carrots[b]					
Uncooked	—[d,e]	500 ± 100	Vac. oven, 6 hours at 70°	”	Blanched, live steam
Cooked	—[d,e]	”	”	”	Cooked
Celery[b]	2.0[c]	500 − 1000	”	”	Unblanched
Garlic[f]	6.5	—	Vac. oven, 16 hours at 60°	—	Unblanched
Onion[g]	4.0	—	”	—	Unblanched
Parsley[a]	4.0[c]	—	”	—	Unblanched
Pepper, Bell[b]					
Green	4.5	1000 − 2500	”	AOAC method **27.078** (Monier-Williams)[k]	Unblanched
Red	7.5	”	”	”	Unblanched
Mixed	4.5	”	”	”	Unblanched
Potato, sweet, instant[j]	3.0	250 ± 50	Vac. oven, 6 hours at 70°	”	—
Potato, white[h,i]					
Diced or sliced, uncooked	6.0	400 ± 100	”	”	Blanched
Mashed, precooked[a]	6.0	300 ± 100	”	”	Cooked and mashed
Tomato powder, juice or paste	3.5	450 max.	Vac. oven, 16 hours at 60° pressure under 100 mm	”	—

[a] Nitrogen-packed in tins. [b] Nitrogen-packed with desiccant bag. [c] At time of packing. [d] Residual moisture shall be absorbed by desiccant. [e] After packing with desiccant. [f] Maximum optical density 0.150, ashed sediment 0.02%, hot water insoluble 10.0%. [g] Francy grade, maximum optical density 0.130 (powder), 0.105 (pieces), hot water insoluble 20.0%, ashed sediment 0.02% (powder only). [h] Reducing sugars, 3.0% ± 0.1% maximum (dry weight). [i] Calcium salts, added, as Ca 0.3% ± 0.1% in uncooked potatoes. [j] Soluble solids by refractometer of 2.0 g/200 ml water, filtered, between 75° and 83° Brix. [k] See *Official Methods of Analysis*, AOAC, 10th ed.

This was adopted as official at the 1966 meeting of AOAC.

Method 17-10. *Moisture in Dehydrated Vegetables, Near-Infrared Spectrophotometric Method.*[21]

This method was adopted in 1966 by AOAC as an official method for moisture in dried vegetables and spices.

APPARATUS AND REAGENTS

a. *Spectrophotometer*—Beckman DK2A recording or equivalent near-infrared instrument.

b. *N,N-dimethylformamide (DMF)*—spectro grade.

PREPARATION OF STANDARD CURVE

Accurately weigh 200, 300, 400, 500 and 600 mg of distilled water into separate dry 125 ml glass-stoppered Erlenmeyer flasks. Pipet 100 ml of dimethylformamide into each flask, Record the absorbances of the solutions in matched 1 cm quartz or silica cells from 1.8 to 2.0 μ, using the same DMF in the reference beam as was used in preparing the standard solutions. Draw a base line between the minima at about 1.82 and 2.0 μ.

The absorbance of each standard solution at its maximum (about 1.92 μ = the total absorbance at maximum—the base line absorbance). Plot the base line corrected absorbance against mg water/100 ml of DMF.

DETERMINATION

Grind samples to pass a 30-mesh sieve and store in tightly sealed containers. Accurately weigh an amount of the ground sample containing 70–100 mg of water into a dry 50 ml glass-stoppered Erlenmeyer flask. Pipet in 20 ml of DMF. Tape the stopper securely to the flask and heat in an oven at 90° for 60 minutes (bell pepper for 40 minutes). Shake the flask mechanically for 10 minutes, then cool to room temperature. Decant the solution into a dry centrifuge tube and centrifuge at 1500 rpm until the solution is clear (about 5 minutes). Record the absorbance of the solution from 1.8 to 2.0 μ, using the same DMF in the reference beam as was used to prepare the sample solutions.

Calculate the absorbance of the solution at its maximum, about 1.92 μ, using the same base line technique as with the standard solutions:

% Moisture in sample

$$= 100\frac{\text{(mg. water from standard curve)}}{5 \times \text{sample weight in mg}}.$$

Bromide Residues in Dehydrated Vegetables Treated With Methyl Bromide

Methyl bromide and ethylene dibromide are used extensively in spices and dehydrated vegetables as fumigants, particularly against insects and insect eggs. The U.S. Food & Drug Administration has established tolerances of varying magnitude for different food products. "In the establishment of (these) tolerances from methyl bromide applications no allowance was made for residues of naturally occurring bromides."[22] Certain vegetables, *e.g.*, paprika and chili pepper grown in coastal regions, contain appreciable amounts of natural bromides. Gentry Corporation, Glendale, California[23], has reported up to 50–60 ppm of bromine attributed to pick-up from the soil in fields where no organic bromides were used.

Several methods have been proposed for determination of bromide residues. The methods we have found satisfactory in our laboratories were developed by chemists of the Dow Chemical Company.[24]

Method 17-11. *Total and Inorganic Bromide Residues in Dehydrated Vegetables and Spices.*

This procedure determines total bromides and inorganic bromides. Methyl bromide residue is calculated as the difference between the two.

REAGENTS

All reagents should be AR or ACS analytical reagent grade. A blank should be run on all reagents, carrying them through the entire procedure, and the thiosulfate titration of the blank run should be subtracted from the final titration.

a. *Digestion solution*—10 g of sodium hydroxide and 35 ml of ethanolamine dissolved in 950 ml of ethanol.

b. 6 *N hydrochloric acid*—Dilute 500 ml of hydrochloric acid with an equal quantity of water and distill, discarding the first and last 100 ml. This distillate will be approximately 6 *N*.

c. *Sodium hypochlorite solution—NF* (4–6%).

d. 50% *sodium formate solution*—50 g made to 100 ml with water.

e. 1% *sodium molybdate solution.*

f. 5% *potassium fluoride solution.*

g. 0.01 *N sodium thiosulfate*—diluted fresh from 0.1 *N* solution.

h. 1% *starch solution*—Make a paste of 1 g of soluble starch and 10 ml of water. Slowly pour into 100 ml of boiling water, boil a few minutes and cool.

i. *Methylene chloride (dichloromethane)*—bromide-free.

Total Bromine Determination

Weigh 5.00 g of ground material into a 100 ml nickel crucible. Add 25 ml of digestion solution (**a**) and let stand for at least 2 hours, preferably overnight. Evaporate to dryness on a steam bath, then dry for at least 1 hour in an oven at 105–110°. Cool and add 10 g of sodium hydroxide pellets, stirring the pellets into the residue with a stiff glass rod. Brush any fragments adhering to the rod into the crucible, using a camel hair brush.

Heat the crucible on a hot plate until bubbling or smoking diminishes (a burner may be used, but care should be observed to prevent excessive foaming or burning.) Now place the crucible at the front of a muffle furnace at 600°, and progressively move it into hotter zones. After volatile gases are evolved, remove the crucible from the furnace and cautiously add a few milligrams of sodium peroxide. After flaring ceases, repeat the addition of sodium peroxide, a few milligrams at a time, to complete the oxidation of organic matter, then return to the furnace for a few minutes. Remove from the furnace. If a white, porous mass remains which refuses to melt, more peroxide must be added and the crucible returned to the furnace. When it is hot, carefully rotate the crucible to wash down any organic matter adhering to the sides, and add a little more peroxide. A few black carbon particles remaining in the melt will not affect the accuracy of results.

Cool the melt and dissolve it in 75 ml of water by placing the crucible on a hot plate, covered with a watch glass, for several minutes. Transfer the solution to a 400 ml beaker and *partially* neutralize by slowly adding 40–60 ml of 6 *N* hydrochloric acid (**b**) (solution must remain alkaline). Boil to destroy peroxides, reducing the volume to 100–125 ml. Filter through a Whatman No. 2 paper, collecting the filtrate and washings in a 500 ml wide-mouthed Erlenmeyer flask. Make slightly acid with 6 *N* hydrochloric acid, using a *small* amount of methyl red indicator solution, then neutralize with sodium hydroxide solution to a methyl red end point. Use as little indicator as possible. The volume here should be about 150 ml.

Add 1.5 to 2 g of crystalline sodium acid phosphate ($NaH_2PO_4.H_2O$) crystals and 5 ml of sodium hypochlorite solution (**c**), and heat the mixture to boiling. After boiling 1 or 2 minutes, add 5 ml of sodium formate solution (**d**). Boil for 2 minutes and cool. Add a few drops of 1% sodium molybdate solution (**e**), 1 ml of 5% potassium fluoride solution (**f**), 0.5 g of potassium iodide and 25 ml of 6 *N* sulfuric acid. Titrate immediately with 0.01 *N* thiosulfate solution (**g**), adding starch solution indicator (**h**) just before the end point. 1 ml = 0.1332 g bromine ion.

Ppm total bromine

$$= \frac{(\text{ml } 0.01\ N \text{ thiosulfate} \times 0.1332)\ 1000}{\text{weight of sample}}.$$

A blank on the reagents should be carried through this procedure.

Inorganic Bromide Determination

Weigh 5.00 g of dry sample into a 100 ml beaker and add 20 ml of methylene chloride (**i**). Filter immediately by suction through a previously prepared Caldwell crucible containing a dry asbestos pad. Rinse with three 5 ml portions of methylene chloride. Do not allow the sample to get so cold by solvent evaporation that moisture condenses on it. Transfer most of the sample back to the beaker without disturbing the asbestos pad. Add 15 ml solvent to the sample and allow it to stand 5 minutes, breaking up any lumps with a glass rod. Filter and rinse as before. Return the sample to the beaker and add 15 ml of solvent. Filter and rinse. Discard all solvent and rinsings. Return the sample and pad to the original beaker, and add 15 ml of solvent, rinsing out the crucible with the solvent and adding the rinsings to the beaker.

Evaporate the contents of the beaker to apparent dryness, stirring to avoid lumping. (Methylene chloride boils at 40°, so avoid a higher temperature.) When dry, lay the beaker on its side in a warm place until solvent odor disappears. Transfer its contents to a 250 ml centrifuge bottle. Rinse the beaker and the Caldwell crucible with water, catching the rinsings in the centrifuge bottle. Add water to about 30 ml volume, mix by swirling, and centrifuge at about 2000 rpm for 5 minutes. Decant into a Gooch crucible fitted with an asbestos pad and filter by suction into a 250 ml beaker. Repeat the extraction with 30 ml of water four times, each time allowing the mixture to stand for 15 minutes in the bottle, then centrifuging and filtering.

Add 3 ml of saturated sodium chloride solution to the combined extracts and evaporate to about 50 ml. Transfer quantitatively to a silica dish and evaporate nearly to dryness. Add 30 ml of 2.5% alcoholic potassium hydroxide, and evaporate the alcohol on a steam bath. Dry the dish and contents in an oven at 105–110° for 15-30 minutes, char in a hood over a flame and continue ashing in a muffle furnace at 400°. After no further flame or smoke appears, bring the furnace to 500–525° and continue ashing for 10 minutes.

Cool, and extract the residue twice with dilute hydrochloric acid (1+50), filtering each time through rapid paper into a 250 ml Erlenmeyer flask. Wash the residue with three 10 ml portions of water. Return the filter paper to the silica dish, treat again with 3 ml of saturated sodium chloride solution and 10 ml of alcoholic potash solution, dry and ignite as before. Extract the residue with 25 ml of dilute hydrochloric acid, wash three times with 10 ml of water, catching the filtrate and washings in the same Erlenmeyer flask.

Neutralize the combined filtrates with 10% sodium hydroxide, using methyl orange as indicator. Evaporate to about 75 ml. Add 1 g of crystalline sodium acid phosphate and 2 ml of sodium hypochlorite solution (**c**). Heat to boiling, and after a minute or so add 2 ml of 50% sodium formate solution (**d**) and boil for 2 minutes. Cool, add 1 drop of molybdate solution (**e**) and 10 ml of sodium fluoride solution (**f**), 0.5 g of potassium iodide and 10 ml of 6 *N* sulfuric acid.

Titrate immediately with the same 0.01 *N* thiosulfate solution used for titration of total bromides, using starch indicator.

CALCULATIONS

$$\frac{(\text{Ml thiosulfate} \times 0.1332)\ 1000}{\text{weight of sample}} = \text{ppm inorganic bromine.}$$

Organically combined bromine as ppm
= total ppm bromine − ppm inorganic bromine.

Ppm bromine × 1.188 = ppm methyl bromide.

Method 17-12. *Bromide Residues in Material Fumigated With Ethylene Dibromide Alone or Combined with Methyl Bromide* (*or whose Type of Organic Bromide Fumigant is Not Known*).

Place a 5 g sample of material in a 125 ml boiling flask fitted with a 24/40 $ joint. Add 20 ml of alcohol and 5 ml of ethanolamine. Attach an upright reflux condenser, and place the flask in a beaker containing hot water so that the water level is about 1″ above the liquid in the flask. Reflux for 3 hours. Wash down the condenser with alcohol, remove it and wash the contents of the flask into a 100 ml nickel crucible. Add about 5 sodium hydroxide pellets, and evaporate the alcohol on a steam bath. Continue drying in an oven at 105° for an hour, add 10 g of sodium hydroxide pellets to the contents of the crubicle, and proceed as in Method 17-11, second paragraph, beginning "Heat the crucible on a hot plate . . ."

NOTE: Refluxing is necessary to hydrolyze ethylene bromide. Methyl bromide hydrolyzes readily on standing in the alkaline mixture.

Sulfur Dioxide

Thrasher[25], in his AOAC referee's report on the classical Monier-Williams method for sulfur dioxide, states that that method as originally written gives high apparent SO_2 values on such dehydrated vegetables as onion, leeks and cabbage (comment by the authors of this book—garlic should also be included) due to the presence of sulfur-containing compounds which distill over and titrate as SO_2.

One of the authors (F.L.H.) has found in his laboratory that Potter's modification[26] of the iodine titration method gives reproducible and fairly accurate results on all vegetables.

Method 17-13. *Sulfur Dioxide in Dried Vegetables, Rapid Method.*

Suspend 8.00 g of ground dehydrated vegetable into each of two 600 ml beakers containing 400 ml of distilled water. Add 5.0 ml of 20% NaOH solution (20 g made up to 100 ml with water) to each beaker. Stir gently, avoiding incorporation of air, and let stand 30 minutes.

To one beaker slowly add 7.0 ml of 5 *N* hydrochloric acid (205 ml of 37% HCl, diluted to 500 ml with water), while stirring to avoid local acid concentration.

Add 10 ml of 1% starch solution and titrate immediately with 0.05 *N* iodine to a definite blue color. It is important that the acidified solution be titrated immediately before recombination occurs. Record the volume of 0.05 *N* iodine consumed as *A*.

To the second beaker add 7.0 ml of 5 *N* hydrochloric acid, 10 ml of 1% starch solution, and 2 ml of 3% hydrogen peroxide to oxidize sulfites to sulfates. Titrate immediately to the same blue endpoint. Report the volume of 0.05 *N* iodine consumed as *B*.

Assuming an 8 g sample and 0.05 *N* iodine, ppm $SO_2 = (A - B)\,200$.

NOTE: *See also* Method 14-2 for the official method for total sulfurous acid in foods.

Packaging Tests, Dehydrated Vegetables

Many dehydrated vegetables are packed in hermetically sealed cans or tins in which the air has been replaced by nitrogen. This is particularly true of products packed in compliance with U.S. Military specifications. Such specifications, *e.g.*, *MIL-P*-43265, *Potatoes, White, Dehydrated*, and *MIL-P*-35090, *Parsley, Dehydrated*, specify the modified Orsat method adopted by the American Dry Milk Institute.

More recent government specifications for nitrogen-packed dehydrated vegetables specify a semi-micro Orsat gas analyzer developed by John J. McMullen and O. J. Stark of the Quartermaster Food and Container Institute for the Armed Forces. This is particularly adaptable to examination of headspace gas in smaller tins, where only 1 to 10 ml of gas (at atmospheric pressure) are available. The original article, containing a diagram of the apparatus, published in *J. Assoc. Offic. Agr. Chemists* **37**:856 (1954) should be consulted. This method is designated as Method No. 237, *Fed. Test Method Std. No.* 101*a* in government specifications. It may be obtained from Federal Supply Service, General Services Administration, Washington, D.C.

Method 17-14. *Oxygen Content of Nitrogen-Packed Dehydrated Vegetables.*[27,28]

REAGENTS

a. *Alkaline pyrogallol solution*—Dissolve 6.5 g of pyrogallic acid, reagent grade, in 25–30 ml of water. Mix this solution with 150 ml of caustic solution, reagent (**b**). This reagent oxidizes readily and should be mixed and introduced into the absorption pipette as quickly as possible. A rubber breather bag should be attached to the open end of the pipette to prevent oxidation. This solution is effective for approximately 200 determinations when the gas examined has an oxygen content of 2 percent or less. Solutions should be changed at least once a month.

b. *Caustic solution*—Dissolve potassium hydroxide, reagent grade, in an equal weight of water. Approximately 175 ml of this solution is required to charge the pipette. Renew at least once a month.

APPARATUS

a. *Gas sampling device*—Any suitable device for puncturing the container and through which a sample of gas in the container may be withdrawn without contamination from the outside atmosphere.*

* A portable device adaptable to all sizes of cans is available from Continental Can Co., Chicago, Ill.

b. *Modified Orsat gas analyzer* (see Fig. 17-1) consisting of:

(1) *Burette*, preferably 100 ml volume, graduated in 0.1 ml divisions;

(2) *Glycerine*, to cover the mercury in the burette for adjustment of vapor pressure;

(3) *Leveling bottle*, for adjusting height of the mercury column in burette;

(4) *Manometer, water*, for equalizing pressure;

(5) *Mercury*, for adjusting gas volume in the burette (about 5 pounds required);

(6) *Pipettes*, gas absorption, with rubber breather bags; and

(7) *Tubing*, rubber, heavy wall (for connections).

DETERMINATION

Arrange the modified Orsat analyzer in a manner similar to that shown in Fig. 17-1. Connect the sample can to stopcock G with a suitable sampling device. Apply enough pressure to the sampling device to give a tight seal with the surface of the container. Do not puncture the container. With stopcocks F, E, and H closed, and stopcocks G and I open, expel all of the air in the burette through stopcock I. Lower the leveling bottle A to the base of the ringstand. Note the level of the mercury column in the burette. If this level remains constant, the apparatus is not leaking. Proceed with the analysis. Apply pressure on the sampling device to puncture the container. Draw 2–3 small samples of gas from the container and expel these through stopcock I before a final sample is taken for analysis.

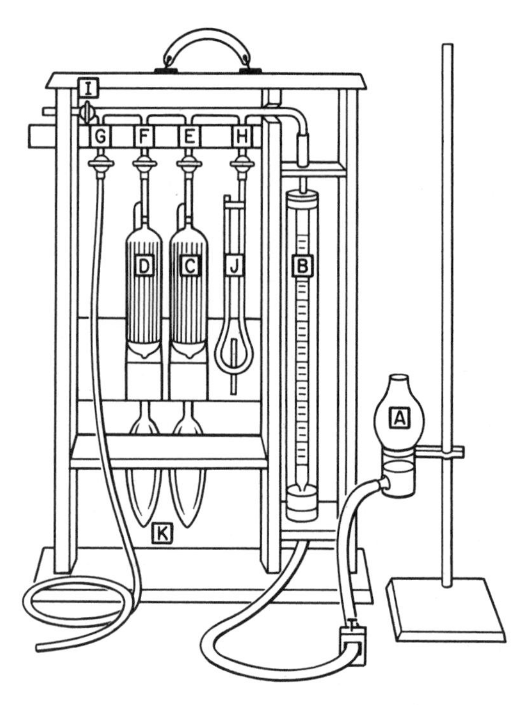

Fig. 17-1. Modified Orsat gas analyzer. A) Leveling bottle; B) Measuring burette; C) Absorption pipette (caustic solution); D) Absorption pipette (alkaline pyrogallol solution); E, F, G, H, I) Two-way stopcocks; J) Manometer; K) Rubber breather bags for absorption pipettes.

Introduce the sample through stopcock G and allow it to displace the mercury in burette B by lowering the leveling bottle A sufficiently to obtain a sample of approximately 100 ml. If the container is small or under partial vacuum, the sample of gas that can be withdrawn into the burette may be considerably less than 100 ml.

Close stopcock G and level the mercury in the burette with that in the leveling bottle by adjusting the height of the bottle. When the levels are approximately equal, open stopcock H and level the water in the manometer by further adjustments of the leveling bottle.

Close stopcock H and determine the volume of the gas sample in the burette without changing the position of the leveling bottle.

Introduce the gas sample into pipette C, containing the caustic solution, through stopcock E by raising the leveling bottle until the mercury level in the burette is at the zero mark at the top of the burette.

Pass the sample back and forth between the pipette and the burette by alternately raising and lowering the leveling bottle until no further change in volume of gas in the pipette is obtained.

Again determine the gas volume in the same manner used for measuring the volume of the original sample. The difference between the original and the last reading is the amount of carbon dioxide in the sample.

The next step in the analysis is to remove the oxygen by passing the sample into pipette D, containing alkaline pyrogallol, through stopcock F.

Again pass the gas back and forth until no further change in the gas volume is noted in the burette. The loss in volume of the gas after passage through pipette D is oxygen, and the residual gas is, for all practical purposes, considered to be nitrogen.

Calculate the percent of oxygen by dividing the milliliters of gas absorbed by the alkaline pyrogallol in pipette D by the volume of the original sample, and multiplying by 100.

If the percent of carbon dioxide is required, it may be calculated by dividing the volume of the gas absorbed in pipette C by the volume of the original sample and multiplying by 100.

If duplicate determinations are desired on the same container, exhaust the residual gas through stopcock I and draw a new sample through stopcock G. The sampling line should remain connected to the container throughout the analysis, for two reasons:

1. It permits the withdrawing of a duplicate sample for analysis, and
2. Prevents the diffusion of air into the sampling line and device.

It is good practice to place a pinch clamp on the tubing immediately following the sampling device before a new container is connected. This will prevent air from entering the tubing if the line is under partial vacuum. After the next container is connected, the residual gas in the burette is allowed to pass through stopcock G and out through the sampling line before puncturing the new container. This residual gas is inert and will tend to prevent the diffusion of air into the sampling line while tapping the new container.

Method 17-15. *Leakage Test For Gas-Filled Cans.*[28]

This test is specified in government specifications for dehydrated vegetables in nitrogen-packed, hermetically sealed tins. The usual tin pack is 401 × 411 (No. $2\frac{1}{2}$ cans) or 603 × 700 (No. 10 cans).

Test

Completely immerse the can under test in water contained in a desiccator, or other vacuum chamber suitable for visual observation. Draw a vacuum and maintain it at 10″ of mercury (atmospheric pressure 29.9″) for at least 30 seconds. A leak is indicated by a steady stream of bubbles. Isolated bubbles caused by trapped air are not considered signs of leakage.

Packaging of Dehydrated Vegetables[28]

Federal or Military Specifications for purchases by the United States government of certain dehydrated vegetables, *e.g.*, *Carrots*, *MIL-C*-839*a*, or *Celery*, *MIL-C*-2408*A*, require packing in a nitrogen atmosphere in hermetically sealed tins containing an envelope of a desiccant, calcined calcium oxide, in each can to absorb residual moisture. The desiccant shall be capable of picking up 28.5 % of its weight in moisture, and shall not increase more than 90% in volume after absorbing 28.5% of its weight in moisture.

Method 17-16. *Activity of Desiccants for Dehydrated Vegetables.*[28,29]

Apparatus and Reagents

a. *One or more 250 mm desiccators* provided with sample supports made of $\frac{1}{4}$-inch mesh wire or similar material (to replace the regular desiccator plate).

b. *500 ml of a saturated solution of pure sodium bromide*, with excess crystals at 24°, in each desiccator.

c. *Two-ounce tin ointment boxes*, 6 cm in diameter and 2 cm in height, or suitable weighing dishes.

d. *Means for maintaining the desiccator at 24° ± 1°C.* throughout the test.

Determination

Weigh 1.9 to 2.0 g of the calcium oxide to the nearest 5 mg in a weighing dish and spread it over the bottom in a uniform layer. Place the weighed dish and sample in the desiccator, using not more than 6 samples in one desiccator, and maintain at 24°. Weigh the dish periodically to determine when the maximum moisture pickup has been reached, and calculate the percentage of water absorbed by the desiccant from its increase in weight. If the moisture pickup has not reached a maximum after 7 days, the test shall be terminated and the results calculated.

Method 17-16A. *Volumetric Expansion of Desiccant.*[29]

Weigh 100 g of the desiccant into a 250 ml

graduated beaker. Tap to compact the desiccant and read the volume.

Place the beaker in a No. 10 can cut to a height of about 3 inches. Add saturated sodium bromide solution to the can around the beaker to a depth of about 1 inch. Place the can in a vacuum desiccator that has no desiccator plate and contains about 1 inch of saturated sodium bromide solution. (The bottom of the can is immersed in the solution in the desiccator to permit rapid dissipation of the heat of hydration.)

Evacuate the desiccator, using a shield to protect against possible implosion.

After hydration is complete or has increased in weight by 28.5%, tap the beaker to compact the hydrated desiccant. Read the volume and calculate the percentage increase. This increase in volume shall not exceed 90%.

Dehydrated Potatoes

The reducing sugars content is a prime factor in the dehydration requirements and storage stability of dehydrated potatoes. Potatoes with less than 1% sugars, dry basis, are considered best for manufacture of potato granules. During storage of raw potatoes at low temperatures (34°–38°), the reducing sugar content increases up to as high as 6% or even 8% (dry basis). This sugar content can be partially reduced by tempering the potatoes at 70°F for 1 or 2 weeks before processing.

The *U.S. Purchasing Specification for Potatoes, White Dehydrated, JJJ-P*-00630 (*Army -GL*), sets maxima of 3% dry weight of reducing sugars and 6% moisture. This specification gives the following analytical method for reducing sugars:

Method 17-17. *Reducing Sugars in Dehydrated Potatoes.*

Preparation of Sample

Grind samples other than granules in a Waring type blender to pass through a 20 mesh, and be retained on a 40 mesh, screen. (Potato granules need not be ground or sieved.)

Note: Raw potatoes may just be ground or blended until a homogenous slurry is obtained.

Determination

Weigh 25 g of granules or ground sieved material into a 250 ml centrifuge bottle. Add 75 ml of water and let stand for one half-hour. (50 g of the ground, raw potatoes shall be used without dilution.) Add 100 ml of 95% ethanol made alkaline with NaOH, plus 1 g of calcium carbonate. Heat in a hot water bath at 70° for 1 hour, stirring frequently.

Centrifuge the sample at 1500 rpm for 10 minutes and decant through a Whatman No. 12 folded filter into a 400 ml beaker. Repeat the extraction 3 more times using 75 ml of alkaline 80% ethanol, heating in the water bath at 70° for 15 minutes, centrifuging at 1500 rpm for 10 minutes, and decanting through a folded filter into the 400 ml beaker. Combine the extracts and evaporate on a steam bath to a volume of 75 to 100 ml. Transfer the extract to a 250 ml volumetric flask with distilled water and bring to the mark.

Determine reducing sugars in a 50 ml aliquot (or a smaller aliquot diluted to 50 ml) by the AOAC Munson-Walker procedure, Method 16-6B. Calculate to percent reducing sugar, dry basis.

Sulfur Dioxide in Dehydrated Potatoes

*Federal Purchasing Specification JJJ-P-*00630 (*Army-GL*) prescribed 450 ± 150 ppm for diced potatoes and potato slices (Types I and III), and 300 ± 100 ppm for mashed precooked granules or flakes (Type II). Sulfur dioxide shall be determined by the Monier-Williams method, as modified by the AOAC. This is Method 14-2.

*Federal Purchasing Specification JJJ-P-*00630 (*Army-GL*) requires that the blanched potatoes shall be given a calcium chloride treatment to firm the potatoes and thus prevent sloughing. This specification sets a limit of 0.3 percent ± 0.1 percent calcium in dehydrated potato. While calcium may be determined by the flame photometer or other convenient method (government specifications permit methods other than those specified if they provide a quality assurance equivalent to those specified), the specification for dehydrated potatoes lists the following volumetric oxalate method:

Method 17-18. Calcium Content of Dehydrated Potatoes.

Weigh 25 g of the blended sample prepared as described in Method 17-17 into a Vycor or platinum 250 ml capacity dish and ash over an open flame until the major portion of the carbon is burned away. Continue ashing in an electric muffle at 550° to a white or nearly white ash. Cool the ash, wet with a little distilled water, then add 2 ml of hydrochloric acid. Warm on a steam bath until the ash dissolves (except for traces of silica), and transfer the contents of the dish to a 250 ml beaker with 100 ml of distilled water. Add 2 or 3 drops of methyl red indicator, adjust to a red color with HCl if necessary, and heat to boiling. If the solution becomes turbid, add concentrated HCl dropwise until clear.

Add 15 ml of saturated ammonium oxalate solution to the gently boiling liquid, and while continuing a gentle boil, stir in 2 heaping tablespoonfuls of dry urea. Cover with a watch glass and continue boiling until the color changes from red to yellow (pH 5). Filter the hot solution through a medium porosity sintered glass filter, and wash the beaker and the precipitate with 15 ml of 2% (by vol.) ammonium hydroxide solution, cooled to 8–10°.

Dissolve the calcium oxalate precipitate on the glass filter by pouring through 100 ml of hot 5% sulfuric acid solution, catching the filtrate in the original beaker. Heat to 60° and immediately titrate with standardized 0.01 *N* or 0.10 *N* potassium permanganate, first adding a few drops and allowing the color to disappear, then continuing the titration. Calculate to percent calcium.

$$1 \text{ ml } 0.1\ N\ KMnO_4 = 0.0020 \text{ g calcium.}$$

The normal calcium content of raw potatoes is about 10 to 20 mg calcium/100 g. This is equivalent to about 45 to 100 mg on dehydrated potatoes.

NOTE: For canned vegetables, mix the contents of the can in a Waring Blendor, weigh 50–100 g into a Vycor or platinum dish and evaporate to dryness, first on the steam bath and finally in a hot-air oven at 105°. Proceed as above, beginning "Ash over an open flame . . ."

Method 17–19. *Soluble Solids in Dehydrated Sweet Potatoes.* This is the method prescribed by *Military Specification MIL-P-43236, Potatoes, Sweet, Instant, Flaked.*

Weigh 20 g of flakes into a 400 ml beaker, add 200 ml of water, and stir the contents with a magnetic stirrer at moderate speed for 10 minutes. Centrifuge for 10 minutes and decant a small portion of the supernatant liquid, or filter through paper, discarding the first few ml.

Determine the refractive index of the decanted or filtered liquid by means of an Abbé or sugar refractometer, and obtain the percent of sucrose from Table 23-3. Report this percent, multiplied by 10, as degrees Brix for the dehydrated flakes. The soluble solids content shall be at least 75° but not over 83° Brix.

This specification comments: "A satisfactory fill may be difficult to obtain if the (soluble solids) content is less than 75 percent. Higher soluble solids may give a gummy product."

Method 17-20. Identification of Synthetic Dyes on Sweet Potatoes.

Market sweet potatoes are at times dyed red to simulate the natural color of red sweet potatoes, thus concealing inferiority. Manganello[32] has devised a thin-layer chromatography technique to separate and identify these dyes. Such dye treatment has since been prohibited by FDA. See *Federal Register*, **33:** 9166 (June 21, 1968).

REAGENTS

a. *Cellulose Powder MN* 300 *G* (Brinkmann Instrument Co., Westbury, N.Y.).

b. *Developing Solvent*—2.5% Sodium citrate/ 25% ammonium hydroxide, 4:1.

c. *Appropriate FD&C dyes*—(Reds #1, 2, 3, 4 and Yellow #6 were used in this study). 0.10–0.15% Aqueous solutions for TLC identification; 15–20 mg/liter Aqueous solutions for visible spectrophotometry.

d. *Nun's veiling* (*wool*).

APPARATUS

a. *Recording spectrophotometer*, with 1 cm cells for visible spectrum.

b. *Desaga/Brinkmann Standard Model Applicator & Mounting Board.*

c. 8 × 8 *inch glass plates.*

d. *Camag Sandwich Chamber TL-521-b*, available from Arthur H. Thomas Co.

e. *Sintered glass filtering funnel*, 25 ml capacity, fine porosity; 50–100 ml suction flask.

f. *Micropipette*, or *Hamilton syringe*, for spotting and streaking TLC plates.

TESTS

A. *Extraction of Dyes from Sweet Potatoes*

Rinse the surfaces of approximately 10 potatoes three times with 100 ml of warm 1% ammonia water. Acidify the ammoniacal dye solution with a few drops of HCl and immerse six 7 square inch pieces of wool (nun's veiling) in the solution. Heat on the steam bath for 20 minutes. (Purification by wool dyeing can be omitted if background impurities for TLC are minimal. Basic dyes transfer to wool from a basic or neutral medium.) After the dye has transferred to the wool, remove the wool from the solution, rinse with warm tap water, place in a porcelain dish containing approximately 25 ml of 1% aqueous ammonia, and heat on a steam bath 20 minutes. After the dye has transferred to the solution, remove the wool and concentrate the solution to approximately 2 ml, driving off the ammonia.

B. *Identification of Dyes by TLC*

1. *Preparation of absorbent layer chromatographic plate*—Blend 15 g of MN 300 G cellulose with 90 ml of water for 1 minute. Spread a 0.25 mm layer on several 8 × 8 glass plates and dry at 100° 20 minutes.

2. *Spotting*—Apply spots equivalent to 1.0–2.0 μg (1–2 ml) of dye standards, individually and in mixture, together with sample dye extract, to the plate, spotting 1.5 cm from the bottom edge of the plate at ½ inch intervals. Contain the spots to a 2 mm diameter, air dry several minutes, and develop in a TLC tank with developing solvent for a distance of 10 cm.

3. *Results—Rf Values of Individually Spotted Dyes:* Compare with tables in next column.

C. *Separation of dyes by preparative TLC with subsequent visible absorption spectra of isolated dyes for confirmation of identity by TLC.*

Prepare a TLC preparative plate 1 mm in thickness, with the same cellulose coating material as used above. Dry for 25 minutes at 100°. Streak approximately ½ ml of the sample dye extract 1.5 cm from the bottom edge of the plate in a narrow band with a syringe, keeping the band as narrow as possible. Air dry for a few minutes and develop in a chromatography tank, advancing the solvent front to the top edge of the plate. After development, air dry and scrape off the heart of each separate color band, and mix the separate scrapings with 10–20 ml each of water. Heat each mixture on the steam bath for 10 minutes and filter through a fine sintered glass filtering funnel with suction. Reduce the colored filtrate to approximately 4 ml, and transfer to a 1 cm spectrophotometer cell. Add a few crystals of ammonium acetate to the solution in the cell, and scan in the visible absorption range. Acidify with 1–2 drops of hydrochloric acid and scan. Make basic with several drops of sodium hydroxide, 1+1, and similarly scan.

For confirmation of previous identification by TLC, compare the sample spectra with those of FD&C dyes similarly treated.

Dye	Rf Value	Apparent Color of Migrated Spot
Individually Spotted FD & C Dye Standards:		
FD&C Red #1	0.20	Rose
2	0.40	Lavender
3	0.04	Reddish Violet
4	0.32	Orangish Red
Yellow #6	0.47	Orange
Synthetic Mixture of Equal Parts of FD & C Standards:		
FD&C Red #2	0.37	Lavender
4	0.30	Orangish Red
Yellow #6	0.47	Orange

Method 17-21. *Optical Index of Dehydrated Onion and Garlic.*

This test is primarily a measure of nonenzymatic browning that may occur during processing. It is an official method of the American Dehydrated Onion and Garlic Association (ADOGA)[33], and is a part of *Federal Purchasing Specification JJJ-O-533, Onions, Dehydrated*, and

Military Specification MIL-G-35008, Garlic, Dehydrated.

DETERMINATION

For products finer than 20 mesh, no preparation is necessary. For flakes, grind under low humidity conditions to pass through a 20-mesh sieve.

Weigh a 2 g sample (1 g for toasted onion) into a beaker. Add 0.5 g of filter-aid (Celite), mix with the dry sample, and add 100 ml of 10% sodium chloride solution, first adding 5–10 ml and stirring to form a slurry, then adding the remainder. Let stand 15 minutes, stirring occasionally. Filter through a folded filter paper, returning the filtrate through the paper until crystal clear.

Determine the percent transmission of the filtrate at 420 mμ, with the instrument standardized at 100% transmission with filtered 10% sodium chloride solution.

Calculate the optical density on the basis of a 1% solution, 420 mμ wave length and 50 mm cell length. Optical density ×1000 equals optical index.

The maximum optical indexes established by ADOGA for dehydrated onion and garlic are:[29,33]

*Grade A Powdered, and Granulated Onion	130
*Grade C Powdered Onion	170
Grade C Chopped Onion	150
All Other White Onion Piece Sizes	105
Toasted Onion Products, Minimum 900	1700
All Grade A Yellow Onion Products	400
All Grade C Yellow Onion Products	450
*Grade A Powdered, and Granulated, Garlic	150
*Grade C Powdered Garlic	200
All other Garlic Products	140

Method 17-22. *Hot-Water Insolubles in Powdered Dehydrated Onion and Garlic.*

This test is a measure of the amount of skin and root base present in the dehydrated product. It is an official ADOGA method[33] and is a part of *Federal Purchasing Specification JJJ-O-533, Onion, Dehydrated,* and *Military Specification MIL-G-35008, Garlic, Dehydrated.* This requirement is of primary value only in fine grinds, *i.e.*, those products passing a 20-mesh sieve.

* These particle sizes will pass through a 20-mesh sieve.

PREPARED GOOCH CRUCIBLE

Wash long-fibre, acid-washed amphibole asbestos with a large quantity of water, and decant to remove fine particles. Prepare a well-packed mat in a Gooch crucible, wash with hot water and dry. Ignite, rewash, and dry at 105°. Repeat the washing, igniting and drying.

DETERMINATION

Transfer an accurately weighed 2 g sample, ground to pass through a 20-mesh sieve, to a 400 ml beaker containing 200 ml of hot distilled water, add 2.000 g of filter-aid, stir, and boil slowly for 5 minutes. (A few drops of antifoam may be added, if necessary, to control foaming.) Filter through the tared, prepared Gooch crucible, wash with about 200 ml of hot distilled water, and dry to constant weight at 105°. Run a blank determination without sample.

% Water-insoluble residue

$$= 100 \times \frac{\text{weight of residue, corrected for blank}}{\text{weight of sample}}.$$

The maximum quantities of hot-water-insoluble material permitted in dehydrated onion and garlic by the ADOGA *Official Standards*[33] and included in Federal and Military Specifications are:

Grade C Powdered Onion	30.0%
All other onion products	20.0%
Grade C Powdered Garlic	20.0%
All other garlic products	10.0%

Method 17-23. *Determination of Pyruvic Acid in Dehydrated Onion.*

As in all natural foods, there are many components in onion and garlic that together make up the sensory factor known as "flavor". Much of the flavor of these vegetables arises from action of the enzyme alliinase upon sulfoxide derivatives, forming volatile odoriferous sulfur compounds. Pyruvic acid and ammonia are by-products of this reaction. Schwimmer and co-workers at USDA Western Regional Research Laboratories have demonstrated that the pyruvic acid thus formed is one measure of the pungency of onion and garlic.[34]

Pyruvic acid reacts with 2,4-dinitrophenylhydrazine to form the corresponding hydrazone, which may be measured colorimetrically.

REAGENTS

a. 0.0125% *w/v 2,4-dinitrophenylhydrazine (DNPH)*—Dissolve 12.5 mg of 2,4-dinitrophenylhydrazine (Eastman 1866) in 100 ml of 2 *N* HCl. Filter through hard paper, and refilter the required amount just before use. This reagent is not stable for more than 6 days.

b. *Sodium pyruvate standard solution*—Dissolve 110 mg of sodium pyruvate (Eastman 8719) in 100 ml of distilled water. This stock solution ≎ 10 micromoles/ml. Dilute 1 ml of the stock solution to 100 ml with recently boiled distilled water. 1 ml of this standard solution ≎ 0.10 micromole/ml. Store both solutions in the refrigerator.

c. *0.6 N NaOH*—Dissolve 12.0 g of reagent grade NaOH in boiled distilled water, and dilute to 500 ml.

d. *40% w/v trichloroacetic acid.*

DETERMINATION

Preparation of Standard Curve: Pipette 0.5, 1.0, 1.5, 2.0 and 2.5 ml of the pyruvate standard solution (**b**) into 5 different test tubes. Add 1 ml of 0.0125% 2.4-dinitrophenylhydrazine (**a**), and distilled water to make a total volume of 5 ml. Let stand 10 minutes at 38°, and continue as described under "Color Development". Plot a curve of absorbance at 420 mμ against concentration of sodium pyruvate. Draw a straight line, passing through the point of origin and parallel to the line obtained above, to obtain the standard curve.

Test Sample: Transfer a 1.000 g sample, ground to pass through a 30–35 mesh sieve, into a small screwcap jar. Add about 50 ml of distilled water. Cap the jar, shake well, and let stand for 10 minutes at 30° to allow enzymatic action to develop. Transfer the mixture quantitatively to a 100 ml volumetric flask containing 5 ml of 40% w/v trichloroacetic acid. Make to volume and let stand for 15–30 minutes. Transfer about 50 ml to a centrifuge tube, and centrifuge for 10 minutes at 300 rpm. Pipette 25 ml of the clear supernatant liquid into a test tube and insert a cork stopper. Mark this test tube "test."

Control Sample: Weigh a 1.000 ground sample into a 150 ml beaker and add about 50 ml of boiling distilled water. Boil for 5 minutes and cool. Transfer quantitatively to a 100 ml volumetric flask containing 5 ml of 40% w/v trichloroacetic acid, make to volume with distilled water, centrifuge about 50 ml as directed under "Test Sample", and pipette 25 ml into a test tube. Mark this "control."

Color Development: Pipette 0.5 ml from the "test" and "control" solutions into each of 2 test tubes, add 3.5 ml of distilled water and 1 ml of 0.0125% DNPH solution, hold at 38° in a water bath for 10 minutes, remove from the bath and cool. Add 5 ml of 0.6 *N* sodium hydroxide to one test tube, and in exactly 2 minutes make the same addition to the second test tube. Let each stand not less than 8 nor more than 10 minutes (use a timer with a second hand), and immediately read their absorbances at 420 mμ against a reference cell containing distilled water.

CALCULATION

The difference in absorbance at 420 mμ between the "control" and "test" samples, as read from the standard curve and multiplied by 200, represents the enzymatically produced pyruvic acid, expressed as micromoles of sodium pyruvate/g of sample.The absorbance of the "test" sample represents the total pyruvic acid developed from sulfoxides.

For determination of enzymatically developed pyruvic acid in dehydrated garlic, proceed as above, using a 0.2 g sample. The general range of enzymatically produced pyruvic acid is in the magnitude of 100–150 micromoles/g, expressed as sodium pyruvate.

INTERPRETATION

Taste-panel pungency tests, compared with analyses for pyruvic acid, in dehydrated onion lead to the conclusion that 25 micromoles of enzymatically produced pyruvates/g represent a highly pungent onion, while 10 micromoles/g or lower represent a mild onion. Commercial production by American dehydrators is usually in the 20 to 35 micromole range.

Enzyme Activity

Vegetables and fruit that have been dried or frozen without heat pretreatment develop an odor and taste in storage usually described as "hay-like". Even after such heat treatment, some dehydrated vegetables develop this odor and taste. The fact that application of heat inhibits or prevents this reaction suggests enzymes as the causative agent.[35]

The usual pretreatment employed by food processors is that of blanching with hot water or live steam. Alternatively, a treatment with sulfur dioxide or its salts, or a combination of heat and sulfiting, is used. Sulfites have been found to be particularly useful in inhibition of peroxidase activity. Further preventive measures are packing in an oxygen-free atmosphere, *i.e.*, a nitrogen pack, and/or in-package desiccation in the case of dehydrated foods.

The adequacy of blanch is usually ascertained on samples drawn immediately after blanching, by testing for residual peroxidase activity, which is inactivated at about 160°F. This is one of the most heat resistant enzymes, and its inactivation assures the destruction of less resistant enzymes. Blanching temperatures are usually in the 170°–190°F range, although occasionally temperatures up to 205°F may be used. Since peroxidase can become reactivated during storage, tests for other enzymes are often made.

Some vegetables cannot be blanched at these high temperatures without adversely affecting their taste and texture. In such cases, catalase is used as the test enzyme for sufficiency of blanch. Off flavors may result, though, even in the absence of catalase, if peroxidase is present.[36]

The enzyme lipoxidase is responsible for the development of certain undesirable flavor changes in vegetables containing considerable fat. This is particularly true of dried beans and peas. Tests for the inactivation of lipoxidase are necessary for such vegetables:[37]

Method 17-24. *Test for Peroxidase Activity.*

This test for peroxidase activity is prescribed in many government specifications (*e.g.*, *MIL-B*-35011, *Beans, Green, Dehydrated*) as an indication of proper blanching, which is one of the in-process quality control requirements for frozen and dehydrated fruits or vegetables:

APPARATUS

a. *Waring Blendor* or equivalent.
b. *Timer* or *stop-watch* with second hand.

REAGENTS

a. 0.5% *ACS* or *USP grade guaiacol in* 50% *ethanol.*
b. 0.08% *hydrogen peroxide solution*, 2.8 ml of 30% H_2O_2/liter. Keep in the refrigerator in a dark bottle; renew each week.

TEST

Weigh out a representative 100 to 200 g sample. Place in the blender with 3 ml water to each gram of vegetable, and grind for 1 minute at moderate or high speed. Filter through a cotton milk filter. Add 2 ml of the filtrate to 20ml of distilled water in a test tube. Prepare a blank by adding 2 ml of filtrate to 22 ml of distilled water in a second test tube, mixing, and using as a color comparison tube. (Do not add any guaiacol or peroxide to this tube.) Add 1 ml of 0.5% guaiacol solution to the first test tube without mixing. Add 1 ml of 0.08% hydrogen peroxide to the same tube, without mixing. Mix the contents thoroughly by inverting, and watch for *development of any color differing from the second tube, regardless of hue*, but of sufficient intensity to show an obvious contrast to this tube. This is a *positive* test, and indicates *inadequate blanching*. If no such color contrast develops in 3½ minutes, consider the test *negative* and the product *adequately blanched.* If color develops after 3½ minutes, it is to be disregarded, and the test shall still be considered negative.

NOTE: To insure adequate blanching, select or include a larger proportion of the bigger units under test, since these are more apt to be underblanched.

Method 17-25. *Control Test for Catalase Activity.*

This is a quick test for use in a vegetable freezing plant, devised by B. E. Proctor.[38]

TEST

Grind 100 g of cooled, blanched vegetable or thawed frozen vegetable through a meat chopper, and quickly transfer 1 g to a mortar containing 1 g of sand and 0.6 g of calcium carbonate. Add 10 ml of water and again grind.

Transfer 4-5 ml of this mixture to a fermentation tube, stopper at once, add 8 ml of 3% hydrogen peroxide (see Note), and shake gently for 3 minutes, avoiding occlusion of air. Compare the volume of gas in the fermentation tube with that given by the same test on the fresh or unblanched vegetable. If the volume of gas is greater than that from the fresh or unblanched vegetable, the test is considered positive. The blank should show gas evolution of less than 0.1 ml.

NOTES

H_2O_2 solution that does not contain a preservative should be used. Dilute 30% H_2O_2 1+9 with water for this solution.

Thompson[39] gives a modification of this method by which the volume of the evolved oxygen is measured and the percent of catalase deactivated calculated by comparison with tests on the unblanched vegetable.

TEST FOR CATALASE ACTIVITY[40]

See AOAC method **17.001–17.004**. This is considerably more sensitive than Method 17-25, since it will show a positive test if 0.5% of the catalase originally present remains after processing. Method 17-25, being rapid, is sufficiently satisfactory for plant control purposes. The AOAC method should be consulted when a more sensitive and precise determination is required.

Method 17-26. *Test for Lipoxidase Activity.*[41]

REAGENTS

a. *Substrate*—Commercial cottonseed oil (0.02–0.04% free-fatty-acid, zero peroxide value, and bland odor and flavor) shall be used as lipoxidase substrate. One gram of cottonseed oil is dissolved in 100 ml of a 50:50 mixture of acetone and 95% ethyl alcohol. This substrate should be made fresh or stored below 0° (32°F).

b. *Buffers*—Mix 8.8 ml of 0.2*M* acetic acid with 41.2 ml of 0.2*M* sodium acetate and dilute to a final volume of 100 ml. The final pH should be 5.5.

c. *Calcium chloride solution*—A solution of 64 mg calcium chloride per ml of distilled water is used to precipitate inactive proteins.

d. *Ferrous ammonium sulfate solution*—Dissolve 0.125 g of ferrous ammonium sulfate (reagent grade) in 100 ml of 3% HCl.

e. *Ammonium thiocyanate*—Dissolve 20 g of fresh ammonium thiocyanate (reagent grade) in 100 ml of distilled water.

f. *Acidified ethyl alcohol*—Add 4ml of concentrated HCl to 500 ml of 95% ethyl alcohol.

NOTE: All water used in test must be distilled and free from ferric iron.

PREPARATION OF SAMPLE

Finely grind the material to be tested to pass a No. 20 mesh sieve. The sample should not be allowed to become warm to the touch during grinding.

TEST

Add 5 g of ground sample to 50 ml of distilled water and extract for 10 minutes with constant stirring at sufficient speed to just keep the solids in suspension. Centrifuge the mixture at approximately 900 times gravity. Add five ml of the calcium chloride solution to the suspended supernatant, mix thoroughly and recentrifuge.

To 100 ml of distilled water add 5 ml of the pH 5.5. acetate buffer and 1 ml of the cottonseed oil substrate. Add one ml of enzyme extract at zero time, and remove 2 ml aliquots from the reaction mixture, which has been held at room temperature (70°F ± 5°), at 8, 16, and 32 minutes to test for peroxide formation. Pipet these aliquots into test tubes containing 25 ml of the acidified ethyl alcohol. Add one ml of the ferrous ammonium sulfate solution, and then 1 ml of the ammonium thiocyanate solution. Swirl the tube to mix the reagents, and watch for color formation within a 3-minute period thereafter.

Prepare the blank by adding 2 ml of the buffered water and substrate mixture to 25 ml of acidified ethyl alcohol. The color, which is developed as before, should be a pale pink. Comparing the color of the blank with the color of the sample being assayed will indicate the amount of lipoxidase activity present in the sample. The colors of the 8, 16 and 32 minute aliquots should not be darker than the blank to indicate 95% or better lipoxidase inactivation. At 90% inactivation, a definitely darker pinkish-red color can be observed in the 32-minute aliquot.

TEXT REFERENCES

1. Lehman, K. R.: *Arch. Hyg. Bakteriol.*, **63**:134 (1907).
2. Kramer, A. and Twigg, B. A.: *Advan. Food Res.*, **9**:153 (1959).
3. Friedman, H. N., *et al.*: *J. Food Sci.*, **28**:390 (1963).
4. Szczesniak, A. S.: *Ibid.*, **28**: 410 (1963).
5. Matz, S. A.: *Food Texture.* Westport: Avi Publishing Company, Inc., (1962).
6. Proctor, B. E., *et al.*: *Food Technol.*, **10**:327 (1956).
7. Kramer, A.: *Canner*, **112**:34, 40 (1951).
8. Bourne, M. C., *et al.*: *Food Technol.*, **20**:522 (1966).
9. *Code of Federal Regulations, Title* 21, *Parts 1 to 129* (Rev. as of 1/1/66). Washington D.C.: Office of the Federal Register (1966).
10. Anon.: *Office Consolidation of the Food & Drugs Act and Food & Drug Regulations, with Amendments to July 14, 1966.* Reg. B.11.002 to B.11.051. Ottawa: Queen's Printer (1966).
11. *Office Consolidation of the Canada Agricultural Products Standards Act and the Processed Fruit and Vegetable Regulations*, pp. 22A to 147. Ottawa: Queen's Printer (1966).
12. *United States Standards for Grades of Processed Fruits and Vegetables.* Fruit and Vegetable, Consumer and Marketing Service, U.S. Department of Agriculture (ask for Standard by name of Processed Fruit or Vegetable.)
13. Ref. 9, Reg. 10.6.
14. Ibid., Regulation 51.3.
15. Lamb, F. C.: *J. Assoc. Offic. Agr. Chemists*, **47**:492 (1964).
16. Ref. 9, Regulation 51.2(b) (5).
17. Ibid., Regulation 51.21(b).
18. Ibid., Regulation 51.11(b).
19. Ibid., Regulation 51.10.
20. Beacham, L. M., Food & Drug Adm.: Private communication (1967).
21. Rader, B. R.: *J. Assoc. Offic. Anal. Chemists*, **49**:726 (1966).
22. Roe, R. S., (Food and Drug Adm.): Private communication (1960).
23. Hart, F. L.: Unpublished work (1960).
24. a. Mapes, D. A. and Shrader, S. A.: *J. Assoc. Offic. Agr. Chemists*, **40**:189 (1957).
    b. Shrader, S. A. and Mapes, D. A.: *Ind. Eng. Chem., Anal. Ed.*, **14**:1 (1942).
25. Thrasher, J. J.: *J. Assoc. Offic. Anal. Chemists*, **49**:834 (1966).
26. Potter, E. F.: *Food Technol.*, **8**:269 (1954).
27. American Dry Milk Institute: *Bull.* **916** (revised). Chicago: American Dry Milk Institute Inc. (1965).
28. *Military Specification MIL-C*-839*a*, *Carrots, Dehydrated.* Washington: General Services Administration (1962).
29. *Military Specification MIL-D*-43266, *Desiccants and Desiccation, Method of: For Packaging Subsistence*, and *MIL-D*-3464*C*, *Desiccants.* Washington, D.C.: General Services Administration.
30. Bernhart, D. N. and Chess, W. B.: *Anal. Chem.*, **31**:1026 (1959).
31. Chess, W. B. (Stauffer Chemical Co.): Private communication (1967).
32. Manganello, S. (Food and Drug Administration, U.S. Dept. Health, Education and Welfare): Private communication (1966).
33. American Dehydrated Onion and Garlic Association: *Official Standards.* San Francisco: American Dehydrated Onion and Garlic Association (1960).
34. a. Schwimmer, S. and Guadagni, D. G.: *J. Food Sci.*, **27**:94 (1962).
    b. Schwimmer, S. and Weston, W. J.: *J. Agr. Food Chem.*, **9**:301 (1961).
35. Selected Reference A, p. 285.
36. Dietrich, W. C., Lindquist, F. E., Bohart, G. S., Morris, H. G. and Nutting, M.-D.: *Food Res.*, **20**:480 (1955).
37. Selected Reference A, p. 288.
38. Proctor, B. E.: *Food Industries*, **14** (11): 51 (1942).
39. Thompson, R. H.: *Ind. Eng. Chem., Anal. Ed.*, **14**:585 (1942).
40. Lineweaver, H.: *J. Assoc. Offic. Agr. Chemists*, **30**:413 (1947).
41. *Military Specification MIL-B*-43193, *Beans, White or Red, Dehydrated.* Washington: General Services Adm. (1964).

SELECTED REFERENCES

A. Acker, L.: "Enzymic Reactions in Low-Moisture Foods." In: *Advan. in Food Res.*, **11**:263. (1962).

B. *Federal Specification, Canned Subsistance Items, Packaging and Packing Of, PPP-C-29a.* Washington: General Services Adm. (1963).

C. Braverman, J. B. S.: "Pretreatment and Preservation, Chemistry and Technology of Vegetable Products." In: *Advan. in Food Res.*, **5**:97, (1954).

D. Howard, F. D., *et al.*: "Nutrient Composition of Fresh California Grown Vegetables," *Bull.* **788.** Davis: Calif. Agricultural Experiment Station. (1962).

E. Huelsen, W. A.: *Sweet Corn.* New York: Interscience Pub. Co. (1954).

F. Jacobs, M. B.: *Chemistry and Technology of Food Products.* New York: Interscience Pub. Co. (1951). II, ch. 27.

G. Joslyn, M. M.: *Food Analysis Applied to Plant Products.* New York: Academic Press (1950).

H. Kirschner, J. G.: "Chemistry of Fruit and Vegetable Flavors." In: *Advan. in Food Res.*, **2**:259, (1949).

I. Lee, F. A.: "The Blanching Process." In: *Advan. In Food Res.*, **8**:63 (1958).

J. Lee, F. A.: "Maturity Tests for Frozen Vegetables." In: *Ind. Eng. Chem., Anal. Ed.*, **13**:38 (1941), **14**:240 (1942).

K. Lynn, G. E. and Vorhes, F. A.: "Symposium on Fumigation Residues." In: *J. Assoc. Offic. Agr. Chemists*, **40**:163 (1957).

L. Talburt, W. F. and Smith, O.: *Potato Processing*, 2nd ed. Westport: Avi Publishing Company Inc. (1967).

M. U.S. Dept. of Agriculture: *Composition of Foods—Raw, Processed, and Canned. Agriculture Handbook, No.* 8 (rev.) (1963).

N. Van Arsdel, W. B. and Copley M. J. (eds.): *Food Dehydration*, I–II. Westport: Avi Publishing Company Inc. (1963, 1964).

O. *Vegetable and Food Fact Sheets (Fruit and Vegetable Facts and Pointers).* Washington: United Fruit and Vegetable Assoc.

CHAPTER 18

Colors

Coloring matters in foods may be either natural ingredients of the foods or substances added thereto to simulate freshness in meats or vegetables, the presence of eggs in noodles or other baked goods, the presence of olive oil in other vegetable oils, high proportions of fruit juice in beverages, etc. Colors may also be added for the wholly innocent purpose of promoting the attractiveness of a food to the eye. Colors naturally present in fruits and vegetables are usually anthocyans, carotenoids or chlorophyll, but occasionally have structures not fitting into these groups. The colors in meats from land-based animals are proteins containing the so-called "heme" iron-pyrrole group (e.g., hemoglobin); in some shellfish, color is due to similar compounds in which copper replaces iron.

Because the coloring ingredients of natural foods are specific for these foods, and their structures have been established in most cases, identification is more a matter of looking up the appropriate reference than it is of running tests in the laboratory. Partly for this reason, the present chapter will be confined to methods for the identification of added colors, both synthetic and those derived from natural sources not indigenous to the foods to which they have been added. Most of the synthetic food colors are azo or triphenylmethane dyes of relatively simple structure.

While a list of the dyes that have at one time or another been used to color foods would be long, the number of such dyes that is legally permitted to be added to foods at the present time is very small. When the present U.S. Food, Drug and Cosmetic Act was adopted, it carried with it a requirement that the Food and Drug Administration establish lists of colors considered sufficiently non-toxic to be safe for use on or in the respective categories of foods, and drugs and cosmetics, and that thereafter no artificial color be permitted to be added unless it came from a lot of a listed color that had been tested by the Food and Drug Administration and certified as suitable. Since the present book is concerned only with food analysis, we shall not discuss the so-called "D & C" and "Ext. D & C" lists. The original list of "FD & C" colors and their lakes permitted for use in foods had by 1965 been reduced to two blues, three greens, three reds (one restricted to coloring oranges), one violet and two yellows.[1] Some of these colors have been removed from the list, which at present includes the following synthetic dyes:[2]

Official Name	Common Name	Restrictions on Use
FD & C Blue No. 1	Brilliant Blue FCF	None
FD & C Blue No. 2	Indigotine	None
FD & C Green No. 3	Fast Green FCF	None
FD & C Red No. 2	Amaranth	None
FD & C Red No. 3	Erythrosine	None
FD & C Red No. 4	Ponçeau SX	150 p.p.m. on Maraschino cherries only
FD & C Violet No. 1	Wool Violet 5BN; Acid Violet 6B	None
FD & C Yellow No. 5	Tartrazine	None
FD & C Yellow No. 6	Sunset Yellow FCF	None
	Citrus Red No. 2	2 ppm on mature oranges only
	Orange B	150 ppm in frankfort and sausage casings

Besides these synthetic dyes, the following natural or inorganic coloring materials are permitted:[3] Algae meal, alkanet (alkanna), annatto, beet juice and powder, bixin and norbixin (from annatto), calcium carbonate, caramel, carbon black, carmine, carminic acid, carotene, carrot oil, charcoal, chlorophyll (and its copper complex), cochineal, corn endosperm oil, cottonseed flour, cudbear, ferric chloride, ferrous gluconate and sulfate, fruit juice, grape skin extract, iron oxides, logwood, paprika (and its oleoresin), riboflavin, safflower, saffron, tagetes meal (Aztec marigold), titanium dioxide, ultramarine blue, vegetable juice, xanthophyll and β-apo-8-carotenal.

British regulations[4] permit use of a larger number of colors, and there is no requirement that such colors be from lots tested by an official agency. The synthetic dyes are the following: Ponçeau MX, Ponçeau 4R, Carmoisine, Amaranth, Red 10B, Erythrosine BS, Red 2G, Red 6B, Ponçeau SX, Ponçeau 3R, Fast Red E, Orange G, Orange RN, Oil Yellow GG, Tartrazine, Naphthol Yellow S, Yellow 2G, Yellow RFS, Yellow RY, Sunset Yellow FCF, Oil Yellow XP, Green S, Blue VRS, Indigo Carmine, Violet BNP, Brown FK, Chocolate Brown FB, Chocolate Brown HT and Black PN. Also permitted are caramel, cochineal, "any coloring material natural to edible fruits or vegetables", alkannet, annatto, carotene, chlorophyll, flavine, indigo, orchil, osage orange, Persian berry, safflower, saffron, sandalwood, turmeric, iron oxide, carbon black, titanium dioxide, ultramarine and aluminum or calcium lakes of the permitted water-soluble colors.

The structures of the synthetic dyes in the above lists (and probably of any dye presently being manufactured) can be found in Selected Reference A.

Methods of Analysis

The food chemist is unlikely to be called on to analyze a food color sample for concentration or the presence of minor impurities. For most colors, the concentration (percentage of "pure dye") is easily determined, if required, by titration of a properly-buffered solution with titanous chloride, using the sample as its own indicator[5]; methods for the determination of minor impurities in the permitted colors may be found in Chapter 35 of AOAC *Methods of Analysis*. If the analyst needs to confirm the identity of a color that he is reasonably certain is not a mixture, the easiest way is to prepare an aqueous solution, run an absorption curve on a recording spectrophotometer, and compare this curve with those of known dyes as given in Selected References E, F, and N. If the color sample contains a mixture of two or more dyes, it will first be necessary to separate these by one of the methods to be discussed herein.

The identification of a truly unknown dyestuff (*i.e.*, one not covered by the following methods and for which a published absorption curve is not available) may require use of resources of structural organic chemistry that are beyond the scope of the present volume. If it is an azo dye and a sufficient quantity is available, reduction of the diazo group(s) with $TiCl_3$ will yield two (or more) amino compounds, which may be identified and so establish the structure of the parent color; but those who wish to try such methods will have to look elsewhere than in this book.

Before the advent of chromatography, dyestuff mixtures found on foods were usually separated by extraction of their aqueous solutions of varying acidities with organic solvents such as *iso*amyl alcohol. One version of this process may be found in AOAC *Methods of Analysis*, Methods **35.002–35.008.** This procedure was largely replaced by column chromatography on cellulose.[6] At the present time, paper or thin-layer[7] chromatography is usually preferred because these techniques are much less time-consuming. Two such methods are given here:

***Method 18-1.** Method of Stanley and Kirk.*[8]

APPARATUS

a. *Chromatographic tube*—20 × 300 mm glass tube with stopcock. (Catalogue No. JC-1505, Size G, of the Scientific Glass Apparatus Co., Inc., 735 Broad St., Bloomfield, N.J., is satisfactory.)

b. *Glass-stoppered graduated cylinders for paper chromatography*—Scientific Glass Apparatus Co. No. C-9830, 100 ml capacity, are satisfactory.

REAGENTS

a. *Activated alumina*—150–250 mesh. Heat 1 hour at 400° before use.

b. *Pyridine-ethyl acetate-water*—1:2:2. Shake a mixture of one volume of pyridine with two volumes each of ethyl acetate and water in a separatory funnel, and allow to separate. Discard the lower layer. (This solvent is best for multicomponent mixtures and mixtures of yellow and orange dyes).

c. *Ammonium hydroxide in isobutyl alcohol*—3.9 *N*. Shake together in a separatory funnel equal volumes of NH_4OH and *iso*butyl alcohol. Allow to separate and discard the lower phase. Pipet 5 ml of the upper phase into 100 ml of water, and titrate to a methyl red endpoint with 1 *N* HCl. On the basis of this titration, adjust the remainder of the upper phase to exactly 3.9 *N* with *iso*butyl alcohol. (This solvent is best for mixtures of blue and green dyes.)

d. *Isoamyl alcohol-ethyl alcohol-ammonium hydroxide-water*—4:4:1:2. Mix these four liquids in the volumetric ratios indicated, and disperse by shaking. (This solvent is best for mixtures of red dyes.)

e. *Aqueous ammonia solution*—1%. Dilute 41 ml of NH_4OH to 1 liter with water.

PREPARATION OF SAMPLE

(The sample should contain about 1 mg of each color present. Normally a 50 g food sample will provide adequate color for subsequent determination.)

a. *Dry gelatin desserts*—Shake the sample with 100 ml of alcohol, and decant with suction through a Büchner funnel. Repeat the extraction until most of the color is removed. Then shake the sample with 100 ml of 1% aqueous NH_3 solution and rapidly filter with suction. Combine the filtrates and make strongly acid to pH indicator paper with acetic acid.

b. *Water-soluble foods* (*including candy, sodas and ices*)—Dissolve solid samples in minimum volumes of 1% NH_3 solution, **e**; use liquid samples undiluted. Filter and make strongly acid with acetic acid.

c. *Oil-soluble foods*—Dissolve in a minimum volume of ether, and extract with 1% NH_3 solution without filtering; acidify the aqueous extract with acetic acid.

d. *Foods insoluble in water and fat solvents*—Cover meat samples with 80% alcohol containing 1% NH_3, grind in a food blender and filter with suction, using Celite Analytical Filter Aid (Fisher No. C-211). Repeat until no more color is extracted.

Treat alimentary pastes and bakery goods in

the same way except to heat the ammoniacal alcohol to boiling prior to blending.

In either case, extract the alcoholic solution with successive portions of *n*-pentane until no further color is extracted. Retain the pentane extract for testing for oil-soluble dyes. Dilute the alcoholic extract with water to about 50% alcohol and acidify with acetic acid.

Separation and Identification of Oil-Soluble Colors

Synthetic oil-soluble food colors are no longer permitted in the United States except for Citrus Red 2 on oranges and Orange B in frankfort and sausage casings, but if a colored pentane extract is obtained in (**d**) the color (or colors) present may be identified by the method of Silk: *J. Assoc. Offic. Agr. Chemists*, **42**:427 (1959).

Preliminary Qualitative Test for Artificial Color

Boil a portion of the prepared sample solution to remove alcohol. Immerse a 2 × 2 cm square of white, defatted wool cloth in this solution and heat 15 minutes on the steam bath. Remove the cloth, rinse with water, cover with 1% aqueous NH_3 solution and heat 15 minutes on the steam bath. Discard the wool, acidify the solution with acetic acid to about pH 2, add a new piece of white wool cloth, and heat 15 minutes on the steam bath. Remove the cloth and rinse well with water. If the cloth is colored, an artificial dye is present. Lack of color indicates absence of acidic artificial dyes.

Column Chromatography

a. *Preparation of column*—Insert a glass-wool plug into the chromatographic tube, fill with the activated alumina to a depth of 150 mm and cover with another glass-wool plug. Wash the packing with HCl (1+9) to remove fines and compact and acidify the column. Pressure may be used to hasten the process.

b. *Isolation of the color*—Introduce the balance of the prepared sample solution into the column and apply pressure or vacuum. Wash the column with an equal volume of water and elute the adsorbed dye with NH_4OH solution (1+9). Two or three column volumes are necessary to elute all of the color, but the total volume should be kept as small as possible.

Separation of the Colors by Paper Chromatography

Separate the colors by paper chromatography of the eluent from the Al_2O_3 column, using the ascending-solvent method. Make three separate paper chromatograms, using the three solvent systems, Reagents (**b**), (**c**) and (**d**). Employ glass-stoppered graduated cylinders so that the chromatograms can be run simultaneously. Chromatograph at room temperature volumes of column-eluate equivalent to about 25 μg of dye on 1 × 12″ strips of Whatman No. 1 or No. 31ET paper. Either fold the strips longitudinally to give them sufficient rigidity to stand alone, or secure them at the top by the stopper, with the end of the paper protruding from the cylinder. (This latter method provides better resolution of colors of low *Rf* value.)

Identification of Colors

Upon removal of the paper strips from the cylinders, note the positions and colors of the spots. When the chromatograms are dry, calculate the *Rf* value for each spot and note the color changes on exposure to HCl and NH_3 fumes. Identify the dyes by comparison with the *Rf* values and color changes of known dyes.

The *Rf* values in Table 18-1 indicate the relative positions of the colored spots, but these values are affected by so many variables that they serve only as a comparative guide. It is desirable to provide an internal standard by adding a trace of known dye (such as erythrosine) to an aliquot of the sample before spotting; the known *Rf* of this dye will facilitate precise identification of the other colors.

Quantitative Determination

Cut the spot from the chromatogram, place in a small beaker, cover with 1–2 ml of 60% alcohol and warm on a steam bath until the paper is colorless. Transfer the cooled solution to a cuvette and compare the transmission with that of a solution of the dye of known concentration. Cells of 2 and 5 cm length and 4 mm bore are satisfactory.

Notes

1. Table 18-2 indicates the percentage of the various dyes present in foods that are recoverable by the above methods.

2. AOAC Method **13.133** gives the following method of sample preparation for egg noodles for the detection of tartrazine (FD & C Yellow No. 5), which is the color usually employed to simulate the presence of egg:

Place 800 ml of cold water and 5 ml of NH_4OH in a liter Erlenmeyer flask and add 200 g of unground sample. Stopper the flask and shake at intervals; 3–4 hours is usually enough time to disintegrate the material. Use a glass rod to dislodge material caking on the bottom. Centrifuge and decant the clear supernatant liquid into a liter flask. Add a solution of 50 g of $MgSO_4 \cdot 7H_2O$ in 100 ml of water, 10 ml of 12% silicotungstic acid solution, and 10 ml of HCl; shake well, and let stand one hour. (This treatment will precipitate almost all of the protein.)

Centrifuge, decant the clear solution into a

TABLE 18-1

Identifying Characteristics of Food Colors

Figures in parentheses indicate secondary (weak) spots which may not be detectable

FDC Designation	Pyridine–Ethyl Acetate *Rf*	IBA–Ammonia *Rf*	IAA–Ethyl Alcohol–Ammonia *Rf*	Neutral Color	Color after exposure to: HCl fumes	NH$_3$ fumes
Red 1	0.45	0.18	0.55 (0.62)	Rose pink	Rose pink	Rose pink
Red 2	0.06	0.01	0.24	Plum	Plum	Plum
Red 3	0.88	0.37	0.70	Pink	Orange	Pink
Red 4	0.50	0.04	0.36	Rose pink	Red	Orange
Yellow 1	0.59	0.22	0.59	Yellow	Decolorizes	Yellow
Yellow 5	0.08	0.00	0.17	Yellow	Yellow	Yellow
Yellow 6	0.43	0.10	0.52 (0.17)	Orange	Orange	Orange
Orange 1	0.71	0.23	0.60	Orange	Purple	Dark red
	(0.58)	(0.27)	(0.69)	Orange-red	Orange	Pale red
Violet 1	0.54 (0.50) (0.58)	0.38[a] (0.21)[a]	0.76 (0.94)	Violet	Yellow or decolorizes	Blue
	(0.72)			Blue	Decolorizes	Blue
Blue 1	0.34 (0.43) (0.53)	0.20	0.59 (0.70)	Blue	Yellow	Blue
Blue 2	0.26 (0.38)	Disappears	0.30[b] (partially decomposes)	Blue	Blue	Blue
Green 1	0.59 (0.79)	0.38[c]	0.75 (0.92)	Aqua green	Orange or decolorizes	Decolorizes[d]
Green 2	0.40[c] (0.54)[c] (0.58)[c]	0.17[c] (0.12)[c] (0.29)[c] (0.40)[c]	0.68 (0.73) (0.78)	Aqua green	Orange or decolorizes	Decolorizes[d]
Green 3	0.42 (0.00)	0.09 (0.00) (0.18)	0.46 (0.62) (0.68)	Aqua green	Orange or decolorizes	Deep blue

[a] Permanently changes to blue. [b] May or may not be detectable, depending upon original concentration of Blue 2. [c] Colorless until dry. [d] Aqua green after ammonia evaporates.

container, and proceed by Method 18-1 or 18-2.

Method 18-2. Mehod of Cox and Pearson.[9]

REAGENTS

a. *Solvent No.* 1—1 ml NH_4OH in 99 ml water.
b. *Solvent No.* 2—2.5% NaCl solution.
c. *Solvent No.* 3—2% NaCl in 50% alcohol.
d. *Solvent No.* 4—*Iso*butanol-alcohol-water (1+2+1).
e. *Solvent No.* 5—*n*-Butanol-water-acetic acid (20+12+5).
f. *Solvent No.* 6—Mix 3 volumes of *iso*butanol with 2 volumes each of alcohol and water; to 99 ml of this mixture add 1 ml of NH_4OH.
g. *Solvent No.* 7—Mix 80 g of phenol with 20 g of water.

PRELIMINARY TREATMENT OF THE FOOD

a. *Non-alcoholic beverages*—As most of these are acidic, they can usually be treated directly with wool. If they are not acid, acidify slightly with acetic acid or $KHSO_4$.

b. *Alcoholic beverages*—Boil to remove the alcohol, and acidify (if necessary) as in (**a**).

c. *Soluble foods* (*e.g., jams, confectionery, icings*)—Dissolve in 30 ml of water and treat as in (**a**).

d. *Starch-based foods*—Grind 10 g of sample very thoroughly with 50 ml of 2% NH_3 in 70% alcohol (700 ml alcohol + 78 ml NH_4OH + 172 ml water). Allow to stand several hours and centrifuge. Pour the separated liquid into an evaporating dish and evaporate on the steam bath. Take up the residue in 30 ml of acidified water and treat as in (**a**).

e. *Candied fruits*—Treat as in (**d**).

f. *Products with a high fat content* (*e.g., sausages and other meat and fish products*)—De-fat with petroleum ether and obtain the color in aqueous solution by treatment with hot water and acidification as in (**a**). *Note that oil-soluble colors will go into and color the petroleum ether.*

g. *Difficult cases*—Treatment with warm, 50–90% acetone or alcohol (which precipitates starch) containing 2% NH_3 is often helpful. Remove the organic solvent before acidifying [*see* (**d**)]. Note that basic colors dye wool in alkaline solution and therefore require treatment with acid instead of NH_3.

EXTRACTION OF THE COLOR FROM THE FOOD

Add a 20 cm strip of white knitting wool (purified by boiling first in very dilute NaOH solution and then in water) to 35 ml of the prepared slightly acidified extract of the sample, and boil. Continue boiling, wash the wool under

TABLE 18-2
Comparison of Recovery Values

FDC Designation	Pyridine–Ethyl Acetate, % Recovery		IBA–Ammonia, % Recovery		Amyl–Ethyl Alcohols, % Recovery	
	Elution	Reflectance	Elution	Reflectance	Elution	Reflectance
Red 1	68	90	82	105[a]		
Red 2	71[a]	80[a]	69	87[a]	97	100
Red 3	62	72	77	76	71	68
Red 4	86	90	102	100	72	85
Yellow 1	60[a]	88[a]	90[a]	94[a]	68[a]	63[a]
Yellow 5	66[a]	100[a]	70[a]	81[a]	83	100
Yellow 6	59[a]	64[a]	75	100	—	—
Orange 1	43	46	—	—	—	—
Violet 1	36	80	—	—	—	—
Blue 1	—	—	69	77	—	—
Blue 2	72	87	—	—	—	—
Green 1	100[a]	89[a]	—	—	—	—
Green 2	—	—	54	76	—	—
Green 3	—	—	62	84	57	71

[a] Estimated from spectral curves of mixed spots.

the cold water faucet, transfer it to a small beaker, and boil gently with dilute NH_4OH. If the color is stripped by the alkali, the presence of an acidic coal-tar dye is indicated. Remove the wool, make the liquid slightly acid, and boil. Add a little wool and continue boiling until all color is taken up. Extract the dye from the wool again with a small volume of dilute NH_4OH, filter through a small plug of cotton wool, and evaporate to a low volume. (This double stripping technique usually gives a purer product, but it is not always necessary. Natural colors may dye the wool during the first treatment, but the color is not usually removed by NH_4OH.)

Basic dyes can be separated by making the food alkaline with NH_4OH, boiling with wool and stripping with acetic acid. Because all of the permitted water-soluble colors (on both the U.S. and British lists) are acidic, an indication of the presence of a basic dye suggests that a non-permitted color is present. (Basic rhodamine dyes were once quite frequently used.)

Separation and Identification of the Extracted Colors

a. *General tests*—While it is always advisable to run chromatograms for final detection of the color, time may often be saved by applying simple qualitative tests to fragments of the dyed wool and comparing the results with those given by known dyes.* For instance, HCl, H_2SO_4 and NH_4OH may produce characteristic color changes. (*See* Selected References L and S.) Some xanthene dyes are extracted by ether from aqueous solution. It is sometimes possible to detect the presence of mixed colors by spotting on filter paper and adding a few drops of water so that separate zones appear.

b. *Chromatographic separations*—Paint a horizontal line of the prepared dye solution about 2 cm from the bottom edge of a piece of Whatman No. 1 chromatographic paper, and dry. Then elute upwards in a covered beaker (or see Method 16-44, "Paper chromatography"), using the most effective solvent to effect a separation. (Solvent No. 5 is useful for general purposes, but it is advisable to study the *R*f values in Table 18-3 first in relation to the probable colors present.) If more than one line of color appears to be present, cut out the separate lines, extract with water or aqueous acetone, evaporate to dryness, and redissolve in a few drops of water. If the separation is indefinite, repeat the process using different solvents.

c. *Identification of the separated colors*—Cut a piece of Whatman No. 1 chromatographic paper so that it can be rolled into a cylinder which fits inside a liter beaker without touching the sides. Draw a line parallel to the bottom of the paper and about 2 cm away from it. Place a spot of the concentrated solution of the unknown dye on the line, together with a series of spots (about 2 cm apart) of aqueous solutions of known dyes of similar color. Dry. Roll the paper into a cylinder and fasten the top corners together with a paper clip so that the vertical edges do not touch. Place the rolled paper into the beaker without touching the sides, and add a 1 cm layer of solvent selected on the basis of the *R*f values in Table 18-3. Cover with a watch-glass and allow the solvent front to rise 10–15 cm. Identify the color by comparison with the spots obtained from the known dyes. Run similar chromatograms using different solvents, chosen in relation to their ability to differentiate between dyes having similar *R*f values on the first run. (*See also* Selected References O and Q). The appearance of the spots under ultraviolet light may be useful for identification purposes.

d. *Confirmation of identity of the dyes with the spectrophotometer*—Dilute the pure neutral dye solution to a suitable color intensity, determine its absorption curve in neutral, acid and alkaline solution, and compare the maxima with those obtained with the known dye. (*See* Table 18-3.) [Values in the table were obtained in neutral (0.02% NH_4 acetate), acid (N/10 HCl) and alkaline (N/10 NaOH) solution.]

If the dye appears to be a non-permitted one, it is particularly important to apply chemical tests. (*See* Selected References I, L and S.)

Oil Soluble Colors

Citrus Red No. 2 and Orange B are the only oil-soluble dyes on the permitted list for the U.S., and these may be used only on oranges and sausages respectively. The British list contains Oil Yellow GG and Oil Yellow XP. Oil Yellow GG can be extracted by NaOH from its ethereal

* Sets of the permitted British colors and many non-permitted ones may be obtained from Solmedia Ltd., London, E. 17, England.

solution, but XP and many of the non-permitted colors cannot. For the isolation of the purified color, the oil should first be extracted from the food with petroleum ether. Then the color can be removed with fairly concentrated acid[10] or by solvent partition as follows:

Dissolve 10 g of oil in 50 ml of petroleum ether and transfer to a separatory funnel. Extract with three 20 ml portions of N,N-dimethylformamide (DMF) and discard the petroleum ether layer. Extract the combined DMF solutions with four 25 ml portions of petroleum ether, back-extracting each time with 5 ml of DMF. Discard the petroleum ether extracts. Dilute the combined DMF solutions with an equal volume of water and extract with 30+10 ml portions of $CHCl_3$. Centrifuge and discard the aqueous DMF layer. Wash the combined $CHCl_3$ extracts with water

TABLE 18-3

_Rf_ Values and Spectrophotometric Peaks of Permitted Water-Soluble Colors[a]

	Class	*Rf* Values						Main Spectrophotometric Peaks
		Solvent No.:						
		1	2	3	4	5	6	mμ
Reds								
Ponçeau MX	Azo	0.47	0.05	0.31	0.40	0.53	0.69	500 in acid
Ponçeau 4R	Azo	0.95	0.36	0.42	0.29	0.33	0.29	505 in acid
Carmoisine	Azo	0.61	0.04	0.56	0.51	0.56	0.28	515 in acid
Amaranth	Azo	0.77	0.06	0.20	0.24	0.19	0.20	520 in acid
Red 10B	Azo	0.59	0.03	0.27	0.32	0.36	0.31	525 in acid
Erythrosine BS	Xanthene	0.23	0.00	0.70	1.00	1.00	0.68	525 in alkali
Red 2G	Azo	0.85	0.17	0.36	0.35	0.45	0.39	530 in acid
Red 6B	Azo	0.66	0.04	0.18	0.27	0.25	0.23	515 in acid
Red FB	Azo	0.10	0.00	0.06	0.32	0.38	0.15	385 and 495 in acid
Ponçeau SX	Azo	0.76	0.04	0.42	0.44	0.56	0.40	500 in acid
Ponçeau 3R	Azo	0.49	0.03	0.32	0.39	0.59	0.50	505 in acid
Fast Red E	Azo	0.45	0.05	0.60	0.45	0.54	0.60	505 in acid
Orange								
Orange G	Azo	1.00	0.59	0.70	0.53	0.52	0.58	475 in acid
Orange RN	Azo	0.33	0.05	0.82	0.84	0.68	0.78	485 in acid
Sunset Yellow FCF	Azo	0.78	0.26	0.65	0.49	0.45	0.56	480 in acid
Yellows[b]								
Tartrazine	Pyrazolone	1.00	0.26	0.30	0.26	0.18	0.22	430 in acid
Naphthol Yellow S	Nitro	0.78	0.31	0.68	0.52	0.53	0.56	370 in acid
Yellow 2G	Pyrazolone	1.00	0.83	1.00	0.74	0.52	0.48	400 in acid
Yellow RFS	Azo	0.94	0.60	0.74	0.47	0.40	0.48	510 in acid
Yellow RY	Azo	0.85	0.21	0.17	0.19	0.11	0.04	435 in acid
Greens, Blues & Violets								
Green S	Triphenylmethane	1.00	0.88	1.00	0.73	0.57	0.50	635 in acid
Blue VRS	Triphenylmethane	1.00	0.85	1.00	0.84	0.62	0.75	415 in acid
Indigo Carmine	Indigoid		0.07	0.30	0.28	0.21	0.27	285 and 615 in acid
Violet BNP	Triphenylmethane	0.40	0.06	1.00	0.80	0.59	0.80	590 in neutral
Browns, Blacks								
Brown FK	Azo	0.24	0.00	0.56	0.45	0.50	0.56	455 in acid
Chocolate Brown FB	Disazo							470 in acid
Chocolate Brown HT	Disazo							465 in acid
Black PN	Disazo	0.31	0.00	0.06	0.10	0.07	0.09	570 in acid

[a] "Separation and Identification of Food Colors Permitted by The Coloring Matters in Food Regulations, 1957," Association of Public Analysts (London), 1960 (see also *J. Chromatography*, 1960, 3, D6–D13).

[b] The permitted list also includes 2 oil-soluble colors, viz. Oil Yellow GG (azo) and Oil Yellow XP (pyrazolone).

to remove dissolved DMF. Evaporate the $CHCl_3$ extract to dryness at room temperature under reduced pressure. Dissolve the residue in 25 ml of DMF, transfer to a separatory funnel, add 25 ml of water, and extract with four 25 ml portions of petroleum ether. Discard the aqueous DMF layer, wash the combined petroleum ether solutions with water to remove the dissolved DMF, and evaporate the petroleum ether solution under reduced pressure at room temperature.

The oil-soluble colors can be separated by reverse-phase paper chromatography. Previously soak the papers in a 10% solution of heavy mineral oil in mixed ethers, drain them and dry under vacuum. Spot with the color dissolved in a suitable solvent (ether, mixed ethers or alcohol). Develop with solvents such as acetone-alcohol (70+30) or dioxane-water (60+40). [*See* Silk: *J. Assoc. Offic. Agr. Chemists*, **42**:427 (1959).]

Oil-soluble colors can also be identified by dissolving the extracted dye in $CHCl_3$ or acetone and obtaining the absorption curve with a recording spectrophotometer. The absorption peaks for the colors on the British permitted list are: Oil Yellow GG, 380 mμ; Oil Yellow XP, 405 mμ.

NOTES

1. It should be borne in mind that the dyeing and stripping process sometimes causes some decomposition of colors and consequently affects the chromatograms and spectrophotometric curves that are obtained.

2. This method has been employed in the laboratory of one of the present authors.

Method 18-3. *Natural Coloring Matters. (AOAC Method* ***35.013-35.015****).*

SEPARATION

a. *By extraction with ether from neutral solutions*—From neutral solutions ether extracts carotene, xanthophyll, coloring matter of tomatoes and paprika, and chlorophyll. The coloring matter remains in the ether solution on shaking with 1 *N* NaOH or 1 *N* HCl. No apparent change takes place, although the substance may be chemically altered more or less by this treatment. Green chlorophyll solutions have a red fluorescence.

b. *By extraction with ether from acid solutions*—From slightly acid solutions ether very readily and completely extracts the coloring matter of alkanet, annatto, turmeric and red dyewoods, sandalwood, camwood and barwood. It extracts in large proportion the flavine coloring matters of fustic, Persian berries and quercitron (after hydrolysis), as well as coloring matter of Brazilwood and green derivatives formed from chlorophyll by alkaline treatment. It extracts in relatively small quantity the coloring matters of logwood, orchil, saffron and cochineal. Coloring matters of this group are readily removed from ether by shaking with alkali solutions, but in most cases they rapidly undergo chemical change.

c. *By extraction with amyl alcohol from acid solutions*—From slightly acid solutions amyl alcohol extracts the major part of the coloring matters of logwood, orchil, saffron and cochineal. Amyl alcohol extracts relatively small proportions of caramel, and the anthocyans which constitute the red coloring matter of most fruits.

IDENTIFICATION

A. *By color changes produced with various reagents.*

Evaporate the ether solutions obtained under (**a**) and (**b**) to dryness, warm the residue with a little alcohol and dilute with water. Dilute the amyl alcohol extract of (**c**) with petroleum ether and extract with water. To portions of these somewhat purified solutions apply reagents as follows:

i. *Hydrochloric acid*—Add to the solution, first one or two drops of HCl and then 3–4 volumes of this acid.

ii. *Sodium or potassium hydroxide*—Make the solution slightly alkaline by adding one drop of 10% NaOH or KOH. (Warning: A red color changing to yellow, especially on warming, may indicate a gallate antioxidant.)

iii. *Sodium hyposulfite*—Add a small crystal of $Na_2S_2O_4$.

iv. *Ferric chloride*—Add a small quantity of freshly-prepared 0.5% $FeCl_3.6H_2O$ solution very carefully, one drop at a time, since colors are not always obtained when an excess is used.

v. *Alum*—Add to the test solution one-fifth its volume of 10% solution of $KAl(SO_4)_2.12H_2O$ or the NH_4 salt.

vi. *Uranium acetate*—Add 5% $UO_2(C_2H_3O_2)_2.2H_2O$ solution drop by drop.

vii. *Sulfuric acid on dry color*—Evaporate a small quantity of the color solution in a porcelain dish. Cool thoroughly and treat the dry residue with 1-2 drops of cold H_2SO_4. The colors formed are sometimes extremely transitory, and observable only the instant the acid wets the residue.

Table 18-4 (p. 450–451) shows the behavior of certain natural coloring matters when treated with the above reagents:

B. *By special tests.*

i. *Chlorophyll*—The "brown phase reaction" may be useful to characterize chlorophyll when it has not been previously treated with alkalies. Treat a green ether or petroleum ether solution of the coloring matter with a small quantity of 10% KOH in methyl alcohol. If the color turns brown and quickly reverts to green, chlorophyll is indicated.

ii. *Annatto*—Pour on a moistened filter an alkaline solution of a color obtained by shaking an oil or melted and filtered fat with warm 2% NaOH. If annatto is present, the paper will absorb the color and remain a straw color after washing with a gentle stream of water. Dry the paper, add a drop of 40% $SnCl_2$ solution, and again dry very carefully. If the paper turns purple, annatto is present.

iii. *Turmeric*—Treat an aqueous or dilute alcoholic solution of the color with HCl until it just begins to turn slightly orange. Divide the mixture into two parts, and add H_3BO_3 (powder or crystals) to one part. If turmeric is present, the portion to which H_3BO_3 has been added becomes markedly redder.

This test may also be performed by dipping a piece of filter paper in the alcoholic solution, drying at 100° and then moistening with a weak H_3BO_3 solution to which a few drops of HCl have been added. On redrying, the paper turns cherry-red if turmeric is present.

iv. *Cochineal*—When cochineal is suspected, acidify the solution with one-third its volume of HCl and shake with amyl alcohol. Wash the amyl alcohol solution 2–4 times with equal volumes of water to remove HCl, then dilute with 1–2 volumes of petroleum ether and shake with a few small volumes of water to remove the color. Divide the combined aqueous extract into two portions and treat as follows:

1. To one portion add, drop by drop, 5% $UO_2(C_2H_3O_2)_2.2H_2O$ solution, shaking thoroughly after each addition. In the presence of cochineal, the solution turns a characteristic emerald green. Uranyl salts do not turn green in the presence of much free acid, so in making the test if a little Na acetate is not added before the U solution, a larger quantity of the U solution is required.

2. To the second portion add one or two drops of NH_4OH; in the presence of cochineal the solution turns violet. (Many fruit colors give nearly identical reactions.) Cochineal differs from orchil in that it is not decolorized by $Na_2S_2O_4$ in acid, neutral or alkaline solution.

v. *Orchil*—This coloring matter may be sulfonated or unsulfonated. Unsulfonated orchil is readily extracted by amyl alcohol from weak acid solution, while extraction of the sulfonated color is incomplete even from a strongly acidified solution. Behavior of the color towards acids and alkalies is similar to that of cochineal, *e.g.*, HCl produces a yellow shade and alkalies a bluish. $Na_2S_2O_4$ reduces orchil, but the color is restored by air oxidation. A characteristic property of orchil is to dye, strip and redye wool readily.

vi. *Caramel*—*See* Method 2-17.

Method 18-4. *Artificial Color in Vegetable Oils.*[11]

APPARATUS

a. *Chromatographic tubes*—22 × 330 mm with coarse fritted glass plate and stopcock.

REAGENTS

a. *Florisil*—Heat at 650° for 2 hours and keep in a closed container in a 130° oven.

b. *Adsorbent for Column* 1—Heat 100–200 g of adsorption alumina (Fisher No. A-540) 1 hour at 400°. Store in a tightly-stoppered bottle in a desiccator.

c. *Adsorbent for Column* 2—Mix equal weights of Sea Sorb 43 (Fisher No. S-120) and Celite 545 (Fisher No. C-212).

d. *Adsorbent for silicic acid column*—Mix equal weights of silicic acid 100 mesh for chromatographic analysis (Mallinckrodt No. 2847) and Celite 545.

e. *Acetonitrile*—Add 1 ml of H_3PO_4 and 30 g of P_2O_5 to a distillation flask containing about 4 liters of acetonitrile, distil, and collect the distillate boiling 81–82°. Saturate with petroleum ether.

PREPARATION OF COLUMNS

a. *Florisil column*—Fill a chromatographic tube with hot Florisil from the 130° oven to a height of 5 inches. Tap, and add 0.5 inch of anhydrous Na_2SO_4. Wash with petroleum ether.

b. *Column 1*—Close the stopcock of a tube. First add 50 ml of petroleum ether, and then sufficient Celite 545 to form a half-inch layer. Add 20 g of the alumina, (**b**), and work a large glass tube or plunger up and down through the alumina to break up lumps and remove air bubbles. Drain the liquid to the level of the alumina.

c. *Column 2*—Prepare in the same manner as Column 1, (**b**), except to use 10 g of the Sea Sorb 43-Celite 545 adsorbent, (**c**), and to compress the adsorbent with slight air pressure.

d. *Silicic acid column*—Add about 4 inches of the silicic acid column adsorbent, (**d**), to a chromatographic column using suction. Gently tap and smooth the surface of the adsorbent with a large glass rod. Pre-wash the column with petroleum ether, using pressure.

DETERMINATION

To 50 ml of an oil sample, add 50 ml of petroleum ether and transfer to the Florisil column. Discard the first (colorless) eluate and collect the colored portion. Continue eluting until no more color is removed. Change receivers and elute with ethyl ether until the eluate is colorless. Change receivers and elute with alcohol-ether (1+3). This removes D & C Violet No. 1 and D & C Yellow No. 11. If a red color remains on the column, elute with sufficient acetonitrile to remove D & C Red No. 35. Evaporate the alcohol-ether and acetonitrile extracts to dryness. Dissolve the residues in $CHCl_3$, dilute to a suitable concentration, record the absorption curves and compare with those of known colors.

Evaporate the ethyl ether extract to dryness, add the petroleum ether extract to this residue, and transfer to Column 1. When all of the solution has entered the alumina, elute with 100 ml of petroleum ether. Discard the eluate, which will contain the color natural to the oil. Wash with two ml portions of $CHCl_3$. If the eluate is green (D & C Green No. 6), continue elution with $CHCl_3$ until all green is removed and reserve the eluate. If it is not green, discard it.

Elute with alcohol-$CHCl_3$ (1+3) until the eluate is colorless. Evaporate to dryness, dissolve in petroleum ether, and add to Column 2. When the solution just passes into the adsorbent, wash the column with 50 ml of petroleum ether and two 10 ml portions of $CHCl_3$. If a trace of green is seen, add to the $CHCl_3$ eluate from Column 1, make to a suitable volume, and determine the spectrophotometric curve to identify D & C Green No. 6.

If D & C Orange No. 4 and/or Ext. D & C Red No. 14 are present they will be eluted with $CHCl_3$ immediately after the D & C Green No. 6. These two dyes cannot be separated from each other, but they can be separated from D & C Green No. 6, and the receiver should be changed as soon as the eluate changes from green to orange-red. Continue to extract with $CHCl_3$ until no more color is removed.

Change receivers and elute with alcohol-$CHCl_3$ (1+3), adding a new receiver whenever the color changes. Several colors may come out at this point. Separately evaporate each color solution, dissolve in $CHCl_3$, and determine its absorption curve.

The silicic acid column will remove the natural color of the oil, and some of the artificial colors can be separated by this column. To use it for this purpose, proceed as follows:

Evaporate a $CHCl_3$ solution of mixed colors obtained from one of the previous columns to dryness, dissolve the residue in petroleum ether, transfer to the silicic acid column, and apply pressure until the solution enters the adsorbent. Then wash with 25 ml of petroleum ether, and elute with petroleum ether-benzene (1+1), collecting each color separately. When the eluate is colorless or the color no longer moves down the column, change to benzene and continue to collect fractions until all color is removed. Petroleum ether-benzene (1+1) will first elute Ext. D & C Yellows Nos. 9 and 10, followed by Ext. D & C Orange No. 4 and Ext. D & C Red No. 14. Benzene will first elute D & C Green No. 6, followed in turn by D & C Reds Nos. 17 and 18 and Ext. D & C Blue No. 5.

An alternative method that gives satisfactory results for some of the colors, and better results for others, is the following:

To 50 ml of oil in a separatory funnel add 150 ml of petroleum ether, and extract with six 50 ml portions of acetonitrile. Prepare Column

TABLE 18-4
Reaction of certain natural coloring matters to common reagents

COLORING MATTER	HYDROCHLORIC ACID	10% SODIUM HYDROXIDE SOLUTION	SODIUM HYPOSULFITE	0.5% FERRIC CHLORIDE SOLUTION	10% ALUM SOLUTION	5% URANIUM ACETATE SOLUTION	SULFURIC ACID ON DRY COLOR
Logwood	Deep red with excess of acid	Violet to violet-blue	Almost decolorized, color returning imperfectly by reoxidation	Dark shades of violet, brown, or black (the first hue often evanescent)	Rose-red (change rather slow)	Violet, quickly fading	Red, changing to yellow
Red woods (brazilwood, sandalwood, camwood, and barwood)	Deep red with excess of acid	Violet-red		Dark shades of violet, brown, or black (the first hue often evanescent)	Rose-red (change rather slow)		
Anthocyans of red fruit colors		Change to green, dull blue, or slate color, usually very quickly becoming browner by oxidation	Anthocyanidins derived by hydrolysis, almost completely decolorized				
Alkanet		Deep blue				Yellowish green	Violet-blue
Orchil	Little or no change	Blue	Decolorized, color returning when shaken with air. Reaction more easily seen in alk. soln				Violet-blue
Cochineal	Little or no change	Violet	No marked change	Slightly darker		Green	
Annatto	Remains orange. Little change		Little affected	No marked change. Perhaps somewhat browner			Blue
Turmeric (soln in ether or alcohol characterized by pure yellow color and light green fluorescence)	Orange-red or carmine-red on addn of several vols of concd acid	Orange-brown	Little affected	No marked change, Perhaps somewhat browner	Little change	Somewhat browner	Red

TABLE 18-4 (continued)

COLORING MATTER	HYDROCHLORIC ACID	10% SODIUM HYDROXIDE SOLUTION	SODIUM HYPOSULFITE	0.5% FERRIC CHLORIDE SOLUTION	10% ALUM SOLUTION	5% URANIUM ACETATE SOLUTION	SULFURIC ACID ON DRY COLOR
Flavone colors of fustic, Persian berries, quercitron, etc.	Becomes intensely yellow with 2–4 vols concd acid	Bright yellow	Little affected	Olive-green or black colorations	More strongly yellow; fustic, developing green fluorescence	Orange colorations	Yellow to orange
Saffron	Little or no change	Remains yellow	Little affected	No marked change. Perhaps somewhat browner	Little change	Not affected	Blue
Carotene and xanthophyll	Little change. Perhaps slightly paler	Little or no change	Little affected	....	....	....	Blue, reaction obtained with difficulty
Chlorophyll	More brownish	'Brown phase reaction"*	....	....	....	....	....
Caramel	Little or no change	Little change or slightly deeper brown	Slightly paler	No change	....	....	....

* AOAC *Methods of Analysis*, 10th ed., **35.015(a)**.

2 using acetonitrile instead of petroleum ether. Add all of the acetonitrile extract to the column, using pressure. Wash with 50 ml of acetonitrile, then with $CHCl_3$ and alcohol-$CHCl_3$ (1+5) as before, collecting color fractions as they come off the column.

This method is especially good for Ext. D & C Yellows Nos. 9 and 10, which are sometimes lost by the previous method. It is also good for Ext. D & C Orange No. 4, Ext. D & C Red No. 14, D & C Reds Nos. 17 and 35 and D & C Yellow No. 11.

NOTE: In examining vegetable oils, the analyst is frequently interested only in determining whether a sample is or is not artificially colored, because the addition of *any* color to a salad oil is considered as making it an imitation olive oil. At a time when Yellow AB (FD & C Yellow No. 3) and Yellow OB (FD & C Yellow No. 4) were almost the only dyes used for coloring such oils, their presence or absence could easily be established by shaking a solution of the oil in 1–2 volumes of gasoline with a 15% (by volume) solution of H_2SO_4 in H_3PO_4; if the acid layer turned red, one or both of these dyes was present.[12] Present-day shifts to the use of dyes not responding to this test have made it necessary to employ more elaborate procedures, such as those outlined above.

TEXT REFERENCES

1. Horowitz, W. (ed.): *Official Methods of Analysis of the Association of Official Agricultural Chemists*, 10th ed. Washington: Association of Official Agricultural Chemists (1965). 680.
2. *Inspection Operations Manual* 10–1–66, *Appendix "A"-63A, Color Additives List*. 1.0.
3. Ibid., 6.0–7.0.
4. Cox, H. E. and Pearson, D.: *The Chemical Analysis of Foods*, 1st American ed. New York: Chemical Publishing Co., Inc. (1962). 99.
5. Ref. 1, Method **35.018–35.020**.
6. Ibid., Method **35.010–35.011**; *see also* Selected Reference H.
7. *See* Selected Reference P.
8. Stanley, R. L. and Kirk, P. L.: *Agr. Food Chem.*, **11**:492 (1963).
9. Ref. 4, pp. 100–104.
10. Ref. 1, Method **35.008 (a)**.
11. Jones, F. B. (New York Station, U.S. Food and Drug Administration): Personal communication (March 1964).
12. Bailey, E. M. (ed.): *Official and Tentative Methods of Analysis of the Association of Official Agricultural Chemists*. 5th ed. Washington: Association of Official Agricultural Chemists (1940). 241.

SELECTED REFERENCES

A. *Colour Index*, 2nd ed. London: Society of Dyers and Colourists, American Association of Textile Chemists and Colorists. (1956–1963). I–IV, Index.

B. Bentley, K. W.: *The Natural Pigments*. New York: Interscience Publishers, Inc. (1960).

C. Furia, T. E. (ed.): *CRC Handbook of Food Additives*, Ch. I. Cleveland: Chemical Rubber Co. (1968).

D. Goodwin, T. W.: *The Comparative Biochemistry of the Carotenoids*. London: Chapman & Hall Ltd. (1952).

E. Freeman, K. A.: *Cosmetics and Color No. 15*. Washington: Food and Drug Administration, Division of Cosmetics (1950). (Contains absorption spectra of all permitted dyes).

F. Gautier, J. A. and Malangeau, P., (eds.): *Mises au Point de Chimie Analytique*, 13me série. Paris: Masson & Cie (1964). (Colored plates, absorption curves.)

G. Gore, T. S., Joshi, B. S., Sunthanker, S. V. and Tilak, B. D.: *Recent Progress in the Chemistry of Natural and Synthetic Colouring Matters and Related Fields*. New York–London: Academic Press (1962).

H. Graichen, C., Sclar, R. M., Ettelstein, N. and Freeman, K. A.: "Isolation and Separation of Coal-Tar Colors in Foods". In: *J. Assoc. Offic. Agr. Chemists*, **38**:792 (1955).

I. Green, A. G.: *The Analysis of Dyestuffs and their*

Identification in Dyed and Colored Materials, Lake Pigments, Foodstuffs, etc., 3rd ed. London: Griffin (1920).

J. Hais, I. M. and Macek, K.: *Handbuch der Papierchromatographie.* Jena: Gustav Fischer Verlag (1958). I, 677–682.

K. Karrer, P. and Jucker, E.: *Carotenoids.* New York: Elsevier Publishing Co. (1950).

L. Leach, A. E. and Winton, A. L.: *Food Inspection and Analysis*, 4th ed. New York: John Wiley & Sons, Inc. (1920). Chapter 17.

M. Palmer, L. S.: *Carotinoids and Related Pigments.* New York: Chemical Publishing Co., Inc. (1922).

N. Peacock, W. H.: "The Application Properties of the Certified 'Coal-Tar' Colorants." In: *American Cyanamid Co., Fine Chemicals Dept., Technical Bull. No.* **715**. (Collection of absorption curves).

O. Sclar, R. N. and Freeman, K. A.: "Chromatographic Procedures for the Separation of Water-Soluble Acid Dye Mixtures". In: *J. Assoc. Offic. Agr. Chemists*, **38**:796 (1955).

P. Stahl, E. (ed.): *Thin-Layer Chromatography, a Laboratory Handbook.* Berlin-Heidelberg-New York: Springer-Verlag (1965). 344–349.

Q. Tilden, D. H.: "Report on Paper Chromatography of Coal-Tar Colors." In: *J. Assoc. Offic. Agr. Chemists*, **35**:423 (1952). (Probably the first systematic study of the paper chromatography of food colors.)

R. Venkataraman, K.: *The Chemistry of Synthetic Dyes.* New York: Academic Press (1952–1970). I–III.

S. Woodman, A. G.: *Food Analysis*, 4th ed. New York: McGraw-Hill Book Co., Inc. (1941). 64–101.

CHAPTER 19

Pesticidal Residues

Until DDT was introduced in Europe by the Army during World War II as a dust applied to the body to kill lice and so prevent the spread of typhus, very few organic pesticides were employed for any purpose. Previously, the number of different pesticides applied to fruits, vegetables, and plants grown as animal feed such as alfalfa, was very small, including few materials beyond lead arsenate, sulphur, lime-sulphur solutions, alkaline copper suspensions (Bordeaux mixture), nicotine, mineral oil, pyrethrum and derris—a list that includes no synthetic organic compounds at all. Of these substances the only one likely to survive as a residue on food reaching the consumer, that might be toxic in the amounts conceivably present, was lead arsenate. Therefore, before World War II the analyst checking foods (particularly fruits) for spray residues could confine his attention to arsenic and lead determinations. It was even possible when running surveys to use only lead determinations for sorting purposes, due to the fact that when a $PbHAsO_4$ deposit weathered in the presence of moisture and atmospheric CO_2, it was converted to $PbCO_3$ and H_3AsO_4, resulting in an insoluble Pb salt that remained on the plant while the As was leached off as soluble arsenate. This meant that if the Pb content of a food did not exceed the established tolerance of 0.05 grain/lb (7 ppm), it was safe to assume that its As_2O_3 content met *its* tolerance of 0.025 grain/lb (3.5 ppm).

The present situation is vastly different. The total number of synthetic organic pesticides that has appeared on the market since World War II is in the hundreds, and a list including only those compounds likely to be encountered on foods would be quite extensive, as is shown by a recent article on a general method of residue-separation in which the authors tested 42 different substances.*

To properly cover the subject would take a whole book. We are therefore confining ourselves to listing a series of references, subdivided into the types of information they furnish.

A few notes may be added:

1. A new method usually appears first (in English) in one of the following journals: *Agricultural and Food Chemistry*, *The Analyst*, and the *Journal of the Association of Official Analytical Chemists*.

2. Until very recently, the official tolerance for all pesticides in milk was zero, which meant, theoretically, that not one molecule of any pesticide could exist in any quantity of

*McLeod, H. A., Mendoza, C., Wales, P., and McKinley, W. P.: *J Assoc. Offic. Anal. Chemists*, **50**:1216 (1967).

milk, however large. This regulation, obviously nonsensical if taken literally, was in practice interpreted to equate "zero" with "any detectable amount", but this interpretation became more and more unsatisfactory as the sensitivity of methods of detection increased with newer developments in gas chromatography. Eventually officialdom succumbed to realities, and on March 14, 1967 the U.S. Food and Drug Administration issued a press release announcing that a tolerance of 0.05 ppm had been set for DDT, DDD and DDE in milk. In *Guide Lines* of July 18, 1967 the following tolerances for other pesticides in milk and butter were announced: Aldrin, BHC, dieldrin, endrin, heptachlor and its epoxide, and lindane, 0.3, and methoxychlor, 1.5, expressed as ppm of the fat present. In a later press release (February 8, 1968), the general tolerance for DDT on fruits and vegetables of 7 ppm was reduced to 3.5 ppm—and to 1 ppm in the case of certain foods such as artichokes, berries, root vegetables, mushrooms, and peanuts.

3. Most of the methods for pesticide residues in foods require the use of large volumes of organic solvents to separate milligram (or microgram) quantities of pesticides from major interferences prior to their final separation and determination by gas chromatography. The ordinary reagent-grade solvents are not of requisite purity, and must be redistilled from glass and perhaps otherwise treated. This can be both a time-consuming and a hazardous chore when large numbers of samples are being routinely analyzed. Fortunately at least two manufacturers now offer grades of solvents especially purified for pesticide residue analysis, which while relatively expensive are probably well worth the price. These are: Burdick and Jackson Laboratories, Inc., Muskegon, Mich.; and Mallinckrodt Chemical Works, St. Louis, Mo.

SELECTED REFERENCES

LISTS OF PESTICIDES

A. Frear, D. E. H.: *A Catalogue of Insecticides and Fungicides*. Waltham: Chronica Botanica Co. (1947–1948). I–II.

B. Frear, D. E. H.: *Pesticide Handbook-Entoma*, 18th ed. State College: College Science Publishers (1966).

C. Martin, H. (ed.): *Guide to Chemicals Used in Crop Production*, 4th ed. Ottawa: Research Branch, Canada Dept. of Agriculture (1965).

D. Rollins, R. Z. (ed.): *Pesticide Chemicals Official Compendium*, 1966 ed. Topeka: Association of American Pesticide Control Officials (1966).

E. Thompson, W. S.: *Agricultural Chemicals*. Davis: Simmons Publishing Co. (1964–1967). I–III.

TOLERANCES AND USES

F. *NAC News and Pesticide Review*, **27**(3) (January-February 1969). Washington: National Agricultural Chemicals Association.

G. *Pesticide Chemical Regulations under the Federal Food, Drug and Cosmetic Act*. Food and Drug Administration, U.S. Dept. of Health, Education and Welfare (December 6, 1962 with subsequent revisions.)

H. "Suggested Guide for the Use of Insecticides to Control Insects Affecting Crops, etc." In: *Agriculture Handbook* 331. Washington: Agricultural Research Service and Forest Service (1967).

I. *USDA Summary of Registered Agricultural Pesticide Chemical Uses*, 2nd ed. (with supplements). Washington: U.S.D.A. Pesticides Regulation Division (1964–1970).

CHEMISTRY

J. Andus, L. J.: *The Physiology and Biochemistry of Herbicides*. London–New York: Academic Press (1964).

K. Anon.: *Pesticides in Tropical Agriculture*. Washington: American Chemical Society (1955).

L. Crafts. A. S.: *The Chemistry and Mode of Action of Herbicides*. New York–London: Interscience Publishing Co. (1961).

M. Frear, D. E. H.: *Chemistry of the Pesticides*, 3rd

ed. Toronto–New York–London: D. Van Nostrand Co., Inc. (1955).

N. Gnadinger, C. C.: *Pyrethrum Flowers*, 2nd ed. and Supplement 1936–1945. Minneapolis: McLaughlin, Gormley, King Co., Inc. (1936 and 1945).

O. Metcalf, R. L.: *Organic Insecticides, their Chemistry and Mode of Action*. New York: Interscience Publishing Co., Inc. (1955). (Still the best general non-analytical text.)

P. Thorn, G. D., and Ludwig, R. A.: *The Dithiocarbamates and Related Compounds*. Amsterdam–New York: Elsevier Publishing Co. (1962).

GENERAL METHODS OF ANALYSIS

Q. Bache, C. A. and Lisk, D. J.: *J. Assoc. Offic. Agr. Chemists*, **37**:1477 (1965); **38**:783, 1757 (1966); **39**:786, 1246 (1967).

R. Barry, H. C., *et al.*: *Pesticide Analytical Manual*. Washington: Food and Drug Administration (1963–1967). I–II.

S. Bowman, M. C. and Beroza, M.: *J. Assoc. Offic. Anal. Chemists*, **50**:1288 (1967).

T. Burchfield, H. P., Johnson, D. E. and Storrs, E. E.: *Guide to the Analysis of Pesticide Residues*. Washington: U.S. Public Health Service (1965). I–II.

U. McLeod, H. A., Mendoza, C., Wales, P., and McKinley, W. P.: *J. Assoc. Offic. Anal. Chemists*, **50**:1216 (1967).

V. Samuel, B. A.: *J. Assoc. Offic. Anal. Chemists*, **49**:346 (1966).

W. Wells, C. E.: *J. Assoc. Offic. Anal. Chemists*, **50**:1205 (1967).

SPECIAL METHODS OF ANALYSIS

X. *Manual of Methods for the Determination of Residues of Shell Pesticides*. New York: Shell Chemical Co. (1964).

Y. "Specifications and Methods of Analysis for Certain Pesticides." In: *Technical Bulletin* No. 1. London: H.M. Stationery Office (1958).

Z. Fisher, H. J.: *Conn. Agr. Exp. Sta. Bull.* **635**:56 (1960). (Method for DDT in milk.)

AA. Gunther, F. A. and Blinn, R. C.: *Analysis of Insecticides and Acaricides*. New York: Interscience Publishing Co., Inc. (1955).

BB. Horwitz, W. A. (ed.): *Official Methods of the Association of Official Agricultural Chemists*, 10th ed. Washington: Association of Official Analytical Chemists (1965).

CC. Zweig, G. (ed.): *Analytical Methods for Pesticides, Plant Growth Regulators and Food Additives*. New York: Academic Press (1963–1967). I–V.

REVIEWS

DD. *Bulletin of Environmental Contamination and Toxicology*. New York: Springer-Verlag New York, Inc. (1966–1969). I–IV.

EE. Gunther, F. A. (ed.): *Residue Reviews*. Berlin–Heidelberg–New York: Springer-Verlag New York, Inc. (1962–1970). I–XXXIV.

CHAPTER 20

Vitamins

It was noted in Chapter 6 that assays for vitamins A and D, thiamine, riboflavin and niacin were required in examining products such as Vitamins A and D Skimmed Milk and Vitamin Mineral Fortified Milk and that a high proportion of the ordinary milk now on the market was fortified with vitamin D only and, therefore, subject to biological assay for that vitamin. The food analyst may also be frequently called on to check the ascorbic acid content of fruit-flavored beverages, and special studies on the dietary properties of natural foods may require assays for practically any known vitamin.

Many of the recognized methods of vitamin assay make use of microbiological techniques, which depend in principle on inoculating a culture of a micro-organism that cannot grow in the absence of the vitamin to be tested for, with an extract of the sample and measuring the amount of growth that takes place. Such methods, because they are biological, would be beyond the scope of the present book in any case. Further, proper coverage of the special field of vitamin assay would require an amount of space that would result in unduly expanding the size of this book. We shall, therefore, confine ourselves to referring those interested in this type of analysis to the volumes listed under "Selected References", which cover this field in great detail.

Two of the vitamins, A and C, are not usually determined by biological methods. Assay for each of these is complicated by structural relationship to other compounds:

Vitamin A, itself, is a colorless primary alcohol of empirical formula $C_{20}H_{30}O$ (*see* Fig. 20–1), whose richest source is the livers of ocean fish. This vitamin does not occur as such in plants, but many vegetables and fodder materials like alfalfa, clover, barley, rye and wheat are relatively high in *carotene*. Carotene is a yellow hydrocarbon of empirical formula $C_{40}H_{56}$ that exists in several isomeric forms, of which the all-*trans*-beta variety is usually encountered in nature (*see* Fig. 20–2). While β-carotene is not itself a vitamin, oxidative

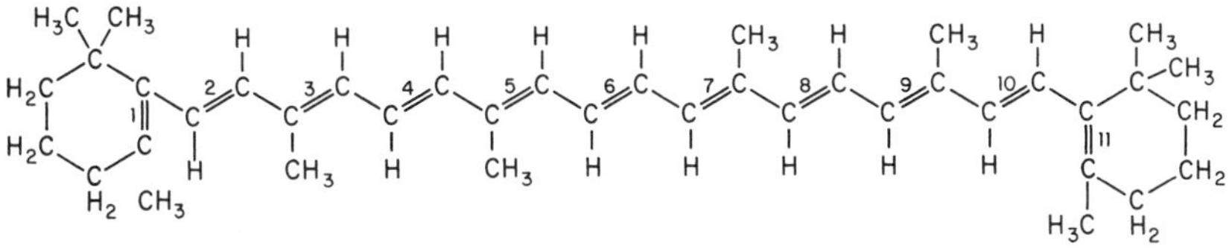

Fig. 20-1. Vitamin A.

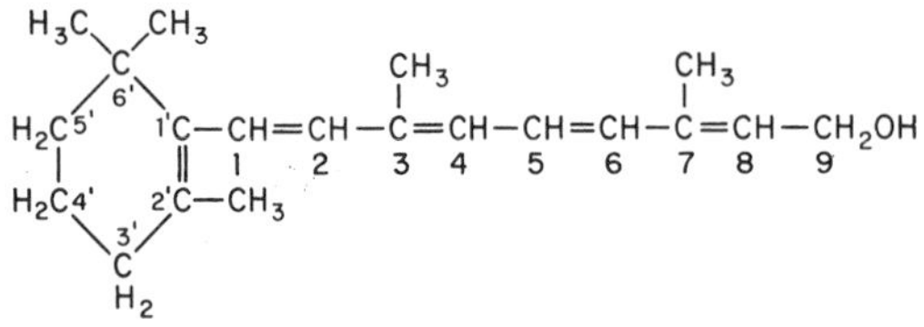

```
 H3C  CH3
    \ /
     C          CH3            CH3
   / 6' \        |              |
H2C5'    1'C—CH=CH—C=CH—CH=CH—C=CH—CH2OH
 |        ||  1   2   3  4   5   6   7  8    9
H2C4'    2'C—CH3
   \ 3' /
     C
     H2
```

Fig. 20-2. All-*trans*-β-carotene.

degradation in the intestines of animals converts each molecule of this compound into two molecules of vitamin A. Since the ultimate physiological effect of β-carotene is at least qualitatively the same as that of vitamin A itself, it is, therefore, necessary in assaying a food for its vitamin A potency to determine both vitamin A and carotene. For further details on this subject the reader is referred to Chapter 1 of Selected Reference G.

In the case of *vitamin C*, assay interferences result from the presence of both physiologically-active and physiologically-inert compounds. Both vitamin C proper (i.e., *l*-ascorbic acid) and the partially oxidized dehydroascorbic acid have vitamin properties, but only the reduced form is determined by the usual routine methods of titration with iodine or 2,6-dichlorophenolindophenol. Such methods are, nevertheless, usually adequate for the analysis of fruit juices, and even extracts of uncooked vegetables, because all of the vitamin C present is in the reduced form. Where there may be reason to believe that dehydroascorbic acid is present, or the matter is in doubt, recourse should be had to methods that will assay both forms of the vitamin. Such a method is the 2,4-dinitrophenylhydrazine method of Roe *et al.**, which will be found on pages 317–327 of Selected Reference A. For the determination of *l*–ascorbic acid alone the authors have found the method of Schmall *et. al.*: *Anal. Chem.*, **25**: 1486 (1953); **26**: 1521 (1954) to be highly satisfactory.

Interference may also arise from d–*iso*ascorbic acid, the so-called *erythorbic acid.* This isomer of ascorbic acid is devoid of vitamin properties, but because it is as good a preservative as ascorbic acid and is cheaper to make, it is not infrequently added to certain foods. The usual titration methods and the colorimetric method of Schmall *et. al.* will not differentiate between the two compounds. If the problem of their separation arises, the reader is referred to the article of Weeks and Deutsch in *J Assoc. Offic. Anal. Chemists*, **50**: 793 (1967).

* Roe, J. H., *et al.*: *J. Biol. Chem.* **147**:399 (1943); **152**:511 (1944).

SELECTED REFERENCES

A. *The Association of Vitamin Chemists, Inc. Methods of Vitamin Assay*, 3rd ed. New York: Interscience Publishers (1966).

B. Booth, V. H.: *Carotene, Its Determination in Biological Materials.* Cambridge: W. Heffer & Sons Ltd. (1957).

C. Deuel, H. J., Jr.: *The Lipids, Their Chemistry and Biochemistry.* New York: Interscience Publishers, Inc. (1951–1957). I–III.

D. Furia, T. E. (ed.): *CRC Handbook of Food Additives.* Cleveland: Chemical Rubber Co. (1968). Chapter 3.

E. Horwitz, W. (ed.): *Official Methods of Analysis of the Association of Official Agricultural Chemists*, 10th ed. Washington D.C.: Association of Official Analytical Chemists (1965). Chapter 39.

F. Klyne, W.: *The Chemistry of the Steroids.* London: Methuen and Co., Ltd. (1957).

G. Sebrell, W. H., Jr. and Harris, R. S.: *The Vitamins: Chemistry, Physiology, Pathology.* New York: Academic Press, Inc. (1954). I–III. 2nd ed. (1967–1968) I–II.

H. Strohecker, R., Henning, H. M. and Libman, D. D.: *Vitamin Assay—Tested Methods.* Weinheim: Verlag Chemie GmbH (1965).

CHAPTER 21

Special Instrumental Methods

Strictly speaking, there is no such thing as a non-instrumental method, since even ordinary volumetric analysis depends ultimately on use of an analytical balance for standardization. This chapter will be restricted to explanations of the functions in analytical chemistry of several instruments that are "special" in that they are not a part of the common equipment of analytical laboratories. These instruments differ in relative availability in rough proportion to their cost. For instance, flame photometers, polarimeters and less expensive types of fluorometers are found in quite a few food laboratories, while mass spectrometers and nuclear magnetic resonance devices are uncommon. In modern times, absorption spectrophotometers for the visual-ultraviolet range such as the Beckman DU have come into almost universal use, so we feel there is no need to discuss these instruments in this chapter.

The 15 instruments whose usefulness for particular types of analytical problems we shall attempt to explain can be divided roughly into two categories: 1) those that are primarily used for qualitative and quantitative analysis of mixtures of compounds; and 2) those that are primarily used for identifying single pure compounds. It should be emphasized that this division is a very rough one; for example, while the infrared spectrophotometer is almost uniquely valuable in its ability to identify an organic compound and the groups that it contains, there are numerous references to its use in quantitative analysis of mixtures of several compounds (more often, however, in the drug than in the food field). Nevertheless, the classification is a useful one, and we have employed it here.

Instruments for Qualitative and Quantitative Analysis of Mixtures

Atomic Absorption Spectrophotometers

It has been pointed out[1] that atomic absorption spectrophotometry "is identical in principle with solution absorption spectrophotometry except that the sample to be examined is in the vapor phase, produced and maintained by a flame." Atomic absorption instruments find their practical use, however, as do flame photometers, primarily as cheaper substitutes for the emission spect-trograph.

Introduction of the atomic absorption spectrophotometer by Walsh in Australia in 1957[2] was rapidly followed by the development of commercial instruments, and this technique has become quite popular in the years that have elapsed since that time. Like the flame photometer, a solution of the sample is sprayed into a flame where it is heated to incandescence. It differs from the flame photometer in this respect: in the flame photometer, the *emitted* light of a particular wavelength, which comes only from the activated (ionized) atoms of an element, is measured; in the atomic absorption spectrophotometer, the vaporized unionized atoms *absorb* some of the light emitted from a hollow cathode lamp whose cathode gives off light of the resonance frequency of a particular element, and it is the amount of this absorption that is measured. One of the claims for the atomic absorption instrument is that because only a minor proportion of the vaporized atoms of an element is ionized, atomic absorption spectrophotometry is more sensitive than flame photometry. Claims for superiority over emission spectrography are also made on the basis that instrument readings can be made directly proportional to concentration. On the other hand, determinations of many elements can be made simultaneously with an emission spectrograph (and it is not even necessary to have prior knowledge of what elements are present), while the atomic absorption spectrophotometer requires a separate hollow cathode lamp, and a separate determination, for each element (or at most each pair of elements) estimated.

As a practical matter: 1) the flame photometer (which is the cheapest of the three instruments) finds its chief use, and is most efficient, in the determination of the alkali metals Na, K, and Li; 2) the atomic absorption spectrophotometer, because its cost is one-fifth as much or even less, is within the means of more laboratories than the emission spectrograph, and in addition is a more sensitive means of determining certain elements such as zinc; and 3) as noted above, the emission spectrograph can determine the widest range of elements in the shortest time, without prior knowledge of what elements are present. Under normal conditions only metallic elements can be determined by any of these instruments, and none determines how the element is combined in the sample.

Method 6–24 is an example of an atomic absorption method.

Manufacturers of atomic absorption spectrophotometers include the following:

Aztec Instruments, Inc., 2 Railroad Place, Westport Conn.

Instrumentation Laboratory, Inc., 113 Hartwell Ave., Lexington, Mass.

Jarrell-Ash., 590 Lincoln St., Waltham, Mass.

Perkin-Elmer Corp., 702G Main Ave., Norwalk, Conn.

Emission Spectographs

The emission spectograph consists of a power source to heat the sample to incandescence, a device (either a prism or a grating) to break up the spectrum of the emitted light, and a means of recording the spectrum (either a photographic plate or film or a bank of photocells). Grating spectrographs have largely replaced prism instruments.

Emission spectrographs find their chief use in food analysis either in the quantitative analysis of plant materials for metallic elements or in semi-quantitative tests for toxic elements such as arsenic, lead, beryllium and thallium. For such purposes the sample is usually ashed, taken up in dilute acid, and applied to a crater in a carbon or graphite electrode. An arc or spark is then formed between this electrode and another of the same material, and the spectrum of the burning sample is recorded and measured. In the metal industries, automatic recording spectrographs employing photocells are used to measure the intensity of the chosen line in each element being determined, but such devices are quite expensive, and in food analysis the cheaper instruments in which the spectrum is recorded on a photographic plate or film are still in use. In photographic recording, after development the film (or plate) is placed in a densitometer which first locates the lines corresponding to the various elements present and then measures their intensities. For semi-quantitative trace element detection, a d-c arc of 5-30 ampere current is the preferred means of excitation because it is the most sensitive, but in quantitative analysis the a-c spark is employed because it yields more uniform spectra.

For a more extended discussion of the details of emission spectography the reader is referred to the article by Scribner and Margoshes in Chapter 64 of Part I, Volume 6 of Kolthoff, Elving and Sandell's *Treatise on Analytical Chemistry* (New York–London–Sydney: Interscience Publishers, 1965). Two things should be noted here, however: 1) The densities of the spectral lines obtained in emission spectography are not strictly proportional to the concentrations of their respective elements; and 2), because it is nearly impossible to take two "shots" under completely identical conditions, it is the common practice in quantitative analysis to correct for small variations in burning, etc., by adding a uniform amount of an element such as cobalt (which is not present in measurable amount in the sample) to both sample and standards. The *ratio* of the density of the line of the element being determined to the density of a chosen cobalt line is then what is calculated, and graphs are constructed relating these density ratios to concentrations of the elements in question.

Examples of emission spectrographic methods especially designed for plant and food analysis are given in Text Reference 3; note also particularly Selected Reference W. The method of the late Waddy Mathis was worked out in the laboratory of one of the authors of this book, and is still in use there. Both of these methods (with slight modification) have been adopted by the AOAC (AOAC Methods **41.001–41.016**).

Manufactures of emission spectrographs include:

Applied Research Laboratories, Inc., 3717 Park Place, Glendale, Calif.

Baird-Atomic, 33 University Road, Cambridge, Mass. and Jarrell-Ash Co., 590 Lincoln St., Waltham, Mass.

Spex Industries, Inc., 3880 Park Ave., Edison, N.J., is a source of spectrographic supplies; graphite electrodes are available from National Spectrographic Laboratories, Inc., 6300 Euclid Ave., Cleveland, Ohio, and Union Carbide Corp., Carbon Products Div., 270 Park Ave., New York, N.Y. Extremely high purity elements and their salts for standardization, manufactured by Johnson-Matthey Co., London, England, are available from Jarrell-Ash Co.

Flame Photometers

Flame photometry is essentially emission spectrography using as excitation source a flame rather than the higher-temperature electric arc or spark. As a matter of fact, Lundegårdh wrote two books describing results obtained with an ordinary spectrograph equipped with an acetylene-air flame source.[4] However, the modern-day flame photometer, besides being a much less expensive device, in our opinion is most useful just when the temperature of its flame, *e.g.*, propane-air, is not high enough to excite most elements, and it consequently can determine the alkali metals with minimum interference. It is possible, by using higher-temperature flames (hydrogen-oxygen at 2900° K or cyanogen-oxygen at 4800° K[5]), and separating the emission lines with a monochromator, to adapt the flame photometer to the determination of numerous other elements, but only at the price of loss in simplicity and freedom from interference, and increase in cost.

An example of the determination of sodium and potassium by flame photometry is Method 1–17.

Relatively inexpensive flame photometers that yield accurate results for Na and K (particularly when using a lithium internal standard) are available from:

Advanced Instruments, Inc., 45 Kenneth St., Newton Highlands, Mass.

Coleman Instruments Corp., 42 West Madison St., Maywood, Ill. and

Jarrell-Ash Co. (See above).

Fluorometers

Disregarding the cause of fluorescence, the visually observed phenomenon is that when a substance is illuminated with light of one wavelength, it emits light of another wavelength; if the emitted light persists after the illumination ceases, the substance is said to phosphoresce. These phenomena have been explained as being due to the release of energy in the form of light by an excited molecule when it reverts to the ground state. A complete explanation would require a knowledge of quantum mechanics, but an understanding of the theory of fluorescence is unnecessary for its use as an analytical tool.

In principle, a fluorometer is a device in which a solution of the substance being analyzed is illuminated by an ultraviolet light source of fixed wavelength, and the intensity of the (usually visible) light emitted by the irradiated solution is measured. The incident radiation of the source must, of course, be separated from the light emitted by the sample; this can most readily be done by measuring the light from the sample at a right angle to the beam of light from the illuminating ultraviolet lamp. For accurate analysis it is necessary to ensure, by the use either of filters or of prisms or gratings, that both the incident and fluorescent light are monochromatic.

Fluorometric methods are valuable where they are applicable (*i.e.*, when the compound being determined is fluorescent or can be converted to a fluorescent derivative) because they are usually highly specific and about 10 times as sensitive as absorption methods; the calibration curves are also usually linear over a 50-fold concentration range. There are pitfalls, however; for instance, other compounds present in a solution of a sample can quench the fluorescence of the fluorescing ingredient, and a practical method must be so designed as to avoid such interferences.

The better grade fluorometers use gratings or prisms and a recorder of some sort to separate and record the fluorescent spectrum of the substance being measured; when this is done, a fluorescence spectrum is obtained, frequently a mirror image of the absorption spectrum, which serves to characterize the substance in addition to determining how much of it is present.[6]

Examples of fluorometric methods of food analysis are those of Watkinson for selenium,[7] McFarland for diethylstilbestrol in meat,[8] and Roberts for cholesterol in egg noodles.[9]

Fluorometers of the quality needed for such methods may be obtained from:

American Instrument Co., Inc., 8030 Georgia Ave., Silver Spring, Md.

Baird-Atomic, 33 University Road, Cambridge, Mass. and

Farrand Optical Co., Inc., 535-A South Fifth Ave., Mt. Vernon, N.Y.

Gas Chromatographs

It can be questioned whether gas chromatography is an "instrumental" method of analysis in the same sense as are the determinations made by means of the other instruments described in this chapter. It is true that commercial gas chromatographs carry various dials and controls on their faces and display their results on x-y recorders, but all that these controls do is adjust temperatures and gas pressures, maintain a detecting device, and record when that device indicates that an ingredient of a sample has come through a column and how much there was of it. In contrast to the other instruments we will describe, the gas chromatograph analyzes a sample not by detecting its ingredients through their characteristic intramolecular properties but by physically separating them. The heart of the gas chromatograph is a tube of glass or metal, of one of two types: either it is of appreciable diameter (about $\frac{1}{4}''$ o.d.) and filled with an inert base coated with an adsorbing medium, or it is a capillary whose inner surface bears a coating of adsorbent. A portion of the sample being analyzed—usually in the form of a solution containing only milligrams or less of the ingredients being determined—is injected onto the heated column, and a stream of inert gas (nitrogen, argon or helium) is passed through the tube and onto the detecting device. The gas carries the sample ingredients through the tube at rates that vary inversely with the tenacity with which they are held by the adsorbent the tube contains; in effect a sort of distillation takes place, and the efficiency of a tube packing is sometimes calculated in terms of so many theoretical plates. As each compound reaches the detector at the end of the tube, it causes the detector to activate a recorder. This recorder plots a height *vs* time curve showing a peak for each time interval that a compound leaves the tube; the height of each peak (or more exactly, the area enclosed by it) is proportional to the quantity of the compound present in the sample aliquot.

The various gas chromatographs differ in control features such as their ability to make rapid temperature changes to permit processing a sample in increasing temperature stages as the more volatile ingredients are removed. The more important developments in gas chromatography, however, have involved the discovery of new detectors that either were more sensitive than previous ones, or had the desirable quality of reacting toward only certain types of compounds. The first and most general type of detector was the thermal conductivity gauge, which was basically an electrically heated fine wire mounted axially in a narrow tube; as the current of gas flowing through the gas chromatograph tube carried a sample ingredient over this wire, its temperature (and consequently its resistance) changed, and a Wheatstone bridge device converted this change to a small e.m.f. that activated the recorder. A later device was the flame temperature detector; this consisted of a hydrogen flame impinging on a thermocouple. As an organic compound emerged from the column, its heat of combustion was added to that of the hydrogen, causing an increase in flame temperature and change in e.m.f. of the thermocouple that was translated into a peak on the recorder.

Nearly all of the later more sensitive detectors, as well as those possessing specificity for particular types of compounds, have been devices involving ionization. One exception is Coulson's microcoulometric detector, in which the effluent compounds are burned and then titrated for their chlorine or sulfur contents. This device has been employed almost exclusively for the detection and determination of organic chlorine or sulfur pesticide residues in foods; for this purpose it has the advantage that non-pesticide residues carried through the column do not affect the recorded curve.

Ionization detectors, in general, possess higher sensitivity than thermal conductivity devices. They also differ in that they do not respond to all compounds. There are several types, but one that was and still is being extensively used in pesticide residue analysis is the electron capture detector (Fig. 21–1):[10]

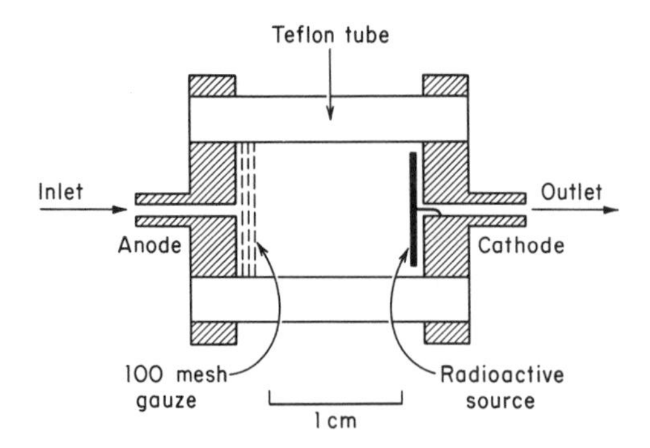

Fig. 21-1. The electron capture detector (after Lovelock).

This device consists of a chamber containing two electrodes and a radioactive ionizing source. A potential is applied across the electrodes just sufficient to collect all ions and electrons. As the carrier gas passes through the chamber it is ionized by the beta particles from the radioactive source, but when a sample ingredient which is electron-capturing emerges from the chromatograph tube and enters the ionization chamber, electron capture occurs to give negative ions which are rapidly removed from the system by recombination. The ion current is thereby reduced, and this drop in current is translated into a peak on the recorder curve by suitable devices.[10] The fact that this detector is exceedingly sensitive to halogen-containing compounds explains its appeal to the pesticide residue analyst.

Until very recently the most common radioactive source was tritium adsorbed on titanium. This source always had the disadvantage that as the temperature of the system rose above 200°, tritium began to evaporate. To avoid this restriction on usable temperature and its associated hazard, sources composed of nickel 63 have come on the market.[12]

Because of its high sensitivity and relative specificity for halogen compounds, the electron capture detector is nearly ideal for the determination of residues of chlorinated organic pesticides, but because it is not very sensitive to most organic phosphorus compounds it is much less satisfactory for the estimation of residues of organic phosphorus pesticides that contain no chlorine atoms. This deficiency was corrected with the discovery by Giuffrida[11] that a detector composed of a hydrogen flame impinging on a platinum coil coated with an alkali metal salt had a sensitivity for phosphorus compounds up to 10,000 times that of a regular flame ionization detector. If the alkali salt used were a halide, the response to halogen compounds was actually depressed. The original coating material was sodium sulfate, but potassium chloride was eventually adopted as the salt giving the best results.

As was pointed out in Chapter 16 (Method 16–11, Note 5), "gas chromatographic methods identify a separated compound only to the extent that they show that, with a particular column and a particular form of detection, the compound gives a peak on a chart at the same place as does the compound it is assumed to be." In order to make more certain identification of gas chromatographic fractions, various authors have worked out means of transferring these fractions to infrared spectrophotometers or mass spectrometers, particularly the latter. This subject is discussed below under "Mass Spectrometers".

For a more detailed yet concise discussion of gas chromatography the reader is referred to Selected Reference U–1.

Gas chromatographic methods described in this book are Methods 2–13, 11–9A, 13–17, 16–11, and 16–28.

There is a rather large number of manufacturers of gas chromatographs in the United States, including:

Barber-Colman Co., Industrial Instruments Div., 1300 Rock St., Rockford, Ill.;

Beckman Instruments, Inc., 2500 Harbor Boulevard, Fullerton Calif.;

F & M Scientific Division of Hewlett-Packard Co., P.O. Box 245, Avondale, Pa.;

Jarrell-Ash Co.;

Packard Instrument Co., Inc. 2200 Warrenville Road, Donners Grove, Ill.;

Perkin-Elmer Corp.; and

Varian Aerograph, 2700 Mitchell Drive, Walnut Creek, Calif.

Manufacturers of gas chromatographic supplies include:

Analabs, Inc., 80 Republic Drive, North Haven, Conn.;

Applied Science Laboratories, State College, Pa.; and

Pierce Chemical Co., P.O. Box 117, Rockford, Ill.

Neutron Activation Analysis Apparatus

Neutron activation analysis depends in principle on the use of a neutron source to convert a nonradioactive isotope to a radioactive isotope, followed by determination of the concentration of that element by measurements of its radioactivity. An example is the determination of nitrogen (and therefore protein) by the $N^{14}(n, 2n)\, N^{13}$ reaction.

This method of analysis is often very sensitive; it can be both rapid and specific, and requires little previous sample preparation. The chief obstacle to its more frequent use has been the initial cost of the apparatus and the shielding for the neutron source.

Manufacturers supplying apparatus of this type include:

General Dynamics Corp., General Atomic Division, P.O. Box 608, San Diego, Calif.;

Kaman Nuclear, 1700 Garden of the Gods Road, Colorado Springs, Colo.

Polarimeters

The polarimeter is of course an instrument for measuring the optical rotation of a solution. Polarimeters calibrated to read directly in terms of per cent sucrose when the rotation of a solution of 26 grams of sample in 100 ml is measured at 20° in a 200 mm tube, using sodium D light, are called *saccharimeters*. While these instruments may be employed in the determination of other optically active compounds than carbohydrates, they find little use in food analysis other than for the estimation of sugars, as in Method 16–6, "Determination" (A). Among the few exceptions are AOAC Method **20.062** for laevo-malic acid (which depends on enhancement of rotation of the acid by uranium salts) and Method 9–14 for lemon and orange oils in extracts.

A discussion of the principles of optical rotation may be found in Text Reference 13.

American manufacturers of better-grade polarimeters are:

Cary Instruments, 2724 South Peck Road, Monrovia, Calif.

O. C. Rudolph Sons, Inc., P.O. Box 446, Caldwell, N.J.; and

Rudolph Instruments Engineering Co., Inc., 61 Stevens Ave., Little Falls, N.J.

Polarographs: Amperometric Titrations

The technique of polarography and the name itself were originated by Jaroslav Heyrovský at the Charles University in Prague about 1929.[14] Polarography is a process of recording the variations in current that take place as a constantly increasing voltage is applied to a system composed of a dropping mercury electrode and a reference electrode immersed in a solution of the sample being analyzed. The mercury electrode consists of a drop of the metal hanging for a few seconds at the surface of a glass capillary from which mercury regularly drops out. Figure 21–2 is a schematic diagram of the polarograph; Figure 21–3 shows the type of curve obtained by plotting current *vs* voltage.

These polarograph curves are completely reproducible, and if experimental conditions are kept constant they depend only on the composition of the electrolyzed solution. In the presence of substances which undergo reduction or oxidation at the surface of the dropping electrode, or substances that catalytically affect the electrode process, or those that form stable compounds with mercury, an increase in cathodic or anodic current is observed. The current rises in a given potential range, and this increase is followed by a region of potentials in which the current has a limiting value. The S-shaped portion of the current-voltage curve is called a *polarographic wave*. The shapes and positions of these waves provide information on both the quantitative and qualitative composition of the electrolyzed solution. The difference between the limiting current and the current before the wave rise, which is

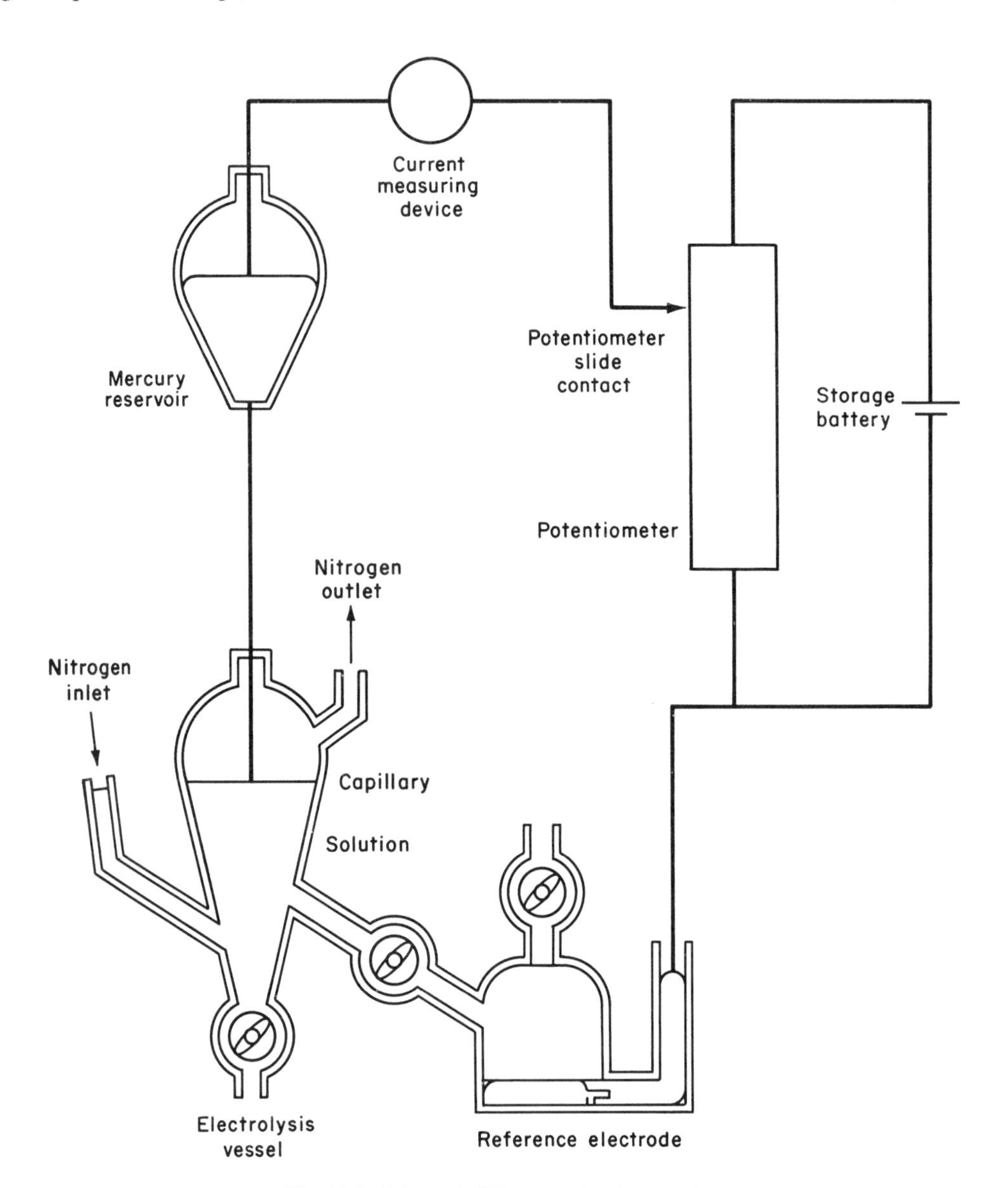

Fig. 21-2. Schematic Diagram of Polarograph.

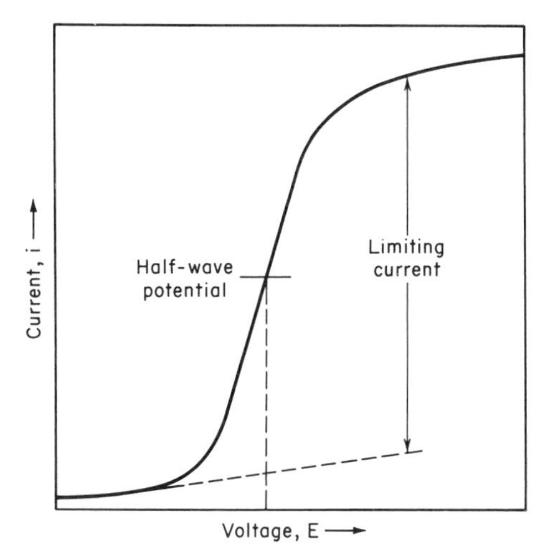

Fig. 21-3. Current–Voltage Relation in Polarography.

called the *wave height*, depends on the concentration of the electroactive substance responsible for the wave, while the *half-wave potential*, i.e., the potential at that point on the curve at which the current reaches half its limiting value, is independent of the concentration but is a function of the nature of the electroactive substance and consequently can be used to identify it qualitatively.[15]

For a more detailed discussion of the principle and practice of polarography, the reader is referred to one of the special texts such as Selected Reference I–2. This technique has been used chiefly to analyze samples for different metallic elements that can be made to yield specific polarographic waves under the proper conditions; the titrations are carried out in a stream of nitrogen to eliminate the oxidation-reduction effects of atmospheric oxygen. The method is however also used to determine organic compounds that yield half-wave potentials. An example of the use of polarography in food analysis is the method of Christian *et al.* for traces of selenium.[16]

Originally polarographic curves were plotted manually, but because this was laborious, Heyrovský and Shikata introduced an instrument in which the movements of a galvanometer lamp were recorded photographically. This was succeeded by various instruments that plotted curves on a chart through an electrically-controlled pen. Derivative polarography has also been employed.

A method of analysis related to polarography is *amperometry*. Without going into detail, in amperometric titrations the current which passes through the titration cell between an indicator electrode and a depolarized reference electrode, at a suitable applied e.m.f., is measured as a function of the volume of the titrating solution. The end point of an amperometric titration is the point of intersection of two lines giving the change in current before and after the equivalence point.[17] Amperometry is still not a common technique, but it did find a certain use in testing for chlorinated organic pesticide residues in the days before gas chromatographic analyses became common: Extracts of the samples were burned, and chloride was determined in solutions of the ash by amperometric titration with silver nitrate.[18]

Manufacturers of polarographic apparatus include:

Fisher Scientific Co., 711 Forbes Ave., Pittsburg, Pa.;

Leeds & Northrup Co., 4901 Stenton Ave., Philadelphia, Pa.;

Melabs Scientific Instruments Operation, 3300 Hillview Ave., Stamford Industrial Park, Palo Alto, Calif.;

E. H. Sargent & Co., 4647 West Foster Ave., Chicago, Ill.

Raman Spectroscopes

The phenomenon of the Raman shift interference effect was discovered independently in 1928 by Raman and Krishnan in India and Landsberg and Madelstam in Russia.[19] When a beam of monochromatic light is passed through a tube containing a liquid sample, and some of the light that is scattered in a direction perpendicular to the incident beam is passed into a spectrograph, the resulting spectrogram shows (in addition to the expected strong line at a position corresponding to the wavelength of the incident light) a pattern of relatively weak lines or narrow bands on the red (long-wavelength) side of the incident line, and possibly a very few weak lines on the blue side as well. These lines are the *Raman spectrum* of the sample; the lines on the long-wavelength side are called the *Stokes Raman lines*, and those on the short-wavelength side, the *anti-Stokes lines*.

If the wavelength of the incident light is changed to larger or smaller values, the *pattern* of Raman lines (*i.e.*, their relative positions to each other) is unaltered, but the group as a whole is shifted to longer or shorter wavelengths. The difference in frequency between a Raman line and the incident light is independent of the frequency of the incident light, and the set of frequency differences or shifts, which is the Raman spectrum, is characteristic of the molecules in the sample. These shifts are commonly reported as wave-numbers.

Because the Raman spectrum of a compound is unique for that compound, and the spectrum of a mixture whose components do not interact is a superposition of the Raman spectra of the components, Raman spectroscopy may be used for the identification of both single compounds and mixtures of as many as six different substances. Since the intensity of each line is proportional to the concentration of the substance giving that line, quantitative analysis is also possible.

The advantages of Raman spectroscopy include: 1) the Raman spectrum of a compound is relatively simple; 2) water may be used as a solvent because the Raman spectrum of water is diffuse and weak; and 3) the relationship between line intensity and concentration is linear. Some disadvantages are: 1) analyses are practically limited to clear, colorless (non-absorbing) liquids; 2) the sample must be completely non-fluorescent; and 3) the sample volume required is larger than for infrared spectroscopy (5–15 ml is normally used). The geatest disadvantage, however, has been the difficulty in obtaining a light source that was sufficiently intense and monochromatic to yield Raman lines that were bright and sharp enough. Until recently almost all instruments utilized the Hg4358A line, but it is now generally expected that the illumination difficulty will be solved by the use of laser beams; in fact one instrument with a laser source, the Cary, is already on the market.

Manufacturers of Raman apparatus include:

Cary Instruments;
Perkin-Elmer Corp.;
Spex Industries, Inc., 3880 Park Ave., Edison, N.J.

X-Ray Fluorescence Spectrographs

There is a difference of opinion over the name for the type of analysis to which we shall refer here. Some authorities prefer the name *X-ray emission spectrography* on the grounds that this designation prevents confusion with ordinary fluorometry as discussed above, and "has a familiar ring to analytical chemists, accustomed as they are to emission spectrography involving radiant energy in the visible and in the ultraviolet."[20] Nevertheless, while such terminology may appear to prevent confusion on the part of technicians knowing little beyond, and interested only in, the practical application of instruments as analytical tools, it (perhaps deliberately) has the effect of completely confusing the distinction between emission and fluorescence phenomena. We have therefore headed this section "X-Ray *Fluorescence* Spectrographs."

When a beam of electrons impinges on a metal target, X-rays characteristic of the particular metal are expelled. This is X-ray *emission.* To use it for analytical purposes would require, however, construction either of separate X-ray tubes for each analysis, or of a single tube so designed that targets could be changed. Because of the physical difficulties in this method of analysis, it has had no practical use, which probably explains in part why some authors in their description of X-ray spectroscopy leave unclear the distinction between emission and fluoroscopic effects.[21]

The X-ray fluorescence spectrometer, on the other hand, is a practical device that has extensive use and may be the instrument of choice for some analyses. It works in principle as follows: an intense beam from an X-ray tube impinges on the sample being analyzed, producing fluorescent X-rays by secondary emission. These rays are dispersed by a crystal of known lattice constant (frequently a bent mica crystal) that serves as a diffraction grating. The diffracted fluorescent X-rays are detected by a Geiger counter attached to a goniometer, and their intensities are automatically recorded as functions of the goniometer angles, which are proportional to the wavelengths or frequencies of the X-rays. Figure 21–4, which was obtained from a chrome-nickel plating on a silver-copper base, illustrates the type of spectrum observed. Each element is identified by its position on the abscissa, while the height of the ordinate peak is proportional to the quantity present in the sample.[21]

X-ray fluorescence analysis has an advantage over emission spectrography in that X-ray spectra never consist of more than four lines, while the transition elements yield many atomic emission lines. In trace metal analysis particularly this plethora of lines can make it difficult to separate a line of the element sought from the lines of a more abundant element such as iron. Another

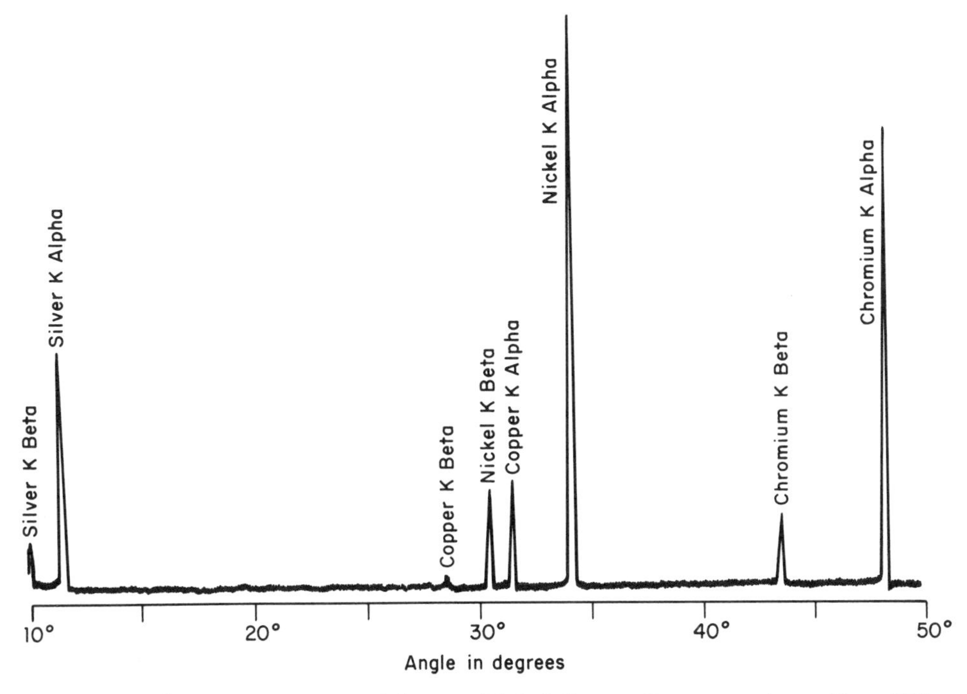

Fig. 21-4. X-ray fluorescence spectrum of chrome–nickel plating on silver–copper brass. [From O'Connor, R. T.: *J. Assoc. Offic. Anal. Chemists*, **51**:244 (1968).]

advantage is that X-ray fluorescence analysis is not restricted to the determination of metallic elements, as emission spectrography is in practice. Its disadvantages are that it is not capable of extreme sensitivity, and cannot be used to determine elements lighter than sodium.[21] Tuchscheerer has applied X-ray fluorescence analysis to the determination of K, P, Cu, Fe, Zn and Mn in foods.[22]

Manufacturers of this type of apparatus include:

Baird-Automic; General Electric Co., X-Ray Dept., 4855 Electric Ave., Milwaukee, Wis.; Instrumentation Division, Engis Equipment Co., 8035 Austin Ave., Morton Grove, Ill. and

Norelco Philips Electronics Instrument Div., Philips Electronics and Pharmaceutical Industries Corp., 750 South Fulton Ave., Mt. Vernon, N.Y.

Instruments Primarily for Identification of Single Compounds

Infrared Spectrophotometers

Despite the fact that they are expensive, infrared spectrophotometers are rapidly on their way to becoming part of the normal equipment of all modern analytical laboratories. These instruments plot the absorption curves of samples in the infrared in essentially the same manner that instruments such as the Beckman DU and the Cary do in the visible and ultraviolet regions of the spectrum. The source of light is different, being a Globar (a bonded SiC rod) or a Nernst glower (a mixture of Zn, Th and Y oxides); the dispersing element of choice is rapidly becoming a grating, but if it is a prism, it must be made of a salt such as NaCl or KBr rather than of quartz; and the detector is usually a thermocouple. These differences in the components of the infrared instrument from those of the visible-ultraviolet apparatus are necessary because the lamps used as sources of visible and ultraviolet light are poor sources of infrared, and glass and even quartz are essentially opaque to the infrared.

We shall not enter into a theoretical discussion of why the infrared absorption curve of a substance tells so much more about that substance than do the visible and ultraviolet spectra. For our purposes, it is sufficient to note that the infrared curve of a compound is a fingerprint that uniquely identifies it, and further, that a specific group within a molecule, *e.g.*, CH_3, C=O, P=O or C=C, absorbs at nearly the same wavelength regardless of the compound in which it is found. Therefore, if an analyst is confronted with a compound of whose structure he is uncertain, he can run its infrared spectrum and compare it with the spectrum of a reference substance he suspects it to be; if the two prove to be the same, identification is certain. If the curve for the unknown does not match that of the postulated compound, it can still be identified if a match can be found with any of the thousands of spectra recorded in the literature. And if even this search fails, the unknown will usually show some characteristic group absorption peaks that will help to point the way toward elucidation of its structure. It should be noted, however, that infrared spectrometry does *not* distinguish between optical isomers.

Infrared spectrophotometry can also be used for the quantitative determination of a single compound or a mixture of several compounds when each of the substances being determined possesses a sharp peak that is not shared with any other substance that is present. This technique has not yet had much use in food analysis, but is currently employed in drug analysis and pesticide formulation analysis.

For accurate work the infrared curve of a compound that is not itself a liquid is preferably obtained from a solution of the substance in an organic liquid that is transparent to infrared over the wavelength range being investigated. Common solvents are carbon disulfide and carbon tetrachloride. For most purposes water cannot be used as a solvent, both because of its opacity to the infrared and because it would dissolve the NaCl or KBr windows of the infrared tubes; the only exception is in working in the near-infrared (*i.e.*, the wavelengths proximate to the red portion of the visible spectrum), where aqueous solutions can be analyzed in tubes with AgCl windows.

If a sample is insoluble in one of the usable solvents, its spectrum is obtained either in a mull with Nujol or a liquid fluorocarbon, or in a pellet made by mixing it with KBr and compressing the mixture.

There is one technique for obtaining infrared spectra that is in a class by itself: Attenuated total reflectance, or "ATR." This method is based on the fact that radiation, when totally internally reflected at the face of a high-index medium such as AgCl or "KRS-5" (thallium bromide-iodide), penetrates a few microns beyond the interface into the adjoining low-index material. By making use of this principle, it is possible with properly-designed and operated equipment to obtain directly spectra of the surfaces of materials such as rubber, plastics and foods. The resulting spectrum is not identical with the absorption spectrum of the sample, but with proper adjustment of the equipment, it resembles the absorption spectrum closely enough to permit direct comparison.[23]

Infrared spectroscopy has been used to identify gas chromatographic fractions, as was noted in Chapter 16. Grasselli and Snavely, for instance, were able to obtain spectra on samples as small as 0.4 microliter in size.[24]

The chief American manufacturers of infrared spectrophotometers are Beckman Instruments, Inc., and Perkin-Elmer Corporation. Barnes Engineering Co., Instrument Division, 30 Commerce Road, Stamford, Conn., and Wilks Scientific Corp., P.O. Box 441, South Norwalk, Conn., are suppliers of infrared equipment. Barnes Engineering Co., Wilks Scientific Corp., and Perkin-Elmer Corp. manufacture ATR aparatus.

Mass Spectrometers

The mass spectrometer is an instrument in which the sample is admitted to an evacuated chamber where it is ionized by bombardment with electrons, and where the ions are then accelerated to a desired kinetic energy, followed by separation by a magnetic field into discrete ion beams according to their respective masses. The ion beams can be detected individually on a collector plate by "sweeping" them past a slit in front of this plate (by varying either the magnetic field or the accelerating voltage); the ion current to the collector plate is amplified and recorded photographically or on a strip-chart recorder. The chart shows a separate peak for each mass represented by an ingredient of the sample, and the peak height is proportional to the relative proportion of that ingredient in the sample.[25]

It should be noted that the chart of a single pure compound will not ordinarily show only one mass, because some of the molecules will have been "cracked" into fragments by the electron bombardment, and the masses of these ionized fragments will be recorded along with that of the ionized intact molecule. This fact makes determination of the mass of the compound less simple than it would otherwise be, but it is nevertheless usually possible to make an unequivocal decision, partly on structural grounds that we cannot go into here.

Probably the greatest value of mass spectrometry to the food chemist is in the identification of gas chromatographic fractions. One illustration of a combination of the two instruments in such fashion that, as each fraction was detected by the gas chromatograph, a portion of it was absorbed and its mass recorded by the mass spectrograph, is given in an article by Gohlke.[26] This author employed a "Bendix" time-of-flight mass spectrometer; for a description of this instrument the reader is referred to Text Reference 27. Teranishi *et al.* describe a specific application of this type of combination to a food analytical problem, namely, the separation and identification of the hydrocarbons present in orange oil.[28] Damico has contributed to the use of the mass spectrometer as a tool for the identification of organic phosphorus pesticides by publishing spectra of many of these compounds.[29]

There are several manufacturers of mass spectrometers, but the time-of-flight instrument employed to identify gas chromatographic fractions is made only by The Bendix Corporation, Cincinnati Division, 3625 Hauck Road, Cincinnati, Ohio. The Perkin-Elmer Corporation has however recently come out with a combination mass spectrometer-gas chromatograph.

Nuclear Magnetic Resonance Apparatus

The explanation for the phenomena measured by NMR instruments lies in the field of quantum mechanics, and for a full understanding the reader should first familiarize himself with that branch of physics.[30] It is possible, however, to obtain NMR spectra and derive structural information with little or no knowledge of what is being measured. With this as an empirical basis, it can be stated (to quote from Bose[31]): "In NMR spectrometry, a liquid or a solution is placed in a strong magnetic field and irradiated with radio frequency waves—the spectrum is a graph (*see* Fig. 21–5) that records an increasing magnetic field along the abscissa and the corresponding intensity of energy absorption along the ordinate.

"The 'chromophore' in proton NMR spectra is the proton, and such a spectrum of a compound will show signals for *nothing* but protons (not even deuterium or tritium) but it will show *every* proton in the molecule.

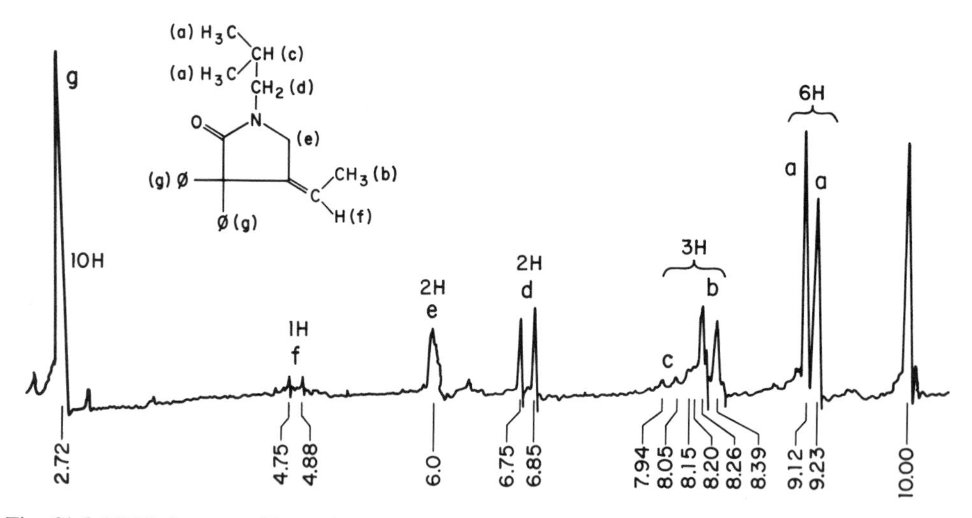

Fig. 21-5. NMR Spectrum [Reproduced from Stanley K. Freeman, ed.: *Interpretive Spectroscopy*, by permission of Ajay K. Bose and Reinhold Book Corporation, a subsidiary of Chapman-Reinhold, Inc., New York 1965).]

Furthermore, there is a quantitative aspect of NMR spectra—the area under a peak or peaks is strictly proportional to the number of protons responsible for the signal."

A specific proton in a molecule will give a single peak if that proton is not interacting with another proton or nucleus; where there is such "coupling", the peak becomes a doublet, triplet or higher multiplet as the nature and extent of the coupling determine. The type of peak and its intensity enable deductions to be made as to the structural position of each proton in the molecule; for a fuller explanation, Text Reference 32 should be consulted.

It is customary to record NMR spectra in terms of the distance of each peak from the signal of a standard compound, usually tetramethylsilane [$(CH_3)_4Si$], that is added to the sample; it is the convention to record the spectrum so that the field increases from left to right, and "TMS" has been chosen because few protons have a higher magnetic field than those in this compound.

The best NMR spectra are obtained on liquid samples[33], and solutions can be employed for solid compounds. A solvent such as chloroform is unsuitable because its molecule contains protons, but deutero-chloroform, in which H is replaced by D, is frequently used.

The primary use of NMR spectroscopy is as an aid in determining the structure of pure compounds, but it has been employed to estimate the total hydrogen content of a sample. Most of the apparatus on the market is limited to the measurement of H^1, but the technique can be expanded to cover F^{19} and even heavier isotopes such as P^{31}, B^{11} and N^{14}.[34]

Probably the chief American manufacturer of NMR apparatus is Varian Associates, 611 Hansen Way, Palo Alto, Calif. Foreign-made instruments may be obtained from:

Analytical Instrument Division, Jeolco USA Inc., 477 Riverside Ave., Medford, Mass.; and

Bruker Scientific, Inc., 1 Westchester Plaza, Cross Westchester Industrial Park, Elmsford, N.Y.

Nuclear Quadrupole Resonance Spectrometers

Nuclear quadrupole resonance spectroscopy has barely come out of the experimental field with the appearance on the market of one commercial instrument. As with NMR, a full understanding of the phenomenon on which it is based requires a knowledge of quantum mechanics. The respective names indicate that both instruments measure the resonance frequencies of certain nuclei. The practical advantage of the nuclear quadrupole resonance spectrometer over the NMR apparatus, for those nuclei that can be detected by both, lies in the fact that the quadrupole instrument requires no magnet and is consequently less expensive. It is, however, applicable only to samples that contain at least one isotope whose nucleus has a quadrupole moment greater than zero and a spin quantum number greater than one half. The samples must also be solid, but this requirement is of course easily met by working at a low enough temperature (77°K is frequently favored because readily obtainable with liquid N_2).

The reason why nuclear quadrupole resonance (NQR) spectrometry does not require an added magnetic field as does NMR is because the frequencies at which NQR signals appear are fixed by the electronic structures of the bonds. There are 130 different isotopes whose nuclei can be detected by the NQR technique, but most work has been done with Cl^{35} and N^{14}. An example of the use of NQR spectrometry in distinguishing between the isomers of hexa-chlorocyclohexane (one of which is of course the pesticide *lindane*) is shown by Brame.[35]

An isotope that shows considerable promise for the use of NQR as an aid in structural placements is N^{14}. When NMR is employed, it is the rare isotope N^{15} that is detected, and since the abundance of this isotope in ordinary nitrogen is only 0.4%, the likelihood of obtaining signals strong enough to be satisfactory in assigning a structure to an N compound should be greater with NQR.

While both NMR and NQR instruments find their chief use as aids in determining the structures of unknown or incompletely characterized compounds, NQR has possibilities for the quantitative analysis of mixtures of compounds—particularly geometrical isomers such as the hexachlorocyclohexanes ("benzene hexachloride").

The only commercial NQR instrument as yet on the market is manufactured by Wilks Scientific Corporation, P.O. Box 441, South Norwalk, Conn.

X-Ray Diffractometers

X-ray fluorescence spectrography was discussed previously because its purpose was such as to place it in a different one of the two classifications used in this chapter. That is, it is a method for the *quantitative determination of the elemental components* of a sample, whereas X-ray diffractometry is a means of *identifying compounds*.

This method originated with the recognition by Von Laue that the wave-lengths of X-rays were the same order of length as the spacings between atoms in a crystal, and that therefore a crystal could act as a diffraction grating for X-rays. Since that time X-ray diffraction has served as an otherwise unavailable tool for determining the three-dimensional structures of many compounds, including that key to heredity deoxyribonucleic acid.

However, X-ray diffractometry finds its everyday use in the analytical laboratory in the identification of much simpler compounds that are more often inorganic than organic. For this purpose the sample (which must, of course, be crystalline) is reduced to a fine powder (about 325 mesh) without segregation. The powder is then usually mixed with a binder and molded into a small cylinder, which is mounted in the X-ray beam. The diffraction pattern is recorded photographically or by some such device as a Geiger counter. For details the reader is referred to Text Reference 36.

The unique feature that distinguishes X-ray diffractometric analysis from chemical analysis is that it detects actual compounds rather than ions. To give one example, the line patterns of a mixture of sodium chloride and potassium bromide would be different from those of a mixture of sodium bromide and potassium chloride. Tables have been built up of the patterns of many compounds, and it is usually possible to identify an unknown compound by reference to these tables[37].

Manufacturers of apparatus for this technique are the same as those who supply X-ray fluorescence spectrometers.

TEXT REFERENCES

ATOMIC ABSORPTION SPECTROPHOTOMETERS

1. Kolthoff, I. M., Elving, P. J. and Sandell, E. B. (eds.): *Treatise on Analytical Chemistry*. New York: John Wiley & Sons (1965). Part I (6), 3526.
2. Walsh, A.: *Spectrochim. Acta*, **7:** 108 (1955); Russell, B. J., Shelton, J. P. and Walsh, A.: *Spectrochim. Acta*, **8**:317 (1957); Walsh A.: U.S. Patent 2,847,899 (August 19, 1958); Allan, J. E.: *Analyst*, **83**:466 (1958).

EMISSION SPECTOGRAPHS

3. Mathis, W. T.: *Anal. Chem.*, **25:**943 (1953). Method of Specht, A. W.: Horticultural Research Branch, Agricultural Research Service, Beltsville, Md.

FLAME PHOTOMETERS

4. Lundegårdh, H.: *The Quantitative Spectral Analysis of the Elements*. Jena: G. Fischer, (1929–1934). I–II.

5. Clark, G. L. (ed.): *The Encyclopedia of Spectroscopy*. New York: Reinhold Publishing Corp. (1960). 330.

FLUOROMETERS

6. Ref. 1, Part I (5), Chapter 59, 3058–3061.
7. Watkinson, J. H.: *Anal. Chem.*, **32**:981 (1960).
8. McFarland, W. B. (Food and Drug Administration): Private communication (January 17, 1962).
9. Roberts, L. A. (Food and Drug Administration): Private communication (March 1968).

GAS CHROMATOGRAPHS

10. Knox, J. H.: *Gas Chromatography*. London: Methuen & Co. Ltd (1962). 78–80.
11. Giuffrida, L. J.: *J. Assoc. Offic. Agr. Chemists*, **47:** 293 (1964);
Giuffrida, L., Ives, N. F. and Bostwich, D. C.: *J. Assoc. Offic. Anal. Chemists*, **49:** 8 (1966);
Ives, N. F. and Giuffrida, L.: *J. Assoc. Offic. Anal. Chemists*, **50:** 1 (1967).
12. Williams, I. H.: *J. Assoc. Offic. Anal. Chemists*, **51:** 64 (1968).

POLARIMETERS

13. Gilman, H. (ed.): *Organic Chemistry, An Advanced Treatise*, 2nd ed. New York: John Wiley & Sons, Inc. (1947). I, Chap. 4.

POLAROGRAPHS; AMPEROMETRIC TITRATIONS

14. Heyrovsky, J.: *Chem. Listy*, **16:** 256 (1922).
15. Zuman, P.: *Chem. and Eng. News*, **46:** 13, 94 (1968). (March 18, 1968.)
16. Christian, G. D., Knoblock, E. C. and Purdy, W. C.: *J. Assoc. Offic. Agr. Chemists*, **43:** 877 (1965).
17. Kolthoff, I. M., and Lingane, J. J.: *Polarography*, 2nd ed. New York: Interscience Publishers (1952). II, 887.
18. Selected Reference I-2, 691–692.

RAMAN SPECTROSCOPES

19. Rosenbaum, C. J.: "Raman Spectroscopy." In: Kolthoff, I. M., *et al.*: (eds.)*: Treatise on Analytical Chemistry*. New York: John Wiley & Sons (1965). Part I (6), Chapter 67, 3746.

X-RAY FLUORESCENCE SPECTROGRAPHS

20. Liebhafsky, H. A., Pfeiffer, H. G. and Winslow, E. H.: In: Ref. 1, Part I (5), Chapter 60, 3138.
21. O'Connor, R. T.: *J. Assoc. Offic. Anal. Chemists*, **51:** 243–244 (1968).
22. Tuchscheerer, Th.: *Z. Lebensm. Unt. Forsch.*, **127:** 13 (1965).

INFRARED SPECTROPHOTOMETERS

23. Ref. I, Part I (6), 3593–3594.
24. Grasselli, J. G. and Snavely, M. K.: *Appl. Spectry*, **16:** 190 (1962).

MASS SPECTROMETERS

25. Leblanc, R. B.: "Mass Spectrometry", In: Clark, G. L. (ed.): *The Encyclopedia of Spectroscopy*. New York: Reinhold Publishing Corp. (1960). 582–583.
26. Gohlke, R. S.: *Anal. Chem.*, **31:** 535 (1958);
Ebert, A. A.: *Anal. Chem.* **33:** 1865 (1961);
Miller, D. O., *Anal. Chem.*, **35:** 2033 (1963).
27. Harrington, D. B.: In: Clark, G. L. (ed.): *The Encyclopedia of Spectroscopy*, New York: Reinhold Publishing Corp. (1960) 630–643. *See also* Cat. TOF8/1066/OZ. *The Time-Of-Flight Mass Spectrometer*. Cincinnati: Cincinnati Division, The Bendix Corporation.
28. Teranishi, R. T., Schultz, T. H., McFadden, W. K., Lundin, R. E. and Black, D. R.: *J. Food Sci.*, **28:** 541 (1963);
Schultz, T. H., Teranishi, R., McFadden, W. H., Kilpatrick, P. W. and Corse, J.: *J. Food Sci.*, **29:** 790 (1965).
29. Damico, J. N.: *J. Assoc. Offic. Anal. Chemists*, **49:** 1027 (1966).

NUCLEAR MAGNETIC RESONANCE APPARATUS

30. Horowitz, S.: *Fundamentals of Quantum Mechanics*. New York: W. A. Benjamin Inc. (1967); *see also* Freeman, S. K. (ed.): *Interpretive Spectroscopy*. New York: Reinhold Publishing Corp. (1965). 282–285.
31. Bose, A. K.: "Protein Nuclear Magnetic Resonance Spectroscopy". In: Freeman, S. K. (ed.): *Interpretive Spectroscopy*. New York: Reinhold Publishing Corp. (1965). 212–213.
32. Ibid., 212–285.
33. Clark, G. L. (ed.): *The Encyclopedia of Spectroscopy*. New York: Reinhold Publishing Corp. (1960). 664.
34. Ibid., 666 and 669–670.

NUCLEAR QUADRUPOLE RESONANCE SPECTROMETERS

35. Brame, E. G., Jr.: *Anal. Chem.*, **39:** 918 (1967).

X-RAY DIFFRACTOMETERS

36. Ref. I, Part I (5), 3108–3123.
37. Hanowalt, J. D., Rinn, H. W. and Frevel, L. K.: *Ind. Eng. Chem., Anal. Ed.*, **10:** 457 (1938); *see also*: File cards of the American Society for Testing and Materials, Philadelphia (1960).

SELECTED REFERENCES

ATOMIC ABSORPTION SPECTROPHOTOMETRY

A. *Analytical Methods for Atomic Absorption Spectrophotometry*. Norwalk: Perkin-Elmer Corp. (1964).

B. *Atomic Absorption Newsletters*. Norwalk: Perkin-Elmer Corp.

C. Capacho-Delgado, L. and LaPaglia, A. J.: *American Laboratory*, (Oct. 1968), 39.

D. Elwell, W. T. and Gridley, J. A. F.: *Atomic Absorption Spectrophotometry*. New York: MacMillen Co. (1962).

E. Menzies, A. C.: *Anal. Chem.*, **22:** 898 (1960); Lewis, L. L.: *Anal. Chem.*, **40:** (No. 12): 28A (October 1968).

F. Neren, E. J. and Wilks, W.: *Bibliography on Atomic Absorption*. Westport: Aztec Instruments, Inc.

G. Ramírez-Muñoz, J.: *Atomic Absorption Spectroscopy*. New York: American Elsevier Publishing Co., Inc. (1968).

H. Robinson, J. W.: *Anal. Chem.*, **32:** 17A (July 1960); *Atomic Absorption Spectroscopy*. New York: Marcel Dekker, Inc. (1966).

EMISSION SPECTROGRAPHY

I. Ahrens, L. H. and Taylor, S. R.: *Spectrochemical Analysis: A Treatise on the D-C Arc Analysis of Geological and Related Materials*, 2nd ed. Reading: Addison-Wesley.

J. *Symposium on Spectrochemical Analysis for Trace Elements*. Special Publication 221. Philadelphia: American Society for Testing Materials (1958).

K. *Symposium on Spectroscopy*. Special Publication 269. Philadelphia: American Society for Testing Materials (1960).

L. Boumans, P. W. J. M.: *Theory of Spectrochemical Excitation*. New York: Plenum Press (1966).

M. Brame, E. G. (ed.): *Applied Spectroscopy Reviews*. New York: Marcel Dekker, Inc. (1968). I.

N. Committee E-2.: *Methods for Emission Spectrochemical Analysis*, 4th ed. Philadelphia: American Society for Testing Materials (1964).

O. Gatterer, A. and Junkes, J.: *Atlas der Restlinien*. Vatican City: Specola Vaticana (1945–1949). I–III.

P. Harrison, G. R.: *Massachusetts Institute of Technology Wavelength Tables*. New York: Technology Press—John Wiley & Sons, Inc. (1939).

Q. Harrison, G. R., Lord, R. C. and Loofburow, J. R.: *Practical Spectroscopy*. New York: Prentice Hall, Inc. (1948).

R. Harvey, C. E.: *A Method of Semi-Quantitative Spectrographic Analysis*. Glendale: Applied Research Laboratories (1947).

S. Harvey, C. E.: *Spectrochemical Procedures*. Glendale: Applied Research Laboratories (1960).

T. Herzeberg, G.: *Atomic Spectra and Atomic Structure*, 2nd ed. New York: Dover Publications (1944).

U. Kroonen, J. and Bader, L.: *Line Interference in Emission Spectrographic Analysis*. Amsterdam–London–New York: Elsevier Publishing Co. (1963).

V. Meggers, W. F., Corliss, C. H. and Scribner, B. F.: *Tables of Spectrochemical Line Intensities*. Washington: Government Printing Office (1961). Parts I–II.

W. Mitchell, R. L.: *The Spectrographic Analysis of Soils, Plants and Related Materials*. Technical Communication No. 44. Harpenden: Commonwealth Bureau of Soil Science (1948).

X. Nachtrieb, N. H.: *Principles and Practice of Spectrochemical Analysis*. New York: McGraw-Hill (1950).

Y. Sawyer, R. A.: *Experimental Spectroscopy*, 3rd ed. New York: Dover Publications (1963).

Z. Thompson, H. W. (ed.): *Advances in Spectroscopy*. New York: Interscience Publishers, Inc. (1959–1961). I–III.

A-1. Various editors: *Developments in Applied Spectroscopy*. New York: Plenum Press (1962–1965). I–IV.

B-1. Zaìdel, A. N., Prokof, V. K. and Raskiì, S. M.: *Tables of Spectrum Lines*, 2nd Int. ed. New York: Pergamon Press (1961).

FLAME PHOTOMETRY

C-1. Burriel-Martí, F., and Ramírez-Muñoz, J.: *Flame Photometry*. Amsterdam–London–New York–Princeton: Elsevier Publishing Co. (1957).

D-1. Dean, J. A.: *Flame Photometry*. New York: McGraw-Hill Book Co. (1960).

E-1. Herrmann, R. I., and Alkemade, C. Th. J.: *Flammenphotometrie*, 2nd ed. Berlin–Göttingen–Heidelberg: Springer-Verlag (1966).

F-1. Margoshes, N. and Vallee, B. L.: *Direct-Reading Flame Spectrometry Principles and Instrumentation*. Tech. Rept. No. PB-111748, Office of Technical Services, U.S. Dept. of Commerce, Washington (1956).

G-1. Mavrodineanu, R. and Boiteux, H.: *Quantitative Spectral Analysis by Flame*. Paris: Masson et Cie. (1954).

H-1. (E-1) Vallee, B. L. and Thiers, R. E.: "Flame Photometry". In: Kolthoff, M. I., *et al.* (eds.): *Treatise on Analytical Chemistry*. New York–London–Sydney: Interscience Publishers (1965). Part I (6), Chapter 65.

FLUOROMETRY

I-1. *Chem. Eng. News*, **45,** (No. 18): 72 (April 24, 1967).

J-1. Clark, G. L. (ed.): *The Encyclopedia of Spectroscopy*. New York: Reinhold Publishing Corp. 1960. 373–376.

K-1. Kolthoff, I. M., Elving, P. J. and Sandell, E. B. (eds.): *Treatise on Analytical Chemistry*. New York–London–Sydney: Interscience Publishers Div., John Wiley & Sons, Inc. (1954). Part I (5), Chapter 59.

L-1. Passwater, R. A.: *Guide to Fluorescence Literature*. New York: Plenum Press (1967).

M-1. Pringsheim, P.: *Fluorescence and Phosphorescence*. New York–London: Interscience Publishers (1949).

N-1. Udenfriend, S.: *Fluorescence Assay in Biology and Medecine*. New York: Academic Press (1962).

GAS CHROMATOGRAPHY

O-1. Berezkin, Viktor G: *Analytical Reaction Gas Chromatography*. New York: Plenum Press (1968).

P-1. Dal Nogare, S. and Juvet, R. S., Jr.: *Gas-Liquid Chromatography Theory and Practice*. New York–London: Interscience Publishers Div., John Wiley & Sons, Inc. (1962).

Q-1. Ettre, L. S: *Open Tubular Columns in Gas Chromatography*. New York: Plenum Press (1965).

R-1. Ettre, L. S. and Zlatkis, A.: *The Practice of Gas Chromatography*. New York: Interscience Publishers Div., John Wiley & Sons, Inc. (1967).

S-1. Goodwin, E. S., Goulden, R. and Reynolds, J. G.: *Analyst*, **86:** 697 (1961).

T-1. Gudzinowicsz, B.: *Gas Chromatographic Analysis of Drugs and Pesticides*. New York: Marcel Dekker, Inc. (1968).

U-1. Knox, J. H.: *Gas Chromatography*. London: Methuen & Co., Ltd. (1962).

V-1. Lovelock, J. E.: *Anal. Chem.*, **35:** 4746 (1963)

W-1. Signeur, A. V.: *Guide to Gas Chromatography Literature*. New York: Plenum Press (1964–1967). I–II.

X-1. Szymanski, H. A.: *Biomedical Applications of Gas Chromatography*. New York: Plenum Press (1964–1968). I–II.

NEUTRON ACTIVATION ANALYSIS

Y-1. Reference J-1. 379–380.

Z-1. O'Connor, R. T.: *J. Assoc. Offic. Anal. Chemists*, **51:** 235–238 (1968).

POLARIMETRY

A-2. Bates, F. J., *et al.*: *Polarimetry, Saccharimetry and the Sugars*. Circular C440. Washington: National Bureau of Standards (1942).

B-2. Crabbé, P.: *Optical Rotatory Dispersion and Circular Dichroism*. San Francisco: Holden Day (1965).

C-2. Djerassi, C.: *Optical Rotatory Dispersion*. New York: McGraw-Hill Book Co. (1960).

D-2. Lowry, M. T.: *Optical Rotatory Power*. New York: Dover Publications (1964).

E-2. Struck, W. S. and Olson, E. C.: *Optical Rotation: Polarimetry*. In Kolthoff, I. M., Elving, P. J. and Sandell, E. V. (eds.): *Treatise on Analytical Chemistry*. New York–London–Sydney: Interscience Publisher Div., John Wiley & Sons, Inc., (1965). Part I(6), chapter 31.

POLAROGRAPHY; AMPEROMETRIC TITRATIONS

F-2. Breyer, B. and Bauer, H. H.: *Alternating Current Polarography and Tensammetry*. New York: Interscience Publishers Div., John Wiley & Sons, Inc. (1963).

G-2. Kambara, T.: *Modern Aspects of Polarography*. New York: Plenum Press (1966).

H-2. Kolthoff, I. M., and Lingane, J. J.: *Polarography*, 2nd ed. New York–London: Interscience Publishers (1952). I–II.

I-2. Meites, L.: *Polarographic Techniques*, 2nd ed. New York: Interscience Publishers Div., John Wiley & Sons, Inc. (1965).

J-2. Milner, G. W. C.: *The Principles and Applications of Polarography and Other Electroanalytical Processes*. London–New York–Toronto: Longmans, Green & Co. (1957).

K-2. Zuman, P.: *Chem. and Eng. News*, **46** (No. 13): 94 (March 16, 1968).

L-2. Zuman, P., *Substituent Effects in Organic Polarography*. New York: Plenum Press (1967).

RAMAN SPECTROSCOPY

M-2. American Petroleum Institute, Research Project 44, *Catalog of Raman Spectral Data*. College Station: A & M College of Texas.

N-2. Clark, G. L. (ed.): *The Encyclopedia of Spectroscopy*, pp. 675–684. New York: Reinhold Publishing Corp. (1960).

O-2. Dudenbostel, B. F., Jr.: *Raman Spectroscopy: Instrumentation and Application Molecular Spectroscopy*, pp. 20–26. London: Institute of Petroleum (1954).

P-2. Hibben, J. H.: "Raman Spectra". In: *Physical Methods in Chemical Analysis.* Berl, W. G. (ed.): New York: Academic Press (1950). I, 405–423.

Q-2. Hibben, J. H.: *The Raman Effect and its Chemical Applications.* New York: Reinhold Publishing Corp. (1939).

R-2. Kohlrausch, K. W. F.: *Der Smekal-Raman Effekt.* Berlin: J. Springer (1931 and Ergänzungsband 1931–37).

S-2. Otting, W.: *Der Raman-Effekt und Seine Analytische Anwendung.* Berlin: Springer-Verlag (1952).

T-2. Rosenbaum, E. J.: "Raman Spectroscopy". In: Kolthoff, I. M., Elving, P. J. and Sandell, E. B. (eds.): *Treatise on Analytical Chemistry.* New York–London–Sydney: Interscience Publishers Div., John Wiley & Sons, Inc. (1965). Part I(6), Chapter 67.

U-2. Stoicheff, B. P.: "High Resolution Raman Spectroscopy". In: Thompson, H. W., (ed.).: *Advances in Spectroscopy*, New York: Interscience Publishers (1950). I, 91–174.

V-2. Szymanski, H. A.: *Raman Spectroscopy, Theory and Practice.* New York: Plenum Press (1967).

X-RAY FLUORESCENCE SPECTROGRAPHY

W-2. Liebhafsky, H. A., Pfeiffer, H. G. and Winslow, E. H. In: Kolthoff, I. M., Elving, P. J. and Sandell, E. B. (eds.): *Treatise on Analytical Chemistry.* New York–London–Sydney: Interscience Publishers Div., John Wiley & Sons, Inc. (1964). Part I (5) Chapter 60, 3138–3171.

X-2. Von Hevesy, G.: *Chemical Analysis by X-Rays and Its Applications.* New York: McGraw-Hill Book Co. (1932).

INFRARED SPECTROPHOTOMETRY

Y-2. Anon.: *Commentary No.* 9. South Norwalk: Wilks Scientific Corp. (1966). (ATR)

Z-2. Anon.: *Internal Reflection Spectroscopy.* South Norwalk: Wilks Scientific Corp. (1965). I. (ATR).

A-3. Barnes, R. B., Gore, R. C., Liddel, U. and Williams, Van Zandt: *Infrared Spectroscopy Industrial Applications and Bibliography.* New York: Reinhold Publishing Corp. (1944).

B-3. Bentley, F. F., Smithson, L. D. and Rozek, A. L.: *Infrared Spectra and Characteristic Frequencies.* New York: Interscience Publishers Div., John Wiley & Sons, Inc. (1967).

C-3. Bellamy, L. J.: *The Infra-Red Spectra of Complex Molecules.* New York: John Wiley & Sons, Inc. (1954).

D-3. Brügel, W.: *An Introduction to Infrared Spectroscopy.* New York: John Wiley & Sons, Inc. (1962).

E-3. Colthup, N. B., Daly, L. H. and Wiberley, S. E.: *Introduction to Infrared and Raman Spectroscopy.* New York–London: Academic Press (1964).

F-3. Conn, G. K. T. and Aveny, D. G.: *Infrared Methods Principles and Applications.* New York–London: Academic Press (1960).

G-3. Cross, A. D.: *An Introduction to Practical Infrared Spectroscopy.* London: Butterworths (1960).

H-3. Dobriner, K., Katzenellenbogen, E. R. and Jones, R. N.: *Infrared Absorption Spectra of Steroids.* New York: Interscience Publishers (1953–1958). I–II.

I-3. Fahrenfort, J.: *Spectrochim. Acta,* **17:** 698 (1961). (ATR).

J-3. Harrick, N. J.: *Internal Reflection Spectroscopy.* New York: Interscience Publishers Div., John Wiley & Sons, Inc. (1967) (ATR).

K-3. Harrick, N. J.: *Ann. N. Y. Acad. Sci.,* **101** (Art. 3): 928 (1963); *J. Opt. Soc.,* **55**(7):351 (1965). (ATR).

L-3. Hershenson, H. M.: *Infrared Absorption Spectra, Indexes for 1945–1957 and 1958–1962.* New York: Academic Press (1959–1964).

M-3. International Union of Pure and Applied Chemistry Commission on Molecular Structure and Spectroscopy: *Table of Wave numbers for the Calibration of Infra-Red Spectrometers.* London: Butterworths, (1961).

N-3. Kirchhoefer, R. D., (U.S. Food and Drug Administration): Private communication (March 1968).

O-3. Kruse, P. W., McGlauchlin, L. D. and McQuiston, R. B.: *Elements of Infrared Technology.* New York–London: John Wiley & Sons, Inc. (1962).

P-3. Lawson, K. E.: *Infrared Absorption of Inorganic Substances.* New York: Reinhold Publishing Corp. (1961).

Q-3. Rao, C. N. R.: *Chemical Applications of Infrared Spectroscopy.* New York–London: Academic Press (1964).

R-3. Silverstein, R. M. and Brooks, G. C.: *Spectrometric Identification of Organic Compounds.* New York–London: John Wiley & Sons, Inc. (1964).

S-3. Smith, A. L. "Infrared Spectroscopy". Chapter 66, pp. 3570–3575 (Tables 66.II–66.V), of Reference K-1, Part I(6). (Bibliographies of works on infrared spectroscopy.)

T-3. Szymanski, H. A., and Erickson, R. E.: *Infrared Band Handbook*, 2nd ed. New York: IFI/Plenum Data Corp. (1970).

U-3. Szymanski, H. A.: *Interpreted Infrared Spectra.* New York: Plenum Press (1964–1967). I–III.

V-3. Szymanski, H. A. (ed.): *Progress in Infrared*

Spectroscopy. New York: Plenum Press (1962–1964). I–II.

W-3. Whetsel, K.: *Chem. and Eng. News*, **46** (No. 6):82 (February 5, 1968).

X-3. White, R. G.: *Handbook of Industrial Infrared Analysis*. New York: Plenum Press (1964).

MASS SPECTROMETRY

Y-3. *Index of Mass Spectral Data*, Special Technical Publication 356. Philadelphia: American Society for Testing Materials.

Z-3. Beynon, J. H.: *Mass Spectrometry and Its Applications to Organic Chemistry*. Amsterdam: Elsevier Publishing Co. (1960).

A-4. Beynon, J. H. and Williams, E. E.: *Mass and Abundance Tables for Use in Mass Spectrometry*. Amsterdam: Elsevier Publishing Co. (1963).

B-4. Biemann, K.: *Mass Spectrometry—Organic Chemical Applications*. New York: McGraw-Hill Publishing Co. (1962).

C-4. Bonarovich, H. A. and Freeman, S. K.: "Mass Spectrometry". In: Freeman, S. K. (ed.): *Interpretive Spectroscopy*. New York: Reinhold Publishing Corp. (1965).

D-4. Budzikiewicz, H., Djerassi, C. and Williams, D. H.: *Interpretation of Mass Spectra of Organic Compounds*. San Francisco: Holden-Day, Inc. (1964).

E-4. Dibeler, V. H.: *Anal. Chem.*, **26**: 58 (1954). (Bibliography).

F-4. McLafferty, F. W.: "Mass Spectrometry". In: Nachod and Phillips (eds.): *Determination of Organic Structure by Physical Methods*. New York: Academic Press, Inc. (1962). II, 63–179.

G-4. McLafferty, F. W. and Pinzelik, J.: *Index and Bibliography of Mass Spectrometry 1962–1965*. New York: Interscience Publishers Div., John Wiley & Sons, Inc. (1967).

H-4. Von Hoene, J. and Loeffler, M.: *Indexed Catalogue of Mass Spectral Data*. Pittsburg: Westinghouse Research Laboratories (1958).

NUCLEAR MAGNETIC RESONANCE SPECTROSCOPY

I-4. Aleksandrov, L.: *The Theory of Nuclear Magnetic Resonance*. New York–London: Academic Press (1966).

J-4. Andrew, E. R.: *Nuclear Magnetic Resonance*. London: Cambridge University Press (1956).

K-4. Bhacca, N. S., Heudert, W. and Röpke, H.: *Atlas of Steroid Spectra*. New York: Springer-Verlag New York, Inc. (1965).

L-4. Bhacca, N. S., *et al.*: *NMR Spectra Catalog*. Palo Alto: Varian Associates, (1962–1963). I–II.

M-4. Bhacca, N. S. and Williams, D. H.: *Applications of NMR Spectroscopy in Organic Chemistry*. San Francisco: Holden-Day, Inc. (1964).

N-4. Bible, R. H.: *Guide to the NMR Empirical Method: A Work-book*. New York: Plenum Press (1967).

O-4. Bible, R. H.: *Interpretation of NMR Spectra*. New York: Plenum Press (1965).

P-4. Bovey, F. A.: *NMR Data Tables for Organic Compounds*. New York: John Wiley & Sons, Inc. (1967).

Q-4. Brownstein, S.: *Chemical Reviews*, **59**: 463 (1959).

R-4. Brügel, W.: *Nuclear Magnetic Resonance Spectra and Chemical Structure*. New York–London: Academic Press, (1968). I.

S-4. Chapman, D. and Magnus, P. D.: *Introduction to Practical High Resolution Nuclear Magnetic Resonance Spectroscopy*. New York–London: Academic Press (1966).

T-4. Corio, P. L.: *Structure of High-Resolution NMR Spectra*. New York–London: Academic Press (1966).

U-4. Crutchfield, M. M., Dungan, C. H., Letchir, J. H., Marsh, V. and Van Wazer, J. R.: *P^{31} Nuclear Magnetic Resonance*. New York: Interscience Publishers Div., John Wiley & Sons, Inc. (1967).

V-4. Emsley, J. W., Freeney, J. and Sutcliffe, L. H. (eds.): *High Resolution Nuclear Magnetic Resonance Spectroscopy*. London: Pergamon Press (1966). I–II.

W-4. Hersehenson, H. M.: *Nuclear Magnetic Resonance and Electron Spin Resonance Spectra. Index for 1958–1963*. New York–London: Academic Press (1963).

X-4. Howell, M. G., Kende, A. S. and Webb, J. S.: *Formula Index to NMR Literature Data*. New York: Plenum Press (1965–1966). I–II.

Y-4. Jackman, L. M.: *Applications of Nuclear Magnetic Resonance Spectroscopy in Organic Chemistry*. New York: Pergamon Press (1959).

Z-4. Mathieson, D. W.: *Nuclear Magnetic Resonance for Organic Chemists*. New York–London: Academic Press (1967).

A-5. Mooney, E. F. (ed.): *Annual Review of NMR Spectroscopy*. New York: Academic Press (1968). I.

B-5. Pesce, B.: *Nuclear Magnetic Resonance in Chemistry*. New York: Academic Press (1965).

C-5. Pople, J. A., Schneider, W. C. and Bernstein, H. J.: *High Resolution Nuclear Magnetic Resonance*. New York: McGraw-Hill Publishing Co., Inc. (1959).

D-5. Roberts, J. D.: *An Introduction to Spin-Spin-Splitting in High-Resolution NMR Spectra*. New York: W. A. Benjamin, Inc. (1961).

E-5. Roberts, J. D.: *Nuclear Magnetic Resonance*. New York: McGraw-Hill Publishing Co., Inc. (1959).

F-5. Szymanski, H. A., and Yelin, R. E.: *NMR Band Handbook*. New York: IFI/Plenum Data Corp. (1968).

G-5. Waugh, J. S. (ed.): *Advances in Magnetic Resonance*. New York–London: Academic Press (1965–1967). I–II.

H-5. Wiberg, K. B. and Nisi, B. J.: *The Interpretation of NMR Spectra*. New York: W. A. Benjamin, Inc. (1962).

NUCLEAR QUADRUPOLE RESONANCE SPECTROMETRY

I-5. Bray, P. J. and Barnes, R. G.: *J. Chem. Phys.*, **27:** 551 (1957).

J-5. Drago, R. S.: *Anal. Chem.*, **38** (No. 4): 31A (1966).

K-5. Fedin, E. I. and Semin, G. K.: *Zh. Strukt. Khim.*, **1**(no. 4): 464 (1960).

L-5. Hooper, H. O. and Bray, P. J.: *J. Chem. Phys.*, **33:** 334 (1960).

M-5. Moross, G. G. and Story, H. S.: *J. Chem. Phys.*, **45:** 2770 (1966).

N-5. Volpicelli, R. J., Rao, B. B. N. and Baldeschwieler, D.: *Rev. Sci. Instr.*, **36:** 150 (1965).

X-RAY DIFFRACTOMETRY

O-5. Brown, J. G.: *X-Rays and Their Applications*. New York: Plenum Press (1960).

P-5. Kaelble, E. F. (ed.): *Handbook of X-Rays*. New York: McGraw-Hill Publishing Co., Inc. (1968).

Q-5. Klug, H. P. and Alexander, L. E.: *X-Ray Diffraction Procedures*. New York: John Wiley & Sons, Inc. (1954).

R-5. Simmons, I. L. and Lublin, P. (eds.): *Progress in Analytical Chemistry*. New York: Consultants Bureau/Plenum Press Divs., Plenum Publishing Corp. (1968).

S-5. Various editors: *Advances in X-Ray Analysis*. New York: Plenum Press (1960–1968). I–XI.

T-5. von Hevesy, G.: *Chemical Analysis by X-Rays and Its Applications*. New York: McGraw-Hill Book Co. (1932).

GENERAL

U-5. Brode, W. R.: *Chemical Spectroscopy*. New York: John Wiley & Sons, Inc. (1943).

V-5. Clark, G. L. (ed.): *The Encyclopedia of Spectroscopy*. New York: Reinhold Publishing Corp. (1960).

W-5. Freeman, S. K. (ed.): *Interpretive Spectroscopy*. New York: Reinhold Publishing Corp. (1965).

X-5. Maier, H. G.: *Leitfaden Moderner Methoden der Lebensmittelanalytik—Optische Methoden*. Darmstadt: Dr. Dietrich Steinkopff Verlag (1966).

Y-5. Mathieson, D. W. (ed.): *Interpretation of Organic Spectra*. New York: Academic Press (1965).

Z-5. Silverstein, R. M. and Bassler, G. C.: *Spectrometric Identification of Organic Compounds*, 2nd ed. New York: John Wiley & Sons, Inc. (1967).

A-6. Various editors: *Developments in Applied Spectroscopy*. New York: Plenum Press (1962–1970). I–XIII.

B-6. West, W. (ed.): *Chemical Applications of Spectroscopy*. New York–London: Interscience Publishers (1965).

C-6. Willard, H. H., Merritt, L., Jr. and Dean, J. A.: *Instrumental Methods of Analysis*. Princeton: D. Van Nostrand & Co. (1958).

CHAPTER 22

Standards and Specifications

Standards and specifications are not new, but as old as commerce itself. As trade grew beyond the barter stage, the need for a common trade language and definition of terms increased. Guilds and trade associations were formed to serve this purpose. Many of our present day standards are nothing more than codified trade practices.

The word "standard" implies some sort of officiality behind its promulgation. It may be by law or by the action of an organized trade association. A "specification" is, or should be, a clear and accurate description of the technical requirements for materials and products sought, including minimum requirements of quality necessary for an acceptable product. It is issued as a guide to processors of the item sought. Compliance with some standards is obligatory. Others are issued as guides to processors, as industry association standards or as definitions of established grades. Compliance with such standards is not obligatory.

Obligatory Standards

Food and Drug Administration Standards

The United States Department of Health, Education and Welfare, through its Food and Drug Administration, has promulgated *Definitions and Standards of Identity*, *Standards of Quality* and *Standards of Fill of Container* under authorization of the *Federal Food, Drug and Cosmetic Act* of 1938, as amended.[1] They are primarily standards for consumer products. These standards are published in the Federal Register as they issue, and copies of individual standards may be obtained from the Food and Drug Administration. To date, standards have been issued for 18 classes of foods; others are in preparation.

These standards have the force and effect of law. All foods for which standards have been promulgated must comply with such standards if they enter into interstate commerce. The standards are exclusive, that is, no ingredient not specifically included in the standard may be used. Certain of the listed ingredients must be included, while others are optional. Many state and local authorities in the United States have included FDA

Standard as parts of their own laws and regulations.

USDA Meat Standards

The United States Department of Agriculture, through its Meat Inspection Division, has established standards for certain meat products made in establishments where inspection is maintained.[2] These, too, are obligatory. The *Meat Inspection Act* requires that a meat inspector be on duty at all times in a plant engaged in slaughtering, processing or otherwise preparing meats for interstate or foreign commerce which are capable of being used as food for man. (The term "meat" is defined as that from cattle, sheep, swine or goats). The inspector has jurisdiction over sanitation, processing, composition and labeling of all products made at the plant and of all ingredients used in processing.

USDA has established by regulation definitions for many meat products. Examples are:

Meat Extract—not more than 25 percent moisture.

Hash—not less than 35 percent cooked and trimmed meat.

Chili-con-carne—not less than 40 percent meat computed on the weight of fresh meat and not more than 8 percent cereal, flour and/or other thickener.

Meat Stew—not less than 25 percent meat computed as fresh meat.

Hamburger—not more than 30 percent fat.

The United States Department of Agriculture through its Poultry Division also enforces the *Poultry Products Inspection Act* (1957). This act defines "Poultry" (chickens, turkeys, ducks, geese and guineas) and "poultry products", and has established obligatory standards for certain of these products. The method of enforcement is very similar to that for meats, viz: production under complete government inspection, with jurisdiction over sanitation, processing, composition and labeling of the output of the establishment. Their regulations[3] define chicken or turkey meat in its natural proportions as consisting of 50–65 percent light meat and 50–35 percent dark meat. Deviations from this must be appropriately labeled. The Poultry Division has also set limits on the minimum amount of poultry meat that must be present in such products as poultry pies, dinners, soups, chop suey, cacciatore, etc.

Canada Meat Standards

Canada Food and Drug Regulations[4] define "meat" as the edible part of the skeletal muscle of an animal that was healthy at the time of slaughter. The word "animal" means any animal, other than marine and fresh water animals, used as food. The *Canada Meat Inspection Act* definition of "meat" is practically the same, but states that it includes muscle of the tongue, diaphragm, heart or esophagus, but does not include muscle of the lips, snout, scalp or ears. The administration of the Canadian Act is very much like that of the U.S. Meat Inspection Act, viz: registration and continuous inspection at the establishment during slaughtering, processing and packing. However, the Canadian Act includes the meat of horses, rabbits, game and poultry (including pigeons), in addition to that of cattle, sheep, swine and goats as named in the United States Act.

The Canada Meat Inspection Act[5] is enforced by the Department of Agriculture, Health and Animals Branch. Their definitions and standards are obligatory.

Canadian Food and Drugs Act Standards

All foods made or sold in Canada are subject to the *Food and Drugs Act and Regulations*, under the jurisdiction of the Department of National Health and Welfare.[4] Inspection and enforcement is administered through that department's Food and Drug Directorate. Standards for foods are prescribed through authority of the Governor in Council, and are promulgated through regulations under the basic act. These standards are obligatory.

The standards define the food, and set certain analytical limits where practicable. The regulations, including these standards should be consulted by analysts and others interested in foods made in, or destined for, Canada. At times they specify "official methods" for analysis. These methods are not included in the text of the regulations. Many of them are identical with those of *Official Methods of Analysis, AOAC.*

Canadian Department of Agriculture Grade Standards

The Production and Marketing Branch of Canada Department of Agriculture enforces the *Canada Agricultural Standards Act.* Regulations respecting "Grading, Packing and Marketing of Processed Fruit and Vegetable Products" have been established.[6] Grades and standards of identity so established are compulsory, and products for which grades have been promulgated must comply with such standards and must bear a declaration of grade when offered for sale in Canada. In this respect Canadian grade standards differ from grade standards set by the United States Department of Agriculture, Consumer and Marketing Service, which are voluntary. These U.S. Standards are described later in this chapter under "Voluntary Standards".

Standards of Identity as listed in the *Processed Fruit and Vegetable Regulations*[6] are in most, but not all, cases identical with standards established under the *Canada Food and Drugs Act*[4], but for all practical purposes they have the same force and effect within the law. The Processed Fruit and Vegetable Regulations include requirements for net and drained weights, can size, quantity of contents declaration, syrup strength, etc., of canned products not covered by the Food and Drugs Act. Regulations issued under both acts should be studied by those interested in Canadian food law.

The regulations issued under the Agricultural Standards Act follow the same philosophy in describing attributes of grade as do their United States counterparts (*q.v.*).

Voluntary Standards

U. S. Department of Agriculture Grade Standards

USDA Consumer & Marketing Service[7] has established voluntary grade standards for some 150 different processed fruit and vegetable products. These are issued primarily as yardsticks of quality, as well as offering a common trading language for voluntary use by the industry. There are three grades available for such use, Grade A or Fancy, Grade B or Choice, and Grade C or standard. All three grades may be in use for certain products; for others only two grades, A and C, may be established. These Grade Standards have been developed in collaboration with the industry. For those products for which the Food and Drug Administration has established *Definitions and Standards of Identity, Quality and Fill of Container* (commonly referred to as FDA Standards), USDA has adopted as their minimum level for Grade C specifications that are at least as high as the mandatory FDA Standards.

Many of the USDA Grade Standards use a weighted scoring system, with a total possible score of 100, in rating certain factors of grade such as color, texture, size uniformity, consistency, *i.e.*, those factors that are usually established by subjective examination.

United States Standards of Grades also use established chemical methods for determining such elements as ash, moisture, acidity, fat, salt, etc. The Standard usually designates the method to be used, either by reference or by inclusion in the Standard. On the other hand the subjective methods of scoring are difficult to describe. Analysts can best obtain proficience in such methods by observing the techniques of an authorized USDA inspector.

U. S. Department of Interior Grade Standards

The Bureau of Commerical Fisheries[8] has issued United States Standards for Grades of certain fishery products. These follow the same grading principles as those used in USDA Standards.

Trade Standards

Some trade associations have established standards of identity and/or quality for specific foods. These are voluntary, but members of the industry association are expected to adhere to them. Examples are "Standards and Grades of Dry Milk, promulgated by the American Dry Milk Institute" (1965) and "Standards of the American Dehydrated Onion and Garlic Association" (1969).

Food Specifications

The United States government issues specifications governing most of the items purchased for official use, including many subsistence items. There are three general classes of U.S. purchasing specifications—Federal, Military and Departmental.[9] All governmental specifications are developed through cooperation of the Federal Agencies with representative segments of the industries that may be interested. Sales to Federal Agencies must comply with all pertinent specifications.

Federal Specifications. These cover standard items of a permanent nature, purchased by two or more Federal Agencies, of which at least one is a civilian agency. Copies of Federal Specifications may be obtained, for a fee, from the Superintendent of Documents, Washington, D.C.

Military Specifications. These are issued by the Department of Defense primarily for use by military agencies. They may be used by other government agencies. Copies of Military Specifications may be obtained from the procuring agency or from the Superintendent of Documents.

Departmental Specifications. These are issued by one Department, for its own use, when no Federal or Military specification exists. They may be used by other agencies and, if interest develops, they may be converted to a Federal Specification. There are relatively few Departmental Specifications other than Military. Some have been issued by the Veterans' Administration and the Post Office Department.

All of these specifications direct methods of sampling and tests to be used on the products. These test methods may be given by literature reference, or details of a test and its interpretation may constitute part of the specification. Such a specification may also list other applicable specifications covering raw materials, packaging and labeling, and any Federal laws that apply to the product.

Trade Specifications. These are specifications established by food manufacturers as a guide to their suppliers. They set up minimum standards of quality and tests applicable to the raw materials and ingredients used by the manufacturers in producing the finished food. Materials shipped to the food manufacturers will be rejected if tests show non-compliance. These trade specifications are often more stringent in their quality requirements than are corresponding government specifications. They also often include tight bacteriological specifications that are usually not included in the government specifications. In this way the processor insures maintenance of high quality of the finished food by tailoring his specifications on raw material and ingredients to fit his own processing needs.

TEXT REFERENCES

1. *Federal Food, Drug and Cosmetic Act and General Regulations for its Enforcement.* Washington: U.S. Department of Health, Education and Welfare (1964).
2. *Regulations Governing the Meat Inspection Act.* Washington: U.S. Department of Agriculture (1965).
3. *Regulations Governing the Inspection of Poultry and Poultry Products.* Washington,: U.S. Department of Agriculture (1965).
4. *Office Consolidation of the Foods and Drugs Act—and Regulations.* Ottawa: Queen's Printer (as of Jan. 27, 1970). B.14.002.
5. *Canada Meat Inspection Act and the Meat Inspection Regulations.* Ottawa: Queen's Printer (1965);
   Canada Agricultural Products Standard Act and Dressed and Eviscerated Poultry Regulations. Ottawa: Queen's Printer (1960).
6. *Office Consolidation of the Canada Agricultural Products Standards Act and the Processed Fruit and Vegetables Regulations.* Ottawa: Queen's Printer (1966).
7. *U.S. Standards for Grades of Processed Fruits, Vegetables and Certain Other Products.* Washington: U.S. Department of Agriculture, Consumer and Marketing Service (1970).
8. *Guide for Buying Fresh and Frozen Fish and Shellfish*, Circular 214. Washington: U.S. Dept. of the Interior, Bureau of Commercial Fisheries (Reprinted 1959). (Lists 14 U.S. Standards for Seafood.)
9. *Guide to Specifications of the Federal Government.* Washington, D.C.: General Services Administration (1965).

SELECTED REFERENCES

A. *Code of Federal Regulations.* Washington: Office of the Federal Register, National Archives and Records Service, General Services Administration (as of Jan. 1, 1970). Title 21, parts 1–119.

B. Allen, R. J. L.: "Food Standards in the United Kingdom". In: *Food Technol.*, **48:**151 (1965).

C. Bartlett, R. P. and Wegener, J. B.: "Sampling Plan Developed by USDA for Inspection of Processed Fruits and Vegetables." In: *Food Technol.*, **11:**526 (1957).

D. Department of Defense: *Sampling Procedures and Tables for Inspection of Attributes*, MIL-STD-105D. Washington: U.S. Government Printing Office (1963).

E. *Requirements of the Food, Drug and Cosmetic Act.* FDA Publication No. 2. (1964).

F. Kramer, A.: "Problem of Developing Grades and Standards of Quality". In: *Food, Drug and Cosmetic Law J.*, **7:**23 (1952).

G. Kramer, A. and Wigg, B. A. T.: *Fundamentals of Quality Control for the Food Industry.* Westport: Avi Publishing Co., Inc. (1962).

H. Somers, R. K.: "Federal Meat Inspection Labeling Program." In: *Assoc. Food & Drug Officials U.S., Quart. Bull.*, **29:**3 (1965).

I. Stefferud, E. (ed.): *Food, The Yearbook of Agriculture.* Washington: U.S. Dept. of Agriculture and U.S. Government Printing Office (1959). Chapter 6.

CHAPTER 23

Appendix (Numerical Tables)

(Tables 23-1 and 23-7 are from the Pharmacopeia of the United States of America, 17th Revision (1965); all others are from Official Methods of Analysis of the Association of Official Agricultural Chemists, 10th Ed. (1965). Reproduced by courtesy of the U.S. Pharmacopeial Convention and the Association of Official Analytical Chemists respectively.)

TABLE 23-1

Atomic Weights

Adopted by the International Union of Pure and Applied Physics (1960) and by the International Union of Pure and Applied Chemistry (1961)

Name	Symbol	Atomic Number	Atomic Weight*	Name	Symbol	Atomic Number	Atomic Weight*
Actinium	Ac	89	(227)	Curium	Cm	96	(247)
Aluminum	Al	13	26.9815	Dysprosium	Dy	66	162.50
Americium	Am	95	(243)	Einsteinium	Es	99	(254)
Antimony	Sb	51	121.75	Erbium	Er	68	167.26
Argon	Ar	18	39.948	Europium	Eu	63	151.96
Arsenic	As	33	74.9216	Fermium	Fm	100	(253)
Astatine	At	85	(210)	Fluorine	F	9	18.9984
Barium	Ba	56	137.34	Francium	Fr	87	(223)
Berkelium	Bk	97	(247)	Gadolinium	Gd	64	157.25
Beryllium	Be	4	9.0122	Gallium	Ga	31	69.72
Bismuth	Bi	83	208.980	Germanium	Ge	32	72.59
Boron	B	5	10.811	Gold	Au	79	196.967
Bromine	Br	35	79.909	Hafnium	Hf	72	178.49
Cadmium	Cd	48	112.40	Helium	He	2	4.0026
Calcium	Ca	20	40.08	Holmium	Ho	67	164.930
Californium	Cf	98	(249)	Hydrogen	H	1	1.00797
Carbon	C	6	12.01115	Indium	In	49	114.82
Cerium	Ce	58	140.12	Iodine	I	53	126.9044
Cesium	Cs	55	132.905	Iridium	Ir	77	192.2
Chlorine	Cl	17	35.453	Iron	Fe	26	55.847
Chromium	Cr	24	51.996	Krypton	Kr	36	83.80
Cobalt	Co	27	58.9332	Lanthanum	La	57	138.91
Copper	Cu	29	63.54	Lead	Pb	82	207.19

* Values in parentheses or brackets denote mass numbers of selected radioactive isotopes. Those in parentheses are for the isotopes of longest known half-life; those in brackets are for isotopes that are better known than the corresponding isotopes of the longest half-life. All atomic weight values are based on the atomic mass of $^{12}C = 12$.

TABLE 23-1 (concluded)
Atomic Weights

Name	Symbol	Atomic Number	Atomic Weight*	Name	Symbol	Atomic Number	Atomic Weight*
Lithium	Li	3	6.939	Rhodium	Rh	45	102.905
Lutetium	Lu	71	174.97	Rubidium	Rb	37	85.47
Magnesium	Mg	12	24.312	Ruthenium	Ru	44	101.07
Manganese	Mn	25	54.9380	Samarium	Sm	62	150.35
Mendelevium	Md	101	(256)	Scandium	Sc	21	44.956
Mercury	Hg	80	200.59	Selenium	Se	34	78.96
Molybdenum	Mo	42	95.94	Silicon	Si	14	28.086
Neodymium	Nd	60	144.24	Silver	Ag	47	107.870
Neon	Ne	10	20.183	Sodium	Na	11	22.9898
Neptunium	Np	93	(237)	Strontium	Sr	38	87.62
Nickel	Ni	28	58.71	Sulfur	S	16	32.064
Niobium	Nb	41	92.906	Tantalum	Ta	73	180.948
Nitrogen	N	7	14.0067	Technetium	Tc	43	[99]
Nobelium	No	102	—	Tellurium	Te	52	127.60
Osmium	Os	76	190.2	Terbium	Tb	65	158.924
Oxygen	O	8	15.9994	Thallium	Tl	81	204.37
Palladium	Pd	46	106.4	Thorium	Th	90	232.038
Phosphorus	P	15	30.9738	Thulium	Tm	69	168.934
Platinum	Pt	78	195.09	Tin	Sn	50	118.69
Plutonium	Pu	94	(242)	Titanium	Ti	22	47.90
Polonium	Po	84	[210]	Tungsten	W	74	183.85
Potassium	K	19	39.102	Uranium	U	92	238.03
Praseodymium	Pr	59	140.907	Vanadium	V	23	50.942
Promethium	Pm	61	[147]	Xenon	Xe	54	131.30
Protactinium	Pa	91	(231)	Ytterbium	Yb	70	173.04
Radium	Ra	88	(226)	Yttrium	Y	39	88.905
Radon	Rn	86	(222)	Zinc	Zn	30	65.37
Rhenium	Re	75	186.2	Zirconium	Zr	40	91.22

TABLE 23-2
Various strength solutions of the common acids, alkalies, and alcohol[a]

(a) *Hydrochloric acid solutions:* Specification requires not $<35\%$ HCl by wt. Sp. gr. = 1.1778 at 15°. Mix with H_2O and dilute to 1 L.

HCL strength desired	Hydrochloric Acid Required		
GRAMS PER LITER	GRAMS	ML	
5	14.29	12.13	
10	28.57	24.26	
15	42.85	36.39	
20	57.14	48.52	
36.46	104.17	88.45	1 *N* solution
50	142.86	121.29	
100	285.71	242.58	
150	428.57	363.88	
200	571.43	485.17	
222.6	636.00	539.99	Constant boiling
278.4	795.43	675.35	Sp. gr. 1.125
300	857.14	727.75	

[a] Prepared by G. C. Spencer and H. J. Fisher.

TABLE 23-2 (continued)

Various strength solutions of common acids, alkalies, and alcohol[a]

(b) *Sulfuric acid solutions:* Specification requires not <94% H_2SO_4 by wt. Sp. gr. = 1.835 at 15°. Pour acid into excess of H_2O and dilute to 1 L.

H_2SO_4 strength desired	Sulfuric Acid Required		
GRAMS PER LITER	GRAMS	ML	
5	5.32	3.0	
12.5	13.29	7.2	For crude fiber
20	21.28	11.6	
30	31.91	17.4	
40	42.55	23.2	
49	52.13	28.4	1 *N* solution
100	106.38	58.0	
150	159.57	87.0	
250	265.96	144.9	
300	319.15	173.9	
400	425.53	231.9	

(c) *Nitric acid solutions:* Specification requires not <68% HNO_3 by wt. Sp. gr. = 1.4146 at 15°. 1 ml conc. HNO_3 contains about 0.96 g HNO_3. Mix with H_2O and dilute to 1 L.

HNO_3 strength desired	Nitric Acid Required		
GRAMS PER LITER	GRAMS	ML	
5	7.35	5.2	
10	14.71	10.4	
20	29.41	20.8	
30	44.12	31.2	
40	58.82	41.6	
50	73.53	52.0	
63	92.65	65.5	1 *N* solution
70	102.94	72.8	
100	147.06	104.0	
150	220.59	156.0	
200	294.12	207.9	
300	441.18	312.9	

(d) *Ammonia solutions:* Specification requires not <27% HN_3 by wt. Sp. gr. = 0.9. Mix and dilute to 1 L.

NH_3 strength desired	Reagent Ammonia Required	
GRAMS PER LITER	GRAMS	ML
5	18.52	20.6
10	37.04	41.1
15	55.55	61.7
20	74.07	82.3
25	92.59	102.9
50	185.18	205.8
75	277.77	308.6
100	370.37	411.5
150	555.55	617.3
200	740.74	823.0

TABLE 23-2 (concluded)

Various strength solutions of common acids, alkalies, and alcohol[a]

(e) *Sodium hydroxide solutions:* Specification requires 95% NaOH in sticks or pellets of caustic soda. Dissolve and dilute to 1 L.

NaOH strength desired	Sodium Hydroxide Required	
GRAMS PER LITER	GRAMS	
12.5	13.16	For crude fiber
30	31.58	
40	42.11	1 *N* solution
50	52.63	
75	78.95	
100	105.26	
150	157.89	
200	210.53	
250	263.16	
300	315.79	

(f) *Alcoholic solutions:*[b] Specification requires 95% C_2H_5OH by vol. Sp. gr. = 0.810 at 25°. Mix and dilute to 1 L.

Alcohol Strength Desired	Alcohol Required	
ML PER LITER	GRAMS	ML
50	42.63	52.6
100	85.26	105.3
150	127.89	157.9
200	170.52	210.5
250	213.16	263.2
300	255.78	315.9
400	341.04	421.1
500	426.32 (proof)	526.3
700	596.84	736.8

[b] Alcohol of any desired strength may be obtained by taking number of ml 95% alcohol equivalent to desired strength and diluting solution to 95 ml. For example: To obtain solution of 70% alcohol, take 70 ml 95% alcohol and dilute to 95 ml.

TABLE 23-3

Refractive indices of sucrose solutions at 20°[a]

International Scale, 1936[b]

REFRACTIVE INDEX AT 20°	SUCROSE, %	REFRACTIVE INDEX AT 20°	SUCROSE, %	REFRACTIVE INDEX AT 20°	SUCROSE, %	REFRACTIVE INDEX AT 20°	SUCROSE, %	REFRACTIVE INDEX AT 20°	SUCROSE, %
1.33299	0.0	.33443	1.0	.33588	2.0	.33733	3.0	.33880	4.0
.33328	0.2	.33472	1.2	.33617	2.2	.33762	3.2	.33909	4.2
.33357	0.4	.33501	1.4	.33646	2.4	.33792	3.4	.33939	4.4
.33385	0.6	.33530	1.6	.33675	2.6	.33821	3.6	.33968	4.6
.33414	0.8	.33559	1.8	.33704	2.8	.33851	3.8	.33998	4.8

[a] The values in this table for the range 0 to 49.8% sucrose are in accordance with the International Scale of Refractive Indices of Sucrose at 20°, 1936 adopted as official at the 1938 meeting of the Association. Values of indices for range 0–24% sucrose are given to five decimal places, those 24.2 to 49.8% to four decimal places. Values for range 50 to 85% are those adopted as official at the 1959 meeting and are given to five decimal places.

[b] *Inter. Sugar J.* **39**:225 (1937).

TABLE 23-3 (continued)

Refractive indices of sucrose solutions at 20°[a]

REFRACTIVE INDEX AT 20°	SUCROSE, %	REFRACTIVE INDEX AT 20°	SUCROSE, %	REFRACTIVE INDEX AT 20°	SUCROSE, %	REFRACTIVE INDEX AT 20°	SUCROSE, %	REFRACTIVE INDEX AT 20°	SUCROSE, %
.34027	5.0	.35250	13.0	.36551	21.0	.3793	29.0	.3939	37.0
.34057	5.2	.35282	13.2	.36585	21.2	.3797	29.2	.3943	37.2
.34087	5.4	.35313	13.4	.36618	21.4	.3800	29.4	.3947	37.4
.34116	5.6	.35345	13.6	.36652	21.6	.3804	29.6	.3950	37.6
.34146	5.8	.35376	13.8	.36685	21.8	.3807	29.8	.3954	37.8
.34176	6.0	.35408	14.0	.36719	22.0	.3811	30.0	.3958	38.0
.34206	6.2	.35440	14.2	.36753	22.2	.3815	30.2	.3962	38.2
.34236	6.4	.35472	14.4	.36787	22.4	.3818	30.4	.3966	38.4
.34266	6.6	.35503	14.6	.36820	22.6	.3822	30.6	.3970	38.6
.34296	6.8	.35535	14.8	.36854	22.8	.3825	30.8	.3974	38.8
.34326	7.0	.35567	15.0	.36888	23.0	.3829	31.0	.3978	39.0
.34356	7.2	.35599	15.2	.36922	23.2	.3833	31.2	.3982	39.2
.34386	7.4	.35631	15.4	.36956	23.4	.3836	31.4	.3986	39.4
.34417	7.6	.35664	15.6	.36991	23.6	.3840	31.6	.3989	39.6
.34447	7.8	.35696	15.8	.37025	23.8	.3843	31.8	.3993	39.8
.34477	8.0	.35728	16.0	.37059	24.0	.3847	32.0	.3997	40.0
.34507	8.2	.35760	16.2	.3709	24.2	.3851	32.2	.4001	40.2
.34538	8.4	.35793	16.4	.3713	24.4	.3854	32.4	.4005	40.4
.34568	8.6	.35825	16.6	.3716	24.6	.3858	32.6	.4008	40.6
.34599	8.8	.35858	16.8	.3720	24.8	.3861	32.8	.4012	40.8
1.34629	9.0	.35890	17.0	.3723	25.0	.3865	33.0	.4016	41.0
.34660	9.2	.35923	17.2	.3726	25.2	.3869	33.2	.4020	41.2
.34691	9.4	.35955	17.4	.3730	25.4	.3872	33.4	.4024	41.4
.34721	9.6	.35988	17.6	.3733	25.6	.3876	33.6	.4028	41.6
.34752	9.8	.36020	17.8	.3737	25.8	.3879	33.8	.4032	41.8
.34783	10.0	1.36053	18.0	.3740	26.0	.3883	34.0	.4036	42.0
.34814	10.2	.36086	18.2	.3744	26.2	.3887	34.2	.4040	42.2
.34845	10.4	.36119	18.4	.3747	26.4	.3891	34.4	.4044	42.4
.34875	10.6	.36152	18.6	.3751	26.6	.3894	34.6	.4048	42.6
.34906	10.8	.36185	18.8	.3754	26.8	.3898	34.8	.4052	42.8
.34937	11.0	.36218	19.0	1.3758	27.0	.3902	35.0	.4056	43.0
.34968	11.2	.36251	19.2	.3761	27.2	.3906	35.2	.4060	43.2
.34999	11.4	.36284	19.4	.3765	27.4	.3909	35.4	.4064	43.4
.35031	11.6	.36318	19.6	.3768	27.6	.3913	35.6	.4068	43.6
.35062	11.8	.36351	19.8	.3772	27.8	.3916	35.8	.4072	43.8
.35093	12.0	.36384	20.0	.3775	28.0	1.3920	36.0	.4076	44.0
.35124	12.2	.36417	20.2	.3779	28.2	.3924	36.2	.4080	44.2
.35156	12.4	.36451	20.4	.3782	28.4	.3928	36.4	.4084	44.4
.35187	12.6	.36484	20.6	.3786	28.6	.3931	36.6	.4088	44.6
.35219	12.8	.36518	20.8	.3789	28.8	.3935	36.8	.4092	44.8

TABLE 23-3 (concluded)

Refractive indices of sucrose solutions at 20°[a]

REFRACTIVE INDEX AT 20°	SUCROSE, %	REFRACTIVE INDEX AT 20°	SUCROSE, %	REFRACTIVE INDEX AT 20°	SUCROSE, %	REFRACTIVE INDEX AT 20°	SUCROSE, %	REFRACTIVE INDEX AT 20°	SUCROSE, %
1.4096	45.0	1.42646	53.0	1.44420	61.0	1.46299	69.0	1.48288	77.0
.4100	45.2	.42689	53.2	.44465	61.2	.46347	69.2	.48339	77.2
.4104	45.4	.42733	53.4	.44511	61.4	.46396	69.4	.48390	77.4
.4109	45.6	.42776	53.6	.44557	61.6	.46444	69.6	.48442	77.6
.4113	45.8	.42819	53.8	.44603	61.8	.46493	69.8	.48493	77.8
.4117	46.0	.42862	54.0	.44649	62.0	.46541	70.0	.48544	78.0
.4121	46.2	.42906	54.2	.44695	62.2	.46590	70.2	.48596	78.2
.4125	46.4	.42949	54.4	.44741	62.4	.46639	70.4	.48648	78.4
.4129	46.6	.42993	54.6	.44787	62.6	.46688	70.6	.48699	78.6
.4133	46.8	.43036	54.8	.44833	62.8	.46737	70.8	.48751	78.8
.4137	47.0	.43080	55.0	.44879	63.0	.46786	71.0	.48803	79.0
.4141	47.2	.43124	55.2	.44926	63.2	.46835	71.2	.48855	79.2
.4145	47.4	.43168	55.4	.44972	63.4	.46884	71.4	.48907	79.4
.4150	47.6	.43211	55.6	.45019	63.6	.46933	71.6	.48959	79.6
.4154	47.8	.43255	55.8	.45065	63.8	.46982	71.8	.49011	79.8
.4158	48.0	.43299	56.0	.45112	64.0	.47032	72.0	.49063	80.0
.4162	48.2	.43343	56.2	.45158	64.2	.47081	72.2	.49115	80.2
.4166	48.4	.43387	56.4	.45205	64.4	.47131	72.4	.49167	80.4
.4171	48.6	.43432	56.6	.45252	64.6	.47180	72.6	.49220	80.6
.4175	48.8	.43476	56.8	.45299	64.8	.47230	72.8	.49272	80.8
.4179	49.0	.43520	57.0	.45346	65.0	.47279	73.0	.49325	81.0
.4183	49.2	.43564	57.2	.45393	65.2	.47329	73.2	.49377	81.2
.4187	49.4	.43609	57.4	.45440	65.4	.47379	73.4	.49430	81.4
.4192	49.6	.43653	57.6	.45487	65.6	.47429	73.6	.49483	81.6
.4196	49.8	.43698	57.8	.45534	65.8	.47479	73.8	.49536	81.8
.42008	50.0	.43742	58.0	.45581	66.0	.47529	74.0	.49589	82.0
.42050	50.2	.43787	58.2	.45629	66.2	.47579	74.2	.49641	82.2
.42092	50.4	.43832	58.4	.45676	66.4	.47629	74.4	.49695	82.4
.42135	50.6	.43877	58.6	.45724	66.6	.47679	74.6	.49748	82.6
.42177	50.8	.43922	58.8	.45771	66.8	.47730	74.8	.49801	82.8
.42219	51.0	.43966	59.0	.45819	67.0	.47780	75.0	.49854	83.0
.42261	51.2	.44011	59.2	.45857	67.2	.47831	75.2	.49907	83.2
.42304	51.4	.44057	59.4	.45914	67.4	.47881	75.4	.49961	83.4
.42347	51.6	.44102	59.6	.45962	67.6	.47932	75.6	.50014	83.6
.42389	51.8	.44147	59.8	.46010	67.8	.47982	75.8	.50068	83.8
.42432	52.0	.44192	60.0	.46058	68.0	.48033	76.0	.50121	84.0
.42475	52.2	.44238	60.2	.46106	68.2	.48084	76.2	.50175	84.2
.42517	52.4	.44283	60.4	.46154	68.4	.48135	76.4	.50229	84.4
.42560	52.6	.44328	60.6	.46202	68.6	.48186	76.6	.50283	86.6
.42603	52.8	.44374	60.8	.46251	68.8	.48237	76.8	.50337	84.8
								.50391	85.0

TABLE 23-4

Corrections for determining % sucrose in sugar solutions by means of either Abbé or immersion refractometer when readings are made at temperatures other than 20°[a]

International Temperature Correction Table, 1936[a]

TEMP. °C	% Sucrose										
	0	5	10	15	20	25	30	40	50	60	70
	Subtract from the % sucrose										
10	0.50	0.54	0.58	0.61	0.64	0.66	0.68	0.72	0.74	0.76	0.79
11	.46	.49	.53	.55	.58	.60	.62	.65	.67	.69	.71
12	.42	.45	.48	.50	.52	.54	.56	.58	.60	.61	.63
13	.37	.40	.42	.44	.46	.48	.49	.51	.53	.54	.55
14	.33	.35	.37	.39	.40	.41	.42	.44	.45	.46	.48
15	.27	.29	.31	.33	.34	.34	.35	.37	.38	.39	.40
16	.22	.24	.25	.26	.27	.28	.28	.30	.30	.31	.32
17	.17	.18	.19	.20	.21	.21	.21	.22	.23	.23	.24
18	.12	.13	.13	.14	.14	.14	.14	.15	.15	.16	.16
19	.06	.06	.07	.07	.07	.07	.07	.08	.08	.08	.08
	Add to the % sucrose										
21	0.06	0.07	0.07	0.07	0.07	0.08	0.08	0.08	0.08	0.08	0.08
22	.13	.13	.14	.14	.15	.15	.15	.15	.16	.16	.16
23	.19	.20	.21	.22	.22	.23	.23	.23	.24	.24	.24
24	.26	.27	.28	.29	.30	.31	.31	.31	.31	.32	.32
25	.33	.35	.36	.37	.38	.38	.39	.40	.40	.40	.40
26	.40	.42	.43	.44	.45	.46	.47	.48	.48	.48	.48
27	.48	.50	.52	.53	.54	.55	.55	.56	.56	.56	.56
28	.56	.57	.60	.61	.62	.63	.63	.64	.64	.64	.64
29	.64	.66	.68	.69	.71	.72	.72	.73	.73	.73	.73
30	.72	.74	.77	.78	.79	.80	.80	.81	.81	.81	.81

[a] *Intern. Sugar J.*, **39:** 24s (1937).

TABLE 23-5

Table for determining % sucrose in sugar solutions from readings of Zeiss immersion refractometer at 20°[a]

Scale reading[b] 20°	n_D^{20}	Sucrose %	Scale reading[b] 20°	n_D^{20}	Sucrose %	Scale reading[b] 20°	n_D^{20}	Sucrose %
14.47	1.33299	0	45	1.34463	7.91	76	1.35606	15.24
15	3320	0.15	46	4500	8.15	77	5642	15.47
16	3358	0.41	47	4537	8.39	78	5678	15.69
17	3397	0.68	48	4575	8.64	79	5714	15.91
18	3435	0.94	49	4612	8.89	80	5750	16.14
19	3474	1.21	50	4650	9.13	81	5786	16.36
20	3513	1.48	51	4687	9.38	82	5822	16.58
21	3551	1.74	52	4724	9.62	83	5858	16.81
22	3590	2.01	53	4761	9.86	84	5894	17.03
23	3628	2.27	54	4798	10.10	85	5930	17.25
24	3667	2.54	55	4836	10.34	86	5966	17.47
25	3705	2.80	56	4873	10.58	87	6002	17.69
26	3743	3.07	57	4910	10.82	88	6038	17.91
27	3781	3.33	58	4947	11.06	89	6074	18.12
28	3820	3.59	59	4984	11.30	90	6109	18.34
29	3858	3.85	60	5021	11.54	91	6145	18.56
30	3896	4.11	61	5058	11.78	92	6181	18.78
31	3934	4.36	62	5095	12.01	93	6127	19.00
32	3972	4.62	63	5132	12.25	94	6252	19.21
33	4010	4.88	64	5169	12.48	95	6287	19.42
34	4048	5.14	65	5205	12.72	96	6323	19.63
35	4086	5.40	66	5242	12.95	97	6359	19.85
36	4124	5.65	67	5279	13.18	98	6394	20.06
37	4162	5.91	68	5316	13.41	99	6429	20.27
38	4199	6.16	69	5352	13.64	100	6464	20.48
39	4237	6.41	70	5388	13.87	101	6500	20.69
40	4275	6.66	71	5425	14.10	102	6535	20.90
41	4313	6.91	72	5461	14.33	103	6570	21.11
42	4350	7.16	73	5497	14.56	104	6605	21.32
43	4388	7.41	74	5533	14.79	105	6640	21.53
44	4426	7.66	75	5569	15.01			

[a] Values in this table were calculated by J. A. Mathews from five-place indices of Schönrock as given by Landt, *Z. Ver. deut. Zucker-Ind.*, **83**:692 (1933).

[b] Scale readings refer only to scale of arbitrary units proposed by Pulfrich, *Z. Angew. Chem.*, p. 1168 (1899). According to this scale 14.5 = 1.33300, 50.0 = 1.34650, and 100.0 = 1.36464. If immersion refractometer used is calibrated according to another arbitrary scale, readings must be converted into refractive indices before this table is used to determine percent sugar.

TABLE 23-6

Revised Hammond table for calculating dextrose, levulose, lactose, invert sugar alone, and invert sugar in presence of sucrose (0.3, 0.4, and 2.0 g of total sugar)[a]

Applicable when Cu is determined by analysis (Expressed in mg)

Copper (Cu)	Dextrose	Levulose	Lactose H_2O	Invert Sugar	Invert Sugar and Sucrose		
					0.3 g of total sugar	0.4 g of total sugar	2.0 g of total sugar
10	4.6	5.1	7.7	5.2	3.2	2.9	....
12	5.6	6.1	9.3	6.2	4.2	3.9	....
14	6.5	7.2	10.8	7.2	5.3	4.9	....
16	7.5	8.3	12.3	8.2	6.3	5.9	....
18	8.5	9.3	13.8	9.2	7.3	6.9	....
20	9.4	10.4	15.4	10.2	8.3	7.9	1.9
22	10.4	11.5	16.9	11.2	9.3	8.9	2.9
24	11.4	12.5	18.4	12.3	10.4	10.0	3.9
26	12.3	13.6	19.9	13.3	11.4	11.0	4.9
28	13.3	14.7	21.5	14.3	12.4	12.0	6.0
30	14.3	15.8	23.0	15.3	13.4	13.0	7.0
32	15.3	16.8	24.5	16.3	14.5	14.1	8.0
34	16.2	17.9	26.1	17.3	15.5	15.1	9.0
36	17.2	19.0	27.6	18.3	16.5	16.1	10.1
38	18.2	20.1	29.1	19.4	17.6	17.1	11.1
40	19.2	21.1	30.6	20.4	18.6	18.2	12.1
42	20.1	22.2	32.2	21.4	19.6	19.2	13.1
44	21.1	23.3	33.7	22.4	20.7	20.2	14.2
46	22.1	24.4	35.2	23.5	21.7	21.3	15.2
48	23.1	25.4	36.8	24.5	22.7	22.3	16.2
50	24.1	26.5	38.3	25.5	23.8	23.3	17.3
52	25.1	27.6	39.8	26.5	24.8	24.3	18.3
54	26.1	28.7	41.4	27.6	25.8	25.4	19.3
56	27.0	29.8	42.9	28.6	26.9	26.4	20.4
58	28.0	30.9	44.4	29.6	27.9	27.5	21.4
60	29.0	31.9	46.0	30.6	28.9	28.5	22.5
62	30.0	33.0	47.5	31.7	30.0	29.5	23.5
64	31.0	34.1	49.0	32.7	31.0	30.6	24.5
66	32.0	35.2	50.6	33.7	32.1	31.6	25.6
68	33.0	36.3	52.1	34.8	33.1	32.7	26.6
70	34.0	37.4	53.6	35.8	34.2	33.7	27.7
72	35.0	38.5	55.2	36.8	35.2	34.7	28.7
74	36.0	39.6	56.7	37.9	36.3	35.8	29.8
76	37.0	40.7	58.2	38.9	37.3	36.8	30.8
78	38.0	41.7	59.8	40.0	38.4	37.9	31.9
80	39.0	42.8	61.3	41.0	39.4	38.9	32.9
82	40.0	43.9	62.8	42.0	40.5	40.0	34.0
84	41.0	45.0	64.4	43.1	41.5	41.0	35.0
86	42.0	46.1	65.9	44.1	42.6	42.1	36.1
88	43.0	47.2	67.4	45.2	43.6	43.1	37.1

[a] *J. Research, NBS*, **24**:589 (1940); **41**:211 (1948).

TABLE 23-6 (continued)

Copper (Cu)	Dextrose	Levulose	Lactose H_2O	Invert Sugar	Invert Sugar and Sucrose		
					0.3 g of total sugar	0.4 g of total sugar	2.0 g of total sugar
90	44.0	48.3	69.0	46.2	44.7	44.2	38.2
92	45.0	49.4	70.5	47.3	45.7	45.2	39.2
94	46.0	50.5	72.1	48.3	46.8	46.3	40.3
96	47.0	51.6	73.6	49.4	47.8	47.4	41.3
98	48.0	52.7	75.1	50.4	48.9	48.4	42.4
100	49.0	53.8	76.7	51.5	50.0	49.5	43.5
102	50.0	54.9	78.2	52.5	51.0	50.5	44.5
104	51.1	56.0	79.7	53.6	52.1	51.6	45.6
106	52.1	57.1	81.3	54.6	53.1	52.7	46.7
108	53.1	58.2	82.8	55.7	54.2	53.7	47.7
110	54.1	59.3	84.4	56.7	55.3	54.8	48.8
112	55.1	60.4	85.9	57.8	56.3	55.8	49.9
114	56.1	61.6	87.4	58.9	57.4	56.9	50.9
116	57.2	62.7	89.0	59.9	58.5	58.0	52.0
118	58.2	63.8	90.5	61.0	59.5	59.0	53.1
120	59.2	64.9	92.1	62.0	60.6	60.1	54.1
122	60.2	66.0	93.6	63.1	61.7	61.2	55.2
124	61.3	67.1	95.2	64.2	62.8	62.3	56.3
126	62.3	68.2	96.7	65.2	63.8	63.3	57.4
128	63.3	69.3	98.2	66.3	64.9	64.4	58.4
130	64.3	70.4	99.8	67.4	66.0	65.5	59.5
132	65.4	71.6	101.3	68.4	67.1	66.6	60.6
134	66.4	72.7	102.9	69.5	68.1	67.6	61.7
136	67.4	73.8	104.4	70.6	69.2	68.7	62.8
138	68.5	74.9	106.0	71.6	70.3	69.8	63.9
140	69.5	76.0	107.5	72.7	71.4	70.9	64.9
142	70.5	77.1	109.0	73.8	72.5	72.0	66.9
144	71.6	78.3	110.6	74.9	73.5	73.0	67.1
146	72.6	79.4	112.1	75.9	74.6	74.1	68.2
148	73.7	80.5	113.7	77.0	75.7	75.2	69.3
150	74.7	81.6	115.2	78.1	76.8	76.3	70.4
152	75.7	82.8	116.8	79.2	77.9	77.4	71.5
154	76.8	83.9	118.3	80.3	79.0	78.5	72.6
156	77.8	85.0	119.9	81.3	80.1	79.6	73.7
158	78.9	86.1	121.4	82.4	81.2	80.6	74.8
160	79.9	87.3	122.9	83.5	82.2	81.7	75.9
162	81.0	88.4	124.5	84.6	83.3	82.8	77.0
164	82.0	89.5	126.0	85.7	84.4	83.9	78.1
166	83.1	90.6	127.6	86.8	85.5	85.0	79.2
168	84.1	91.8	129.1	87.8	86.6	86.1	80.3
170	85.2	92.9	130.7	88.9	87.7	87.2	81.4
172	86.2	94.0	132.2	90.0	88.8	88.3	82.5
174	87.3	95.2	133.8	91.1	89.9	89.4	83.6
176	88.3	96.3	135.3	92.2	91.0	90.5	84.7
178	89.4	97.4	136.9	93.3	92.1	91.6	85.8

TABLE 23-6 (continued)

Copper (Cu)	Dextrose	Levulose	Lactose H_2O	Invert Sugar	Invert Sugar and Sucrose		
					0.3 g of total sugar	0.4 g of total sugar	2.0 g of total sugar
180	90.4	98.6	138.4	94.4	93.2	92.7	86.9
182	91.5	99.7	140.0	95.5	94.3	93.8	88.0
184	92.6	100.9	141.5	96.6	95.4	94.9	89.1
186	93.6	102.0	143.1	97.7	96.5	96.0	90.2
188	94.7	103.1	144.6	98.8	97.6	97.1	91.3
190	95.7	104.3	146.2	99.9	98.7	98.2	92.4
192	96.8	105.4	147.7	101.0	99.9	99.4	93.6
194	97.9	106.6	149.3	102.1	101.0	100.5	94.7
196	98.9	107.7	150.8	103.2	102.1	101.6	95.8
198	100.0	108.8	152.4	104.3	103.2	102.7	96.9
200	101.1	110.0	153.9	105.4	104.3	103.8	98.0
202	102.2	111.1	155.5	106.5	105.4	104.9	99.2
204	103.2	112.3	157.0	107.6	106.5	106.0	100.3
206	104.3	113.4	158.6	108.7	107.6	107.2	101.4
208	105.4	114.6	160.2	109.8	108.8	108.3	102.5
210	106.5	115.7	161.7	110.9	109.9	109.4	103.7
212	107.5	116.9	163.3	112.1	111.0	110.5	104.8
214	108.6	118.0	164.8	113.2	112.1	111.6	105.9
216	109.7	119.2	166.4	114.3	113.2	112.8	107.1
218	110.8	120.3	167.9	115.4	114.4	113.9	108.2
220	111.9	121.5	169.5	116.5	115.5	115.0	109.3
222	112.9	122.6	171.0	117.6	116.6	116.1	110.5
224	114.0	123.8	172.6	118.8	117.7	117.3	111.6
226	115.1	125.0	174.2	119.9	118.9	118.4	112.7
228	116.2	126.1	175.7	121.0	120.0	119.5	113.9
230	117.3	127.3	177.3	122.1	121.1	120.7	115.0
232	118.4	128.4	178.8	123.3	122.3	121.8	116.2
234	119.5	129.6	180.4	124.4	123.4	122.9	117.3
236	120.6	130.8	181.9	125.5	124.5	124.1	118.4
238	121.7	131.9	183.5	126.6	125.7	125.2	119.6
240	122.7	133.1	185.1	127.8	126.8	126.3	120.7
242	123.8	134.2	186.6	128.9	127.9	127.5	121.9
244	124.9	135.4	188.2	130.0	129.1	128.6	123.0
246	126.0	136.6	189.7	131.2	130.2	129.8	124.2
248	127.1	137.7	191.3	132.3	131.3	130.9	125.3
250	128.2	138.9	192.9	133.4	132.5	132.0	126.5
252	129.3	140.1	194.4	134.6	133.6	133.2	127.6
254	130.5	141.3	196.0	135.7	134.8	134.3	128.8
256	131.6	142.4	197.5	136.8	135.9	135.5	130.0
258	132.7	143.6	199.1	138.0	137.1	136.6	131.1
260	133.8	144.8	200.7	139.1	138.2	137.8	132.3
262	134.9	145.9	202.2	140.3	139.4	138.9	133.4
264	136.0	147.1	203.8	141.4	140.5	140.1	134.6
266	137.1	148.3	205.3	142.6	141.7	141.2	135.8
268	138.2	149.5	206.9	143.7	142.8	142.4	136.9

TABLE 23-6 (continued)

Copper (Cu)	Dextrose	Levulose	Lactose H_2O	Invert Sugar	Invert Sugar and Sucrose		
					0.3 g of total sugar	0.4 g of total sugar	2.0 g of total sugar
270	139.3	150.6	208.5	144.8	144.0	143.5	138.1
272	140.4	151.8	210.0	146.0	145.1	144.7	139.3
274	141.6	153.0	211.6	147.1	146.3	145.9	140.4
276	142.7	154.2	213.2	148.3	147.4	147.0	141.6
278	143.8	155.4	214.7	149.4	148.6	148.2	142.8
280	144.9	156.5	216.3	150.6	149.7	149.3	143.9
282	146.0	157.7	217.9	151.8	150.9	150.5	145.1
284	147.2	158.9	219.4	152.9	152.1	151.7	146.3
286	148.3	160.1	221.0	154.1	153.2	152.8	147.5
288	149.4	161.3	222.6	155.2	154.4	154.0	148.6
290	150.5	162.5	224.1	156.4	155.5	155.2	149.8
292	151.7	163.7	225.7	157.5	156.7	156.3	151.0
294	152.8	164.9	227.3	158.7	157.9	157.5	152.2
296	153.9	166.0	228.8	159.9	159.0	158.7	153.4
298	155.1	167.2	230.4	161.0	160.2	159.9	154.6
300	156.2	168.4	232.0	162.2	161.4	161.0	155.7
302	157.3	169.6	233.5	163.4	162.5	162.2	156.9
304	158.5	170.8	235.1	164.5	163.7	163.4	158.1
306	159.6	172.0	236.7	165.7	164.9	164.6	159.3
308	160.7	173.2	238.2	166.9	166.1	165.7	160.5
310	161.9	174.4	239.8	168.0	167.2	166.9	161.7
312	163.0	175.6	241.4	169.2	168.4	168.1	162.9
314	164.2	176.8	243.0	170.4	169.6	169.3	164.1
316	165.3	178.0	244.5	171.6	170.8	170.5	165.3
318	166.5	179.2	246.1	172.8	172.0	171.7	166.5
320	167.6	180.4	247.7	173.9	173.1	172.8	167.7
322	168.8	181.6	249.2	175.1	174.3	174.0	168.9
324	169.9	182.8	250.8	176.3	175.5	175.2	170.1
326	171.1	184.0	252.4	177.5	176.7	176.4	171.3
328	172.2	185.2	253.9	178.7	177.9	177.6	172.5
330	173.4	186.4	255.5	179.8	179.1	178.8	173.7
332	174.5	187.6	257.1	181.0	180.3	180.0	174.9
334	175.7	188.8	258.7	182.2	181.5	181.2	176.1
336	176.8	190.1	260.2	183.4	182.6	182.4	177.3
338	178.0	191.3	261.8	184.6	183.8	183.6	178.6
340	179.2	192.5	263.4	185.8	185.0	184.8	179.8
342	180.3	193.7	265.0	187.0	186.2	186.0	181.0
344	181.5	194.9	266.6	188.2	187.4	187.2	182.2
346	182.7	196.1	268.1	189.4	188.6	188.4	183.4
348	183.8	197.3	269.7	190.6	189.8	189.6	184.6
350	185.0	198.5	271.3	191.8	191.0	190.8	185.9
352	186.2	199.8	272.9	193.0	192.2	192.0	187.1
354	187.3	201.0	274.4	194.2	193.4	193.2	188.3
356	188.5	202.2	276.0	195.4	194.6	194.4	189.5
358	189.7	203.4	277.6	196.6	195.8	195.7	190.8

TABLE 23-6 (concluded)

Copper (Cu)	Dextrose	Levulose	Lactose H_2O	Invert Sugar	Invert Sugar and Sucrose		
					0.3 g of total sugar	0.4 g of total sugar	2.0 g of total sugar
360	190.9	204.7	279.2	197.8	197.1	196.9	192.0
362	192.0	205.9	280.8	199.0	198.3	198.1	193.2
364	193.2	207.1	282.4	200.2	199.5	199.3	194.5
366	194.4	208.3	284.0	201.4	200.7	200.5	195.7
368	195.6	209.6	285.6	202.6	201.9	201.7	196.9
370	196.8	210.8	287.1	203.8	203.1	203.0	198.2
372	198.0	212.0	288.7	205.0	204.3	204.2	199.4
374	199.1	213.3	290.3	206.3	205.6	205.4	200.7
376	200.3	214.5	291.9	207.5	206.8	206.6	201.9
378	201.5	215.7	293.5	208.7	208.0	207.9	203.1
380	202.7	217.0	295.0	209.9	209.2	209.1	204.4
382	203.9	218.2	296.6	211.1	210.4	210.3	205.6
384	205.1	219.5	298.2	212.4	211.7	211.6	206.9
386	206.3	220.7	299.8	213.6	212.9	212.8	208.1
388	207.5	221.9	301.4	214.8	214.1	214.0	209.4
390	208.7	223.2	303.0	216.0	215.4	215.3	210.6
392	209.9	224.4	304.6	217.3	216.6	216.5	211.9
394	211.1	225.7	306.2	218.5	217.8	217.8	213.2
396	212.3	226.9	307.8	219.8	219.1	219.0	214.4
398	213.5	228.2	309.4	221.0	220.3	220.3	215.7
400	214.7	229.4	311.0	222.2	221.5	221.5	217.0
402	215.9	230.7	312.6	223.5	222.8	222.8	218.2
404	217.1	232.0	314.2	224.7	224.0	224.0	219.5
406	218.4	233.2	315.9	226.0	225.3	225.3	220.8
408	219.6	234.5	317.5	227.2	226.6	226.5	222.0
410	220.8	235.8	319.1	228.5	227.8	227.8	223.3
412	222.0	237.1	320.7	229.7	229.1	229.1	224.6
414	223.3	238.4	322.4	231.0	230.4	230.4	225.9
416	224.5	239.7	324.0	232.3	231.6	231.7	227.2
418	225.7	241.0	325.7	233.6	232.9	232.9	228.5
420	227.0	242.2	327.4	234.8	234.2	234.2	229.8
422	228.2	243.6	329.1	236.1	235.5	235.5	231.1
424	229.5	244.9	330.8	237.5	236.8	236.9	232.4
426	230.7	246.3	332.6	238.8	238.2	238.2	233.8
428	232.0	247.8	334.4	240.2	239.5	239.6	235.1
430	233.3	249.2	336.3	241.5	240.9	241.0	236.5
432	234.7	250.8	338.3	243.0	242.4	242.5	238.0
434	236.1	252.7	340.7	244.7	244.1	244.2	239.6

TABLE 23-7
Alcoholometric Table*

Percentage of C_2H_5OH		Specific gravity in air		Percentage of C_2H_5OH		Specific gravity in air	
By volume at 15.56° C.	By weight	at $\frac{25°}{25°}$	at $\frac{15.56°}{15.56°}$	By weight	By volume at 15.56° C.	at $\frac{25°}{25°}$	at $\frac{15.56°}{15.56°}$
0	0.00	1.0000	1.0000	0	0.00	1.0000	1.0000
1	0.80	0.9985	0.9985	1	1.26	0.9981	0.9981
2	1.59	0.9970	0.9970	2	2.51	0.9963	0.9963
3	2.39	0.9956	0.9956	3	3.76	0.9945	0.9945
4	3.19	0.9941	0.9942	4	5.00	0.9927	0.9928
5	4.00	0.9927	0.9928	5	6.24	0.9911	0.9912
6	4.80	0.9914	0.9915	6	7.48	0.9894	0.9896
7	5.61	0.9901	0.9902	7	8.71	0.9879	0.9881
8	6.42	0.9888	0.9890	8	9.94	0.9863	0.9867
9	7.23	0.9875	0.9878	9	11.17	0.9848	0.9852
10	8.05	0.9862	0.9866	10	12.39	0.9833	0.9839
11	8.86	0.9850	0.9854	11	13.61	0.9818	0.9825
12	9.68	0.9838	0.9843	12	14.83	0.9804	0.9812
13	10.50	0.9826	0.9832	13	16.05	0.9789	0.9799
14	11.32	0.9814	0.9821	14	17.26	0.9776	0.9787
15	12.14	0.9802	0.9810	15	18.47	0.9762	0.9774
16	12.96	0.9790	0.9800	16	19.68	0.9748	0.9763
17	13.79	0.9778	0.9789	17	20.88	0.9734	0.9751
18	14.61	0.9767	0.9779	18	22.08	0.9720	0.9738
19	15.44	0.9756	0.9769	19	23.28	0.9706	0.9726
20	16.27	0.9744	0.9759	20	24.47	0.9692	0.9714
21	17.10	0.9733	0.9749	21	25.66	0.9677	0.9701
22	17.93	0.9721	0.9739	22	26.85	0.9663	0.9688
23	18.77	0.9710	0.9729	23	28.03	0.9648	0.9675
24	19.60	0.9698	0.9719	24	29.21	0.9633	0.9662
25	20.44	0.9685	0.9708	25	30.39	0.9617	0.9648
26	21.29	0.9673	0.9697	26	31.56	0.9601	0.9635
27	22.13	0.9661	0.9687	27	32.72	0.9585	0.9620
28	22.97	0.9648	0.9676	28	33.88	0.9568	0.9605
29	23.82	0.9635	0.9664	29	35.03	0.9551	0.9590
30	24.67	0.9622	0.9653	30	36.18	0.9534	0.9574
31	25.52	0.9609	0.9641	31	37.32	0.9516	0.9558
32	26.38	0.9595	0.9629	32	38.46	0.9498	0.9541
33	27.24	0.9581	0.9617	33	39.59	0.9480	0.9524
34	28.10	0.9567	0.9604	34	40.72	0.9461	0.9506
35	28.97	0.9552	0.9590	35	41.83	0.9442	0.9488
36	29.84	0.9537	0.9576	36	42.94	0.9422	0.9470
37	30.72	0.9521	0.9562	37	44.05	0.9402	0.9451
38	31.60	0.9506	0.9548	38	45.15	0.9832	0.9432
39	32.48	0.9489	0.9533	39	46.24	0.9362	0.9412
40	33.36	0.9473	0.9517	40	47.33	0.9341	0.9392
41	34.25	0.9456	0.9501	41	48.41	0.9320	0.9372
42	35.15	0.9439	0.9485	42	49.48	0.9299	0.9352
43	36.05	0.9421	0.9469	43	50.55	0.9278	0.9331
44	36.96	0.9403	0.9452	44	51.61	0.9256	0.9310
45	37.87	0.9385	0.9434	45	52.66	0.9235	0.9289

* Based upon data appearing in the *National Bureau of Standards Bulletin*, **9**:424–5.

TABLE 23-7 (concluded)

46	38.78	0.9366	0.9417	46	53.71	0.9213	0.9268
47	39.70	0.9348	0.9399	47	54.75	0.9191	0.9246
48	40.62	0.9328	0.9380	48	55.78	0.9169	0.9225
49	41.55	0.9309	0.9361	49	56.81	0.9147	0.9203
50	42.49	0.9289	0.9342	50	57.83	0.9124	0.9181
51	43.43	0.9269	0.9322	51	58.84	0.9102	0.9159
52	44.37	0.9248	0.9302	52	59.85	0.9079	0.9137
53	45.33	0.9228	0.9282	53	60.85	0.9056	0.9114
54	46.28	0.9207	0.9262	54	61.85	0.9033	0.9092
55	47.25	0.9185	0.9241	55	62.84	0.9010	0.9069
56	48.21	0.9164	0.9220	56	63.82	0.8987	0.9046
57	49.19	0.9142	0.9199	57	64.80	0.8964	0.9024
58	50.17	0.9120	0.9117	58	65.77	0.8941	0.9001
59	51.15	0.9098	0.9155	59	66.73	0.8918	0.8978
60	52.15	0.9076	0.9133	60	67.69	0.8895	0.8955
61	53.15	0.9053	0.9111	61	68.64	0.8871	0.8932
62	54.15	0.9030	0.9088	62	69.59	0.8848	0.8909
63	55.17	0.9006	0.9065	63	70.52	0.8824	0.8886
64	56.18	0.8983	0.9042	64	71.46	0.8801	0.8862
65	57.21	0.8959	0.9019	65	72.38	0.8777	0.8839
66	58.24	0.8936	0.8995	66	73.30	0.8753	0.8815
67	59.28	0.8911	0.8972	67	74.21	0.8729	0.8792
68	60.33	0.8887	0.8948	68	75.12	0.8706	0.8768
69	61.38	0.8862	0.8923	69	76.02	0.8682	0.8745
70	62.44	0.8837	0.8899	70	76.91	0.8658	0.8721
71	63.51	0.8812	0.8874	71	77.79	0.8634	0.8697
72	64.59	0.8787	0.8848	72	78.67	0.8609	0.8673
73	65.67	0.8761	0.8823	73	79.54	0.8585	0.8649
74	66.77	0.8735	0.8797	74	80.41	0.8561	0.8625
75	67.87	0.8709	0.8771	75	81.27	0.8537	0.8601
76	68.98	0.8682	0.8745	76	82.12	0.8512	0.8576
77	70.10	0.8655	0.8718	77	82.97	0.8488	0.8552
78	71.23	0.8628	0.8691	78	83.81	0.8463	0.8528
79	72.38	0.8600	0.8664	79	84.64	0.8439	0.8503
80	73.53	0.8572	0.8636	80	85.46	0.8414	0.8479
81	74.69	0.8544	0.8608	81	86.28	0.8389	0.8454
82	75.86	0.8516	0.8580	82	87.08	0.8364	0.8429
83	77.04	0.8487	0.8551	83	87.89	0.8339	0.8404
84	78.23	0.8458	0.8522	84	88.68	0.8314	0.8379
85	79.44	0.8428	0.8493	85	89.46	0.8288	0.8353
86	80.66	0.8397	0.8462	86	90.24	0.8263	0.8328
87	81.90	0.8367	0.8432	87	91.01	0.8237	0.8303
88	83.14	0.8335	0.8401	88	91.77	0.8211	0.8276
89	84.41	0.8303	0.8369	89	92.52	0.8184	0.8250
90	85.69	0.8271	0.8336	90	93.25	0.8158	0.8224
91	86.99	0.8237	0.8303	91	93.98	0.8131	0.8197
92	88.31	0.8202	0.8268	92	94.70	0.8104	0.8170
93	89.65	0.8167	0.8233	93	95.41	0.8076	0.8142
94	91.03	0.8130	0.8196	94	96.10	0.8048	0.8114
95	92.42	0.8092	0.8158	95	96.79	0.8020	0.8086
96	93.85	0.8053	0.8118	96	97.46	0.7992	0.8057
97	95.32	0.8011	0.8077	97	98.12	0.7962	0.8028
98	96.82	0.7968	0.8033	98	98.76	0.7932	0.7988
99	98.38	0.7921	0.7986	99	99.39	0.7902	0.7967
100	100.00	0.7871	0.7936	100	100.00	0.7871	0.7936

TABLE 23-8

Alcohol table for calculating percentages of alcohol by volume at 15.56°C (60°F) in mixtures of ethyl alcohol and H_2O from their Zeiss immersion refractometer readings and indices of refraction at 17.5–25°C[a]

SCALE READING[b]	INDEX OF REFRACTION	17.5° C	18° C	19° C	20° C	21° C	22° C	23° C	24° C	25° C
13.2	1.33250	......	......	......	......	......	......	......	......	0.00
13.4	3257	......	......	......	......	......	......	......	......	0.18
13.6	3265	......	......	......	......	......	......	......	0.14	0.35
13.8	3273	......	......	......	......	......	......	0.10	0.31	0.53
14.0	3281	......	......	......	......	......	0.08	0.28	0.49	0.70
14.2	3288	......	......	......	......	0.04	0.24	0.45	0.67	0.88
14.4	3296	......	......	......	......	0.21	0.41	0.63	0.84	1.06
14.6	3304	......	......	......	0.16	0.38	0.59	0.80	1.02	1.24
14.8	3312	......	......	0.14	0.34	0.55	0.77	0.98	1.19	1.40
15.0	3319	0.00	0.10	0.31	0.52	0.73	0.94	1.16	1.36	1.55
15.2	3327	0.17	0.27	0.48	0.69	0.91	1.12	1.32	1.51	1.71
15.4	3335	0.34	0.44	0.65	0.85	1.07	1.29	1.47	1.66	1.86
15.6	3343	0.52	0.60	0.82	1.03	1.24	1.44	1.62	1.82	2.01
15.8	3350	0.68	0.78	0.99	1.21	1.40	1.60	1.77	1.97	2.17
16.0	3358	0.84	0.94	1.17	1.36	1.55	1.75	1.92	2.12	2.33
16.2	3366	1.02	1.12	1.32	1.51	1.70	1.90	2.08	2.27	2.48
16.4	3374	1.18	1.29	1.47	1.66	1.85	2.05	2.24	2.43	2.62
16.6	3381	1.34	1.43	1.62	1.81	2.00	2.20	2.39	2.57	2.77
16.8	3389	1.49	1.57	1.77	1.96	2.15	2.35	2.53	2.72	2.92
17.0	3397	1.63	1.72	1.92	2.11	2.30	2.50	2.69	2.87	3.06
17.2	3405	1.77	1.87	2.06	2.26	2.45	2.65	2.82	3.02	3.21
17.4	3412	1.92	2.01	2.21	2.41	2.59	2.79	2.97	3.17	3.36
17.6	3420	2.07	2.16	2.36	2.56	2.74	2.94	3.12	3.32	3.51
17.8	3428	2.21	2.31	2.51	2.70	2.89	3.09	3.27	3.46	3.66
18.0	3435	2.36	2.45	2.66	2.85	3.04	3.23	3.42	3.61	3.81
18.2	3443	2.50	2.60	2.81	3.00	3.19	3.37	3.57	3.76	3.96
18.4	3451	2.65	2.75	2.96	3.15	3.34	3.52	3.71	3.91	4.11
18.6	3459	2.80	2.90	3.10	3.30	3.48	3.66	3.86	4.06	4.26
18.8	3466	2.95	3.05	3.25	3.45	3.63	3.81	4.01	4.21	4.41
19.0	3474	3.10	3.19	3.40	3.59	3.77	3.96	4.16	4.36	4.56
19.2	3482	3.25	3.34	3.55	3.73	3.92	4.11	4.31	4.51	4.70
19.4	3489	3.39	3.48	3.70	3.88	4.07	4.26	4.46	4.65	4.85
19.6	3497	3.53	3.63	3.84	4.03	4.22	4.41	4.61	4.80	5.00
19.8	3505	3.68	3.78	3.98	4.17	4.37	4.56	4.75	4.95	5.15

[a] Rearranged from table of B. H. St. John, which is based upon data of Doroschevskii and Dovrzhanchik, *J. Russ. Phys. Chem. Soc.*, **40**:101 (1908). Scale readings were converted into refractive indices by using formula $n_D = 1.327338 + 0.00039347X - 0.00000020446X^2$.

[b] Scale readings refer only to scale of arbitrary units proposed by Pulfrich, *Z. Angew. Chem.*, p. 1168 (1899). According to this scale, 14.5 = 1.33300, 50.0 = 1.34650, and 100.0 = 1.36464. If immersion refractometer used is calibrated to another arbitrary scale, readings must be converted into refractive indices before table is used to determine % alcohol.

TABLE 23-8 (continued)

SCALE READING	INDEX OF REFRACTION	17.5° C	18° C	19° C	20° C	21° C	22° C	23° C	24° C	25° C
20.0	3513	3.83	3.93	4.13	4.32	4.52	4.72	4.90	5.10	5.29
20.2	3520	3.97	4.07	4.27	4.47	4.66	4.87	5.05	5.24	5.44
20.4	3528	4.12	4.22	4.42	4.61	4.82	5.01	5.20	5.38	5.58
20.6	3536	4.26	4.36	4.56	4.75	4.96	5.15	5.34	5.52	5.72
20.8	3543	4.41	4.51	4.70	4.90	5.10	5.29	5.48	5.67	5.87
21.0	3551	4.56	4.65	4.85	5.04	5.24	5.44	5.62	5.82	6.02
21.2	3559	4.70	4.80	4.99	5.19	5.39	5.58	5.77	5.96	6.16
21.4	3566	4.84	4.94	5.14	5.33	5.53	5.72	5.91	6.11	6.30
21.6	3574	4.99	5.09	5.28	5.47	5.67	5.87	6.06	6.25	6.44
21.8	3582	5.13	5.23	5.43	5.61	5.82	6.01	6.20	6.39	6.59
22.0	1.33590	5.27	5.37	5.57	5.76	5.96	6.15	6.34	6.54	6.73
22.2	3597	5.41	5.51	5.71	5.90	6.11	6.29	6.49	6.68	6.87
22.4	3605	5.56	5.65	5.85	6.05	6.25	6.43	6.63	6.82	7.01
22.6	3613	5.70	5.80	6.00	6.19	6.39	6.57	6.77	6.96	7.16
22.8	3620	5.85	5.94	6.14	6.33	6.53	6.71	6.91	7.10	7.31
23.0	3628	5.99	6.08	6.28	6.47	6.67	6.86	7.06	7.24	7.45
23.2	3636	6.13	6.22	6.42	6.61	6.81	7.00	7.20	7.39	7.59
23.4	3643	6.27	6.36	6.56	6.75	6.95	7.14	7.34	7.53	7.73
23.6	3651	6.41	6.50	6.70	6.90	7.09	7.28	7.48	7.67	7.87
23.8	3659	6.55	6.64	6.85	7.04	7.23	7.42	7.62	7.81	8.00
24.0	3666	6.69	6.78	6.99	7.18	7.38	7.56	7.76	7.95	8.14
24.2	3674	6.83	6.93	7.13	7.32	7.52	7.70	7.90	8.09	8.28
24.4	3682	6.97	7.06	7.27	7.46	7.66	7.84	8.04	8.23	8.42
24.6	3689	7.11	7.20	7.41	7.60	7.80	7.98	8.17	8.37	8.55
24.8	3697	7.25	7.35	7.55	7.74	7.93	8.12	8.31	8.51	8.69
25.0	3705	7.39	7.49	7.68	7.88	8.06	8.26	8.45	8.64	8.84
25.2	3712	7.53	7.63	7.82	8.01	8.20	8.40	8.59	8.78	8.98
25.4	3720	7.66	7.76	7.95	8.14	8.34	8.54	8.73	8.92	9.12
25.6	3728	7.80	7.90	8.09	8.28	8.48	8.68	8.86	9.06	9.26
25.8	3735	7.94	8.03	8.22	8.42	8.62	8.82	9.00	9.20	9.39
26.0	3743	8.07	8.16	8.36	8.55	8.75	8.95	9.14	9.34	9.53
26.2	3751	8.21	8.30	8.50	8.69	8.89	9.09	9.28	9.48	9.67
26.4	3758	8.34	8.44	8.63	8.82	9.03	9.22	9.42	9.61	9.81
26.6	3766	8.48	8.57	8.77	8.96	9.16	9.36	9.55	9.75	9.95
26.8	3774	8.62	8.71	8.91	9.10	9.30	9.49	9.69	9.89	10.09
27.0	3781	8.75	8.85	9.05	9.23	9.44	9.63	9.83	10.03	10.23
27.2	3789	8.89	8.98	9.18	9.37	9.58	9.76	9.97	10.17	10.37
27.4	3796	9.02	9.12	9.32	9.51	9.71	9.90	10.10	10.31	10.51
27.6	3804	9.16	9.26	9.45	9.65	9.85	10.03	10.24	10.45	10.65
27.8	3812	9.29	9.39	9.59	9.79	9.98	10.17	10.38	10.58	10.79
28.0	3820	9.43	9.53	9.72	9.92	10.12	10.31	10.51	10.72	10.93
28.2	3827	9.57	9.66	9.86	10.06	10.25	10.45	10.65	10.86	11.06
28.4	3835	9.70	9.80	9.99	10.19	10.39	10.59	10.79	11.00	11.20
28.6	3842	9.84	9.93	10.13	10.32	10.52	10.72	10.93	11.13	11.33
28.8	3850	9.97	10.07	10.26	10.46	10.66	10.86	11.06	11.27	11.47

TABLE 23-8 (continued)

SCALE READING	INDEX OF REFRACTION	17.5° C	18° C	19° C	20° C	21° C	22° C	23° C	24° C	25° C
29.0	3858	10.10	10.19	10.40	10.59	10.79	11.00	11.20	11.40	11.61
29.2	3865	10.24	10.33	10.52	10.73	10.93	11.13	11.33	11.54	11.75
29.4	3873	10.36	10.46	10.66	10.86	11.06	11.27	11.47	11.67	11.88
29.6	3881	10.50	10.59	1.079	10.99	11.20	11.39	11.60	11.81	12.01
29.8	3888	10.63	10.72	10.93	11.12	11.33	11.53	11.74	11.94	12.15
30.0	3896	10.76	10.86	11.05	11.26	11.46	11.66	11.87	12.08	12.29
30.2	3904	10.89	10.99	11.18	11.38	11.59	11.79	12.00	12.21	12.42
30.4	3911	11.02	11.12	11.31	11.51	11.72	11.93	12.13	12.34	12.56
30.6	3919	11.15	11.25	11.44	11.64	11.85	12.06	12.27	12.48	12.70
30.8	3926	11.28	11.38	11.58	11.78	11.99	12.19	12.40	12.61	12.84
31.0	1.33934	11.41	11.51	11.71	11.91	12.12	12.32	12.54	12.75	12.97
31.2	3942	11.54	11.64	11.84	12.04	12.25	12.46	12.67	12.89	13.11
31.4	3949	11.66	11.77	11.97	12.17	12.38	12.59	12.81	13.02	13.24
31.6	3957	11.79	11.90	12.10	12.30	12.51	12.72	12.94	13.15	13.37
31.8	3964	11.92	12.03	12.23	12.43	12.64	12.85	13.07	13.29	1351.
32.0	3972	12.05	12.15	12.36	12.57	12.78	12.99	13.20	13.42	13.64
32.2	3980	12.18	12.28	12.49	12.70	12.91	13.12	13.34	13.55	13.77
32.4	3987	12.31	12.40	12.62	12.83	13.04	13.25	13.47	13.69	13.91
32.6	3995	12.43	12.54	12.75	12.96	13.17	13.38	13.60	13.82	14.04
32.8	4002	12.56	12.67	12.88	13.09	13.30	13.51	13.73	13.95	14.17
33.0	4010	12.69	12.79	13.01	13.22	13.43	13.64	13.86	14.09	14.31
33.2	4018	12.82	12.92	13.13	13.35	13.56	13.78	13.99	14.22	14.44
33.4	4025	12.95	13.05	13.26	13.48	13.69	13.91	14.13	14.35	14.58
33.6	4033	13.08	13.18	13.39	13.61	13.82	14.04	14.26	14.48	14.71
33.8	4040	13.20	13.30	13.52	13.74	13.95	14.17	14.39	14.62	14.85
34.0	4048	13.33	13.43	13.64	13.86	14.08	14.30	14.52	14.75	14.98
34.2	4056	13.45	13.56	13.77	13.99	14.21	14.43	14.65	14.88	15.11
34.4	4063	13.58	13.68	13.90	14.12	14.34	14.57	14.78	15.01	15.25
34.6	4071	13.70	13.81	14.02	14.25	14.47	14.70	14.91	15.14	15.38
34.8	4078	13.83	13.94	14.14	14.37	14.59	14.83	15.05	15.28	15.51
35.0	4086	13.96	14.06	14.27	14.50	14.72	14.96	15.18	15.41	15.65
35.2	4094	14.08	14.19	14.39	14.62	14.85	15.09	15.31	15.54	15.78
35.4	4101	14.21	14.31	14.52	14.75	14.97	15.22	15.44	15.67	15.91
35.6	4109	14.33	14.44	14.65	14.87	15.10	15.34	15.56	15.80	16.05
35.8	4116	14.46	14.56	14.78	15.00	15.23	15.47	15.69	15.93	16.18
36.0	4124	14.58	14.69	14.90	15.13	15.35	15.59	15.82	16.06	16.31
36.2	4131	14.71	14.81	15.03	15.25	15.48	15.72	15.95	16.19	16.44
36.4	4139	14.83	14.94	15.16	15.38	15.61	15.85	16.08	16.32	16.56
36.6	4146	14.96	15.06	15.28	15.51	15.73	15.97	16.21	16.45	16.69
36.8	4154	15.08	15.19	15.41	15.63	15.86	16.10	16.34	16.58	16.82
37.0	4162	15.20	15.31	15.53	15.76	15.99	16.23	16.47	16.71	16.95
37.2	4169	15.33	15.44	15.66	15.89	16.11	16.35	16.60	16.84	17.08
37.4	4177	15.45	15.56	15.79	16.01	16.24	16.48	16.72	16.97	17.21
37.6	4184	15.57	15.69	15.91	16.14	16.37	16.61	16.85	17.09	17.34
37.8	4192	15.70	15.81	16.04	16.26	16.49	16.73	16.98	17.22	17.46

TABLE 23-8 (continued)

SCALE READING	INDEX OF REFRACTION	17.5° C	18° C	19° C	20° C	21° C	22° C	23° C	24° C	25° C
38.0	4199	15.82	15.94	16.16	16.39	16.62	16.86	17.11	17.35	17.59
38.2	4207	15.94	16.06	16.29	16.51	16.75	16.99	17.23	17.47	17.72
38.4	4215	16.07	16.18	16.41	16.64	16.87	17.11	17.36	17.60	17.85
38.6	4222	16.19	16.31	16.53	16.76	17.00	17.24	17.48	17.73	17.97
39.8	4230	16.31	16.43	16.66	16.89	17.13	17.36	17.61	17.85	18.10
39.0	4237	16.44	16.55	16.78	17.01	17.25	17.49	17.74	17.98	18.23
39.2	4245	16.56	16.67	16.91	17.14	17.38	17.62	17.86	18.11	18.35
39.4	4252	16.68	16.80	17.03	17.26	17.50	17.74	17.99	18.23	18.48
39.6	4260	16.80	16.92	17.15	17.39	17.63	17.87	18.11	18.36	18.61
39.8	4267	16.93	17.04	17.28	17.51	17.75	17.99	18.24	18.48	18.73
40.0	1.34275	17.05	17.16	17.40	17.63	17.88	18.12	18.36	18.61	18.86
40.2	4282	17.17	17.29	17.52	17.76	18.00	18.24	18.49	18.74	18.99
40.4	4290	17.29	17.41	17.64	17.88	18.12	18.37	18.61	18.86	19.11
40.6	4298	17.41	17.53	17.77	18.01	18.25	18.49	18.74	18.99	19.24
40.8	4305	17.54	17.65	17.89	18.13	18.37	18.61	18.86	19.11	19.37
41.0	4313	17.66	17.77	18.01	18.25	18.49	18.74	18.99	19.24	19.49
41.2	4320	17.78	17.90	18.13	18.37	18.62	18.86	19.11	19.36	19.62
41.4	4328	17.90	18.03	18.26	18.50	18.74	18.99	19.24	19.49	19.75
41.6	4335	18.02	18.14	18.38	18.62	18.86	19.11	19.36	19.61	19.87
41.8	4343	18.14	18.26	18.50	18.74	18.99	19.23	19.48	19.74	20.00
42.0	4350	18.27	18.38	18.62	18.87	19.11	19.36	19.61	19.86	20.13
42.2	4358	18.39	18.50	18.74	18.99	19.23	19.48	19.73	19.99	20.25
42.4	4365	18.51	18.62	18.87	19.11	19.36	19.60	19.86	20.11	20.38
42.6	4373	18.63	18.75	18.99	19.23	19.48	19.72	19.98	20.24	20.50
42.8	4380	18.75	18.87	19.11	19.36	19.60	19.85	20.10	20.36	20.63
43.0	4388	18.87	18.99	19.23	19.48	19.72	19.97	20.23	20.49	20.75
43.2	4395	18.99	19.11	19.35	19.60	19.85	20.09	20.35	20.61	20.88
43.4	4403	19.11	19.23	19.47	19.72	19.97	20.21	20.47	20.74	21.01
43.6	4410	19.23	19.35	19.59	19.85	20.09	20.34	20.60	20.86	21.13
43.8	4418	19.35	19.47	19.72	19.97	20.21	20.46	20.72	20.99	21.25
44.0	4426	19.46	19.59	19.84	20.09	20.34	20.58	20.84	21.11	21.38
44.2	4433	19.58	19.71	19.96	20.21	20.46	20.71	20.96	21.23	21.50
44.4	4440	19.70	19.83	20.08	20.33	20.58	20.83	21.09	21.36	21.63
44.6	4448	19.82	19.95	20.20	20.45	20.70	20.95	21.21	21.48	21.75
44.8	4456	19.94	20.07	20.32	20.58	20.82	21.07	21.33	21.60	21.88
45.0	4463	20.06	20.18	20.44	20.70	20.95	21.19	21.45	21.73	22.00
45.2	4470	20.18	20.30	20.56	20.82	21.07	21.31	21.58	21.85	22.13
45.4	4478	20.29	20.42	20.68	20.94	21.19	21.43	21.70	21.98	22.25
45.6	4486	20.41	20.54	20.80	21.06	21.31	21.55	21.82	22.10	22.38
45.8	4493	20.53	20.66	20.92	21.18	21.43	21.67	21.94	22.23	22.51
46.0	4500	20.65	20.78	21.04	21.30	21.54	21.79	22.07	22.35	22.64
46.2	4508	20.76	20.89	21.16	21.42	21.66	21.91	22.19	22.48	22.76
46.4	4516	20.88	21.01	21.28	21.54	21.78	22.03	22.32	22.61	22.89
46.6	4523	21.00	21.13	21.40	21.66	21.90	22.16	22.44	22.73	23.02
46.8	4530	21.12	21.25	21.52	21.78	22.02	22.28	22.57	22.86	23.15

TABLE 23-8 (continued)

SCALE READING	INDEX OF REFRACTION	17.5° C	18° C	19° C	20 °C	21° C	22° C	23 °C	24° C	25 °C
47.0	4538	21.24	21.37	21.64	21.90	22.15	22.41	22.69	22.99	23.28
47.2	4545	21.36	21.49	21.76	22.02	22.27	22.53	22.82	23.12	23.41
47.4	4553	21.48	21.61	21.88	22.15	22.39	22.66	22.94	23.24	23.54
47.6	4560	21.60	21.73	22.00	22.27	22.51	22.78	23.07	23.37	23.67
47.8	4568	21.72	21.85	22.12	22.39	22.64	22.91	23.20	23.50	23.80
48.0	4575	21.84	21.97	22.24	22.51	22.76	23.03	23.32	23.63	23.93
48.2	4583	21.96	22.09	22.36	22.63	22.88	23.16	23.45	23.76	24.06
48.4	4590	22.08	22.21	22.48	22.75	23.01	23.28	23.58	23.89	24.19
48.6	4598	22.20	22.33	22.60	22.87	23.13	23.41	23.71	24.02	24.32
48.8	4605	22.32	22.45	22.72	22.99	23.26	23.54	23.83	24.14	24.45
49.0	1.34613	22.44	22.57	22.84	23.12	23.38	23.66	23.96	24.27	24.59
49.2	4620	22.56	22.69	22.96	23.24	23.51	23.79	24.09	24.40	24.72
49.4	4628	22.68	22.81	23.08	23.36	23.63	23.92	24.22	24.53	24.85
49.6	4635	22.80	22.93	23.21	23.48	23.76	24.04	24.35	24.66	24.98
49.8	4643	22.92	23.05	23.33	23.61	23.88	24.17	24.48	24.79	25.11
50.0	4650	23.04	23.17	23.45	23.73	24.01	24.30	24.61	24.92	25.25
50.2	4658	23.16	23.30	23.57	23.85	24.13	24.23	24.74	25.05	25.38
50.4	4665	23.28	23.42	23.69	23.98	24.26	24.56	24.86	25.18	25.51
50.6	4672	23.40	23.54	23.81	24.10	24.38	24.69	24.99	25.32	25.65
50.8	4680	23.51	23.66	23.93	24.22	24.51	24.81	25.12	25.45	25.78
51.0	4687	23.63	23.78	24.05	24.35	24.64	24.94	25.25	25.58	25.91
51.2	4695	23.75	23.90	24.18	24.47	24.76	25.07	25.38	25.71	26.05
51.4	4702	23.87	24.02	24.30	24.59	24.89	25.20	25.51	25.84	26.18
51.6	4710	23.99	24.14	24.42	24.72	25.01	25.33	25.64	25.97	26.32
51.8	4717	24.11	24.26	24.54	24.84	25.14	25.46	25.77	26.11	26.45
52.0	4724	24.23	24.38	24.66	24.96	25.27	25.58	25.90	26.24	26.59
52.2	4732	24.36	24.50	24.79	25.09	25.39	25.71	26.03	26.37	27.72
52.4	4740	24.48	24.62	24.91	25.21	25.52	25.84	26.16	26.51	26.86
52.6	4747	24.60	24.74	25.03	25.34	25.65	25.97	26.29	26.64	26.99
52.8	4754	24.72	24.86	25.15	25.46	25.77	26.10	26.42	26.77	27.13
53.0	4762	24.84	24.98	25.28	25.59	25.90	26.23	26.56	26.91	27.27
53.2	4769	24.96	25.10	25.40	25.71	26.03	26.35	26.69	27.04	27.40
53.4	4777	25.08	25.23	25.52	25.84	26.15	26.48	26.82	27.17	27.54
53.6	4784	25.20	25.35	25.65	25.96	26.28	26.61	26.95	27.31	27.67
53.8	4792	25.32	25.47	25.77	26.09	26.41	26.74	27.08	27.44	27.81
54.0	4799	25.44	25.59	25.90	26.22	26.54	26.87	27.21	27.58	27.95
54.2	4806	25.56	25.71	26.02	26.34	26.67	27.00	27.35	27.71	28.08
54.4	4814	25.68	25.84	26.14	26.47	26.79	27.13	27.48	27.85	28.22
54.6	4821	25.81	25.96	26.27	26.59	26.92	27.26	27.61	27.98	28.36
54.8	4829	25.93	26.08	26.39	26.72	27.05	27.39	27.75	28.11	28.49
55.0	4836	26.05	26.20	26.52	26.85	27.18	27.52	27.88	28.25	28.63
55.2	4844	26.17	26.32	26.64	26.97	27.31	27.65	28.01	28.38	28.77
55.4	4851	26.29	26.45	26.76	27.10	27.43	27.78	28.15	28.52	28.90
55.6	4858	26.41	26.57	26.89	27.23	27.55	27.92	28.28	28.65	29.04
55.8	4866	26.53	26.69	27.01	27.35	27.69	28.05	28.41	28.78	29.18

TABLE 23-8 (continued)

SCALE READING	INDEX OF REFRACTION	17.5° C	18° C	19° C	20° C	21° C	22° C	23° C	24° C	25° C
56.0	4873	26.65	26.81	27.14	27.48	27.82	28.18	28.54	28.92	29.31
56.2	4880	26.78	26.93	27.26	27.60	27.94	28.31	28.68	29.05	29.45
56.4	4888	26.90	27.05	27.38	27.73	28.07	28.44	28.81	29.19	29.58
56.6	4895	27.02	27.18	27.51	27.85	28.20	28.56	28.94	29.32	29.72
56.8	4903	27.14	27.30	27.63	27.98	28.33	28.69	29.07	29.46	29.86
57.0	4910	27.26	27.42	27.75	28.10	28.46	28.82	29.20	29.59	29.99
57.2	4918	27.38	27.54	27.88	28.23	28.59	28.95	29.34	29.73	30.13
57.4	4925	27.50	27.66	28.00	28.35	28.72	29.08	29.47	29.86	30.27
57.6	4932	27.62	27.79	28.13	28.48	28.85	29.21	29.60	30.00	30.41
57.8	4940	27.75	27.91	28.25	28.60	28.97	29.34	29.73	30.14	30.55
58.0	1.34947	27.87	28.03	28.38	28.73	29.10	29.47	29.87	30.27	30.69
58.2	4954	27.99	28.15	28.50	28.86	29.23	29.60	29.99	30.41	30.83
58.4	4962	28.11	28.28	28.62	28.98	29.36	29.73	30.13	30.54	30.97
58.6	4969	28.23	28.40	28.75	29.11	29.48	29.86	30.26	30.68	31.11
58.8	4977	28.35	28.52	28.88	29.23	29.61	29.99	30.40	30.82	31.25
59.0	4984	28.47	28.64	29.00	29.36	29.74	30.13	30.53	30.95	31.40
59.2	4991	28.59	28.77	29.12	29.49	29.87	30.26	30.67	31.09	31.54
59.4	4999	28.71	28.89	29.25	29.61	29.99	30.39	30.81	31.23	31.68
59.6	5006	28.84	29.01	29.37	29.74	30.13	30.53	30.94	31.38	31.83
59.8	5014	28.96	29.13	29.50	29.87	30.26	30.66	31.08	31.52	31.97
60.0	5021	29.08	29.26	29.62	29.99	30.39	30.79	31.22	31.66	32.12
60.2	5028	29.20	29.38	29.74	30.12	30.52	30.93	31.36	31.80	32.27
60.4	5036	29.32	29.50	29.87	30.25	30.65	31.06	31.50	31.94	32.41
60.6	5043	29.45	29.63	29.99	30.38	30.78	31.20	31.64	32.09	32.56
60.8	5050	29.57	29.75	30.12	30.51	30.91	31.33	31.78	32.32.	32.71
61.0	5058	29.69	29.87	30.25	30.64	31.05	31.47	31.92	32.38	32.86
61.2	5065	29.81	29.99	30.38	30.77	31.18	31.61	32.06	32.52	33.01
61.4	5073	29.93	30.12	30.50	30.90	31.32	31.74	32.20	32.67	33.16
61.6	5080	30.06	30.25	30.63	31.03	31.45	31.88	32.34	32.81	33.31
61.8	5087	30.18	30.37	30.76	31.16	31.59	32.01	32.49	32.96	33.46
62.0	5095	30.31	30.50	30.89	31.29	31.72	32.16	32.63	33.10	33.60
62.2	5102	30.43	30.63	31.01	31.43	31.86	32.30	32.77	33.25	33.75
62.4	5110	30.56	30.75	31.14	31.56	31.99	32.44	32.91	33.40	33.90
62.6	5117	30.69	30.88	31.28	31.69	32.13	32.58	33.06	33.55	34.05
62.8	5124	30.81	31.01	31.41	31.83	32.27	32.72	33.20	33.70	34.21
63.0	5132	30.94	31.14	31.54	31.96	32.41	32.87	33.35	33.84	34.36
63.2	5139	31.06	31.26	31.67	32.10	32.55	33.01	33.50	33.99	34.52
63.4	5146	31.19	31.39	31.80	32.23	32.69	33.15	33.64	34.15	34.67
63.6	5154	31.32	31.52	31.93	32.37	32.83	33.30	33.79	34.30	34.83
63.8	5161	31.45	31.65	32.07	32.51	32.97	33.44	33.93	34.45	34.98
64.0	5168	31.58	31.78	32.20	32.65	33.11	33.59	34.08	34.61	35.15
64.2	5176	31.70	31.91	32.34	32.79	33.25	33.73	34.23	34.76	35.31
64.4	5183	31.83	32.04	32.47	32.92	33.39	33.88	34.39	34.92	35.48
64.6	5190	31.96	32.17	32.60	33.06	33.53	34.02	34.54	35.07	35.64
64.8	5198	32.09	32.30	32.74	33.20	33.67	34.17	34.69	35.23	35.80

TABLE 23-8 (continued)

SCALE READING	INDEX OF REFRACTION	17.5° C	18° C	19° C	20° C	21° C	22° C	23° C	24° C	25° C
65.0	5205	32.22	32.43	32.87	33.34	33.82	34.32	34.84	35.39	35.97
65.2	5212	32.35	32.57	33.01	33.48	33.96	34.47	34.99	35.55	36.13
65.4	5220	32.48	32.70	33.15	33.62	34.10	34.61	35.15	35.71	36.30
65.6	5227	32.61	32.83	33.28	33.76	34.25	34.76	35.30	35.87	36.64
65.8	5234	32.75	32.96	33.42	33.90	34.40	34.91	35.46	36.02	36.63
66.0	5242	32.88	33.10	33.56	34.04	34.54	35.06	35.62	36.19	36.79
66.2	5249	33.01	33.23	33.70	34.18	34.69	35.22	35.77	36.35	36.96
66.4	5256	33.14	33.37	33.84	34.33	34.84	35.38	35.93	36.52	37.13
66.6	5264	33.28	33.51	33.98	34.47	34.99	35.53	36.09	36.68	37.30
66.8	5271	33.41	33.65	34.12	34.62	35.14	35.69	36.25	36.84	37.48
67.0	1.35278	33.55	33.79	34.26	34.76	35.29	35.84	36.41	37.01	37.65
67.2	5286	33.69	33.92	34.41	34.91	35.44	36.00	36.57	37.18	37.83
67.4	5293	33.82	34.06	34.55	35.05	35.60	36.16	36.73	37.35	38.00
67.6	5300	33.96	34.20	34.69	35.20	35.75	36.32	36.90	37.52	38.18
67.8	5308	34.09	34.34	34.84	35.35	35.90	36.48	37.06	37.69	38.35
68.0	5315	34.23	34.48	34.98	35.50	36.05	36.36	37.23	37.86	38.53
68.2	5322	34.36	34.62	35.13	35.65	36.21	36.79	37.39	38.03	38.70
68.4	5329	34.50	34.76	35.27	35.80	36.37	36.95	37.56	38.21	38.88
68.6	5337	34.64	34.90	35.42	35.95	36.52	37.12	37.73	38.38	39.06
68.8	5344	34.77	35.04	35.57	36.10	36.68	37.28	37.90	38.56	39.24
69.0	5351	34.91	35.19	35.71	36.25	36.84	37.45	38.07	38.73	39.43
69.2	5359	35.04	35.33	35.86	36.41	36.99	37.61	38.24	38.90	39.61
69.4	5366	35.19	35.47	36.01	36.56	37.15	37.78	38.41	39.08	39.80
69.6	5373	35.34	35.62	36.16	36.72	37.32	37.94	38.58	39.26	39.98
69.8	5381	35.49	35.76	36.31	36.87	37.48	38.11	38.75	39.45	40.17
70.0	5388	35.64	35.91	36.46	37.02	37.64	38.28	38.92	39.63	40.35
70.2	5395	35.78	36.05	36.61	37.19	37.80	38.45	39.10	39.81	40.53
70.4	5403	35.93	36.20	36.76	37.35	37.97	38.61	39.28	39.99	40.72
70.6	5410	36.08	36.35	36.92	37.51	38.13	38.78	39.46	40.17	40.90
70.8	5417	36.23	36.50	37.07	37.67	38.30	38.95	39.64	40.35	41.08
71.0	5424	36.38	36.65	37.23	37.83	38.47	39.12	39.82	40.54	41.27
71.2	5432	36.53	36.80	37.39	37.99	38.63	39.30	40.00	40.72	40.46
71.4	5439	36.68	36.95	37.55	38.16	38.80	39.48	40.18	40.90	41.64
71.6	5446	36.83	37.11	37.71	38.32	38.97	39.65	40.36	41.08	41.83
71.8	5454	36.98	37.27	37.87	38.49	39.14	39.83	40.54	41.27	42.02
72.0	5461	37.13	37.42	38.02	38.65	39.31	40.01	40.72	41.45	42.21
72.2	5468	37.29	37.58	38.19	38.82	39.49	40.18	40.90	41.64	42.40
72.4	5475	37.44	37.73	38.35	38.98	39.66	40.36	41.08	41.82	42.58
72.6	5483	37.60	37.89	38.51	39.16	39.83	40.54	41.26	42.01	42.77
72.8	5490	37.75	38.05	38.67	39.33	40.01	40.71	41.45	42.19	42.96
73.0	5497	37.91	38.21	38.84	39.50	40.18	40.88	41.63	42.38	43.15
73.2	5504	38.06	38.37	39.00	39.67	40.36	41.06	41.81	42.56	43.33
73.4	5512	38.22	38.53	39.17	39.84	40.53	41.24	41.99	42.75	43.52
73.6	5519	38.38	38.69	39.34	40.02	40.70	41.42	42.17	42.93	43.70
73.8	5526	38.54	38.85	39.50	40.19	40.88	41.60	42.36	43.12	43.89

TABLE 23-8 (concluded)

SCALE READING	INDEX OF REFRACTION	17.5° C	18° C	19° C	20° C	21° C	22° C	23° C	24° C	25° C
74.0	5533	38.70	39.01	39.67	40.36	41.05	41.78	42.54	43.31	44.08
74.2	5541	38.86	39.18	39.84	40.53	41.23	41.96	42.72	43.49	44.28
74.4	5548	39.02	39.34	40.01	40.71	41.41	42.15	42.91	43.68	44.48
74.6	5555	39.18	39.51	40.18	40.88	41.59	42.33	43.09	43.86	44.67
74.8	5563	39.35	39.68	40.35	41.05	41.77	42.51	43.28	44.05	44.87
75.0	5570	39.51	39.84	40.53	41.23	41.95	42.70	43.46	44.25	45.07
75.2	5577	39.68	40.01	40.70	41.41	42.13	42.88	43.65	44.44	45.29
75.4	5584	39.84	40.18	40.87	41.58	42.31	43.07	43.83	44.63	45.50
75.6	5592	40.01	40.35	41.04	41.76	42.49	43.25	44.02	44.83	45.71
75.8	5599	40.18	40.53	41.22	41.94	42.67	43.44	44.21	45.03	45.92
76.0	1.35606	40.35	40.70	41.40	42.12	42.85	43.63	44.41	45.24	46.12
76.2	5613	40.53	40.87	41.57	42.30	43.04	43.81	44.60	45.44	46.34
76.4	5621	40.70	41.04	41.75	42.48	43.22	44.00	44.80	45.65	46.56
76.6	5628	40.87	41.22	41.92	42.66	43.41	44.19	44.99	45.86	46.78
76.8	5635	41.04	41.39	42.10	42.84	43.60	44.38	45.19	46.07	47.00
77.0	5642	41.22	41.57	42.28	43.02	43.79	44.57	45.40	46.29	47.23
77.2	5650	41.39	41.74	42.46	43.20	43.97	44.76	45.60	46.51	47.45
77.4	5657	41.57	41.91	42.63	43.39	44.16	44.95	45.81	46.73	47.68
77.6	5664	41.75	42.09	42.81	43.57	44.35	45.15	46.01	46.95	47.91
77.8	5671	41.92	42.26	42.99	43.76	44.54	45.35	46.23	47.17	48.14
78.0	5678	42.09	42.43	43.17	43.94	44.73	45.56	46.45	47.40	48.37
78.2	5686	42.26	42.61	43.36	44.13	44.92	45.76	46.67	47.63	48.60
78.4	5693	42.44	42.78	43.54	44.32	45.12	45.96	46.89	47.85	48.84
78.6	5700	42.61	42.96	43.72	44.51	45.32	46.17	47.11	48.08	49.07
78.8	5707	42.78	43.14	43.91	44.70	45.52	46.39	47.34	48.31	49.31
79.0	5715	42.95	43.32	44.09	44.89	45.72	46.61	47.56	48.53	49.54
79.2	5722	43.13	43.50	44.28	45.08	45.92	46.83	47.79	48.76	49.77
79.4	5729	43.31	43.68	44.47	45.28	46.13	47.04	48.01	48.99	50.01
79.6	5736	43.49	43.86	44.65	45.48	46.34	47.26	48.23	49.22	50.24
79.8	5744	43.67	44.05	44.84	45.68	46.56	47.48	48.46	49.45	50.48
80.0	5751	43.85	44.24	45.04	45.88	46.77	47.70	48.68	49.68	50.71

TABLE 23-9

Correction table for specific gravity of milk (Quévenne degrees)*

Lactometer	Temperature (F) 51	52	53	54	55	56	57	58	59	60	61	62	63	64	65	66	67	68	69	70
20	19.3	19.4	19.4	19.5	19.6	19.7	19.8	19.9	19.9	20.0	20.1	20.2	20.2	20.3	20.4	20.5	20.6	20.7	20.9	21.0
21	20.3	20.3	20.4	20.5	20.6	20.7	20.8	20.9	20.9	21.0	21.1	21.2	21.3	21.4	21.5	21.6	21.7	21.8	22.0	22.1
22	21.3	21.3	21.4	21.5	21.6	21.7	21.8	21.9	21.9	22.0	22.1	22.2	22.3	22.4	22.5	22.6	22.7	22.8	23.0	23.1
23	22.3	22.3	22.4	22.5	22.6	22.7	22.8	22.8	22.9	23.0	23.1	23.2	23.3	23.4	23.5	23.6	23.7	23.8	24.0	24.1
24	23.3	23.3	23.4	23.5	23.6	23.6	23.7	23.8	23.9	24.0	24.1	24.2	24.3	24.4	24.5	24.6	24.7	24.9	25.0	25.1
25	24.2	24.3	24.4	24.5	24.6	24.6	24.7	24.8	24.9	25.0	25.1	25.2	25.3	25.4	25.5	25.6	25.7	25.9	26.0	26.1
26	25.2	25.2	25.3	25.4	25.5	25.6	25.7	25.8	25.9	26.0	26.1	26.2	26.3	26.5	26.6	26.7	26.8	27.0	27.1	27.2
27	26.2	26.2	26.3	26.4	26.5	26.6	26.7	26.8	26.9	27.0	27.1	27.3	27.4	27.5	27.6	27.7	27.8	28.0	28.1	28.2
28	27.1	27.2	27.3	27.4	27.5	27.6	27.7	27.8	27.9	28.0	28.1	28.3	28.4	28.5	28.6	28.7	28.8	29.0	29.1	29.2
29	28.1	28.2	28.3	28.4	28.5	28.6	28.7	28.8	28.9	29.0	29.1	29.3	29.4	29.5	29.6	29.7	29.9	30.1	30.2	30.3
30	29.1	29.1	29.2	29.3	29.4	29.6	29.7	29.8	29.9	30.0	30.1	30.3	30.4	30.5	30.7	30.8	30.9	31.1	31.2	31.3
31	30.0	30.1	30.2	30.3	30.4	30.5	30.6	30.8	30.9	31.0	31.2	31.3	31.4	31.5	31.7	31.8	31.9	32.1	32.2	32.4
32	31.0	31.1	31.2	31.3	31.4	31.5	31.6	31.7	31.9	32.0	32.2	32.3	32.5	32.6	32.7	32.9	33.0	33.2	33.3	33.4
33	31.9	32.0	32.1	32.3	32.4	32.5	32.6	32.7	32.9	33.0	33.2	33.3	33.5	33.6	33.8	33.9	34.0	34.2	34.3	34.5
34	32.9	33.0	33.1	33.2	33.3	33.5	33.6	33.7	33.9	34.0	34.2	34.3	34.5	34.6	34.8	34.9	35.0	35.2	35.3	35.5
35	33.8	33.9	34.0	34.2	34.3	34.5	34.6	34.7	34.9	35.0	35.2	35.3	35.5	35.6	35.8	35.9	36.1	36.2	36.4	36.5

* Paul G. Heineman, "Milk," p. 144. W. B. Saunders Co. (1921).

TABLE 23-10

Table for determining total solids in milk from any given specific gravity and percentage of fat (Shaw and Eckles)

Results expressed as total solids, percent

Percentage of fat	Lactometer reading at 60° F. (Quévenne degrees).										
	26	27	28	29	30	31	32	33	34	35	36
2.00	8.90	9.15	9.40	9.65	9.90	10.15	10.40	10.66	10.91	11.16	11.41
2.05	8.96	9.21	9.46	9.71	9.96	10.21	10.46	10.72	10.97	11.22	11.47
2.10	9.02	9.27	9.52	9.77	10.02	10.27	10.52	10.78	11.03	11.28	11.53
2.15	9.08	9.33	9.58	9.83	10.08	10.33	10.58	10.84	11.09	11.34	11.59
2.20	9.14	9.39	9.64	9.89	10.14	10.39	10.64	10.90	11.15	11.40	11.65
2.25	9.20	9.45	9.70	9.95	10.20	10.45	10.70	10.96	11.21	11.46	11.71
2.30	9.26	9.51	9.76	10.01	10.26	10.51	10.76	11.02	11.27	11.52	11.77
2.35	9.32	9.57	9.82	10.07	10.32	10.57	10.82	11.08	11.33	11.58	11.83
2.40	9.38	9.63	9.88	10.13	10.38	10.63	10.88	11.14	11.39	11.64	11.89
2.45	9.44	9.69	9.94	10.19	10.44	10.69	10.94	11.20	11.45	11.70	11.95
2.50	9.50	9.75	10.00	10.25	10.50	10.75	11.00	11.26	11.51	11.76	12.01
2.55	9.56	9.81	10.06	10.31	10.56	10.81	11.06	11.32	11.57	11.82	12.07
2.60	9.62	9.87	10.12	10.37	10.62	10.87	11.12	11.38	11.63	11.88	12.13
2.65	9.68	9.93	10.18	10.43	10.68	10.93	11.18	11.44	11.69	11.94	12.19
2.70	9.74	9.99	10.24	10.49	10.74	19.09	11.24	11.50	11.75	12.00	12.25
2.75	9.80	10.05	10.30	10.55	10.80	11.05	11.31	11.56	11.81	12.06	12.31
2.80	9.86	10.11	10.36	10.61	10.86	11.11	11.37	11.62	11.87	12.12	12.37
2.85	9.92	10.17	10.42	10.67	10.92	11.17	11.43	11.68	11.93	12.18	12.43
2.90	9.98	10.23	10.48	10.73	10.98	11.23	11.49	11.74	11.99	12.24	12.49
2.95	10.04	10.29	10.54	10.79	11.04	11.30	11.55	11.80	12.05	12.30	12.55
3.00	10.10	10.35	10.60	10.85	11.10	11.36	11.61	11.86	12.11	12.36	12.61
3.05	10.16	10.41	10.66	10.91	11.17	11.42	11.67	11.92	12.17	12.42	12.68
3.10	10.22	10.47	10.72	10.97	11.23	11.48	11.73	11.98	12.23	12.48	12.74
3.15	10.28	10.53	10.78	11.03	11.29	11.54	11.79	12.04	12.29	12.55	12.80
3.20	10.34	10.59	10.84	11.09	11.35	11.60	11.85	12.10	12.35	12.61	12.86
3.25	10.40	10.65	10.90	11.16	11.41	11.66	11.91	12.16	12.42	12.67	12.92
3.30	10.46	10.71	10.96	11.22	11.47	11.72	11.97	12.22	12.48	12.73	12.98
3.35	10.52	10.77	11.03	11.28	11.53	11.78	12.03	12.28	12.54	12.79	13.04
3.40	10.58	10.83	11.09	11.34	11.59	11.84	12.09	12.34	12.60	12.85	13.10
3.45	10.64	10.89	11.15	11.40	11.65	11.90	12.15	12.40	12.66	12.91	13.16
3.50	10.70	10.95	11.21	11.46	11.71	11.96	12.21	12.46	12.72	12.97	13.22
3.55	10.76	11.02	11.27	11.52	11.77	12.02	12.27	12.52	12.78	13.03	13.28
3.60	10.82	11.08	11.33	11.58	11.83	12.08	12.33	12.58	12.84	13.09	13.34
3.65	10.88	11.14	11.39	11.64	11.89	12.14	12.39	12.64	12.90	13.15	13.40
3.70	10.94	11.20	11.45	11.70	11.95	12.20	12.45	12.70	12.96	13.21	13.46
3.75	11.00	11.26	11.51	11.76	12.01	12.26	12.51	12.76	13.02	13.27	13.52
3.80	11.06	11.32	11.57	11.82	12.07	12.32	12.57	12.82	13.08	13.33	13.58
3.85	11.12	11.38	11.63	11.88	12.13	12.38	12.63	12.88	13.14	13.39	13.64
3.90	11.18	11.44	11.69	11.94	12.19	12.44	12.69	12.94	13.20	13.45	13.70
3.95	11.24	11.50	11.75	12.00	12.25	12.50	12.75	13.00	13.26	13.51	13.77
4.00	11.30	11.56	11.81	12.06	12.31	12.56	12.81	13.06	13.32	13.57	13.83
4.05	11.36	11.62	11.87	12.12	12.37	12.62	12.87	13.12	13.38	13.63	13.89
4.10	11.42	11.68	11.93	12.18	12.43	12.68	12.93	13.18	13.44	13.69	13.95
4.15	11.48	11.74	11.99	12.24	12.49	12.74	12.99	13.25	13.50	13.76	14.01
4.20	11.54	11.80	12.05	12.30	12.55	12.80	13.05	13.31	13.56	13.82	14.07

TABLE 23-10 (continued)

Per-centage of fat	Lactometer reading at 60° F. (Quévenne degrees).										
	26	27	28	29	30	31	32	33	34	35	36
4.25	11.60	11.86	12.11	12.36	12.61	12.86	13.12	13.37	13.62	13.88	14.13
4.30	11.66	11.92	12.17	12.42	12.67	12.92	13.18	13.43	13.68	13.94	14.19
4.35	11.72	11.98	12.23	12.48	12.73	12.98	13.24	13.49	13.74	14.00	14.25
4.40	11.78	12.04	12.29	12.54	12.79	13.04	13.30	13.55	13.80	14.06	14.31
4.45	11.84	12.10	12.35	12.60	12.85	13.10	13.36	13.61	13.86	14.12	14.37
4.50	11.90	12.16	12.41	12.66	12.91	13.16	13.42	13.67	13.92	14.18	14.43
4.55	11.97	12.22	12.47	12.72	12.97	13.22	13.48	13.73	13.98	14.24	14.49
4.60	12.03	12.28	12.53	12.78	13.03	13.28	13.54	13.79	14.04	14.30	14.55
4.65	12.09	12.34	12.59	12.84	13.09	13.34	13.60	13.85	14.10	14.36	14.61
4.70	12.15	12.40	12.65	12.90	13.15	13.40	13.66	13.91	14.16	14.42	14.67
4.75	12.21	12.46	12.71	12.96	13.21	13.46	13.72	13.97	14.22	14.48	14.73
4.80	12.27	12.52	12.77	13.02	13.27	13.52	13.78	14.03	14.28	14.54	14.79
4.85	12.33	12.58	12.83	13.08	13.33	13.58	13.84	14.09	14.34	14.60	14.85
4.90	12.39	12.64	12.89	13.14	13.39	13.64	13.90	14.15	14.40	14.66	14.91
4.95	12.45	12.70	12.95	13.20	13.45	13.70	13.96	14.21	14.46	14.72	14.97
5.00	12.51	12.76	13.01	13.26	13.51	13.76	14.02	14.27	14.52	14.78	15.03
5.05	12.57	12.82	13.07	13.32	13.57	13.83	14.08	14.33	14.58	14.84	15.09
5.10	12.63	12.88	13.13	13.38	13.63	13.89	14.14	14.39	14.64	14.90	15.15
5.15	12.69	12.94	13.19	13.44	13.69	13.95	14.20	14.45	14.70	14.96	15.21
5.20	12.75	13.00	13.25	13.50	13.75	14.01	14.26	14.51	14.76	15.02	15.27
5.25	12.81	13.06	13.31	13.56	13.81	14.07	14.32	14.57	14.82	15.08	15.33
5.30	12.87	13.12	13.37	13.62	13.87	14.13	14.38	14.63	14.88	15.14	15.39
5.35	12.93	13.18	13.43	13.68	13.93	14.19	14.44	14.70	14.95	15.20	15.45
5.40	12.99	13.24	13.49	13.74	14.00	14.25	14.50	14.76	15.01	15.26	15.51
5.45	13.05	13.30	13.55	13.80	14.06	14.31	14.56	14.82	15.07	15.32	15.57
5.50	13.11	13.36	13.61	13.86	14.12	14.37	14.62	14.88	15.13	15.38	15.63
5.55	13.17	13.42	13.67	13.93	14.18	14.43	14.69	14.94	15.19	15.44	15.69
5.60	13.23	13.48	13.73	13.99	14.24	14.49	14.75	15.00	15.25	15.50	15.75
5.65	13.29	13.54	13.79	14.05	14.30	14.55	14.81	15.06	15.31	15.56	15.81
5.70	13.35	13.60	13.85	14.11	14.36	14.61	14.87	15.12	15.37	15.62	15.87
5.75	13.41	13.66	13.91	14.17	14.42	14.68	14.93	15.18	15.43	15.68	15.93
5.80	13.47	13.72	13.97	14.23	14.48	14.74	14.99	15.24	15.49	15.74	15.99
5.85	13.53	13.78	14.04	14.29	14.54	14.80	15.05	15.30	15.55	15.80	16.06
5.90	13.59	13.84	14.10	14.35	14.60	14.86	15.11	15.36	15.61	15.86	16.12
5.95	13.65	13.90	14.16	14.41	14.66	14.92	15.17	15.42	15.67	15.92	16.18
6.00	13.71	13.96	14.22	14.47	14.72	14.98	15.23	15.48	15.73	15.98	16.24
6.05	13.77	14.02	14.28	14.53	14.78	15.04	15.29	15.54	15.79	16.04	16.30
6.10	13.83	14.08	13.44	14.59	14.84	15.10	15.35	15.60	15.85	16.10	16.35
6.15	13.89	14.14	14.40	14.65	14.90	15.16	15.41	15.66	15.91	16.16	16.42
6.20	13.95	14.20	14.46	14.71	14.96	15.22	15.47	15.72	15.97	16.22	16.48
6.25	14.01	14.26	14.52	14.77	15.02	15.28	15.53	15.78	16.03	16.28	16.54
6.30	14.07	14.32	14.58	14.83	15.08	15.34	15.59	15.84	16.09	16.34	16.60
6.35	14.13	14.38	14.64	14.90	15.14	15.40	15.65	15.90	16.15	16.40	16.66
6.40	14.19	14.44	14.70	14.96	15.20	15.46	15.71	15.96	16.21	16.46	16.72
6.45	14.25	14.50	14.76	15.02	15.26	15.52	15.77	16.02	16.27	16.52	16.78
6.50	14.31	14.56	14.82	15.08	15.32	15.58	15.83	16.08	16.33	16.58	16.84

TABLE 23-10 (continued)

Percentage of fat	Lactometer reading at 60° F. (Quévenne degrees)										
	26	27	28	29	30	31	32	33	34	35	36
6.55	14.37	14.62	14.88	15.14	15.38	15.64	15.89	16.14	16.39	16.64	16.90
6.60	14.43	14.68	14.94	15.20	15.44	15.70	15.95	16.20	16.45	16.70	16.96
6.65	14.49	14.74	15.00	15.26	15.50	15.76	16.01	16.26	16.51	16.76	17.02
6.70	14.55	14.80	15.06	15.32	15.56	15.82	16.07	16.32	16.57	16.82	17.08
6.75	14.61	14.86	15.12	15.38	15.62	15.88	16.13	16.38	16.63	16.88	17.14
6.80	14.67	14.92	15.18	15.44	15.68	15.94	16.19	16.44	16.69	16.94	17.20
6.85	14.73	14.98	15.24	15.50	15.74	16.00	16.25	16.50	16.75	17.00	17.26
6.90	14.79	15.04	15.30	15.56	15.80	16.06	16.31	16.56	16.81	17.06	17.32
6.95	14.85	15.10	15.36	15.62	15.86	16.12	16.37	16.62	16.87	17.12	17.38

TABLE 23-10 (concluded)

Proportional Parts*

Lactometer fraction	Fraction to be added to total solids	Lactometer fraction	Fraction to be added to total solids	Lactometer fraction	Fraction to be added to total solids
0.1	0.03	0.4	0.10	0.7	0.18
.2	.05	.5	.13	.8	.20
.3	.08	.6	.15	.9	.23

* Table giving Proportional Parts shows amount to be added when lactometer readings are in whole numbers and decimals.

Index